Computer Arithmetic

Volume II

Computer Arithmetic

Volume II

Editor

Earl E Swartzlander

University of Texas at Austin, USA

World Scientific

NEW JERSEY • LONDON • SINGAPORE • BEIJING • SHANGHAI • HONG KONG • TAIPEI • CHENNAI

Published by

World Scientific Publishing Co. Pte. Ltd.
5 Toh Tuck Link, Singapore 596224
USA office: 27 Warren Street, Suite 401-402, Hackensack, NJ 07601
UK office: 57 Shelton Street, Covent Garden, London WC2H 9HE

Library of Congress Cataloging-in-Publication Data
Computer arithmetic / [edited by] Earl Swartzlander, The University of Texas at Austin, USA. -- New edition.
3 volumes cm
Includes bibliographical references and indexes.
ISBN 978-9814641463 (hardbound : alk. paper) -- ISBN 9814641464 (hardbound : alk. paper) --
ISBN 978-9814651134 (hardbound : alk. paper) -- ISBN 9814651133 (hardbound : alk. paper) --
ISBN 978-9814651561 (hardbound : alk. paper) -- ISBN 9814651567 (hardbound : alk. paper)
1. Computer programming. 2. Floating-point arithmetic. 3. Computer arithmetic.
I. Swartzlander, Earl E., Jr., 1945– editor.
QA76.6.C633 2014
005.1--dc23
2014038310

British Library Cataloguing-in-Publication Data
A catalogue record for this book is available from the British Library.

Typeset by Stallion Press
Email: enquiries@stallionpress.com

Printed in Singapore

About the Editor

Earl E. Swartzlander, Jr. is a Professor of Electrical and Computer Engineering at the University of Texas at Austin. In this position, he and his students conduct research in computer engineering with emphasis on application specific processor design, including high-speed computer arithmetic, systolic signal processor architecture, VLSI technology, and QCA technology (an emerging form of nanotechnology). Dr. Swartzlander holds degrees in Electrical Engineering from Purdue University, the University of Colorado, and the University of Southern California.

Prof. Swartzlander is a Fellow of the IEEE. He has been honored with the IEEE Third Millennium Medal, the Distinguished Engineering Alumnus Award from the University of Colorado, the Outstanding Electrical Engineer and Distinguished Engineering Alumnus awards from Purdue University and the IEEE Computer Society Golden Core Award. He was the Editor-in-Chief of the *IEEE Transactions on Computers* from 1990–1994 and was the founding Editor-in-Chief of the *Journal of VLSI Signal Processing.* In addition, he has served as an associate editor for the *IEEE Transactions on Computers*, the *IEEE Transactions on Parallel and Distributed Systems*, and the *IEEE Journal of Solid-State Circuits.*

He has been a member of the Board of Governors of the IEEE Computer Society (1987–1991), the IEEE Signal Processing Society (1992–1994) and the IEEE Solid-State Circuits Council/Society (1986–1991). He has been a member of the IEEE History Committee (1996–2004), the IEEE Fellows Committee (2000–2003), the IEEE James H. Mulligan, Jr. Education Medal Committee (2006–2011).

Preface

This is the second printing of *Computer Arithmetic* Volume II (Los Alamitos, CA: IEEE Computer Society Press, 1990). That book is a sequel to the earlier collection of "Benchmark" papers on computer arithmetic *(Computer Arithmetic,* Stroudsburg, PA: Dowden, Hutchinson, & Ross, 1980). Both this volume and that volume are being reprinted as two parts of the present three volume collection on computer arithmetic, with reprint permission from AT&T, IEEE, Elsevier, ACM, and IPSJ.

Volume I examined the basic arithmetic operations of add, subtract, multiply, divide and their application to the computation of logarithms, functions, and floating point arithmetic. With the tremendous progress in computing and microelectronic technology it was appropriate to extend the material.

As with Volume I, this volume is addressed to researchers in computer arithmetic and students in graduate courses that need to easily access a cross section of the key papers in the field as well as practicing engineers who need access to practical material on the design and application of arithmetic elements.

This volume examines theoretical aspects of arithmetic (with Sections on number representation and on-line arithmetic), design aspects (with Sections on error detection and correction), and implementation aspects (with chapters on VLSI Adders, VLSI Multipliers, and Floating-Point VLSI Chips). The bibliography is updated to 1990 and greatly expanded from that of *Computer Arithmetic* Volume I.

Selecting topics and papers for a book such as this is an exercise in dynamic optimization. The topics that are included are intended to complement the coverage of the previous book. The papers on chips overlap the sections on arithmetic operations in the predecessor (and serve to validate the importance of much of the previous material). As the size of the bibliography shows, this field is quite active. Accordingly, selecting forty papers has forced the omission of many fine papers. Some of these are included in Volume III.

The book begins with an overview of basic computer arithmetic. This material explains the basic concepts and the basic operations of add, subtract, multiply, and divide.

Part I examines error detection/correction. With the high density of current VLSI (and projections into the future for WSI and ULSI!) error correction has become increasingly important. The chapter begins with a paper describing the Hamming code (which is used to protect against errors that arise in storage and communication) and continues with five papers on arithmetic error correction.

Part II introduces and summarizes the current status of digit on-line arithmetic. This is an approach where digits of the operands are processed serially beginning with the most significant. Although not widely used, this approach is quite promising for high speed special purpose processors.

The next focus is on the implementation of adders with VLSI in Part III. Papers are included that present alternatives to the widely used carry lookahead adder in addition to examining the implementation of carry lookahead adders. This chapter concludes with a paper on the design of an error detecting adder.

VLSI multiplier implementation is covered in the next part. The first paper describes the first commercial single-chip parallel multiplier. The four succeeding papers describe techniques for achieving better performance. The final paper describes a state-of-the-art multiplier implemented with sub-micron CMOS.

Part V describes floating-point chips. The chips range from coprocessors that implement a complete repertoire of arithmetic operations to specialized chips that implement a few operations. Word sizes range from 32 bits to 80 bits.

The next part explores advanced topics in number representation. These papers discuss various aspects that arise in the design of floating-point number systems and present a variety of techniques for accurately representing rational numbers.

The final part concerns implementations with papers on distributed arithmetic, low power design concepts, and a logarithmic processor that provides better performance than floating-point processors implemented with similar technology.

Earl E. Swartzlander, Jr.

Contents by Author

Contents

Part II: ON-LINE ARITHMETIC

Part III: VLSI ADDER IMPLEMENTATIONS

Part IV: VLSI MULTIPLIER IMPLEMENTATIONS

Part V: FLOATING-POINT VLSI CHIPS

Part VI: NUMBER REPRESENTATION

Overview of Computer Arithmetic

1 Introduction

This introduction begins with a discussion of binary number systems in Section 2. Section 3 provides examples of fixed point implementations of the four basic arithmetic operations with two's complement binary arithmetic. Section 4 very briefly describes the implementation of floating-point arithmetic operations. Throughout, there are many references to material reprinted in Swartzlander, Jr. (1980), which includes several papers that provide more detail on the material in this overview. Regarding notation, capital letters represent digital numbers (i.e., words), while lower case letters represent bits of the corresponding word. The subscripts range from n − 1 to 0 to indicate the bit position within the word (x_{n-1} is the most significant bit of X, x_0 is the least significant bit of X, etc.). The logic designs presented in this overview are based on positive logic with AND, OR, and invert operations. Depending on the technology used for implementation, different logical operations (such as NAND and NOR) or direct transistor realizations may be used, but the basic concepts do not change significantly.

2 Binary Number Systems

Most arithmetic is performed with fixed point numbers that have constant scaling (i.e., the position of the binary point is fixed). In most cases the numbers represent fractions, although integers and mixed fixed point numbers are occasionally used for special applications. Pairs of fixed point numbers may be used to create floating-point numbers, as discussed in Section 4.

At the present time, fixed point binary numbers are generally represented using the two's complement number system. This choice has prevailed over the sign magnitude and other number systems, because the frequently performed operations of addition and subtraction are easily performed on two's complement numbers.

Fixed point number systems represent numbers, for example, A, by n bits: a sign bit and n − 1 data bits. By convention, the most significant bit, a_{n-1}, is the sign bit, which is a 1 for negative numbers and a 0 for positive numbers. The n − 1 data bits are $a_{n-2}, a_{n-3}, \ldots, a_1, a_0$. In the following material, fixed point fractions will be described for the two's complement and sign magnitude systems.

Two's Complement Number System

In the two's complement fractional number system, the value of a number is the sum of n − 1 positive binary fractional bits and a sign bit which has a weight of −1:

$$A = -a_{n-1} + \sum_{i=0}^{n-2} a_i 2^{i-n+1} \tag{1}$$

Two's complement numbers are negated by complementing all bits and adding a 1 to the least significant bit (lsb) position. For example, using 3-bit numbers to form $-3/4$:

$$\begin{aligned} +3/4 &= 011 \\ \text{invert all bits} &= 100 \\ \text{add 1 lsb}\quad & \underline{001} \\ & 101 = -3/4 \\ \text{Check: invert all bits} &= 010 \\ \text{Add 1 lsb}\quad & \underline{001} \\ & 011 = 3/4 \end{aligned}$$

Sign Magnitude Number System

Sign magnitude numbers consist of a sign bit and n – 1 bits that express the magnitude of the number,

$$A = (1 - 2a_{n-1}) + \sum_{i=0}^{n-2} a_i 2^{i-n+1}$$

Sign magnitude numbers are negated by complementing the sign bit. For example, to form $-3/4$,

$$\begin{aligned} +3/4 &= 011 \\ \text{invert sign bit} &= 111 = -3/4 \\ \text{Check: invert sign bit} &= 011 = 3/4 \end{aligned}$$

Table 1 compares 3-bit fractional fixed point numbers in the two number systems. Note that the sign magnitude number system has two zeros (i.e., positive zero and negative zero) and that only two's complement is capable of representing -1. For positive numbers, both number systems use identical representations.

A significant difference between the two number systems is their behavior under truncation. Figure 1 shows the effect of truncating high-precision fixed point fractions X, to form three bit fractions T(X). Truncation of two's complement numbers never increases the value of the number (i.e., the truncated numbers have values that are unchanged or shift toward negative infinity), as can be seen from Equation (1) since any truncated bits have positive weight. This bias can cause an accumulation of errors for computations that involve summing many truncated numbers (which may occur in scientific and signal processing applications). In the sign magnitude number system, truncated numbers are unchanged or shifted toward zero, so that if approximately half of the numbers are positive and half are negative, the errors will tend to cancel.

TABLE 1 Example of 3-bit Fractional Fixed Point Numbers.

Number	Two's Complement	Sign Magnitude
+3/4	011	011
+1/2	010	010
+1/4	001	001
+0	000	000
−0	N/A	100
−1/4	111	101
−1/2	110	110
−3/4	101	111
−1	100	N/A

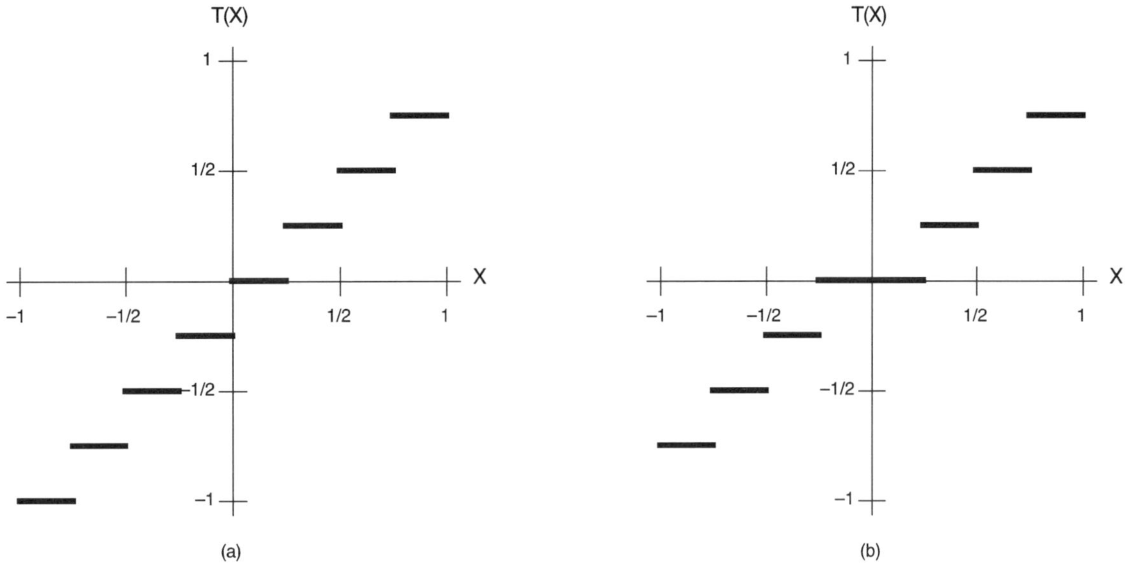

FIGURE 1 Behavior of fixed point fractions under truncation: (a) two's complement and (b) sign magnitude.

TABLE 2 Full Adder Truth Table.

Inputs			Outputs	
a_k	b_k	c_k	c_{k+1}	s_k
0	0	0	0	0
0	0	1	0	1
0	1	0	0	1
0	1	1	1	0
1	0	0	0	1
1	0	1	1	0
1	1	0	1	0
1	1	1	1	1

3 Fixed-Point Arithmetic Algorithms

This section presents an assortment of typical fixed point algorithms for addition, subtraction, multiplication, and division.

Fixed-Point Addition and Subtraction

Addition is performed by summing the corresponding bits of the two n-bit numbers, including the sign bit. Subtraction is performed by summing the corresponding bits of the minuend and the two's complement of the subtrahend. Overflow is detected in a two's complement adder by comparing the carry signals into and out of the most significant adder stage (i.e., the stage that computes the sign bit). If the carries differ, an overflow has occurred and the result is invalid.

Full Adder: The full adder is the fundamental building block of most arithmetic circuits. Its operation is defined by the truth table shown in Table 2. The sum and carry outputs are described by the following equations:

$$s_k = c_{k+1} = a_k\overline{b_k}\overline{c_k} + \overline{a_k}b_k\overline{c_k} + \overline{a_k}\overline{b_k}c_k + a_kb_kc_k = a_k \oplus b_k \oplus c_k$$
$$c_{k+1} = \overline{a_k}b_kc_k + a_k\overline{b_k}c_k + a_kb_k\overline{c_k} + a_kb_kc_k = a_kb_k + a_kc_k + b_kc_k \tag{2}$$

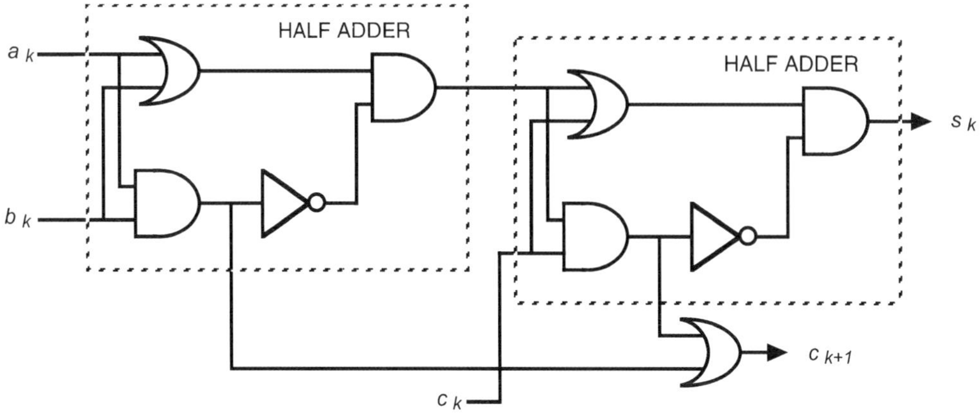

FIGURE 2 Nine gate full adder.

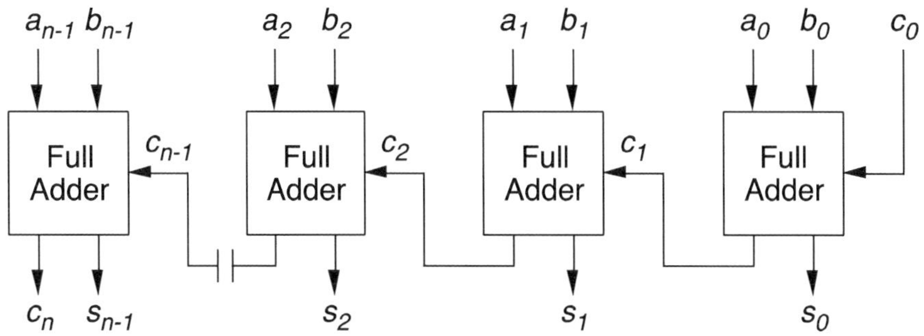

FIGURE 3 Ripple carry adder.

where: $\mathrm{a_k}$, $\mathrm{b_k}$, and $\mathrm{c_k}$ are the inputs to the k-th full adder stage, and $\mathrm{s_k}$ and $\mathrm{c_{k+1}}$ are the sum and carry outputs, respectively and $\oplus$ denotes the Exclusive-OR logic operation.

In evaluating the relative complexity of implementations it is often convenient to assume a nine gate realization of the full adder, as shown in Figure 2. For this implementation, the delay from either $\mathrm{a_k}$ or $\mathrm{b_k}$ to $\mathrm{s_k}$ is six gate delays and the delay from $\mathrm{c_k}$ to $\mathrm{c_{k+1}}$ is two gate delays. Some technologies, such as CMOS, form inverting gates (e.g., NAND and NOR gates) more efficiently than the non-inverting gates that are assumed in this overview. Circuits with equivalent speed and complexity can be constructed with inverting gates.

Ripple Carry Adder: A ripple carry adder for n-bit numbers can be implemented by concatenating n full adders, as shown in Figure 3. At the kth bit position, bits a_k and b_k of operands A and B and the carry signal from the preceding adder stage, c_k, are used to generate the kth bit of the sum, s_k, and the carry to the next stage, $\mathrm{c_{k+1}}$. This is called a ripple carry adder, since the carry signals "ripple" from the least significant bit position to the most significant.

If the ripple carry adder is implemented by concatenating n of the nine gate full adders, like those shown in Figure 2, an n-bit ripple carry adder requires $2n+4$ gate delays to produce the most significant sum bit and $2n+3$ gate delays to produce the carry output. A total of $9n$ logic gates are required to implement the n-bit ripple carry adder.

In comparing the delay and complexity of adders, the delay from data input to most significant sum output denoted by DELAY and the gate count denoted by GATES will be used. The DELAY and GATES values are subscripted by RCA to indicate that they are for a ripple carry adder. Although these simple metrics are suitable for first order comparisons, more accurate comparisons require more exact modeling since the implementations may be effected with transistor networks (as opposed to gates), which will have different delay and complexity characteristics.

$$DELAY_{RCA} = 2n + 4$$
$$GATES_{RCA} = 9n$$

Carry Lookahead Adder: Another popular adder approach is the carry lookahead adder (Weinberger and Smith, 1958, MacSorley, 1961). Here specialized logic computes the carries in parallel. The carry lookahead adder uses MFAs, or modified full adders (modified in the sense that a carry is not formed) for each bit position and lookahead modules, that form the carries. The lookahead modules form individual carries and also block generate and block propagate outputs that indicate that a carry is generated within the module or that an incoming carry would propagate across the module, respectively. Rewriting Equation (2) with $g_k = a_k b_k$ and $p_k = a_k + b_k$.

$$c_{k+1} = g_k + p_k c_k \tag{3}$$

This explains the concept of carry generation and carry propagation. At a given stage, a carry is "generated" if g_k is true (i.e., both a_k and b_k are 1), and a carry is "propagated" from its input to its output if p_k is true (i.e., either a_k or b_k is a 1). The nine gate full adder shown in Figure 2 has AND and OR gates that produce a_k and b_k with no additional complexity. Because the carry out is produced by the lookahead logic, the OR gate that produces c_{k+1} can be eliminated. The result is an eight gate modified full adder (MFA). Extending Equation (3) to a second stage,

$$\begin{aligned} c_{k+2} &= g_{k+1} + p_{k+1} c_{k+1} \\ &= g_{k+1} + p_{k+1}(g_k + p_k c_k) \\ &= g_{k+1} + p_{k+1} g_k + p_{k+1} p_k c_k \end{aligned} \tag{4}$$

Equation (4) results from evaluating Equation (3) for the $(k+1)$-th stage and substituting c_{k+1} from Equation (3). Carry c_{k+2} exits from stage $k+1$ if: (1) a carry is generated there or (2) a carry is generated in stage k and propagates across stage $k+1$ or (3) a carry enters stage k and propagates across both stages k and $k+1$, etc. Extending to a third stage:

$$\begin{aligned} c_{k+3} &= g_{k+2} + p_{k+2} c_{k+2} \\ &= g_{k+2} + p_{k+2}(g_{k+1} + p_{k+1} g_k + p_{k+1} p_k c_k) \\ &= g_{k+2} + p_{k+2} g_{k+1} + p_{k+2} p_{k+1} g_k + p_{k+2} p_{k+1} p_k c_k \end{aligned} \tag{5}$$

Although it would be possible to continue this process indefinitely, each additional stage increases the size (i.e., the number of inputs) of the logic gates. Four inputs required to implement Equation (5) are frequently the maximum number of inputs per gate for current technologies. To continue the process, block generate and block propagate signals ($g_{k+3:k}$ and $p_{k+3:k}$, respectively are defined over 4-bit blocks (stages k to $k+3$):

$$g_{k+3:k} = g_{k+3} + p_{k+3} g_{k+2} + p_{k+3} p_{k+2} g_{k+1} + p_{k+3} p_{k+2} p_{k+1} g_k \tag{6}$$

and

$$p_{k+3:k} = p_{k+3} p_{k+2} p_{k+1} p_k \tag{7}$$

Equation (3) can be expressed in terms of the 4-bit block generate and propagate signals:

$$c_{k+4} = g_{k+3:k} + p_{k+3:k} c_k \tag{8}$$

Thus, the carry out from a 4-bit wide block can be computed in only four gate delays (the first to compute p_i and g_i for $i = k$ through $k+3$, the second to evaluate $p_{k+3:k}$, the second and third to evaluate $g_{k+3:k}$, and the third and fourth to evaluate c_{k+4} using Equation (8).

An n-bit carry lookahead adder implemented with r-bit blocks requires $\lfloor(n-1)/(r-1)\rfloor$ lookahead logic blocks. A 4-bit lookahead logic block is a direct implementation of Equations (3) through (7), requiring 14 logic gates. In general, an r-bit lookahead logic block requires $\frac{1}{2}\left(3r+r^2\right)$ logic gates. The Manchester carry chain Kilburn, Edwards, and Aspinall (1960) is an alternative switch-based technique for the implementation of a lookahead logic block.

Figure 4 shows the interconnection of 16 modified full adders and five 4-bit lookahead logic blocks to realize a 16-bit carry lookahead adder. The sequence of events that occur during an add operation, is as follows: (1) apply A, B, and carry in signals: (2) each adder computes P and G; (3) first level lookahead logic computes the 4-bit propagate and generate signals; (4) second level lookahead logic computes c_4, c_8, and c_{12}; (5) first level lookahead logic computes the individual carries; and (6) each modified full adder computes the sum outputs. This process may be extended to larger adders by subdividing the large adder into 16-bit blocks and using additional levels of carry lookahead (e.g., a 64-bit adder requires three levels).

The delay of carry lookahead adders is evaluated by recognizing that an adder with a single level of carry lookahead (for 4-bit words if 4-bit blocks are used) has six gate delays, and that each additional level of lookahead increases the maximum word size by a factor of four and adds four gate delays. More generally (Waser and Flynn, 1982, pp. 83–88), the number of lookahead levels for an n-bit adder is $\lceil \log_r n \rceil$ where r is the maximum number of inputs per gate. Since an r-bit carry lookahead adder has six gate delays and there are four additional gate delays per carry lookahead level after the first,

$$DELAY_{CLA} = 2 + 4\lceil \log_r n \rceil \tag{9}$$

The complexity of an n-bit carry lookahead adder implemented with r-bit lookahead logic blocks is n modified full adders (each of which requires eight gates) and $\lceil(n-1)/(r-1)\rceil$ lookahead logic blocks (each of which requires $\frac{1}{2}\left(3r+r^2\right)$ gates). In addition 2 gates are used to calculate the carry out from the adder, c_n, from $p_{n-1:0}$ and $g_{n-1:0}$.

$$GATES_{CLA} = 8n + \frac{1}{2}(3r + r^2)\left\lceil \frac{n-1}{r-1} \right\rceil$$

For the case of $r = 4$

$$GATES_{CLA} \approx 12\frac{2}{3}n - 4\frac{2}{3}$$

The carry lookahead approach reduces the delay of adders from increasing in proportion to the word size (as is the case for ripple carry adders) to increasing in proportion to the logarithm of the word size. As with ripple carry adders, the carry lookahead adder complexity grows linearly with the word size (for $r = 4$, the complexity of a carry lookahead adder is about 40% greater than the complexity of a ripple carry adder). It is important to realize that most carry lookahead adders require gates with up to 4 inputs while ripple carry adders use only inverters and 2 input gates.

Carry Skip Adder: The carry skip adder Lehman and Burla (1961) divides the words to be added into blocks (like the carry lookahead adder). The basic structure of a 16-bit carry skip adder implemented with five 3-bit blocks and one 1-bit-wide block is shown in Figure 5. Within each block, a ripple carry adder is used to produce the sum bits and the carry (which is used as a block generate). In addition, an AND gate is used to form the block propagate signal. These signals are combined using Equation (8) to produce a fast carry signal.

For example, with $k = 4$, Equation (8) yields $c_8 = g_{7:4} + p_{7:4}\, c_4$. The carry out of the second ripple carry adder is a block generate signal if it is evaluated when carries generated by the data inputs (i.e., $a_{7:4}$ and $b_{7:4}$ in Figure 5) are valid, but before the carry that results from c_4. Normally, these two types of carries coincide in time, but in the carry skip adder the c_4 signal is produced by a 4-bit ripple carry adder, so the carry output is a block generate from 11 gate delays after application of A and B until it becomes c_8 at 19 gate delays after the application of A and B.

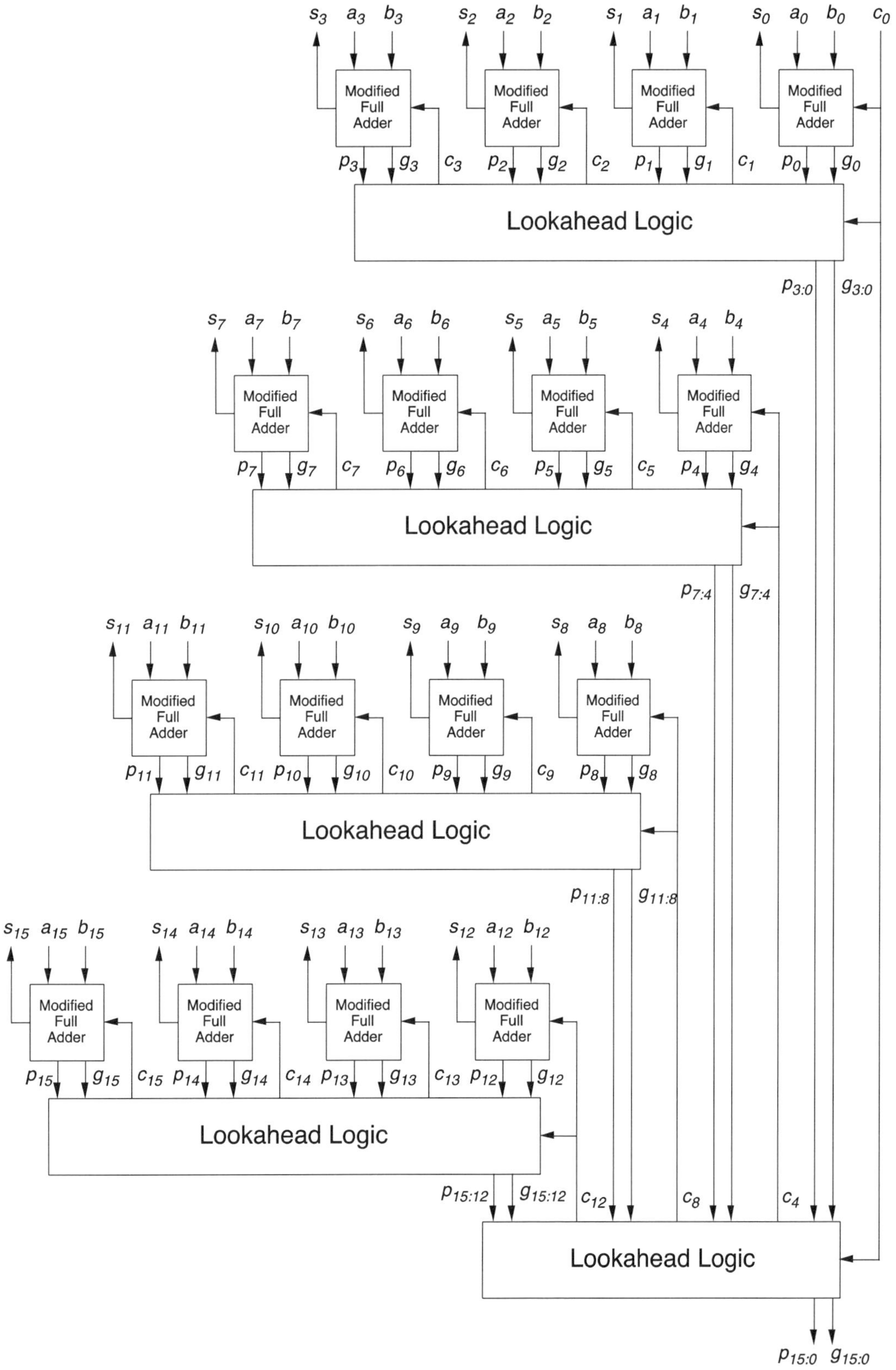

FIGURE 4 16-bit carry lookahead adder.

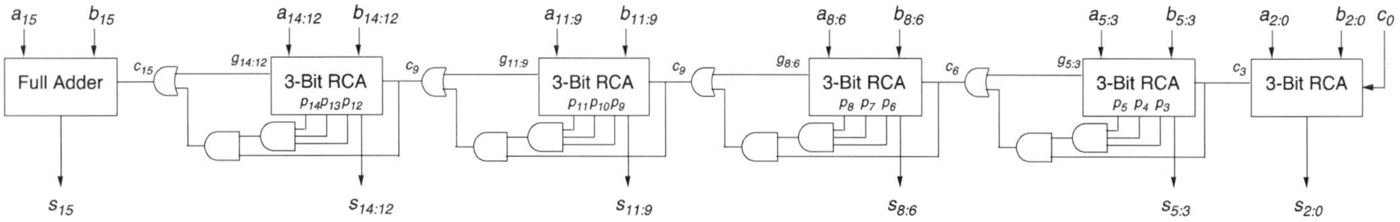

FIGURE 5 16-bit carry skip adder.

In the carry skip adder, the first and last blocks are simple ripple carry adders, whereas the $\lceil n/k \rceil - 2$ intermediate blocks are ripple carry adders augmented with three gates. The delay of a carry skip adder is the sum of $2k + 3$ gate delays to produce the carry in the first block, 2 gate delays through each of the intermediate blocks, and $2k + 1$ gate delays to produce the most significant sum bit in the last block.

To simplify the analysis, the ceiling function in the count of intermediate blocks is ignored. If the block width is k,

$$DELAY_{SKIP} = 2k + 3 + 2\left(\frac{n}{k} - 2\right) + 2k + 1$$
$$DELAY_{SKIP} = 4k + 2\frac{n}{k}$$

where: $DELAY_{SKIP}$ is the total delay of the carry skip adder with a single level of k-bit-wide blocks. The optimum block size is determined by taking the derivative of $DELAY_{SKIP}$ with respect to k, setting it to zero, and solving for k. The resulting optimum values for k and $DELAY_{SKIP}$ are

$$k = \sqrt{n/2}$$
$$DELAY_{SKIP} = 4\sqrt{2n}$$

Better results can be obtained by varying the block width so that the first and last blocks are smaller while the intermediate blocks are larger and also by using multiple levels of carry skip (Chen, and Schlag, 1990, Turrini, 1989).

The complexity of the carry skip adder is only slightly greater than that of a ripple carry adder because the first and last blocks are standard ripple carry adders while the intermediate blocks are ripple carry adders with three gates added for carry skipping.

$$GATES_{SKIP} = 9n + 3\left(\left\lceil\frac{n}{k}\right\rceil - 2\right)$$

Carry Select Adder: The carry select adder Bedrij (1962) divides the words to be added into blocks and forms two sums for each block in parallel (one with a carry in of 0 and the other with a carry in of 1). As shown for a 16-bit carry select adder in Figure 6, the carry out from the previous block controls a multiplexer that selects the appropriate sum. The carry out is computed using Equation (8), because the group propagate signal is the carry out of an adder with a carry input of 1 and the group generate signal is the carry out of an adder with a carry input of 0.

If a constant block width of k is used, there will be $\lceil n/k \rceil$blocks and the delay to generate the sum is $2k + 3$ gate delays to form the carry out of the first block, two gate delays for each of the $\lceil n/k \rceil - 2$ intermediate blocks, and three gate delays (for the multiplexer) in the final block. To simplify the analysis, the ceiling function in the count of intermediate blocks is ignored. The total delay is thus

$$DELAY_{C-SEL} = 2k + 2n/k + 2$$

where $\mathrm{DELAY}_{\mathrm{C-SEL}}$ is the total delay. The optimum block size is determined by taking the derivative of $\mathrm{DELAY}_{\mathrm{C-SEL}}$ with respect to k, setting it to zero and solving for k. The result is

$$k = \sqrt{n}$$
$$DELAY_{C-SEL} = 2 + 4\sqrt{n}$$

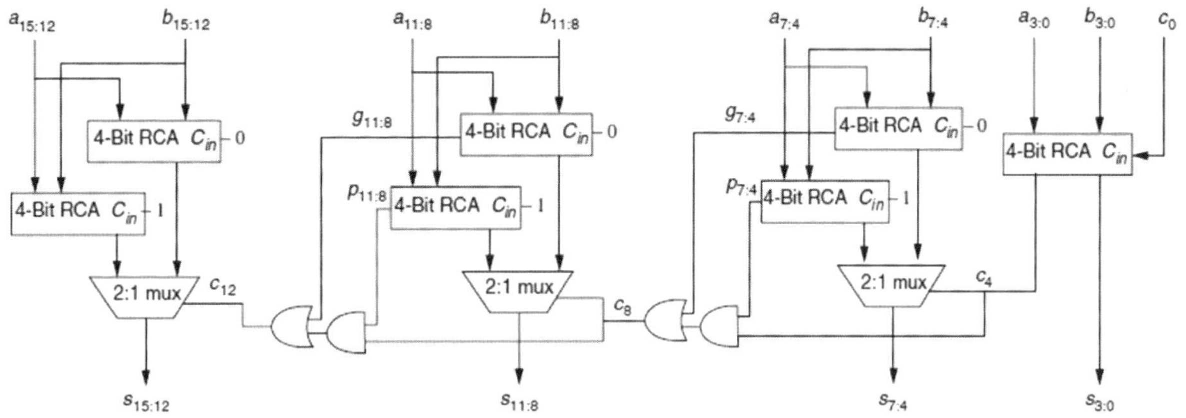

FIGURE 6 16-bit carry select adder.

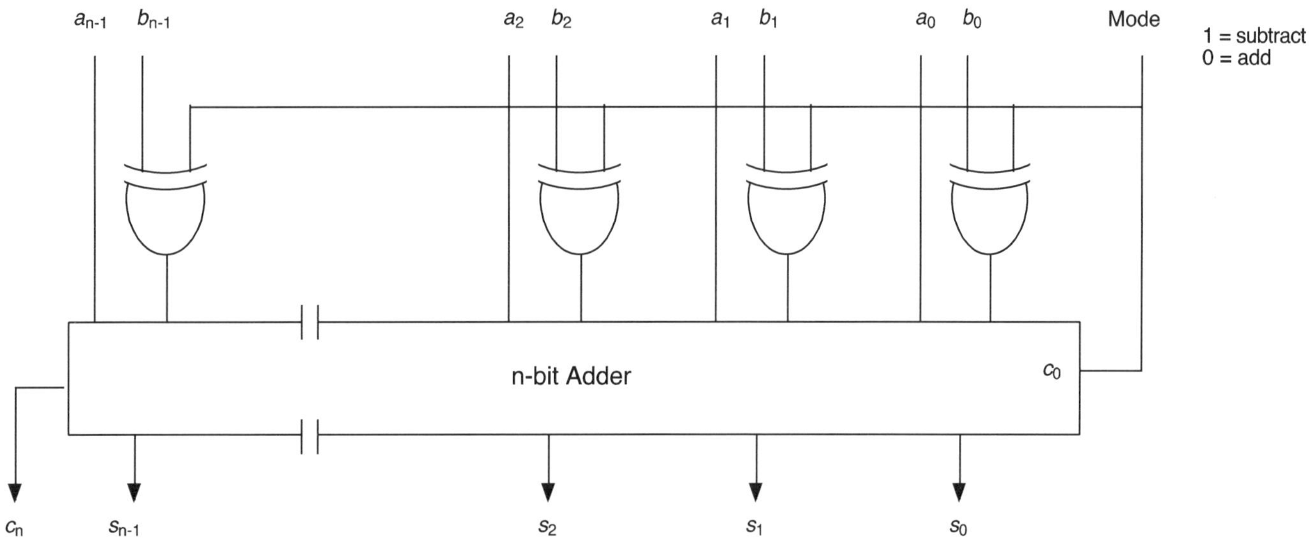

FIGURE 7 Adder/subtracter.

As for the carry skip adder, better results can be obtained by varying the width of the blocks. In this case, the optimum is to make the two least significant blocks the same size and each successively more significant block 1 bit larger. In this configuration, the delay for each block's most significant sum bit will equal the delay to the multiplexer control signal (Goldberg, 1990, p. A-38).

The complexity of the carry select adder is $2n - k$ ripple carry adder stages, the intermediate carry logic and $(\lceil n/k \rceil - 1)k$-bit-wide 2:1 multiplexers:

$$GATES_{C-SEL} = 9(2n-k) + 2\left(\left\lceil \frac{n}{k} \right\rceil - 2\right) + 3(n-k) + \left\lceil \frac{n}{k} \right\rceil - 1$$

$$GATES_{C-SEL} = 21n - 12k + 3\left\lceil \frac{n}{k} \right\rceil - 5$$

This is slightly more than twice the complexity of a ripple carry adder of the same size.

Subtraction: To produce an adder/subtracter, the adder is modified as shown in Figure 7 by including Exclusive-OR gates to complement operand B when performing subtraction. It forms either A + B or A − B. In the case of A + B, the mode selector is set to logic 0, which causes the Exclusive-OR gates to pass operand B through unchanged to the adder. The carry into the least significant adder stage is also set to ZERO, so standard addition occurs. Subtraction is implemented by setting the mode selector to logic ONE, which causes the Exclusive-OR gates to complement the bits of B; formation of the two's complement of B is completed by setting the carry into the least significant adder stage to ONE.

Fixed-Point Multiplication

Multiplication is generally implemented by direct logic implementations. Array multipliers have a very regular layout, but are slower than "Fast" multipliers.

Array Multiplier: A direct approach to multiplication involves the combinational generation of all bit products and their summation with an array of full adders. The block diagram for an unsigned 8-by-8 array multiplier is shown in Figure 8. It has an 9-by-8 array of cells (15 G cells, 7 HA cells and 42 FA cells) in the upper portion of the diagram and a 7-bit ripple carry adder (constructed with 1 HA cell and 6 FA cells) along the bottom row. The upper portion uses an 8-by-8 array of AND gates to form the bit products (some of the AND gates are in the G cells and the remainder are contained within the FA and HA blocks). The output of the upper portion of the circuit is a pair of 7-bit words whose sum is the most significant 8-bits of the product $p_8 - p_{15}$. The complete product is available after 14 adder delays.

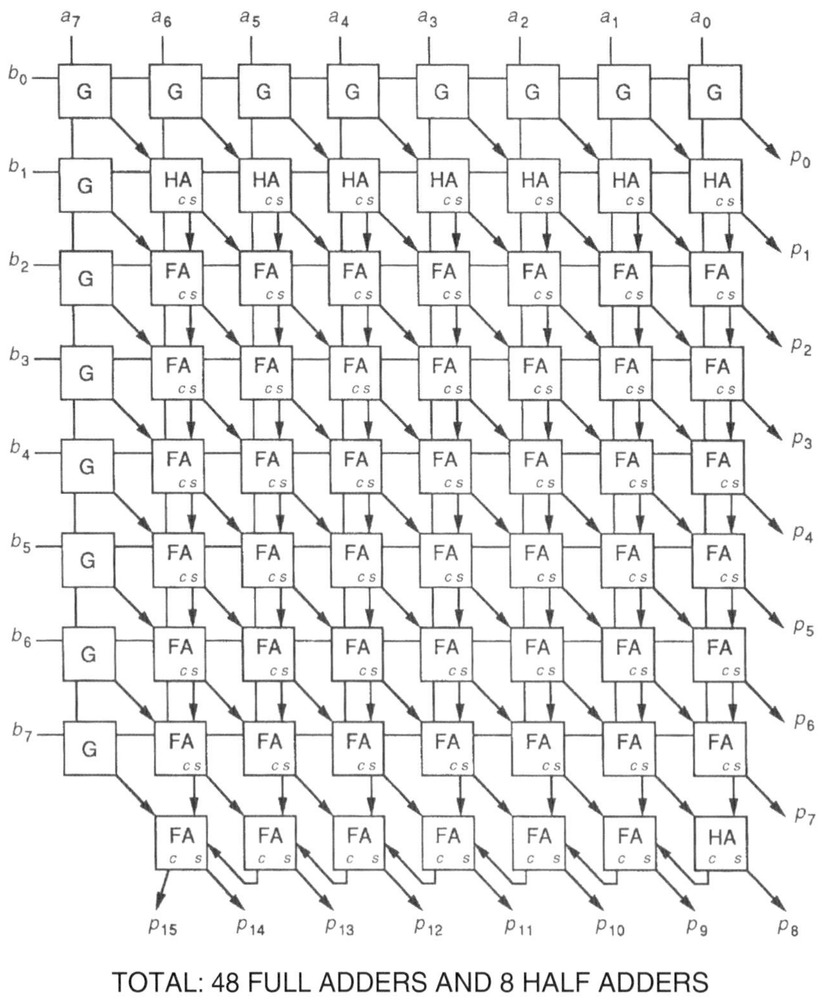

FIGURE 8 Unsigned 8-bit by 8-bit array multiplier.

Modification of the array multiplier to multiply two's complement numbers requires using NAND gates instead of AND gates to form the bit products in the most significant row (the b_7 row) and column (the a_7 column) (an AND gate is used for the a_7b_7 bit product) and also adding two correction terms (Parhami, 2010, p. 223).

The delay of the array multiplier is evaluated by following the pathways from the inputs to the outputs. The longest path starts at the upper left corner, progresses to the lower right corner, and then across the bottom to the lower left corner. If it is assumed that the delay from any adder input (for either half or full adders) to any adder output is k gate delays then the total delay of an n-bit by n-bit array multiplier is:

$$DELAY_{ARRAY\ MPY} = k(2n - 2) + 1$$

The complexity of the array multiplier is n^2 AND gates, n half adders, and $n^2 - 2n$ full adders. If a half adder comprises four gates and a full adder comprises nine gates, the total complexity of an n-bit by n-bit array multiplier is

$$GATES_{ARRAY\ MPY} = 10n^2 - 14n$$

Array multipliers are easily laid out in a cellular fashion, making them attractive for VLSI implementation, where minimizing the design effort may be more important than maximizing the speed.

Fast Multipliers: A method for fast multiplication was developed by Wallace (1964) and refined by Dadda (1965). With this method, a three-step process is used to multiply two numbers: (1) the bit products are formed; (2) the bit product matrix is reduced to a two-row matrix in which the sum of the rows equals the sum of the bit products; and (3) the two numbers are summed using a fast adder to produce the product. Although this may seem to be a complex process, it yields multipliers with delay proportional to the logarithm of the operand word size, which is faster than array multipliers that have delay that is proportional to the word size.

The second step in the fast multiplication process is shown for an unsigned 8-by-8 Dadda multiplier in Figure 9. An input 8-by-8 matrix of dots (each dot represents a bit product) is shown as Matrix 0. Columns having more than six dots (or that will grow to more than six dots due to carries) are reduced by the use of half-adders (each half-adder takes in two dots from a given column and outputs one dot in the same column and one dot in the next more significant column) and full adders (each full adder takes in three dots from a given column and outputs one in the same column and one in the next more significant column) so that no column in Matrix 1 will have more than six dots. Half-adders are shown by two dots connected with a crossed line in the succeeding matrix and full adders are shown by two dots connected with a line in the succeeding matrix. In each case, the rightmost dot of the pair connected by a line is in the column from which the inputs were taken for the adder. In the succeeding steps, reduction to Matrix 2 with no more than four dots per column, Matrix 3 with no more than three dots per column, and finally Matrix 4 with no more than two dots per column is performed. The height of the matrices is determined by working back from the final (two-row) matrix and limiting the height of each matrix to the largest integer that is no more than 1.5 times the height of its successor (i.e., 2, 3, 4, 6, 9, 13, 19, 28, 42, etc.). Each matrix is produced from its predecessor in one adder delay. Because the number of matrices is logarithmically related to the number of bits in the words to be multiplied, the delay of the matrix reduction process is proportional to log n. Because the adder that reduces the final two-row matrix can be implemented as a carry lookahead adder (which also has logarithmic delay), the total delay for this multiplier is proportional to the logarithm of the word size.

The delay of a Dadda multiplier is evaluated by following the pathways from the inputs to the outputs. The longest path starts at the center column of bit products (which require 1 gate delay to be formed), progresses through the successive reduction matrices (which requires approximately $\text{Log}_{1.44}(n) - 2$ full

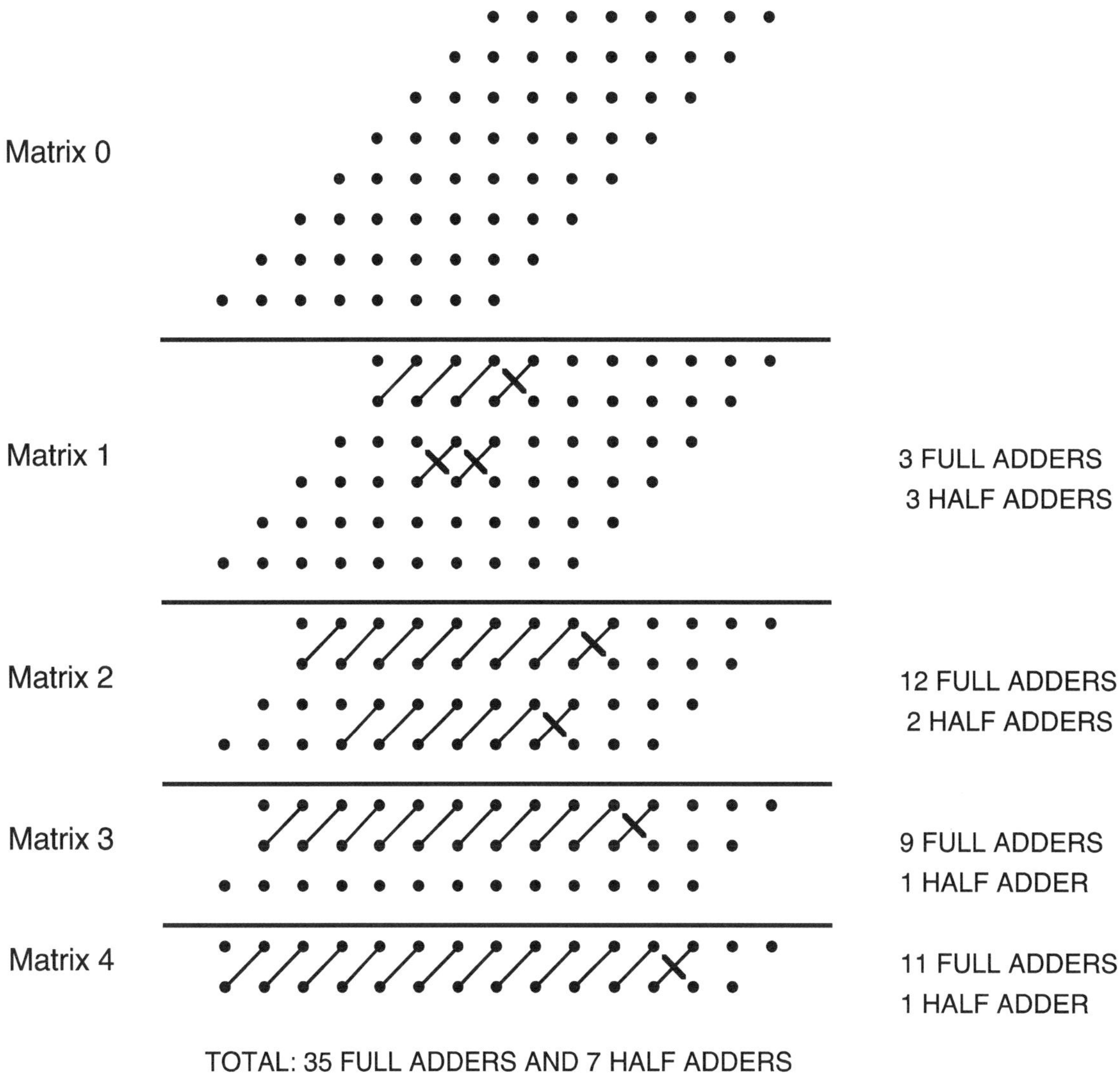

FIGURE 9 Unsigned 8-bit by 8-bit Dadda multiplier.

adder delays) and finally through the $2n-2$-bit carry propagate adder. If the delay from any adder input (for either half or full adders) to any adder output is k gate delays and if the carry propagate adder is realized with a carry lookahead adder implemented with 4-bit lookahead logic blocks (with delay given by Equation (9)), the total delay (in gate delays) of an n-bit by n-bit Dadda multiplier is:

$$DELAY_{DADDAMPY} = 1 + k(\log_{1.44}(n) - 2) + 2 + 4\lceil \log_r(2n-2)\rceil$$

The complexity of a Dadda multiplier is determined by evaluating the complexity of its parts. There are n^2 gates ($2n-2$ are NAND gates, the rest are AND gates) to form the bit product matrix, $(n-2)^2$ full adders, $n-1$ half adders, and a $2n-2$-bit carry propagate adder for the addition of the final two row matrix. If the carry propagate adder is realized with a carry lookahead adder (implemented with 4-bit lookahead logic blocks) and if the complexity of a full adder is 9 gates and the complexity of a half adder is 4 gates, then the total complexity is:

$$GATES_{DADDA\ MPY} = 10n^2 - 62/3n - 26$$

The Wallace multiplier is very similar to the Dadda multiplier, except that it does more reduction in the first stages of the reduction process, it uses more half adders and it uses a slightly smaller carry

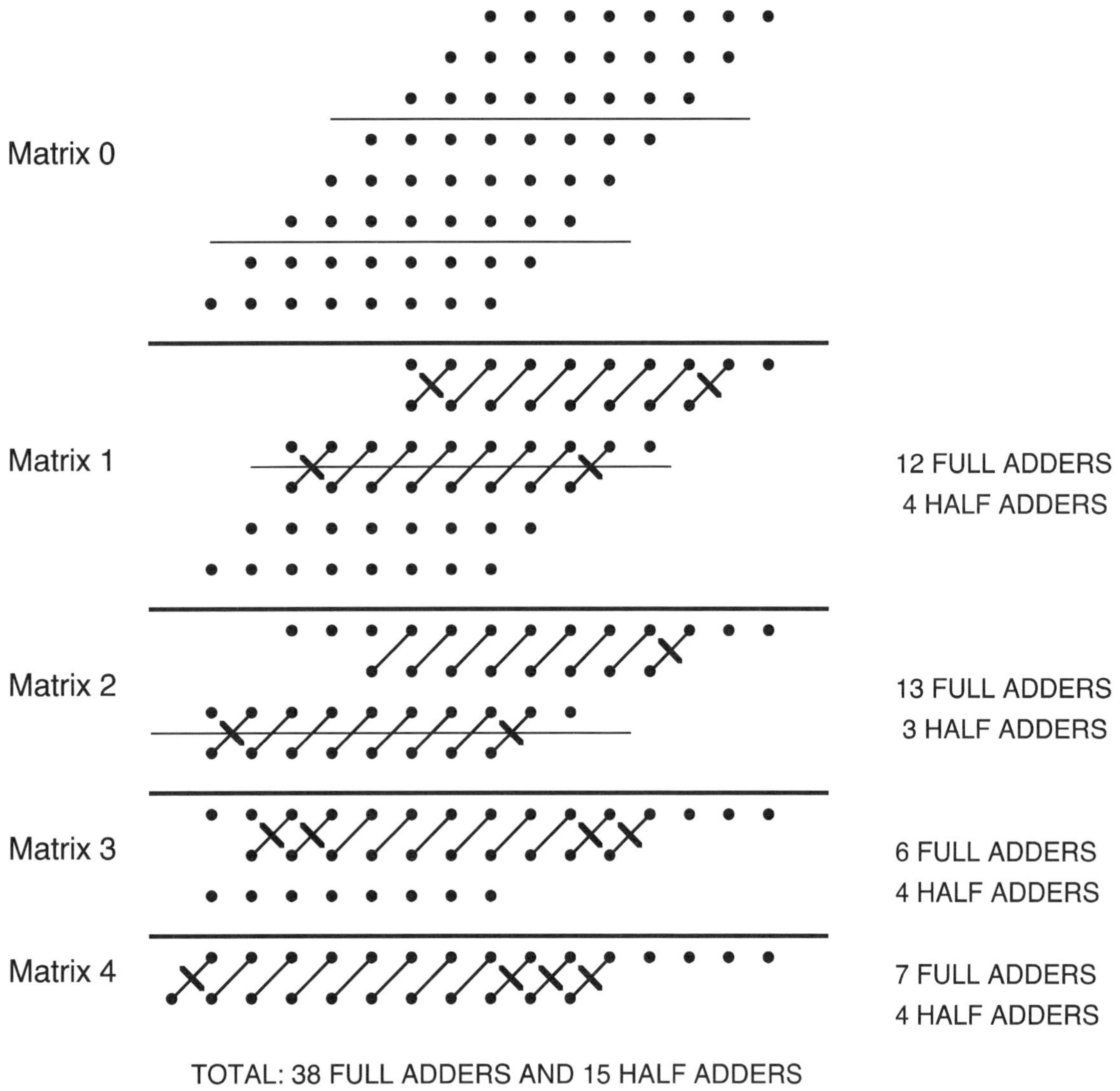

FIGURE 10 Unsigned 8-bit by 8-bit Wallace multiplier.

propagating adder. A dot diagram for an unsigned 8-bit by 8-bit Wallace multiplier is shown on Figure 10. This reduction (which like the Dadda reduction requires 4 full adder delays) is followed by an 11-bit carry propagating adder. The total complexity of the Wallace multiplier is slightly greater than the total complexity of the Dadda multiplier. In most cases the Wallace and Dadda multipliers have the same delay.

Fixed-Point Division

There are two types of division algorithms in common use: digit recurrence and convergence methods. The digit recurrence approach computes the quotient on a digit-by-digit basis, hence they have a delay proportional to the precision of the quotient. In contrast the convergence methods compute an approximation that converges to the value of the quotient. For the common algorithms, the convergence is quadratic, meaning that the number of accurate bits approximately doubles on each iteration.

The digit recurrence methods that use a sequence of shift, add or subtract, and compare operations are relatively simple to implement. On the other hand, the convergence methods use multiplication on each cycle. This means higher hardware complexity, but if a fast multiplier is available, potentially higher speed.

Digit Recurrent Division: The digit recurrent algorithms Robertson (1958) based on selecting digits of the quotient Q (where $Q = N/D$) to satisfy the following equation:

$$P_{k+1} = rP_k - q_{n-k-1}D \quad \text{for } k = 1, 2, \ldots, n-1$$

where P_k is the partial remainder after the selection of the k-th quotient digit, $P_0 = N$ (subject to the constraint $|P_0| < |D|$), r is the radix, q_{n-k-1} is the k-th quotient digit to the right of the binary point, and D is the divisor. In this subsection, it is assumed that both N and D are positive, see Ercegovac and Lang (1994) for details on handling the general case.

Binary SRT Divider: The binary SRT division process (also known as radix-2 SRT division) selects the quotient from three candidate quotient digits $\{\pm 1, 0\}$. The divisor is restricted to $0.5 \leq D =< 1$. A flowchart of the basic binary SRT scheme is shown in Figure 11. Block 1 initializes the algorithm. In steps 3 and 5, 2 P_k is used to select the quotient digit. In steps 4 and 6, $\mathrm{P_{k+1}} = 2\ \mathrm{P_k} - \mathrm{q\ D}$. Step 8 tests whether all bits of the quotient have been formed and goes to step 2 if more need to be computed. Each pass through steps 2–8 forms one digit of the quotient. The result upon exiting from step 8 is a collection of n signed binary digits.

Step 9 converts the n digit signed digit number into an n-bit two's complement number by subtracting, N, which has a 1 for each bit position where $\mathrm{q}_i = -1$ and 0 elsewhere from, P, which has a 1 for each bit position where $\mathrm{q}_i = 1$ and 0 elsewhere. For example:

$$\begin{aligned}
\mathrm{Q} &= 0\ .\ 1\ 1 - 1\ 0\ 1\ = 21/32 \\
\mathrm{P} &= 0\ .\ 1\ 1\ 0\ 0\ 1 \\
\mathrm{N} &= 0\ .\ 0\ 0\ 1\ 0\ 0 \\
\mathrm{Q} &= 0\ .\ 1\ 1\ 0\ 0\ 1 - 0\ .\ 0\ 0\ 1\ 0\ 0 \\
\mathrm{Q} &= 0\ .\ 1\ 1\ 0\ 0\ 1\ + 1\ .\ 1\ 1\ 1\ 0\ 0 \\
\mathrm{Q} &= 0\ .\ 1\ 0\ 1\ 0\ 1\ = 21/32
\end{aligned}$$

The selection of the quotient digit can be visualized with a P-D Plot such as the one shown in Figure 12. The plot shows the divisor along the x-axis and the shifted partial remainder (in this case 2 $\mathrm{P_k}$) along the y-axis. In the area where $0.5 \leq D < 1$, values of the quotient digit are shown as a function of the value of the shifted partial remainder. In this case the relations are especially simple. The digit selection and resulting partial remainder are given for the k-th iteration by the following relations:

$$\begin{aligned}
&\text{If } P_k \geq .5, && q_{n-k-1} = 1 \quad \text{and} \quad P_{k+1} = 2P_k - D \\
&\text{If } -.5 < P_k < .5, && q_{n-k-1} = 0 \quad \text{and} \quad P_{k+1} = 2P_k \\
&\text{If } P_k \leq -.5, && q_{n-k-1} = -1 \quad \text{and} \quad P_{k+1} = 2P_k + D
\end{aligned}$$

Computing an n-bit quotient will involve selecting n quotient digits and up to $n + 1$ additions.

Radix-4 SRT Divider: The higher radix SRT division process is similar to the binary SRT algorithm. Radix 4 is the most common higher radix SRT division algorithm with either a minimally redundant digit set of $\{\pm 2, \pm 1, 0\}$ or the maximally redundant digit set of $\{\pm 3, \pm 2, \pm 1, 0\}$. The operation of the algorithm is similar to the binary SRT algorithm shown on Figure 11, except that in steps 3 and 5, 4 P_k and D are used to determine the quotient digit. A P-D Plot is shown on Figure 13 for the maximum redundancy version of radix-4 SRT division. There are seven possible values for the quotient digit at each stage. The

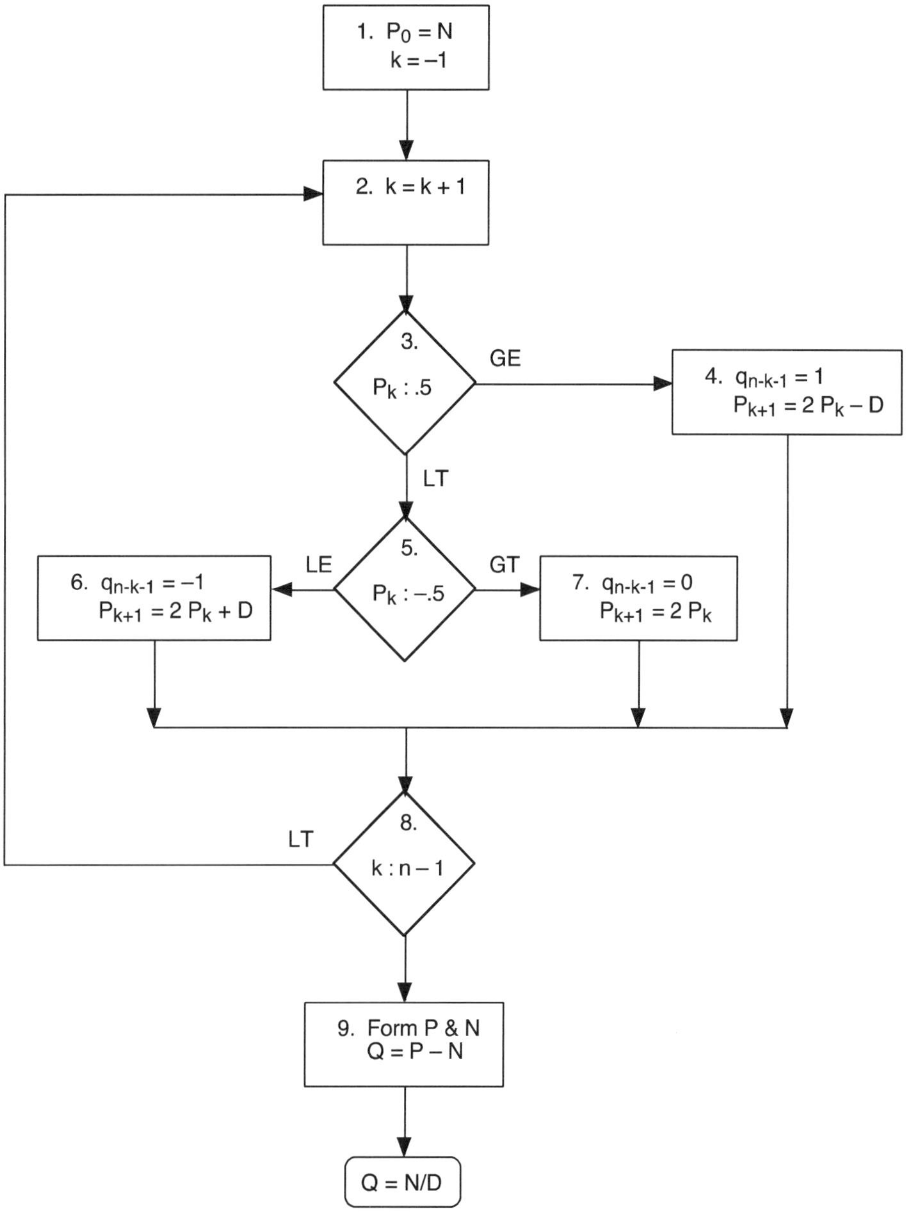

FIGURE 11 Flowchart of binary SRT division.

test for completion in step 8 becomes k: $\frac{n}{2} - 1$. Also the conversion to two's complement in step 9 is modified slightly since each quotient digit occupies two bits of the P and N numbers that are used to form the two's complement number.

Goldschmidt Divider: The second category of division techniques uses a multiplication based iteration to compute a quadratically convergent approximation to the quotient. In systems that include a fast multiplier, this process may be faster than the digit recurrent methods. The Goldschmidt divider Anderson *et al.* (1967) treats the division problem as a fraction whose numerator and denominator are multiplied by a succession of numbers so that the final denominator is approximately 1 and the final numerator is approximately equal to the quotient.

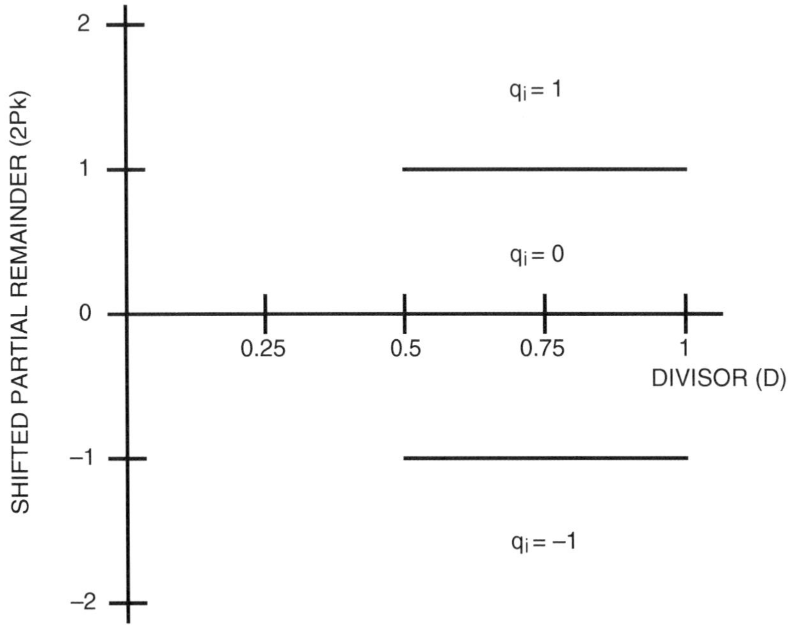

FIGURE 12 P-D Plot for binary SRT division.

The steps to compute a quotient using the Goldschmidt method are:

1. Scale the numerator, N_0, and the denominator, D_0, so that the value of the denominator is "close" to 1 (e.g. $3/4 < D_0 < 1\frac{1}{4}$). Use a table to find a factor, F_1, that will make $D_0^* F_1 \approx 1$. Set $i = 0$
2. Multiply N_i and D_i by F_{i+1} forming N_{i+1} and D_{i+1}, respectively.
3. Increment i, Calculate $F_{i+1} = 2 - D_i$.
   Repeat steps 2 and 3 until $D_n = 1$ within the desired precision.

This process is illustrated by the decimal example shown in Figure 14.

The value of N_k quadratically approaches the value of the quotient. If F_1 in step 1 results in $D_1 \approx 1$ within a maximum error, the number of iterations to achieve a specific accuracy of the quotient can be predetermined. In the example, D_1 is within 2.5% of 1, D_2 is within 0.06%, etc. Thus each iteration (which requires 1 subtract and 2 multiplies) doubles the number of accurate bits to which N_k approximates the quotient. If the initial accuracy is known, the final multiplication that forms D_k is not needed.

4 Floating-Point Arithmetic Algorithms

Recent advances in VLSI have increased the feasibility of hardware implementations of floating-point arithmetic units. The main advantage of floating-point arithmetic is that its wide dynamic range virtually eliminates overflow for most applications.

Floating-Point Format

A floating-point number, A, consists of a significand (or mantissa), S_a, and an exponent, E_a. The value of a number, A, is given by the equation:

$$A = S_a r^{Ea}$$

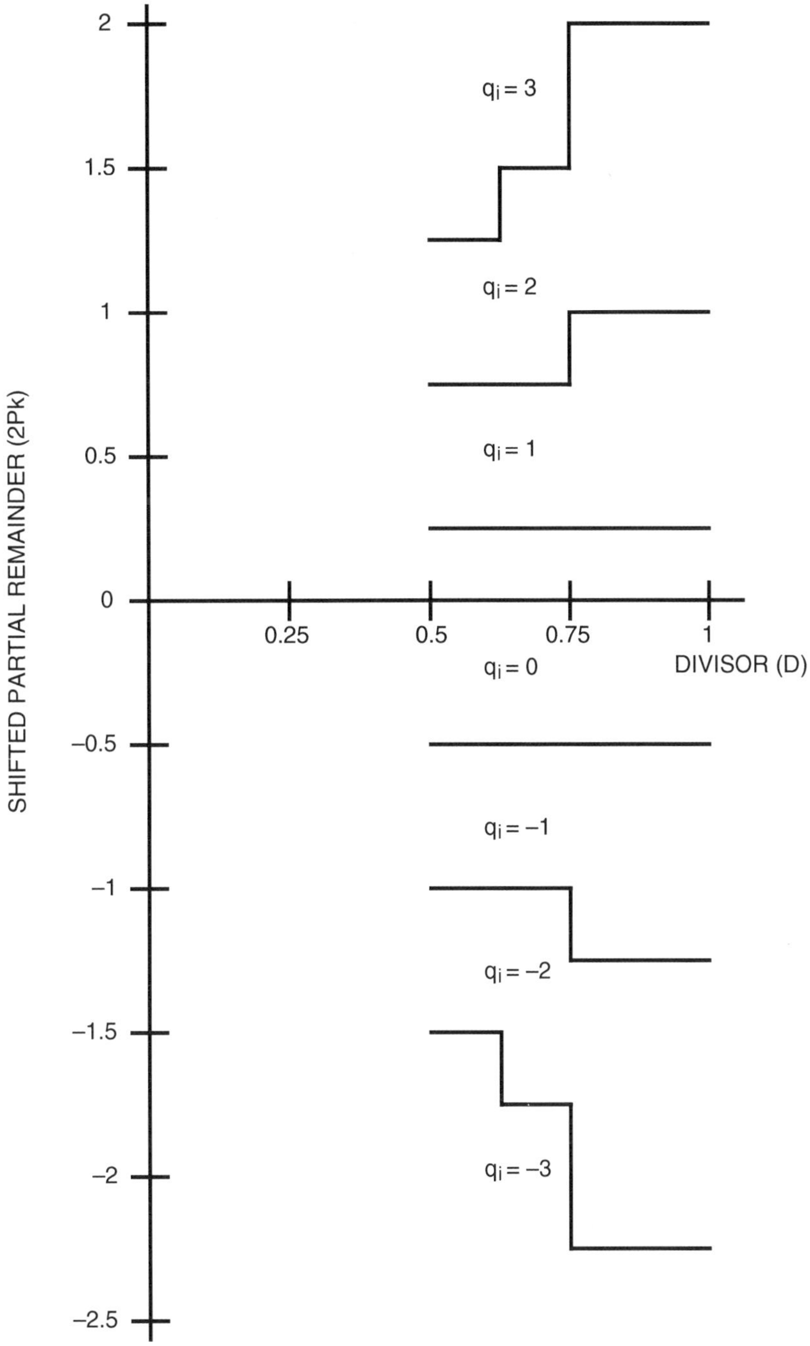

FIGURE 13 P-D Plot for radix-4 maximally redundant SRT division.

where r is the radix (or base) of the number system. The significand is generally normalized by requiring that the most significant digit be nonzero. Use of the binary radix (i.e., r = 2) gives maximum accuracy but may require more frequent normalization than higher radices.

The IEEE Standard 754 single precision (32-bit) floating-point format, which is widely implemented, has an 8-bit biased integer exponent that ranges between 0 and 255 IEEE Standard for Floating-Point Arithmetic (2008). The exponent is expressed in excess 127 code so that its effective value is determined by subtracting 127 from the stored value. Thus, the range of effective values of the exponent is −127 to

$N_0 = N = 0.625$	
$D_0 = D = 0.75$	
$F_1 = 1.3$	1 Table Look Up
$N_1 = N_0 \cdot F_1 = 0.625 \cdot 1.3 = 0.8125$	1 Multiply
$D_1 = D_0 \cdot F_1 = 0.75 \cdot 1.3 = 0.975$	1 Multiply
$F_2 = 2 - D_1 = 2 - 0.975 = 1.025$	1 Subtract
$N_2 = N_1 \cdot F_2 = 0.8125 \cdot 1.025 = 0.8328125$	1 Multiply
$D_2 = D_1 \cdot F_2 = 0.975 \cdot 1.025 = 0.999375$	1 Multiply
$F_3 = 2 - D_2 = 2 - 0.999375 = 1.000625$	1 Subtract
$N_3 = N_2 \cdot F_3 = 0.8328125 \cdot 1.000625 = 0.83333300781$	1 Multiply
$D_3 = D_2 \cdot F_3 = 0.999375 \cdot 1.000625 = 0.99999960937$	1 Multiply
$Q = N_3 = 0.83333300781$	

FIGURE 14 Example of Goldschmidt division.

128, corresponding to stored values of 0 to 255, respectively. A stored exponent value of 0 ($E_{\min}$) serves as a flag indicating that the value of the number is 0 if the significand is 0 and for denormalized numbers if the significand is non-zero. A stored exponent value of 255 ($E_{\max}$) serves as a flag indicating that the value of the number is infinity if the significand is 0 or "not a number" if the significand is non-zero. The significand is a 25-bit sign magnitude mixed number (the binary point is to the right of the most significant bit). The leading bit of the significand is always a 1 (except for denormalized numbers). As a result when numbers are stored, the leading bit can be omitted, giving one extra bit of precision. More details on floating-point formats and on the various considerations that arise in the implementation of floating-point arithmetic units are given in (Goldberg, 1990, Parhami, 2010, Gosling, 1980, Flynn and Oberman, 2001, Koren, 2002).

Floating-Point Addition

A flow chart for the functions required for floating-point addition is shown in Figure 15. For this flowchart, the operands are assumed to be IEEE Std. 754 single precision numbers that have been "unpacked" and normalized with significand magnitudes in the range [1, 2) and exponents in the range [−126, 127].

On the flow chart, the operands are (E_a, S_a) and (E_b, S_b), the result is (E_s, S_s), and the radix is 2. In step 1 the operand exponents are compared; if they are unequal, the significand of the number with the smaller exponent is shifted right in step 3 or 4 by the difference in the exponent values to properly align the significands. For example, to add the decimal operands 0.867×10^5 and 0.512×10^4, the latter would be shifted right by one digit and 0.867 added to 0.0512 to give a sum of 0.9182×10^5. The addition of the significands is performed in step 5. Step 6 tests for overflow and step 7 corrects, if necessary, by shifting the significand one position to the right and incrementing the exponent. Step 9 tests for a zero significand. The loop of steps 10–11 scales unnormalized (but non-zero) significands upward to normalize the result. Step 12 tests for underflow.

It is important to recognize that the flowchart shows the basic functions that are required, but not necessarily the best way to perform those functions. For example, the loop in steps 10–11 is a really poor

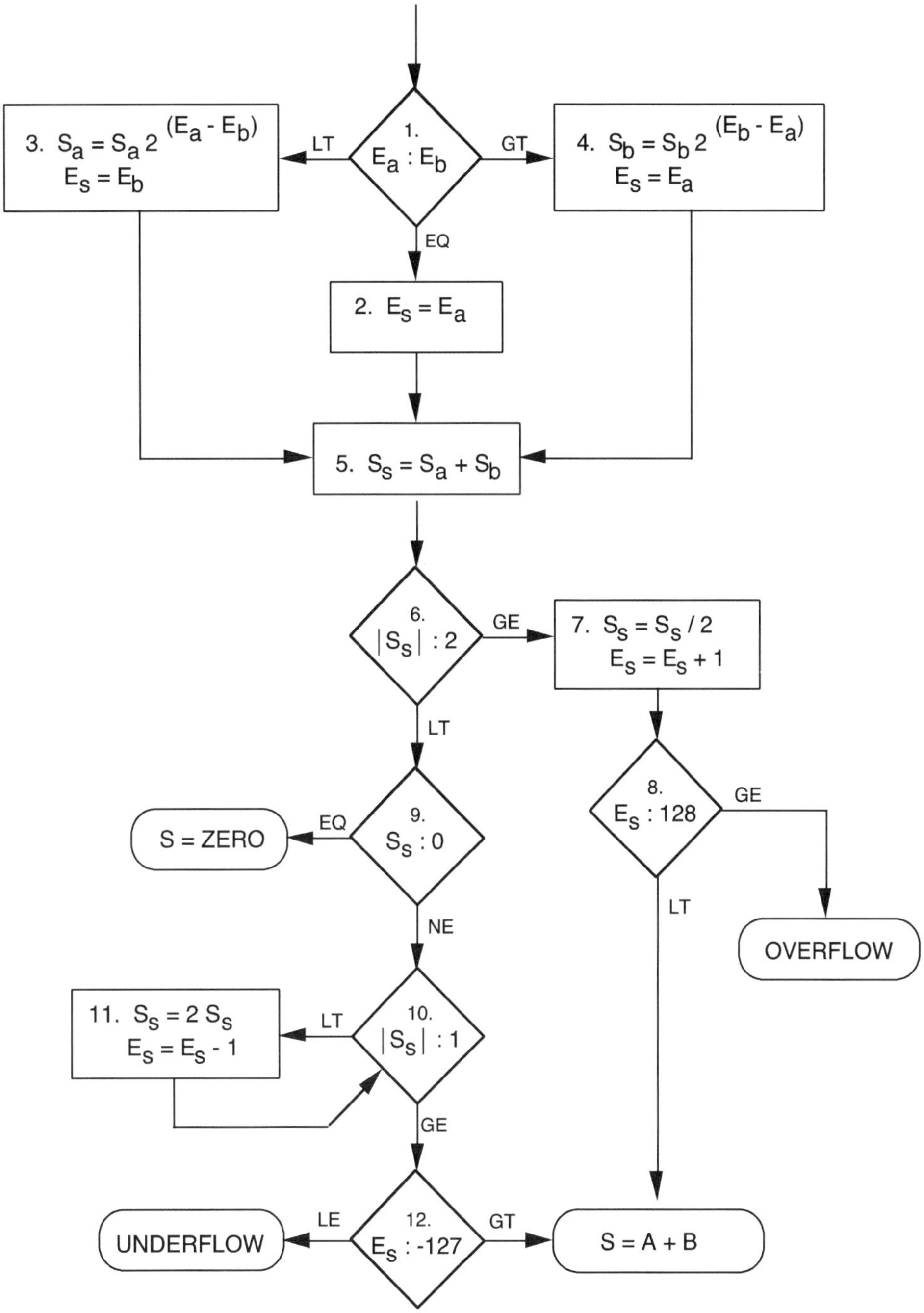

FIGURE 15 Floating-point addition.

way to normalize the significand of the sum. It is a serial loop which is executed a variable number of times depending on the data value. A much better way is to determine the number of leading zeros in the significand and then use a barrel shifter to do the shift in a single step in hardware.

Floating-point subtraction is implemented with a similar algorithm. Many refinements are possible to improve the speed of the addition and subtraction algorithms, but floating-point addition and subtraction will, in general, be much slower than fixed-point addition as a result of the need for operand alignment and result normalization and rounding.

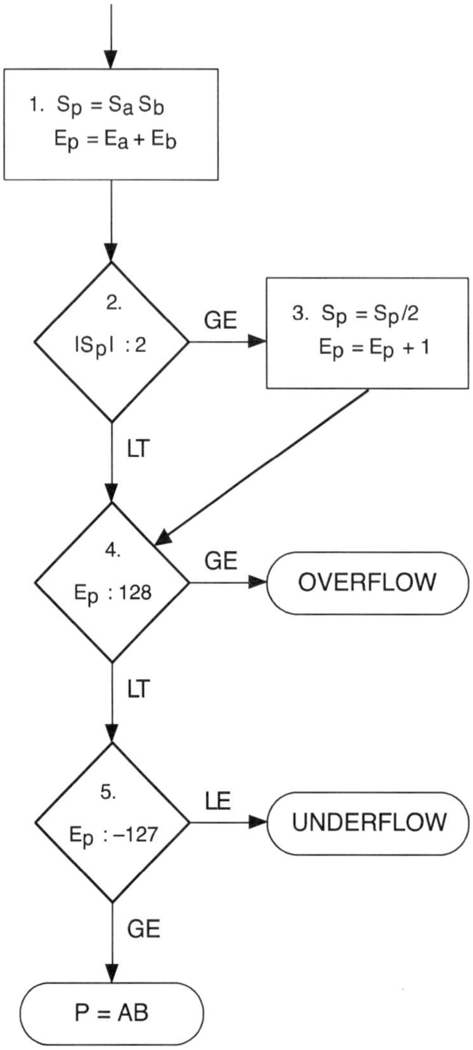

FIGURE 16 Floating-point multiplier.

Floating-Point Multiplication

The algorithm for floating-point multiplication is shown in the flowchart of Figure 16. In step 1, the product of the operand significands and the sum of the operand exponents are computed. Steps 2 and 3 normalize the significand if necessary. For radix-2 floating-point numbers, if the operands are normalized, at most a single shift is required to normalize the product. Step 4 tests the exponent for overflow. Finally, step 5 tests for underflow.

Floating-Point Division

The floating-point division algorithm is shown in the flowchart of Figure 17. The quotient of the significands and the difference of the exponents are computed in step 1. The quotient is normalized (if necessary) in steps 2 and 3 by shifting the quotient significand while the quotient exponent is adjusted appropriately. For radix-2, if the operands are normalized, only a single shift is required to normalize the quotient. The computed exponent is tested for underflow in step 4. Finally, the fifth step tests for overflow.

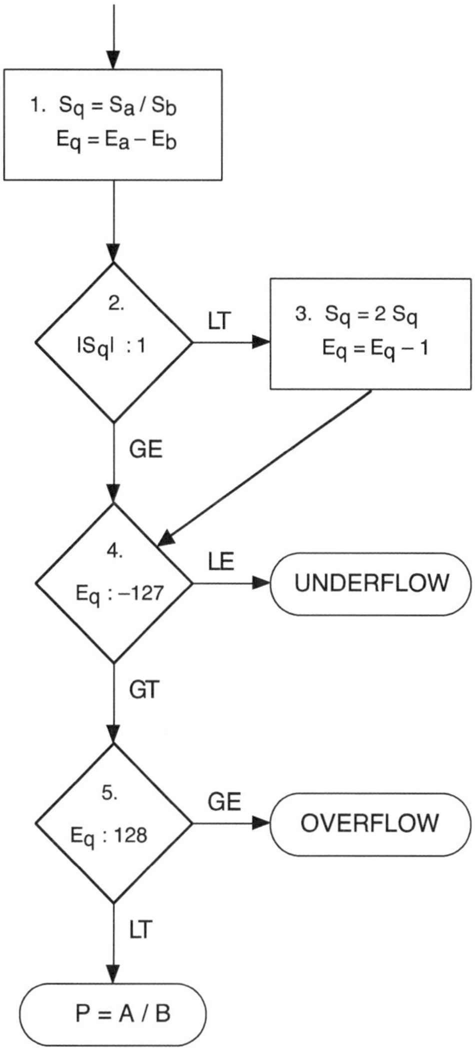

FIGURE 17 Floating-point divider.

Floating-Point Rounding

All floating-point algorithms may require rounding to produce a result in the correct format. A variety of alternative rounding schemes have been developed for specific applications. Round to nearest with ties to even, round to nearest with ties away from 0, round toward ∞, round toward $-\infty$, and round toward 0 are all available in implementations of the IEEE floating-point standard IEEE Standard for Floating-Point Arithmetic (2008).

5 Conclusions

This overview has presented an overview of the two's complement number system, algorithms for the basic integer arithmetic operations of addition, subtraction, multiplication, and division, and a very brief discussion of floating-point operations. When implementing arithmetic units there is often an opportunity to optimize the performance and the complexity to match the requirements of the specific application. In general, faster algorithms require more area or more complex control; it is often useful to use the fastest algorithm that will fit the available area.

Acknowledgment

This is a revision of chapter originally presented in Swartzlander, Jr. (1992).

References

S. F. Anderson, J. G. Earle, R. E Goldschmidt, and D. M Powers, "The IBM System/360 Model 91: Floating-Point Execution Unit," *IBM Journal of Research and Development*, 11, 1967, pp. 34–53.

O. J., Bedrij, "Carry Select Adders," *IEEE Transactions on Electronic Computers*, vol. EC-11, 1962, pp. 340–346.

P. K. Chen, and M. D. F. Schlag, "Analysis and Design of CMOS Manchester Adders with Variable Carry Skip," *IEEE Transactions on Computers*, vol. 39, 1990, pp. 983–992.

L. Dadda, "Some Schemes for Parallel Multipliers," *Alta Frequenza*, vol. 34, 1965, pp. 349–356.

M. D. Ercegovac and T. Lang, *Division and Square Root: Digit-Recurrence Algorithms and Their Implementations*, Boston, MA: Kluwer Academic Publishers, 1994.

M. J. Flynn and S. F. Oberman, *Advanced Computer Arithmetic Design*, New York: John Wiley & Sons, 2001.

D. Goldberg, "Computer Arithmetic" Appendix A in *Computer Architecture: A Quantitative Approach*, In D. A. Patterson and J. L. Hennessy, Editors, San Mateo, CA: Morgan Kauffmann, 1990.

J. B. Gosling, *Design of Arithmetic Units for Digital Computers*, New York: Macmillan, 1980.

IEEE Standard for Floating-Point Arithmetic, ANSI/IEEE Std.754-2008, New York: The Institute of Electrical and Electronic Engineers, Inc., August 2008.

T. Kilburn, D. B. G. Edwards, and D. Aspinall, "A Parallel Arithmetic Unit Using a Saturated Transistor Fast-Carry Circuit," *Proceedings of the Institution of Electrical Engineers*, vol. 107, Pt. B, 1960, pp. 573–584.

I. Koren, *Computer Arithmetic Algorithms*, Second Edition, Natick, MA: A.K. Peters, 2002.

M. Lehman and N. Burla, "Skip Technique for High Speed Carry Propagation in Binary Arithmetic Units," *IEEE Transactions on Electronic Computers*, vol. EC-10, 1961, pp. 691– 698.

O. L. MacSorley, "High-Speed Arithmetic in Binary Computers," *Proceedings of the IRE*, vol. 49, 1961, pp. 67–91.

B. Parhami, *Computer Arithmetic: Algorithms and Hardware Designs*, Second Edition, New York: Oxford University Press, 2010.

J. E. Robertson, "A New Class of Digital Division Methods," *IEEE Transactions on Electronic Computers*, vol. EC-7, 1958, pp. 218–222.

E. E. Swartzlander, Jr., *Computer Arithmetic*, Stroudsburg, PA: Dowden, Hutchinson & Ross, Inc., 1980.

E. E. Swartzlander, Jr., "Computer Arithmetic," Chapter 4 in C. H. Chen, Editor, *Computer Engineering Handbook*, NewYork: McGraw–Hill, 1992, pp. 4-1–4-20.

S. Turrini, Optimal Group Distribution in Carry-Skip Adders," *Proceedings of the 9th Symposium on Computer Arithmetic*, 1989, pp. 96–103.

C. S. Wallace, "A Suggestion for a Fast Multiplier," *IEEE Transactions on Electronic Computers*, vol. EC-13, 1964, pp. 14–17.

S. Waser, and M. J. Flynn, *Introduction to Arithmetic for Digital Systems Designers*, New York: Holt, Rinehart and Winston, 1982.

A. Weinberger and J. L. Smith, "A Logic for High-Speed Addition," *National Bureau of Standards, Circular 591*, Washington, DC: National Bureau of Standards, 1958, pp. 3–12.

Part I:
ERROR TOLERANT ARITHMETIC

Editor's Comments on Papers 1 Through 5

1. **Hamming**
   Error Detecting and Error Correcting Codes
2. **Garner**
   Generalized Parity Checking
3. **Rao**
   Error-Checking Logic for Arithmetic-Type Operations of a Processor
4. **Avizienis**
   Arithmetic Algorithms for Error-Coded Operands
5. **Robertson**
   Error Detection and Correction for Addition and Subtraction, Through Higher Radix Extensions of Hamming Codes

Interest in error detection and correction has been a cyclic phenomenon with peaks in the early 1950s as the first generation of computers was constructed using failure prone vacuum tubes, again in the late 1960s for the space program where computers needed to operate correctly even though individual components may fail in the harsh environment of space, and at present where error correction allows the development of circuits that would be infeasible otherwise. Additionally there are many requirements for computing where failures simply cannot be permitted (e.g., flight control systems for commercial airplanes, nuclear reactor control, etc.).

Currently, VLSI circuits with hundreds of thousands of logic gates can be realized on a single chip. Such high densities require very small devices which are sensitive to upset due to natural radiation. With such a large number of devices, error tolerant communication and arithmetic is of great importance. Wafer Scale Integration, which is in the research phase is another area that demands error tolerance (Jesshope and Moore, 1986, Saucier and Thilhe, 1986, Lea, 1988, Swartzlander and Brewer, 1989, Swartzlander, 1989, Tewksbury, 1989, Brewer and Little, 1990).

This part begins with Hamming's classic paper on error detecting and correcting codes. Hamming describes codes that can (1) detect single errors, (2) correct single errors, or (3) correct single errors and detect double errors. These systematic codes add parity bits computed on groups of bits in each data word. The Hamming code is useful in checking the storage and transmission of data (not in the checking of arithmetic operations). It is easily understood and encoders and decoders for it are easy to implement.

Paper 2 by Garner, extends the notion of parity, which may be viewed as the congruence Mod 2 (or the residue Mod 2) to general bases. The paper introduces generalized parity, derives pertinent properties, and then shows how diminished and augmented base parity may be used to check the correctness of numerical operations.

The next paper, by Rao, develops residue based error checking algorithms for add, complement, shift, and rotate operations. Rao describes the residue check process, provides error checking formulas for the various operations, and concludes with the design of a residue checked processor that has an overhead of under 400 gates for a 25 bit ALU.

In the fourth paper, Avizienis gives a detailed treatment of the nonseparate AN code. The algorithms and their application to the design of the Jet Propulsion Laboratory STAR (Self-Testing And Repairing) Computer are presented. The paper includes a comparison between AN and inverse residue coding for implementation of the STAR Computer.

Finally, Robertson in Paper 5 shows how the concept of parity checking as embodied in the Hamming code can be extended to arithmetic, specifically addition modulo 2r. The paper discusses modulo k addition, extension of the Hamming code, and optimal representation of digits for the radix 4 adder.

References

Joe Brewer and Michael J. Little, eds., *1990 Proceedings of the International Conference on Wafer Scale Integration*, Los Alamitos, CA: IEEE Computer Society Press, 1990.

Chris Jesshope and Will Moore, eds., *Wafer Scale Integration*, Boston: Adam Hilger, 1986.

R. M. Lea, ed., *Wafer Scale Integration, II*, New York: North-Holland, 1988.

Gabrièle Saucier and Jacques Thilhe, eds., *Wafer Scale Integration*, New York: North-Holland, 1986.

Earl Swartzlander and Joe Brewer, eds., *Proceedings of the International Conference on Wafer Scale Integration*, Washington: IEEE Computer Society Press, 1989.

Earl E. Swartzlander, Jr., ed., *Wafer Scale Integration*, Boston: Kluwer Academic Publishers, 1989.

Stuart K. Tewksbury, *Wafer-Level Integrated Systems: Implementation Issues*, Boston: Kluwer Academic Publishers, 1989.

The Bell System Technical Journal

Vol. XXIX *April, 1950* *No. 2*

Error Detecting and Error Correcting Codes

By R. W. HAMMING

1. Introduction

THE author was led to the study given in this paper from a consideration of large scale computing machines in which a large number of operations must be performed without a single error in the end result. This problem of "doing things right" on a large scale is not essentially new; in a telephone central office, for example, a very large number of operations are performed while the errors leading to wrong numbers are kept well under control, though they have not been completely eliminated. This has been achieved, in part, through the use of self-checking circuits. The occasional failure that escapes routine checking is still detected by the customer and will, if it persists, result in customer complaint, while if it is transient it will produce only occasional wrong numbers. At the same time the rest of the central office functions satisfactorily. In a digital computer, on the other hand, a single failure usually means the complete failure, in the sense that if it is detected no more computing can be done until the failure is located and corrected, while if it escapes detection then it invalidates all subsequent operations of the machine. Put in other words, in a telephone central office there are a number of parallel paths which are more or less independent of each other; in a digital machine there is usually a single long path which passes through the same piece of equipment many, many times before the answer is obtained.

In transmitting information from one place to another digital machines use codes which are simply sets of symbols to which meanings or values are attached. Examples of codes which were designed to detect isolated errors are numerous; among them are the highly developed 2 out of 5 codes used extensively in common control switching systems and in the Bell Relay

Computers,[1] the 3 out of 7 code used for radio telegraphy,[2] and the word count sent at the end of telegrams.

In some situations self checking is not enough. For example, in the Model 5 Relay Computers built by Bell Telephone Laboratories for the Aberdeen Proving Grounds,[1] observations in the early period indicated about two or three relay failures per day in the 8900 relays of the two computers, representing about one failure per two to three million relay operations. The self-checking feature meant that these failures did not introduce undetected errors. Since the machines were run on an unattended basis over nights and week-ends, however, the errors meant that frequently the computations came to a halt although often the machines took up new problems. The present trend is toward electronic speeds in digital computers where the basic elements are somewhat more reliable per operation than relays. However, the incidence of isolated failures, even when detected, may seriously interfere with the normal use of such machines. Thus it appears desirable to examine the next step beyond error detection, namely error correction.

We shall assume that the transmitting equipment handles information in the binary form of a sequence of 0's and 1's. This assumption is made both for mathematical convenience and because the binary system is the natural form for representing the open and closed relays, flip-flop circuits, dots and dashes, and perforated tapes that are used in many forms of communication. Thus each code symbol will be represented by a sequence of 0's and 1's.

The codes used in this paper are called *systematic* codes. Systematic codes may be defined[3] as codes in which each code symbol has exactly n binary digits, where m digits are associated with the information while the other $k = n - m$ digits are used for error detection and correction. This produces a *redundancy* R defined as the ratio of the number of binary digits used to the minimum number necessary to convey the same information, that is,

$$R = n/m.$$

This serves to measure the efficiency of the code as far as the transmission of information is concerned, and is the only aspect of the problem discussed in any detail here. The redundancy may be said to lower the effective channel capacity for sending information.

The need for error correction having assumed importance only recently, very little is known about the economics of the matter. It is clear that in

[1] Franz Alt, "A Bell Telephone Laboratories' Computing Machine"—I, II. Mathematical Tables and Other Aids to Computation, Vol. 3, pp. 1–13 and 60–84, Jan. and Apr. 1948.

[2] S. Sparks, and R. G. Kreer, "Tape Relay System for Radio Telegraph Operation," *R.C.A. Review*, Vol. 8, pp. 393–426, (especially p. 417), 1947.

[3] In Section 7 this is shown to be equivalent to a much weaker appearing definition.

using such codes there will be extra equipment for encoding and correcting errors as well as the lowered effective channel capacity referred to above. Because of these considerations applications of these codes may be expected to occur first only under extreme conditions. Some typical situations seem to be:

a. unattended operation over long periods of time with the minimum of standby equipment.

b. extremely large and tightly interrelated systems where a single failure incapacitates the entire installation.

c. signaling in the presence of noise where it is either impossible or uneconomical to reduce the effect of the noise on the signal.

These situations are occurring more and more often. The first two are particularly true of large scale digital computing machines, while the third occurs, among other places, in "jamming" situations.

The principles for designing error detecting and correcting codes in the cases most likely to be applied first are given in this paper. Circuits for implementing these principles may be designed by the application of well-known techniques, but the problem is not discussed here. Part I of the paper shows how to construct special minimum redundancy codes in the following cases:

a. single error detecting codes

b. single error correcting codes

c. single error correcting plus double error detecting codes.

Part II discusses the general theory of such codes and proves that under the assumptions made the codes of Part I are the "best" possible.

PART I

SPECIAL CODES

2. Single Error Detecting Codes

We may construct a single error detecting code having n binary digits in the following manner: In the first $n-1$ positions we put $n-1$ digits of information. In the n-th position we place either 0 or 1, so that the entire n positions have an even number of 1's. This is clearly a single error detecting code since any single error in transmission would leave an odd number of 1's in a code symbol.

The redundancy of these codes is, since $m = n - 1$,

$$R = \frac{n}{n-1} = 1 + \frac{1}{n-1}.$$

It might appear that to gain a low redundancy we should let n become very large. However, by increasing n, the probability of at least one error in a symbol increases; and the risk of a double error, which would pass undetected, also increases. For example, if $p \ll 1$ is the probability of any error, then for n so large as $1/p$, the probability of a correct symbol is approximately $1/e = 0.3679\ldots$, while a double error has probability $1/2e = 0.1839\ldots$.

The type of check used above to determine whether or not the symbol has any single error will be used throughout the paper and will be called a *parity check*. The above was an *even* parity check; had we used an odd number of 1's to determine the setting of the check position it would have been an odd parity check. Furthermore, a parity check need not always involve all the positions of the symbol but may be a check over selected positions only.

3. Single Error Correcting Codes

To construct a single error correcting code we first assign m of the n available positions as information positions. We shall regard the m as fixed, but the specific positions are left to a later determination. We next assign the k remaining positions as check positions. The values in these k positions are to be determined in the encoding process by even parity checks over selected information positions.

Let us imagine for the moment that we have received a code symbol, with or without an error. Let us apply the k parity checks, in order, and for each time the parity check assigns the value observed in its check position we write a 0, while for each time the assigned and observed values disagree we write a 1. When written from right to left in a line this sequence of k 0's and 1's (to be distinguished from the values assigned by the parity checks) may be regarded as a binary number and will be called the *checking number*. We shall require that this checking number give the position of any single error, with the zero value meaning no error in the symbol. Thus the check number must describe $m + k + 1$ different things, so that

$$2^k \geq m + k + 1$$

is a condition on k. Writing $n = m + k$ we find

$$2^m \leq \frac{2^n}{n+1}.$$

Using this inequality we may calculate Table I, which gives the maximum m for a given n, or, what is the same thing, the minimum n for a given m.

We now determine the positions over which each of the various parity checks is to be applied. The checking number is obtained digit by digit, from right to left, by applying the parity checks in order and writing down the corresponding 0 or 1 as the case may be. Since the checking number is

TABLE I

n	m	Corresponding k
1	0	1
2	0	2
3	1	2
4	1	3
5	2	3
6	3	3
7	4	3
8	4	4
9	5	4
10	6	4
11	7	4
12	8	4
13	9	4
14	10	4
15	11	4
16	11	5
	Etc.	

to give the position of any error in a code symbol, any position which has a 1 on the right of its binary representation must cause the first check to fail. Examining the binary form of the various integers we find

$$\begin{aligned} 1 &= 1 \\ 3 &= 11 \\ 5 &= 101 \\ 7 &= 111 \\ 9 &= 1001 \\ &\text{Etc.} \end{aligned}$$

have a 1 on the extreme right. Thus the first parity check must use positions

$$1, 3, 5, 7, 9, \cdots .$$

In an exactly similar fashion we find that the second parity check must use those positions which have 1's for the second digit from the right of their binary representation,

$$\begin{aligned} 2 &= 10 \\ 3 &= 11 \\ 6 &= 110 \\ 7 &= 111 \\ 10 &= 1010 \\ 11 &= 1011 \\ &\text{Etc.,} \end{aligned}$$

the third parity check

$$\begin{aligned} 4 &= 100 \\ 5 &= 101 \\ 6 &= 110 \\ 7 &= 111 \\ 12 &= 1100 \\ 13 &= 1101 \\ 14 &= 1110 \\ 15 &= 1111 \\ 20 &= 10100 \end{aligned}$$

Etc.

It remains to decide for each parity check which positions are to contain information and which the check. The choice of the positions 1, 2, 4, 8, $\cdots$ for check positions, as given in the following table, has the advantage of making the setting of the check positions independent of each other. All other positions are information positions. Thus we obtain Table II.

TABLE II

Check Number	Check Positions	Positions Checked
1	1	1, 3, 5, 7, 9, 11, 13, 15, 17,$\cdots$
2	2	2, 3, 6, 7, 10, 11, 14, 15, 18,$\cdots$
3	4	4, 5, 6, 7, 12, 13, 14, 15, 20,$\cdots$
4	8	8, 9, 10, 11, 12, 13, 14, 15, 24,$\cdots$
.	.	.
.	.	.
.	.	.

As an illustration of the above theory we apply it to the case of a seven-position code. From Table I we find for $n = 7$, $m = 4$ and $k = 3$. From Table II we find that the first parity check involves positions 1, 3, 5, 7 and is used to determine the value in the first position; the second parity check, positions 2, 3, 6, 7, and determines the value in the second position; and the third parity check, positions 4, 5, 6, 7, and determines the value in position four. This leaves positions 3, 5, 6, 7 as information positions. The results of writing down all possible binary numbers using positions 3, 5, 6, 7, and then calculating the values in the check positions 1, 2, 4, are shown in Table III.

Thus a seven-position single error correcting code admits of 16 code symbols. There are, of course, $2^7 - 16 = 112$ meaningless symbols. In some applications it may be desirable to drop the first symbol from the code to avoid the all zero combination as either a code symbol or a code symbol plus a single error, since this might be confused with no message. This would still leave 15 useful code symbols.

Table III

Position							Decimal Value of Symbol
1	2	3	4	5	6	7	
0	0	0	0	0	0	0	0
1	1	0	1	0	0	1	1
0	1	0	1	0	1	0	2
1	0	0	0	0	1	1	3
1	0	0	1	1	0	0	4
0	1	0	0	1	0	1	5
1	1	0	0	1	1	0	6
0	0	0	1	1	1	1	7
1	1	1	0	0	0	0	8
0	0	1	1	0	0	1	9
1	0	1	1	0	1	0	10
0	1	1	0	0	1	1	11
0	1	1	1	1	0	0	12
1	0	1	0	1	0	1	13
0	0	1	0	1	1	0	14
1	1	1	1	1	1	1	15

As an illustration of how this code "works" let us take the symbol 0 1 1 1 1 0 0 corresponding to the decimal value 12 and change the 1 in the fifth position to a 0. We now examine the new symbol

0 1 1 1 0 0 0

by the methods of this section to see how the error is located. From Table II the first parity check is over positions 1, 3, 5, 7 and predicts a 1 for the first position while we find a 0 there; hence we write a

1 .

The second parity check is over positions 2, 3, 6, 7, and predicts the second position correctly; hence we write a 0 to the left of the 1, obtaining

0 1 .

The third parity check is over positions 4, 5, 6, 7 and predicts wrongly; hence we write a 1 to the left of the 0 1, obtaining

1 0 1 .

This sequence of 0's and 1's regarded as a binary number is the number 5; hence the error is in the fifth position. The correct symbol is therefore obtained by changing the 0 in the fifth position to a 1.

4. Single Error Correcting Plus Double Error Detecting Codes

To construct a single error correcting plus double error detecting code we begin with a single error correcting code. To this code we add one more position for checking all the previous positions, using an even parity check. To see the operation of this code we have to examine a number of cases:

1. No errors. All parity checks, including the last, are satisfied.
2. Single error. The last parity check fails in all such situations whether the error be in the information, the original check positions, or the last check position. The original checking number gives the position of the error, where now the zero value means the last check position.
3. Two errors. In all such situations the last parity check is satisfied, and the checking number indicates some kind of error.

As an illustration let us construct an eight-position code from the previous seven-position code. To do this we add an eighth position which is chosen so that there are an even number of 1's in the eight positions. Thus we add an eighth column to Table III which has:

TABLE IV

0
0
1
1
1
1
0
0
1
1
0
0
0
0
1
1

PART II

GENERAL THEORY

5. A Geometrical Model

When examining various problems connected with error detecting and correcting codes it is often convenient to introduce a geometric model. The model used here consists in identifying the various sequences of 0's and 1's which are the symbols of a code with vertices of a unit n-dimensional cube. The code points, labelled x, y, z, $\cdots$, form a subset of the set of all vertices of the cube.

Into this space of 2^n points we introduce a *distance*, or, as it is usually called, a *metric*, $D(x, y)$. The definition of the metric is based on the observation that a single error in a code point changes one coordinate, two errors, two coordinates, and in general d errors produce a difference in d coordinates.

Thus we define the distance $D(x, y)$ between two points x and y as the number of coordinates for which x and y are different. This is the same as the least number of edges which must be traversed in going from x to y. This distance function satisfies the usual three conditions for a metric, namely,

$$D(x, y) = 0 \quad \text{if and only if } x = y$$

$$D(x, y) = D(y, x) > 0 \quad \text{if } x \neq y$$

$$D(z, y) + D(y, z) \geq \mathrm{D}(x, z) \text{ (triangle inequality).}$$

As an example we note that each of the following code points in the three-dimensional cube is two units away from the others,

0 0 1
0 1 0
1 0 0
1 1 1 .

To continue the geometric language, a sphere of radius r about a point x is defined as all points which are at a distance r from the point x. Thus, in the above example, the first three code points are on a sphere of radius 2 about the point (1, 1, 1). In fact, in this example any one code point may be chosen as the center and the other three will lie on the surface of a sphere of radius 2.

If all the code points are at a distance of at least 2 from each other, then it follows that any single error will carry a code point over to a point that is *not* a code point, and hence is a meaningless symbol. This in turn means that any single error is detectable. If the minimum distance between code points is at least three units then any single error will leave the point nearer to the correct code point than to any other code point, and this means that any single error will be correctable. This type of information is summarized in the following table:

TABLE V

Minimum Distance	Meaning
1	uniqueness
2	single error detection
3	single error correction
4	single error correction plus double error detection
5	double error correction
	Etc.

Conversely, it is evident that, if we are to effect the detection and correction listed, then all the distances between code points must equal or exceed the minimum distance listed. Thus the problem of finding suitable codes is the same as that of finding subsets of points in the space which maintain at least the minimum distance condition. The special codes in sections 2, 3, and 4 were merely descriptions of how to choose a particular subset of points for minimum distances 2, 3, and 4 respectively.

It should perhaps be noted that, at a given minimum distance, some of the correctability may be exchanged for more detectability. For example, a subset with minimum distance 5 may be used for:

a. double error correction, (with, of course, double error detection).

b. single error correction plus triple error detection.

c. quadruple error detection.

Returning for the moment to the particular codes constructed in Part I we note that any interchanges of positions in a code do not change the code in any essential way. Neither does interchanging the 0's and 1's in any position, a process usually called complementing. This idea is made more precise in the following definition:

Definition. Two codes are said to be *equivalent* to each other if, by a finite number of the following operations, one can be transformed into the other:

1. The interchange of any two positions in the code symbols.
2. The complementing of the values in any position in the code symbols.

This is a formal equivalence relation ($\sim$) since $A \sim A$; $A \sim B$ implies $B \sim A$; and $A \sim B$, $B \sim C$ implies $A \sim C$. Thus we can reduce the study of a class of codes to the study of typical members of each equivalence class.

In terms of the geometric model, equivalence transformations amount to rotations and reflections of the unit cube.

6. Single Error Detecting Codes

The problem studied in this section is that of packing the maximum number of points in a unit n-dimensional cube such that no two points are closer than 2 units from each other. We shall show that, as in section 2, 2^{n-1} points can be so packed, and, further, that any such optimal packing is equivalent to that used in section 2.

To prove these statements we first observe that the vertices of the n-dimensional cube are composed of those of two $(n-1)$-dimensional cubes. Let A be the maximum number of points packed in the original cube. Then one of the two $(n-1)$-dimensional cubes has at least $A/2$ points. This cube being again decomposed into two lower dimensional cubes, we find that one of them has at least $A/2^2$ points. Continuing in this way we come to a two-dimensional cube having $A/2^{n-2}$ points. We now observe that a square can have at most two points separated by at least two units; hence the original n-dimensional cube had at most 2^{n-1} points not less than two units apart.

To prove the equivalence of any two optimal packings we note that, if the packing is optimal, then each of the two sub-cubes has half the points. Calling this the first coordinate we see that half the points have a 0 and half have a 1. The next subdivision will again divide these into two equal groups having 0's and 1's respectively. After $(n - 1)$ such stages we have, upon reordering the assigned values if there be any, exactly the first $n - 1$ positions of the code devised in section 2. To each sequence of the first $n - 1$ coordinates there exist $n - 1$ other sequences which differ from it by one coordinate. Once we fix the n-th coordinate of some one point, say the origin which has all 0's, then to maintain the known minimum distance of two units between code points the n-th coordinate is uniquely determined for all other code points. Thus the last coordinate is determined within a complementation so that any optimal code is equivalent to that given in section 2.

It is interesting to note that in these two proofs we have used only the assumption that the code symbols are all of length n.

7. Single Error Correcting Codes

It has probably been noted by the reader that, in the particular codes of Part I, a distinction was made between information and check positions, while, in the geometric model, there is no real distinction between the various coordinates. To bring the two treatments more in line with each other we redefine a *systematic* code as a code whose symbol lengths are all equal and

1. The positions checked are independent of the information contained in the symbol.
2. The checks are independent of each other.
3. We use parity checks.

This is equivalent to the earlier definition. To show this we form a matrix whose i-th row has 1's in the positions of the i-th parity check and 0's elsewhere. By assumption 1 the matrix is fixed and does not change from code symbol to code symbol. From 2 the rank of the matrix is k. This in turn means that the system can be solved for k of the positions expressed in terms of the other $n - k$ positions. Assumption 3 indicates that in this solving we use the arithmetic in which $1 + 1 = 0$.

There exist non-systematic codes, but so far none have been found which for a given n and minimum distance d have more code symbols than a systematic code. Section 9 gives an example of a non-systematic code.

Turning to the main problem of this section we find from Table V that a single error correcting code has code points at least three units from each other. Thus each point may be surrounded by a sphere of radius 1 with no two spheres having a point in common. Each sphere has a center point and

n points on its surface, a total of $n + 1$ points. Thus the space of 2^n points can have at most:

$$\frac{2^n}{n+1}$$

spheres. This is exactly the bound we found before in section 3.

While we have shown that the special single error correcting code constructed in section 3 is of minimum redundancy, we cannot show that all optimal codes are equivalent, since the following trivial example shows that this is not so. For $n = 4$ we find from Table I that $m = 1$ and $k = 3$. Thus there are at most two code symbols in a four-position code. The following two optimal codes are clearly not equivalent:

0 0 0 0	and	0 0 0 0
1 1 1 1		0 1 1 1

8. Single Error Correcting Plus Double Error Detecting Codes

In this section we shall prove that the codes constructed in section 4 are of minimum redundancy. We have already shown in section 4 how, for a minimum redundancy code of $n - 1$ dimensions with a minimum distance of 3, we can construct an n dimensional code having the same number of code symbols but with a minimum distance of 4. If this were not of minimum redundancy there would exist a code having more code symbols but with the same n and the same minimum distance 4 between them. Taking this code we remove the last coordinate. This reduces the dimension from n to $n - 1$ and the minimum distance between code symbols by, at most, one unit, while leaving the number of code symbols the same. This contradicts the assumption that the code we began our construction with was of minimum reduncancy. Thus the codes of section 4 are of minimum redundancy.

This is a special case of the following general theorem: To any minimum redundancy code of N points in $n - 1$ dimensions and having a minimum distance of $2k - 1$ there corresponds a minimum redundancy code of N points in n dimensions having a minimum distance of $2k$, and conversely. To construct the n dimensional code from the $n - 1$ dimensional code we simply add a single n-th coordinate which is fixed by an even parity check over the n positions. This also increases the minimum distance by 1 for the following reason: Any two points which, in the $n - 1$ dimensional code, were at a distance $2k - 1$ from each other had an odd number of differences between their coordinates. Thus the parity check was set oppositely for the two points, increasing the distance between them to $2k$. The additional coordinate could not decrease any distances, so that all points in the code are now at a minimum distance of $2k$. To go in the reverse direction we simply drop one coordinate from the n dimensional code. This reduces the minimum distance of $2k$ to $2k - 1$ while leaving N the same. It is clear that if one code is of minimum redundancy then the other is, too.

9. Miscellaneous Observations

For the next case, minimum distance of five units, one can surround each code point by a sphere of radius 2. Each sphere will contain

$$1 + C(n, 1) + C(n, 2)$$

points, where $C(n, k)$ is the binomial coefficient, so that an upper bound on the number of code points in a systematic code is

$$\frac{2^n}{1 + C(n, 1) + C(n, 2)} = \frac{2^{n+1}}{n^2 + n + 2} \geq 2^m.$$

This bound is too high. For example, in the case of $n = 7$, we find that $m = 2$ so that there should be a code with four code points. The maximum possible, as can be easily found by trial and error, is two.

In a similar fashion a bound on the number of code points may be found whenever the minimum distance between code points is an odd number. A bound on the even cases can then be found by use of the general theorem of the preceding section. These bounds are, in general, too high, as the above example shows.

If we write the bound on the number of code points in a unit cube of dimension n and with minimum distance d between them as $B(n, d)$, then the information of this type in the present paper may be summarized as follows:

$$B(n, 1) = 2^n$$

$$B(n, 2) = 2^{n-1}$$

$$B(n, 3) = 2^m \leq \frac{2^n}{n + 1}$$

$$B(n, 4) = 2^m \leq \frac{2^{n-1}}{n}$$

$$B(n - 1, 2k - 1) = B(n, 2k)$$

$$B(n, 2k - 1) = 2^m \leq \frac{2^n}{1 + C(n, 1) + \cdots + C(n, k - 1)}.$$

While these bounds have been attained for certain cases, no general methods have yet been found for contructing optimal codes when the minimum distance between code points exceeds four units. nor is it known whether the bound is or is not attainable by systematic codes.

We have dealt mainly with systematic codes. The existence of non-systematic codes is proved by the following example of a single error correcting code with $n = 6$.

$$
\begin{array}{cccccc}
0 & 0 & 0 & 0 & 0 & 0 \\
0 & 1 & 0 & 1 & 0 & 1 \\
1 & 0 & 0 & 1 & 1 & 0 \\
1 & 1 & 1 & 0 & 0 & 0 \\
0 & 0 & 1 & 0 & 1 & 1 \\
1 & 1 & 1 & 1 & 1 & 1
\end{array}\ .
$$

The all 0 symbol indicates that any parity check must be an even one. The all 1 symbol indicates that each parity check must involve an even number of positions. A direct comparison indicates that since no two columns are the same the even parity checks must involve four or six positions. An examination of the second symbol, which has three 1's in it, indicates that no six-position parity check can exist. Trying now the four-position parity checks we find that

$$
\begin{array}{cccccc}
1 & 2 & & & 5 & 6 \\
& 2 & 3 & 4 & 5 &
\end{array}
$$

are two independent parity checks and that no third one is independent of these two. Two parity checks can at most locate four positions, and, since there are six positions in the code, these two parity checks are not enough to locate any single error. The code is, however, single error correcting since it satisfies the minimum distance condition of three units.

The only previous work in the field of error correction that has appeared in print, so far as the author is aware, is that of M. J. E. Golay.[4]

[4] M. J. E. Golay, Correspondence, Notes on Digital Coding, *Proceedings of the I.R.E.*, Vol. 37, p. 657, June 1949.

Generalized Parity Checking*

HARVEY L. GARNER†

Summary—**The usual definition given for the parity check is unwieldy and not particularly suited for the analysis or the study of the arithmetic properties of parity checking. The definition of parity by means of congruences provides a convenient mathematical basis for the concepts of the parity check. In this paper congruence notation is used to generalize the concepts of parity to include nonbase two number systems. Consideration is given to the cases where the check base is equal to the number base, and where it is not equal to the number base. The arithmetic properties of each case are considered by means of congruences.**

Introduction

THE CONCEPT of the parity check[1] is widely known and has frequently been used to detect and in some cases to correct errors in digital systems. The parity check is obtained by associating with each representation of information an auxiliary digit or group of digits known as the parity check digit(s). The proper choice of the relationship between the check digits and the information digits permits the establishment of mathematical operations, which may indicate erroneous digits or provide correction data. Parity checking usually is defined for the binary system in terms of the odd or evenness of the number of ONE digits in a specified block of binary digits. If the number of ONE digits is even the parity check digit is a zero. If the number of ONE digits is odd the parity check digit is a one. This definition is sufficient for most practical applications of the parity concept, but gives no insight into the fundamental principles of parity. A restatement of parity in terms of linear congruences provides a mathematical basis for the parity concept. In terms of congruences, the parity check may be defined for the general case where the check base does not equal the number base. The parity check which employs a check base equal to the number base (as in the usual definition given above) is a special case of the general parity check. The congruence notation also facilitates the study of the arithmetic properties of the parity check. The properties of the parity check with regard to arithmetic operations has received little attention in the literature of error correcting and detecting codes. Previous research in this area is due to Block *et al.*[2,3] and Robertson.[4] Robertson studied the application of a Hamming error correcting code in a general-purpose digital computer. The papers by Block describe a particular error detecting scheme employed in the RAYDAC computer.

The primary contribution of this paper is the unification of the parity check concepts within the framework of congruences.

Congruences

The remainder of this paper makes considerable use of congruence notation. A short presentation of pertinent concepts and relationships is included at this point. Additional material on the subject may be found in any standard text on number theory.[5]

The congruence

$$A \equiv \alpha \text{ Mod } b \tag{1}$$

is read "A is congruent to α modulo b." The congruence relationship states that

$$A = \alpha + bt \tag{2}$$

is valid for some value of t, where A, α, b and t are integers. α is called the residue and b the base or modulus of the number A.

As examples of congruences, consider

$$10 \equiv 7 \text{ Mod } 3 \tag{3}$$

$$10 \equiv 4 \text{ Mod } 3 \tag{4}$$

$$10 \equiv 1 \text{ Mod } 3. \tag{5}$$

In these examples the integers 7, 4, and 1 form a residue class of 10 Mod 3. Of particular importance is the least positive residue of the class which in this example is one. The least positive residue is that residue for which $0 \leq \alpha \leq b$.[6]

Unless otherwise specified, the term residue as used in this paper means the least positive residue.

For the most part, congruences may be treated in the same manner as the equality sign in ordinary arithmetic. The main exception pertains to division. The properties of congruences used here are presented below without proof.

1) Congruences to the same modulus may be added and the result is a valid congruence.

$$\sum_{i=1}^{n} A_i \equiv \left(\sum_{i=1}^{n} \alpha_i \right) \text{ Mod } b. \tag{6}$$

* Manuscript received by the PGEC, March 10, 1958; revised manuscript received, June 23, 1958.

† University of Michigan, Ann Arbor, Mich.

[1] R. W. Hamming, "Error detecting and correcting codes," *Bell Sys. Tech. J.*, vol. 26, pp. 147–160; April, 1950.

[2] R. M. Block, R. V. D. Cambell, and M. Ellis, "The logical design of the raytheon computer," *Math. Tables and Aids to Comp.*, vol. 3, pp. 286–296, 317–323; October, 1948.

[3] R. Block, U. S. Patent No. 2,634,052.

[4] J. E. Robertson, "Error Detection and Correction in Binary Parallel Digital Computers," Digital Computer Lab., University of Illinois, Urbana, Ill., "Electronic Digital Computer" Internal Rep. No. 37, pp. 70–81; 1952.

[5] G. H. Hardy, and E. M. Wright, "An Introduction to the Theory of Numbers," Oxford University Press, London, Eng.; 1956.

[6] The equality sign may exist on only one side of the inequality.

2) Congruences to the same modulus may be multiplied and the result is a valid congruence.

$$\prod_{i=1}^{n} A_i \equiv \left(\prod_{i=1}^{n} \alpha_i \right) \text{Mod } b. \quad (7)$$

3) Congruences are transitive. If $A \equiv B$ and $B \equiv C$ then $A \equiv C$.

4) The following congruence relationships provide a basis for numerical checking procedures

$$\sigma b^n \equiv \sigma \text{ Mod } (b-1) \quad (8)$$

$$\sigma b^n \equiv + \sigma \text{ Mod } (b+1) \quad n \text{ even} \quad (9)$$

$$\equiv - \sigma \text{ Mod } (b+1) \quad n \text{ odd.} \quad (10)$$

Generalization of Parity

A parity check need not necessarily include every digit of the number. In the general case, multiple parity checks may be employed, each check concerning only specific digits. However, for purposes of simplification, only the case of a single parity check over all digits is considered here. The parity of a number from a consistently weighted number system is defined now. Consider the number

$$\sigma_n \sigma_{n-1} \cdots \sigma_2 \sigma_1 \quad (11)$$

This is the usual shorthand notation for the polynomial representation of a number N in a consistently weighted number system, base b:

$$N = \sigma_n b^{n-1} + \sigma_{n-1} b^{n-2} + \cdots + \sigma_2 b^1 + \sigma_1 b^0. \quad (12)$$

A general parity check digit p associated with number N is defined by

$$F(N) \equiv p \text{ Mod } m \quad (13)$$

where p is the least positive residue.

This paper considers two different functions of N. The first function is

$$F(N) = N \quad (14)$$

and the parity check digit p is given by

$$N \equiv p \text{ Mod } m \qquad m \neq b. \quad (15)$$

This type of check is based directly on the numerical properties of linear congruences and the check digit is some function of the magnitude of the number N. This type of check is called a numerical parity check.

If the check base m is identical to the number base b, then

$$N \equiv \sigma_1 \text{ Mod } b. \quad (16)$$

The check is unsatisfactory because it is a function of only the low order digit of the number representation. This is due to the fact that

$$\sigma_i b^n \equiv 0 \text{ Mod } b \qquad n > 0. \quad (17)$$

A check for the case $m = b$ is obtained if

$$F(N) = \sigma_n + \sigma_{n-1} + \cdots + \sigma_2 + \sigma_1. \quad (18)$$

The parity check digit p for this type of check, which is termed digital checking, is defined by the relation

$$\sum_{i=1}^{n} \sigma_i \equiv p \text{ Mod } b. \quad (19)$$

An equivalent definition may be given in terms of ring addition. Ring addition modulo b is specified by the symbol $\oplus$. Ring addition is simply normal addition without carry, for example,

$$(b-1) \oplus 1 = 0. \quad (20)$$

In terms of ring addition modulo b the parity digit p is given as

$$p = \sigma_n \oplus \sigma_{n-1} \oplus \cdots \oplus \sigma_2 \oplus \sigma_1. \quad (21)$$

If the number system is binary, and the check base is two, then the previous definition for the digital check provides a parity digit which makes the total number of ONE digits of the number plus check representation even in correspondence with the usual parity procedures. The definition is extended easily to an odd parity check. For this case the parity check digit is given by $\bar{p}$, the diminished radix complement of p defined as

$$\bar{p} \oplus p = b - 1. \quad (22)$$

As an example of the parity definition for $m = b$, consider the following.

Given 12894, a number in base 10, then

$$(1 + 2 + 8 + 9 + 4) \equiv p \text{ Mod } 10 \quad (23)$$

or

$$p = 1 \oplus 2 \oplus 8 \oplus 9 \oplus 4 = 4. \quad (24)$$

Both types of parity checks are restricted in application to the detection of certain classes of errors in the number representation. Error correction is only possible if multiple parity checks are employed.[7,8] The digital parity check, $m = b$, detects all alterations except the alterations in which the sum of the digit perturbations is congruent to 0 modulo b. Let the decimal number of the previous example be corrupted by $+5$ in the first column and by -5 in the third column. The result is an erroneous representation but the error is obviously undetectable.

$$1 \oplus 2 \oplus 3 \oplus 9 \oplus 9 = p = 4. \quad (25)$$

A numerical parity check Mod m, $m \neq b$, is insensitive to an error if the magnitude of the error is congruent to zero Mod m. Thus the error is detectable only if the check base m is not a divisor of the magnitude of the error. The magnitude of the error in the previous example is 495. The error is not detected by Mod 3, 5, 9 or 11 checks.

Parity checks are particularly effective in detecting

[7] M. J. E. Golay, "Notes on digital coding," Proc. IRE, vol. 37, p. 657; June, 1949.
[8] M. J. E. Golay, "Binary coding," IRE Trans. on Information Theory, No. PGIT-4, pp. 23–28; September, 1954.

errors for which only one digit has been modified. For this class of errors a parity check $m=b$ is always effective. If m is less than b then perturbations congruent to zero modulo m are undetectable.

Arithmetic Invariance of the Digital Parity Check

In the arithmetic system employed in most digital machines, a one-to-one correspondence is established between the machine digit representation and a set of positive and negative fractions. The rules of machine arithmetic operations are set up in such a way as to preserve this one-to-one correspondence under the operation of addition. Addition is the fundamental arithmetic operation of the machine and the other arithmetic operations, subtraction, multiplication and division, are defined in terms of addition and the auxiliary operations, shift and complement.

It is obvious that the digital parity check is invariant to the shift operation if no digits are deleted or added to the representation. This is true because the digital parity check is a function of the modulo sum of the individual digits rather than a function of the magnitude of the representation. Also, the parity check is independent of the position of the radix point. For this reason the remaining discussion does not explicitly consider whether the representations are in correspondence with fractions or integers.

Consider now the addition of two consistently weighted numbers in base b representation. The check base, of course, is also equal to b.

$$A + D = S$$

or

$$\begin{array}{r} a_n \cdots a_1 \\ + \quad d_n \cdots d_1 \\ \hline c_{n+1}s_n \cdots s_1 \end{array} .$$

Now

$$s_i = a_i \oplus d_i \oplus c_i \tag{26}$$

and

$$(a_i + d_i + c_i) = s_i + c_{i+1}b \tag{27}$$

$$c_1 = 0 \tag{28}$$

$$c_i = 0 \text{ or } c_i = 1 \text{ if } i \neq 1; \tag{29}$$

c_i is the ith carry digit from the ith minus one column.

The parity check digit p for the sum is given as

$$c_{n+1} \oplus s_n \oplus s_{n-1} \oplus \cdots \oplus s_1 = p_S \tag{30}$$

or

$$c_{n+1} \oplus a_n \oplus d_n \oplus c_n \oplus \cdots \oplus a_1 \oplus d_1 = p_S, \tag{31}$$

but

$$a_n \oplus \cdots \oplus a_1 = p_A \tag{32}$$

and

$$d_n \oplus \cdots \oplus d_1 = p_D; \tag{33}$$

therefore,

$$c_{n+1} \oplus \cdots c_2 \oplus p_A \oplus p_D = p_S. \tag{34}$$

The last expression provides a means of checking the addition operation in terms of the parity check digits of the sum, the addend, and the augend and the generated carries. Alternatively, the parity check may be in terms of p', the radix complement or additive inverse of p defined as

$$p' \oplus p = 0. \tag{35}$$

If p' rather than p is used as the check digit, the check procedure for addition is given as

$$c_n \oplus \cdots \oplus c_1 \oplus p_S' = p_A' \oplus p_D'. \tag{36}$$

The class of addition errors detected by the digital parity check is the same as the class of representation errors defined in the previous section. However, it must be remembered that the logic of the usual addition circuits is such that the various carry digits are not independent. The parity checking procedure is valid only if the carry digits are correct.

Consider the decimal addition of the following numbers:

					p
6	6	9	4	1	6
3	2	4	5	1	5
9	9	3	9	2	2

Due to the occurrence of one carry, the check of addition is

$$6 \oplus 5 \oplus 1 = 2$$

$$2 = 2. \tag{37}$$

The check of the complement operation is straightforward. The diminished radix complement of a representation A is obtained by substituting for each digit a_i the diminished radix complement, $\bar{a}_i$.

$$a_i \oplus \bar{a}_i = b - 1 \tag{38}$$

given

$$A = a_n \cdots a_1 \quad \text{and} \quad \bar{A} = \bar{a}_n \cdots \bar{a}_1 \tag{39}$$

$$p_A = a_n \oplus \cdots \oplus a_1 \quad \text{and} \quad p_{\bar{A}} = \bar{a}_n \oplus \cdots \oplus \bar{a}_1 \tag{40}$$

and

$$a_n \oplus \bar{a}_n \oplus \cdots \oplus a_1 \oplus \bar{a}_1 = p_A \oplus p_{\bar{A}} = f. \tag{41}$$

Then

$$f \equiv n(b-1) \text{ Mod } b \tag{42}$$

$$(f+n) \equiv nb \text{ Mod } b \tag{43}$$

$$(f+n) \equiv 0 \text{ Mod } b \tag{44}$$

and

$$p_{\bar{A}} + p_{\bar{A}} + n \equiv 0 \text{ Mod } b, \tag{45}$$

or

$$p_{\bar{A}} \oplus p_A \oplus n \text{ Mod } b = 0. \tag{46}$$

A check of the diminished radix complement consists of the modulo b sum of the parity check of the normal representation, the parity check of the complemented representation and the number of digits comprising the representation. In the special case where $b=2$, the check of the complemented representation is equal to the check of the normal representation if n is even, and opposite if n is odd.

A check of the radix complement operation is somewhat more complex than that required for the diminished radix complement. The complications are due to the definition of the radix complement rather than to the nature of the parity check procedure. Two checking procedures are possible. The first is based on the fact that the radix complement A' is obtained from the diminished radix complement $\bar{A}$ by the addition of one to the lowest order digit. This method requires cognizance of the carries produced when the one is added to the lowest order digit. The check is given by the relationship

$$p_{A'} = 1 \oplus p_{\bar{A}} \oplus \sum_{i=2}^{n+1} c_i \text{ Mod } b. \tag{47}$$

The second procedure for obtaining the radix complement check is based upon the well-known rule for obtaining the digitwise radix complement: "Starting with the low order digit, radix complement the first non-zero digit; the remaining higher order digits are changed by diminished radix complementation." This method requires a knowledge of the number of zeros preceding the first nonzero digit. Let this number be q.

Given

$$A = a_n \cdots a_{q+2}a_{q+1}a_q \cdots a_1 \tag{48}$$

$$A' = \bar{a}_n \cdots \bar{a}_{q+2}a'_{q+1}a_q \cdots a_1 \tag{49}$$

$$a_i = 0 \text{ if } i \leqq q \tag{50}$$

so

$$p_A = a_n \oplus \cdots \oplus a_{q+2} \oplus a_{q+1} \tag{51}$$

$$p_{A'} = \bar{a}_n \oplus \cdots \oplus \bar{a}_{q+2} \oplus a'_{q+1} \tag{52}$$

$$p_A \oplus p_{A'} = a_n \oplus \bar{a}_n \oplus \cdots \oplus a_{q+2} \oplus \bar{a}_{q+2} \oplus a_{q+1} \oplus a'_{q+1}, \tag{53}$$

$$p_A \oplus p_{A'} = [(n-q-1)(b-1)+b] \text{ Mod } b \tag{54}$$

or

$$p_A \oplus p_{A'} \oplus (n-q) \text{ Mod } b = 1. \tag{55}$$

Multiplication consists of repetitive addition with appropriate shift operations. The result is a double-length product. The digital parity of the double-length product is found from consideration of the addition operations since parity is not affected by the shift operation. The linear congruence is multiplicative. The check procedure for the product $P = A \times D$ is

$$p_P = (p_A \odot p_D) \oplus \sum c_i \text{ Mod } b \tag{56}$$

or

$$(p_A \times p_D) + \sum c_i \equiv p_P \text{ Mod } b. \tag{57}$$

The symbol $\odot$ indicates multiplication, modulo b. It is observed that the digital parity check provides no protection against faulty shift operations. Parity protection against this type of error is obtained if multiple checks are employed.

The division algorithm commonly employed is easily explained in terms of the remainder theorem.

Consider

$$\frac{X}{Y} = Q + \frac{R}{Y} \tag{58}$$

or

$$X = QY + R \tag{59}$$

$$R = X - QY \tag{60}$$

where

$$Q = q_n b^{n-1} + \cdots + q_j b^{j-1}. \tag{61}$$

Then

$$R = X - (q_n b^{n-1} + \cdots + q_j b^{j-1})Y. \tag{62}$$

The subtraction of Y from X, q_i times is accomplished by the addition of $\bar{Y}$, q_i times. Carries which are generated must be considered. The parity check expression is given as

$$p_R = p_X \oplus (p_Q \odot p_{\bar{Y}}) \oplus \sum c_i \text{ Mod } b. \tag{63}$$

This check may be employed for each division operation or only as a check of the over-all division process. In any case the check is insensitive to shift errors.

Examples of Arithmetic Checking ($m=b=10$)

Addition

	3	4	6	③
+	4	5	8	⑦
	8	0	4	②

○ the circled digits are the check digits

Check

$$p_A \oplus p_D \oplus \sum c_i \text{ Mod } b = p_s. \tag{64}$$

$$3 \oplus 7 \oplus 2 = 2 \tag{65}$$

Complement

nines complement

$A = 0345$②

$A = 9654$④

Check

$$p_{\bar{A}} \oplus p_A \oplus n \text{ Mod } b = 0 \tag{66}$$

$$4 \oplus 2 \oplus 4 = 0 \tag{67}$$

ten's complement

$A = 034500$ ②

$A' = 965500$ ⑤

Check

$$p_{A'} \oplus p_A \oplus (n - q) \text{ Mod } b = 1 \tag{68}$$

$$5 \oplus 2 \oplus 4 = 1 \tag{69}$$

Product

346 ③	
213 ⑥	
346	
346	
692	
346	
1038	5 carries generated
346	
4498	
346	
39098	
346	
73698 ③	

Check

$$p_p = (p_A \odot p_D) \oplus \sum c_i \text{ Mod } b \tag{70}$$

$$3 = (3 \odot 6) \oplus 5 \tag{71}$$

$$3 = 8 \oplus 5 = 3$$

Division

$$Q = \frac{073699 \text{ ④}}{0346 \quad \text{③}} = \frac{X}{Y}$$

the ten's complement of 0346 ③ is 9654 ④

	Carries
073699	2
9654	
039099	2
9654	
004499	
9654	
969899	
04499	3
9654	
01039	
9654	
97579	
1039	1
9654	
0693	2
9654	

	Carries
0347	3
9654	
0001	
9654	
9655	
001	

$Q = 213$ ⑥

$R = 1$

$\sum c_i = 13$

Check

$$p_R = p_X \oplus (p_Q \odot p_{\bar{Y}}) \oplus \sum c_i \text{ Mod } b$$

$$1 = 4 \oplus (6 \odot 4) \oplus 3 = 4 \oplus 4 \oplus 3 = 1. \tag{72}$$

Numerical Checking, Mod $m (m \neq b)$, of Arithmetic Operations

The modulus of the numerical parity check is not the same magnitude as the radix of the number representation. Congruence relationships are valid for any modulus. However, only certain moduli yield procedures which permit straightforward computation of the parity digits. There exist in particular two interesting and useful checks based on procedures congruent modulo $(b-1)$ and congruent modulo $(b+1)$. These checks are called the diminished base numerical check and the augmented base numerical check, respectively. The simplest checks of the numerical class depend directly on certain congruence relationships. Numerical checking procedures have one very desirable feature not associated with digital checking procedures; they do not require a tally or cognizance of the carries generated by the addition process.

The Diminished Base Numerical Check

The diminished base numerical check is dependent on the following properties of linear congruences:

$$\sigma_i b^n \equiv \sigma_i \text{ Mod } (b - 1). \tag{73}$$

This relationship permits parity calculation independent of the digit position or weight. Consider the sum

$$X + Y = S$$

$$X = x_n b^{n-1} + \cdots + x_1 b^0 \tag{74}$$

$$Y = y_n b^{n-1} + \cdots + y_1 b^0 \tag{75}$$

$$S = c_{n+1} b^n + s_n b^{n-1} + \cdots + s_1 b^0. \tag{76}$$

Since congruences are additive,

$$X = \sum_{i=1}^{n} x_i b^{i-1} \equiv \sum_{i=1}^{n} x_i \text{ Mod } (b - 1). \tag{77}$$

The parity check digit p_X associated with the representation X is defined by the term

$$\sum_{i=1}^{n} x_i$$

reduced to the least positive residue of the class. The parity check digit p_X also may be defined by

$$p_X = x_n \oplus \cdots \oplus x_1 \tag{78}$$

for either definition

$$X = \sum_{i=1}^{n} x_i b^{i-1} \equiv p_X \text{ Mod } (b - 1). \tag{79}$$

The check digit p_Y is given as

$$Y = \sum_{i=1}^{n} y_i b^{i-1} \equiv p_X \text{ Mod } (b - 1). \tag{80}$$

Linear congruences are additive, therefore,

$$X + Y = \sum_{i=1}^{n} (x_i + y_i)b^{i-1} \equiv (p_X + p_Y) \text{ Mod } (b-1) \quad (81)$$

and

$$p_S = p_X \oplus p_Y. \quad (82)$$

The modulo $(b-1)$ check is numerical and hence does not require cognizance of the generated carries. Nevertheless, the check is sensitive to errors in carry generation and propagation. The check detects all digit alterations for which the modulo $(b-1)$ sum of the changes does not equal zero. This is equivalent to the statement that an alteration is detectable only if $b-1$ is not a factor of the magnitude of the arithmetic error. A single carry failure causes an alteration of minus one and is therefore detectable. However, the error is not correctable on the basis of the parity information.

A check of the complement operation is obtained from the basic definition of the complemented representation. The definitions which follow are for representations of $n+m+1$ digits. The representation has n digits to the left of the radix point, m digits to the right of the radix point and one sign digit. The diminished radix complement of X is $\overline{X}$ where

$$X + \overline{X} = b^{n+1} - b^{-m}. \quad (83)$$

The radix complement X' is given as

$$X + X' = b^{n+1}. \quad (84)$$

The usual congruence relationships are defined for integer values. It is, therefore, convenient to consider the radix point on the extreme right of the representation. The check is obtained by observing that

$$b^m(b^{n+1} - b^{-m}) \equiv 0 \text{ Mod } (b-1) \quad (85)$$

and

$$b^m b^{n+1} \equiv 1 \text{ Mod } (b-1). \quad (86)$$

Then the check for the diminished radix complement is

$$p_X \oplus p_{\bar{X}} = 0 \quad (87)$$

and the check for the radix complement is

$$p_X \oplus p_{X'} = 1. \quad (88)$$

Congruences are multiplicative. Therefore, a check of the product

$$P = R \times S$$

is given simply as

$$p_P = p_R \odot P_S \quad (89)$$

with no cognizance of carries required.

A check involving the division of congruences is not practical since the correct division procedure is dependent upon whether the divisor and $(b-1)$ have any common factors. However, this is of little consequence since machine division is usually obtained by a process consisting of repetitive add and shift operations.

$$\frac{X}{Y} = Q + \frac{R}{Y}. \quad (90)$$

$$X = QY + R. \quad (91)$$

The check of the division process is independent of carries and must be true at every step in the division process. The check is

$$p_X = (p_Q \odot p_Y) \oplus p_R. \quad (92)$$

The Augmented Base Numerical Check

The diminished base numerical check avoids the need for carry cognizance and for that reason offers a simpler check procedure than is obtainable by means of digital checking.

However, for the binary number system the diminished base check is meaningless. One solution is the augmented base check. This system is also carry independent, and parity check digits are relatively easy to obtain. The augmented base check is dependent on the following property of congruences.

$$\sigma_i b^j \equiv + \sigma_i \text{ Mod } (b+1) \text{ if } j \text{ is even} \quad (93)$$

$$\equiv - \sigma_i \text{ Mod } (b+1) \text{ if } j \text{ is odd.} \quad (94)$$

The check digit for representation A is given as

$$(\cdots + a_5 - a_4 + a_3 - a_2 + a_1) \equiv p \text{ Mod } (b+1) \quad (95)$$

where

$$A = \cdots + a_5 b^4 + a_4 b^3 + a_3 b^2 + a_2 b^1 + a_1 b^0. \quad (96)$$

The checking procedures employed for the arithmetic processes are similar to the procedures for the diminished radix except that ring addition $\oplus$ and ring multiplication $\odot$ are defined modulo $(b+1)$.

In the binary case the number base is two and the check base is three. The parity of a representation A is given as

$$(\cdots + a_5 + 2a_4 + a_3 + 2a_2 + a_1) \equiv p \text{ Mod } 3 \quad (97)$$

or

$$\cdots \oplus a_5 \oplus 2a_4 \oplus a_3 \oplus 2a_2 \oplus a_1 = p. \quad (98)$$

Direct subtraction is avoided by using the radix complement of one with respect to the base three.

Note that for the binary case the augmented base check is identical to the diminished base check, base four.

Examples of the Augmented Base Check

Let $b=2$; then the check is modulo 3.

Addition $X+Y=S$

0 0 0 1 1 1 0 1	(10)	Check
0 1 0 1 0 1 0 1	(01)	$p_X \oplus p_Y = p_S$
		$10 \oplus 01 = 00$
0 1 1 1 0 0 1 0	(00)	

Complement

$A = 00011101.$ (10) $n=7$

Diminished Radix Complement

$\bar{A} = 11100010.$ (01)

Check

$P_A \oplus P_{\bar{A}}$

$= (b^{n+1}-1) \text{ Mod } (b+1)$

since $n+1$ is even

$10 \oplus 01 = 00$

Radix Complement

$A' = 11100011.$ (10)

Check

$p_A \oplus p_{A'} = b^{n+1} \text{ Mod } (b+1)$

since $n+1$ is even

$10 \oplus 10 = 01$

Multiplication $R \times S = P$

1011	(10)
×1001	(00)
1011	
101100	
1100011	(00)

Check

$p_R \odot p_S = p_P$

$10 \odot 00 = 00$

Division $X/Y = Q + V_R/Y$

0	110011	$-X$
1	0101	$-Y$
0	000111	$-R$

↑ sign column

Check

$p_X = (p_Q \odot p_Y) \oplus p_R$

$00 = (01 \odot 10) \oplus 01$

$= 10 \oplus 10 = 00$

The above is a check of the first step in the division process.

Conclusion

The congruence notation used provides a fundamental basis for the concept of both the digital and the numerical type of parity check. Congruence notation also provides the mathematical tools needed to consider the arithmetic properties of the parity checks and the extension of the parity check concept to nonbinary systems.

In the final analysis, the effectiveness of a parity check is dependent upon the error structure of the carry process. It is necessary to consider in detail the mechanism of binary addition before the effectiveness of the parity check may be evaluated. A subsequent paper by the author considers the error structure of the binary adder and the effectiveness of the various parity checks.

Error-Checking Logic for Arithmetic-Type Operations of a Processor

THAMMAVARAPU R. N. RAO, MEMBER, IEEE

Abstract—A method of extending the usefulness of residue coding (or congruence checking) to check for errors in operations such as complement, shift, and rotate (or cycle) is presented. The checking logic and a practical method for its implementation are derived. The cost of check circuitry is only of the order of 30 to 40 percent of the part of the processor that is subjected to checking. In another paper presently under review, a method for single error correction using a new code called "bi-residue code" is presented.

Index Terms—Congruence checking, error-checking circuitry, residue coding, residue generator, self-checking processors.

1. Introduction

ERROR checking of the internal operations of a processor, such as addition, complementation, shifting, and rotation, is an important step in the design of self-checking processors. Parity-based checks, such as Hamming checks, are used for checking errors in operations involving transmission of digital information. Residue coding [1], [2], on the other hand, is known to be useful in checking for errors in both arithmetic and transmission operations.

We develop here a method of extending the usefulness of residue coding to check operations such as shift and rotate which are not strictly arithmetic operations, but are classified here for convenience as "arithmetic-type" operations. The complexity of error-checking logic is dependent to a large extent on the processor configuration, the type of arithmetic logic used, the size of arithmetic registers, and, above all, on the residue base used.

The checking logic presented here is suitable for diminished binary complement arithmetic [3]. The method is derived for a processor having a configuration as shown in Fig. 1. However, the logic can be easily modified for use in other processors with different configurations or arithmetic logic. Next, in this section, we introduce the processor configuration, the diminished complement arithmetic, and the notion of residue checking. In Section 2, the checking logic for add, complement, shift, and rotate operations is derived. In Section 3, the implementation of the checking logic is discussed. Finally, the cost of check circuitry and other practical aspects are briefly discussed. In another paper presently under review, a method for single error correction using "a bi-residue code" is presented.

Manuscript received July 24, 1967. Part of this work was carried out while the author was with the Bell Telephone Laboratories, Inc., Holmdel, N. J.
The author is with the Department of Electrical Engineering, University of Maryland, College Park, Md.

1.1 *Accumulator, Adder, and Other Circuitry*

Fig. 1 schematically describes the accumulator register and its adder. Shift-rotate logic circuitry is also indicated. The communication between these registers and the rest of the processor is by means of the two buses which could be called input and output buses. The control and clock signals to various registers are also indicated in Fig. 1.

1.2 *Arithmetic Logic*

Each of the registers (accumulator, augend, and addend) are n binary bits in length. Since the negative numbers are represented in 1's complement form, the range of numbers M that can be represented is equal to 2^n-1. Specifically, those represented are $-2^{n-1}+1$, $-2^{n-1}+2, \cdots, -1, 0, 1, 2, \cdots$, and $2^{n-1}-1$. Zero will have two representations, one with all n 0's and the other, its complement, with all n 1's. (Sometimes the latter is referred to as the negative zero.) The addition is done with an end-around carry (EAC), that is, a carry from the most significant bit (or sign bit) to the least significant bit.

1.3 *Residue Checking*

Residue checking can be employed in different ways. Our method is as follows. We use a check register (CR) in which we shall maintain a quantity $|A|_r$ where A is the integer represented in the accumulator at that instant of time, r is a positive integer called the residue

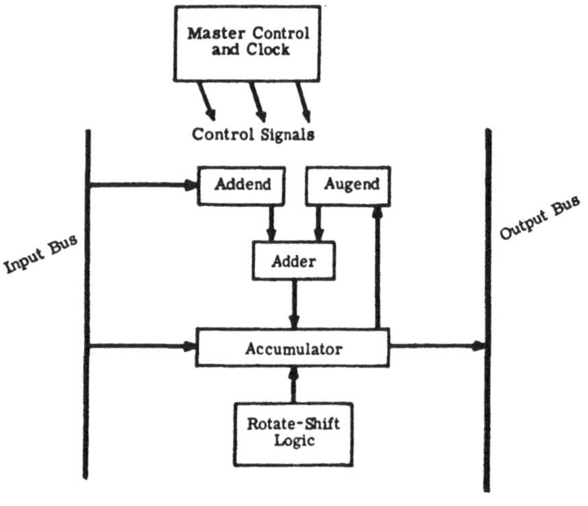

Fig. 1. Accumulator and adder schematic.

base, and $|A|_r$ is the least non-negative integer congruent to A modulo r ($|A|_r$ is also said to be the residue of A modulo r).

Consider an operation ϕ (where ϕ can be any one of the operations add, shift, complement, load accumulator, and rotate) which modifies the contents of the accumulator from A to A'. We denote this as $\phi: A \to A'$ or as $\phi(A) \to A'$. This means that A' is the result of operation ϕ on A. A' is not necessarily a function of A. We now desire to find a parallel operation ϕ_r (corresponding to ϕ), which modifies the check-register contents $|A|_r$ to $|\phi(A)|_r = |A'|_r$. This we may denote as ϕ_r: $|A|_r \to |\phi(A)|_r$. Once again ϕ_r may operate on either, both, or neither of A and $|A|_r$. If a ϕ_r can be devised for each given ϕ, then error checking is simply reduced to checking the relationship between accumulator and check-register contents periodically. Any violation of this relationship (i.e., the computed residue of the accumulator contents disagrees with CR contents) can be arranged to generate a maintenance alarm or program interrupt signal. Consider, for example, the operation of complementation of an n-bit accumulator register. $\phi(A) = M - A$ where $M = 2^n - 1$, and hence the ϕ_r operation consists of obtaining $|M - A|_r = ||M|_r - |A|_r|_r$ and storing that in the check register.

Thus, after each such operation, if both ϕ and ϕ_r are correctly executed, the check register will contain the residue of the accumulator contents. If r is chosen to be one of 3, 7, 15, and 31, etc. ($r = 2^k - 1$ for $k = 2, 3, 4, \cdots$), there are a number of advantages: 1) obtaining residue modulo r can be accomplished without actual division, 2) the addition and other manipulation of residues can be carried out using conventional binary logic, and 3) if k divides n, $2^k - 1$ would divide $2^n - 1$ so that $|M|_r = 0$, which simplifies to a large extent the checking logic. The formulas for error-checking logic are derived in the next section for the case $r = 3$, using a 2-bit check register.

1.4 *The Type of Errors Detected By Residue Coding*

It has been shown [2] that when the $r = 2^k - 1$ any burst error of length less than k can be detected. In general, unless the error pattern is such that its magnitude is divisible by r, it can be detected. For $r = 3$, all single errors and nearly 2/3 of all other types of errors combined are detectable.

2. Formulas for Error Checking

Once again we use n for the number of bits in the accumulator register and denote the range by $M = 2^n - 1$. A residue base of 3 is assumed.

2.1 *Addition*

Let $0 \leq a$, $b < M$ and then the addition (modulo M) of a and b is

$$|a + b|_M = a + b \quad \text{if there is no EAC}$$
$$= a + b - M \quad \text{if there is an EAC.}$$

Therefore, the residue check for the sum is $||a+b|_M|_3$, which is equal to $||a|_3 + |b|_3 + X|_3$, where X is equal to 0 if either $|M|_3 = 0$ or there is no EAC generated in the addition. Otherwise, X is equal to 2. ($|M|_3$ can be shown to be equal to 0 for all even values of n and is equal to 1 for odd values. $|-M|_3$ has a value 2 for n odd.)

2.2 *Complement* (1's *Complement*)

If the residue check for a is $|a|_3$, then since the complement of a, denoted by $\bar{a}$, is congruent to $M - a$ (modulo M) its residue will be $|M - a|_3$. Hence

$$|M - a|_3 = |-a|_3 = |\bar{a}|_3 \quad \text{for } |M|_3 = 0 \ (n \text{ even})$$
$$= |M - a|_3 = |1 + \overline{|a_3|}|_3 \quad \text{for } |M|_3 = 1 \ (n \text{ odd}).$$

2.3 *Left Shift* (*by c places for* $0 \leq c \leq n$)

Let A be the contents of the accumulator before shifting, and X be the number represented by the rightmost $n - c$ bits (see Fig. 2). Since in a left shift by c places the leftmost c bits are lost, the result in the accumulator after shifting is equal to $X \cdot 2^c$. Therefore, the check register must contain

$$|X \cdot 2^c|_3 = ||X|_3 |2^c|_3|_3 = |X|_3 \quad \text{for } c \text{ even}$$
$$= |2|X|_3|_3 = \overline{|X|}_3 \quad \text{for } c \text{ odd.}$$

2.4 *Right Shift* (*by c places for* $0 \leq c \leq n$)

As before, let A be the contents before shifting. Let Y denote the number obtained by resetting the rightmost c places to 0. The result after shifting is trivially $Y2^{-c}$ (see Fig. 3). The residue must be

$$|Y2^{-c}|_3 = |Y|_3 |2^{-c}|_3 = |Y|_3 \quad \text{for } c \text{ even}$$
$$= |2|Y|_3|_3 = \overline{|Y|}_3 \quad \text{for } c \text{ odd.}$$

Thus, for a left (or right) shift by c places, the first step is to obtain the residue of the number obtained by

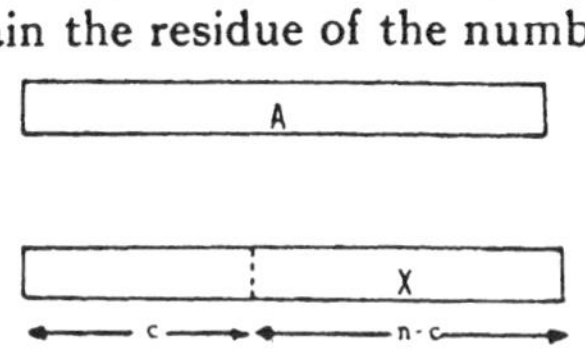

Fig. 2.

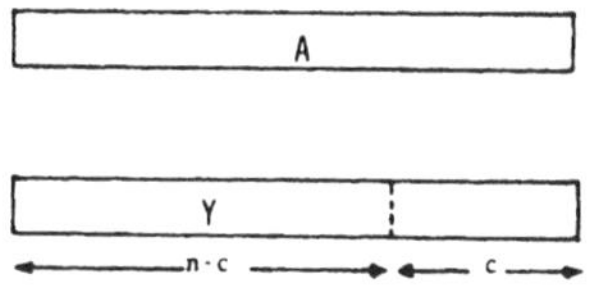

Fig. 3.

TABLE I

$\vert a\vert_3$	$2\vert a\vert_3$	Complement of $\vert a\vert_3$	Exchanging of Bits
00	00	11	00
01	10	10	10
10	01	01	01
11	11	00	11

resetting the leftmost (or rightmost) c places to 0, and then multiplying by 2 if c is odd. In this connection it must be said that multiplying by 2, modulo 3 of $|a|_3$ is equivalent to exchanging the two bits, which is also the same as complementing if one views 11 as equal to 00 (see Table I).

2.5 *Rotate*

Since error checking for left or right rotate operations will be quite simple if $|M|_3=0$, we shall treat that case first.

2.5.1 Case of $|M|_3=0$

Left rotate by c places (or right rotate by $n-c$ *places)* $0\leq c\leq n$: Left rotate by c places is equivalent to multiplying 2^c modulo M. Hence, if accumulator contents are A before rotating, they would be $|A\cdot 2^c|_M$ after rotating. Hence, the check register should contain $||A\cdot 2^c|_M|_3$.

Since $|M|_3=0$, the above is equal to

$$|A\cdot 2^c|_3 = |A|_3|2^c|_3 = |A|_3 \quad \text{for } c \text{ even}$$

$$= 2|A|_3 = \overline{|A_3|} \quad \text{for } c \text{ odd.}$$

Since right rotate by c places is, in effect, the same as left rotate by $n-c$ places, and $n-c$ is odd whenever c is odd (as n is even for the case $|M|_3=0$), there needs to be no distinction between the right or left rotate as far as the check bit manipulation is concerned.

2.5.2 Case of $|M|_3=1$

For this case, left rotate and right rotate are considered separately.

2.5.2.1 Left rotate by c places: Let A as before represent the number in the accumulator before shifting, and X be the number obtained by resetting the leftmost c places to 0 (Fig. 2). The number in the accumulator after rotation is $|A\cdot 2^c|_M$.

Hence, the residue (that is, the check-register contents) must be $||A\cdot 2^c|_M|_3$. Since A can be broken up into parts as X and $A-X$, we could write the residue as

$$R = ||X\cdot 2^c|_M|_3 + ||(A-X)2^c|_M|_3.$$

Since $X\cdot 2^c<M$, we have

$$R = ||X\cdot 2^c|_3| + ||(A-X)2^c|_M|_3.$$

$|(A-X)\cdot 2^c|_M$ is also equal to $|(A-X)2^{-(n-c)}|_M$, which corresponds to right-rotating the c leftmost bits by $n-c$ places. Since $(A-X)2^{-(n-c)}<M$, modulo M quotes can be eliminated. Hence

$$R = |X\cdot 2^c|_3 + |(A-X)2^{-(n-c)}|_3.$$

Since we have assumed $|M|_3=1$, $|2^n|_3=|M+1|_3=2^1$, and thus we have

$$\begin{aligned} R &= |X\cdot 2^c|_3 + |(A-X)2^{c-1}|_3 \\ &= |X\cdot 2^c + (A-X)2^{c-1}|_3 = |(X+A)2^{c-1}|_3 \\ &= |(|X|_3 + |A|_3)|2^{c-1}|_3|_3 \\ &= ||X|_3 + |A|_3|_3 \quad \text{for } c \text{ odd} \\ &= |\overline{|X|_3 + |A|_3}|_3 \quad \text{for } c \text{ even.} \end{aligned}$$

Thus, the residue $|X|_3$ is obtained as in the case of left shift, but this time it should be added to the contents of the check register, and then complemented if c is odd. (Note that for left shift, $|X|_3$ is complemented only if c is even.)

2.5.2.2 Right Rotate by c Places: As before, A represents the content before rotating and Y the number obtained by resetting the c rightmost bits (see Fig. 3). By a derivation similar to that in Section 2.6.1 one could show the residue to be

$$\begin{aligned} R &= ||A|_3 + |Y|_3|_3 \quad \text{for } c \text{ odd} \\ &= |\overline{|A|_3 + |Y|_3}|_3 \quad \text{for } c \text{ even.} \end{aligned}$$

3. Error-Checking Circuitry

We consider here the implementation of the error-checking logic described earlier. The error checking will consist of: 1) the residue generator (RG), 2) the residue manipulator (RM), 3) the check register (CR), 4) the shift, rotate check control (SRCC). Communications between these parts and the part of the processor whose operations they are expected to check are indicated in Fig. 4. In the succeeding sections, we describe the functions of these parts individually and collectively.

3.1 *Residue Generator*

The residue generator consists of an n-bit register (RGR), the residue generation logic, the reset logic, and a 2-bit register called residue register (RR) as shown in Fig. 5.

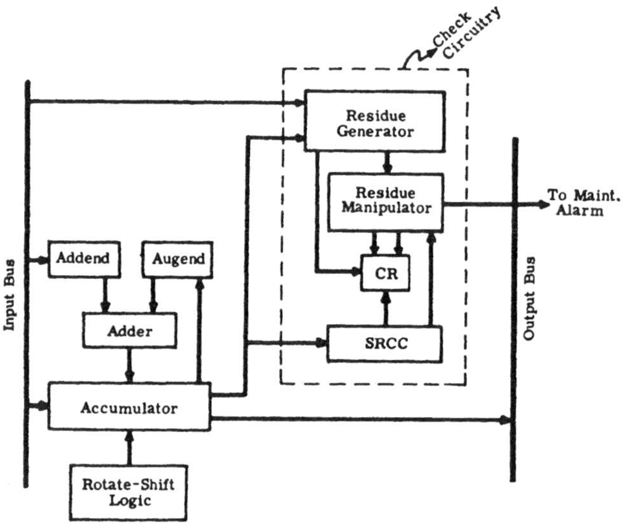

Fig. 4. Processor and check circuitry schematic.

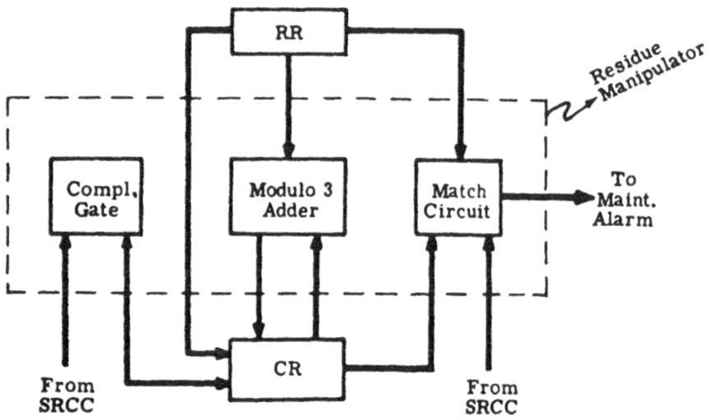

Fig. 6. Residue manipulator schematic.

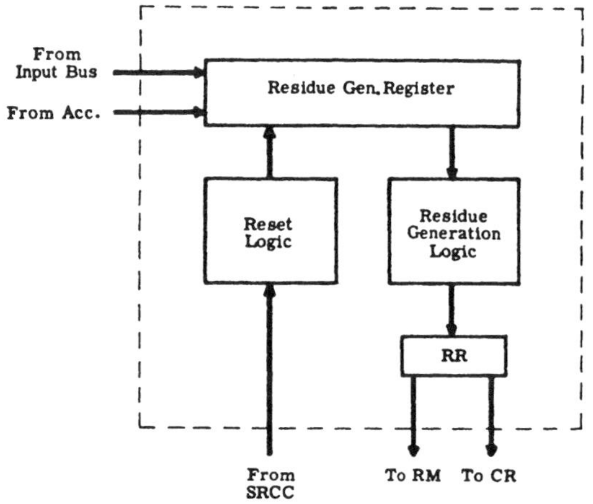

Fig. 5. Residue generator schematic.

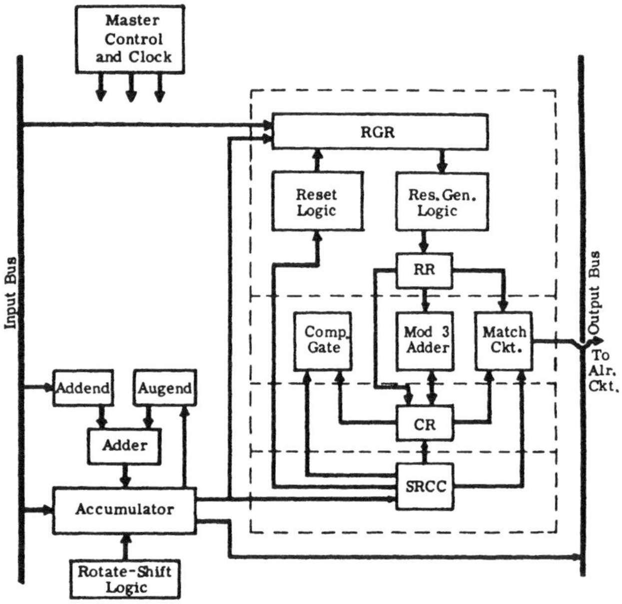

Fig. 7. Processor and its check circuitry.

The number whose residue is to be obtained is gated into the RGR. Depending on the operation, this comes from either the accumulator or the input bus. For operations such as shift (or rotate if n is odd) by c places, it is required that the residue of a part of the contents of the accumulator be reset to 0. This is implemented by use of the reset logic circuit. The number c and the direction of shift (or rotate) are obtained from the SRCC.

The residue generation logic consists of treating the n bits as $n/2$ (for n even) or $(n+1)/2$ (for n odd) number of bit pairs, and then combining two adjacent pairs into a single new pair by using a modulo 3 adder. This process of combining two pairs into one is carried out in stages until they are reduced to a single pair. This pair is then stored in the residue register.

3.2 *Residue Manipulator*

Residue manipulator functions are 1) to match the CR and RR, and in case of mismatch, to send an error-indication signal to the maintenance alarm or maintenance interrupt circuits as the case may be; 2) to provide for modulo 3 addition of the contents of CR and RR and to store the result in CR; and 3) to provide a complementing circuit for the contents of CR. These functions are indicated in Fig. 6.

3.3 *Shift, Rotate Check Control (SRCC)*

The function of SRCC is to modify the check bits appropriately for the shift or rotate instructions. This is done by controlling the reset logic circuit of the residue generator, the complement gate, and the modulo 3 adder of the residue manipulator. In addition, SRCC works as the control unit for the entire checking circuitry. It receives the decoded instruction from the master control, and distributes signals and clock pulses to activate and coordinate the functions of RG and RM. The instructions executed by the processor are treated by SRCC as two distinct types: 1) those that *do not* alter the contents of the accumulator and 2) those that *do* alter the contents of the accumulator.

For the first type of instructions, the check bits (contents of CR) are not modified. However, a routine check for the agreement between the accumulator and CR is made. This is done by generating the residue of the contents of the accumulator and comparing it with the contents of CR by means of the match circuit. A mismatch generates an error signal which will be transmitted automatically to the maintenance alarm circuit or interrupt program. The processor and its check circuitry are shown in block diagram in Fig. 7.

For the instructions which alter the contents of accumulator, the modification of check bits would be necessary to maintain the agreement as mentioned earlier. For shift (and rotate if n is odd) instruction, the accumulator contents are gated into RGR. For others in this same category, such as "add to the contents of accumulator," the same word that comes into addend register also enters RGR parallel from the input bus. Similarly, for "store accumulator" instructions, the word that comes into accumulator comes into RGR from the input bus. This gating is controlled by SRCC. For shift (and rotate only if n is odd) operation, the reset logic circuit is activated and it is supplied with the information such as the number of places to be reset and the direction of the shift.

4. The Cost of Checking

The relative sizes (in terms of NAND gates) of the check circuitry and the part of the processor that is being checked will be of importance to the designer. The processor part, exclusive of the master control, shown in Fig. 1 is of the order of 1000 gates (for 25-bit registers). The check circuitry shown in Fig. 5 appears to be realizable for less than 400 gates. Thus, the check circuitry increases the hardware cost by 40 percent. However, this increase is dependent on the size of processor registers. An increase in the size of processor registers will proportionally increase the size of the processor, while the check circuitry may be less affected. Thus, for the processors with larger registers, the increase of cost due to checking is expected to be of the order of 30 to 35 percent.

4.1 *Other Arithmetic Operations*

The check circuitry is derived to check add, complement, shift, and rotate operations. Checking other arithmetic operations, namely, subtract, multiply, and divide, follows naturally from the checks on add, complement, and shift operations. Since the arithmetic functions constitute a significant part of a processor, the error checking of these operations can be considered as an important step in the design of self-checking processors.

References

[1] H. L. Garner, "Generalized parity checking," *IRE Trans. Electronic Computers*, vol. EC-7, pp. 207–213, September 1958.

[2] D. S. Henderson, "Residue class error checking codes," *Proc. 16th Nat'l Meeting of the Association for Computing Machinery*, Los Angeles, Calif., September 5–8, 1961.

[3] N. R. Scott, *Analog and Digital Computer Technology*. New York: McGraw-Hill, 1960, pp. 227–229.

Arithmetic Algorithms for Error-Coded Operands

ALGIRDAS AVIŽIENIS

Reprinted from *IEEE Transactions on Computers*, Volume C-22, Number 6, June 1973, pages 567-572.

***Abstract*—A set of arithmetic algorithms is described for operands that are encoded in the "*AN*" error-detecting code with the low-cost check modulus $A = 2^a - 1$. The set includes addition, additive inverse (complementation), multiplication, division, roundoff, and two auxiliary algorithms: "multiply by $2^a - 1$," and "divide by $2^a - 1$." The design of a serial radix-16 processor is presented in which these algorithms are implemented for the low-cost *AN* code with $A = 15$. This processor has been constructed for the Jet Propulsion Laboratory STAR computer. The adaptation of "two's complement" arithmetic for an inverse-residue code is also described.**

***Index Terms*—*AN* codes, arithmetic algorithms, arithmetic processor design, error-detecting codes, low-cost arithmetic error codes, residue codes.**

I. Introduction

Arithmetic error codes are of interest to computer designers because they are preserved during arithmetic operations in the computer, and therefore can be applied to all computer words in order to attain concurrent error detection during the execution of programs. In addition to the theoretical properties of arithmetic error codes [1], [2] the designer must consider two additional aspects: the cost versus effectiveness tradeoffs of their application [3] and the implementation of the entire set of arithmetic algorithms of the computer using coded operands. This paper considers arithmetic algorithms for both major classes of arithmetic error codes: the *separate* (residue) and *nonseparate* (*AN*) codes [2]. Attention is concentrated on the low-cost radix-2 codes of both classes [3].

Algorithms for the residue codes have been developed as early as 1948 in the Raydac computer, which employed a variant of modulo 31 residue checking for its arithmetic unit [4]. Later, modulo 3 residue checking has been mentioned (without details) in descriptions of the IBM 7030 (Stretch) and the Univac III computers. Sets of algorithms were discussed by Garner [5], Davis [6], Sellers *et al.* [7], and Rao [8].

The first set of algorithms for the *AN* codes (with the low-cost check modulus $A = 2^a - 1$) has been devised for the STAR computer developed at the Jet Propulsion Laboratory [9], [10]. A complete 4-bit byte-organized arithmetic processor with $A = 15$ has been designed and constructed [11]. Experience with this processor led to the follow-on design of a second arithmetic processor that employed the separate modulo 15 inverse residue code [3], [12].

The nonseparate code considered in this paper is the *AN* code [1], which is formed when an uncoded operand X is multiplied by the *check modulus* A to give the coded operand AX. The separate codes are the *residue* code [1], and the *inverse-residue* code [3] which has significant advantages in fault detection of repeated-use faults. The modulo A inverse-residue encoding for a number X attaches a check symbol X'' to form the pair (X, X''). The value of X'' is given by

$$X'' = A - (A|X)$$

where $A|X$ means "the modulo A residue of X." $A|X$ is the value of the check symbol X' employed in modulo A residue encoding (X, X'). The inverse-residue code is a separate code, since it has no arithmetic interaction between X and X'', and therefore differs from the nonseparate systematic subcodes of *AN* codes [2].

The following sections present a set of algorithms for the low-cost (modulo $2^a - 1$) *AN*-coded operands and a discussion of their implementation in a byte-organized arithmetic processor. The last section discusses the algorithms for inverse-residue coded operands of the STAR computer.

II. Algorithms for *AN*-Coded Operands $(2^a - 1)X$

The entire set of binary arithmetical algorithms is conveniently adaptable to *AN* codes with the check modulus $2^a - 1$ and word length of ka bits [9], [11]. A fundamental constraint on the choice of A for binary n-bit *AN*-coded numbers $Z = AX$ is the requirement that the code should be complementable with respect to K(the values 2^n or $2^n - 1$) in order to implement the additive inverse algorithm. Assuming that the uncoded numbers X are complemented with respect to the integer value M, we have the requirement

$$K - AX = A(M - X)$$

which gives

$$K = AM,$$

that is, A must be a factor of K. Since A is an odd integer, this immediately excludes the choice $K = 2^n$. Consequently, a *complementable AN code* can be obtained if and only if

$$2^n - 1 = AM$$

is satisfied. All possible choices of A are available from a table of the prime factors of $2^n - 1$ [9]. The "$An + B$" code [1] allows other choices of A, but the algorithms are cumbersome to

Manuscript received January 31, 1973. This work was supported in part by the National Aeronautics and Space Administration under Contract NAS 7-100, and was performed at the Jet Propulsion Laboratory, California Institute of Technology, Pasadena, Calif.
The author is with the Department of Computer Science, University of California, Los Angeles, Calif. 90024, and the Jet Propulsion Laboratory, Pasadena, Calif. 91103.

implement. All codes with $n = ka$ and $A = 2^a - 1$ are complementable; for these codes the well-known "one's complement" algorithms apply directly, including complementation, sign detection, range extension, range contraction, left and right arithmetical shifting, and addition. In range extension the next allowed word length (after ka bits) is $(k + 1)a$ bits and range is extended by prefixing a bits identical to the leftmost bit. Similarly, range can be contracted by dropping a leftmost bits when $a + 1$ leftmost bits are identical. The existence of range extension and range contraction algorithms makes error-coded arithmetic with variable-length operands possible when a byte-organized processor is employed.

Multiplication and division of the AN-coded number Z by the check modulus $2^a - 1$ are needed in computing coded products and quotients. The "multiply by $2^a - 1$" algorithm requires one n-bit parallel modulo $2^n - 1$ addition. The $n - a$ bits-long coded operand Z is extended by a bits, shifted a positions left to form 2^aZ, and then added modulo $2^n - 1$ to its own (a bits extended) one's complement $\overline{Z}$ to form the n-bit result

$$(2^n - 1)|(2^aZ + \overline{Z}) = (2^a - 1)Z.$$

The algorithm is illustrated in Example 1.

Example 1–Multiply by $(2^a - 1) = 15$:

$$\begin{aligned} 1 &= C_0 \\ 16X &= 0011\ \ 0111\ \ 0000 \\ \overline{X} &= \underline{1111\ \ 1100\ \ 1000} \\ 15X &= 0011\ \ 0011\ \ 1001. \end{aligned}$$

The inverse operation, i.e., the "divide by $2^a - 1$" algorithm, generates the result $(2^a - 1)Z \div (2^a - 1)$ in bytes of a bits per step, employing an a-bit parallel adder to implement

$$Z = \overline{(2^n - 1)|\{(2^a - 1)Z + \overline{2^aZ}\}}.$$

The overbar designates one's complement and the addition is modulo $2^n - 1$, where n is the length of operand $(2^a - 1)Z$. The algorithm can be initiated because the "end-around" carry C_0 and the rightmost four bits (W_3, W_2, W_1, W_0) of $\overline{2^aZ}$ are deduced from the most significant bit Y_{n-1} of the operand $Y = (2^a - 1)Z$. The rule is as follows.

1) If $Y_{n-1} = 0$, then $C_0 = 0$ and $W_3W_2W_1W_0 = 1111$.
2) If $Y_{n-1} = 1$, then $C_0 = 1$ and $W_3W_2W_1W_0 = 0000$.

The algorithm is illustrated in Example 2.

Example 2–Divide by $(2^a - 1) = 15$:

$Y = 15Z =$	0000	0010	0111	0100	0110	1011	
$W = \overline{16Z} =$	1111	1101	0110	0001	1010	[1111]	$= W_3W_2W_1W_0$
	0	0	0	1	1	0	$= C_0$
$\overline{Z} =$	$\underline{0}1111$	$\underline{0}1111$	$\underline{0}1101$	$\underline{0}0110$	$\underline{1}0001$	$\underline{1}1010$	
$Z =$	0000	0000	0010	1001	1110	0101.	

A double-length coded product $(2^a - 1)XY$ of the n-bit operands $(2^a - 1)X$ and $(2^a - 1)Y$ is obtained by forming the one's complement binary product $(2^a - 1)^2XY$ and then dividing it by $2^a - 1$. The length of the result is $2n - a$ bits. A roundoff algorithm must be employed to obtain a single-length (n-bit) product.

Roundoff by truncation will not yield an AN-coded number; the algorithm must be modified. To round off the rightmost $n - a$ bits of the coded operand $(2^a - 1)XY$ we add the constant $(2^a - 1)G$, such that in $(2^a - 1)(XY + G)$ the rightmost $n - a$ bits are identical to the leftmost ("sign") bit and may be dropped. The "divide by $2^a - 1$" algorithm is applied to $(2^a - 1)XY$ to obtain the $n - a$ rightmost bits of XY. These bits are inspected and the value of G is selected to implement either up-rounding or down-rounding. The roundoff algorithm is illustrated in Example 3.

Example 3–Roundoff of Eight Bits: The eight rightmost bits of $XY = K = 1100\ 1011$. Roundup: add $15G$ where $G = 2^8 - K = 0011\ 0101$.

$$\begin{aligned} 15XY &= 0000\ \ 0010\ \ 1001\ \ 1110\ \ 0101 \\ 15G &= \underline{0000\ \ 0000\ \ 0011\ \ 0001\ \ 1011} \\ 15(XY + G) &= 0000\ \ 0010\ \ 1101\ \ 0000\ \ 0000. \end{aligned}$$

The division algorithm must produce two coded results: the quotient and the remainder. Given the $(2n - a)$-bit dividend $(2^a - 1)X$ and n-bit divisor $(2^a - 1)Y$, the n-bit quotient $(2^a - 1)Q$ is obtained by first forming $(2^a - 1)^2X$ and then executing binary division which yields the coded quotient $(2^a - 1)Q$ and the remainder $(2^a - 1)^2R$ satisfying

$$(2^a - 1)^2X = (2^a - 1)Q(2^a - 1)Y + (2^a - 1)^2R.$$

The coded remainder $(2^a - 1)R$ is obtained using the "divide by $2^a - 1$" algorithm. An uncoded quotient Q would result without the initial "multiply by $2^a - 1$" algorithm for the dividend. This initial algorithm may cause a special condition which is analyzed in the Appendix.

In summary, we note that the AN codes with the check modulus $2^a - 1$ display an exceptional adaptability to binary arithmetic and have a low-cost checking algorithm when the length of operands is chosen to be some multiple ka of the check length a. Byte-serial variable-length operand arithmetic can be implemented in bytes of a, $2a$, etc., bits. An experimental byte-serial processor using the above algorithms with $n = 32$, $a = 4$, and $2^a - 1 = 15$ has been constructed and tested as part of the STAR computer [11], [12]. It is described in the next section.

III. Design of a Processor for AN-Coded Operands

The operand precision of 28 bits was selected for the expected class of problems, and the check modulus $A = 15$ was chosen in order to provide 100 percent detection of single determinate repeated-use faults when binary numbers are transmitted and added in 4-bit bytes [3]. The length of coded numbers $15X$ is 32 bits. The choice of $A = 15$ provides 100 percent detection of such repeated-use faults for binary coded numbers up to 56 bits in length [3]. The STAR computer is a replacement system, and information is transmitted between its subsystems in 4-bit bytes in order to reduce the size of the data bus.

The block diagram of the AN-code arithmetical processor for the Star computer is shown in Fig. 1. The processor accepts five operation codes: clear add, add, subtract, multiply, and divide. One processor cycle consists of 10 byte times (clock

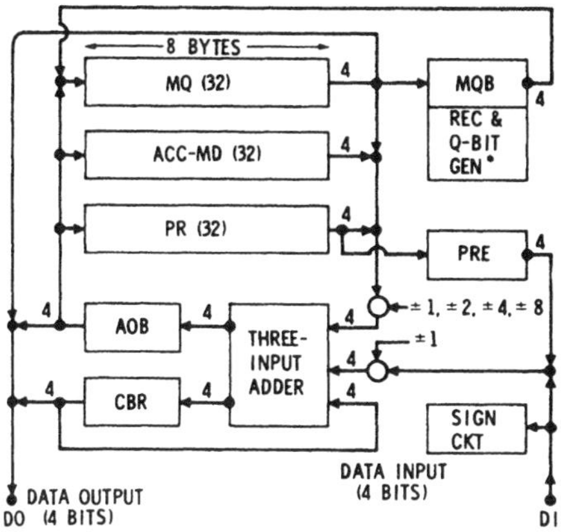

Fig. 1. *AN*-code arithmetic processor.

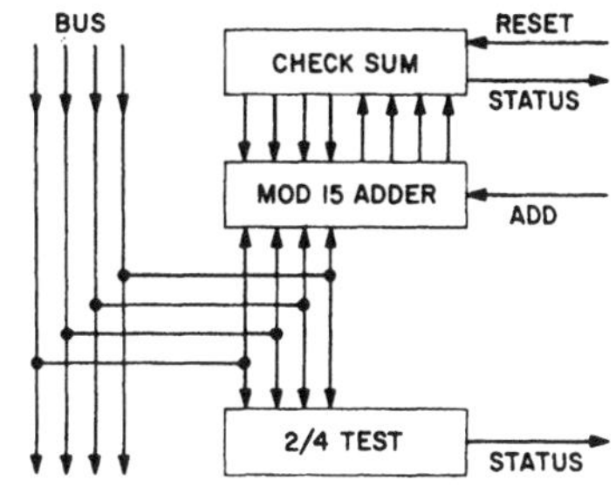

Fig. 2. Checker element of the STAR computer.

times). Operands are received on four data-input lines (DI), and results (both partial and final) are delivered on four data-output lines (DO). Partial results (partial sums, products, remainders) may be up to 10 bytes long, and the last byte is accompanied by a "perform check" signal on a control line. The final results (sum, difference, product, quotient) are always 8 bytes long and also have the perform check signal along with the eighth byte. They are followed by one nonnumerical condition-code byte, which employs a two-out-of-four encoding. There are three *singularity* codes: sum overflow, quotient overflow, and zero divisor, and three *result* codes: positive, zero, and negative. The latter three serve as branching information for "jump" class instructions. All partial and final results are sent to the *Checker*, which consists of a modulo 15 adder and accumulator, as shown in Fig. 2. Upon receipt of the perform check signal, the 4-bit check-sum accumulator is inspected for the "1111" ("all ones") check result. Any other check sum indicates that the result was not a properly *AN*-coded number, and a fault warning is issued. The Checker is outside the processor—there is one copy each in the three majority-voted test-and-repair processors [12].

Internally, the processor contains three 8-byte double-ranked shift registers (Fig. 1): the accumulator/multiplicand/divisor (ACC-MD) register, the product-remainder (PR) register, and the multiplier-quotient (MQ) register. There also are four 1-byte registers: adder output buffer (AOB), multiplier-quotient buffer (MQB), PR extension (PRE), and the carry-byte register (CBR). The *adder* is a three-input parallel adder that generates a *sum byte* (SB) and a *carry byte* (CB). The CB is stored in the 4-bit double-ranked CBR. The 8-bit output exists because the ACC-MD input to the adder supplies the multiples ±1, ±2, ±4, and ±8 of ACC-MD contents as operands for radix-16 multiplication and division. In this design all movement of numerical information takes place *four bits at once*, in the series parallel mode. At no point do two or more bits of the same byte pass through the same logic element; consequently, a fault will not damage more than one bit in one byte, and always the same bit position will be subject to damage when a sequence of bytes passes through the faulty element. The bit channels of the processor are physically isolated; for example, the ACC-MD, PR, and MQ shift registers consist of four 8-bit serial shift registers each. All data lines are 4 bits wide in Fig. 1. Multiplication and division are radix-16 operations; therefore, only 4-bit shifts are employed.

The *AN*-coded binary operands are 32 bits long. The one's complement form is used to represent negative numbers, and the binary point is assumed to be at the left end of the encoded 32-bit number, giving the range $-1/2 < Z < 1/2$ for any coded operand Z, and the range $-1/30 < X < 1/30$ for the uncoded operands X. This range was most convenient for the experimental model; other ranges can be readily employed by moving the binary point and appropriately adjusting the algorithms. However, the range $-1 < X < 1$ is not available for uncoded operands, since moving the binary point will scale the range of X by 2^j, and the jump in range will be from 16/30 to 32/30. The location of the binary point must be referred to the coded numbers in order to obtain simple sign detection and overflow detection algorithms. Zero is represented by a string of 32 ones; the all-zero number is not a properly coded number.

The *clear-add* algorithm needs one cycle for its execution. It replaces the contents of the ACC-MD by the input operand, which is added to zero and then stored in the ACC-MD. At the same time this sum is also returned on the output lines to the external Checker.

The *add* and *subtract* algorithms add the input operand or its digitwise complement, respectively, to the operand in the ACC-MD, store the sum in the ACC-MD, and also send it to the Checker. Additions that do not generate an end-around carry (eac) are completed in one cycle. If an eac is detected, the carry byte is sent to the Checker, a check is performed on this incomplete result, and a second cycle is employed to add the eac to the contents of ACC-MD. In case of overflow the "additive overflow" singularity code follows the result on the output lines.

The *multiply* algorithm requires that the multiplicand $15X$ should be already in the ACC-MD register; it is placed there by clear-add command, or is left there by the preceding operation. One cycle is employed to load the multiplier $15Y$ into the MQ register and to derive sign information for the product in a duplexed sign-test circuit. The multiplier is then sensed one byte at a time; consequently, there are eight radix-16 steps, each one consisting of from one to three cycles. The multiplier byte is recoded into a form possessing not more than two non-zero (±1) bits, and appropriate multiples (±1, ±2, ±4, ±8) of the multiplicand are added to the partial product in PR. Every partial result (nine bytes, plus the eac byte) is also delivered to the Checker to check the addition. None, one, or two cycles may be required. Each step is concluded with a contraction-and-shift cycle for the 9-byte partial result P^* of the preceding additions. This cycle rounds the partial result P^* to a 9-byte result P (which is a multiple of 16) by subtracting a constant $15N_i$ to get

$$P = P^* - 15N_i = 15(16T)$$

since

$$P^* = 15(16T + N_i).$$

Simultaneously, P is shifted one byte right (divided by 16) and the value of N_i $(0 \leqslant N_i \leqslant 15)$ is stored in one byte of MQ, which has delivered a multiplier byte for decoding and thus can accommodate the "correction byte" N_i. At the end of the eighth step, the PR contains the result

$$P^{(8)} = (15X)(15Y) - 15N = 15(15XY - N).$$

The terminal step is now performed to get an 8-byte rounded product P as a result. One cycle is used to divide $P^{(8)}$ by 15, and the 8-byte result

$$P^{(8)} \div 15 = 15XY - N$$

(which is usually not a multiple of 15) is returned to PR and the Checker. However, the value of N has been stored in the MQ and is now available to form the 16-byte result by attaching MQ at the right end of PR:

$$P_{\text{long}} = (15XY - N) + N.$$

During the next cycle, N is sent to the Checker and added to the check sum of $15XY - N$, which was not tested. Internally, the roundoff constant G is computed during this cycle from the bytes of N. The algorithm is concluded by adding the coded roundoff constant $15G$ to get the final 8-byte rounded result

$$P_{\text{rnd}} = (15XY - N) + N \pm 15G = 15(XY \pm G)$$

where G is the up- or down-rounding constant that would have been used for an uncoded product XY. Although one's complement is being used for subtractions, only nine bytes of the partial products were needed for the steps of the algorithm. The contract-and-shift cycle permits this time-saving variation of one's complement multiplication. The multiplication time for nonzero operands is

$$t_m = 1 + \sum_{i=0}^{7} (k_i + 1) + 3 \text{ cycles}$$

where k_i is the count of nonzero digits in the recoded ith multiplier byte. Since a multiplier can have as many as 16 and as few as two nonzero digits (one +1 and one −1) in its recoded form, the variation of t_m is

$$14 \leqslant t_m \leqslant 28 \text{ cycles.}$$

Zero operands are detected upon arrival and a zero result is delivered immediately, taking only two cycles.

The *divide* algorithm requires that the divisor $15Y$ should be already in the ACC-MD register. One cycle is used to load the single-length (8-byte) dividend $15X$ into the PR register and to record the signs in the duplex sign circuits. Tests for zero dividend (giving an immediate zero result) and zero divisor (a singularity) are performed during this cycle, and in either case only one additional cycle is needed to complete the algorithm. The divisor is returned to the Checker in case of a "zero divisor" singularity. In the nonzero cases the next cycle is used to form $15(15X)$, that is, to multiply the dividend by 15 in order to get a properly coded quotient $15Q$, such that

$$15^2X = (15Y)(15Q) + 15^2R$$

is satisfied, where R is the remainder in uncoded division. The quotient $15Q$ is then generated in eight radix-16 steps, one byte of the quotient at a time. The first cycle of each step is employed to finish restoring the remainder (if required by the preceding step) and to shift the remainder one byte left (multiplying by 16). Four cycles of quotient generation follow. First, the magnitude of the remainder is diminished by eight times the divisor. The sign of this result selects the leftmost bit of the new quotient byte and decides whether +4 (restoration) or −4 (further decrease) will be the next multiple of the divisor. If restoration is called for in the last cycle, it will be performed during the shift cycle of the next step. All partial results are returned to the PR register and also are sent to the Checker. Two cycles are needed for the terminal step: one completes the storage of the quotient into MQ, the other is used to move the quotient into ACC-MD and to deliver it to the Checker. The total time for division is

$$t_d = 2 + 5 \times 8 + 2 = 44 \text{ cycles.}$$

The "quotient overflow" singularity occurs whenever the allowed operand range $-1/30 < Q < 1/30$ is exceeded. This implies $|15Q| \lessapprox 1/2$ as a requirement, and the test is an attempt (during the first step, second cycle) to generate a bit of value 1 for the leftmost position of the coded-quotient magnitude; if it succeeds, the quotient will overflow. This condition leads to an immediate termination of the algorithm (in one cycle), and an output of a "zero" value pseudoresult and an appropriate singularity code. The remainder of division is not made accessible in this experimental model of the processor, although it can be obtained from the contents of PR.

IV. Conversion to an Inverse-Residue Code Processor

The AN-coded arithmetic processor of the STAR computer displays two disadvantages of the AN codes.

Disadvantage 1: One's complement arithmetic must be employed which is less convenient than two's complement, due to eac, especially in a byte-serial processor.

Disadvantage 2: The use of the nonseparate AN code makes the implementation of multiple-precision and floating-point operations relatively cumbersome.

In view of these limitations, the decision was made to design an inverse-residue code processor for two's complement arithmetic as a replacement for the original AN-code processor. The algorithms for residue code processors were known to exist [4]-[8]. The principal difficulty was posed by the adaptation of two's complement representation of negative numbers.

In the residue code processor (Fig. 3), the operands (X, Y) enter the Main Processor, while the check symbols $(X', Y'$, or $X'', Y'')$ enter the Check Processor. The operations of the two processors must be independent. Such independence is readily attainable for one's complement, following the argument used in the discussion of AN codes. In a two's complement processor a special condition occurs when the two's complement (modulo 2^n) Main Processor addition discards a left-end carry-out C_n. This reduction of the sum by 2^n requires the modulo A reduction of the residue sum in the Check Processor by the constant $A|2^n$. This constant has the value 1 for $A = 2^a - 1$ and $n = ka$. (The inverse-residue sum must be increased by 1 modulo $2^a - 1$.) The "correction signal" $C_n = 1$ that is conveyed from the Main Processor to the Check Processor makes them interdependent and raises the concern about the possibility of compensating errors. To prove that such compensating errors cannot occur we consider two cases for which the "correction signal" value is in error.

Case 1–"Correction Signal" $C_n = 1$ is Caused by an Error: An incorrect carry-out $C_n = 1$ will occur only due to the addition of an error value 2^j which will cause a carry $C_{j+1} = 1$ when the adder positions to the left $(j + 1, \cdots, n - 1)$ all have the carry-propagate condition (input bit pairs 0, 1 or 1, 0). The error value is $-2^n + 2^j$, the correction signal compensates only for the -2^n part, and the $+2^j$ error value remains detectable.

Case 2–Correction Signal $C_n = 1$ is Inhibited by an Error: This case is symmetric to the previous one. An error value of -2^j has prevented the carry-out $C_{j+1} = 1$, and this in turn has inhibited the signal $C_n = 1$ which should have been sent. The result $X + Y$ of the Main Processor differs by $2^n - 2^j$ from the correct result, and $2^n - 2^j$ is a multiple of $2^a - 1$ for the values $j = 0, a, 2a, \cdots, (k - 1)a$ when $n = ka$. This is an undetectable error for $A = 2^a - 1$ in general; however, it will be detected for all j because of the following. Since $C_n = 0$ is sent, the check processor computes the check sum $(2^n - 1)|(X'' + Y'')$, while the main processor sum is $X + Y - 2^j$ due to the error. The error value -2^j has not been compensated for in the check sum, and it remains detectable.

In both cases neither the Main Processor result nor the Check Processor result are equal to the correct values, but it has been shown that the check algorithm will always produce a disagreement, and the occurrence of the error will be detected. This proof overcomes the principal objection to two's complement arithmetic. The algorithms of the inverse-residue code processor in general follow those developed for one's complement arithmetic. The preceding two cases are illustrated in Examples 4 and 5.

Example 4–Case 1:

$X'' = 1011$	$X = 1111\ \ 0100$	
$Y'' = \underline{0101}$	$Y = \underline{0000\ \ 1010}$	
$15|(X''+Y'') = 0001$	$0\ 1111\ \ 1110$	correct sum S
correct $C_n = \underline{\ \ 0}$ ←	$\underline{\ \ \ \ 1\ \ \ \ }$	error adds 2^4
correct $S'' = 0001$	$1\ 0000\ \ 1110$	incorrect sum S^*
incorrect $C_n = \underline{\ \ 1}$ ←		
incorrect $S'' = 0010$	the error is detected.	

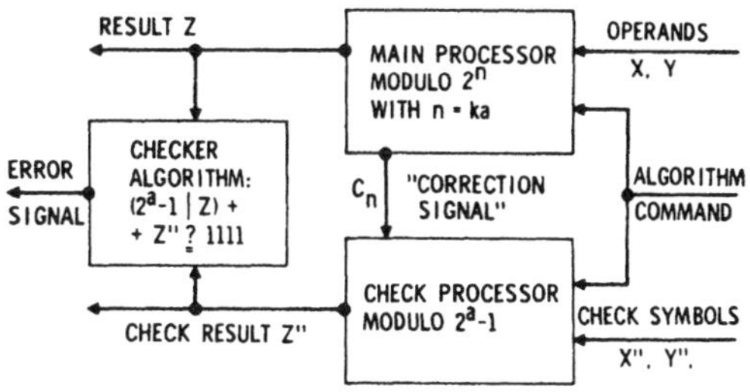

Fig. 3. Inverse residue code arithmetic processor.

Example 5–Case 2:

$X'' = 1011$	$X = 1111\ \ 0100$	
$Y'' = \underline{0100}$	$Y = \underline{0001\ \ 1010}$	
$15|(X'' + Y'') = 1111$	$1\ 0000\ \ 1110$	correct sum S
correct $C_n = \underline{\ \ +1}$ ←	$\underline{\ \ \ -1\ \ \ }$	error adds -2^4
correct $S'' = 0001$	$0\ 1111\ \ 1110$	incorrect sum
incorrect $C_n = \ \ 0$ ←		
incorrect $S'' = 1111$	the error is detected.	

APPENDIX

THE DIVISION ANOMALY

Given the dividend AX and the divisor AY, the restoring division algorithm will yield the results Q and AR such that

$$AX = Q \times AY + AR, \quad 0 \leqslant |AR| < |AY|, \text{or } 0 \leqslant |R| < |Y|$$

is satisfied, and the sign of AR is the same as the sign of AX. The dividend A^2X and divisor AY should yield AQ and A^2R such that

$$A^2X = AQ \times AY + A^2R, \quad 0 \leqslant |A^2R| < |A^2Y|,$$

$$\text{or } 0 \leqslant |R| < |Y|$$

is the required condition. However, the direct application of the restoring algorithm will give

$$A^2X = Q^* \times AY + R^*, \quad 0 \leqslant |R^*| < |AY|.$$

The largest quotient Q^* that is computed may not be the required multiple AQ of A, but exceed it by the value K:

$$Q^* = AQ + K.$$

This, in turn, will make R^* less than the required multiple $A^2R = A^2X - AQ \times AY$ of A^2:

$$R^* = A^2X - AQ \times AY - K \times AY$$

such that

$$A^2R = R^* + K \times AY, \quad 0 \leqslant K < A.$$

In radix 2^a division with $A = 2^a - 1$ we have

$$0 \leqslant K \leqslant 2^a - 2$$

as the value of the "quotient overshoot" K.

The correction of the quotient consists of choosing the last quotient digit q_0 such that the quotient has the value AQ and the remainder A^2R has the same sign as the dividend A^2X. This means that q_0 can be in the range

$$-(2^a - 2) \leqslant q_0 \leqslant 1.$$

The only positive value of q_0 can be $q_0 = 1$, since the division criterion gives the last partial remainder $R(1)$ such that

$$0 \leqslant |R(1)| < |AY| \text{ or } 0 \leqslant 2^a \times |R(1)| < 2^a \times |AY|.$$

When $A = 2^a - 1$, we need to satisfy $|A^2R| < |A^2Y|$ and this can be attained with $q_0 \leqslant 1$ because of what preceded.

In the STAR computer implementation of division, the value of q_0 is obtained by computing the modulo $A = 15$ residue N of all preceding radix 16 quotient digits:

$$N = 15|(q_7 + \cdots + q_1).$$

The value $N = 0$ indicates $q_0 = 0$. When $1 \leqslant N \leqslant 13$, it is the value of the "quotient overshoot" and $q_0 = -N$ is selected. When $N = 14$, the test is made for $q_0 = 1$; if this changes the sign of the remainder, then $q_0 = -14$ is selected. A negative value of q_0 requires a subtraction to generate the final quotient AQ.

ACKNOWLEDGMENT

The logic design of the processors was performed and construction of the AN-code processor was directed by D. A. Rennels and A. D. Weeks from the Spacecraft Computers Section of the Jet Propulsion Laboratory. The author wishes to thank them for numerous stimulating discussions, as well as D. K. Rubin and J. J. Wedel of the JPL who also have contributed valuable advice and constructive criticism.

REFERENCES

[1] W. W. Peterson, *Error Correcting Codes*. New York: Wiley, 1961, pp. 236–244.
[2] H. L. Garner, "Error codes for arithmetic operations," *IEEE Trans. Electron. Comput.*, vol. EC-15, pp. 763–770, Oct. 1966.
[3] A. Avižienis, "Arithmetic error codes: Cost and effectiveness studies for application in digital system design," *IEEE Trans. Comput.*, vol. C-20, pp. 1322–1331, Nov. 1971.
[4] R. M. Bloch, R. V. D. Campbell, and M. Ellis, "The logical design of the Raytheon computer," *Math. Tables and Other Aids to Computation*, vol. 3, pp. 286–295, 317–323, Oct. 1948.
[5] H. L. Garner, "Generalized parity checking," *IRE Trans. Electron. Comput.*, vol. EC-7, pp. 207–213, Sept. 1958.
[6] R. A. Davis, "A checking arithmetic unit," in *1965 Fall Joint Comput. Conf., AFIPS Conf. Proc.*, vol. 27. Montvale, N.J.: AFIPS Press, 1965, pp. 705–713.
[7] F. F. Sellers, Jr., M.-Y. Hsiao, and L. L. Bearnson, *Error Detecting Logic for Digital Computers*. New York: McGraw-Hill, 1968, ch. 8, 10.
[8] T. R. N. Rao, "Error-checking logic for arithmetic-type operations of a processor," *IEEE Trans. Comput.*, vol. C-17, pp. 845–849, Sept. 1968.
[9] A. Avižienis, "A set of algorithms for a diagnosable arithmetic unit," Jet Propulsion Lab., Pasadena, Calif., Tech. Rep. 32-546, Mar. 1964.
[10] —, "An experimental self-repairing computer," in *Information Processing '68, Proc. IFIP Congr. 1968*, vol. 2, A. J. H. Morrell, Ed. Amsterdam: North-Holland, 1969, pp. 872–877.
[11] —, "The diagnosable arithmetic processor," *Space Programs Summary No. 37-37* (Jet Propulsion Lab., Pasadena, Calif.), vol. 4, pp. 76–80, Feb. 1966.
[12] A. Avižienis, G. C. Gilley, F. P. Mathur, D. A. Rennels, J. A. Rohr, and D. K. Rubin, "The STAR (self-testing-and-repairing) computer: An investigation of the theory and practice of fault-tolerant computer design," *IEEE Trans. Comput.*, vol. C-20, pp. 1312–1321, Nov. 1971.

Algirdas Avižienis (S'55–M'56–F'73) was born in Kaunas, Lithuania, on July 8, 1932. He received the B.S., M.S., and Ph.D. degrees in electrical engineering from the University of Illinois, Urbana, in 1954, 1955, and 1960, respectively.

From 1955 to 1956 he was a Research Engineer at the Jet Propulsion Laboratory, California Institute of Technology, Pasadena. During graduate studies at the Digital Computer Laboratory, University of Illinois, he was a Fellow in 1954–1955 and 1956–1958, and a Research Assistant in 1959–1960, participating in the design of the Illiac II computer. In 1960 he rejoined the Jet Propulsion Laboratory and initiated the JPL self-testing and repairing (Star) computer research project. In 1962 he joined the faculty of the School of Engineering and Applied Science at the University of California, Los Angeles. He is currently Professor and Vice Chairman in the Department of Computer Science and conducts research in digital system design and fault tolerance. He has also remained associated with the Jet Propulsion Laboratory as principal investigator of the STAR computer research project.

Dr. Avižienis is a member of Sigma Xi, Tau Beta Pi, Eta Kappa Nu, and the Association for Computing Machinery, and is the Chairman of the Technical Committee on Fault-Tolerant Computing and a member of the Governing Board of the IEEE Computer Society. He has served as Chairman of the First International Symposium on Fault-Tolerant Computing (1971) and as Program Chairman of COMPCON 1972. In 1968 he received the Computer Society Honor Roll award for organizing and chairing the first Workshop on the Organization of Reliable Automata. He received the NASA Apollo Achievement Award in 1969.

ERROR DETECTION AND CORRECTION FOR ADDITION AND SUBTRACTION, THROUGH USE OF HIGHER RADIX EXTENSIONS OF HAMMING CODES

by
James E. Robertson

Department of Computer Science
University of Illinois at Urbana-Champaign
Urbana, Illinois 61801

Abstract

The properties of Hamming codes for error detection and correction can be extended from the binary parity check to addition, modulo 2r. Malfunctions in hardware during addition, modulo 2r, can be detected and corrected. Since carry-save and signed-digit addition, radix r, are included in addition, modulo 2r, this extension of Hamming codes makes possible new techniques for detection and correction of hardware malfunctions during signed-digit and carry-save addition.

1. Introduction

Binary Hamming codes are currently limited almost exclusively to detection and correction of errors during transmission or storage of information. Little or no use is made of the fact that binary Hamming codes can be used for detection and correction of errors resulting from hardware malfunctions for the Boolean operations of EXCLUSIVE OR and complementation.

It is the purpose of this paper to indicate that errors resulting from malfunctions in hardware for performing addition and subtraction in radix r, can be detected and corrected through use of Hamming codes extended to addition, modulo 2r. The first observation is the obvious one that the parity check, the Boolean EXCLUSIVE OR, addition modulo 2, and subtraction modulo 2, are identical operations. The generalization used here is therefore from addition, modulo 2 to addition, modulo 2r. The second observation is that the fact that binary Hamming codes can be used for detection and correction of hardware malfunctions during addition modulo 2 can be generalized to extended Hamming codes being used for detection and correction of hardware malfunctions during addition, modulo 2r. The third observation is that the operations of carry-save and signed-digit arithmetic, radix r, are subsets of addition, modulo 2r.

The details of the generalization for Hamming codes are given in Section 2. In particular, it is shown that the generalization depends only on the associativity of addition, modulo k.

In section 3, the discussion is particularized in several ways. The Hamming codes are extended to addition, modulo 4, for checking binary arithmetic. The eight digit single error correcting, double error detecting code is described in detail. In section 4, a specific numerical example is discussed for the radix 4 single error correcting, double error detecting extended Hamming code.

In section 5, the considerations determining the representation of a radix 4 digit by two binary digits are given. This lends to a specific logical design. Section 6 summarizes the problems encountered when negative digital values are employed.

2. Extension to Addition modulo k.

Let X represent an n-tuple of digits X_i (i=0,1,2,...,n-1) in radix k. One or more of the digits X_i, say X_j, is not independent, but is related to the other digits by the expression:

$X_j = \sum_{s_j} X_i$ (modulo k), where s_j designates a subset of all possible values of i, excluding j. The n-tuple Y and Z are similarly defined, with digits Y_j and Z_j related by addition (modulo k) of similarly designated subsets of Y and Z, respectively,

Consider the transformation Z = X + Y modulo k, which means $Z_i = X_i + Y_i$ modulo k for every i. In particular, Z_j is formed as $X_j + Y_j$, modulo k. Therefore $Z_j = X_j + Y_j$ (modulo k) $= \sum_{s_j} X_i + \sum_{s_j} Y_i$ (modulo k). By associativity of addition modulo k, $Z_j = \sum_{s_j} (X_i + Y_i)$ (modulo k) $= \sum_{s_j} Z_i$ (modulo k). Therefore, in the absence of error, the error protection properties associated with X_j and Y_j apply also to Z_j.

A check digit C_j is generated as $C_j = -Z_j + \sum_{s_j} Z_i$. In the absence of an error in the digits Z_i (i=j, or $i \epsilon s_j$), the check digit C_j is zero.

All of the equations in this section reduce to the binary Hamming codes if k = 2 and if it is recognized that addition modulo 2 and subtraction modulo 2 are the same.

3. Extension to Addition (modulo 4) of The Single Error Correcting, Double Error Detecting Hamming Code, With n = 8.

For the single error correcting, double error detecting Hamming code, the equations of the previous section for the extension to addition (modulo 4) become: n=8, i=0,1,2,3,4,5,6,7 j=0,1,2,4 $\quad s_0 = 1,2,3,4,5,6,7$

$s_1 = 3,5,7 \quad s_2 = 3,6,7 \quad s_4 = 5,6,7$

$$X_0 = \sum_{i=1}^{7} X_i \text{ modulo } 4 \qquad \text{Eqn 3.1a}$$

$$X_1 = X_3 + X_5 + X_7 \text{ modulo } 4 \qquad \text{Eqn 3.1b}$$

$$X_2 = X_3 + X_6 + X_7 \text{ modulo } 4 \qquad \text{Eqn 3.1c}$$

$$X_4 = X_5 + X_6 + X_7 \text{ modulo } 4. \qquad \text{Eqn 3.1d}$$

Similar expressions are applicable for the Y_i

If each Z_i (i=0,1,2,...,7) is defined as $Z_i = X_i + Y_i$ (modulo 4), then, for example,

$Z_1 = X_1 + Y_1$ (modulo 4)

$= X_3 + X_5 + X_7 + Y_3 + Y_5 + Y_7$ (modulo 4)

$= (X_3 + Y_3) + (X_5 + Y_5) + (X_7 + Y_7)$ (modulo 4)

$= Z_3 + Z_5 + Z_7$ (modulo 4)

The associated check digit C_1 is

$C_1 = -Z_1 + Z_3 + Z_5 + Z_7$ (modulo 4)

For error correction and detection, assume that the n–tuples X and Y have been checked and are free of error. An error E_i ($E_i \neq 0$, modulo 4) is introduced for one value of i during the addition $Z_i = X_i + Y_i$ modulo 4, so that $Z_i = X_i + Y_i + E_i$ (modulo 4). When the check digits C_j are generated from the Z_i, there are two cases:

1) i=3,5,6, or 7

$C_0 = E_i$				Eqn 3.2.1a
$C_1 = E_i$	(i = 3,5,7)	$C_1 = 0$	(i=6)	Eqn 3.2.2a
$C_2 = E_i$	(i=3,6,7)	$C_2 = 0$	(i=5)	Eqn 3.2.3a
$C_4 = E_i$	(i = 5,6,7)	$C_4 = 0$	(i = 3)	Eqn 3.2.4a

2) i= 0,1,2,4

$C_0 = E_i$	(i = 1,2,4)	$C_0 = -E_0$	Eqn 3.2.1b
$C_1 = 0$	(i = 0,2,4)	$C_1 = -E_1$	Eqn 3.2.2.b
$C_2 = 0$	(i = 0,1,4)	$C_2 = -E_2$	Eqn 3.2.3b
$C_4 = 0$	(i = 0,1,2)	$C_4 = -E_4$	Eqn 3.2.4b

In the first case (i = 3,5,6,7), a single error is indicated by the value of the error E_i in C_0, and by the same value E_i in two or more of C_1, C_2, and C_4 in a pattern indicative of the value of i, and hence the location of the error. In the second case (i=0,1,2,4) a single error is indicated by the value $-E_i$ in C_i. For i=0, the only error indication is $C_0 = -E_0$; for i=1,2 or 4, the error is indicated by $C_0 = E_i$ and $C_i = -E_i$.

4. A numerical example.

A numerical example of the binary addition 7+5+3 is given in table 4.1. For addition purposes, column 7 is assigned binary weight 8, column 6 is weight 4, column 5 is weight 2, and column 3 is weight 1. Row 1 is therefore the binary operand 7, row 2 is the operand 5, and row 9 is the operand 3. For these rows, the values in columns 4,2,1, and 0 are obtained by addition, modulo 4, of appropriate subsets of columns 7,6,5, and 3, in accordance with equations 3.1. It is assumed that all digits in rows 1,2,4 and 0 have been checked and are free of error, within the limitations of the error correction procedure in use.

A single binary addition requires the use of two simpler binary operations; the half-adder $b^0 \leftarrow a^0 + a^0$, followed by the carry generator $2a^0 + a^0 \leftarrow b^0 + a^0$ (reference 9.2). Theoretically, each digit in row 3 could be obtained from the corresponding digits of rows 1 and 2 by addition modulo 4; in practice, the arithmetic digits of columns 7,6,5, and 3 would be obtained using the simple hardware of the half adder $b^0 \leftarrow a^0 + a^0$. Malfunctions in the hardware used to generate the digits Z_i, of row 3, should be detectable and correctable (equations 3.2), within the limitations of the error correction method in use.

The X_i of row 4 are the Z_i of row 3, and the Y_i of row 5 are zeroes. The addition of the operands 7 and 5 is completed by combining rows 4 and 5 to form row 6, using carry generators for the arithmetic digits of columns 7,6,5, and 3, and using addition, modulo 4, for the check digits of columns 4,2,1, and 0. The sum, 12, is represented by the digits in row 6, as 8x0+4x2+2x1+1x2.

It is imperative that the digits in row 6 be checked and made free of error. Each of the radix 4 digits in columns 7,6,5, and 3 of row 6 are split into two binary digits, one a carry which is shifted left to appear in row 7, and the other a binary sum which appears in row 8. For example, the 2 in row 6, column 6, becomes the carry of 1 in row 7, column 7, and the binary sum digit 0 in row 8, column 6. Thus the value 12, represented as binary-weighted radix 4 digits in row 6, becomes two conventional binary numbers whose values are 10 in row 7, from the shifted carries, and 2 in row 8, the binary sum digits. Because of the shift of the carries, it is necessary at this point to calculate, using equations 3.1, the digits of columns 4,2,1, and 0, of rows 7 and 8.

Row / i	7	6	5	3	4	2	1	0
1 X_i	0	1	1	1	2	2	2	1
2 Y_i	0	1	0	1	1	2	1	2
	$b^0 \leftarrow a^0 + a^0$				$c^0 = c^0 + c^0$ (mod. 4)			
3 Z_i	0	2	1	2	3	0	3	3
4 X_i	0	2	1	2	3	0	3	3
5 Y_i	0	0	0	0	0	0	0	0
	$2a^0 + a^0 \leftarrow a^0 + b^0$				$c^0 = c^0 + c^0$ (mod. 4)			
6 Z_i	0	2	1	2	3	0	3	3
7 K_i	1	0	1	0	2	1	2	3
8 X_i	0	0	1	0	1	0	1	3
9 Y_i	0	0	1	1	1	1	2	2
	$b^0 \leftarrow a^0 + a^0$				$c^0 = c^0 + c^0$ (mod. 4)			
10 Z_i	0	0	2	1	2	1	3	1
11 X_i	0	0	2	1	2	1	3	1
12 Y_i	1	0	1	0	2	1	2	3
	$2a^0 + a^0 \leftarrow a^0 + b^0$				$c^0 = c^0 + c^0$ (mod. 4)			
13 Z_i	1	0	3	1	0	2	1	0
14 K_i	0	1	0	0	1	1	0	3
15 X_i	1	0	1	1	2	2	3	2

Table 4.1 A numerical example of binary addition.

Thereafter, the addition of the third operand, of value 3, is performed in a manner similar to that of rows 1 through 8. The sum digits in row 8, of numerical value 2, are added to the third operand of value 3 in row 9, using a half adder, to generate the sum of value 5 in row 10. This sum, copied without change to become row 11, is added to the carry digits of value 10 in row 12, copied from row 7. The sum, whose value is 15 is represented as binary-weighted radix 4 digits in row 13, and collectively by the two binary numbers, the shifted carries of value 4 in row 14, and the binary sum digits of value 11 in row 15.

5. Representation of digits and logical design.

The problem of representation is one of establishing a correspondence between the four values of the radix 4 digit set $c^0 = \{0,1,2,3\}$ and the four states 00,01,10 and 11, of two binary variables η and y. Clearly, there are $4! = 24$ possible correspondences, which fall into 3 groups of 8 each under permutation and negation of η and y. For addition, it is reasonable to impose the condition that the value 0 of digit set c^0 shall be represented by state 00. This reduces the number of correspondences to the six shown in Table 5.1, which form three pairs of correspondences under permutation of η and y.

η	y	c_a^0	c_b^0	c_c^0	c_d^0	c_e^0	c_f^0
0	0	0	0	0	0	0	0
0	1	1	3	1	2	2	3
1	0	2	2	3	1	3	1
1	1	3	1	2	3	1	2

Table 5.1
Correspondencies between four states and four values.

Under permutation of η and y, $c_a^0=c_d^0, c_b^0=c_e^0$, and $c_e^0=c_f^0$. Since permutation reduces to a matter of nomenclature, further consideration is given only to c_a^0, c_b^0, and c_c^0.

The final consideration which determines the correspondence is that of separability of the radix 4 digit c^0 into two binary digits, a carry and a sum, as exemplified by the generation of rows 7 and 8 from row 6, (and of rows 14 and 15 from row 13) in Table 4.1 Separability requires the relationship $c^0=2\eta+y$, which indicates that the correspondence indicated by column c_a^0 in Table 5.1 should be used.

The other correspondences c_b^0 and c_c^0 can be made separable with very little hardware (an EXCLUSIVE OR gate in each case), but this hardware would have to be used at a particularly critical point. With reference to Table 4.1, the digits c^0 in row 6 (or row 13) can be checked and corrected, implying that the individual binary digits are also correct. If the correspondence is separable, it is reasonable to assume that the binary digits in rows 7 and 8 (or rows 14 and 15) are correct, and that it is safe to compute the check digits of columns 4,2,1, and 0 for these rows.

The logical design for a radix four adder (mod. 4) is completely determined once the correspondence c_a^0 of Table 5.1 is chosen. In terms of states, the addition table is that of Table 5.2.

	y_i (η_i, y_i)				
x_i (ξ_i, x_i)	00	01	10	11	
00	00	01	10	11	sum z_i
01	01	10	11	00	
10	10	11	00	01	(ζ_i, z_i)
11	11	00	01	10	

Table 5.2
Addition table for $c_a^0=C_a^0+c_a^0$ (mod.4)

Table 5.2, in turn, can be converted directly into the truth tables necessary for logical design, and leads to the design equations:

$$\zeta_i = \xi \oplus \eta_i \oplus x_i y_i \qquad \text{Eqn. 5.1}$$

$$z_i = x_i \oplus y_i \qquad \text{Eqn. 5.2}$$

The notation for these equations is given by $Z_i=X_i+Y_i$, modulo 4, $Z_i=2\zeta_i+z_i$, $X_i=2\xi_i+x_i$, and $Y_i=2\eta_i+y_i$.

All of the additions (modulo 4) implicit in the operations of Table 4.1 are of the type given by Equations 5.1 and 5.2.

6. Extension to negative values and subtraction.

If all variants of half adders and carry generators are to be used, consideration must be given to digit sets with negative values, as exemplified by those in Tables 6.1 and 6.2.

η	y	c_a^0	c_a^1	c_a^2	c_a^3
0	0	0	0	0	0
0	1	1	1	1	$\bar{3}$
1	0	2	2	$\bar{2}$	$\bar{2}$
1	1	3	$\bar{1}$	$\bar{1}$	$\bar{1}$

Table 6.1. Format a for four-valued digit sets.

η	y	c_b^0	c_b^1	c_b^2	c_b^3
0	0	0	0	0	0
0	1	3	$\bar{1}$	$\bar{1}$	$\bar{1}$
1	0	2	2	$\bar{2}$	$\bar{2}$
1	1	1	1	1	$\bar{3}$

Table 6.2. Format b for four-valued digit sets.

In the previous section, it was noted that separability requires that the format choice $c_a^0=2\eta + y$ be made. Similarly, for digit sets with negative values, separability necessitates the unique choices $c_b^1=2\eta - y$ $c_a^2=-2\eta + y$, and $c_b^3=-2\eta - y$. To summarize, format a must be used for even values of offset and format b must be used when the offset is odd.

Consider Table 5.2, the addition table for $z_i = x_i + y_i$, modulo 4, when the digit sets for z_i, x_i, and y_i, are all of type c_a^0. It can be shown that when z_i, x_i and y_i are all of format a, regardless of offset, then table 5.2 applies, and therefore the same logical design of equations 5.1 and 5.2 should be used. Similiarly, table 5.2 and equations 5.1 and 5.2 are valid if z_i, x_i, and y_i are all represented using format b. If the formats of z_i, x_i, and y_i are mixed, then a different addition table, and hence a different logical design, will result.

It is proposed that all adders, modulo 4, used shall be os the same logical design. This choice necessitates that the inputs to a carry generator must be of the same format as

b^0	α	a	c_a^0	η	y	c_b^1	η	y
0	0	0	0	0	0	0	0	0
1	0	1	1	0	1	1	1	1
2	1	0	2	1	0	$\underline{2}$	1	0
			3	1	1	1	0	1

Table 6.3 Two interpretations of digit set b^0.

that of imposed by the output. Table 6.3 is an example to illustrate that this is possible. For format a, $b^0 \epsilon c_a^0$, and the adder inputs are $\eta = \alpha$ and y = a. For format b, $b^0 \epsilon c_b^1$, and the adder inputs are $\eta = \alpha \vee$a and y = a. The fact that all inputs to half adders and carry generators are either two-valued or three-valued indicates that an interpretation for either format choice can always be made.

7. Comments on the Radix 4 Example

Of the two bits which comprise a radix 4 digit, the least significant, or sum bit, is also a parity bit in the sense of the binary Hamming code, for the particular representations chosen in the example. Thus the properties of the binary Hamming code apply, independent of the most significant, or carry, bit.

With proper account taken of the negative values in equations 3.2, it seems possible to identify both the amount and the location of any single error E_i, and hence to correct it. It also seems likely that a single sum error $E_i = 1$ and a single carry error $E_j = 2$, can both be corrected, whether or not i=j. When $E_i = 1$ and $E_j = 1$, with i≠j, the consequent carry indications provide additional information not available for the binary Hamming code. This case warrants further investigation.

The hardware cost of generating the carry digits in the adders modulo 4 is roughly twice as great as the binary parity check. Another disadvantage is that, for generality, two different interpretations of the same digit set are necessary, as exemplified by table 6.3.

8. Conclusions

The previous sections represent a preliminary investigation into higher radix extensions of Hamming codes. Many directions for further investigation suggest themselves.

It is also evident that all the Boolean operations, when performed in parallel, can be checked and corrected, since the AND operation is implicit in the half adder, and the NOT operation is a special case of the EXCLUSIVE OR.

The approach here also suggests that the binary parity check be considered a count modulo 2 of the errors that have occurred, and suggests that the effect of counting the errors to a greater value than two be investigated.

The extensive research inspired by Hamming's original discoveries should be reviewed with higher radix applications to arithmetic in mind.

9. Bibliography

9.1 Hamming, R. W.; Error Detecting and Error Correcting Codes, Bell System Technical Journal, Volume XXIX (April, 1950) pp. 147-160.

9.2 Robertson, J.E. A Systematic Approach to the Design of Structures of Arithmetic, Proc. 5th Symposium on Computer Arithmetic, May 1981.

Acknolwedgement:

This research was supported in part by the National Science Foundation under grant NSF MCS 83-08576.

Part II:
ON-LINE ARITHMETIC

Editor's Comments on Papers 6 Through 13

6. **Avizienis**
   Signed-Digit Number Representations for Fast Parallel Arithmetic

7. **Ercegovac and Lang**
   On-Line Arithmetic: A Design Methodology and Applications in Digital Signal Processing

8. **Trividi and Ercegovac**
   On-Line Algorithms for Division and Multiplication

9. **Watanuki and Ercegovac**
   Error Analysis of Certain Floating-Point On-Line Algorithms

10. **Zaccone and Barlow**
    Improved Normalization Results for Digit On-Line Arithmetic

11. **Irwin and Owens**
    Fully Digit On-Line Networks

12. **Ercegovac and Lang**
    On-Line Scheme for Computing Rotation Factors

13. **Ercegovac and Lang**
    On-the-Fly Conversion of Redundant into Conventional Representations

Digit on-line arithmetic (frequently called on-line arithmetic) is a process for performing arithmetic on a serial (by digit) basis. All on-line arithmetic processors accept inputs and generate outputs in a most significant digit first format, since the on-line implementation of division requires that data be received with the most significant digit first. Redundant number representations are used for the digits to avoid carry propagation in addition and subtraction.

With their serial data flow, on-line arithmetic processors can be pipelined (i.e., cascaded) directly to perform complicated calculations. Additionally, they may be internally pipelined at rates limited only by the time required to calculate a single digit. Elimination of carry propagation allows on-line operations to be overlapped. Application specific systems benefit especially because arithmetic operations can be overlapped by starting operations as soon as digits become available from previous operations (i.e., it is not necessary to wait until the previous operations have been completed).

The first paper describes the use of signed-digit arithmetic to implement high speed arithmetic processors. By reducing carry propagation to one digit position, signed-digit arithmetic forms the basis for cascadable on-line arithmetic algorithms. Avizienis presents the properties of signed-digit numbers, gives algorithms for addition, subtraction, multiplication, division, and rounding, and concludes with a discussion of a signed-digit adder.

The second paper presents an overview of on-line arithmetic and a methodology for the design of special purpose on-line processors. This paper by Ercegovac and Lang provides a description of the properties of on-line arithmetic, describes example algorithms for fixed point addition, multiplication, and division, shows how elementary operations are combined to implement composite calculations, and ends with an IIR filter example that shows that on-line processors can achieve significant speedup factors relative to conventional processors.

In the next paper, Trivedi and Ercegovac present on-line algorithms for division and multiplication. On-line division algorithms for nonredundant and redundant operands are presented as is an on-line multiplication algorithm for redundant operands.

The fourth paper, by Watanuki and Ercegovac, presents floating-point on-line algorithms for addition and multiplication. The floating-point numbers use conventional binary exponents and signed digit significands (mantissas). The algorithms produce quasi-normalized results, where at least one of the two most significant digits is non zero. The paper presents the algorithms, derives bounds on the absolute and maximum relative representational errors, and presents simulations of the error arising in repeated addition and the Jacobi two dimensional boundary value problem.

The fifth paper examines normalization requirements and techniques in greater depth. Zaccone and Barlow develop several normalization schemes and apply them to on-line addition, multiplication, and division. They also present a post normalization algorithm that can be used to normalize numbers in an on-line system.

Irwin and Owens examine an on-line processor architecture in the next paper. The paper presents the requirements for an on-line primitive processing component, shows how these components are concatenated to perform larger algorithms, and concludes with a comparison of two on-line systems, one based on a systolic array and the other based on a pipeline network architecture.

In the seventh paper, Ercegovac and Lang demonstrate the application of on-line methods to the computation of rotation factors for matrix transformation (e.g., Given's rotation for QR decomposition). This computation is a good example as it requires multiplication, addition, square rooting, and division. The on-line implementation is roughly three times faster than a conventional implementation.

The final paper, by Ercegovac and Lang, provides a method for converting on-line results to conventional (two's complement) binary numbers. This capability is necessary if on-line computation elements are to be integrated into larger systems that use conventional binary numbers.

Signed-Digit Number Representations for Fast Parallel Arithmetic*

ALGIRDAS AVIZIENIS†, MEMBER, IRE

Summary—**This paper describes a class of number representations which are called *signed-digit* representations. Signed-digit representations limit carry-propagation to one position to the left during the operations of addition and subtraction in digital computers. Carry-propagation chains are eliminated by the use of redundant representations for the operands. Redundancy in the number representation allows a method of fast addition and subtraction in which each sum (or difference) digit is the function only of the digits in two adjacent digital positions of the operands. The addition time for signed-digit numbers of any length is equal to the addition time for two digits. The paper discusses the properties of signed-digit representations and arithmetic operations with signed-digit numbers: addition, subtraction, multiplication, division and roundoff. A brief discussion of logical design problems for a signed-digit adder concludes the presentation.**

I. Introduction

THIS PAPER describes a class of number representations which are called *signed-digit* representations. Signed-digit representations limit carry propagation to one position to the left during the operations of addition and subtraction in digital computers. Carry-propagation chains are eliminated by the use of redundant representations for the operands. In a *conventional number representation* with an integer radix $r>1$ each digit is allowed to assume exactly r values: $0, 1, \cdots, r-1$. In a *redundant representation* with the same radix r each digit is allowed to assume more than r values.

Previous methods of carry elimination formed redundant representations by the combination of explicitly identified digits, such as stored carries or borrows, with conventional number representations.[1–6] A difficulty in redundant representations of the stored-carry type is the lack of unique representation for the zero algebraic value of a number; the former sign digit (now indicator digit) does not always indicate the true sign of a number and overflow detection becomes more complicated. Furthermore, the number of states used to represent the values of one digital position is doubled by the binary stored-carry digit; this amount of redundancy is excessive.

In the method described here, each digit of a positional constant radix number representation with an integer radix r is allowed to assume q values

$$r + 2 \leq q \leq 2r - 1,$$

that is, more than the r values allowed in the conventional representation.[7] Both positive and negative digit values are allowed for this purpose. Redundancy in the number representation allows a method of fast addition and subtraction in which each sum (or difference) digit is the function only of the digits in two adjacent digital positions of the operands. These operations are called *totally-parallel* addition and subtraction. The requirement for totally-parallel addition and subtraction determines the minimum redundancy ($r+2$ values) which is necessary in the representation of one digit. The upper limit for the redundancy of digit values results from the requirement for a unique representation of the zero algebraic value of a number. To satisfy this requirement the magnitude of allowed digit values may not exceed $r-1$.

Requirements of totally-parallel addition and subtraction and of a unique representation for the zero value are satisfied by a class of redundant representations for radices $r>2$ which are called *signed-digit* representations. The digits of a signed-digit representation individually assume both positive and negative integer values and contain the sign information for the number; no special sign digit is necessary. This property leads to the name of "signed-digit" representations. The number of digit values in a radix $r>2$ representation ranges from a required minimum of $r+2$ to an allowable maximum of $2r-1$.

The following sections of this paper describe the properties of individual digits and of complete numbers in signed-digit representations. After development of these properties, arithmetic operations for signed-digit representations are discussed. Examples are given of

* Received by the PGEC, March 23, 1961; revised manuscript received, May 22, 1961.

† Jet Propulsion Lab., California Inst. Tech., Pasadena, Calif.

[1] A. W. Burks, *et al.*, "Preliminary Discussion of the Logical Design of an Electronic Computing Instrument," Institute for Advanced Study, Princeton, N. J.; pt. 1, vol. I, 2nd ed. revised; September, 1947.

[2] J. E. Robertson, "Preliminary Design of an Arithmetic Unit for Use with a Self-Checking Binary Parallel Digital Computer," Digital Computer Lab., University of Illinois, Urbana, Ill., Internal Rept. No. 19; June, 1950.

[3] Staff of the Digital Computer Lab., "On the Design of a Very High-Speed Computer," University of Illinois, Urbana, Ill., Rept. No. 80, ch. 8; October, 1957.

[4] G. Metze, "A Study of Parallel One's Complement Arithmetic Units with Separate Carry or Borrow Storage," Ph.D. dissertation, University of Illinois, Urbana, Ill., 1958; Digital Computer Lab., University of Illinois, Rept. No. 81; November 11, 1957.

[5] G. Metze and J. E. Robertson, "Elimination of carry propagation in digital computers," *Proc. Internatl. Conf. on Information Processing,* UNESCO, Paris, France, June 15–20, 1959; pp. 389–396.

[6] J. E. Robertson, "Redundant number systems for digital computer arithmetic," notes for the University of Michigan Engrg. Summer Conf., "Topics in the Design of Digital Computing Machines," Ann Arbor, Mich.; July 6–10, 1959.

[7] A. Avizienis, "A Study of Redundant Number Representations for Parallel Digital Computers," Ph.D. dissertation, University of Illinois, Urbana, Ill., 1960; Digital Computer Laboratory, University of Illinois, Rept. No. 101; May 20, 1960.

signed-digit addition, subtraction, multiplication and division. The paper concludes with some logical design considerations for a signed-digit adder.

II. Properties of Signed-Digit Representations

In this section the class of signed-digit representations is derived according to four requirements which are postulated as necessary for number representations in fast parallel arithmetic.

A. Requirements for Signed-Digit Representations

The purpose of signed-digit representations is to allow addition and subtraction of two numbers in which no serial signal propagation is required along the adder; that is, the time duration of the operation is independent of the length of the operands and is equal to the time required for the addition or subtraction of two digits. Furthermore, the signed-digit representations must have a unique representation of zero algebraic value of a number. The desired principal properties of signed-digit representations are specified by the following description of a signed-digit number:

A *signed-digit number* is represented by $n+m+1$ digits z_i ($i=-n, \cdots, -1, 0, 1, \cdots, m$) and has the algebraic value

$$Z = \sum_{i=-n}^{m} z_i r^{-i},$$

where the values of r and z_i are such that the following requirements are satisfied:

1) The radix r is a positive integer.
2) The algebraic value $Z=0$ has a unique representation.
3) There exist transformations between the conventional representation and the signed-digit representation for every algebraic value Z within a specified range.
4) Totally-parallel addition and subtraction is possible for all digits in corresponding positions of two representations.

The arithmetic operations of totally-parallel addition and subtraction of two digits z_i and y_i from the corresponding ith positions of the representations of numbers Z and Y are defined as follows:

Definition 1: Addition of digits z_i and y_i is totally-parallel if the following two conditions are satisfied:

1) The *sum digit* s_i (ith digit of the sum $S=Z+Y$) is a function only of the *augend digit* z_i, *addend digit* y_i and the *transfer*[8] *digit* t_i from the $(i+1)$th position on the right: $s_i=f(z_i, y_i, t_i)$.

2) The transfer digit t_{i-1} to the $(i-1)$th position on the left is a function only of the augend digit z_i and the addend digit y_i: $t_{i-1}=f(z_i, y_i)$.

Definition 2: Totally-parallel subtraction of the *subtrahend digit* y_i from the *minuend digit* z_i is performed as the totally-parallel addition of the additive inverse of y_i, *i.e.*, $z_i-y_i=z_i+(-y_i)$.

Definitions 1 and 2 impose limiting conditions on the values which the digits z_i and y_i may assume. The addition of two digits is performed in two successive steps. First, an outgoing transfer digit t_{i-1} and an *interim sum digit* w_i are formed:

$$z_i + y_i = rt_{i-1} + w_i. \tag{1}$$

Then the sum digit s_i is formed:

$$s_i = w_i + t_i. \tag{2}$$

Definition 1 will be satisfied if the range of values which s_i may assume in (2) does not exceed the allowed range of values for the digits z_i and y_i in (1). The block diagram of a totally-parallel adder for signed-digit representations is shown in Fig. 1.

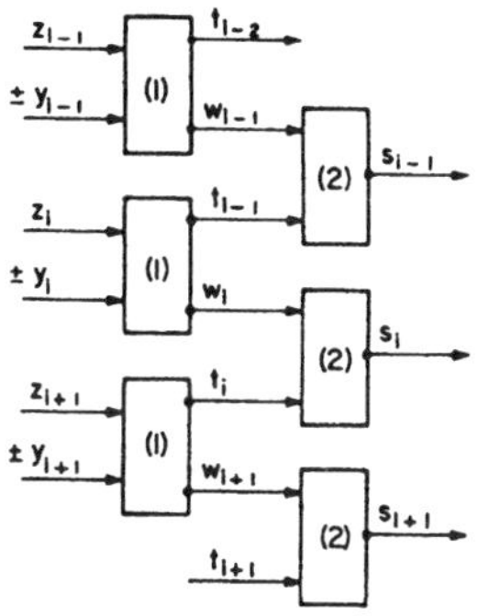

Fig. 1—Section of a totally-parallel adder for signed-digit representation.

Definition 2 will be satisfied if, for every allowed nonzero value of the digit y_i, there exists an additive inverse in the set of all allowed values of y_i;

for every $y_i=a$, there exists $y_i=-a$ such that $a+(-a)=0$. (3)

The requirement for unique representation of the zero value of a number will be satisfied by the condition

$$|z_i| \leq r - 1. \tag{4}$$

The requirement for conversion from the conventional sign and magnitude, m-digit representation to signed-digit representation (m digits long), will be satisfied if the procedure of totally-parallel addition, applied to the nonsign digits x_i of the conventional representation,

$$x_i = rt_{i-1} + w_i \tag{5}$$

$$z_i = w_i + t_i, \tag{6}$$

[8] The term "transfer digit" is used here instead of the commonly used terms "carry" or "borrow" for two reasons:

1) the transfer digit may assume both positive and negative values in either addition or subtraction;
2) unlike the "carry" or "borrow" of conventional addition or subtraction, the transfer digit is never propagated past the first adder position on the left.

will yield an allowed value of digit z_i for every positive and negative value of the digit x_i ($x_i=0, 1, \cdots, r-1$) in the conventional radix r representation. Conversion from signed-digit to conventional representation is always possible, since condition (4) allows the interpretation of any signed-digit representation as the sum of two conventional sign and magnitude representations, one consisting of the positive and one of the negative digits of the signed-digit representation.

B. Required and Allowed Values of Digits

The conditions (1)–(6) of the preceding section determine the required and allowed values of one digit z_i for an arbitrary integer radix $r \geq 2$. As the first step, the required and allowed values of t_i and w_i in (1) and (2) must be established. The conditions (1)–(4) establish[9] the set of values

$$t_i = -1, 0, 1 \tag{7}$$

as the least sufficient set of values for the transfer digit t_i, and the condition

$$|w_i| \leq r - 2 \tag{8}$$

as the upper limit for the magnitude of the interim sum w_i if t_i is limited by (7). Values of $|t_i| > 1$ have no practical application for two-digit operations, but they are necessary for simultaneous totally-parallel addition and subtraction of several digits which will be discussed later. An immediate result of (8) is the restriction of allowed values of radix r to the integer values

$$r > 2 \tag{9}$$

because, for $r=2$, the only allowed value of $w_i=0$ does not satisfy (5) for the value $x_i=1$.

Requirements (5), (7) and (8) establish that at least r values of w_i are required if all $2r-1$ positive and negative values of x_i are to be transformed according to (5), and that the values $w_i = -1, 0, 1$ should be included for all $r>2$. The relationship between the greatest value w_{max} and smallest value w_{min} of w_i is:

$$w_{max} - w_{min} \geq r - 1, \tag{10}$$

where the equality sign applies if only r values of w_i are chosen. *The required set of values of w_i therefore must* consist of r integers, which include $-1, 0, 1$ and satisfy (8) and (5) for all $2r-1$ values of x_i. More than one such set may exist, but the choice of a sequence symmetric around zero is preferable in further development. *The set of all allowed values of w_i* is unique and consists of $2r-3$ integers from $-(r-2)$ to $r-2$.

The values of digits w_i and t_i and the totally-parallel subtraction requirement stated in (3) determine which values of z_i are required and how many values of z_i are allowed for a digit z_i of a radix $r \geq 3$ signed-digit representation. The digit z_i assumes the minimum number of values when the required set of w_i is chosen as a sequence of r integers $\{w_{min}, \cdots, -1, 0, 1, \cdots, w_{max}\}$ in which additive inverses exist for all (when r is odd) or all but one (when r is even) nonzero values of w_i. In both cases the required values of the sum digit s_i in (2) consist of a sequence of $r+2$ integers: $\{w_{min}-1, w_{min}, \cdots, -1, 0, 1, \cdots, w_{max}, w_{max}+1\}$.

For an odd radix $r_0 \geq 3$ we choose $w_{max} = -w_{min} = \frac{1}{2}(r_0-1)$; here the additive inverse exists for every value of w_i and s_i, and *the required (minimum) set of values for digits z_i or y_i* in (1) consists of the sequence of r_0+2 integers

$$\{-\tfrac{1}{2}(r_0+1), \cdots, -1, 0, 1, \cdots, \tfrac{1}{2}(r_0+1)\}. \tag{11}$$

For an even radix $r_e \geq 4$, either $(w_{min}-1)$ or $(w_{max}+1)$ in the set of the required values of s_i has no additive inverse. This additive inverse is required as a subtrahend input to the adder (y_i) in order to satisfy (3). *The required (minimum) set of values for the subtrahend digit y_i* consists of the sequence of r_e+3 integers

$$\{-(\tfrac{1}{2}r_e+1), \cdots, -1, 0, 1, \cdots, \tfrac{1}{2}r_e+1\}. \tag{12}$$

The minimum set of values for the sum digit s_i and the augend (or minuend) digit z_i requires only r_e+2 integers and either $\frac{1}{2}r_e+1$ or $-(\frac{1}{2}r_e+1)$ is omitted from (12) to give this set. Only r_e+2 digit values require storage, but the adder must accept r_e+3 values of the subtrahend digit y_i. For instance, given $r=4$, the values $-2, -1, 0, 1, 2, 3$ are sufficient to represent any sum or difference, but the adder must also accept the subtrahend digit value $y_i=-3$ as the additive inverse of $y_i=3$ during a subtraction.

C. Types of Signed-Digit Representations

The sets (11) and (12) are the required (minimum) sets of digit values which satisfy the requirements for signed-digit representations. They are the only allowed sets for radix 3 (values $-2, -1, 0, 1, 2$) and radix 4 (values $-3, -2, -1, 0, 1, 2, 3$). For all $r>4$ there exists more than one set of allowed digit values. Since the maximum allowed absolute value of a digit is $r-1$ according to (4), all sequences of integers,

$$\{-a, -(a-1), \cdots, -1, 0, 1, \cdots, a-1, a\}$$
$$\tfrac{1}{2}(r_0+1) \leq a \leq r_0 - 1 \text{ or } \tfrac{1}{2}r_e + 1 \leq a \leq r_e - 1; \tag{13}$$

r_0 is an odd integer $r_0 \geq 3$,

and r_e is an even integer $r_e \geq 4$,

will satisfy the requirements for signed-digit representation. *The maximum allowed set of values* occurs when $a=r-1$ and consists of $2r-1$ integers. For instance, two sets exist for radix 5, one with 7 values (-3 to 3) and one with 9 values (-4 to 4). Four sets exist for radix 10, from 13 values (-6 to 6) to 19 values (-9 to 9).

All signed-digit representations, *i.e.*, representations which satisfy the requirements 1) to 4) of Section II-A, may now be described in terms of the allowed values of radix r and digits z_i as follows:

[9] Derivations of the results presented here appear in the Appendix

Definition 3: The *class of signed-digit representations* consists of all positional, constant radix representations of algebraic values

$$Z = \sum_{i=-n}^{m} z_i r^{-i}$$

in which the digits z_i assume one of the allowed sets of values given by (13) and the radix is a positive integer $r>2$. The *redundancy* of a signed-digit representation is *minimal* when $a=\frac{1}{2}(r_0+1)$ or $a=\frac{1}{2}r_e+1$, and the redundancy is *maximal* when $a=r_0-1$ or $a=r_e-1$ in (13).

A consequence of restriction (4), $|z_i| \leq r-1$, is an upper limit to the maximum weighted sum of all digits z_i $(i=k+1, \cdots, m)$ to the right of an arbitrary digit z_k in the signed-digit representation of the algebraic value Z, which is

$$\sum_{i=k+1}^{m} |z_i|_{\max} r^{-i} \leq r^{-k}(1 - r^{-m+k}) < r^{-k}. \quad (14)$$

Important consequences of (14) are as follows:

1) The sign of the algebraic value Z is indicated by the sign of the most significant nonzero digit.
2) The algebraic value Z is zero if, and only if, all digits of its signed-digit representation have the value $z_i=0$.
3) Given a signed-digit representation of the algebraic value Z, the signed-digit representation of $-Z$ is formed by changing the sign of the value of every nonzero digit z_i.

By repetitive formation and propagation of transfer digits, a signed-digit representation may be converted to a *canonical form* in which the values of all digits z_i are in the chosen set of the values of w_i; that is, no more transfer digits can be formed. A maximum of m additions of the number to zero may be required to put an m-digit number into the canonical form. If the set of values of w_i is the minimal (required) set, we obtain the *minimal canonical form*, which is nonredundant, since its digits assume only r values.

D. Range of Numbers and Overflow Indication

Generally, the range for algebraic values Z of fixed-point numbers (or mantissas in floating point arithmetic) is chosen as $1>Z>-1$ (or 1 or -1 may be included) in conventional number representations. Two considerations determine the choice of a convenient range for signed-digit representations: 1) the entire conventional range $1 \geq Z \geq -1$ should be covered, and 2) a simple, immediate method of overflow indication should exist.

These two requirements will be satisfied if the absolute range limits for the algebraic value

$$Z = \sum_{i=0}^{m} z_i r^{-i}$$

of a signed-digit representation are defined to be

$$1 + |z_i|_{\max} \sum_{i=2}^{m} r^{-i} \geq Z \geq -1 - |z_i|_{\max} \sum_{i=2}^{m} r^{-i}. \quad (15)$$

Overflow is detected from the inspection of the two most significant digits z_0 and z_1 as follows:

positive overflow occurs if $z_0 = 1$ and

$$z_1 \geq 1, \text{ or if } z_0 > 1, \quad (16)$$

negative overflow occurs if $z_0 = -1$ and

$$z_1 \leq -1, \text{ or if } z_0 < -1. \quad (17)$$

In a redundant representation one algebraic value may be represented in more than one way. For this reason overflow will be indicated for some representations of numbers near the limits, but within the allowed maximum range, of (15). For the maximum allowed range of (15), and the overflow detection rules (16) and (17), there exist three ranges of overflow indication:

1) *Certain overflow indication*, irrespective of representation, will occur if Z exceeds the maximum range of (15)

$$|Z| > 1 + \frac{r^{-1} - r^{-m}}{r-1} |z_i|_{\max}. \quad (18)$$

2) *Potential overflow indication*, dependent on the specific representation of Z, covers the range

$$1 + \frac{r^{-1} - r^{-m}}{r-1} |z_i|_{\max} \geq |Z| \geq 1 + \frac{1}{r} - \frac{r^{-1} - r^{-m}}{r-1} |z_i|_{\max}. \quad (19)$$

3) *No overflow indication* for any representation of Z will occur in the range

$$|Z| < 1 + \frac{1}{r} - \frac{r^{-1} - r^{-m}}{r-1} |z_i|_{\max}. \quad (20)$$

The no-overflow range is minimum when $|z_i|_{\max} = r-1$. In this case, the range of (20) reduces to

$$|Z| < 1 + r^{-m}. \quad (21)$$

The range $1 \geq Z \geq -1$ is therefore covered by every signed-digit representation and is considered as the allowed range for the scaling of numbers. It must be noted that an overflow indication may not occur until the algebraic value of a number reaches the certain overflow indication range of (18). Other ranges and more precise overflow detection may be chosen in an analogous manner.

E. Conversion Procedures

The conversion of a conventional (sign and magnitude) representation to a signed-digit representation of the same radix $r>2$ occurs according to (5) and (6).

Given the conventional representation of

$$Z = (1 - 2x_s) \sum_{i=-n}^{m} x_i r^{-i},$$

in which the digits x_i range over r values $0, 1, \cdots, r-1$, and $x_s = 0$ if Z is positive and $x_s = 1$ if Z is negative, the conversion to signed-digit representation is performed as follows:

1) Choose the set of allowed values $(w_{\min}, \cdots, -1, 0, 1, \cdots, w_{\max})$ of the interim sum w_i.
2) Interpret each digit x_i $(i = -n, \cdots, -1, 0, 1, \cdots, m)$ to be negative if the sign digit $x_s = 1$.
3) Consider each digit x_i as the sum of two digits in signed-digit representation and recode according to (5)

$$w_i = x_i - r t_{i-1},$$

where

$$t_{i-1} = 0 \text{ if } w_{\min} \leq x_i \leq w_{\max}$$

$$t_{i-1} = 1 \text{ if } x_i > w_{\max}$$

$$t_{i-1} = -1 \text{ if } x_i < w_{\min}.$$

4) Form the digits z_i of the signed-digit representation: $z_i = w_i + t_i$ for $i = -n, \cdots, -1, 0, 1, \cdots, m$, and $z_{-n-1} = t_{-n-1}$.

The conversion from signed-digit to a conventional representation may be performed by several methods. A direct approach is to consider the signed-digit number as the sum of two numbers in conventional representation of the same length, one of which is positive and the other negative. Negative and positive digits are separated to form these two numbers, which then can be added in an adder for conventional representations to yield the desired conventional representation. Conversion can also be executed in a serial manner, starting with the least significant digit of the signed-digit representation.[7]

F. Simultaneous Addition of Several Digits

Previous discussion of totally-parallel addition and subtraction was limited to simultaneous addition of two digits and therefore to two numbers only. However, rules (1) and (2) may be extended to simultaneous totally-parallel addition of an arbitrary number n of digits if values of the transfer digit $|t_i| > 1$ are allowed and the radix r is sufficiently large.

The conditions (1), (2), (4) and (10) yield the upper and lower limits for the greatest absolute value of t_i as follows:

$$\frac{n-1}{r-1} |z_i|_{\max} \leq |t_i|_{\max} \leq |z_i|_{\max} - |w_i|_{\max}. \quad (22)$$

The values of t_i are chosen as a sequence of integers ranging from $-|t_i|_{\max}$ to $|t_i|_{\max}$, and the values of w_i must satisfy (2) and (4); *i.e.*, $|w_i + t_i| \leq r - 1$ for all values of t_i and w_i. Consequently, the number n of digits which may be added simultaneously is a function of the radix r and of the maximum allowed magnitudes of digits z_i and w_i

$$n \leq r - \frac{|w_i|_{\max}}{|z_i|_{\max}} (r - 1). \quad (23)$$

If the greatest $|z_i|_{\max}$ and the least $|w_i|_{\max}$ (see Appendix) are used,

$$n \leq \tfrac{1}{2}(r_0 + 1) \quad \text{and} \quad n \leq \tfrac{1}{2} r_e \quad (24)$$

is the upper limit for n. For instance, a maximum of two digits may be added simultaneously for radix 4, three digits for radix 5, and five digits for radix 10.

A total of n numbers in signed-digit representation may be simultaneously added (or subtracted) in the totally-parallel mode and will yield a sum or difference in signed-digit representation if (22) and (23) are satisfied.

G. Modified Signed-Digit Representations

The requirement of totally-parallel addition and subtraction restricts signed-digit representations to radices $r > 2$. Furthermore, at least $r+2$ values are required for the sum digit s_i in (2), and the subtraction requirement increases the required number of subtrahend digit values to $r+3$ for even radices, as shown in (12). The digit addition rules (1) and (2) may be modified to allow the propagation of the transfer digit over two digital positions to the left.[10] If this type of *two-transfer* addition is allowed, the radix $r = 2$ may be used and only $r+1$ values are required for the sum digit. Two-transfer addition is executed in three successive steps (the adder diagram is shown in Fig. 2):

$$1) \quad z_i' + y_i' = r t_{i-1}' + w_i' \quad (25)$$

$$2) \quad w_i' + t_i' = r t_{i-1}'' + w_i'' \quad (26)$$

$$3) \quad s_i' = w_i'' + t_i''. \quad (27)$$

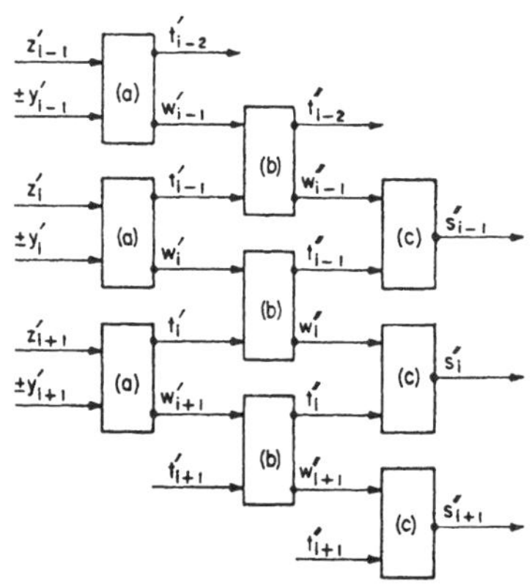

Fig. 2—Section of a two-transfer adder for modified signed-digit representation.

[10] J. E. Robertson, private communication.

The digits z_i', y_i' and s_i' are digits of a *modified* signed-digit representation.

For any radix $r \geq 2$, w_i' assumes $r+1$ values. For any *even* radix r_e, the values of w_i' range from $\frac{1}{2}r_e$ to $-\frac{1}{2}r_e$. In step 1) a transfer digit is generated whenever possible, *i.e.*, whenever $|z_i'+y_i'| \geq \frac{1}{2}r_e$, $|t_{i-1}'| = 1$; in step 2) a transfer digit is generated only if it is mandatory, *i.e.*, $|t_{i-1}''| = 1$ only if $|w_i'+t_i'| = \frac{1}{2}r_e+1$. Under these conditions t_i' and t_i'' cannot be both $+1$ or both -1, and s_i' assumes only r_e+1 values which are the same as the values of w_i'

$$\{-\tfrac{1}{2}r_e, \cdots, -1, 0, 1, \cdots, \tfrac{1}{2}r_e\}. \tag{28}$$

These r_e+1 values are the required values of one digit z_i' or y_i' for any even radix $r_e \geq 2$ if addition is performed according to (25)–(27). For instance, for the radix $r=2$ the required values are $-1, 0, 1$; for $r=4$ the required values range from -2 to 2; and for $r=10$ from -5 to 5.

For any *odd* radix $r_0 \geq 3$ the same procedure of addition applies if r_0+1 values of w_i' are chosen, ranging from $\pm\frac{1}{2}(r_0+1)$ to $\mp\frac{1}{2}(r_0-1)$. Only r_0+1 values of s_i' are required, which are the same as the values of w_i'. However, the principal advantage of two-transfer addition is lost because one more value of a digit must be introduced to provide the additive inverse of $\pm\frac{1}{2}(r_0+1)$, and r_0+2 values of a subtrahend digit y_i' are required at the input of the adder.

The potential advantages of two-transfer addition and of the resulting modified signed-digit representations are the use of radix $r=2$ (with digit values -1, 0, 1) and the decreased redundancy requirement (only $r+1$ values of a sum digit are required), especially for even radices. The disadvantages are the more complicated process of addition and the greater length of transfer digit propagation. It is interesting to observe that the procedure of two-transfer addition corresponds to the method of addition used for the "coincident carry-borrow storage" representation.[4]

Generally, it may be concluded that the lower limit on the required redundancy of one digit is a function of the number of digital positions over which a signal is allowed to propagate. If no redundancy exists and each sum digit assumes only r values, a sum digit s_i is the function of all the addend digits z_i and augend digits y_i to the right, *i.e.*, $s_i = f(z_i, y_i, z_{i+1}, \cdots, z_m, y_m)$. If each sum digit assumes $r+1$ values, we have $s_i = f(z_i, y_i, z_{i+1}, y_{i+1}, z_{i+2}, y_{i+2})$, and the operation is two-transfer addition. If each sum digit assumes $r+2$ values or more, we have $s_i = f(z_i, y_i, z_{i+1}, y_{i+1})$ with the restriction $r \geq 3$, and the addition is totally-parallel.

III. Signed-Digit Arithmetic

The preceding section presented the properties of a class of number representations which satisfy the four requirements 1) to 4) postulated in Section II-A. Given the class of signed-digit representations described by Definition 3, this section will describe the execution of addition, subtraction, shifting, multiplication and division for signed-digit numbers.

A. Elementary Arithmetical Operations

Addition, subtraction, left shifting and right shifting are the elementary operations for a signed-digit arithmetic unit. Multiplication and division are executed as sequences of additions or subtractions and shifts.

The addition of two signed-digit numbers is performed as the totally-parallel addition of all corresponding digits z_i and y_i according to (1) and (2). The block diagram of a signed-digit adder is shown in Fig. 1. To perform subtraction, first the sign of the subtrahend is changed by inverting the signs of all nonzero digits, then totally-parallel addition of corresponding digits is performed. Given the allowed values of w_i as the sequence: $w_{\min}, \cdots, -1, 0, 1, \cdots, w_{\max}$; the rules for forming w_i, t_{i-1} and s_i are as follows:

$$w_i = (z_i + y_i) - rt_{i-1}, \tag{29}$$

where

$$\begin{aligned} t_{i-1} &= 0 \text{ if } w_{\min} \leq z_i + y_i \leq w_{\max}, \\ t_{i-1} &= 1 \text{ if } z_i + y_i > w_{\max}, \\ t_{i-1} &= -1 \text{ if } z_i + y_i < w_{\min}; \\ s_i &= w_i + t_i. \end{aligned} \tag{30}$$

It is interesting to note that serial addition or subtraction of signed-digit numbers may be initiated with the most significant digits. Example 1 demonstrates the addition of two radix 10 signed-digit numbers.

Example 1—Signed-Digit Addition (Radix 10): The allowed digit values are: for w_i: 5, 4, 3, 2, 1, 0, $\bar{1}$, $\bar{2}$, $\bar{3}$, $\bar{4}$, $\bar{5}$;

for t_i: 1, 0, $\bar{1}$;

for s_i, z_i, y_i: 6, 5, 4, 3, 2, 1, 0, $\bar{1}$, $\bar{2}$, $\bar{3}$, $\bar{4}$, $\bar{5}$, $\bar{6}$.

Negative values are identified by a bar above the integer. The radix 10 signed-digit operands are:

augend z: $1.\bar{3}65\bar{1}\bar{4}$, algebraic value $Z=0.76486$
addend y: $0.\bar{4}053\bar{1}$, algebraic value $Y=-0.39471$.

The procedure of addition is as follows:

augend z:	1.	$\bar{3}$	6	5	$\bar{1}$	$\bar{4}$
addend y:	0.	$\bar{4}$	0	5	3	$\bar{1}$
Step 1):	0+1.	$\bar{1}\bar{0}$+3	10+$\bar{4}$	10+0	0+2	0+$\bar{5}$
Step 2):	$\bar{1}$	1	1	0	0	
sum s:	0.	4	$\bar{3}$	0	2	$\bar{5}$

The sum is s: $0.4\bar{3}02\bar{5}$, algebraic value $S=0.37015$.

For the absolute range limits of (15), the limits for the sum or difference of two numbers $S = Z \pm Y$ are

$$|S| \leq 2 + 2\frac{r^{-1} - r^{-m}}{r - 1}|z_i|_{\max}. \tag{31}$$

The overflow detection rules (16) and (17) assure that $|s_0| \leq 2$ will always hold and that $|t_{-1}| = 1$ may exist only for radix $r=3$, since $|w_0| = 2$ is allowed for all radices $r>3$. The interim sum digits w_0, w_1 and transfer digits t_{-1}, t_0, t_1 contain all the necessary overflow information before s_0 and s_1 have been formed, and the overflow indication may be generated simultaneously with the sum digits. Immediate standardization may be performed in floating-point addition.

Shifting (left and right) of signed-digit numbers is performed in the conventional manner. If the digits z_i of a number pass through the adder logic during the shifting operation, transfer digits t_{i-1} will be generated when $z_i > w_{max}$ or $z_i < w_{min}$ and added to the interim sum w_{i-1} at the left according to (29) and (30). During the right shift of one position, $z_0=0$ is inserted as the most significant digit of the shifted number. If this shift includes transfer generation, the least significant digit z_m in the shifting register will retain the property $|z_m| \leq |z_i|_{max}-1$, which is useful in multiple precision operations. Before the left shift of one position, the inspection of digits z_1 and z_2 will predict overflow after the shift. If the left shift includes transfer generation, digit z_3 must be inspected for a potential transfer digit generation.

B. Multiplication

Signed-digit multiplication is performed as a sequence of additions or subtractions and right shifts. Given the multiplicand z and the $m+1$ digits-long-multiplier y

$$\left(\text{algebraic value } Y = \sum_{i=0}^{m} y_i r^{-i}\right)$$

in radix $r \geq 3$ signed-digit representations, the product $p_{m+1} = zy + p_0 r^{-m}$ is formed by the following recursive process:

$$p_{m+1} = p_m + zy_0 \tag{32}$$

$$p_{j+1} = \frac{1}{r}(p_j + zy_{m-j}) \quad \text{for } j = 0, 1, 2, \cdots, m-1. \tag{33}$$

Here the p_{j+1} are partial products, p_0 is the initial number in the accumulator register and y_{m-j} is the multiplier digit sensed during the jth step of multiplication.

Temporary overflow may occur in the sum $p_j + zy_{m-j}$ before the right shift which forms p_{j+1}. If p_j and z are in range, $(|p_j| \leq |z|_{max}, |z| \leq |z|_{max})$, we have

$$|p_j + zy_{m-j}| \leq |z|_{max}(|y_{m-j}|_{max} + 1), \tag{34}$$

where $|z|_{max}$ is the maximum allowed magnitude of a number. Since $|y_{m-j}|_{max} \leq r-1$ always holds, p_{j+1} will also be in range. To store the temporary sum (before right shift) one more storage position (s_{-1}) must be allowed.

It is necessary to note that although both z and y do not indicate overflow, the product zy may indicate overflow if either z or y or both were in the potential overflow indication range given by (19). A test for overflow is necessary at the end of multiplication. If the temporary overflow storage position s_{-1} is available, the correct value of the product is retained in case of overflow.

Example 2 demonstrates signed-digit multiplication for radix 10 numbers.

Example 2—Signed-Digit Multiplication (Radix 10): The allowed digit values are given in Example 1. The radix 10 signed-digit operands are

multiplicand z: $0.\bar{4}62$, algebraic value $Z=-0.338$
multiplier y: $1.\bar{3}1\bar{5}$, algebraic value $Y=0.705$.

The procedure of multiplication is as follows:

step	digit y_{m-j}	quantity	representation	next adder operation
$j=0$	$y_3=\bar{5}$	p_0	0.000	
		zy_3	$2.\bar{3}\bar{1}0$	add
		p_0+zy_3	$2.\bar{3}\bar{1}0$	shift right
		p_1	$0.2\bar{3}\bar{1}0$	
$j=1$	$y_2=1$	zy_2	$0.\bar{4}62$	add
		p_1+zy_2	$0.\bar{2}310$	shift right
		p_2	$0.0\bar{2}310$	
$j=2$	$y_1=\bar{3}$	zy_1	1.014	add
		p_2+zy_1	$1.00\bar{3}10$	shift right
		p_3	$0.100\bar{3}10$	
$j=3$	$y_0=1$	zy_0	$0.\bar{4}62$	add
		p_4	$0.\bar{2}\bar{4}2\bar{3}10$	end

The product $zy=p_4$ is $0.\bar{2}\bar{4}2\bar{3}10$, algebraic value $ZY=-0.238290$.

It should be noted that the greatest absolute value of the multiplier digit y_{m-j} in (33) is $\frac{1}{2}r_e+1$ for even radices and $\frac{1}{2}(r_0+1)$ for odd radices when the multiplier y is in the minimal redundancy representation. By serial recoding of y (before sensing of y_{m-j}) to the minimal canonical (nonredundant) form, the greatest absolute value of y_{m-j} is reduced to $\frac{1}{2}r_e$ for even radices and $\frac{1}{2}(r_0-1)$ for odd radices. In this case, for the radix 3, the values of y_{m-j} are -1, 0, 1; for radix 4 they are -1, 0, 1 and 2 or -2; for radix 10 the ten required values are -4 to 4 and 5 or -5. Since y_{m-j} has values from 0 to $r-1$ in conventional representation, fewer additions (or fewer multiple-generating circuits) will be necessary on the average for one multiplication in signed-digit representation.

Eq. (24) shows that $n \leq \frac{1}{2}(r_0+1)$ or $n \leq \frac{1}{2}r_e$ digits can be simultaneously added for signed-digit numbers. If the multiplier y is recoded to the minimal canonical form, only one addition per step of (33) is sufficient for any odd radix $r_0 \geq 3$ when each stage of the adder is designed to accept up to $\frac{1}{2}(r_0-1)$ inputs of the multiplicand digit z_i. The resulting product will be in the maximal redundancy form. For any even radix $r_e \geq 4$, two additions per step of (33) will be needed when $|y_{m-j}| = \frac{1}{2}r_e$. For all smaller magnitudes of y_{m-j} one addition per step is sufficient. This shows that the use of higher redundancy in the representation and of greater adder complexity permits very fast signed-digit multiplication.

C. Division

Signed-digit division is performed as a sequence of additions or subtractions and left shifts. The method of division which is most readily applicable to signed-digit representations is due to Robertson.[11] The representation of the quotient digits in this method may be redundant; however, the redundancy allows an inexact selection of quotient digits. In a signed-digit representation the sign of the dividend or a partial remainder is not readily available for inspection if several most significant digits are zero. The magnitude, however, may be estimated, with a limited uncertainty, from the inspection of a few most significant digits. The Robertson division method which requires such an estimate is therefore preferable to restoring or nonrestoring division, both of which require knowledge of the sign for exact selection of quotient digits.

Given the dividend z and the divisor d, the quotient digits q_j are generated by the following recursive process:

$$p_{j+1} = rp_j - dq_{j+1} \quad \text{for} \quad j = 0, 1, 2, \cdots, m-1 \tag{35}$$

$$p_0 = z - dq_0. \tag{36}$$

Here the p_j are partial remainders, p_m is the remainder and q is the quotient ($m+1$ digits long) with the algebraic value

$$Q = \sum_{i=0}^{m} q_i r^{-i}.$$

During each step of division the quotient digit q_{j+1} must be chosen which has a value such that the next partial remainder p_{j+1} in (35) is within the same allowed range as p_j. This range is a function of the magnitude of divisor d. If $|z|$ is not in this range, the choice of $|q_0| = 1$ in (36) must bring p_0 into allowed range; otherwise the quotient overflows. Every quotient digit q_{j+1} in (35) is assigned the value which satisfies the condition

$$|rp_j - dq_{j+1}| \leq c|d|, \tag{37}$$

where c is the *range test constant* whose allowed range is determined by the choice of the allowed values of the quotient digit. A *comparison circuit* must be constructed which performs (37); the value $c = \frac{1}{2}$ is the most practical choice within the allowed range.[12] If the representation of the quotient digits is redundant (q_{j+1} assumes more than r values), the comparison (37) may be inexact, that is, it is sufficient to perform the comparison between truncated values of $|rp_j - dq_{j+1}|$ and $|d|$. To facilitate the comparison (37) the divisor d must be standardized before division; otherwise, very great precision of comparison is required when $|d|$ is small.

The choice of allowed values for the quotient digits q_j is governed by two considerations. The least number of additions (or multiple-generating circuits) will be required for one division when the digits q_j assume the least possible number of values. However, at least minimal redundancy in representation of q_j is necessary to allow an inexact selection of quotient digits. In case of minimal redundancy the values of q_j range from $-\frac{1}{2}r_e$ to $\frac{1}{2}r_e$ (a total of $r+1$ values) for even radices $r_e \geq 4$, and from $-\frac{1}{2}(r_0+1)$ to $\frac{1}{2}(r_0+1)$ (a total of $r+2$ values) for odd radices $r_0 \geq 3$. For these values of q_j, it is sufficient to compare the first four digits of the test quantities in (37).

In the simplest mechanization of (35) and (36), the standardized divisor d is repetitively added to (if signs of d and rp_j disagree) or subtracted from (if signs of d and rp_j agree) the shifted partial remainder rp_j, until the condition

$$\left|\sum_{i=0}^{3} z_i r^{-i}\right| \leq \left|\frac{1}{2}\sum_{i=0}^{3} d_i r^{-i}\right| \tag{38}$$

is detected by the comparison circuit, indicating that p_{j+1} has been generated. Here d_i ($i = 0, 1, 2, 3$) are the first four digits of d and z_i ($i = 0, 1, 2, 3$) are the first four digits in the accumulator register, which contains the dividend z at the start of division, and later—the partial remainders and incomplete partial remainders. The value of q_{j+1} is equal to the number of additions (sign of q_{j+1} is minus) or the number of subtractions (sign of q_{j+1} is plus) required to generate p_{j+1}. If (38) is satisfied immediately after the left shift of p_j, $q_{j+1} = 0$ is the required value. In any case, p_{j+1} is now shifted left and the procedure repeated to generate q_{j+2} and p_{j+2} until $m+1$ quotient digits have been generated. A numerical example of this process of division is given in Example 3.

The above-presented (binary) version of Robertson division requires the least amount of special circuitry. Quotient digits q_j may be generated during one addition cycle if more comparator circuits and circuits which generate multiples of the divisor are added to the arithmetic unit.[11]

Example 3—Signed-Digit Division (Radix 10): The allowed digit values are given in Example 1. The values of quotient digits q_j are 5 to $\bar{5}$ inclusive. The radix 10 signed-digit operands are

dividend z: $0.\bar{2}\bar{3}6\bar{4}15$, algebraic value $Z = -0.224385$
divisor d: $1.\bar{3}1\bar{5}$, algebraic value $D = 0.705$.

The test quantity is

$$\left|\frac{1}{2}\sum_{i=0}^{3} d_i r^{-i}\right|,$$

algebraic value $T = 0.3525$.

[11] J. E. Robertson, "A new class of digital division methods," IRE Trans. on Electronic Computers, vol. EC-7, pp. 218–222; September, 1958.

[12] Derivations of the range test constant, the precision requirement in comparison, and other results presented here appear in the Appendix.

The procedure of division is as follows:

step	quantity	representation	test	next adder operation	q_{j+1}
$j=-1$	z	$0.\bar{2}\bar{3}6\bar{4}15$	$0.224<T$	shift	$q_0=0$
$j=0$	rp_0 d	$\bar{2}.\bar{3}6\bar{4}150$ $1.\bar{3}1\bar{5}$	$2.244>T$	add d	
	rp_0+d d	$\bar{2}.5\bar{4}1$ $1.\bar{3}1\bar{5}$	$1.539>T$	add d	
	rp_0+2d d	$\bar{1}.2\bar{3}\bar{4}$ $1.\bar{3}1\bar{5}$	$0.834>T$	add d	
	p_1	$0.\bar{1}\bar{3}1$	$0.129<T$	shift	$q_1=\bar{3}$
$j=1$	rp_1 d	$\bar{1}.\bar{3}11500$ $1.\bar{3}1\bar{5}$	$1.289>T$	add d	
	rp_1+d d	$\bar{1}.42\bar{4}$ $1.\bar{3}1\bar{5}$	$0.584>T$	add d	
	p_2	0.121	$0.121<T$	shift	$q_2=\bar{2}$
$j=2$	rp_2 $-d$	1.215000 $\bar{1}.3\bar{1}5$	$1.215>T$	add $-d$	
	rp_2-d $-d$	0.510 $\bar{1}.3\bar{1}5$	$0.510>T$	add $-d$	
	p_3	$0.\bar{2}05000$	$0.195<T$	end	$q_3=2$

The quotient q is $0.\bar{3}\bar{2}2$, algebraic value $Q=-0.318$; the remainder p_3 is $0.\bar{2}05$, algebraic value $P_3=-0.195\times10^{-3}$.

D. Roundoff and Multiple Precision Operations

Addition and subtraction of numbers whose length is a multiple k of the $m+1$ digits long adder is performed as a sequence of k ordinary additions or subtractions in signed-digit arithmetic. The least significant parts are brought up first, and if a transfer digit t_{-1} is generated out of the most significant adder stage, it is applied as an input transfer digit t_m during the following operation on the two next higher-significance parts of the operands. This is always possible because $t_m=0$ for any two single-length operands. It should, furthermore, be noted that a single-length number may be added to or subtracted from the more significant half of a double-length number without affecting the less significant half in any manner. This is due to the general property of signed-digit representations that every digit contains its own sign information and therefore any section of a representation is a complete signed-digit representation in its own right.

The operation of roundoff is often required to reduce a number to the standard length. Roundoff for signed-digit numbers is performed by truncation. If the allowed values of a digit z_i are chosen symmetrically around zero $(-a, \cdots, -1, 0, 1, \cdots, a)$ and we assume that every one of these values will occur with the same probability, then the average error which is introduced by truncation is zero and the roundoff is without bias.

The knowledge of the number of significant digits in a fixed-length representation may be desirable in digital computer arithmetic, especially in floating-point operations. Because roundoff is performed by truncation for signed-digit numbers, it is possible to maintain a continuous indication of nonsignificant digits during addition, subtraction and left shifts. A new digit value $z_i=0'$ called the *space-zero* (distinct from the *value-zero* $z_i=0$) may be introduced to identify nonsignificant positions in a signed-digit representation. The following rule applies to the space-zero value $0'$:

$$z_i \pm 0' = 0' \text{ for all values of } z_i. \tag{39}$$

In the first step of totally-parallel digit addition ($z_i \pm 0' = rt_{i-1}+w_i$), $t_{i-1}=0$ and $w_i=0'$ are formed; the second step yields the sum digit $s_i=w_i+t_i=0'+t_i=0'$. Space-zero value digits may appear only as a continuous string of digits extending to the left from the least significant position of a register.

The nonsignificant positions in input numbers are represented by space-zero digits. During left shifts and during standardization in floating-point arithmetic space-zero digits are placed in the least significant position of the shifting register. When two signed-digit numbers are added or subtracted, the space-zero digits in the number with fewer significant digits will round off the sum or difference to the lesser precision. The space-zero digits will maintain a continuous indication of precision for all signed-digit numbers in the machine.

IV. Logical Design of Signed-Digit Adders

One problem of practical interest is the logical design of an adder to perform totally-parallel addition and subtraction with numbers in signed-digit representation. A parallel adder for signed-digit numbers of m-digit length consists of m identical *digit-adders* (also stage-adders[7]) which are linked to their immediate neighbors by transfer digit lines, an output to the left and an input at the right. Fig. 1 shows a block diagram with the required interconnections, inputs and outputs of digit-adders. All transfer digits t_i and interim sum digits w_i are formed simultaneously, thus the addition time for numbers of any length is equal to the addition time for one digit-adder. The objective of a digit-adder is to execute the operations described by (29) and (30) for digits of a given signed-digit representation, which is defined by the choice of the radix r and the allowed values of digits z_i, y_i, w_i and t_i. The allowed ranges of choice have been derived in the preceding sections. Minimal-redundancy representations require the least storage capacity for the values of a digit and therefore are preferable to representations with higher redundancy. Furthermore, less complicated digit-adder logic may be expected when the least possible number of digit values is employed.

The choice of the radix r depends on the desired balance between the increase in storage requirements and the logical complexity of one digit-adder. The relative increase in storage capacity requirements diminishes when r is large; however, one digit-adder must accept more values of a digit and the logical circuits become more complex. Few digit values are also preferable in multiplication and division. For instance, radix 4 requires seven values (-3 to 3) for the digits of a subtrahend at the input to the digit-adder, at least six of these values (-3 or 3 may be avoided) for the sum digit s_i, and in storage, five values of the quotient digits (-2 to 2) and at least four of these values (-2 or 2 may be avoided) for the multiplier digits. Three binary storage elements per digit are required, compared to two for conventional radix 4 representations. Three binary storage elements per digit are sufficient for all radices $3 \leq r \leq 6$, four elements for radices $7 \leq r \leq 14$. The radix 10 (with 13 values) requires four elements per digit, the same as for conventional representations; the required values of quotient digits are -5 to 5. The space-zero value and additional redundancy may be introduced for radix 10 and many other radices without increasing the storage requirements.

The choice of the binary representation in storage of the allowed values for one digit affects the complexity of logical circuits which perform (29) and (30). A one digit-adder may be treated as a multiple-input and multiple-output switching circuit which is specified by these rules of addition. Alternatively, a digit-adder may be designed as a small conventional adder which consists of basic building blocks (such as half-adders) and has appropriate modifications to generate t_{i-1} and w_i according to the specifications of (29) and (30). Digit values are represented by binary digits corresponding to the storage elements, and conventional methods are used to represent negative values. The minimization of the logic for a digit-adder as a function of the specific storage representation of digit values, of the redundancy of digit representation, and of the choice of basic logic circuits remains as an interesting unsolved program for further investigation.

As an example, logical design has been performed for a radix 4 digit-adder.[13] The "building-block" approach was used in this design and conventional adder design techniques were employed. The seven values of a radix 4 digit (-3, -2, -1, 0, 1, 2, 3), were represented by three binary digits weighted -4, 2, and 1. The radix 4 digit-adder required the equivalent logical circuitry of approximately 12 half-adders of the conventional type: this is 2.5 to 3 times more than required by one position of a conventional radix 4 adder with serial carry propagation. The comparison is more favorable when the auxiliary circuitry of the conventional adders (carry acceleration, carry-completion sensing, conditional corrections) is considered. An investigation of a radix 10 digit-adder showed that it was about two times more complex than one stage of a conventional excess-three code adder. The experimental designs were intended to serve as an existence demonstration and an indication of the order of complexity of a digit-adder. No claim is made for minimal complexity of these designs; furthermore, design problems which are not directly affected by the totally-parallel addition requirements, such as gating, synchronization or timing, and arithmetic control were not considered.

[13] Avizienis, *op. cit.*, pp. 63–70.

V. Conclusion

A class of redundant number representations which permit the elimination of carry-propagation chains in addition and subtraction has been demonstrated. Unique representation exists for the zero value of a number, and all sign information is contained by the individual digits. No special sign digit is required; this eliminates preliminary and terminal corrections caused by sign digits.

Procedures for arithmetical operations have been described for the class of signed-digit representations. The time required for addition or subtraction is independent of the length of the operands; also, simultaneous addition of several numbers is possible. Convenient procedures exist for overflow detection, roundoff, and multiple-precision operations. Multiplier and quotient digits assume values which require fewer adder operations than an equivalent multiplication or division in conventional representation. A number in signed-digit representation requires more allowed values and more storage capacity per digit than the same number in conventional representation. The logical circuits are more complex for a signed-digit adder. The minimization of digit-adders remains a problem for further investigation.

A signed-digit number of arbitrary length consists of a positionally-ordered collection of complete one-digit long representations; this property leads to useful simplifications in arithmetical operations. Partition of a signed-digit number at any position yields two complete signed-digit numbers whose sum is equal to the original number. In a conventional parallel adder all iterative adder circuits are bound together by the requirement for carry propagation. In a signed-digit parallel adder, each iterative digit-adder depends only on its own inputs and the information generated by its immediate neighbor to complete one cycle of addition. It may be concluded that signed-digit representations offer the means for a completely parallel execution of arithmetical operations in a digital computer.

Appendix

The Division Comparison

The maximum allowed magnitude of a partial remainder p_j in (35) and (36) is a function of $|d|$, *i.e.*, $|p_j|_{\max} = k|d|$.

Two requirements are to be satisfied:

Requirement 1: The greatest partial remainder should be returnable into range after a shift by means of the greatest quotient digit $|q_j|_{\max}$

$$rk|d| - |d| \, |q_j|_{\max} \leq k|d|. \qquad (40)$$

Eq. (40) yields the upper bound of k as

$$k \leq \frac{|q_j|_{\max}}{r-1}.$$

Requirement 2: If the shifted partial remainder is outside the range ($|rp_j| = k|d| + \delta > k|d|$) by the least detectable difference δ, a single reduction of rp_j by d should yield a partial remainder within range, *i.e.*,

$$k|d| + \delta - |d| \geq -k|d|. \tag{41}$$

Eq. (41) yields the lower limit of k as

$$k \geq \frac{1}{2} - \frac{1}{2}\frac{\delta}{|d|} \approx \frac{1}{2},$$

where the approximation $k = \frac{1}{2}$ is satisfactory since $|d|_{\max} \geq 1$ always holds.

The division requirements are satisfied for any value of k in the range limited by the above requirements, that is, for

$$\frac{1}{2} \leq k \leq \frac{|q_j|_{\max}}{r-1}. \tag{42}$$

The value of k may be allowed to vary within the range of (42); therefore, an *inexact* comparison is possible in (37). The allowed error of truncation must be determined. The quantities which are compared in (37) are the divisor $d = d' + d''$ and the quantity $p = p' + p''$. Here p is first the dividend z, then a shifted partial remainder rp_j or a diminished shifted partial remainder.

The test is to detect the condition $|p'| \leq c|d'|^*$ where c is the *range test constant.* An uncertainty of comparison occurs because of the uncompared lower-significance parts d'' and p''. When both p and d are truncated behind the position $i = h$, we denote $|p''|_{\max} = |d''|_{\max} = \Delta$, and determine the maximum value of Δ which satisfies $|p_j|_{\max} = k|d|$. In the test $|p'| \leq c|d'|^*$ the precision of $c|d'|^*$ is the same as of $|p'|$, that is, $h+1$ digits. Therefore $c|d'| = c|d'|^* + e$, where $0 \leq e < r^{-h}$.

Two conditions must be satisfied:

Condition 1: When $|p'| \leq c|d'|^*$ is indicated, $|p| = k_{\max}|d|$ must hold under the worst conditions, which occur when $|p| = |p'| + \Delta$, $|d| = |d'| - \Delta$ and $|p'| = c|d'|^*$ with $e = 0$. This yields

$$\Delta_1 = |d| \frac{k_{\max} - c}{1 + c} \tag{43}$$

as the maximum allowable error of truncation.

Condition 2: When $|p'| > c|d'|^*$ is indicated, one more reduction of $|p|$ by $|d|$ should retain $|p| - |d|$ in range, that is, $|p| - |d| = -k_{\max}|d|$ must hold under the worst conditions which occur when $|p| = |p'| - \Delta$, $|d| = |d'| + \Delta$ and $|p'| = c|d'|^* + r^{-h}$ with $e = e_{\max}$. This yields

$$\Delta = |d| \frac{k_{\max} + c - 1}{1 + c} + \frac{r^{-h} - e_{\max}}{1 + c}, \tag{44}$$

and since $(r^{-h} - e_{\max})$ is a small positive quantity, a good approximation is

$$\Delta_2 = |d| \frac{k_{\max} + c - 1}{1 + c}. \tag{45}$$

From the above results, the condition $\Delta \geq 0$ applied to (43) and (45) yields the range

$$1 - k_{\max} \leq c \leq k_{\max}, \tag{46}$$

in which no truncation uncertainty is allowed ($\Delta = 0$) for $c = k_{\max}$ or $c = 1 - k_{\max}$, and an uncertainty given by (43) and (45) may exist for $1 - k_{\max} < c < k_{\max}$, where

$$k_{\max} = \frac{|q_j|_{\max}}{r-1} \text{ from (42).}$$

Setting $\Delta_1 = \Delta_2$, we get $c = \frac{1}{2}$ and

$$\Delta_{\max} = \tfrac{1}{3}(2k_{\max} - 1)|d| \tag{47}$$

as the greatest value of Δ as a function of c, since Δ_1 decreases and Δ_2 increases as c ranges from $1 - k_{\max}$ to $k_{\max}$.

When the divisor d is standardized, its least value is

$$|d|_{\min} = \frac{1}{r} + r^{-2}\left(1 - \frac{|z_i|_{\max}}{r-1}\right) + r^{-m}\frac{|z_i|_{\max}}{r-1} \tag{48}$$

for $r \geq 3$, the range of (15) and overflow rules (16) and (17). Condition (48) occurs when $d_0 = 0$, $d_1 = d_2 = 1$ and $d_i = -|z_i|_{\max}$ for $i = 3, 4, \cdots, m$. Therefore $|d|_{\min} \geq 1/r$ holds for all signed-digit representations, and for standardized divisors we have

$$\Delta_{\max} = \frac{1}{3r}\left(2\frac{|q_j|_{\max}}{r-1} - 1\right). \tag{49}$$

This equation indicates the required precision of comparison for any choice of $|q_j|_{\max}$. For example, if $|q_j|_{\max} = r - 1$ (maximum redundancy case), we have $\Delta_{\max} = 1/3r$, and it is sufficient to compare positions $i = 0, 1, 2$. For minimal redundancy and even radices

$$r_e \geq 4, \quad |q_j|_{\max} = \tfrac{1}{2}r_e \quad \text{and} \quad \Delta_{\max} = \frac{1}{3r_e(r_e - 1)};$$

here the positions $i = 0, 1, 2, 3$ must be compared. For minimal redundancy and odd radices

$$r_0 \geq 3, \quad |q_j|_{\max} = \tfrac{1}{2}(r_0 + 1) \quad \text{and} \quad \Delta_{\max} = \frac{2}{3r_0(r_0 - 1)};$$

here the positions $i = 0, 1, 2, 3$ must be compared except for $r = 3$, where $i = 0, 1, 2$ are sufficient.

Derivation of Allowed and Required Digit Values

The conditions (1)–(4) establish the restrictions

$$2|z_i|_{\max} \leq r|t_i|_{\max} + |w_i|_{\max} \tag{50}$$

$$|t_i|_{\max} + |w_i|_{\max} \leq |z_i|_{\max} \tag{51}$$

on the maximum absolute values of t_i, w_i and z_i (or s_i and y_i). Given $|z_i|_{max}=r-1$, we have the limiting conditions

$$2(r-1) = r|t_i|_{max} + |w_i|_{max} \quad (52)$$

$$r-1 = |t_i|_{max} + |w_i|_{max} \quad (53)$$

which are simultaneously satisfied by $|t_i|_{max}=1$ and $|w_i|_{max}=r-2$, leading to (7)-(9).

In the conversion procedure described by (5) and (6), the $2r-1$ values of x_i (from $-r+1$ to $r-1$) are to be recoded as $x_i=rt_{i-1}+w_i$. The values $x_i=-1, 0, 1$ require corresponding values $w_i=-1, 0, 1$, because (8) does not allow their recoding with $|t_i|>0$. At least r values of w_i are needed to recode all $2r-1$ values of x_i. In this case, $w_i=a$ is generated for each pair of values $x_i=a$ and $x_i=-(r-a)$, except for $w_i=0$ which is generated for $x_i=0$ only. Furthermore, in this case $w_{max}+1$ is recoded as $r+w_{min}$, where w_{max} is the greatest allowed value and w_{min} is the least allowed value of w_i, giving

$$w_{max} - w_{min} = r - 1, \quad (54)$$

which is the limiting case of (10). The value $|w_i|_{max}$ is least when we choose $w_{max}=|w_i|_{max}=\frac{1}{2}(r_0-1)$ for odd radices r_0 and $w_{max}=|w_i|_{max}=\frac{1}{2}r_e$ for even radices r_e.

In simultaneous totally-parallel addition of n digits, the restrictions are

$$|z_i|_{max} = |w_i|_{max} + |t_i|_{max} \quad (55)$$

$$n|z_i|_{max} \leq r|t_i|_{max} + |w_i|_{max}, \quad (56)$$

which give the lower limit of $|t_i|_{max}$ as

$$|t_i|_{max} \geq \frac{n-1}{r-1}|z_i|_{max}. \quad (57)$$

The upper limit of $|t_i|_{max}$ is given by (55) as

$$|t_i|_{max} \leq |z_i|_{max} - |w_i|_{max}. \quad (58)$$

The greatest range of $|t_i|_{max}$ is obtained when the greatest allowed value of $|z_i|_{max}=r-1$ and the least allowed value of $|w_i|_{max}=\frac{1}{2}r_e$ or $|w_i|_{max}=\frac{1}{2}(r_0-1)$ is used in (22). These choices also lead to the maximum values of n given in (24).

ACKNOWLEDGMENT

The author wishes to thank Prof. J. E. Robertson of the Digital Computer Laboratory, University of Illinois, Urbana, Ill., for his comments on this paper and for guidance and critical advice during the writing of the doctoral dissertation[7] on which this paper is based.

CHAPTER 24

On-Line Arithmetic: A Design Methodology and Applications in Digital Signal Processing

Miloš D. Ercegovac and Tomas Lang

Computer Science Department
School of Engineering and Applied Science
University of California, Los Angeles
Los Angeles, CA 90024

Abstract. A method to design on-line algorithms is presented and its application illustrated for addition, multiplication, and division. It is shown that the resulting implementations have a canonical structure and the implementation for division is given. Multioperation algorithms and integrated composite implementations are useful in reducing the overall area and on-line delay; we illustrate the approach by a sum-of-squares algorithm and the implementation of floating point $(A^2 + B^2)^{1/2}$, which is part of a unit to compute the rotation factors for Givens rotation. As an example of the use of multioperation algorithms in signal processing, the on-line implementation of a recursive IIR filter is given and compared with a conventional approach.

1. Introduction

Recent developments in VLSI technologies have revived the need for further research and development in fast and efficient arithmetic algorithms and implementations. In many arithmetic-intensive applications, such as signal processing, the use of traditional arithmetic algorithms is not always satisfactory since these algorithms were developed for the constraints imposed by SSI/MSI technologies, which are quite different than those encountered in VLSI. The capabilities of VLSI technology in terms of density and speed can only be achieved if new algorithms are developed to satisfy the specific requirements. Of particular importance are algorithms that consider (i) reduction of on-chip and off-chip communication, (ii) modularity of primitive operators, iii) elimination of time consuming carry-propagate addition, and iv) parallelism between several operations and, in particular, efficient implementation of sequences of dependent operation, such as those occurring in recursive structures.

To deal with problems such as on-chip signal routing, inter-chip communications, and modular partitioning, digit-serial least-significant-digit-first (LSDF) arithmetic has been often suggested [5,18]. The potential advantages of the LSDF approach include simplicity of the primitive operations and their implementation, serial communications, ease of pipelining, and a potential to overlap several LSDF operations with one-digit time delays. There are, however, several disadvantages of the LSDF approach. First, operation such as division, square root, and some recursive structures produce the outputs in most-significant-digit-first (MSDF) form. Consequently, a sequence involving these operations cannot be performed without large delays between successive operations to transform these outputs to LSDF form. Second, multiplication in the LSDF mode produces the least significant half of the result first, which may not be used in subsequent operations and adds a significant delay. Third, LSDF operations make floating-point implementations slow since normalization of the result depends on the most significant digits.

An attractive alternative to the LSDF serial approach is to perform all operations in a most-significant-digit-first manner on digit-serial operands. Such a systematic method of performing *all* operations in a MSDF manner, ***on-line arithmetic***, has been introduced [6,19] and investigated in several application areas such as function evaluation, solving linear systems, matrix triangularization, and FIR filter implementation [6,7,8,9,11,12,16]. At this time, only a few prototype chips have been produced for some elementary operations such as $y = ax + b$ [20] and some others of a more complex nature are in development [4]. These have shown the potential benefits of the approach. As a complement to using the on-line approach, it is possible to develop integrated algorithms and implementations for a specific application, in such a way that the interfaces between modules are optimized. We have used this approach to applications in the area of matrix computations, such as matrix triangularization and singular value decomposition [8,11], with good results.

The paper is organized as follows. After briefly presenting the basic principles and properties of on-line arithmetic we describe a systematic method to derive on-line algorithms and implementations, illustrating the method with addition, multiplication, and division. Next we discuss the implementation choices for computations which involve complicated arithmetic expressions and show three alternative approaches. We finish with the discussion of alternative designs for a second-order IIR filter.

2. Principles and Properties of On-Line Arithmetic

We summarize the most important principles and properties of on-line arithmetic.

MSDF mode and on-line delay. On-line arithmetic algorithms operate in a digit-serial manner, beginning with the most significant digit [6,7,12,13,15,19]. To generate the first digit of the result, δ digits of the input operands are needed. That is, after the δ most significant digits of the operands are received, the most significant digit of the result is generated. Thereafter, for each new digit of the operands, an extra digit of the result is obtained as illustrated below. The *on-line* delay δ is typically a small integer, e.g., 1 to 4.

Input x	x_1	x_2	x_3	x_4	x_5	x_6	x_7
Input y	y_1	y_2	y_3	y_4	y_5	y_6	y_7
Output z	$\leftarrow$	δ	$\rightarrow$	z_1	z_2	z_3	z_4

Overlapping dependent operations. Since on-line algorithms effectively produce the output during the digit-serial input of the operands, successive (dependent) operations execute in an overlapped manner - they can be chained with a latency of δ steps. In contrast, conventional bit-parallel and bit-serial approaches require, in general, completion of an operation before continuing with the dependent successor operations. The more nesting levels there are in the sequence of operations, the more advantageous the on-line approach is. This is particularly suitable in implementing recursive algorithms.

Low-bandwidth Communication. Since the result is produced in digit-serial manner, the on-line approach has low-bandwidth communication requirements. This is especially attractive for highly concurrent special-purpose VLSI designs, where the interconnections are very significant in determining the area and time.

Redundant representation. The left-to-right mode of computation requires a flexibility in computing digits on the basis of partial information about inputs. This is achieved by the use of *redundancy* in the number representation system [1,2,17]. A radix-r weighted number system is redundant if its digit set is redundant, that is, it has more than r values. This allows several representations of a given value. Consequently, there is a flexibility in choosing an output digit at a given step so that, if necessary, a compensation can be introduced.

Examples of radix-4 redundant digit sets are: {-1,0,1,2,3} (asymmetric, minimally-redundant), {-2,-1,0,1,2} (symmetric, minimally-redundant), and {-5,-4,-3,-2,-1,0,1,2,3,4,5} (symmetric, over-redundant). Similarly, useful radix-2 redundant sets are {-1,0,1} and {0,1,2}

Redundant digit sets have the disadvantage that their representation requires more bits than the nonredundant counterpart. For example, to represent the radix-2 set {-1,0,1} two bits are required, while the nonredundant set {0,1} requires just one bit. Since different redundant representations are possible, an integrated approach for a complex computation can use heterogeneous representations to optimize the implementation.

Conversion to conventional representation. In some cases, conversion from redundant to conventional representation is needed; this conversion in parallel arithmetic requires a carry-propagate addition. In the on-line approach, the conversion can be performed efficiently without carry-propagate addition using an on-the-fly conversion method [10].

Form of the algorithm and basic components. On-line algorithms are of the digit-recurrence type, in which a step consists of computing one digit of the result and a new value of a residual (internal state) function. That is,

$$c_j = F(W[j-1], A[j], B[j], C[j-1]) \quad W[j] = G(W[j-1], A[j], B[j], C[j])$$

where $A[j] = \sum_{i=0}^{j+\delta} a_i r^{-i}$ and $B[j] = \sum_{i=0}^{j+\delta} b_i r^{-i}$ are the operands and $C[j] = \sum_{i=0}^{j} c_i r^{-i}$ is the result.

Due to this similar structure of the algorithms, they are all implemented using the same basic components, such as the following: (i) registers to store operands, results, and residual vectors; (ii) append units to append a new digit to a vector; (iii) Two operand and multioperand redundant adders, such as signed digit adders, 3-2 carry-save adders and their generalization to 4-2 and 5-2 reductions; (iv) converters from redundant representations (i.e., signed digit and carry save) to conventional representations; (v) carry-propagate adders of limited precision (3 to 6 bits) to produce estimates of the residual functions; and (vi) digit-selection schemes to obtain output digits. Figure 1 shows the general structure of an on-line implementation.

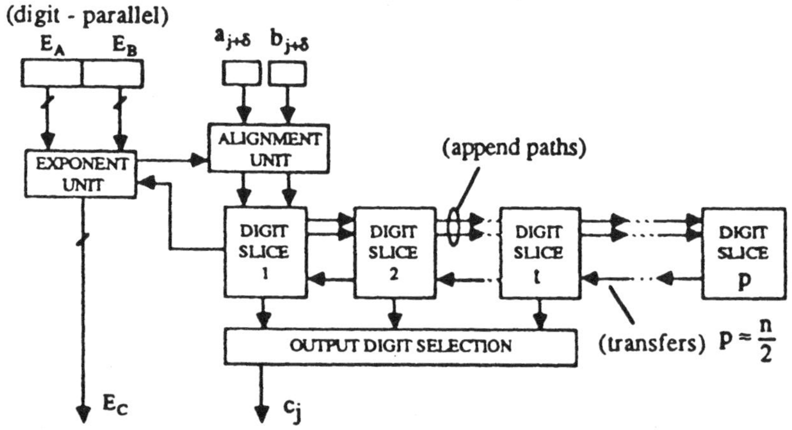

Figure 1. General Structure of On-Line Implementation

Precision and number of bit-slices. Because of the on-line nature of the operands (and a initial value zero for the residual), in general the width of the residual increases as the digits of the operands arrive. Moreover, because of the carry-free redundant additions used and the fact that the output selection function operates on a limited precision estimate of the residual, for an output precision of n digits, only about $n/2$ bit-slices are required in the computation of the residual. This is illustrated in Figure 1, and contrasts with the implementation of parallel algorithms, which require about n slices.

Algorithms and implementations. Algorithms have been developed for most of the basic arithmetic operations, as summarized in [7,13], as well as for certain composite operations. Of significance is the larger set of operations for which an on-line implementation is possible, in contrast with the LSDF approach. Implementations of these algorithms are in a development stage.

3. Systematic method for developing algorithms and implementations

We now describe a systematic method to develop on-line algorithms and implementations. We then illustrate the method by the operations of addition, multiplication, and division. Other operations, such as square root, sum of squares, and maximum, can be developed in the same manner. The method consists of the following steps:

Step 1: Describe the on-line operation by the bound on the error after j digits have been computed. For a binary operation with operands A and B and result C, this bound has the form

$$|f(A[j], B[j]) - C[j]| < r^{-j} \tag{3.1}$$

Step 2: Transform expression (3.1) so that it can be used to develop a recurrence that uses only multiplication by r (shift), addition/subtraction, and multiplication by a single digit. The form of the resulting expression is

$$G(-r^{-j}) < G(f(A[j], B[j]) - C[j]) < G(r^{-j}) \tag{3.2}$$

where G represents the required transformation.

Step 3: Define a scaled residual (called in the sequel just residual) as follows

$$W[j] = r^j(G(f(A[j], B[j] - C[j])) \tag{3.3}$$

with the bound $r^j G(-r^{-j}) \le \omega^- < W[j] < \omega^+ \le r^j G(r^{-j})$ and initial condition $W[0] = G(f(A[0], B[0]) - C[0])$.

The scaling is done to have a bound that is not multiplied by r^{-j}. The values ω^- and ω^+ are the actual bounds to be determined in Step 6.

Step 4: Determine a recurrence on $W[j]$ of the form

$$W[j] = rW[j-1] + r^j(G(f(A[j], B[j]) - C[j]) - G(f(A[j-1], B[j-1]) - C[j-1])) \tag{3.4}$$

Step 5: For purposes of selection of the result digit c_j, decompose the recurrence as follows so that H_1 is independent of c_j

$$W[j] = rW[j-1] + H_1 + H_2(c_j) = z[j-1] + H_2(c_j) \tag{3.5}$$

Step 6: Determine the limits of the selection intervals of $z[j-1]$. That is, determine L_k and U_k

so that

$$\omega^{+} > U_k + H_2(k) \quad \omega^{-} < L_k + H_2(k) \tag{3.6}$$

Step 7: Determine the value of ω^{+} and ω^{-} using the containment condition, that is,

$$U_\rho \geq \max(z[j-1]) = \max(rW[j-1] + H_1) = r\omega^{+} + \max(H_1) \tag{3.7}$$

$$L_{-\rho} \leq \min(z[j-1]) = \min(rW[j-1] + H_1) = r\omega^{-} + \min(H_1) \tag{3.8}$$

Step 8: Determine minimum overlap between adjacent intervals to use estimate of $z[j-1]$. Compute bound on on-line delay δ.

Since the computation of the recurrence is performed in a redundant form, the selection is based on an estimate of $z[j-1]$ with an error such that

$$e_{neg} \leq z[j-1] - \hat{z}[j-1] \leq e_{pos} \tag{3.9}$$

If the estimate is obtained by assimilating t fractional bits of $z[j-1]$, then we have the following relations

$$m(k) = M(k-1) + 2^{-t}, \quad M(k-1) + e_{pos} \leq U_{k-1}, \quad m(k) - |e_{neg}| \geq L_k \tag{3.10}$$

where $[m(k), M(k)]$ is the *actual* selection interval for the value k.

This results in a minimum overlap between adjacent regions

$$\Delta(k-1,k) = U_{k-1} - L_k \geq e_{pos} + |e_{neg}| - 2^{-t} \tag{3.11}$$

From Step 6 we obtain

$$U_{k-1} - L_k \leq \omega^{+} - \omega^{-} + H_2(k-1) - H_2(k) \tag{3.12}$$

Therefore,

$$\omega^{+} - \omega^{-} + H_2(k-1) - H_2(k) \geq e_{pos} + |e_{neg}| - 2^{-t} \tag{3.13}$$

The values of the errors depend on the type of redundant adder used. For a carry-save adder, for example, $e_{pos} = 2\times 2^{-t}$ and $e_{neg} = 0$. For this case then,

$$\omega^{+} - \omega^{-} + H_2(k-1) - H_2(k) \geq 2^{-t} \tag{3.14}$$

From the worst case value of this expression we get a relation between δ and t, with, typically, a minimum value of δ chosen.

Step 9: Determine the c-digit selection function and value of t.

The actual selection intervals $[m(k), M(k)]$ of $\hat{z}[j-1]$ (an estimate of $z[j-1]$ with t fractional bits) satisfy the following conditions

$$m(k) = I(k)2^{-t}, \ I(k) \text{ integer}, \ m(k) \geq L_k + |e_{neg}|, \tag{3.15}$$

$$M(k-1) = m(k) - 2^{-t} \leq U_{k-1} - e_{pos}$$

Since, in general, L_k and U_{k-1} can depend on $A[j]$, $B[j]$, $C[j-1]$, and δ, the selection intervals can also depend on these values. The range of $A[j]$, $B[j]$, and $C[j-1]$ are divided into intervals so that just a few most-significant bits of the values are used in determining the selection intervals.

As an illustration of the development method, we show in Table 3.1 the outcomes of Steps 1 - 9 in deriving on-line algorithms for addition, multiplication, and division.

Table 3.1 The Development Steps, Equations, and Parameters for Addition, Multiplication, and Division

	Step	Addition/Multiplication		Division
1	Description	$\lvert A[j]\ op\ B[j] - C[j]\rvert < r^{-j}$		$\lvert A[j]/B[j] - C[j]\rvert < r^{-j}$
2	Transform	$\lvert A[j]\ op\ B[j] - C[j]\rvert < r^{-j}$		$\lvert A[j] - C[j]B[j]\rvert < r^{-j}B[j]$
3	Residual			
	$W[j] =$	$r^j(A[j]\ op\ B[j] - C[j])$		$r^j(A[j] - C[j]B[j])$
	bound	$\lvert W[j]\rvert < \omega \le 1$		$\lvert W[j]\rvert < \omega \le B[j]$
	initialization	$W[0]=A[0]\ op\ B[0]$		$W[0] = A[0]$
4	Recurrence			
	$W[j] - rW[j-1] =$	$(a_{j+\delta} + b_{j+\delta})r^{-\delta} - c_j$ (*add*)		$a_{j+\delta} - C[j-1]b_{j+\delta} - B[j]c_j$
	$W[j] - rW[j-1] =$	$(A[j-1]b_{j+\delta} + B[j]a_{j+\delta})r^{-\delta} - c_j$ (*mult*)		
5	Decompose			
	$z[j-1] - rW[j-1] =$	$(a_{j+\delta} + b_{j+\delta})r^{-\delta}$ (*add*)		$a_{j+\delta}r^{-\delta} - C[j-1]b_{j+\delta}r^{-\delta}$
	$z[j-1] - rW[j-1] =$	$(A[j-1]b_{j+\delta} + B[j]a_{j+\delta})r^{-\delta}$ (*mult*)		
	$W[j]=$	$z[j-1] - c_j$		$z[j-1] - B[j]c_j$
6	Limits			
	U_k	$\omega + k$		$\omega + B[j]k$
	L_k	$-\omega + k$		$-\omega + B[j]k$
7	Containment			
	$U_\rho \ge max(z[j-1])$	$\omega + \rho \ge r\omega + 2\rho r^{-\delta}$		$\omega + B[j]\rho \ge r\omega + 2\rho r^{-\delta}$
	ω	$\omega \le K(1 - 2r^{-\delta})$		$\omega \le K(B[j] - 2r^{-\delta})$
8	Value of δ			
	$\Delta(k-1,k) =$	$2K(1 - 2r^{-\delta}) - 1$		$2K(B[j] - 2r^{-\delta}) - B[j]$
		Addition	Multiplication	
	$r^{-\delta} \le$	$\frac{1}{2} - \frac{1}{4K}$	$\frac{(2K-1) - 2^{-t}}{4K}$	$\frac{(2K-1)B[j] - 2^{-t}}{4K}$
	$\delta\ (r=2) \ge$	2	3	4
	$\delta\ (r>2) \ge$	1	2	3
9	Selection (for $r=2$)			
	$\delta\ (\omega)$	2 $(\frac{1}{2})$	3 $(\frac{3}{4})$	4 $(B[j] - 2^{-3})$
	U_1, L_1	3/2, 1/2	7/4, 1/4	$2B[j] - 2^{-3}, 2^{-3}$
	U_0, L_0	1/2, −1/2	3/4, −3/4	$B[j] - 2^{-3}, -B[j] + 2^{-3}$
	U_{-1}, L_{-1}	−1/2, −3/2	−1/4, −7/4	$-2^{-3}, -B[j] + 2^{-3}$
	(carry-save) $t=$		2	3
	$m(1)$	1/2	1/2	1/4
	$m(0)$	−1/2	−1/2	−3/8

As an example of an on-line algorithm and its implementation, we present a radix-2 on-line division $C = A/B$, implemented using a carry-save adder. For a minimum on-line delay of 4, the accuracy of the estimate $\hat{z}[j-1]$ is 3 fractional bits. The corresponding quotient-selection function is

$$c_j = select(\hat{z}[j-1]) = \begin{cases} 1 & \text{if } \hat{z}[j-1] \geq 1/4 \\ 0 & \text{if } -3/8 \leq \hat{z}[j-1] \leq 1/8 \\ -1 & \text{if } \hat{z}[j-1] \leq -4/8 \end{cases}$$

In summary, the radix-2 algorithm is

Initialization

$$A[0] = \sum_{i=1}^{4} a_i 2^{-i};\ B[0] = \sum_{i=1}^{4} b_i 2^{-i};\ W[0] = A[0];\ Q[0] = 0$$

Recurrence

for $j = 1,2,\ldots,m$ **do**

$$z[j-1] = 2W[j-1] + a_{j+4}2^{-4} - C[j-1]b_{j+4}2^{-4}$$
$$c_j = select(\hat{z}[j-1])$$

$$W[j] \leftarrow z[j-1] - c_j B[j]$$

$$B[j] \leftarrow convert(B[j-1], b_{j+4});\ C[j] \leftarrow convert(C[j-1], c_j)$$

End

The corresponding implementation is shown in Figure 2. In comparison with conventional redundant division, the on-line division is somewhat more complex and has a longer cycle time. For example, while the SRT division uses a 3-to-2 adder, 1 fractional bit estimate of the partial remainder, one 2-input multiplexer, and 5 registers (1 for non-redundant divisor, 2 for the redundant partial remainder, and 2 for the redundant quotient), the on-line algorithm uses a 4-to-2 adder, a 3-bit (fractional) estimate, 6 registers (2 for the redundant divisor, 2 for the redundant partial remainder, and 2 for the redundant quotient), two 2-input multiplexers, two on-the-fly converters (using the divisor and the quotient registers), and two appending networks. We estimate that the critical path delay corresponds to two multiplexers, one 4-bit CPA and one CSA.

4. Multi-Operation and Composite On-Line Algorithms

To reduce the overall on-line delay of a group of operations, it is often advantageous and feasible to combine several operations into a single *multi-operation* on-line algorithm. As an example, described fully in [11], below we show an on-line algorithm for sum of squares which has an on-line delay $\delta = 0$, in contrast with the delay of five of the corresponding network consisting of two on-line multipliers and one adder. This reduction in delay is partially due to the overredundant output digit.

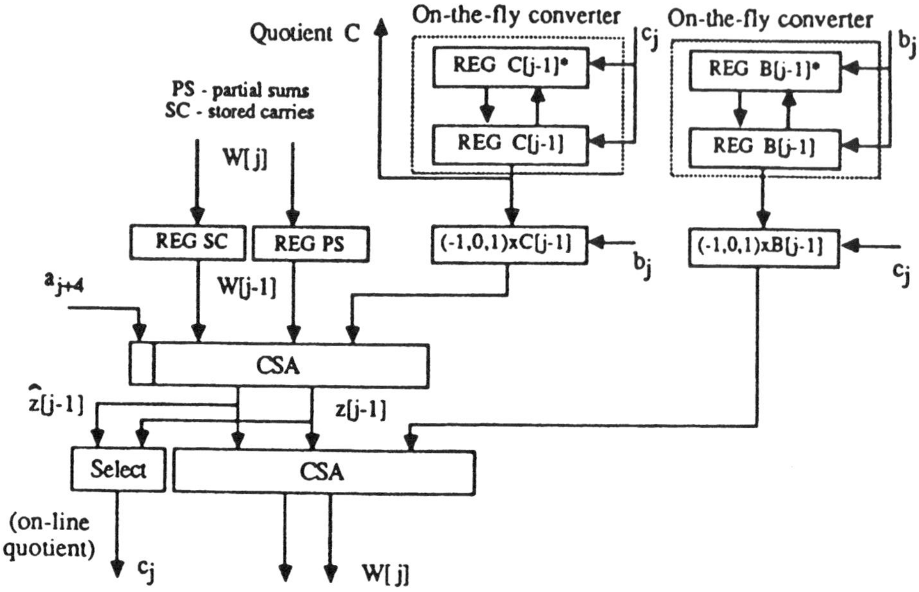

Figure 2. Implementation of On-Line Division

Initialization

$W[0] \leftarrow 0;\quad A[0] \leftarrow 0;\quad B[0] \leftarrow 0;$

Recurrence

for $j=1,2,\ldots,n+1$

$\{W[j] \leftarrow \mathit{fract}\{2W[j-1] + (2A[j-1]+a_j 2^{-j})a_j + (2B[j-1]+b_j 2^{-j})b_j\};$

$c_j \leftarrow \mathit{int}\{2W[j-1] + (2A[j-1]+A_j 2^{-j})a_j + (2B[j-1]+b_j 2^{-j})b_j\};$

$A[j] \leftarrow \mathit{append}\ (A[j-1], a_j);\quad B[j] \leftarrow \mathit{append}\ (B[j-1], b_j)\ \}$

end *Sum of Squares*

For more complicated algorithms that cannot be implemented by a single on-line module, an interconnection of modules is required. A modular approach would be to use the standard on-line implementation of primitive operations as components. However, this might lead to suboptimal implementations with respect to area and on-line delay. An alternative is to develop an integrated approach and develop one *composite* algorithm. As an illustration we show in Figure 3 the use of an alignment module, the sum-of-squares with overredundant output digit, and an appropriate square root module to perform floating-point $(A^2 + B^2)^{1/2}$, with an overall on-line delay of 4. This is part of an on-line unit to compute the Givens' rotation factors, as described in [11]. A network of standard on-line modules would have an on-line delay of 7.

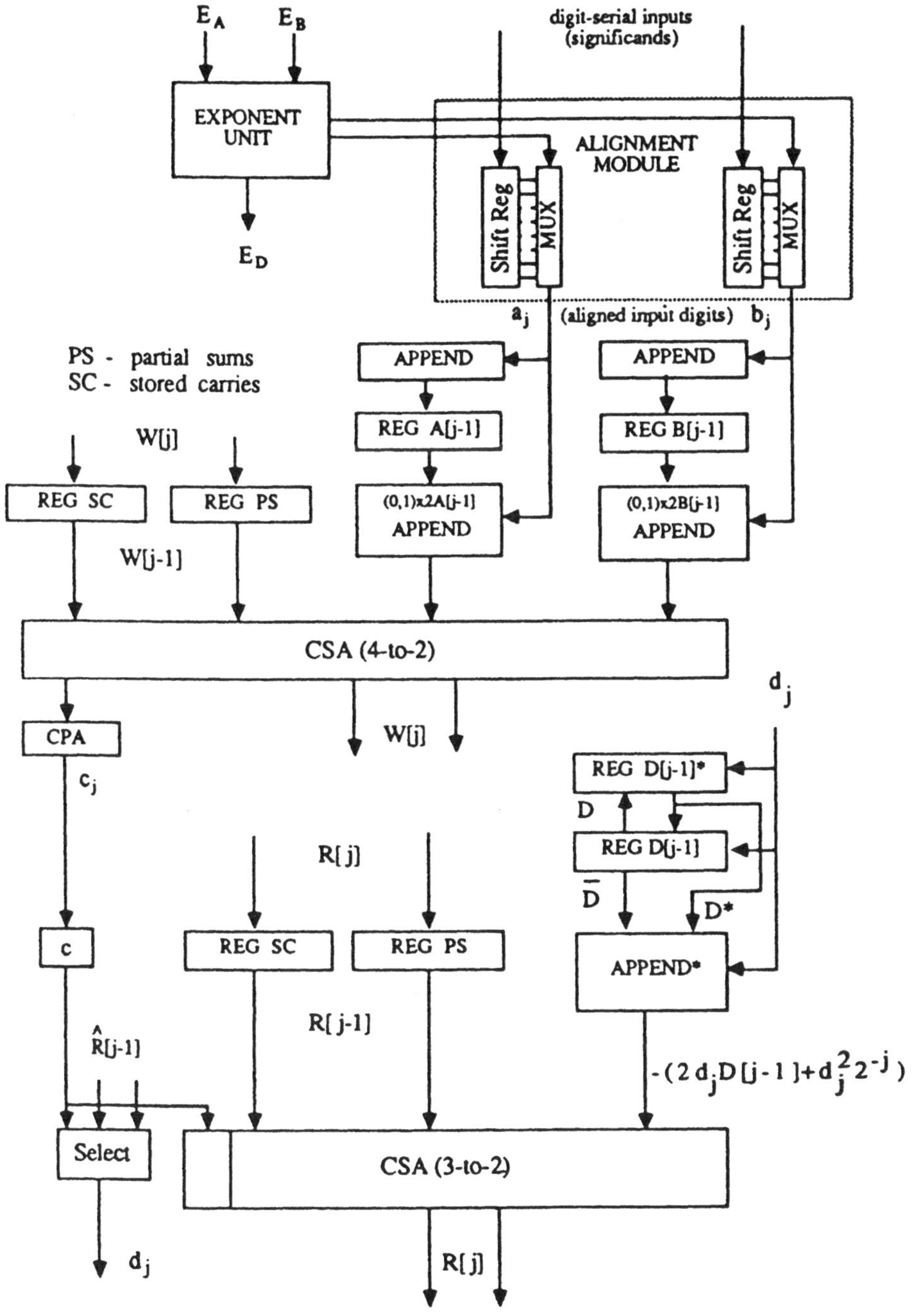

Figure 3. Implementation of On-Line Algorithm for Computing $D = \sqrt{A^2+B^2}$

5. Example: Implementations of Second-Order IIR Filter

As an illustration of potential benefits of on-line arithmetic in signal processing, we consider a second-order IIR filter, described by [14]. The output expression is

$$y(n) = a_1 y(n-1) + a_2 y(n-2) + bx(n)$$

Without attempting any optimizations, a direct form I of filter realization is analyzed, assuming a fixed-point format with m fractional bits and the following implementation alternatives:

(i) **Conventional parallel arithmetic** implementation may, for example, use a two-stage CSA array pipelined multiplier for right and left halves of the product (MR and ML), and a two-stage 6-to-2 carry-save adder, and a CPA adder (Fig.4). As indicated in the timing diagram, the time to obtain $y(n)$ is $T_{CONV} = 5t_{stage}$, where the stage delay $t_{stage} \approx kt_{FA}$ corresponds to an m-bit carry-propagate adder (t_{FA} is the delay of full adder). For precision $m \leq 32$, we assume that $k = 6$ so that $T_{CONV} \approx 30t_{FA}$. The next output cannot be computed before completing the previous one. Thus, the rate of filter computation is $1/T_{CONV}$.

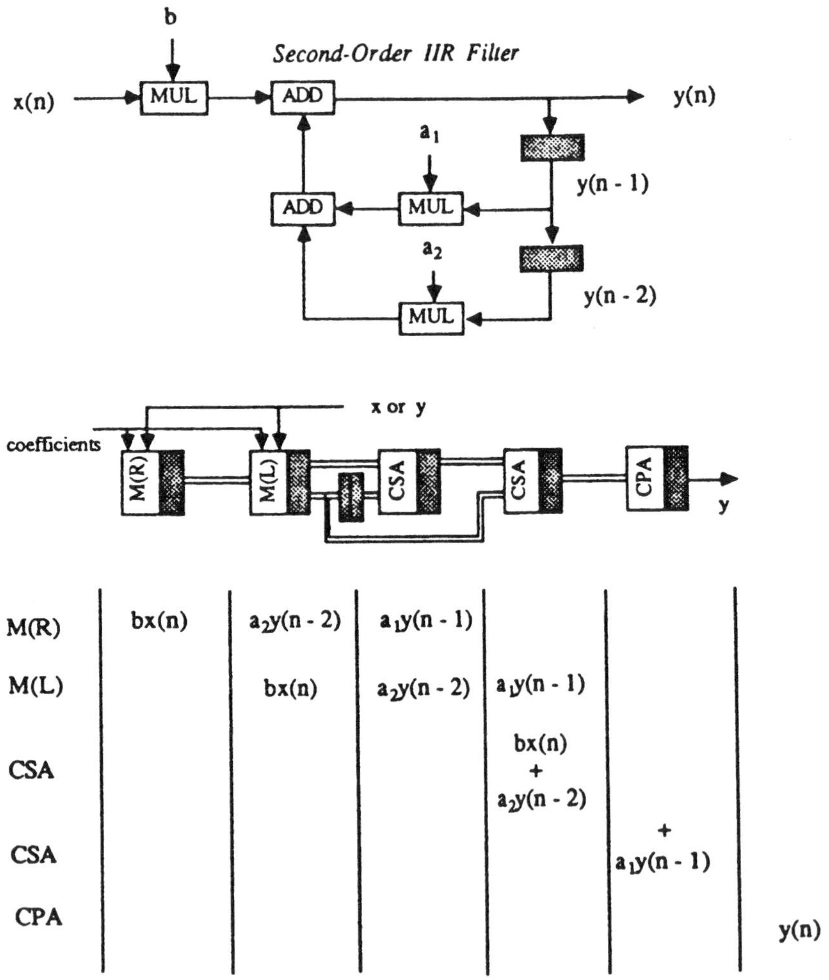

Figure 4. Conventional Parallel Implementation of Second-Order Filter

(ii) **Least Significant Digit First (LSDF) serial conventional arithmetic** implementation uses three serial multipliers and two serial adders. Since it takes m clocks to begin producing the most significant half of the product, the time to produce $y(n)$ is $T_{LSDF} \approx mt_{FA}$. The rate of generating outputs in this alternative is $1/T_{LSDF}$.

(iii) **On-line arithmetic** implementation (Fig.5) with three multioperation modules of the type $vu + w$, with an on-line delay of 4, requires time $T_{OL} = (12 + m)t_{OL-stage}$ to produce $y(n)$, where $t_{OL-stage} = 2t_{FA}$. Since on-line algorithms operate in the MSDF mode, several successive computations can be overlapped. In this case, the next recursive step may begin producing $y(n+1)$ after only $4t_{OL-stage}$. That is, the throughput rate is $\approx 1/(8t_{FA})$, which is significantly better than rates achievable with approaches (i) and (ii). The throughput is independent of the working precision but not the number of on-line units. To achieve the maximum throughput, there should be about $3\lceil m/4 \rceil$ multioperation operators. The implementation cost would be comparable to the cost of that many conventional serial-parallel carry-save multipliers (without carry-propagate adders). Note that the number of bit-slices in on-line units is about $m/2$.

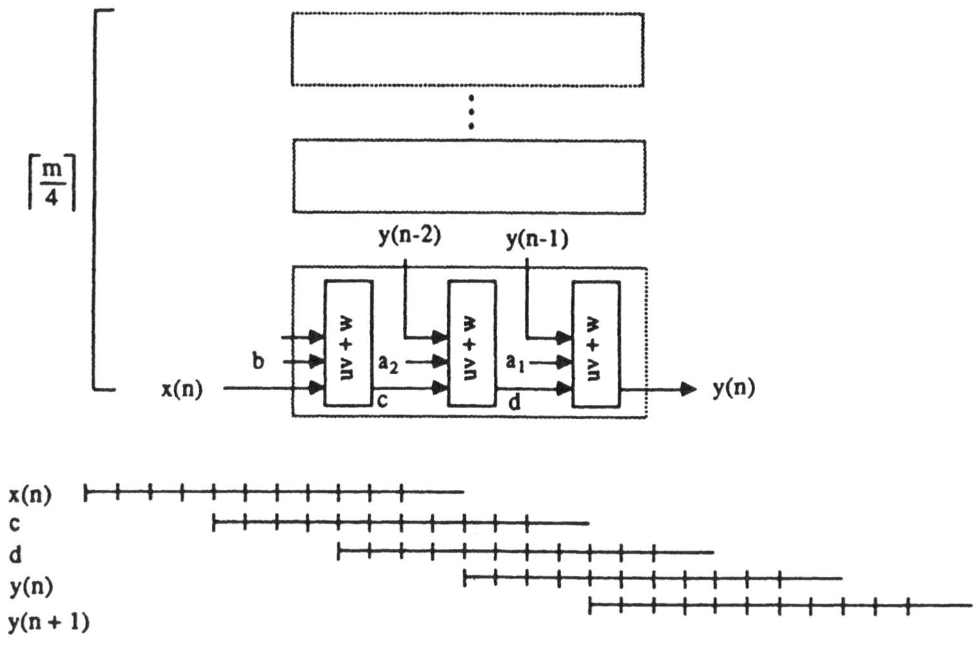

Figure 5. On-Line Implementation of Second-Order Filter

This example indicates that significant speedups might be possible using the on-line alternative. A detailed study of the on-line design and VLSI implementation of recursive filters, including the effects of error behavior and floating-point representation, is in progress [3].

Acknowledgments This research has been supported in part by the State of California/Rockwell International MICRO Grant *On-Line Arithmetic Algorithms for VLSI Design of Recursive Filters.*

References

[1] Avizienis, A. "Signed Digit Number Representation for Fast Parallel Arithmetic", IRE Trans. Electron. Comput., Vol. EC-10, pp. 389-400, 1961.

[2] Atkins, D., "Introduction to the Role of Redundancy in Computer Arithmetic", Computer, June 1975, pp.74-75.

[3] Brackert, R., "Design and Implementation of Recursive Filters Using On-Line Arithmetic", Ph.D. Dissertation (in progress), Electrical Engineering Department, UCLA, 1988.

[4] VLSI Class Project, "On-Line Rotation Chip", CS 258, UCLA Computer Science Department, 87/88.

[5] Denyer, P. and Renshaw, D., *VLSI Signal Processing: A Bit-Serial Approach*, Addison-Wesley, 1985.

[6] Ercegovac, M.D., "A General Method for Evaluation of Functions and Computations in a Digital Computer", Ph.D. Thesis, Report No. 750, Department of Computer Science, University of Illinois, Urbana, August 1975.

[7] Ercegovac, M.D., "On-Line Arithmetic: An Overview", Proc. SPIE Conference on Real-Time Signal Processing, 1984.

[8] Ercegovac, M.D. and T. Lang, "On-Line Schemes for Computing Rotation Angles for SVDs", Proc. SPIE Conference on Real-Time Signal Processing, San Diego, August 1987.

[9] Ercegovac, M.D. and T. Lang, "Fast Radix-4 Multiplication Without Carry-Propagate Addition", Proc. ICCD '87 Conference, New York, 1987.

[10] Ercegovac, M.D. and T. Lang, "On-the-Fly Conversion of Redundant into Conventional Representations", IEEE Transactions on Computers, Vol. C-36, No.7, July 1987, pp.895-897.

[11] Ercegovac, M.D. and T. Lang, "On-line Scheme for Computing Rotation Factors", Journal of Parallel and Distributed Computing, Vol.5, 1988, pp.209-227.

[12] Irwin, M.J., "An Arithmetic Unit for On-Line Computation", report No. UIUCDCS-R-77-873.

[13] Irwin, M.J. and R.M. Owens, "Digit-Pipelined Arithmetic as Illustrated by the Paste-Up System: A Tutorial", Computer, April 1987, pp. 61-73.

[14] Oppenheim, A. and R. Shafer, *Digital Signal Processing*, Prentice-Hall, 1975.

[15] Owens, R.M., "Digit On-Line Algorithms for Pipeline Architectures," Ph.D. thesis, Tech. Report CS-80-21, Dept. of Computer Science, Penn State University, August 1980.

[16] Owens, R.M. and J. JaJa, "A VLSI Chip for the Winograd/Prime Factor Algorithm to Compute the DFT," IEEE Transactions on Acoustics, Speech, and Signal Processing, vol. ASSP-34, no. 4, August 1986, pp. 979-988.

[17] Robertson, J.E., "A New Class of Digital Division Methods," IRE Transactions on Electronic Computers, vol. EC-7, September 1958, pp. 88-92.

[18] Smith, S.G. and Denyer, P., *Serial-Data Computation*, Kluwer Academic Publishers, 1988.

[19] Trivedi, K.S. and Ercegovac, M.D., "On-Line Algorithms for Division and Multiplication" IEEE Trans. on Comput., Vol. C-26, No.7, July 1977, pp. 681-687.

[20] Tullsen, D.M. and M.D. Ercegovac, "Design and Implementation of an On-Line Algorithm", SPIE Vol. 698 Real Time Signal Processing IX (1986), pp. 92-99.

On-Line Algorithms for Division and Multiplication

KISHOR S. TRIVEDI AND MILOŠ D. ERCEGOVAC

Reprinted from *IEEE Transactions on Computers*, Volume C-26, Number 7, July 1977, pages 681-687.

***Abstract*—In this paper, on-line algorithms for division and multiplication are developed. It is assumed that the operands as well as the result flow through the arithmetic unit in a digit-by-digit, most significant digit first fashion. The use of a redundant digit set, at least for the digits of the result, is required.**

***Index Terms*—Computer arithmetic, division, multiplication, on-line algorithms, pipelining, radix, redundancy.**

I. Introduction

IN THIS PAPER we consider problems of division and multiplication in a computational environment in which all basic arithmetic algorithms satisfy the "on-line" property. This implies that to generate the jth digit of the result, it is necessary and sufficient to have the operands available up to the $(j + \delta)$th digit, where the index difference δ is a small positive constant. It is necessary to accumulate δ initial digits of the operands in order to produce the first digit of the result. Subsequently, one digit of the result is produced upon receiving one digit of each of the operands. Thus δ is the on-line delay. Such algorithms can be used to speed up an arithmetic multiprocessor structure due to their potential to perform a sequence of operations in an overlapped fashion. Another possible application is in performing variable precision arithmetic. The on-line property implies a left-to-right digit-by-digit type of algorithm and consequently, a redundant representation, at least, of the results. For addition and subtraction, algorithms satisfying the on-line property can be easily specified. Multiplication requires a somewhat more elaborate approach and there are several possible ways of defining an on-line algorithm. However, the existence of an on-line division algorithm is not obvious and its analysis appears interesting.

Note that it is possible to define the on-line property with respect to a subset of the operands and/or the result. For example, conventional division has the on-line property with respect to the quotient. Similarly, conventional multiplication has the on-line property with respect to the multiplier. Several authors have extended the on-line property of multiplication to the product digits as well [1], [2]. It is also possible to define the on-line property with respect to a right-to-left mode of operation. Addition and subtraction operations always satisfy this property. Atrubin [3] has developed a right-to-left on-line algorithm for multiplication. In this paper, we will retain our earlier definition of the on-line property where all of the operands as well as the result digits flow through the arithmetic unit in a left-to-right digit-by-digit fashion.

Consider an m-digit radix r number $N = \Sigma_{i=1}^{m} n_i r^{-i}$. In the conventional representation, each digit n_i can take any value from the digit set $\{0,1,\cdots,r-1\}$. Such representations, which allow only r values in the digit set, are nonredundant since there is a unique representation for each (representable) number. By contrast, number systems that allow more than r values in the digit set are redundant. Redundant number representations are often useful in speeding up arithmetic operations [4], [5]. It is not difficult to see that the use of redundant number representation is mandatory for on-line algorithms. If we were to use a nonredundant number system, then even for simple operations like addition and subtraction, there is an on-line delay $\delta = m$ due to carry propagation. If we allow redundancy in the number representation, then it is possible to limit carry propagation to one digital position. A well-known example is the totally parallel addition in the signed digit representation [4]. This gives us an on-line algorithm for addition (and subtraction) with $\delta = 1$. Campeau [6] has developed an on-line algorithm for multiplication with $\delta = 1$. In this paper, we develop a new on-line algorithm for division where the value of δ depends upon the radix and other properties of the number system employed. In particular, we will present an on-line binary division algorithm with $\delta = 4$ (i.e., a 4-bit on-line delay). We will also show that an on-line decimal division algorithm can be obtained with $\delta = 4$ (i.e., a 4-digit on-line delay). We also develop a compatible on-line multiplication algorithm with $\delta = 1$.

After introductory remarks, an analysis of the on-line division problem is given in Section II. The radix-2 case with nonredundant operands is considered first and a feasible on-line algorithm is defined. Later, a generalization of this on-line algorithm for redundantly represented operands and higher radices is given. In Section III, a compatible on-line multiplication algorithm is considered.

II. Division

Let the radix r representations of the positive dividend, divisor, and quotient be denoted by N, D, and Q, respectively, such that

$$N = \sum_{i=1}^{m} n_i r^{-i}, \quad D = \sum_{i=1}^{m} d_i r^{-i}, \quad Q = \sum_{i=1}^{m} q_i r^{-i},$$

Manuscript received January 20, 1976; revised January 5, 1977. This work was supported in part by the National Science Foundation under Grant US NSF DCR 73-07998.

K. S. Trivedi was with the University of Illinois at Urbana-Champaign, Urbana, IL 61801. He is now with the Department of Computer Science, Duke University, Durham, NC 27706.

M. D. Ercegovac was with the Department of Computer Science, University of Illinois at Urbana-Champaign, Urbana, IL 61801. He is now with the Department of Computer Science, School of Engineering and Applied Science, University of California, Los Angeles, CA 90024.

and

$$Q = N/D \text{ to } m \text{ digit precision.}$$

If all the digits of the operands N and D are known in advance then the division can be carried out by the following well-known algorithm $D1$.

Algorithm D1:
Step 1 [Initialize]: $P_0 \leftarrow N; j \leftarrow 0;$
Step 2 [Selection]: $q_{j+1} \leftarrow$ Select (rP_j, D);
Step 3 [Basic Recursion]: $P_{j+1} \leftarrow rP_j - q_{j+1}D;$
Step 4 [Test]: If $j < m - 1$ then $j \leftarrow j + 1$, and go to Step 2;
End $D1$;

In this algorithm, P_j is known as the jth partial remainder and rP_j is known as the jth shifted partial remainder. The selection procedure of Step 2 obtains the new quotient digit q_{j+1} such that P_{j+1} satisfies certain range restrictions [5]. This process requires the comparison of rP_j against some constant multiples of D. The use of redundancy in the representation of each digital position of the quotient allows us to select q_{j+1} based on the inspection of a limited number of leading digits of rP_j and D [5]. Furthermore, these methods can also be extended to the case when both the partial remainder and the divisor are in redundant form [7]. It is clear that without the use of redundancy it is not possible to avoid a set of *full* precision comparisons.

The present problem is that the digits of the dividend and the divisor are not known in advance but are available on-line, digit-by-digit, most significant digit first. It should be clear at the outset that in the absence of redundancy the problem cannot be solved.

Let us assume that the first digit (q_1) of the quotient can be obtained after δ leading digits of the dividend and the divisor are known. Thereafter, one new digit of the quotient can be obtained upon receiving one digit each of the dividend and the divisor. We can then specify the following algorithm $D2$.

Algorithm D2:

Step 1 [Initialize]: $\hat{P}_0 \leftarrow \sum_{i=1}^{\delta} n_i r^{-i};$

$\hat{D}_0 \leftarrow \sum_{i=1}^{\delta} d_i r^{-i}; j \leftarrow 0;$

Step 2 [Selection]: $q_{j+1} \leftarrow$ Select $(r\hat{P}_j, \hat{D}_j)$
Step 3 [Basic Recursions]:

$$\hat{D}_{j+1} \leftarrow \hat{D}_j + d_{j+\delta+1} r^{-j-\delta-1}$$
$$\hat{P}_{j+1} \leftarrow r\hat{P}_j + n_{j+\delta+1} r^{-\delta} - q_{j+1}\hat{D}_{j+1} - \left(\sum_{i=1}^{j} q_i r^{-i}\right) \cdot d_{j+\delta+1} r^{-\delta}$$

Step 4 [Test]: If $j < m - 1$ then $j \leftarrow j + 1$ and go to Step 2;
End $D2$;

Note that we have assumed that the dividend and the divisor are padded with δ zero digits to the right. The basic recursion of algorithm $D2$ is slightly more complex than that in algorithm $D1$ due to the corrective action necessitated by the operand digits that arrive at the jth step.

The convergence of algorithm $D2$ can be established as follows. From the recursions of algorithm $D2$, the following expression for $\hat{P}_j$ is easily established by induction on j.

$$\hat{P}_j = r^j \left[\sum_{i=1}^{j+\delta} n_i r^{-i} - \left(\sum_{i=1}^{j} q_i r^{-i}\right)\left(\sum_{i=1}^{j+\delta} d_i r^{-i}\right)\right], \tag{2.1}$$

which implies that

$$\hat{P}_m = r^m[N - QD]$$

and

$$Q = N/D - \hat{P}_m r^{-m}/D. \tag{2.2}$$

Therefore, if we can establish a range restriction

$$|\hat{P}_j| \leq \alpha D \tag{2.3}$$

with $\frac{1}{2} \leq \alpha \leq 1$, then from (2.2) we deduce that Q is the required quotient to m-digit precision [5]. Note that the value of α is determined by the properties of the number system employed [5]. For $j = 0$, (2.3) can be satisfied by appropriately preshifting the dividend. Assume that there is a selection procedure that generates the quotient digit q_{j+1} so as to guarantee $|\hat{P}_{j+1}| \leq \alpha D$ given that $|\hat{P}_j| \leq \alpha D$. Then by induction, the range restriction (2.3) will hold for all values of j. Therefore, we will now derive such a selection procedure. Instead of proceeding directly, we will first establish a bound on $|P_j - \hat{P}_j|$ and then a selection procedure will be developed that guarantees $|P_j| \leq \alpha' D$. This in turn gives us a bound on $\hat{P}_j$. We will first discuss a simple case in which the radix is two and the divisor and the partial remainders are in nonredundant forms. Later, we will generalize to an arbitrary radix with all operands in redundant form.

A. Binary Division with Nonredundant Operands

From the recursions of algorithm $D1$, the following equation can be derived by induction.

$$P_j = r^j \left[\sum_{i=1}^{m} n_i r^{-i} - \left(\sum_{i=1}^{j} q_i r^{-i}\right)\left(\sum_{i=1}^{m} d_i r^{-i}\right)\right]. \tag{2.4}$$

From (2.1) and (2.4) we have

$$P_j - \hat{P}_j = r^j \left[\sum_{i=j+\delta+1}^{m} n_i r^{-i} - \left(\sum_{i=1}^{j} q_i r^{-i}\right)\left(\sum_{i=j+\delta+1}^{m} d_i r^{-i}\right)\right]. \tag{2.5}$$

Recall that a binary ($r = 2$) nonredundant form of the dividend and the divisor implies that $n_i, d_i \in \{0,1\}$. Since the quotient is required to be in a redundant form, we let $q_i \in \{0,1,-1\}$. Therefore, from (2.5) we can get the bounds on $P_j - \hat{P}_j$,

$$P_j - \hat{P}_j \leq 2^j \left[\sum_{i=j+\delta+1}^{m} 2^{-i} + \left(\sum_{i=1}^{j} 2^{-i}\right)\left(\sum_{i=j+\delta+1}^{m} 2^{-i}\right)\right]$$

and

$$P_j - \hat{P}_j \geq -2^j \left[\left(\sum_{i=1}^{j} 2^{-i} \right) \left(\sum_{i=j+\delta+1}^{m} 2^{-i} \right) \right],$$

from which we have

$$-2^{-\delta} \leq P_j - \hat{P}_j \leq 2 \cdot 2^{-\delta}. \tag{2.6}$$

Assuming that the divisor is the fractional part of a normalized floating-point number, we have $\frac{1}{2} \leq D < 1$. From the digit set $\{0,1,-1\}$ of quotient digits, we obtain the range restriction on P_j [5],

$$-D \leq P_j \leq D. \tag{2.7}$$

From (2.6) and (2.7) we get opaque

$$-D + 2^{-\delta} \leq \hat{P}_j \leq D - 2 \cdot 2^{-\delta}. \tag{2.8}$$

From (2.7), we get

$$-2D \leq 2P_j \leq 2D.$$

For a given value of q_{j+1}, there is a range of $2P_j$ values such that the selection of this value of q_{j+1} in step 2 of algorithm $D1$ will force the subsequent value of P_{j+1} to satisfy the range restriction (2.7). These ranges can be specified as follows:

If $0 \leq 2P_j \leq 2D$, then $q_{j+1} = 1$,

if $-D \leq 2P_j \leq D$, then $q_{j+1} = 0$, and

if $-2D \leq 2P_j \leq 0$, then $q_{j+1} = -1$.

Corresponding ranges of $2\hat{P}_j$ values can be obtained by using the inequality (2.6).

If $2 \cdot 2^{-\delta} \leq 2\hat{P}_j \leq 2D - 4 \cdot 2^{-\delta}$, then $q_{j+1} = 1$,

if $-D + 2 \cdot 2^{-\delta} \leq 2\hat{P}_j \leq D - 4 \cdot 2^{-\delta}$, then $q_{j+1} = 0$, and

if $-2D + 2 \cdot 2^{-\delta} \leq 2\hat{P}_j \leq 0 - 4 \cdot 2^{-\delta}$, then $q_{j+1} = -1$.

(2.9)

It is not possible to use these rules for quotient digit selection in algorithm $D2$ since the knowledge of all the bits of the divisor D is implied above. However, the redundancy in the representation of the quotient allows us to make the selection, independent of the value of D.

Each inequality of (2.9) determines a wedge-shaped area on the $2\hat{P}_j$ versus D plane as shown in Fig. 1. This area together with its associated value of $q_{j+1} = i$ will be called the i-selection region. Let the intersection of the i-selection region with $(i+1)$-selection region be called $(i, i+1)$-selection overlap region. The existence of nonnull selection overlap regions is due to the redundancy in quotient digits. The $(-1,0)$ and $(0,1)$-selection overlap regions are respectively given by

$$-D + 2 \cdot 2^{-\delta} \leq 2\hat{P}_j \leq -4 \cdot 2^{-\delta}$$

and

$$2 \cdot 2^{-\delta} \leq 2\hat{P}_j \leq D - 4 \cdot 2^{-\delta}.$$

Now we require that there be nonzero-selection overlap for all values of $\frac{1}{2} \leq D < 1$. Since the worst case occurs when $D = \frac{1}{2}$, we get the condition

$$2 \cdot 2^{-\delta} < \tfrac{1}{2} - 4 \cdot 2^{-\delta}.$$

Since δ is required to be an integer, we get $\delta \geq 4$. Next we select two constants C_1 and C_2 such that the line $2\hat{P}_j = C_1$ is entirely within the $(0,1)$-selection overlap region. A similar condition holds for C_2 with respect to the $(-1,0)$-selection overlap region. These two lines can be used as the selection lines to obtain the following selection procedure.

If $2\hat{P}_j > C_1$, then $q_{j+1} = 1$,

if $2\hat{P}_j < C_2$, then $q_{j+1} = -1$

otherwise $\quad q_{j+1} = 0$. (2.10)

The development of the selection rules (2.10) was based on range restriction (2.8). Therefore, using (2.2), we see that algorithm $D2$ will converge. We choose the following values of the constants C_1 and C_2 from Fig. 1: $C_1 = \frac{1}{4}$ and $C_2 = -\frac{1}{4}$.

If we substitute the selection rules (2.10) in algorithm $D2$ together with $\delta = 4$, $r = 2$, $C_1 = \frac{1}{4}$, and $C_2 = -\frac{1}{4}$, then we have a binary on-line division algorithm. In a similar fashion, we can also obtain on-line division algorithms for higher radices.

An example of binary on-line division now follows.

Let $m = 24$,

$N = 0.101000110110101110010101$, and

$D = 0.111101100101100011110011$.

$\hat{P}_j$	q_{j+1}
0.101000000000000000000000	1
0.010100000000000000000000	1
1.101001000000000000000000	−1
0.010000100000000000000000	1
1.100111100000000000000000	−1
0.001100100000000000000000	1
1.011100110100000000000000	−1
1.111011001100000000000000	0
1.110011101110000000000000	−1
0.100110010111100000000000	1
0.001111001001100000000000	1
1.100100101101100000000000	−1
0.001011000000100000000000	1
1.011001110001101010000000	−1
1.101110011111000000100000	−1
0.010111111100111000010000	1
1.110011100100000100011000	−1
0.100100101101110101010000	1
0.001111110110000110110000	1
1.011111011100110101000001	−1
1.111101110101011001001000	0
1.111011101010110010010010	0
1.110111010101100100100100	−1
0.101100010000101100111010	1

We note that negative values of $\hat{P}_j$ are represented in two's complement notation.

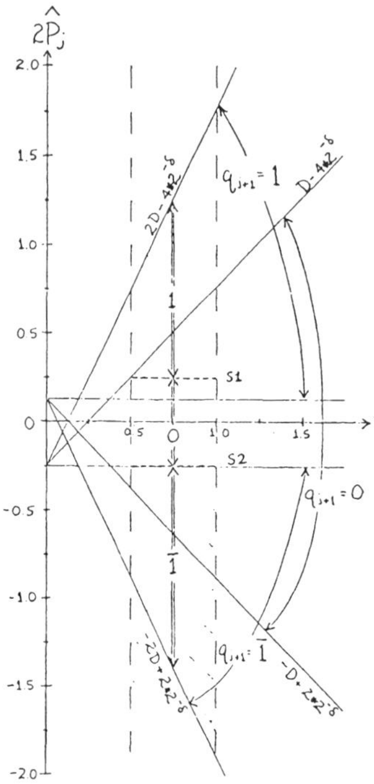

Fig. 1. The selection regions for binary on-line division ($\delta = 4$).

B. Division with Redundant Operands

In Section A, we required that the operands and the partial remainder be in nonredundant form. As a result, we cannot use carry-free addition (or subtraction) in Step 3 of algorithm $D2$. In this section we generalize to the case where the dividend, the divisor, and the partial remainder are in a redundant form. As a result, carry-free addition in Step 3 of algorithm $D2$ can be used. For a discussion of carry-free addition see [4].

We assume that the digits of the dividend, the divisor, and the quotient are all chosen from the symmetric redundant digit set

$$D_\rho = \{-\rho, \cdots -1, 0, 1, \cdots \rho\} \text{ and } \frac{r}{2} \leq \rho \leq r - 1.$$

$K = \rho/(r-1)$ is known as the degree of redundancy. It can be shown that the appropriate range restriction on P_j is given by [5], [7]

$$-KD \leq P_j \leq KD. \tag{2.11}$$

From (2.5) and the fact that $|n_i|, |d_i|, |q_i| \leq \rho$, we have

$$|P_j - \hat{P}_j| \leq r^j \left[\rho \cdot \frac{r^{-j-\delta-1} - r^{-m-1}}{1 - r^{-1}} + \rho \cdot \left(\frac{r^{-1} - r^{-j-1}}{1 - r^{-1}} \right) \cdot \rho \cdot \left(\frac{r^{-j-\delta-1} - r^{-m-1}}{1 - r^{-1}} \right) \right]$$

$$= (K + K^2) r^{-\delta} - (K + K^2) r^{-m+j} - K^2 r^{-j-\delta} + K^2 r^{-m},$$

which implies that

$$-(K + K^2) r^{-\delta} \leq P_j - \hat{P}_j \leq (K + K^2) r^{-\delta}. \tag{2.12}$$

From (2.11) and (2.12), we get the range restriction on $\hat{P}_j$:

$$-KD + (K + K^2) r^{-\delta} \leq \hat{P}_j \leq KD - (K + K^2) r^{-\delta}. \tag{2.13}$$

For a given value of $-\rho \leq i \leq \rho$, there exists a range of values of rP_j such that the selection $q_{j+1} = i$ guarantees the range restriction (2.11) on P_{j+1}. The range of rP_j values for $q_{j+1} = i$ is given by

$$(-K + i)D \leq rP_j \leq (K + i)D. \tag{2.14}$$

The corresponding i-selection region for $r\hat{P}_j$ is obtained using (2.12) and (2.14) as

$$(-K + i)D + (K + K^2) r^{-\delta+1} \leq r\hat{P}_j \leq (K + i)D - (K + K^2) r^{-\delta+1}. \tag{2.15}$$

In Fig. 2, we have shown these selection regions on a graph of $r\hat{P}_j$ versus D. Such a graph is known as a P-D plot in the literature [7]. But for obvious reasons, we call it a $\hat{P}$-D plot. The use of inequality (2.15) in a selection procedure will imply a set of full precision comparisons between $r\hat{P}_j$ with several multiples of D. However, only δ digits of D are known initially. The existence of ranges of $r\hat{P}_j$ values, where more than one value of q_{j+1} is possible, allows us to make a selection with the limited information on D [6], [7]. Recall that the redundancy in number representation is the cause of such selection overlap regions. Quotient digit selection can now be made from the estimate R_j of $\hat{P}_j$ and the estimate $\hat{D}$ of D. Assume that δ most significant digits of $\hat{P}_j$ are used as its estimate R_j, and β most significant digits of the divisor D are used as its estimate $\hat{D}$. Since only δ digits of the divisor are available initially, $\beta \leq \delta$ must hold. Let Δp denote the upper bound on the error made in estimating $r\hat{P}_j$ by rR_j, i.e., $|r\hat{P}_j - rR_j| \leq \Delta p$. Similarly, let $|D - \hat{D}| \leq \Delta d$. The rest of the analysis for determining the appropriate selection rules and determining the values of δ and β is very similar to the analysis in Atkins' Ph.D. dissertation [7]. The difference is that his P-D plot is that of rP_j versus D, whereas ours is that of $r\hat{P}_j$ versus D. The main point is that the errors of estimation (Δp and Δd) should be small compared to the latitude of choice allowed by the selection overlap regions. It may be deduced from (2.15) that the smallest selection overlap region occurs near the minimal value, $D_{\min}$, of the divisor [7]. Therefore, we require that the rectangle of height $2\Delta p$, width $2\Delta d$, and with its center at $\hat{D}_{\min}$, should be completely contained in the $(i-1, i)$-selection overlap region (see Fig. 2). The above condition is equivalent to the condition [7] opaque

$$\Delta p + \tfrac{1}{2}(2i - 1)\Delta d + (K + K^2) r^{-\delta+1} \leq \tfrac{1}{2}(2K - 1)\hat{D}_{\min}.$$

Note that the worst case occurs when $i = \rho$, therefore, we get the condition

$$\Delta p + (\rho - \tfrac{1}{2})\Delta d + (K + K^2)r^{-\delta+1} \le (K - \tfrac{1}{2})\hat{D}_{\min}. \quad (2.16)$$

To obtain a bound on Δd, we note that it consists of the trailing digits of D starting from the $(\beta + 1)$st digit. The largest value of each of these digits is ρ.

Therefore,

$$\Delta d \le \sum_{i=\beta+1}^{m} \rho r^{-i} \le \frac{\rho}{r-1} r^{-\beta} = Kr^{-\beta}.$$

Similarly,

$$\Delta p \le \frac{\rho}{r-1} r^{-\delta+1} = Kr^{-\delta+1}.$$

Therefore, the inequality (2.16) reduces to

$$Kr^{-\delta+1} + (\rho - \tfrac{1}{2})Kr^{-\beta} + (K + K^2)r^{-\delta+1} \le (K - \tfrac{1}{2})\hat{D}_{\min}$$

$$(2K + K^2)r^{-\delta+1} + K(\rho - \tfrac{1}{2})r^{-\beta} \le (K - \tfrac{1}{2})\hat{D}_{\min}. \quad (2.17)$$

If we assume that the divisor D is standardized [7], i.e., $d_1 \neq 0$, then since $|d_i| \le \rho$, we have

$$\hat{D}_{\min} = \frac{1}{r} - \sum_{i=2}^{\beta} \rho r^{-i}$$

or

$$\hat{D}_{\min} = \frac{1}{r} - \frac{\rho}{r^2}\frac{1 - r^{-\beta+1}}{1 - r^{-1}} = \frac{1}{r} - \frac{\rho}{r}\frac{1 - r^{-\beta+1}}{r - 1} = \frac{1}{r}(1 - K(1 - r^{-\beta+1})).$$

Therefore, the condition (2.17) reduces to

$$(2K + K^2)r^{-\delta+1} + K(\rho - \tfrac{1}{2})r^{-\beta} \le \frac{1}{r}(K - \tfrac{1}{2})[1 - K(1 - r^{-\beta+1})]$$

or

$$(2K + K^2)r^{-\delta+2} + K(\rho - K)r^{-\beta+1} \le (1 - K)(K - \tfrac{1}{2}). \quad (2.18)$$

Once r and K are decided, we can determine β and δ from this condition. From condition (2.18) we infer that number systems with $K = \tfrac{1}{2}$ (i.e., a nonredundant system) and $K = 1$ (i.e., a maximally redundant system) are not permissible. As an example, let $r = 10$, $\rho = 6$ and $K = \tfrac{2}{3}$. Then from (2.18), we have

$$(4/3 + 4/9)10^{-\delta+2} + 2/3(16/3)10^{-\beta+1} \le \frac{1}{3}\cdot\frac{1}{6},$$

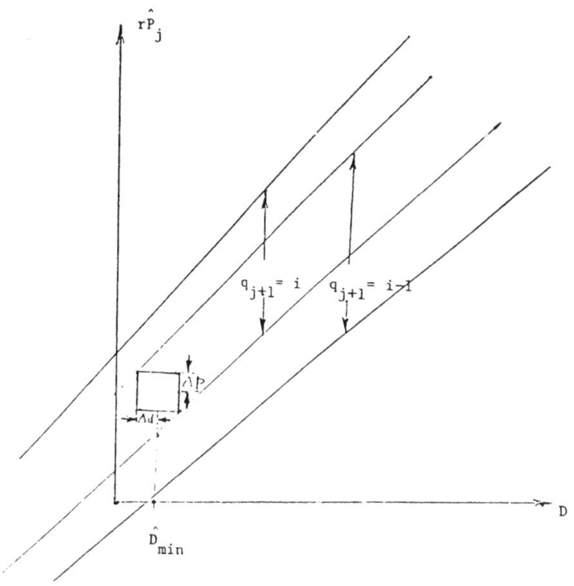

Fig. 2. $\hat{P}$-D plot for higher radix on-line division.

or

$$(3200)10^{-\delta} + (640)10^{-\beta} \le 1.$$

If we let $\delta = \beta + 1$, then

$$(320 + 640)10^{-\beta} \le 1$$

or

$$\beta \ge 3, \delta \ge 4.$$

Thus, we can define a decimal on-line division algorithm with a four digit on-line delay. The selection rules may be obtained from the $\hat{P}$-D plot following the procedure in [7], but the details are omitted here.

III. Multiplication

An on-line algorithm for multiplication, compatible with the previously considered on-line division algorithm, can be conveniently derived following the well-known technique of incremental multiplication, as used in the digital differential analyzers [6], [8], combined with the use of redundant number of systems.

Let

$$X = \sum_{i=1}^{m} x_i \cdot r^{-i}$$

$$Y = \sum_{i=1}^{m} y_i \cdot r^{-i} \quad (3.1)$$

be the radix r representations of the positive multiplicand and the multiplier, respectively. Define

$$X_j = \sum_{i=1}^{j} x_i \cdot r^{-i} = X_{j-1} + x_j \cdot r^{-j}$$

$$Y_j = \sum_{i=1}^{j} y_i \cdot r^{-i} = Y_{j-1} + y_j \cdot r^{-j} \quad (3.2)$$

to be the j digit representations of the operands X and Y, available at the jth step by definition of an on-line algorithm. The corresponding partial product is, then,

$$X_j \cdot Y_j = X_{j-1} \cdot Y_{j-1} + (X_j \cdot y_j + Y_{j-1} \cdot x_j)r^{-j}. \tag{3.3}$$

Let P_j be the scaled partial product, i.e.,

$$P_j = X_j \cdot Y_j \cdot r^j \tag{3.4}$$

so that the recursion of the multiplication algorithm can be expressed as follows:

$$P_j = rP_{j-1} + X_j \cdot y_j + Y_{j-1} \cdot x_j. \tag{3.5}$$

Defining $P_0 = 0$, the scaled product $P_m = X \cdot Y \cdot r^m$ can be generated in m steps (3.5). If a nonredundant number system is used in representing the partial products, the digits of the desired product appear in a right-to-left fashion, as determined by the conventional carry propagation requirements. If, however, a redundant number representation is adopted, the desired left-to-right generation of the product digits, as required by the on-line property, can be easily provided. Moreover, the redundancy in number representation can make the time required to perform the recursive step independent of the operand precision (i.e., a carry-free addition is possible).

We will use a symmetric redundant digit set

$$D_\rho = \{-\rho, -(\rho-1), \cdots, -1, 0, 1, \cdots, \rho-1, \rho\} \tag{3.6}$$

where·

$$\frac{r}{2} \leq \rho \leq r-1.$$

Following the general computational method described in [9], the basic recursion (3.5) of the multiplication algorithm is redefined in the following way:

$$w_j = r(w_{j-1} - d_{j-1}) + X_j \cdot y_j + Y_{j-1} \cdot x_j \tag{3.7}$$

where the digits $d_j \in D_\rho$, can be determined by the following selection function:

$$d_j = S(w_j) = \text{sign}\, w_j \cdot \lfloor |w_j| + 1/2 \rfloor \tag{3.8}$$

which clearly corresponds to a rounding procedure.

Then, from (3.5) and (3.7), the following relation can be obtained by induction

$$w_j = P_j - \sum_{i=1}^{j-1} d_i \cdot r^{(j-i)}. \tag{3.9}$$

Substituting $j = m$ in (3.9) and rearranging, we have

$$P_m = X \cdot Y \cdot r^m = r^m \sum_{i=1}^{m-1} d_i r^{-i} + w_m \tag{3.10}$$

or

$$X \cdot Y = \sum_{i=1}^{m} d_i r^{-i} + (w_m - d_m) r^{-m}.$$

By definition of the selection function $S(w_j)$, $|w_m - d_m| \leq \frac{1}{2}$, so that $\Sigma_{i=1}^{m} d_i r^{-i}$ is indeed the redundant representation of the most significant half of the product $X \cdot Y$. Thus, the algorithm will converge.

We have assumed above that the selection function (3.8) will produce the digit d_i such that $|d_i| \leq \rho$. This condition will be satisfied provided

$$|w_j| < \rho + \tfrac{1}{2} \tag{3.11}$$

holds for $j = 1, 2, \cdots, m$. We will now obtain an upper bound M, on the values of the operands X and Y such that (3.11) holds. Let $|X|, |Y| \leq M$. Noting that $|w_{j-1} - d_{j-1}| \leq \frac{1}{2}$ and $|y_j|, |x_j| \leq \rho$, from (3.7) we get

$$|w_j| \leq \frac{r}{2} + 2M\rho.$$

Now using (3.11) and rearranging we get

$$M < \frac{1}{2} - \frac{r-1}{4\rho}.$$

Thus, for a minimally redundant system, defined by $\rho = r/2$, the required operand bound is

$$|X|, |Y| < \frac{1}{2r}$$

and for a maximally redundant system, defined by $\rho = r - 1$, the required operand bound is

$$|X|, |Y| < \frac{1}{4}.$$

On closer inspection, we are often able to improve upon these bounds. For example, $|X|, |Y| \leq \frac{1}{2}$ suffices for $r = 2$.

As discussed in detail elsewhere [9], it is a simple matter to make the time required for the computation of w_j independent of the precision of the corresponding operands. Thus, by allowing

$$\tfrac{1}{2} < |w_j - d_j| < 1$$

a carry propagation free addition can be utilized in (3.7).

The following example illustrates the on-line multiplication algorithm for $r = 2$.

$X = 0.01101001$
$Y = 0.01110011$

j	$X_j y_j + Y_{j-1} x_j$	w_j	d_j	$2(w_j - d_j)$
1	0.0	0.0	0	0.0
2	0.01	0.01	0	0.1
3	0.101	1.001	1	0.01
4	0.0110	0.1010	1	−0.11
5	0.0111	−0.0101	0	−0.101
6	0.0	−0.101	−1	0.11
7	0.0110100	1.0010100	1	0.010100
8	0.11011011	1.00101011	1	0.0101011

$$\sum_{j=1}^{8} d_j 2^{-j} + 2^{-8}(w_8 - d_8) = 0.0010111100101011$$

$$= X \cdot Y. \square$$

IV. Concluding Remarks

Two compatible algorithms for on-line division and multiplication, based on the redundant number systems, have been presented. These on-line algorithms provide an effective way of speeding up the execution of sequences of the basic arithmetic operations by minimizing the delay between successive operations in an overlapped mode of operation. In real-time applications, where the inputs are serially generated by an analog-to-digital conversion process beginning with the most significant digits, the on-line algorithms can be used to increase the overall speed by overlapping the computation with the conversion. The described algorithms can be seen to have rather simple implementation requirements and properties which are compatible with the desirable modularity in implementation and variable precision operations.

At the present time, computer systems employing a pipelined arithmetic unit for floating-point computation consider fraction arithmetic as a single stage of the pipeline [10], [11]. The on-line algorithms presented in this paper will allow further decomposition of this stage so that a digitally pipelined arithmetic unit will be possible.

References

[1] A. Avizienis, "On a flexible implementation of digital computer arithmetic," in *Proc. IFIP*, pp. 664–668, 1962.

[2] M. J. Pisterzi, "A limited connection arithmetic unit," Ph.D. dissertation, Dep. Comput. Sci., Univ. of Illinois, Urbana-Champaign, IL, June 1970.

[3] A. J. Atrubin, "A one-dimensional real-time iterative multiplier," *IEEE Trans. Comput.*, vol. C-14, pp. 394–399, 1965.

[4] A. Avizienis, "Signed digit number representation for fast parallel arithmetic," *IRE Trans. Electron. Comput.*, vol. EC-10, pp. 389–400, 1961.

[5] J. E. Robertson, "A new class of digital division methods," *IRE Trans. Electron. Comput.*, vol. EC-7, pp. 218–222, Sept. 1958.

[6] J. O. Campeau, "Communication and sequential problems in the parallel processor," in *Parallel Processor Systems, Technologies and Applications.* New York: Spartan, 1970.

[7] D. E. Atkins, "A study of methods for selection of quotient digits during digital division," Ph.D. dissertation, Dep. of Comput. Sci., Univ. of Illinois, Urbana-Champaign, IL, June 1970.

[8] E. L. Braum, *Digital Computer Design—Logic, Circuitry, and Synthesis.* New York: Academic Press, 1963.

[9] M. D. Ercegovac, "A general method for evaluation of functions and computations in a digital computer," Ph.D. dissertation, Dep. Comput. Sci., University of Illinois, Urbana-Champaign, IL, July 1975.

[10] C. Stephenson, "Case study of the pipelined arithmetic unit for the TI advanced scientific computer," in *IEEE Proc. 3rd Symp. on Computer Arithmetic*, Dallas, TX, Nov. 1975.

[11] S. F. Anderson *et al.*, "The system/360 model 91 floating-point execution unit," *IBM Syst. J.*, vol. 11, p. 34, 1967.

Kishor S. Trivedi was born in Bhavnagar, India, on August 20, 1946. He received the Ph.D. degree in computer science from the University of Illinois, Urbana-Champaign, IL, in 1974.

From 1974–1975 he was a Visiting Research Associate at the University of Illinois. In the summer of 1976, he was a Visiting Scientist at the Institute for Computer Applications to Science and Engineering, NASA Langley Research Center, Hampton, VA. He is currently an Assistant Professor of Computer Science at Duke University, Durham, NC. His current interests include computer architecture and operating systems.

Miloš D. Ercegovac, for a photograph and biography please see this issue, p. 680.

Error Analysis of Certain Floating-Point On-Line Algorithms

OSAAKI WATANUKI, MEMBER, IEEE, AND MILOŠ D. ERCEGOVAC

Reprinted from *IEEE Transactions on Computers*, Volume C-32, Number 4, April 1983, pages 352-358.

***Abstract*—The properties of redundant number system in significand (mantissa) representation are studied and the range of redundant significand is derived. From the range of the redundant significand and the absolute error of on-line operations, the MRRE (maximum relative representation error) is defined and analyzed for floating-point on-line addition and multiplication.**

***Index Terms*—Absolute error, floating-point on-line algorithm, maximum relative representation error, pseudo-normalized significand, redundant significand.**

I. INTRODUCTION

In designing a highly parallel processor it is important to consider features such as modularity, reconfigurability, LSI realizability, intermodule interconnection bandwidth, etc. [1]. For this reason we may consider using on-line arithmetic [8], [9], [14], [17], [18], because on-line arithmetic is advantageous from the standpoints of 1) high concurrency, 2) low interconnection bandwidth, 3) modularity, and 4) VLSI realizability [11], [12], [17].

The on-line arithmetic algorithms have the following properties: 1) the computation is performed digit by digit starting with the most significant digit (msd); 2) the jth digit of the result can be computed as soon as $j + \delta$ digits of the operands are available, where δ is a small integer called on-line delay.

The totally parallel addition [2], [3] can be easily adapted for on-line addition [14]. On-line algorithms for multiplication and division were developed in [18]. The on-line multiplication requires $\delta = 1$ and division $\delta = 3$ to 5. An on-line square-rooting algorithm with $\delta = 1$ was described in [9], and [17]. The on-line approach can also be applied for implementing more complex operations [8], [12].

In this paper we discuss the range of the on-line significand (mantissa) and analyze the representation errors for floating-point on-line addition and multiplication. Section II presents preliminaries of the on-line approach and illustrates the use of on-line arithmetic. In Section III we derive the range of the on-line (redundant) significand representation and show that the pseudonormalized significands do not occur for $r > 2$. Section IV presents an analysis of the representation errors for floating-point on-line addition and multiplication. These results are illustrated by simulation examples in Section V.

Manuscript received April 7, 1982; revised November 20, 1982. This work was supported in part by the Office of Naval Research under Contract N00014-79-C-0866.

O. Watanuki was with the Department of Computer Science, University of California, Los Angeles, CA 90024. He is now with the IBM Japan Science Institute, Tokyo, Japan.

M. D. Ercegovac is with the Department of Computer Science, University of California, Los Angeles, CA 90024.

II. PRELIMINARIES

Floating-point algorithms are required in order to apply on-line arithmetic effectively to practical numerical computations. By introducing the notion of the quasi-normalization, floating-point algorithms for on-line addition and multiplication are presented in [19], and [20]. In the floating-point on-line (FLPOL) algorithms, fixed-point on-line addition [14] and multiplication [18] are modified and used to compute the result significand. The exponent processing is similar to the method used in the conventional floating-point algorithms [7], [22]. However, special care must be taken with timing so that operand alignment, treatment of significand overflow and postnormalization are performed in an on-line fashion.

In order to represent the significand of redundant floating-point numbers, signed-digits [2], [3] are used, while the exponent is represented in the conventional form.

Definition 2.1: The k-digit redundant floating-point representation of number x is

$$x = r^{x_e} \sum_{j=1}^{k} x_j r^{-j} \tag{2.1}$$

where r is the radix, x_e is the exponent represented in the conventional form, and the signed-digits of significand $x_j \in \{-\rho, \cdots, -1, 0, 1, \cdots, \rho\}$ where $\rho = r/2$ for minimal redundancy and $\rho = r - 1$ for maximal redundancy.

In applying the signed-digit sets to on-line operations, maximally redundant representation is preferable to minimally redundant representation because the operands represented using a conventional number system can be directly input to on-line operations. Furthermore, the on-line arithmetic algorithms with maximally redundant digit sets usually require smaller on-line delays, and a faster execution is expected.

However, if a maximally redundant representation is used, it may happen that a redundant significand x_f has only one significant digit even though all the digit positions of the significand have nonzero digits (the pseudo-normalization problem).

Definition 2.2: A nonzero redundant floating-point number $x = x_f r^{x_e}$ with k digits of significand represented by a maximally redundant digit set is said to be

1) normalized if $r^{-1} \le |x_f| < 1$;
2) quasi-normalized if $r^{-2} \le |x_f| < 1$;
3) pseudo-normalized if $r^{-k} \le |x_f| < 1$.

Example: For $r = 10$, $\rho = 9$ and $m = 4$,

1) normalized $0.11\bar{8}3 = 0.1023$;
2) quasi-normalized $0.1\bar{9}23 = 0.0123$;
3) pseudo-normalized $0.1\overline{996} = 0.0004$.

As one of the main results in this paper, we show that the FLPOL arithmetic algorithms [19] yield quasi-normalized results for $r > 2$. In order to avoid loss of significance due to quasi-normalization, $m + 1$ digits of redundant significand must be retained if m significant digits are required. Pseudo-normalized numbers can be only artificially generated, and would not appear in the intermediate results of on-line operations. Hereafter we shall use the term "normalized" to refer to quasi-normalized redundant floating-point numbers. The proposed normalized fixed-precision FLPOL algorithms can be easily modified into variable precision algorithms.

Time Requirements

In the following discussion, we assume that the exponent processing time is equal to the digit computation time and that a maximally redundant digit set and a higher radix are used. FLPOL addition requires the on-line delay δ_a:

$$0 \le \delta_a \le m + 3 \tag{2.2}$$

where $\delta_a = 0$ if at least one of the operands is zero, $\delta_a = 1$ or 2 for addition of the operands with similar signs, and $\delta_a = 2$ to $m + 3$ for subtraction.

Using the above results, the total number of time steps to obtain a result of $m + 1$ digits can be given. If at least one of the operands is zero, the total number of steps is $m + 1$. For on-line addition of operands with similar signs, the $m + 1$-digit result is obtained in $m + 2$ to $m + 3$ steps, and for on-line subtraction $m + 3$ to $2m + 4$ steps. If we assume that alignment shift or normalization shift of one digit position in conventional floating-point arithmetic takes one time step, then conventional floating-point addition and subtraction take $m + 1$ to $2m + 2$ steps including one time step for the exponent processing. The excessive delay which occurs in subtraction can be improved by using unnormalized arithmetic or significance arithmetic [5], [16].

Since the absolute values of quasi-normalized operands are in the interval $[r^{-2}, 1)$, the absolute value of the product should lie in the range $[r^{-4}, 1)$. That is, the number of leading zeros is at most three. Thus the total on-line delay δ_m of FLPOL multiplication is bounded by the following inequality:

$$0 \le \delta_m \le 4 \tag{2.3}$$

where $\delta_m = 0$ if at least one of the operands is zero.

Thus the total number of steps to yield $m + 1$-digit product becomes $m + 2$ to $m + 5$. In particular, if at least one of the operands is zero, the result is obtained in $m + 1$ steps.

As stated previously, the exponent and the msd (most significant digit) of the result significand are obtained at the same time in on-line arithmetic, and the subsequent operations can be initiated as soon as both of the msd's of the operands are available. We shall illustrate this process using an example shown in Fig. 1.

Consider the expression

$$z = a + (bx + cy) \tag{2.4}$$

where a, b and c are constants represented in the conventional form, and x and y are variables represented in the redundant form. The msd's of bx and cy are produced after a small number of delays. Then addition $bx + cy$ can be initiated as soon as the msd's of both bx and cy become available. Thus the computation is carried out in a digit-serial fashion, starting with the msd of the significand. The computing time of expression (2.4) requires 14 digit steps using four FLPOL arithmetic processors.

```
R=10, ρ=9, M=8,

  Z = A + (BX + CY),  WHERE
A=0.32751613E+00, B=0 14752103E+00, C=0.71253192E-01,
X=0.604̄1152̄15̄E-01. Y=0.14̄33012̄5̄3E+01.

TIME    1 2 3 4 5 6 7 8 9 0 1 2 3 4 5
B_E     0
B_J     1 4 7 5 2 1 0 3 0
X_E    -1
X_J     6 0 4̄ 1 1 5 2̄ 1 5̄
BX       -1
          1 1̄ 2̄ 1̄ 4 1̄ 5 3̄ 0
C_E    -1
C_J     7 1 2 5 3 1 9 2 0
Y_E     1
Y_J     1 4̄ 3 3 0 1 2̄ 5̄ 3
CY         -1
            4 5 1 0 4 2̄ 1 3̄ 1
BX+CY         0
              1 5̄ 4 1̄ 0 2̄ 2̄ 5̄ 4
A_E     0
A_J     3 2 7 5 1 6 1 3 0
Z_E               0
Z_J               4 2̄ 1 4 1 4 1̄ 2̄ 4
```

Fig. 1. Example of floating-point on-line computation.

In conventional normalized floating-point operations, the exponent must be adjusted after the msd of the significand is obtained. Therefore, floating-point addition cannot be initiated until the preceding operations are completed. If we assume that the exponent processing time and the digit computation time of conventional arithmetic are equal to the digit computation time of on-line arithmetic, it takes 27 digit steps to yield the same result using two conventional floating-point processors. Since the exponent and the msd of the significand are obtained at the same time in FLPOL operations, the subsequent operations can be initiated as soon as the msd's of the both operands become available.

III. Range of Redundant Significand

Due to the nature of the digit selection function used in the on-line algorithms, the result digits are always chosen so that the computed result of an FLPOL operation is within a certain error range from the true result. This property is briefly stated in the following theorem.

Theorem 1: The absolute error incurred by terminating the FLPOL algorithms for addition, subtraction and multiplication after the kth digit of the result significand has been obtained satisfies the inequality

$$|\epsilon_k| < \left[\frac{1+\Delta}{2} + 2\rho\frac{r^{-\delta}}{r-1}\right]r^{-k+z_e} \quad (3.1)$$

where ϵ_k is the absolute error due to truncating the computed result at the kth digit, r is the radix, ρ is the maximum digit that could appear in the operand significand, Δ is the overlap between the digit selection intervals and z_e is the exponent of the result.

Proof: Let two redundant operands be $x = x_f r^{x_e}$ and $y = y_f r^{y_e}$.

A. Addition/Subtraction

Without loss of generality, we may assume $x_e \geq y_e$. Then the operand y is rewritten in the aligned form

$$y = r^{x_e}\sum_{j=1}^{m+1} y_j r^{-(j+i_d)} \quad (3.2)$$

where i_d is the difference of the exponents, i.e., $0 \leq i_d = x_e - y_e < m+1$.

The absolute error committed at the kth digit is given by

$$\epsilon_k = \hat{z} - Z_k \quad (3.3)$$

where $\hat{z}$ is the true sum and Z_k denotes the result computed up to the kth digit. The true sum is written in the form

$$\hat{z} = r^{x_e}\sum_{j=1}^{m+1+i_d}(x_j + y_{j-i_d})r^{-j} \quad (3.4)$$

where $x_j = 0$ for $j > m+1$ and $y_j = 0$ for $j \leq 0$.

We must note that the FLPOL addition algorithm produces and discards $\delta_a = \delta + L$ digits of zeros in the beginning, where δ is the delay of fixed-point on-line addition, and L is the number of canceled digits in case the operands have opposite signs. For an overflow in the result significand, $L = -1$. The exponent is initially set to x_e, and is adjusted to $z_e = x_e - L$ when the most significant digit of the result significand is obtained. If we stop the recursion at the $(k+\delta_a)$th step, then we have obtained the partial remainder $w(k+\delta_a)$ and the result digit $d_{k+\delta_a}$, but the operand digits $x_{k+\delta_a+1}$ and $y_{k+\delta_a+1-i_d}$ have not been included in the recursion. We define the partial remainder at the $(k+\delta_a+1)$st recursive step which has not included the operand digits as $\hat{w}(k+\delta_a+1)$.

Consider the recursion for on-line addition:

begin OLADD

$w(0) \leftarrow 0;\ d_0 \leftarrow 0;$

for $j = 1$ **step** 1 **until** $m + \delta + 1$ **do**

begin

$w(j) \leftarrow r[w(j-1) - d_{j-1}] + r^{-\delta}(x_j + y_j);$

$d_j \leftarrow \text{sign}(w(j))\lfloor |w(j)| + 0.5\rfloor$

end

end OLADD

$/* z = x + y = \sum_{j=1}^{m+\delta+1} d_j r^{-j} */.$

From the on-line addition recursion we obtain the following relation:

$$r^{-(k+1+L)}\hat{w}(k+\delta_a+1) = \sum_{j=1}^{k+\delta_a}(x_j + y_{j-i_d})r^{-j} - r^{\delta_a - L}\sum_{j=1}^{k+\delta_a} d_j r^{-j} \quad (3.5)$$

where $y_j = 0$ for $j \leq 0$. From (3.5) and the fact that initial δ_a zeros are discarded and the exponent is adjusted to $x_e - L$, we obtain

$$\begin{aligned} Z_k &= r^{z_e}\sum_{i=1}^{k} z_i r^{-i} = \left(r^{\delta_a}\sum_{j=1}^{k+\delta_a} d_j r^{-j}\right)r^{x_e - L} \\ &= r^{x_e}\left[\sum_{j=1}^{k+\delta_a}(x_j + y_{j-i_d})r^{-j} - r^{-(k+1+L)}\hat{w}(k+\delta_a+1)\right]. \end{aligned} \quad (3.6)$$

From (3.3), (3.4) and (3.6),

$$\begin{aligned} |\epsilon_k| &= r^{x_e}\left|\sum_{j=k+\delta_a+1}^{m+1+i_d}(x_j + y_{j-i_d})r^{-j} + r^{-(k+1+L)}\hat{w}(k+\delta_a+1)\right| \\ &\leq \left|\hat{w}(k+\delta_a+1)r^{-(k+1+L)} + 2\rho\sum_{j=k+\delta_a+1}^{m+1+i_d} r^{-j}\right| r^{x_e}. \end{aligned} \quad (3.7)$$

Since $|\hat{w}(k+\delta_a+1)| \leq r\frac{1+\Delta}{2}$ and $\sum_{j=1}^{k} r^{-j} < \frac{1}{r-1}$, the result (3.1) follows. If an infinite precision of the operand representation is allowed, the smallest absolute value of a redundant significand approaches the righthand side of (3.1). In practice, the precision of the operands is limited and the magnitude of significand is greater than this value.

B. Multiplication

In case of multiplication, the problem is simpler than in addition because the significand alignment is not needed. Let us assume that the significand of the product has L leading zeros.

For FLPOL multiplication, we have for the true product

$$\hat{z} = r^{x_e + y_e}\sum_{j=1}^{m+1}(x_j Y_j + y_j X_{j-1})r^{-j} \quad (3.8)$$

where

$$X_{j-1} = \sum_{n=1}^{j-1} x_n r^{-n}, \text{ and } Y_j = \sum_{n=1}^{j} y_n r^{-n}.$$

The algorithm discards initial $\delta_m = \delta + L$ digits of zeros and the exponent is adjusted to $x_e + y_e - L$. Then it follows that:

$$\begin{aligned} Z_k &= r^{z_e}\sum_{i=1}^{k} z_i r^{-i} = \left(r^{\delta_m}\sum_{j=1}^{k+\delta_m} d_j r^{-j}\right)r^{x_e + y_e - L} \\ &= \left[\sum_{j=1}^{k+\delta_m}(x_j Y_j + y_j X_{j-1})r^{-j} - r^{-(k+1+L)}\hat{w}(k+\delta_m+1)\right]r^{x_e + y_e}. \end{aligned} \quad (3.9)$$

Hence, the absolute error becomes

$$|\epsilon_k| = \left| r^{-(k+1+L)}\hat{w}(k+\delta_m+1) + \sum_{j=k+\delta_m+1}^{m+1} (x_j Y_j + y_j X_{j-1}) r^{-j} \right| r^{x_e+y_e}. \quad (3.10)$$

Since $|w(k+\delta_m+1)| \le r\dfrac{1+\Delta}{2}$ and $|X_j|, |Y_j| < \dfrac{\rho}{r-1}$, we obtain

$$|\epsilon_k| < \left[\frac{1+\Delta}{2} + 2\rho^2 \frac{r^{-\delta}}{(r-1)^2}\right] r^{-k+x_e+y_e-L}$$
$$\le \left[\frac{1+\Delta}{2} + 2\rho \frac{r^{-\delta}}{r-1}\right] r^{-k+z_e}.$$

Hence, Theorem 1 is proved. □

This theorem is of basic importance, because it leads to the derivation of the redundant significand with the smallest magnitude and provides the proof that the results of the FLPOL operations are quasi-normalized for higher radices.

With regard to the significand of FLPOL operations, the following theorem is provided.

Theorem 2: The absolute value of the redundant significand is no less than $[(1-\Delta)/2 - 2\rho r^{-\delta}/(r-1)]\, r^{-1}$ if the significand is produced by the FLPOL algorithms. In other words,

$$|z_f| > \left[\frac{1-\Delta}{2} - \frac{2\rho r^{-\delta}}{r-1}\right] r^{-1} = z_{f\min}. \quad (3.11)$$

Proof: Obviously, the initial $\delta + L$ digits of zeros are removed by the FLPOL algorithms, and we have a digit "1" or "−1" at the $(\delta + L + 1)$st recursive step. From Theorem 1, this partially computed result with $k = 1$ has an absolute error of magnitude

$$|\epsilon_1| < \left[\frac{1+\Delta}{2} + \frac{2\rho r^{-\delta}}{r-1}\right] r^{-1+z_e}.$$

Then, the minimum bound of the result is given by the inequality

$$|z| > r^{-1+z_e} - |\epsilon_1|$$
$$= \left[\frac{1-\Delta}{2} - \frac{2\rho r^{-\delta}}{r-1}\right] r^{-1+z_e}.$$

Hence, (3.11) follows. □

For multiplication we obtain the range of the significand in a similar manner

$$|z_f| > \left[\frac{1-\Delta}{2} - \frac{2\rho^2 r^{-\delta}}{(r-1)^2}\right] r^{-1} \quad (3.12)$$
$$\ge z_{f\min}.$$

$z_{f\min}$ is the smallest bound of the redundant significand if the operands are represented with an infinite precision. For instance, in case $r = \rho - 1$, $\delta = 1$ and $\Delta = 0$, (3.11) yields $z_{f\min} = 1/2r^{-1} - 2r^{-2}$. This value can be derived also by an inductive method [20].

Corollary: The FLPOL arithmetic operations guarantee a result with quasi-normalized significand for any radix higher than 2 by choosing an appropriate value of delay δ and a sufficient number β of comparison digits.

Proof: Let $\Delta = 2r^{1-\beta}$. Then, from (3.12), it follows that

$$z_{f\min}/r^{-2} \ge r/2 - r^{2-\beta} - 2r^{1-\delta}. \quad (3.14)$$

In order for the results to be quasi-normalized, the right-hand side of (3.14) must be no less than unity. For $r > 6$, it possible to satisfy $z_{f\min} \ge r^{-2}$ by setting $\beta = 2$ and $\delta = 1$, and radix 6 requires $\beta = \delta = 2$. In the case of radix 4, the necessary condition is $\beta = 3$ and $\delta = 2$. □

Although quasi-normalization is not guaranteed for $r = 2$, it is possible to satisfy $z_{f\min} \ge r^{-3}$ by choosing $\beta = \delta = 4$. More details on the range of redundant significands are discussed in [20].

IV. Representation Errors

With regard to error analysis, a number of measures have been proposed to evaluate the performance of computer arithmetic. Wilkinson [21] used the maximum relative representation error (MRRE) for worst case error analyses. McKeeman [15] proposed the average relative representation error (ARRE) and compared floating-point number systems with different radices. Brent [6] suggested using the root mean square (RMS) error criterion to discuss the choice of a radix for floating-point numbers of fixed length.

The roundoff methods used in conventional arithmetic are not required in the on-line algorithms, since simple truncation produces no bias [3]. We study in this paper the errors resulting from the FLPOL algorithms for addition/subtraction and multiplication in terms of the absolute error and the MRRE. The logarithmic distribution [13] is generally used as an approximation to the distribution of significand for conventional floating-point arithmetic. However, the statistical properties of redundant floating-point numbers have not been established, and therefore the ARRE or the RMS criterion cannot be discussed at this moment.

A. The Absolute Error

The maximum bound of the absolute error is easily derived from Theorem 1 by setting $k = m + 1$ in (3.1). However, the maximum error bound can be improved if we study the FLPOL arithmetic operations carefully. For static error analysis, we study the effect of truncating the result at $(m + 1)$ st digit assuming the operands are free of errors.

1) *Addition:* Summarizing the cases discussed in the Appendix and considering that $\Delta = 0$ for addition, the absolute error e_a of FLPOL addition is bounded by the following inequality.

$$|e_a| < \left[\frac{1}{2} + \frac{\rho r^{-\delta}}{r-1}\right] r^{-(m+1)+z_e}. \quad (4.1)$$

2) *Multiplication:* After $m + \delta_m + 1$ recursive steps, all the digits of the operands are included in the recursion, since $\delta \ge 1$ and $L \ge -1$. From (3.10), the absolute error e_m of FLPOL multiplication satisfies the following inequality:

$$|e_m| \leq \frac{1+\Delta}{2} r^{-(L+m+1)+x_e+y_e}$$
$$= \frac{1+\Delta}{2} r^{-(m+1)+z_e}, \tag{4.2}$$

where

$$z_e = x_e + y_e - L.$$

We see from (4.1) that the absolute error of FLPOL addition is nearly rounded for higher radices or a large value of on-line delay. (4.2) also shows that the absolute error of FLPOL multiplication is nearly rounded if the number of comparison digits is large enough. These results are quite natural because the the digit selection function is a rounding procedure.

B. Maximum Relative Representation Error

The maximum relative representation error (MRRE) for redundant floating-point numbers is defined as

$$\text{MRRE} = \sup \frac{|\hat{z}_f - z_f|}{|\hat{z}_{f}f|} \tag{4.3}$$

where $\hat{z} = r^{z_e}\hat{z}_f$ is the true result of an arithmetic operation and $z = r^{z_e}z_f$ is the computed value.

1) *Addition:* From (3.11) and (4.1) we obtain the expression for the MRRE of FLPOL addition as

$$E_a = \frac{\frac{1}{2} + \frac{\rho r^{-\delta}}{r-1}}{\frac{1}{2} - \frac{2\rho r^{-\sigma}}{r-1}} r^{-m}. \tag{4.4}$$

2) *Multiplication:* Similarly, from (3.12) and (4.2), it follows that

$$E_m = \frac{\frac{1+\Delta}{2}}{\frac{1-\Delta}{2} - \frac{2\rho^2 r^{-\delta}}{(r-1)^2}} r^{-m}. \tag{4.5}$$

The values of the MRRE for addition and multiplication are shown in Table I for different combination of r, ρ and δ.

The MRRE of redundant floating-point numbers can be used to compare the characteristics of different floating-point number systems [15], and also it may be used for worst case error analysis [21]. As we see in Table I, a larger radix with smaller ρ yields better figures of the MRRE.

TABLE I
MAXIMUM RELATIVE REPRESENTATION ERROR

			MRRE ($\times R^{-M}$)	
			ADDITION	MULTIPLICATION
R	ρ	δ	(Δ=0)	(Δ =2R^{-2})
8	4	2	1.0555556	1.0874243
	5	1	1.8333333	1.4450402
	6	1	2.125	1.7147402
	7	1	2.5	2.2
10	5	2	1.0340909	1.0540954
	6	1	1.5454545	1.2714681
	7	1	1.6774193	1.3820676
	9	1	2.0	1.7586207
16	8	2	1.0126050	1.0203185
	9	1	1.2647059	1.1170765
	10	1	1.3	1.1438424
	15	1	1.5	1.3578947
32	16	2	1.0030303	1.0049617
	17	1	1.1103896	1.0432058
	18	1	1.1173913	1.0481742
	31	1	1.2142857	1.1476510

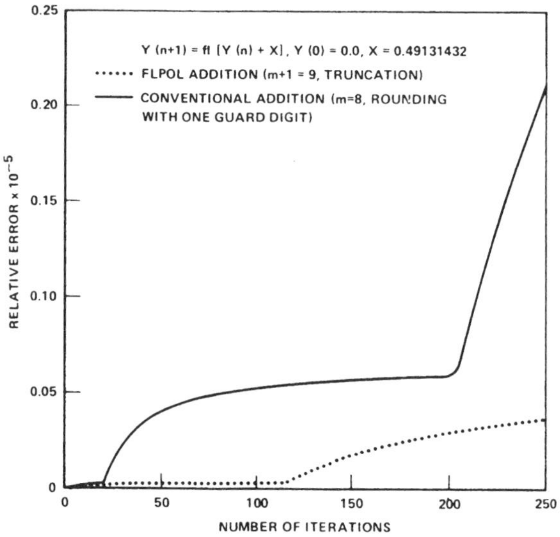

Fig. 2. Error accumulation in repeated addition (r = 10).

V. SIMULATION RESULTS

A result of simulation is shown in Fig. 2. A constant is added repeatedly, and the error growth is measured for both on-line and conventional arithmetic. In conventional rounding arithmetic, the MRRE is given by $1/2r^{1-m}$. For instance, in radix-10 8-digit number representation systems the MRRE is 5.0×10^{-8}. In FLPOL arithmetic, one additional digit is computed to prevent loss of significant digit in the result significand because of quasi-normalization. As a result, the MRRE of FLPOL arithmetic is maintained smaller than that of conventional arithmetic. In the radix-10 9-digit maximally redundant representation, the MRRE for FLPOL addition is 2.0×10^{-8} as shown in Table I. As expected from the comparison of the MRRE's, we observe in Fig. 2 that the error growth of FLPOL addition is smaller than that of conventional addition. The relative error of FLPOL addition grows relatively rapidly near the points where the number of repetition is approximately 130. This is due to quasi-normalization of the result which causes loss of one significant digit.

The worst case absolute error of

$$Y(n+1) = [Y(n) + X](1 + E_a), \qquad n = 0, 1, 2, \cdots \quad (5.1)$$

with $Y(0) = 0$ is approximately given by $E_a n(n-1)X/2$ neglecting higher order terms. Since the true value of $Y(n)$ is nX, the relative error is $e_r = E_a(n-1)/2$. If we substitute the MRRE of FLPOL addition $E_a = 2.0 \times 10^{-8}$, and the number of iterations $n = 201$, we obtain $e_r = 2 \times 10^{-6}$. In Fig. 2, the relative error of on-line arithmetic at $n = 200$ is approximately one-seventh of the worst case error.

Jacobi's iterative algorithm for solving two-dimensional boundary value problems is written as

$$U_{i,j}^{n+1} \leftarrow 0.25\,(U_{i-1,j}^{n} + U_{i,j-1}^{n} + U_{i+1,j}^{n} + U_{i,j+1}^{n}) + b_{i,j} \qquad (5.2)$$

where $U_{i,j}^n$ is the approximate solution at the point (i, j) in the domain at the nth iterative step.

In Fig. 3, error propagation in the Jacobi algorithm is shown, assuming a square mesh with 8×8 internal grid points. The boundary values are $U_{i,0}^n = 0.1875$, $i = 0, 1, 2, \cdots, 9$, $U_{9,j}^n = 0.5000$, $j = 1, 2, \cdots, 8$, $U_{i,9}^n = 0.8750$, $i = 1, 2, \cdots, 9$, $U_{0,j}^n = 0.4375$, $j = 1, 2, \cdots, 9$, and the initial conditions are $Ui, j^0 = 0.0$, $1 \le i, j \le 8$. The constant $b_{i,j} = -0.48828125 \times 10^{-3}$, $1 \le i, j \le 8$. Radix 16 and 6 digits are used for both on-line and conventional arithmetic. The largest absolute error of $U_{i,j}^n$ in each iterative step is plotted up to 100 iterations.

In this type of iterative methods, accumulation of errors does not occur if unbiased arithmetic is used [4], since the magnitude of eigenvalues of the constant matrix used in the iteration is no larger than unity. As shown in Fig. 3, the errors of on-line arithmetic always stay below a constant level which is approximately 1×10^{-6}. On the other hand, in the case of conventional arithmetic using truncation, the error grows as the number of iteration increases. This result supports the discussion on the bias of on-line arithmetic that the symmetric signed-digits representation statistically cancels the bias [3], even though truncation scheme is used.

VI. Conclusion

The error characteristics of the FLPOL arithmetic operations have been discussed. General properties of the redundant significand, required by on-line arithmetic, have been studied. By using a method of error analysis, the range of the redundant significand has been derived for general redundant representation. The range can be also obtained by an inductive method. As a result, the FLPOL operations are proved to yield quasi-normalized results. The MRRE has been defined for redundant floating-point numbers using the absolute error and the range of the redundant significand. The MRRE is improved if a higher radix with smaller ρ is used. The MRRE of FLPOL operations can be used to compare the characteristics of different redundant representations and can be used for the worst case error analysis of computations using FLPOL arithmetic also.

Appendix

The cases of the absolute error analysis for the FLPOL addition are as follows.

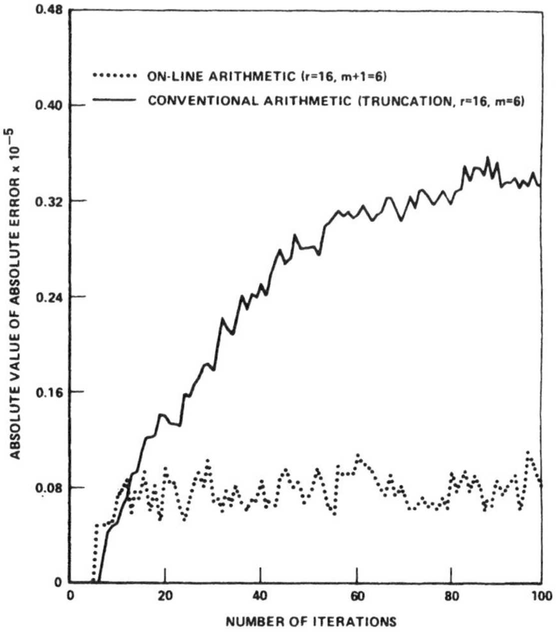

Fig. 3. Error propagation in the Jacobi algorithm.

1) Overflow in the result significand (i.e., $L = -1$)

i) $i_d \le \delta_a - 1$

After $m + 1 + \delta_a$ steps, all the digits of the operand are included in the recursion, and (3.7) leads to

$$|\epsilon_{m+1}| = |r^{-(m+1)}\hat{w}(m + \delta + 1)|r^{x_e} \le \frac{1+\Delta}{2} r^{-(m+1)+z_e} \qquad (A.1)$$

where

$$z_e = x_e + 1.$$

ii) $i_d > \delta_a - 1$.

Similarly from (3.7),

$$|\epsilon_{m+1}| = r^{x_e}|y_{m+\delta-i_d+1}r^{-(m+\delta+1)} + \cdots + y_{m+1}r^{-(m+i_d+1)} + r^{-(m+1)}\hat{w}(m+\delta+1)| < \left[\frac{1+\Delta}{2} + \frac{\rho r^{-\delta}}{r-1}\right] r^{-(m+1)+z_e} \qquad (A.2)$$

where

$$z_e = x_e + 1.$$

2) No overflow in the result significand (i.e., $L \ge 0$)

i) $i_d \le \delta_a$

After $m + \delta_a + 1$ recursive steps, all the operand digits are included in the recursion, and (3.7) leads to

$$|\epsilon_{m+1}| \le |r^{-(m+2+L)}\hat{w}(m + \delta_a + 2)|r^{x_e} = \frac{1+\Delta}{2} r^{-(m+1)+z_e} \qquad (A.3)$$

where

$$z_e = x_e - L.$$

ii) $i_d > \delta_a$.

Similarly from (3.7),

$$|\epsilon_{m+1}| \leq r^{x_e}|y_{m+\delta_a+2-i_d}r^{-(m+\delta_a+2)} + \cdots + y_{m+1}r^{-(m+i_d+1)} + r^{-(m+2+L)}\hat{w}(m+\delta_a+2)| < \left[\frac{1+\Delta}{2} + \frac{\rho r^{-\delta}}{r-1}\right]r^{-(m+1)+z_e}, \qquad (A.4)$$

where

$$z_e = x_e - L.$$

References

[1] A. Avizienis, M. D. Ercegovac, T. Lang, P. Sylvain, and A. Thomasian, "An investigation of fault-tolerant architecture for large-scale numerical computing," in *High Speed Computer and Algorithm Organization*, D. J. Kuck *et al.*, Eds. New York: Academic, 1977.

[2] A. Avizienis, "Signed-digit number representation for fast parallel arithmetic," *IRE Trans. Electron. Comput.*, vol. EC-10, pp. 389–400, Sept. 1961.

[3] A. Avizienis, "On flexible implementation of digital computer arithmetic," in *Proc. IFIP Congress 1962*, Amsterdam, The Netherlands: North-Holland, 1963, pp. 664–678.

[4] N. F. Benschop and H. C. Ratz, "A mean square estimate of the generated roundoff error in constant matrix iterative processes," *J. Ass. Comput. Mach.*, vol. 18, pp. 48–62, Jan. 1971.

[5] R. L. Bivins and N. Metropolis, "Significance arithmetic: application to a partial differential equation," *IEEE Trans. Comput.*, vol. C-26, July 1977.

[6] R. P. Brent, "On the precision attainable with various floating-point number systems," *IEEE Trans. Comput.*, vol. C-22, pp. 601–607, June 1973.

[7] J. T. Coonen, "An implementation guide to a proposed standard for floating-point arithmetic," *Computer*, vol. 13, pp. 68–79, Jan. 1980.

[8] M. D. Ercegovac, "A general hardware-oriented method for evaluation of functions and computations in a digital computer," *IEEE Trans. Comput.*, vol. C-26, pp. 667–680, July 1977.

[9] ——, "An on-line square rooting algorithm," in *Proc. 4th Symp. Comput. Arith.*, Oct. 1978, pp. 183–189.

[10] M. D. Ercegovac and A.L. Grnarov, "On the performance of on-line arithmetic," in *Proc. 1980 Int. Conf. Parallel Process.*, Aug. 1980, pp. 55–61.

[11] A. Gorji-Sinaki and M. D. Ercegovac, "Design of a digit-slice on-line arithmetic unit," in *Proc. 5th Symp. Comput. Arith.*, May 1981, pp. 72–80.

[12] A. L. Grnarov and M. D. Ercegovac, "VLSI-oriented iterative networks for array computations," in *Proc. IEEE Int. Conf. Circuits Comput.*, 1980.

[13] R. W. Hamming, "On the distribution of numbers," *Bell. Syst. Tech. J.*, vol. 49, pp. 1609–1625, Oct. 1970.

[14] M. J. Irwin, "An arithmetic unit for on-line computation," Ph.D. dissertation, Dep. Comput. Sci., Univ. Illinois, Urbana-Champaign, Tech. Rep. 873, May 1977.

[15] W. M. McKeeman, "Representation error for real numbers in binary computer arithmetic," *IEEE Trans. Electron. Comput.*, vol. EC-16, Oct. 1967.

[16] N. Metropolis and R. L. Ashenhurst, "Significant digit arithmetic," *IRE Trans. Electron. Comput.*, vol. EC-7, pp. 265–267, Dec. 1958.

[17] V. G. Oklobdzija and M. D. Ercegovac, "An on-line square root algorithm," *IEEE Trans. Comput.*, vol. C-31, pp. 70–75, Jan. 1982.

[18] K. S. Trivedi and M. D. Ercegovac, "On-line algorithms for division and multiplication," *IEEE Trans. Comput.*, vol. C-26, pp. 681–687, July 1977.

[19] O. Watanuki and M. D. Ercegovac, "Floating-point on-line arithmetic: algorithms," in *Proc. 5th Symp. Comput. Arith.*, May 1981, pp. 81–86.

[20] O. Watanuki, "Floating-point on-line arithmetic for highly concurrent digit-serial computation: application to mesh problems," Ph.D. dissertation, Dep. Comput. Sci. Univ California, Los Angeles, CA, Tech. Rep. UCLA-ENG-8107, May 1981.

[21] J. H. Wilkinson, *Rounding Errors in Algebraic Processes*. New York: Prentice-Hall, 1963.

[22] J. M. Yohe, "Roundings in floating-point arithmetic," *IEEE Trans. Comput.*, vol. C-22, pp. 577–586, June 1973.

Osaaki Watanuki (S'79-M'79-S'79-M'81) was born in Tokyo, Japan. He received the B.S. degree in electrical engineering from Tokyo Institute of Technology in 1970, and the M.S. and Ph.D. degrees in computer science from the University of California, Los Angeles, in 1976 and 1981, respectively.

He was with the Mitsubishi Co. from 1970 to 1974 as a design engineer and computer analyst. From 1979 to 1981 he was a Research Assistant in the Department of Computer Science at the University of California, Los Angeles. He is presently a researcher at the IBM Japan Science Institute, Toyko, Japan. His research interests are in the areas of parallel numerical computations, computer arithmetic, digital signal processing, and speech recognition.

Miloš D. Ercegovac was born in Belgrade, Yugoslavia. He received the B.S.E.E. degree in 1965 from the University of Belgrade, and the M.S. and Ph.D. degrees in computer science from the University of Illinois, Urbana-Champaign, in 1972 and 1975, respectively.

From 1966 to 1970 he was associated with the Institute for Automation and Telecommunications "Milhailo Pupin" in Belgrade, Yugoslavia, and from 1970 to 1975 with the Department of Computer Science, University of Illinois, Urbana-Champaign, as a Research Assistant. He is currently an Associate Professor and Vice-Chairman with the Department of Computer Science, University of California, Los Angeles. He specializes in research and teaching in the areas of computer system design and architecture, and theory and practice of computer arithmetic. In particular, his research interests include fast arithmetic and hardware-oriented algorithms, high-speed computer architectures, functional (applicative) languages and data-driven machines, and fault-tolerant building blocks. blocks.

Dr. Ercegovac is a member of the Association for Computing Machinery and the IEEE Computer Society.

Improved Normalization Results for Digit On-line Arithmetic

Richard J. Zaccone[1] and Jesse L. Barlow[2]

Department of Computer Science
The Pennsylvania State University
University Park, PA 16802

In digit on-line arithmetic, operands are introduced a digit at a time. After the first few operand digits have been introduced, the result begins to appear a digit at a time. This feature of digit on-line arithmetic allows a significant amount of overlapping of arithmetic operations.

Digit on-line arithmetic can sometimes produce unnormalized results. This can present a problem for the divide and square root algorithms. If the divisor and radicand are highly unnormalized, these algorithms will not produce the correct results. Two advances in overcoming this problem are presented. First, several techniques for producing results that are closer to being normalized are developed. Second, it is shown that normalized results are not necessary for divide and square root to work properly. Combining these results yields algorithms that will always give the correct results.

1. Introduction

Digit on-line algorithms for performing computer arithmetic are a recent development [6, 8, 13, 14]. These algorithms are a radical departure from conventional techniques for performing computer arithmetic.

Digit on-line algorithms are iterative algorithms based on the continued sums/products algorithms of DeLugish [4]. What distinguishes these algorithms from conventional algorithms is that the operands are introduced a digit at a time, and the result is generated a digit at a time. At each iteration new digits of the operands are introduced. After a brief delay, result digits are generated, one per iteration.

With digit on-line arithmetic, one can construct pipeline architecures for solving recurrence relations. Digit on-line arithmetic allows these pipelines to be constructed with a high degree of modularity, and simple interconnections [9]. Also, a speedup factor between 2 and 16 over conventional arithmetic can be expected [7].

The algorithms for digit on-line arithmetic can sometimes produce unnormalized results. This has seriously limited their usefulness, since the divide and square root algorithms, as originally presented [9], require a normalized divisor and radicand. In this paper, it will be shown that a normalized divisor and radicand are not necessary for divide and square root to work properly. A condition which is much easier to satisfy will be shown to be sufficient. In the cases where this new condition is not met, methods for conditioning the operands so that it will be met are presented. The result is that the normalization problem in digit on-line arithmetic can always be avoided.

[1]Work supported by the Office of Naval Research under contract No. N0014-80-0517.

[2]Work supported by the National Science Foundation under contract No. DCR-8402363.

2. Notation

All of the digit on-line algorithms will use a redundant base 8 number system.[3] Thus all digits used to represent a number are from the digit set D, where

$$D = \{\bar{7}, \bar{6}, \bar{5}, \bar{4}, \bar{3}, \bar{2}, \bar{1}, 0, 1, 2, 3, 4, 5, 6, 7\},$$

and $\bar{7} = -7$, $\bar{6} = -6$, etc. (X_C, X_M) and (Y_C, Y_M) will be used to represent the floating point operands of an operation. X_C and Y_C are the characteristic portions of the floating point number, and X_M and Y_M are the mantissas. (Z_C, Z_M) will represent the floating point result. The following symbols will also be used.

Symbols Used

symbol	meaning
m_C	number of characteristic digits
m_M	number of mantissa digits
$\hat{x}_i$	i^{th} digit of X_C
x_i	i^{th} digit of X_M
$\hat{y}_i$	i^{th} digit of Y_C
y_i	i^{th} digit of Y_M

Then

$$X_C = \hat{x}_1\hat{x}_2 \cdots \hat{x}_{m_C}$$

$$X_M = .x_1x_2 \cdots x_{m_M}$$

$$Y_C = \hat{y}_1\hat{y}_2 \cdots \hat{y}_{m_C}$$

$$Y_M = .y_1y_2 \cdots y_{m_M}$$

where $\hat{x}_i, x_i, \hat{y}_i, y_i \in D$. Thus,

$$\text{value of } X_C = \sum_{i=0}^{m_C-1} \hat{x}_{m_C-i} 8^i,$$

and

$$\text{value of } X_M = \sum_{i=1}^{m_M} x_i 8^{-i}.$$

Also, X_M^i will be used to represent the first i digits of X_M. I.e.

$$X_M^i = .x_1 \cdots x_i.$$

A floating point number (X_C, X_M) is ***normalized*** if

$$.1_8 \le |X_M| < 1.$$

[3]Other bases such as 2, 4 and 16 are possible, and the results presented here can easily be extended to these bases.

Note that because of the signed digits, it is not sufficient to check the first digit of the mantissa to determine if it is normalized. For example .1$\bar{2}$ is not normalized. In the worst case, it is not possible to determine if a number is unnormalized until the last mantissa digit has been examined.

Digit on-line algorithms are iterative, and at each iteration, they expect to be provided with the next digit of the operands. Thus, if an algorithm requires two operands, then $\hat{x}_j$ and $\hat{y}_j$ must be provided at the jth characteristic iteration. Likewise, x_j and y_j must be provided during the jth mantissa iteration.

3. Floating Point Division - RDIV

Let (Y_C, Y_M) be the floating point dividend, (X_C, X_M) the floating point divisor, and (Z_C, Z_M) the floating point quotient. Algorithm RDIV computes successive approximations to the floating point quotient.

Following is the algorithm RDIV for computing the floating point quotient.

Characteristic Initialization

$$A_{-3} = 0$$

Characteristic Iterations $j = 1, 2, \ldots, m_C$

$$A_{j-3} = A_{j-4} + 8^{m_C - j}(\hat{y}_j - \hat{x}_j)$$

Characteristic Trailing Iteration $j = m_C + 1$

$$A_{j-3} = A_{j-4} + 1$$

Characteristic Trailing Iterations $j = m_C + 2, m_C + 3$

$$A_{j-3} = A_{j-4}$$

Mantissa Initialization

$$U_0 = 0$$

$$Q_{-3} = 0$$

$$S_0 = 1$$

Mantissa Iteration $j = 1$

$$U_j = 8^1 U_{j-1} + 8^{-2} S_{j-1} x_j$$

$$Q_{j-3} = Q_{j-4} + 8^{-1} y_j$$

$$S_1 = S_0$$

Mantissa Iteration $j = 2$

$$t_j = 8^1 U_{j-1} + 8^{-2} S_{j-1} x_j$$

$$s_j = \mathrm{DIV}_j(t_j)$$

$$U_j = t_j s_j - 1$$

$$Q_{j-3} = (Q_{j-4} + 8^{-j} y_j) s_j$$

$$S_j = s_j S_{j-1}$$

Mantissa Iterations $j = 3, 4, \ldots, m_M$

$$t_j = 8^1 U_{j-1} + 8^{-2} S_{j-1} x_j$$

$$s_j = \mathrm{DIV}_j(t_j)$$

$$U_j = t_j(1 + 8^{2-j} s_j) + s_j$$

$$Q_{j-3} = (Q_{j-4} + 8^{-j} S_{j-1} y_j)(1 + 8^{2-j} s_j)$$

$$S_j = S_{j-1}(1 + 8^{2-j} s_j)$$

Mantissa Trailing Iterations $j = m_M + 1, m_M + 2, m_M + 3$

$$t_j = 8^1 U_{j-1}$$

$$s_j = \mathrm{DIV}_j(t_j)$$

$$U_j = t_j(1 + 8^{2-j} s_j) + s_j$$

$$Q_{j-3} = Q_{j-4}(1 + 8^{2-j} s_j)$$

$$S_j = S_{j-1}(1 + 8^{2-j} s_j)$$

The definitions for DIV_j, $j = 1, \ldots, m_M+3$ can be found in [16].

4. One Step Normalization

Owens [9] describes an algorithm called "one step normalization" that can be incorporated into each floating point operation to bring the result one step closer to being normalized. It does this by waiting until the first approximation to the result mantissa has been generated, before producing the last characteristic digit. If the mantissa can be shifted left one place (i.e. if the first mantissa digit produced would have been a zero), the last characteristic digit is decreased by one, and the mantissa is shifted.

For example, in the multiplication algorithm, let P_1 be the first mantissa approximation. P_1 is an approximation to the product which has been produced after examining only the first two digits of the operands. One step normalization says that the mantissa may be shifted left and the characteristic adjusted if $|P_1| < .04_8$. If $|P_1| \geq .04_8$, the result is produced as usual.

The are many instances where $|P_1| \geq .04_8$ but it is possible to shift the mantissa anyway. One step normalization will not detect these cases, and the following sections suggest some improvements that allow many more shifts to occur.

5. Multiplication

Let (Z_C, Z_M) be the result of a floating point multiplication, where (X_C, X_M) and (Y_C, Y_M) are the operands. By definition,

$$Z_M = X_M Y_M. \tag{1}$$

If one step normalization is used to determine if a shift is possible, the on-line delay increases by one, and a shift left is performed if

$$|X_M^2 Y_M^2| < .04_8. \tag{2}$$

This test will miss many of the instances when a shift can be performed. The following discussion will show how it can be improved.

From (1) it follows that

$$\begin{aligned} |Z_M| &< (|X_M^1| + 8^{-1})(|Y_M^1| + 8^{-1}) \\ &= |X_M^1 Y_M^1| + 8^{-1}(|X_M^1| + |Y_M^1|) + 8^{-2}. \end{aligned}$$

Therefore, it is possible to shift Z_M left one place after the first digit of each operand has been introduced if

$$|X_M^1 Y_M^1| + 8^{-1}(|X_M^1| + |y_M^1|) + 8^{-2} \leq .1_8, \tag{3}$$

or equivalently if

$$|X_M^1 Y_M^1| + 8^{-1}(|X_M^1| + |Y_M^1|) \leq .07_8. \tag{4}$$

This test is easy to compute, and since only the first digit of each operand is needed, it does not increase the on-line delay.

Another benefit is that this method can also allow a shift of two places. If the right hand side of (3) is replaced with $.01_8$, then this gives the test necessary to check if a shift of two is possible. A shift of two is possible if

$$X_M^1 = Y_M^1 = 0. \tag{5}$$

For purposes of comparison, consider the best that can be done after examining two digits of each operand. From (1),

$$|Z_M| < (|X_M^2| + 8^{-2})(|Y_M^2| + 8^{-2}),$$

and therefore, a shift left is possible if

$$(|X_M^2| + 8^{-2})(|Y_M^2| + 8^{-2}) \leq .1_8,$$

or equivalently if

$$|X_M^2 Y_M^2| + 8^{-2}(|X_M^2| + |Y_M^2|) \leq .0777_8. \tag{6}$$

This test will increase the on-line delay by one, and gives optimal results if only two digits are being examined. However, the number of bits involved in the comparison (6) make this test prohibitively expensive and beyond current technology. Therefore, this test will serve to gauge the effectiveness of the other two methods presented in this section.

The following table summarizes the performance of one-step normalization (2), test (4), and test (6). Each of these tests was applied to all positive combinations for X_M^2 and Y_M^2 (i.e. all combinations of X_M^2 and Y_M^2 were assumed to be equally likely). "Percent Allowed" shows how often a particular method allowed a shift.

Improvements to One Step Normalization for Multiply

Test	Percent Allowed	On-line Delay Increases?
One step normalization	25%	Yes
Test (4)	31%	No
Test (6)	37%	Yes

Test (4) is clearly superior to one step normalization, since it can detect when a shift is possible more often, and it does not increase the on-line delay. It has the added benefit of allowing two shifts in a special case (5). Although test (6) is not feasible to implement, it shows that by looking at only one operand digit, test (4) can perform almost as well as the best possible case when looking at two operand digits.

6. Division

Let (Z_C, Z_M) be the result of a floating point division, where (Y_C, Y_M) is the numerator and (X_C, X_M) is the denominator.

The divide algorithm generates its first approximation to the quotient after the fourth operand digits have been introduced. Call this approximation Q_1. According to one step normalization, a shift is possible if

$$|Q_1| < .4_8. \tag{7}$$

The rest of this section will describe how the test (7) can be improved.

By definition,

$$|Z_M| = \frac{|Y_M|}{|X_M|} 8^{-1}, \tag{8}$$

and therefore

$$|Z_M| < \frac{|Y_M^2| + 8^{-2}}{|X_M^2| - 8^{-2}} 8^{-1}$$

Thus, after the first two operand digits have been introduced, Z_M can be shifted left one place if

$$\frac{|Y_M^2| + 8^{-2}}{|X_M^2| - 8^{-2}} 8^{-1} \leq .1_8,$$

or equivalently if

$$|X_M^2| - |Y_M^2| \geq .02_8. \tag{9}$$

Also, Z_M can be shifted left two places if

$$\frac{|Y_M^2| + 8^{-2}}{|X_M^2| - 8^{-2}} 8^{-1} \leq .01_8,$$

or equivalently if

$$8^{-1}|X_M^2| - |Y_M^2| \geq .011_8. \tag{10}$$

Combining (9) and (10) yields the following shift algorithm for divisions.

New Shift Algorithm for Division

if $8^{-1}|X_M^2| - |Y_M^2| \geq .011_8$ **then**
shift = 2
else if $|X_M^2| - |Y_M^2| \geq .02_8$ **then**
shift = 1
else
shift = 0

For comparison, consider using a similar technique after the first four operand digits have been introduced. Using (8),

$$|Z_M| < \frac{|Y_M^4| + 8^{-4}}{|X_M^4| - 8^{-4}} 8^{-1},$$

and so a shift can be performed if

$$|X_M^4| - |Y_M^4| \geq .0002_8. \tag{11}$$

Similarly, a shift of two is possible if

$$8^{-1}|X_M^4| - |Y_M^4| \geq .00011_8. \tag{12}$$

This method is too expensive to practically compute, but serves as a good benchmark since no better than this can be done after the first four operand digits have been introduced.

The following table summarizes the performance of one step normalization (7), new shift algorithm for division, and the tests given by (11) and (12). "Percent Allowed" indicates how often a shift of one or more was allowed "Two Shifts" indicates how often two shifts were allowed. The observations were made by considering all positive possibilities for X_M^4 and Y_M^4. The decisions for the new shift algorithm for division were made by considering only the first 2 digits of X_M^4 and Y_M^4.

Improvements to One Step Normalization for Divide

Test	Percent Allowed	Two Shifts	On-line Delay Increases?
One step normalization	28%	0%	Yes
New shift algorithm for division	54%	6%	No
Tests (11) and (12)	56%	6%	Yes

The new shift algorithm for division clearly performs better than one step normalization. Also, by only looking at the first two operand digits, it performs almost as well as the optimal algorithm which considers four operand digits, and it does not increase the on-line delay.

7. Addition

Let (Z_C, Z_M) be the result of a floating point addition, where (X_C, X_M) and (Y_C, Y_M) are the operands. $\tilde{X}_M$ and $\tilde{Y}_M$ are equal to X_M and Y_M after they have been shifted so that X_C and Y_C are equal. If one step normalization is used, the on-line delay increases by one, and a shift is possible when

$$|\tilde{X}_M^2 + \tilde{Y}_M^2| < .4_8. \quad (13)$$

The following analysis will show how (13) can be improved.

By definition,

$$Z_M = 8^{-1}(\tilde{X}_M + \tilde{Y}_M),$$

and therefore,

$$|Z_M| < 8^{-1}(|\tilde{X}_M^1 + \tilde{Y}_M^1| + 2\cdot 8^{-1}).$$

This suggests that Z_M can be shifted left if

$$8^{-1}(|\tilde{X}_M^1 + \tilde{Y}_M^1| + 2\cdot 8^{-1}) \le .1_8,$$

or equivalently if

$$|\tilde{X}_M^1 + \tilde{Y}_M^1| \le .6_8. \quad (14)$$

Since (14) requires only the first digit of each operand to decide if a shift is possible, the on-line delay does not increase.

If however, an increase of one in the on-line delay is tolerable, the following technique can be used. Note that

$$|Z_M| < 8^{-1}(|\tilde{X}_M^2 + \tilde{Y}_M^2| + 2\cdot 8^{-2}),$$

and a shift is therefore possible if

$$8^{-1}(|\tilde{X}_M^2 + \tilde{Y}_M^2| + 2\cdot 8^{-2}) \le .1_8,$$

which reduces to

$$|\tilde{X}_M^2 + \tilde{Y}_M^2| \le .76_8. \quad (15)$$

The following table summarizes the performance of one step normalization (13), test (14), and test (15). To obtain these results, all possible values for $\tilde{X}_M^2$ and $\tilde{Y}_M^2$ were tested to see when a shift was possible. The "Percent Allowed" is the percent of the total number of observations where a shift was permitted.

Improvements to One Step Normalization for Add

Test	Percent Allowed	On-line Delay Increases?
One step normalization (13)	47%	Yes
Test (14)	68%	No
Test (15)	77%	Yes

The performance of test (14) and test (15) exceeds that of one step normalization. Note that although test (14) performs slightly worse than test (15), it has the advantage of not increasing the on-line delay.

8. Post Normalization

This section describes an algorithm which can be appended to any one of the digit on-line algorithms, and which will bring the result one step closer to being normalized. The advantage of this algorithm is that it can be appended more than once to a digit on-line algorithm and thus bring a result even closer to being completely normalized.

Let (Z_C, Z_M) be the result of a floating point operation, and let $\hat{z}_i$ be the ith characteristic digit, and z_i be the ith mantissa digit. If the post normalization algorithm is supplied with the digits of (Z_C, Z_M) in an on-line fashion, it will produce (Z_C', Z_M'), where (Z_C', Z_M') is one step closer to being normalized.

The post normalization algorithm examines z_1 to decide if it is possible to decrease Z_C by one and then shift Z_M left one place. If $z_1 = 0$, then the shift and decrement takes place. If $z_1 \neq 0$, then no shift or decrement takes place. Note that because of the redundant arithmetic being used, there are instances where $|Z_M| < .1_8$ and it is not possible to shift Z_M. For instance, suppose that $Z_M^2 = .1\bar{7}_8$. The post normalization algorithm will not shift Z_M left since $z_1 = 1$. However, since $Z_M < .1_8$, it is not normalized. The post normalization algorithm handles this case by changing Z_M^2 to $.01_8$. Then, if it is applied again, it will perform the shift. Note that each application of the post normalization algorithm increases the on-line delay by one.

Post Normalization

Characteristic Iteration $j = 1$

$$E_j = 8^{-1}\hat{z}_j$$

Characteristic Iterations $j = 2, \ldots, m_C$

$$t_j = 8E_{j-1} + 8^{-1}\hat{z}_j$$

$$s_j = \mathrm{DIS}(t_j)$$

$$E_j = t_j - s_j$$

$$\hat{z}'_{j-1} = s_j$$

Mantissa Iteration $j = 1$

$$e = \begin{cases} 1 & \text{if } z_j = 0 \\ 0 & \text{otherwise} \end{cases}$$

$$t_j = 8E_{m_C} - e$$

$$s_j = \mathrm{DIS}(t_j)$$

$$\hat{z}'_{m_C} = s_j$$

$$F_j = 8^{-1}z_j$$

Mantissa Iterations $j = 2, \ldots, m_M$

$$t_j = 8^e\,(8F_{j-1} + 8^{-1} z_j)$$

$$s_j = \mathrm{NORM}_j(t_j)$$

$$z'_{j-1} = s_j$$

$$F_j = t_j - s_j$$

Mantissa Trailing Iteration $j = m_M + 1$

$$t_j = 8^{e+1} F_{j-1}$$

$$s_j = \mathrm{NORM}_j(t_j)$$

$$z'_{j-1} = s_j$$

$$\mathrm{NORM}_2(x) = \begin{cases} 7 & 7 \le x \\ 6 & 6 \le x < 7 \\ 5 & 5 \le x < 6 \\ 4 & 4 \le x < 5 \\ 3 & 3 \le x < 4 \\ 2 & 2 \le x < 3 \\ 1 & 1 \le x < 2 \\ 0 & -1 < x < 1 \\ \bar{1} & -2 < x \le -1 \\ \bar{2} & -3 < x \le -2 \\ \bar{3} & -4 < x \le -3 \\ \bar{4} & -5 < x \le -4 \\ \bar{5} & -6 < x \le -5 \\ \bar{6} & -7 < x \le -6 \\ \bar{7} & -10_8 < x \end{cases}$$

$$\begin{matrix}\mathrm{NORM}_j(x) \\ j = 3, \ldots, m_M+1\end{matrix} = \begin{cases} 7 & 7 \le x \\ 6 & 6 \le x < 7 \\ 5 & 5 \le x < 6 \\ 4 & 4 \le x < 5 \\ 3 & 3 \le x < 4 \\ 2 & 2 \le x < 3 \\ 1 & 0 < x < 2 \\ 0 & x = 0 \\ \bar{1} & -2 < x < 0 \\ \bar{2} & -3 < x \le -2 \\ \bar{3} & -4 < x \le -3 \\ \bar{4} & -5 < x \le -4 \\ \bar{5} & -6 < x \le -5 \\ \bar{6} & -7 < x \le -6 \\ \bar{7} & -10_8 < x \end{cases}$$

Since the only function of the post normalization algorithm is to try and normalize its input, it is necessary to verify that it has no other unexpected effects. In particular, one would like to verify that (Z_C, Z_M) and (Z'_C, Z'_M) will always have the same value. It is fairly easy to see that the characteristic iterations of the post normalization algorithm work correctly. The following lemma and theorem will show that the post normalization algorithm does in fact preserve the value of (Z_C, Z_M).

Lemma 16. Let $z_1, z_2, \ldots, z_{m_M}$ be the mantissa digits which are input to the post normalization algorithm, and $z'_1, z'_2, \ldots, z'_{m_M}$ be the mantissa digits output by the post normalization algorithm. Let $a_j = 8F_j$. For $j = 2, \ldots, m_M-1$, where F_j is as defined in the mantissa iterations of the post normalization algorithm. Then $|a_j| \le 7$, and $.z_1 z_2 \cdots z_j = .z'_1 z'_2 \cdots z'_{j-1} a_j$.

Proof. By induction on j. Assume that $z_1 \ne 0$ and hence $e = 0$. The proof for when $z_1 = 0$ is similar.
Base $j = 2$.
By definition, $t_2 = z_1.z_2$. If z_1 and z_2 are of the same sign, then from the definition of NORM_2, $z'_1 = z_1$. Hence $F_2 = .z_2$, $a_2 = z_2$, and $|a_2| \le 7$. So, $.z_1 z_2 = .z'_1 a_2$.

Suppose that z_1 and z_2 have different signs. Without loss of generality, assume $z_1 \ge 1$ and $z_2 \le -1$. By definition, $t_2 = z_1.z_2$, and from the definition of NORM_2, $z'_1 = z_1 - 1$. Therefore,

$$\begin{aligned} F_2 &= z_1.z_2 - (z_1 - 1) \\ &= .z_2 + 1. \end{aligned}$$

Thus $a_2 = 8 + z_2$, which implies $|a_2| \le 7$. Since

$$z_1 8^{-1} + z_2 8^{-2} = (z_1 - 1)8^{-1} + (8 + z_2)8^{-2},$$

then $.z_1 z_2 = z'_1 a_2$.

Induction Hypothesis

$$.z_1 z_2 \cdots z_j = .z'_1 z'_2 \cdots z'_{j-1} a_j \qquad \text{where } |a_j| \le 7.$$

Induction Step $j = 3, \ldots, m_M+1$
Suppose that either $a_j \ge 1$ and $z_{j+1} \ge 1$, or $a_j \le -1$ and $z_{j+1} \le -1$ By definition $t_{j+1} = a_j.z_{j+1}$. Therefore, from the definition of NORM_j $z'_j = a_j$, and thus by the induction hypothesis

$$.z_1 \cdots z_j = .z'_1 \cdots z'_j. \tag{17}$$

Also, from the definition of F_{j+1}, $F_{j+1} = .z_{j+1}$, and therefore,

$$a_{j+1} = z_{j+1} \qquad \text{where } |a_{j+1}| \le 7. \tag{18}$$

Thus, using (17) and (18)

$$.z_1 \cdots z_{j+1} = .z'_1 \cdots z'_j a_{j+1}.$$

Suppose $a_j = 0$ and $z_{j+1} \ge 1$. Then $t_{j+1} = 0.z_{j+1}$. Then from the definition of NORM_j, $z'_j = 1$ and from the induction hypothesis

$$.z_1 \cdots z_j = .z'_1 \cdots z'_j - 8^{-j}. \tag{19}$$

From the definition of F_{j+1} $F_{j+1} = 0.z_{j+1} - 1$, and therefore

$$a_{j+1} = z_{j+1} - 8 \qquad \text{where } |a_{j+1}| \le 7. \tag{20}$$

Using (19) and (20) yields

$$.z_1 z_2 \cdots z_{j+1} - 8^{-j} = .z'_1 \cdots z'_j - 8^{-j} + 8^{-j-1} a_{j+1},$$

and hence

$$.z_1 z_2 \cdots z_{j+1} = .z'_1 \cdots z'_j a_{j+1}.$$

The cases for $a_j = 0$ and $z_{j+1} \le -1$, $a_j = 0$ and $z_{j+1} = 0$, $a_j \le -1$ and $z_{j+1} = 0$, $a_j \le -1$ and $z_{j+1} \ge 1$, $a_j \ge 1$ and $z_{j+1} \le -1$, and $a_j \ge 1$ and $z_{j+1} = 0$ are similar. □

Let $z_1, z_2, \ldots, z_{m_M}$ be the mantissa digits which are input to the post normalization algorithm, and $z'_1, z'_2, \ldots, z'_{m_M}$ be the mantissa digits output by the post normalization algorithm. The following theorem shows that the value of the mantissa is not altered by the post normalization algorithm.

Theorem 21.

$$.z_1 \cdots z_{m_M} = .z'_1 \cdots z'_{m_M}.$$

Proof. By lemma 16,

$$.z_1 z_2 \cdots z_{m_M} = .z'_1 z'_2 \cdots z'_{m_M-1} a_{m_M}.$$

But, by definition

$$\begin{aligned} z'_{m_M} &= \text{NORM}_j(t_{m_M}) \\ &= \text{NORM}_j(8F_{m_M}) \\ &= \text{NORM}_j(a_{m_M}) \\ &= a_{m_M}, \end{aligned}$$

where $|a_{m_M}| \le 7$. □

The improvements suggested for multiply, add, and divide work significantly better than one step normalization. Also, they are easy to implement and they do not increase the on-line delay. Therefore, they should probably be incorporated into the formulation of the digit online algorithms.

9. A Closer Look at Division

Owens [9] says that a normalized divisor is a sufficient condition for RDIV to give correct results. This condition can be relaxed considerably. It will be shown that X_M^2 (the first two digits of the mantissa of the divisor) must only satisfy

$$.10_8 \le |X_M^2| \le .77_8$$

so that RDIV gives correct results. Since the divisor does not need to be normalized, it is no longer necessary to wait until the last mantissa digit has been seen to determine if a divide will be successful. This can now be determined after the second mantissa digit.

The characteristic iterations are fairly straightforward since they compute $Y_C - X_C + 1$. However, the mantissa iterations are a little bit more complicated. Here, the object is to compute $Z_M = Y_M / X_M$. (To make the discussion simpler, the 8^{-1} factor will be ignored). If d_j, $j = 2, \ldots, m_M+3$ are chosen such that

$$X_M \prod_{j=2}^{m_M+3} d_j \approx 1,$$

then

$$\frac{Y_M}{X_M} = \frac{Y_M \prod_{j=2}^{m_M+3} d_j}{X_M \prod_{j=2}^{m_M+3} d_j} \approx Y_M \prod_{j=2}^{m_M+3} d_j.$$

The following lemma shows that S_M+3 is the same as $\prod_{j=2}^{m_M+3} d_j$ in the above discussion.

Lemma 22. Let s_j and S_j be as in algorithm RDIV. By definition, $s_j = \text{DIV}_j(t_j)$, $j = 2, \ldots, m_M+3$. Let $d_j = (1 + 8^{2-j} s_j)$, $j = 3, \ldots, m_M+3$ and $d_2 = s_2$. Then

$$\begin{aligned} S_j &= \prod_{i=2}^{j} d_i \\ &= S_{j-1} d_j, \end{aligned}$$

where $S_1 = 1$.

Proof. $S_2 = d_2$ since by hypothesis $s_2 = d_2$ and by definition $S_1 = 1$. Also by definition $S_j = S_{j-1}(1 + 8^{2-j} s_j)$. □

Lemma 23 establishes the definition of U_j. A direct result of this lemma is that if U_j is "small" (this will be defined later), $S_j \approx 1/X_M^j$.

Thus, the value of U_j can be used as a measure of whether S_j is a good approximation to $1/X_M^j$. Note also that t_j is an updated version of $8^1 U_{j-1}$ when $2 \le j \le m_M$.

Lemma 23. For $2 \le j \le m_M+3$, $U_j = 8^{j-2}(S_j X_M^j - 1)$, where $X_M^j = X_M$ if $j \ge m_M$. Also, for $3 \le j \le m_M$, $t_j = 8^{j-2}(S_{j-1} X_M^j - 1)$, where $X_M^j = X_M$ if $j \ge m_M$.

Proof. By induction on j.
Base. $j = 2$. By definition $U_2 = t_2 s_2 - 1$. It can easily be seen that $t_2 = X_M^2$ and $s_2 = S_2$. Thus $U_2 = S_2 X_M^2 - 1$.

Induction Step. By definition

$$\begin{aligned} t_j &= 8^1 U_{j-1} + 8^{-2} S_{j-1} x_j \\ &= 8^{j-2}(S_{j-1} X_M^{j-1} - 1) + 8^{-2} S_{j-1} x_j \\ &= 8^{j-2} S_{j-1}(X_M^{j-1} + 8^{-j} x_j) - 8^{j-2} \\ &= 8^{j-2}(S_{j-1} X_M^j - 1). \end{aligned}$$

The proof for U_j is similar.

$$\begin{aligned} U_j &= t_j(1 + 8^{2-j} s_j) + s_j \\ &= 8^{j-2}(S_{j-1} X_M^j - 1)(1 + 8^{2-j} s_j) + s_j \\ &= 8^{j-2} X_M^j S_{j-1}(1 + 8^{2-j} s_j) - 8^{j-2} \\ &= 8^{j-2}(X_M^j S_j - 1). \end{aligned}$$

□

Lemma 24. For $2 \le j \le m_M+3$ $Q_{j-3} = Y_M^j S_j$, where $Y_M^j = Y_M$ for $j \ge m_M$.

Proof. By induction on j.
Base. $j = 2$. By definition,

$$Q_{-1} = (Q_{-2} + 8^{-2} y_2) s_2.$$

But $Q_{-2} = Y_M^1$ and $s_2 = S_2$. Therefore, $Q_{-1} = Y_M^2 S_2$.

Induction Step. By definition,

$$\begin{aligned} Q_{j-3} &= (Q_{j-4} + 8^{-j} S_{j-1} y_j)(1 + 8^{2-j} s_j) \\ &= (Y_M^{j-1} S_{j-1} + 8^{-j} S_{j-1} y_j)(1 + 8^{2-j} s_j) \\ &= (Y_M^{j-1} + 8^{-j} y_j) S_{j-1}(1 + 8^{2-j} s_j) \\ &= Y_M^j S_j. \end{aligned}$$

□

By Lemma 23, if U_j is "small", $S_j \approx 1/X_M^1$, and therefore by Lemma 24 $Q_{m_M} \approx Y_M/X_M$. This will be formalized later. The following three lemmas can easily be verified by writing a small program. These lemmas help to establish bounds on U_j (Remember that U_j can be used to estimate how close S_j is to the reciprocal of X_M^1). Lemma 28 uses these results to establish a bound for U_j when X_M^2 (the first two digits of the mantissa of the divisor) satisfies (29). t_j, S_j, and s_j are as defined in algorithm RDIV.

Lemma 25. By definition, $t_2 = X_M^2$, $s_2 = \mathrm{DIV}_2(t_2)$ and $U_2 = t_2 s_2 - 1$. If $.10_8 \le |t_2| \le .77_8$, then $|U_2| \le .25_8$.

Lemma 26. By definition, $s_3 = \mathrm{DIV}_3(t_3)$ and $U_3 = t_3(1 + 8^{-1}s_3) + s_3$. If $|t_3| \le 3.31_8$, then $|U_3| \le .5_8$.

Lemma 27. By definition, if $4 \le j \le m_M+3$, then $s_j = \mathrm{DIV}_j(t_j)$ and $U_j = t_j(1 + 8^{2-j}s_j) + s_j$. If $|t_j| \le 5.7_8$, then $|U_j| \le .441_8$.

Lemma 28. If $.10_8 \le |X_M^2| \le .77_8$, ($X_M^2$ is the first two digits of the mantissa of the divisor) then

$$|U_j| < 0.5_8 \quad j = 2, \ldots, m_M+3. \tag{29}$$

Thus, if X_M^2 satisfies (29), S_j will be a good approximation to $1/X_M^1$.

Proof. By induction on j.
Base. $j = 2$. If $.10_8 \le |X_M^2| \le .77_8$, then $.10_8 \le |t_2| \le .77_8$, since by definition $t_2 = X_M^2$. By lemma 25, $|U_2| \le .25_8$.
For $j = 3$, $t_3 = 8^1 U_2 + 8^{-2} S_2 x_3$, which implies that $|t_3| \le 3.4_8$ since $U_2 \le .25_8$, $|S_2| \le 7$, and $|x_3| \le 7$. By lemma 26, $|U_3| \le .5_8$.

Induction Step. By definition,

$$t_j = 8^1 U_{j-1} + 8^{-2} S_{j-1} x_j,$$

and hence

$$\begin{aligned} |t_j| &\le |8^1 U_{j-1}| + 8^{-2}|S_{j-1}x_j| \\ &\le 5 + 8^{-2}|S_{j-1}x_j| \\ &\le 5.7_8. \end{aligned}$$

Therefore, by lemma 27, $|U_j| \le .441_8$. □

The following theorem establishes a condition on X_M (the mantissa of the divisor) which is sufficient to cause the divide algorithm to produce the correct quotient. Note that this is a much stronger result than just requiring that X_M be normalized. Condition (31) requires that the first two digits of the mantissa must fall within a certain range. This is stronger than previous results [9] since normalized divisors are a small subset of the values allowed by (31). Previously there were many instances when every mantissa digit of the divisor had to be examined so it could be determined if it was normalized.

Theorem 30. If

$$.10_8 \le |X_M^2| \le .77_8, \tag{31}$$

then

$$\left|\frac{Y_M}{X_M} - Q_{m_M}\right| < 6.2_8 \times 8^{-(m_M+1)},$$

where Q_{m_M} is the final approximation to $\frac{Y_M}{X_M}$ produced by algorithm RDIV. Therefore, if the first two digits of the mantissa of the divisor satisfy (31), then RDIV produces an accurate approximation the the true quotient $\frac{Y_M}{X_M}$.

Proof. Using lemma 24,

$$\begin{aligned} \left|\frac{Y_M}{X_M} - Q_{m_M}\right| &= \left|\frac{Y_M}{X_M} - Y_M S_{m_M+3}\right| \\ &\le |Y_M| \left|\frac{1}{X_M} - S_{m_M+3}\right|. \end{aligned}$$

Then by lemma 23,

$$\left|\frac{Y_M}{X_M} - Q_{m_M}\right| \le |Y_M| \left|\frac{1}{X_M} U_{m_M+3} 8^{-(m_M+1)}\right|.$$

By definition, $|Y_M| < 1$, $\frac{1}{X_M} < 12_8$, and by lemma 28, $|U_{m_M+3}| < .5_8$. Thus

$$\left|\frac{Y_M}{X_M} - Q_{m_M}\right| < 6.2_8 \times 8^{-(m_M+1)}.$$ □

The following result can also be proven for the square root algorithm. The proof is precisely analogous to that for the division algorithm. It is given in [17].

Theorem 32. If

$$.10_8 \le X_M^2 \le .77_8, \tag{33}$$

then

$$|\sqrt{X_M} - R_{m_M}| < 5.62_8 \times 8^{-(m_M+1)}$$

where R_{m_M} is the final approximation to $\sqrt{X_M}$ produced by algorithm RSQR from [9]. Therefore, if the first two digits of the mantissa of the radicand satisfy (33), then RSQR produces an accurate approximation the the true square root $\sqrt{X_M}$.

Note that this is a much stronger result than that given by Owens [9], since theorem 32 does not require that X_M be normalized, but only that (33) be satisfied.

10. Conclusion

The improvements suggested for multiply, add, and divide work significantly better than one step normalization. Also, they are easy to implement and they do not increase the on-line delay. Therefore, they should probably be incorporated into the formulation of the digit on-line algorithms.

According to Owens [9], a normalized divisor is a sufficient condition for RDIV to give correct results. Therefore, it was necessary to make sure that the divisor was normalized to insure a correct result. In the worst case, this means that if the mantissa of the divisor is $.10 \cdots 01$, where the 1 occurs in the last mantissa digit, the divide algorithm won't know that it has an unnormalized divisor until it has seen every digit. Thus a number that was only slightly unnormalized has completely disrupted a divide operation. The only way that the problem could be avoided with certainty was to increase the on-line delay to m_M (the number of mantissa digits). However, all advantages over conventional arithmetic have then been lost.

With the results presented in the preceding sections, this problem can be eliminated. In order to do so, it is necessary to use the post normalization algorithm that was presented in an earlier section of this paper. This algorithm, which has an on-line delay of one, takes the result of a floating point operation, (Z_C, Z_M), as its input. It produces (Z_C', Z_M') where (Z_C', Z_M') is one step closer to being normalized.

The post normalization algorithm and the results presented in the previous two sections can now be combined to produce a new approach to the normalization problem.

This approach involves a pre-conditioning for all divisors and radicands. As shown in the previous two sections, it is only necessary to look at the first two digits of the mantissas of these values. If conditions (31) for divide and (33) for square root are satisfied, then the operation can proceed without any pre-conditioning, since it has been shown that these are sufficient conditions for divide and square root to work properly. If these conditions are not satisfied, then the post normalization algorithm can be used as many times as necessary.

The fact that the post normalization algorithm is digit on-line can be used to make this easier. The result digits of one application of the algorithm can be fed back into the algorithm as they are produced. The net effect is that a new version of the operand is available at each time step. That is, at each time step the operand is brought closer to satisfying either (31) or (33). As soon as the proper condition is satisfied, the divide or square root can continue.

Since the divide and square root algorithms will now have variable on-line delays, this will affect other operations in the pipeline. Results will need to be buffered and some timing problems may result. However, these are implementation details that shouldn't be difficult to solve.

11. References

1. R. L. Ashenhurst and N. Metropolis, Unnormalized Floating Point Arithmetic, *J. ACM 6*, (1959), pp. 415-428.
2. R. L. Ashenhurst, Experimental Invesigation of Unnormalized Arithmetic, in *Error in Digital Computation*, vol. 2, L. B. Rall (ed.), John Wiley & Sons, New York, NY, 1965, pp. 3-37.
3. D. E. Atkins, Introduction to the Role of Redundancy in Computer Arithmetic, *Computer 8*, 6 (June 1975), pp. 76-84.
4. B. G. DeLugish, A Class of Algorithms for Automatic Evaluation of Certain Elementary Functions in a Binary Computer, Ph.D. Diss., Report 399, Dept. of Computer Science, University of Illinois, Urbana, IL, 1970.
5. M. D. Ercegovac, Radix-16 Evaluation of Certain Elementary Functions, *IEEE Trans. on Computers C-22*, 6 (June 1973), pp. 561-566.
6. M. D. Ercegovac, An On-line Square Rooting Algorithm, *Proc. of fourth Symp. on Computer Arithmetic*, Santa Monica, CA, Oct. 1978.
7. M. D. Ercegovac and A. L. Grnarov, On the Performance of On-line Arithmetic, *Proc. of the 1980 Inter. Conf. on Parallel Processing*, , 1980, pp. 55-61.
8. A. L. Grnarov and M. D. Ercegovac, An Algorithm for On-Line Normalization, Quarterly Report, UCLA Computer Science Dept., July 1979.
9. R. M. Owens, *Digit On-Line Algorithms for Pipeline Architectures*, Ph.D. Diss., Dept. of Computer Science, The Pennsylvania State Univ., University Park, PA, 1980.
10. R. M. Owens, Compound Algorithms for Digit On-Line Arithmetic, *Proc. 5th Symp on Computer Arithmetic*, Ann Arbor, MI, May 1981, pp. 64-71.
11. P. H. Sterbenz, *Floating-Point Computation*, Prentice Hall, Englewood Cliffs, NJ, 1974.
12. K. S. Trivedi and J. G. Rusnak, Higher Radix On-line Division, *Proc. of the Fourth Symp. on Computer Arithmetic*, Santa Monica, CA, Oct. 1978.
13. O. Watanuki and M. D. Ercegovac, Floating-Point On-Line Arithmetic: Error Analysis, *Proc. 5th Symp. on Computer Arithmetic*, Ann Arbor, MI, May 1981, pp. 87-91.
14. O. Watanuki and M. D. Ercegovac, Floating-Point On-Line Arithmetic: Algorithms, *Proc. 5th Symp. on Computer Arithmetic*, Ann Arbor, MI, May 1981, pp. 81-86.
15. J. H. Wilkinson, *Rounding Errors in Algebraic Processes*, Prentice Hall, Englewood Cliffs, NJ, 1963.
16. R. J. Zaccone and J. L. Barlow, Improved Normalization Results for Digit Online Arithmetic, CS-84-16, Dept. of Computer Science, The Pennsylvania State Univ., University Park, PA, Oct. 1984.
17. R. J. Zaccone, *Numerical Properties of Digit On-line Arithmetic*, Ph.D. Diss., Dept. of Computer Science, The Pennsylvania State Univ., University Park, PA, 1984.

Reprinted from *IEEE Transactions on Computers*, Volume C-32, Number 4, April 1983, pages 402-406.

Fully Digit On-Line Networks

MARY JANE IRWIN AND ROBERT MICHAEL OWENS

***Abstract*—Research in computer architecture in the last decade has been driven largely by the motivation to overcome the "von Neumann" bottleneck. This paper describes the design and use of one such architecture—fully digit on-line networks. First, digit on-line algorithms and processing are defined. The key advantage to digit on-line processing is that it allows a digit serial, most significant digit first, type of data flow. Processing of the most significant operand digits starts immediately and generation of the most significant result digits soon follows. The minimum set of primitive logic operations required to implement a digit on-line processing component in VLSI are outlined. Then, digit on-line networks consisting of many of these digit on-line components are examined. Finally, two different network configurations are discussed and compared.**

***Index Terms*—Digit on-line algorithms, digit on-line architectures, linear recurrences, pipelining, redundant digit sets, systolic arrays.**

I. AN INTRODUCTION TO DIGIT ON-LINE ALGORITHMS

Since the definition of "on-line processing" as it relates to computer arithmetic by Trivedi and Ercegovac [1], digit on-line algorithms and their corresponding computer architectures have undergone investigation [2]-[8]. Fully digit on-line arithmetic algorithms generate the jth digit of the result after having been supplied with up to only the $(j + k)$th digits of the input operand(s) where k is a small predetermined on-line delay. Digit on-line algorithms can be derived which input digits starting at the least significant end of the operand(s) (left-directed algorithms) or at the most significant end of the operand(s) (right-directed algorithms). It is this latter digit flow—right-directed—upon which this paper is based. In digit on-line processing, the generation of result digits proceeds on-line with the input of operand digits. Full precision of the operands is not required in order for the evaluation of the arithmetic function to begin; the payoff can be a decrease in execution time. Algorithms have been developed [6] for the right-directed fully digit on-line evaluation of floating-point addition, multiplication, division, square root, and the trigonometric functions of sine and cosine. Fixed-point algorithms for the logarithm and exponential functions also exist. This family of digit on-line algorithms is based upon the use of continued sums and products [9] with the result digits being selected from a redundant digit set [10].

In fully digit on-line right-directed floating-point algorithms, the operand digits are input starting with the integer exponent and working through the fraction mantissa. Exponent first-ordering is mandatory in the case of addition and subtraction, and so is imposed on all operations. The exponent is processed in a digit on-line fashion, not accumulated and then processed in a more conventional manner as is the case in other proposed floating-point quasi-digit on-line schemes [4], [8]. Fully digit on-line right-directed algorithms for the various functions can be derived in a two-step process. First, the appropriate continued sums or products algorithm [9] can be converted in a straightforward manner to an algorithm which is digit on-line with respect to the input operand(s). These algorithms operate by using one more digit of the input operand(s) per iteration to generate closer and closer approximations to the results. Since the last (and best) approximation to the result is not available until the final iteration, this algorithm is not fully digit on-line. There are, however, several discretization algorithms which, when given these approximations, generate the result in a digit on-line manner. One such algorithm uses a limited precision rounding on those approximations to generate one digit of the result per iteration. The equations given in Table I for the fully digit on-line right-directed floating-point multiplication algorithm have been reformulated after using this two-step approach to produce a more computationally efficient algorithm.

Manuscript received April 19, 1982; revised October 15, 1982 and November 29, 1982. This work was supported, in part, by the National Science Foundation under Grant US-NSF-MCS-7809126A01, and by the Office of Naval Research under Grant N00014-80-C-0517.

The authors are with the Department of Computer Science, Pennsylvania State University, University Park, PA 16802.

The two important properties of digit on-line algorithms are evident in the equations of Table I.

1) Operand digits are input in a digit on-line fashion with an initial on-line delay of $k = 1$, i.e., only one digit of both input operands is required initially.

2) Result digits are produced in a digit on-line fashion; the jth operand digits are supplied at step j while the $(j - k)$th result digit is produced.

The discretization algorithm (DIS) is just a limited precision rounding on the temporary partials T_j's generated by the approximation recursions, i.e., a rounding which is based on an inspection of the first few most significant digits of T_j. Thus, this discretization scheme can be implemented as a limited precision table look-up operation. Table II gives the discretization scheme (DIS) when the radix is octal.

The recursive equations for fully digit on-line right-directed floating-point addition ($k = 1$) are similar to those for multiplication, whereas more complex but still similar recursive equations exist for division ($k = 3$) and square root ($k = 3$). The next section addresses the minimum set of primitive logic operations required to implement a digit on-line component. Then, various digit on-line networks consisting of many of these digit on-line components and how they work together with the central system are examined. Finally, a comparison of the networks configuration is given.

TABLE I
FULLY DIGIT ON-LINE FLOATING-POINT MULTIPLICATION

Solve: $(z_c . z_m) = (y_c . y_m) \cdot (x_c . x_m)$

CHARACTERISTIC EQUATIONS

Initialization

$$A_0 = r^{(-1)}((y_c)_1 + (x_c)_1)$$

$$(y_c)_{m_c+1}, (x_c)_{m_c+1} = 0$$

Recursions for j = 2, 3, . . . , m_c, m_c+1

$$T_j = rA_{j-2} + r^{(-1)}((y_c)_j + (x_c)_j)$$

$$(z_c)_{j-1} = DIS(T_j)$$

$$A_{j-1} = T_j - (z_c)_{j-1}$$

MANTISSA EQUATIONS

Initialization

$$Y_1 = r^{(-1)}(y_m)_1, \quad X_1 = r^{(-1)}(x_m)_1$$

$$P_0 = r^{(-1)}((x_m)_1 Y_1)$$

$$(y_m)_{m_m+1}, (x_m)_{m_m+1} = 0$$

Recursions for j = 2, 3, . . . , m_m, m_m+1

$$Y_j = Y_{j-1} + r^{(-j)}(y_m)_j, \quad X_j = X_{j-1} + r^{(-j)}(x_m)_j$$

$$T_j = rP_{j-2} + r^{(-1)}((x_m)_j Y_j + X_{j-1}(y_m)_j)$$

$$(z_m)_{j-1} = DIS(T_j)$$

$$P_{j-1} = T_j - (z_m)_{j-1}$$

WHERE

r is the radix and $r^{(-1)}$ is 1/r,

$(y_c)_j, (x_c)_j, (z_c)_j, (y_m)_j, (x_m)_j$, and $(z_m)_j$ are the jth digits of $y_c, x_c, z_c, y_m, x_m, z_m$, and

m_c and m_m are the lengths in digits of the integer characteristics and normalized fractional mantissas.

II. Logic Requirements of a Digit On-Line Processing Component

The minimum set of primitive logic operations which must be provided in hardware to build a fully digit on-line processing component (for floating-point add, subtract, multiply, divide, and square root) as determined by inspection of the family of unified fully digit on-line algorithms are the following.

1) The addition or subtraction of two fixed-point values which can be implemented in four gate delays using redundant adders with very regular structure and interconnections.

2) The multiplication of a fixed-point value by a single digit value which can be implemented in six gate delays using redundants adders with some front-end combinational logic.

3) The shifting of a fixed-point value by a constant integer value which can be implemented in zero gate delays by appropriately off-setting the interconnections.

4) Result digit selection, determined on-line by applying a discretization function, like DIS, to a truncated version of the appropriate T_j in two gate delays using table look-up.

Once these primitive operations have been specified, the logic design of a fully digit on-line component can be considered. A block diagram of the mantissa processor for the fully digit on-line multiplication algorithm given in Table I is shown in Fig. 1. With only minor gating and control modifications, this multiply component could be designed to also do fully digit on-line floating-point additions of up to four operands.

TABLE II
Result Digit Selection Function ($r = 8$)

T_j	DIS(T_j)	Encoding
$T_j \geq 6.4_8$	7	0111
$6.4_8 > T_j \geq 5.4_8$	6	0110
$5.4_8 > T_j \geq 4.4_8$	5	0101
$4.4_8 > T_j \geq 3.4_8$	4	0100
$3.4_8 > T_j \geq 2.4_8$	3	0011
$2.4_8 > T_j \geq 1.4_8$	2	0010
$1.4_8 > T_j \geq 0.4_8$	1	0001
$0.4_8 > T_j \geq -0.4_8$	0	0000
$-0.4_8 > T_j \geq -1.4_8$	-1	1111
$-1.4_8 > T_j \geq -2.4_8$	-2	1110
$-2.4_8 > T_j \geq -3.4_8$	-3	1101
$-3.4_8 > T_j \geq -4.4_8$	-4	1100
$-4.4_8 > T_j \geq -5.4_8$	-5	1011
$-5.4_8 > T_j \geq -6.5_8$	-6	1010
$-6.4_8 > T_j$	-7	1001

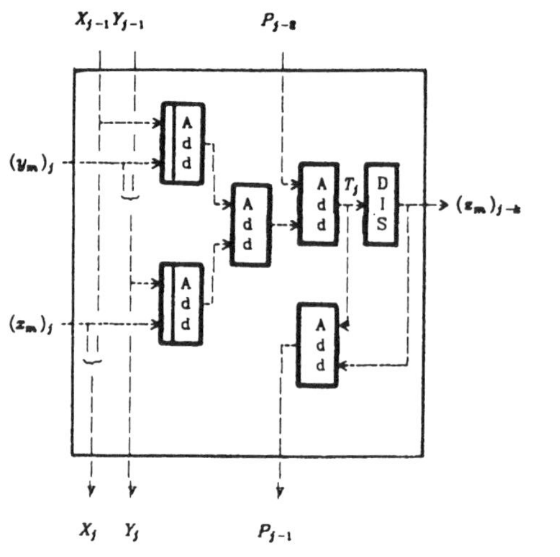

Fig. 1. Block diagram of a multiply component.

Once the worst case gate delays for the minimum set of primitive operations have been analyzed, and the block diagram for a digit on-line processing component has been specified, the worst case processing speed can be determined. Sixteen gate delays are needed to produce the next product digit $(z_m)_{j-1}$, while 20 gate delays are required to accomplish one complete iteration of the fully digit on-line multiplication algorithm using the logic of Fig. 1. This does not include time for register loading and control functions. A similar analysis of the other floating-point algorithms indicates that addition requires ten gate delays, while division and square root both require 28 gate delays. This iteration step time (i.e., the time to obtain one result digit) is independent of the length of the operands.

III. Digit On-Line Networks

Various computer architectures using these digit on-line components will now be investigated. Since the algorithms are defined by a set of very similar recursive equations, a single multifunctional hardware component could be built which, in one time step, inputs one digit of the operand(s) and outputs the previously computed digit of the result while cycling through one recursion of the algorithm as shown in Fig. 2. This single component could be employed as a coprocessor in a cost-effective manner by a small computer system to implement elementary functions. In this situation, the component would be used recursively with the residual produced by each iteration reused by that unit to complete the next iteration. (The residual R_{j-1} would correspond to either A_{j-1} or P_{j-1} in the multiplication algorithm of Table I.)

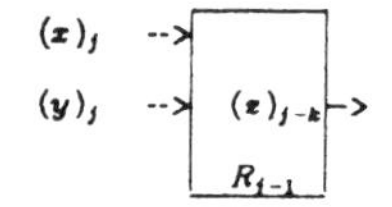

Fig. 2. Digit on-line uniprocessor component.

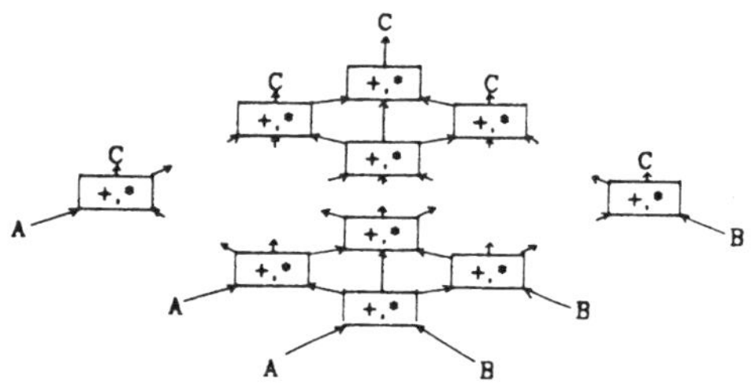

Fig. 3. Conventional systolic array.

For applications where matrix and vector operations are prevalent and where the need for high speed may override cost considerations, a multiprocessor architecture may be more appropriate than this uniprocessor approach. Several digit on-line processing components working together may be interconnected to make a digit on-line network. In many cases, a digit on-line network can be formed by simply replacing the components of a conventional network with their equivalent digit on-line counterparts. For example, consider the well known systolic array [11] for computing the product C of two n diagonal matrices A and B as shown in Fig. 3.

Fig. 4 illustrates a fully digit on-line network for matrix multiplication created by substituting the add/multiply components of the systolic array with their digit on-line counterparts. In this mode, the digit on-line components are each functioning as uniprocessors. The skewing/deskewing components (S and D) augmenting the input-output paths are needed to maintain input-output compatibility with the conventional systolic array. While the digit on-line systolic array shares the same structure as its nondigit on-line counterpart, it avoids, because of the digit on-line property of its components, several implementation problems.

The operation of either array involves moving data between the processors of the array and performing the necessary arithmetic operations within each processor. In a conventional array these two functions can not be overlapped if the arithmetic operation to be performed can not be started before the full precision of all operands required by that operation has been transferred to the processor. Hence, such arrays have two distinct phases. In the first phase, the links of the array are used to move data between the processors of the array. In the next phase, the processors of the array are used to perform the necessary arithmetic operations. Data can be transferred over links in either a parallel or serial fashion. If data are transferred in a totally parallel fashion (all the bits are transferred at once over separate bit level interconnections), the time required to perform the actual transfer is minimal. However, the number of bit level interconnections is maximized. If the array is to be realized in VLSI, this can present quite a problem. On the other hand, if data are transferred in a totally serial fashion, the number of interconnections is minimized. However, the time required to transfer each piece of data, the time required for the data movement phase of operations, and hence, the running time of the array, is increased.

In a digit on-line network, these two functions can be overlapped. This is the direct result of being able to start an arithmetic operation before all the data required by that operation has been totally transferred to the processor. Hence, the time dedicated to the transfer of data does not suffer because of a decrease in the number of interconnections. Therefore, the designer can attain both high speed and a minimal number of interconnections.

Others have also recognized this problem and have proposed left-directed digit on-line algorithms as a solution. The drawback with left-directed algorithms is that the set of arithmetic operations for which there exists a left-directed digit on-line algorithm is quite small (basically integer addition and multiplication.) While this set is sufficient for many applications, it is far too limited for many other applications.

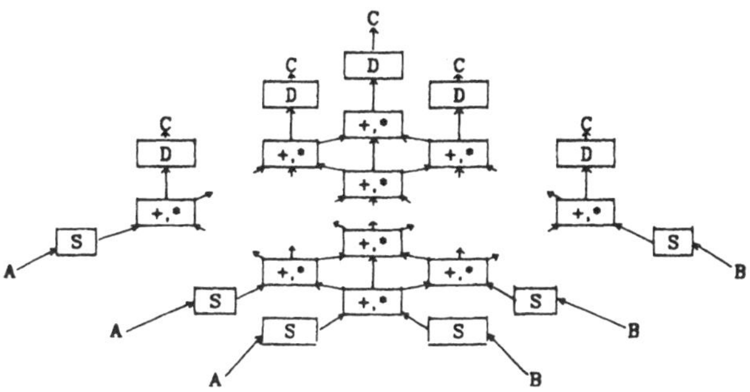

Fig. 4. Digit on-line systolic array.

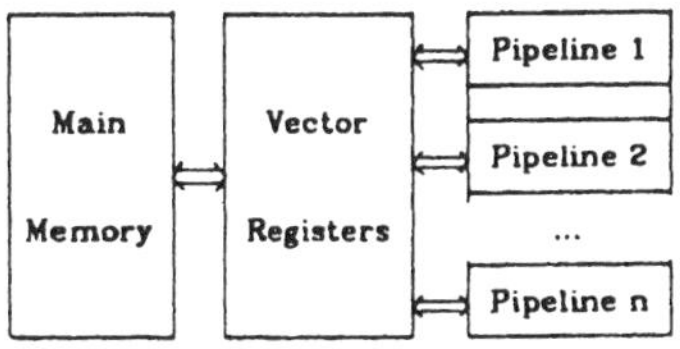

Fig. 5. Conventional pipeline computer.

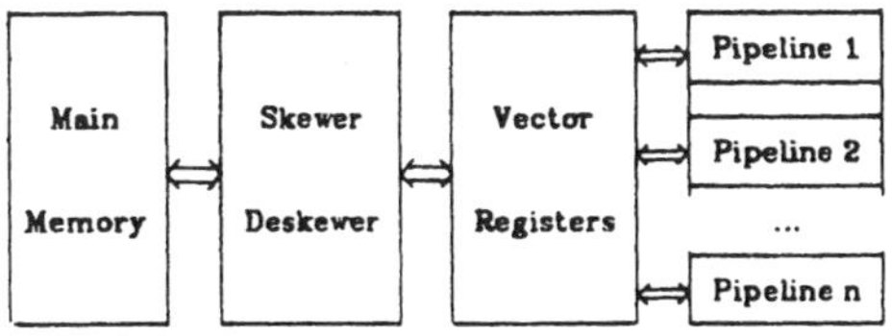

Fig. 6. Digit on-line pipeline computer.

Now consider Fig. 5 which illustrates a fragment of a vector computer. Fig. 6 illustrates a fully digit on-line vector computer created by substituting the pipelines and vector registers of the vector computer with their digit on-line counterparts. In contrast to the digit on-line systolic array, the segments of each pipeline do not function in a uniprocessor mode, but instead the residuals are passed from segment to segment. The skewing/deskewing components augmenting the input-output paths are needed to maintain input-output compatibility with the conventional main memory. While the digit on-line vector computer shares the same structure as its nondigit on-line counterpart, it too avoids, because of the digit on-line property of its components, several implementation problems.

The operation of either pipeline involves moving data down the segments of the pipelines. When designing a pipeline for a given arithmetic operation, the designer can elect to use either a small number of relatively complex segments or a larger number of less complex segments. However, in the case of a conventional pipeline, a large number of segments implies a large fill/flush time. Likewise, a complex set of segments implies a low throughput. Hence, no matter what the designer does, the pipeline is suboptimal for either short or long vectors. Usually the choice has been to make the pipeline suboptimal for short vectors. The problem of dealing effectively with algorithms which use short vectors is left to the user.

However, in the case of a fully digit on-line pipeline, it is possible to have both a large number of segments (which affords a high throughput) and low fill/flush times as described in the next section. Hence, the digit on-line vector computer can have high throughput while also being suitable for algorithms which use short vectors.

One obvious concern with the digit on-line pipeline approach is that of pin-out. As shown in Fig. 1, several full precision values must be passed from segment to segment. If one segment is implemented on a chip, pin-out may be the limiting factor. If, however, the chip is partitioned according to data flow instead of by segment delineation, some of the pin-out concerns are alleviated. Most pin-out concerns are avoided altogether with the uniprocessor approach, since the residual is reused internally in each chip and only the digits of the operands and the result must enter and leave the chip. The next section investigates other advantages and disadvantages of these two alternative digit on-line architectures.

IV. Comparison of Two Digit On-Line Networks

Pipelined machines, like the CRAY-1, are very efficient in handling problems where the computations required involve componentwise vector operations, especially when such operations can be chained. Chaining in the CRAY-1 consists of "connecting" pipelines end to end to improve the processing speed of independent operation. However, these machines are usually intended for a market of complex scientific computations, typically characterized by the numerical solution of partial differential equations which, in turn, call for the solution of various linear and nonlinear recurrences. For obvious reasons, conventional vector pipelining is not directly amenable to the solution of such recurrences. Thus, one must either resort to fast scalar processing, a reformulation of the problem to one which can be solved more efficiently on existing pipeline machines [12], [13] or the design of unconventional pipeline processors which would provide directly for the efficient solution of linear recurrences [5], [14], [15].

Digit on-line networks are one such unconventional processor design. Two alternative configurations will be compared. One is based upon the use of multiple, digit on-line uniprocessors (Fig. 4) connected as a linear systolic array. The other configuration is based upon the use of digit on-line pipelines (Fig. 6). Using digit on-line pipelines, a technique similar to chaining can be used to increase processing speed. With this approach, the digit on-line pipelines are strung side to side with a *chaining of digits* between pipes, creating a pipeline network. Each of these two approaches has its advantages and limitations; however, in either case, improvements in processing speed can be dramatic, especially when solving recurrences.

Consider a simple first-order linear recurrence, characterized by two streams of input vectors $\vec{a}$ and $\vec{b}$ and one output vector stream $\vec{x}$ with a recycling of the result of the previous computation:

$$x^i = \begin{cases} b^i & i = 1 \\ a^i x^{i-1} + b^i & i = 2, 3, \cdots n. \end{cases}$$

For simplicity it is assumed that $p = m_c + m_m = 4$ (i.e., operands have four digits of precision). Two digit on-line uniprocessors as shown in Fig. 7, each with a unit processing time of t_d, can be connected in a digit on-line linear array to solve this problem. In this array, output digits from one unit stream into the other unit. Because of the inherent on-line delay of the add and multiply algorithms, additional buffering must be inserted into the array as shown. Using this two unit approach, all of the digits of x^n have been generated after $(np + nk + 1)t_d$ unit delays.

A digit on-line pipeline network constructed of two digit on-line pipelines for computing the simple linear recurrence is shown in Fig. 8. Each digit on-line pipeline consists of $(s = m_c + m_m + k = 5)$ segments with a fixed segment processing time of t_d. In the steady state, the pipeline is computing up to $m_c + m_m$ different results, the $(k + 1)$st stage producing the first digit of the ith result and the last stage producing the last digit of the $(i - s + k)$th result.

Nine consecutive time steps are shown in Fig. 8, with the digits of the result x^{i-1} being recycled back into the network immediately after generation. For example, during time step t_6, the output of the fifth segment of the add pipe, digit x_4^1, is immediately fed back as input to the fourth segment of the multiply pipe along with a_4^2. This information is used to compute the intermediate results y_3^2, which in turn is fed into the add pipe along with b_3^2 at time step t_7. This overlapping of computation of the leading digits of y^i and the trailing digits of x^{i-1} gives the desired speedup for recursive type computations. Using the pipeline network approach, all the digits of x^n have been generated after $(4n + p - 2)t_d$ time delays.

Comparing these two digit on-line configurations gives insight into their relative advantages and disadvantages. The issue of pin-out has been previously discussed and would probably lead the designer to choose the linear network. The processing speed would, on the other hand, indicate that the pipeline network is the faster of the two for $n > 1$ and $p > 3$. Of course, the linear network requires only two digit on-line uniprocessors, while the pipeline network takes ten digit on-line segments. However, the linear network requires additional buffering of operand digits which is variable and dependent on the on-line delay of the pipelines, while the pipeline network does not require buffering.

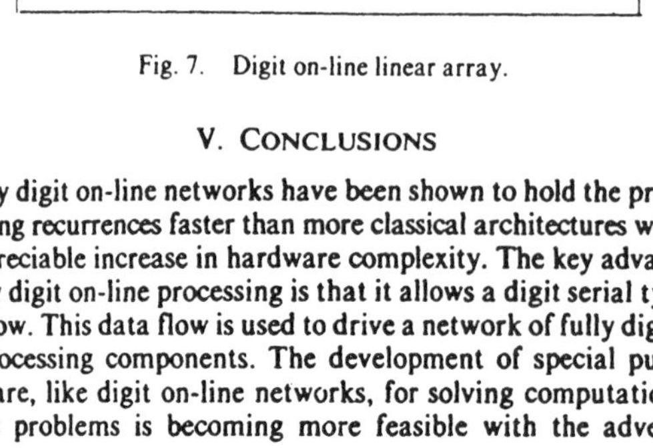

Fig. 7. Digit on-line linear array.

V. Conclusions

Fully digit on-line networks have been shown to hold the promise of solving recurrences faster than more classical architectures without an appreciable increase in hardware complexity. The key advantage to fully digit on-line processing is that it allows a digit serial type of data flow. This data flow is used to drive a network of fully digit on-line processing components. The development of special purpose hardware, like digit on-line networks, for solving computationally intense problems is becoming more feasible with the advent of VLSI.

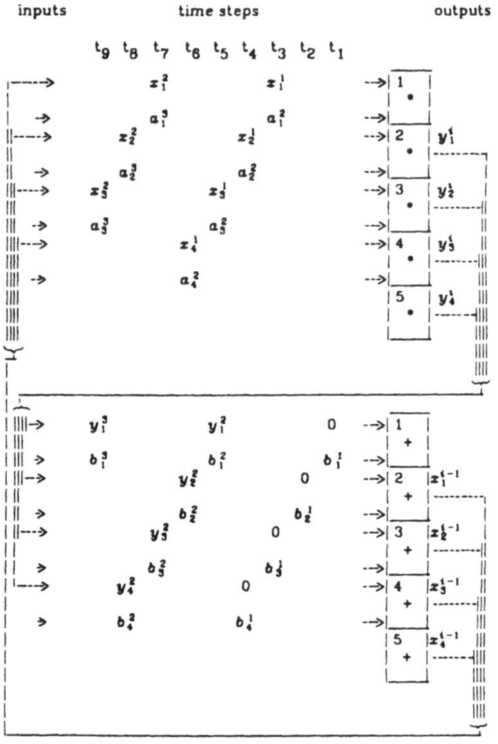

Fig. 8. A digit on-line pipeline network.

References

[1] K. S. Trivedi and M. D. Ercegovac, "On-line algorithms for division and multiplication," *IEEE Trans. Comput.*, vol. C-26, no. 7, pp. 681-687, July 1977.

[2] M. J. Irwin, "A pipelined processing unit for on-line division," in *Proc. 5th Symp. Comput. Arch.*, Palo Alto, CA, Apr. 1978.

[3] R. M. Owens and M. J. Irwin, "On-line algorithms for the design of pipeline architectures," in *Proc. 6th Symp. Comput. Arch.*, Philadelphia, PA, Apr. 1979, pp. 12-19.

[4] M. D. Ercegovac and A. L. Grnarov, "On the performance of on-line arithmetic," in *Proc. 1980 Conf. Parallel Processing*, Aug. 1980, pp. 55-62.

[5] M. J. Irwin and D. Heller, "On-line pipeline systems for recursive numeric computation," *Proc. 7th Symp. Comput. Arch.*, pp. 292-299, LaBaule, France, May 1980, pp. 292-299.

[6] R. M. Owens, "Compound algorithms for digit on-line arithmetic," *Proc. 5th Symp. Comput. Arith.*, Ann Arbor, MI, May 1981, pp. 64-71.

[7] ——, "Error analysis of unnormalized arithmetic," Rep. CS-81-16, Dep. Comput. Sci., Penn State Univ., Univ. Park, Aug. 1981.

[8] V. G. Oklobdzija and M. D. Ercegovac, "An on-line square root algorithm," *IEEE Trans. Comput.*, vol C-31, no. 1, pp. 70-75, Jan. 1982.

[9] B. G. DeLugish, "A class of algorithms for automatic evaluation of certain elementary functions on a binary computer," Ph.D. dissertation, Dep. Comput. Sci., Univ. Illinois, Urbana, Rep. 399, June 1970.

[10] D. E. Atkins, "Introduction to the role of redundancy in computer arithmetic," *Comput.*, vol. 8, no. 6, pp. 74-76, June 1975.

[11] H. T. Kung and C. E. Leiserson, "Systolic arrays for VLSI," in *Introduction to VLSI Systems*, C. A. Mead and L. A. Conway, Eds. Reading, MA: Addison-Wesley, 1980.

[12] D. D. Gajski, "An algorithm for solving linear recurrence systems on parallel and pipelined machines," *IEEE Trans. Comput.*, vol. C-30, pp. 190-206, Mar. 1981.

[13] D. Heller, "A survey of parallal algorithms in numerical linear algebra," *SIAM Rev.*, vol. 20, pp. 740-777, Oct. 1978.

[14] S. Y. Kung and D. V. B. Rao, "Highly parallel architectures for solving linear equations," in *Proc. IEEE Conf. Acoust., Speech, Signal Processing*, 1981, pp. 39-42.

[15] D. Gannon and L. Snyder, "Linear recurrence systems for VLSI: The configurable, highly parallel approach," in *Proc. 1981 Conf. Parallel Processing*, Aug. 1981, pp. 259-260.

On-Line Scheme for Computing Rotation Factors*

MILOŠ D. ERCEGOVAC AND TOMAS LANG

UCLA Computer Science Department, University of California, Los Angeles, California 90024

Received August 19, 1987

An integrated radix-2 on-line algorithm for computing rotation factors for matrix transformations is presented. The inputs are in sign-and-magnitude, floating-point representation and the outputs can be used in on-line signed-digit or in parallel form. The exponents are computed using conventional arithmetic while the significands are processed using on-line algorithms. The conventional result is obtained by using an on-the-fly conversion scheme. The rotation factors are computed in $10 + n$ clock cycles for n-bit significands. The clock period is kept small by the use of redundant adder schemes and low-precision estimates. The implementation and performance of the algorithm are discussed and compared with other approaches.

1. INTRODUCTION

The rotation factors that we propose to compute are an essential part of some matrix transformations such as matrix triangularization methods [7] and the *QR* decomposition [9]. In particular, in systolic arrays in which the rotations are done in parallel, the time of computation of the rotation factors is critical since it determines the step time of the array. We present an implementation using on-line arithmetic [4, 8, 11, 12], since this approach is potentially advantageous in evaluating arithmetic expressions consisting of sequentially dependent operations. The on-line scheme for the computation of rotation factors was proposed previously in [2], where it is shown that it can produce a significant speed-up with respect to the use of conventional algorithms. However, the description is given at a high level and using previously developed algorithms for the operations of multiplication, division, and square root [3, 10, 12]. In this paper we develop an *integrated* algorithm for the rotation factors, which consists of modifications of the algorithms of

* This research has been supported in part by the ONR under Contract N00014-85-K-0159, "On-Line Arithmetic Algorithms and Structures for VLSI."

the primitive operations to provide for suitable interfaces with their predecessors and successors in the computation. In this way, both the on-line delay and the complexity of the implementation are reduced. Moreover, the scheme is for floating-point representations.

The clock period is kept small by the use of redundant techniques, such as redundant adders (i.e., carry-save or signed-digit) and selection functions that depend on low-precision estimates. The conversion of the result from redundant to conventional representation is done using an on-the-fly scheme [5]. In principle, it is possible to select for each component operation the most suitable redundant representation. However, preliminary analysis has indicated that there are no clear differences between the use of carry-save and signed-digit approaches. Therefore, we illustrate the scheme by an implementation with carry-save adders throughout. This has the advantage of making the bit slices of all component operations the same.

We propose to compute the rotation factors of the Givens matrix transformation [7], defined as

$$C = \frac{G}{(G^2 + H^2)^{1/2}}, \qquad S = \frac{H}{(G^2 + H^2)^{1/2}}. \tag{1.1}$$

We assume that $G = g \cdot 2^{e_G}$, $H = h \cdot 2^{e_H}$, $C = c \cdot 2^{e_C}$, and $S = s \cdot 2^{e_S}$ are normalized floating-point numbers with n-bit fractions in sign-and-magnitude. In order to reduce the on-line delay, the algorithm for division uses the dividend in bit-parallel form. Since division begins after several clock periods, it should be possible to load operands byte-serially. If this is not acceptable, the division algorithm can be easily modified to accept the dividend in on-line form. The results are produced in both the bit-parallel and the on-line forms. The exponents are represented and processed in a conventional, bit-parallel manner. The special cases $G = 0$ (producing $C = 0$ and $S = 1$) and $H = 0$ (producing $C = 1$ and $S = 0$) are omitted from further discussion.

The scheme, shown in Fig. 1a, performs the following functions:

1. Computation of exponents
2. Computation of aligned fractions $x = align(g)$ and $y = align(h)$
3. Computation of $z = x^2 + y^2$ $(0.25 \leqslant z < 2)$
4. Computation of $d = z^{1/2}$ $(0.5 \leqslant d < 2^{1/2})$
5. Computation of $c = g/d$ and $s = h/d$ $(0.5 \leqslant c, s < 1)$.

Figure 1b shows the timing diagram indicating the on-line delays to be determined later. We now describe each component separately and then comment on the whole system.

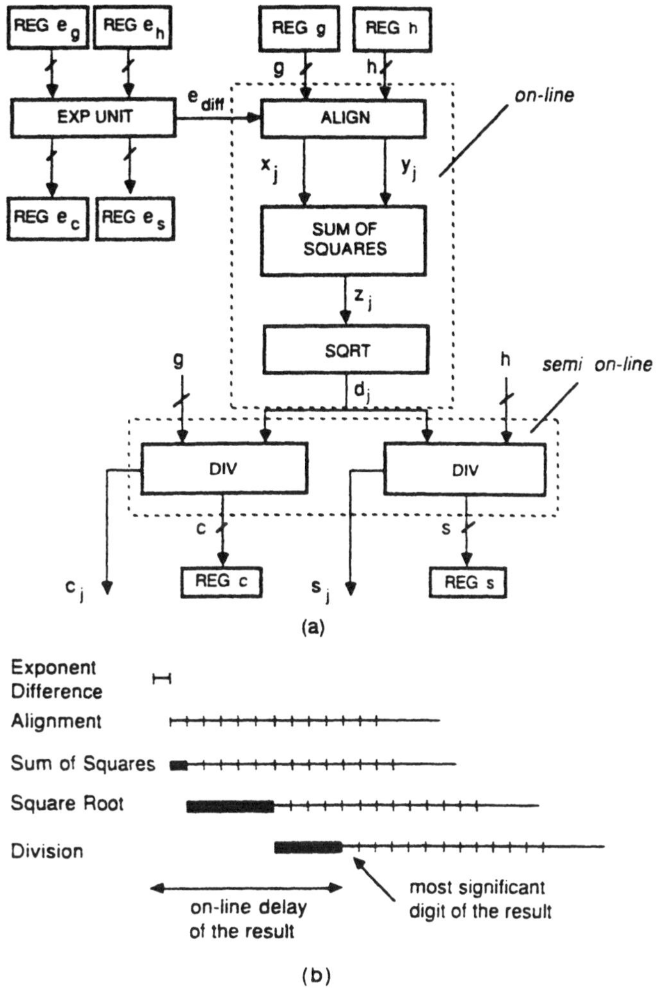

FIG. 1. (a) General scheme. (b) Overall timing.

2. PROCESSING OF EXPONENTS

The exponents of the result can be obtained directly, instead of going through the intermediate steps. Since the sequence of operations is sum of squares, square root, and division, we get *exponent of sum of squares* = $2e$ (where $e = \max(e_G, e_H)$) and *exponent of square root* = e. Therefore, $e_C = e_G - e$ and $e_S = e_H - e$. By defining $e_{diff} = e_G - e_H$ we finally get

$$e_C = \begin{cases} 0 & \textbf{if} \quad e_{diff} \geqslant 0 \\ e_{diff} & \textbf{otherwise} \end{cases} \qquad e_s = \begin{cases} -e_{diff} & \textbf{if} \quad e_{diff} \geqslant 0 \\ 0 & \textbf{otherwise.} \end{cases}$$

3. ALIGNMENT

The alignment of the input fractions is performed in an on-line manner during the parallel-to-serial conversion according to following algorithm:

Algorithm *Align*

/* Delay operand with smaller exponent */
for $i = 1,2,\ldots,n$
    if $e_{diff} \geqslant 0$ **then**
        $\{x_i = g_i;$
        if $i \leqslant e_{diff}$ **then** $y_i = 0;$
        else $y_i = h_{(i-|e_{diff}|)}\}$
    else
        $\{y_i = h_i;$
        if $i \leqslant |e_{diff}|$ **then** $x_i = 0;$
        else $x_i = g_{(i-|e_{diff}|)}\}$
end *Align*

The alignment scheme is shown in Fig. 2. The exponent difference is obtained in one clock cycle. Thereafter, the bits of the aligned operands are obtained one per cycle. The on-line delay $p_{align} = 1$ due to the calculation of exponent difference.

4. On-Line Sum of Squares Algorithm

In this section we present an algorithm and implementation for on-line sum of squares that integrates well in the computation of the rotation factors.

To incorporate a possible on-line delay p we compute

$$z^* = 2^{-p}z = 2^{-p}(x^2 + y^2). \tag{4.1}$$

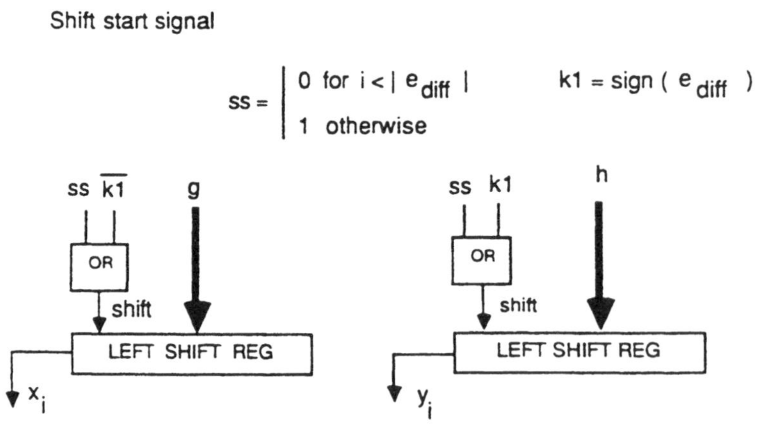

Fig. 2. Alignment scheme.

For the on-line forms of x, y, and z^*,

$$X[j] = X[j-1] + x_j 2^{-j}, \qquad X[0] = 0$$

$$Y[j] = Y[j-1] + y_j 2^{-j}, \qquad Y[0] = 0$$

$$Z[j] = Z[j-1] + z_j^* 2^{-j}, \qquad Z[0] = 0, \tag{4.2}$$

we define the residual function

$$W[j] = 2^j(2^{-p}X[j]^2 + 2^{-p}Y[j]^2 - Z[j]). \tag{4.3}$$

To obtain the correct result, the residual is bounded so that $W[j] < 1$ which results in

$$(2^{-p}X[n]^2 + 2^{-p}Y[n]^2 - Z[n]) < 2^{-n}. \tag{4.4}$$

From the residual expression we get the recurrence

$$W[j] = 2W[j-1] + (2X[j-1]x_j + x_j^2 2^{-j})2^{-p} + (2Y[j-1]y_j + y_j^2 2^{-j})2^{-p} - z_j^* \tag{4.5}$$

with the initial condition

$$W[0] = (X[0]^2 + Y[0]^2)2^{-p} - Z[0] = 0. \tag{4.6}$$

To keep $W[j]$ bounded we make the selection

$$z_j^* = integer\{2W[j-1] + (2X[j-1]x_j + x_j^2 2^{-j})2^{-p} + (2Y[j-1]y_j + y_j^2 2^{-j})2^{-p}\} \tag{4.7}$$

which transforms the recurrence into

$$W[j] = fraction\{2W[j-1] + (2X[j-1]x_j + x_j^2 2^{-j})2^{-p} + (2Y[j-1]y_j + y_j^2 2^{-j})2^{-p}\}. \tag{4.8}$$

To have a short step time, this recurrence is implemented using a redundant addition (e.g., carry-save, signed-digit). As mentioned in the Introduction, here we use the carry-save adder variant. In this case, the exact integer and fraction parts cannot be determined (without propagating carries). Consequently, the previous expressions are transformed into

$$z_j^* = csint\{2W[j-1] + (2X[j-1]x_j + x_j^2 2^{-j})2^{-p} + (2Y[j-1]y_j + y_j^2 2^{-j})2^{-p}\} \quad (4.9)$$

$$W[j] = csfrac\{2W[j-1] + (2X[j-1]x_j + x_j^2 2^{-j})2^{-p} + (2Y[j-1]y_j + y_j^2 2^{-j})2^{-p}\}, \quad (4.10)$$

where *csfrac* produces a value in the range [0, 2).

The corresponding carry-save operation is shown in Fig. 3a. From it we see that the bounds on z_j^* are (for $x, y < 1$)

$$0 \leqslant z_j^* \leqslant 6 \quad \textbf{if} \quad p = 0, \qquad 0 \leqslant z_j^* \leqslant 4 \quad \textbf{if} \quad p = 1. \quad (4.11)$$

These values of z_j^* are over-redundant (that is, they are larger than 1 for a radix-2 representation). We choose this alternative because it simplifies the selection, reduces the on-line delay, and is suitable for the interface with the square root. We choose to implement the case with $p = 0$ to minimize the overall on-line delay. The result is in the range $0.25 \leqslant z < 2$. The algorithm is summarized next.

Algorithm *Sum of Squares*

/* Initialization */
$W[0] \leftarrow 0$; $z_0 \leftarrow 0$;
$X[0] \leftarrow 0$; $Y[0] \leftarrow 0$;

/* Recurrence */
for $j = 1,2,...,n+1$
$\{W[j] \leftarrow csfract\{2W[j-1] + (2X[j-1]+x_j 2^{-j})x_j + (2Y[j-1]+y_j 2^{-j})y_j\}$;
$z_j \leftarrow \mathrm{csint}\{2W[j-1] + (2X[j-1]+x_j 2^{-j})x_j + (2Y[j-1]+y_j 2^{-j})y_j\}$;
$X[j] \leftarrow append(X[j-1], x_j)$;
$Y[j] \leftarrow append(Y[j-1], y_j)\}$

end *Sum of Squares*

Implementation

The scheme is shown in Fig. 3b. Note that its implementation complexity is similar to that of an on-line multiplier. The on-line delay is $p_{ss} = 0$. The critical path consists of two multiplexers, one 4-to-2 carry-save adder, and a 3-bit carry-propagate adder.

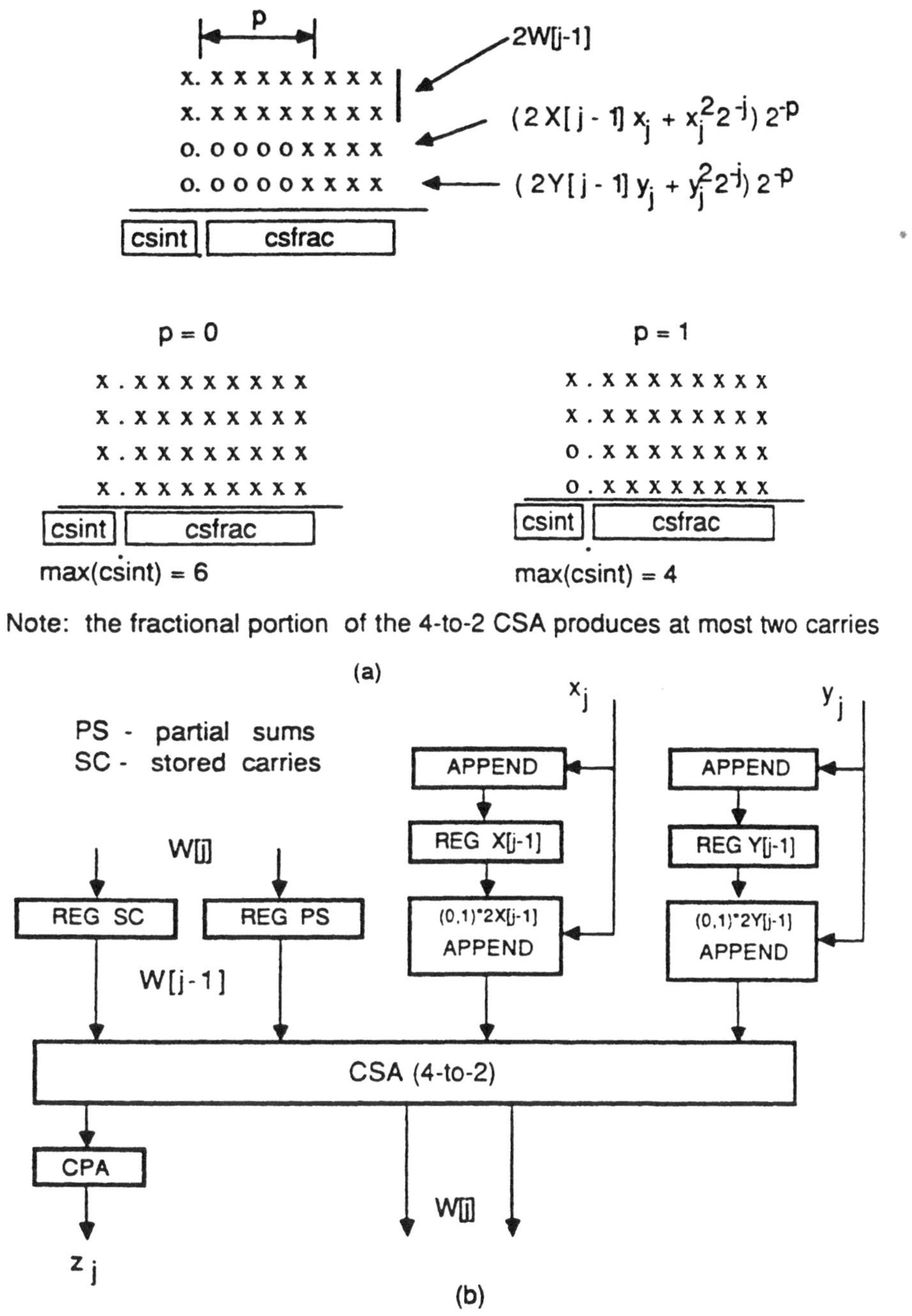

FIG. 3. (a) Carry-save operation in sum of squares. (b) Scheme.

5. ON-LINE SQUARE ROOT

We now describe an on-line square root algorithm to interface with the sum of squares and the division, for the computation of rotation factors. This algorithm allows an over-redundant input digit set, differing in this respect from previous on-line square root algorithms [3, 10].

Let the on-line forms of the argument z and of the square root d be

$$Z[j] = Z[j-1] + z_j 2^{-j}, \qquad -b \leqslant z_j \leqslant a$$
$$D[j] = D[j-1] + d_j 2^{-j}, \qquad d_j \in \{-1, 0, 1\}. \tag{5.1}$$

To bound the error of the computation we make

$$(D[j] - 2^{-j})^2 + b2^{-p-j} \leqslant Z[j+p] \leqslant (D[j] + 2^{-j})^2 - a2^{-p-j}, \tag{5.2}$$

where p is the on-line delay determined in Appendix A and, as indicated, the argument digits satisfy $z_j \in \{-b, \ldots, 0, \ldots, a\}$.

We define a residual $R[j]$ such that

$$R[j] = 2^j(Z[j+p] - D[j]^2), \qquad R[0] = Z[p]. \tag{5.3}$$

From (4.2) the residual should satisfy the bounds $C_{-1}[j] \leqslant R[j] \leqslant C_1[j]$ where

$$C_{-1}[j] = (-2D[j] + 2^{-j} + b2^{-p}), \qquad C_1[j] = (2D[j] + 2^{-j} - a2^{-p}). \tag{5.4}$$

The resulting recurrence is

$$R[j] = 2R[j-1] + z_{j+p}2^{-p} - 2d_jD[j-1] - d_j^2 2^{-j}. \tag{5.5}$$

As stated before, to have a short step time, we use a redundant addition and a selection function that depends on a low-precision estimate. The corresponding selection function *dsel* is derived in Appendix A. Using a carry-save adder, $t = 3$ fractional bits in the estimate of the remainder, and the on-line delay $p_{sqr} = 4$, we obtain the selection function

$$d_j = dsel(\hat{R}^*[j-1]) = \begin{cases} 1 & \text{if} \quad \hat{R}^*[j-1] \geqslant 0 \\ 0 & \text{if} \quad \hat{R}^*[j-1] = -\frac{1}{4} \\ -1 & \text{if} \quad \hat{R}^*[j-1] \leqslant -\frac{1}{2}, \end{cases}$$

where $\hat{R}^*[j-1]$ is an estimate with three fractional bits of $R^*[j-1] = R[j-1] + z_{j+p}2^{-p-1}$.

The corresponding algorithm for on-line square root is summarized next.

Algorithm *Sqrt*

/* Initialization */

$R[-5] \leftarrow 0$; $D[-1] \leftarrow 0$;

/* Accumulate 4 input digits */

for $j = -4,-3,-2,-1$

$\{R[j] \leftarrow 2R[j-1] + z_{4+j}2^{-4}\}$

/*Recurrence*/

for $j = 0,1,2,\ldots,n$

$\{\hat{R}^*[j-1] = \hat{R}[j-1] + z_{4+j}2^{-5}$;
$d_j = dsel(\hat{R}^*[j-1])$;
$R[j] \leftarrow 2R[j-1] + z_{4+j}2^{-4} - 2d_jD[j-1] - d_j^2 2^{-j}$;
$D[j] \leftarrow convert(D[j-1], d_j)\}$

end *Sqrt*

Note that $z_k = 0$ for $k > n$. The result is in the range $[\frac{1}{2}, 2^{1/2})$.

Implementation

The square root scheme is shown in Fig. 4a. The internal arithmetic is performed in 2's complement system. The signed-digit operand is converted to 2's complement using the on-the-fly conversion method [5]. The operation performed by APPEND* module is discussed in Appendix C. The critical path (Fig. 4b), consists of a 5-bit CPA, one multiplexer, and a 3-to-2 carry-save adder.

6. On-Line Division Algorithm and Implementation

We now present the algorithm for the division operation. We consider the computation of s (that of c is similar). To incorporate the necessary on-line delay p we redefine the division operation as

$$s = 2^{-p}\frac{h}{d}, \qquad h \in \left[\frac{1}{2}, 1\right), \qquad d \in \left[\frac{1}{2}, 2^{1/2}\right), \tag{6.1}$$

where s is obtained without delay. The real quotient is obtained by shifting left s by p positions. Consequently, the real quotient is obtained with an on-line delay of p.

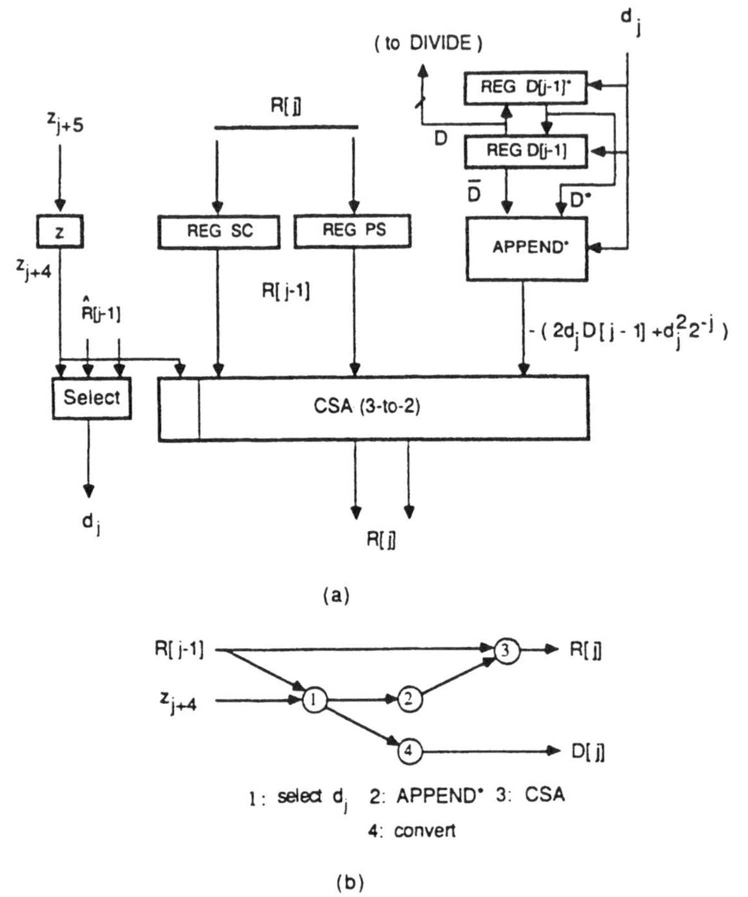

FIG. 4. (a) Square root scheme. (b) Critical paths.

Since we are developing an algorithm that is specifically suited to the computation of the rotation factors, we assume that h (the dividend) is known in parallel form (not on-line), while d is on-line and s is also produced on-line (but converted on-the-fly to conventional representation). However, this approach requires that the dividend be communicated to the division unit in parallel. Because of this, it might be more convenient for a particular implementation to use an algorithm which also uses the dividend in an online fashion. This would require minor modifications to the algorithm. The formulation using the dividend off-line leads to the following recurrence. Let

$$P[j] = 2^j(2^{-p}h - S[j] \times D[j]) \tag{6.2}$$

be a partial remainder, satisfying the bound $|P[j]| < |D[j]|$.

The corresponding recurrence is

$$P[j] = 2P[j-1] - S[j-1]d_j - D[j]s_j, \qquad P[0] = 2^{-p}h, \qquad S[0] = 0. \tag{6.3}$$

The recurrence can be decomposed into the two steps

$$W[j] = 2P[j-1] - S[j-1]d_j \tag{6.4}$$

$$P[j] = W[j] - D[j]s_j \tag{6.5}$$

so that the selection function can depend on W.

As for the other algorithms, to have a short clock period, a redundant adder is used, and the selection is done on a low-precision estimate of W. The selection function *qsel* is determined in Appendix B. For an estimate of $t = 3$ fractional bits and an on-line delay of $p_{div} = 3$, and using carry-save adder, the selection function is

$$s_j = qsel(\hat{W}[j]) = \begin{cases} 1 & \textbf{if} \quad \hat{W}[j] \geq \frac{2}{8} \\ 0 & \textbf{if} \quad -\frac{2}{8} \leq \hat{W}[j] \leq \frac{1}{8} \\ -1 & \textbf{if} \quad \hat{W}[j] \leq -\frac{3}{8}. \end{cases}$$

The corresponding algorithm is given next.

Algorithm *Division*

/* Initialization */
$P[0] \leftarrow h\times 2^{-3}$; $D[0] \leftarrow d_0$; $S[0] \leftarrow 0$; $s_0 \leftarrow 0$;

/* Recurrence */

for $j = 1,2,\ldots,n+3$:

{ $W[j] = 2P[j-1] - S[j-1]d_j$;

/* Note that $S[j-1] < 2^{-p} = 2^{-3}$ */

$\hat{W}[j] = (2P[j-1])_4 - (1-sign(d_j))*2^{-3}$;

$s_j = qsel(\hat{W}[j])$;
$P[j] \leftarrow W[j] - D[j]s_j$;
$S[j] \leftarrow convert\,(S[j-1], s_j)$ }

end *Division*

Note that $(A)_4$ denotes the four most significant bits of A and $d_k = 0$ for $k > n$.

Implementation

The implementation is shown in Fig. 5a. Note that the additions/subtractions are done using carry-save adders in 2's complement system; for this

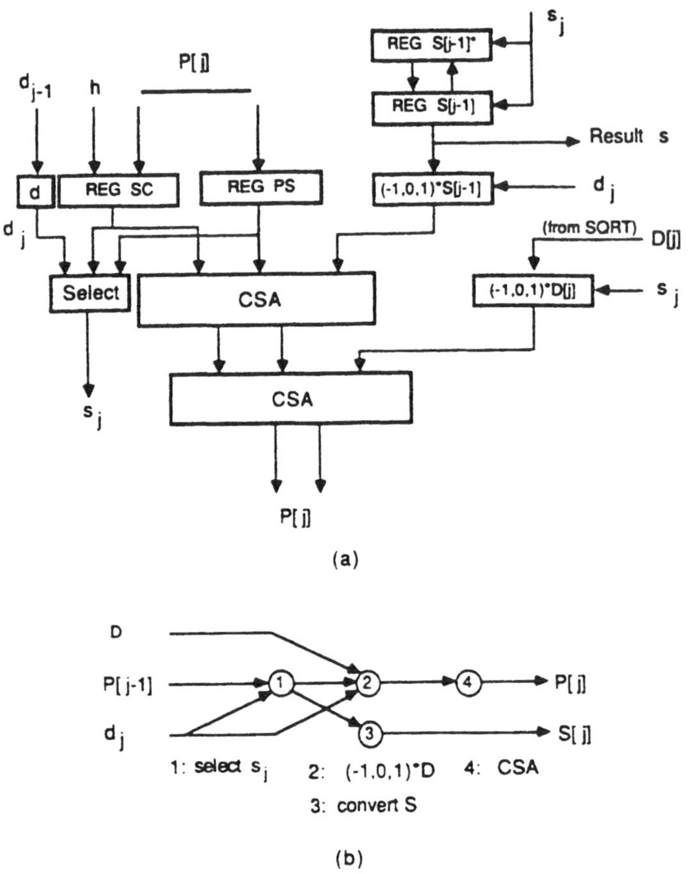

FIG. 5. (a) Division scheme. (b) Critical paths.

reason, the quotient is converted on-the-fly [5] into conventional form to be used as operands for the additions. The critical path is shown in Fig. 5b. We estimate that its delay corresponds to two multiplexers, one 4-bit CPA and one CSA.

7. Summary of the Proposed Scheme and Comparisons

An integrated scheme for computing rotation factors using on-line arithmetic algorithms has been presented. The scheme has the following characteristics:

–The inputs and outputs are in sign-and-magnitude floating-point format; the inputs may be loaded byte-serially if appropriate. The outputs can also be used in on-line signed-digit form.

–The on-line delay of the scheme, defined as the number of clock cycles between loading of the operands and the appearance of the most significant digit of the result, is

$$P_{rot} = (p_{align} + 1) + (p_{ss} + 1) + (p_{sqr} + 1) + (p_{div} + 1)$$
$$= 1 + 1 + 5 + 4 = 11.$$

-The latency, defined as the total time to generate the result, is $T = p_{rot} + n - 1 = 10 + n$ for n-bit significands (Fig. 1b).

-The cycle time is determined roughly by a multiplexer, a 4-to-2 carry-save adder, and a 5-bit carry-propagate adder.

-Since it takes roughly k cycles to propagate the kth bit of an on-line unit to influence the selection function, the number of bit slices required to implement the carry-save adders is approximately $n/2$.

Comparison with Alternative Schemes

We now compare the execution time of the presented on-line implementation with that of other schemes proposed for the computation of the rotation factors. We just do a rough comparison since a more accurate one would require many assumptions on the implementation of these other schemes as well as on the technology. To remain independent of the particular technology, we make the comparison in terms of number of clock cycles, instead of using actual time. Even so, we must make some assumptions, since not all clock cycles are the same. Therefore, we define a *basic* clock cycle and relate all others to this basic one. The basic clock cycle has a delay of roughly a multiplexer, a 4-to-2 (carry-save) adder, a selection function composed of a short carry-propagate adder and some logic, and the loading of a register. This basic clock cycle is suitable for the on-line implementation presented in this paper.

As shown in Table I, we compare with the following schemes, assuming 32-bit significands and single-chip implementations:

(a) Using only one floating-point multiplier (array multiplier) and one floating-point adder. For this scheme, the divisions and the square root use an iterative multiplicative approach. We estimate the number of floating-point operations to be about 20 (1 addition + 2 multiplications + 1 square

TABLE I
COMPARISON WITH OTHER SCHEMES

Scheme	Basic cycles	Speed-up
Multiplier + adder	120	1
MULT + ADD + SQR + DIV	$5n$ = 160	0.75
2MULT + ADD + SQR + 2DIV	$3n$ = 96	1.25
CORDIC	$7.5n$ = 226	0.55
Redundant/on-line CORDIC	$2n$ = 64	1.9
On-line (this paper)	$10+n$ = 42	3

root + 1 reciprocal + 2 multiplications). We estimate that the time for each of these floating-point operations is of 6 basic clock cycles. This results in a total of 120 basic clock cycles.

(b) Using one adder, one multiplier, one divider, and one square root unit. In this case we assume that the multiplier is sequential (to fit all units in one chip). Each unit takes n cycles (except addition, which is neglected) for a total of $5n$ clock cycles. Fast algorithms (with carry-save addition) can be used so that the clock cycle corresponds roughly to the basic cycle.

(c) Using one adder, two multipliers, two dividers, and one square root unit. In this case the parallelism in the computation can be exploited to achieve a time of $3n$ clock cycles.

(d) It is possible to use a CORDIC approach to calculate the rotation angle and also to perform the actual rotation [1]. The operation consists in performing roughly $1.25n$ iterations of the CORDIC recurrence. The clock cycle for this recurrence is larger than the basic clock cycle because it involves a variable shift and a carry-propagate addition. We estimate that the CORDIC cycle is about six times the basic cycle, resulting in a time of $7.5n$ basic cycles.

(e) The speed of the CORDIC recurrence can be increased by using a redundant adder and performing the operation on-line to eliminate the shifter [6]. This results in an execution time of about $2n$ cycles.

The speed-up with respect to case (a) is shown in the table. As can be seen, the scheme presented here is the fastest and produces a speed-up of about 3. The next fastest is the redundant and on-line CORDIC, which might have the advantage of smaller area.

It is difficult to compare the schemes with respect to implementation complexity without doing detailed designs. However, we can make the following general statements:

(i) The general-purpose approach of using one multiplier and one adder is probably inefficient in area since it requires fast modules to be competitive in speed.

(ii) The on-line approach is comparable in complexity with respect to the specialized non-on-line case (c) and provides better performance. Moreover, the on-line approach has the advantage of reduced interconnection among modules. It also uses only about $n/2$ bit slices per operation.

(iii) The CORDIC approach is promising because of the complex operation performed by the single CORDIC module. However, to compete in speed it is necessary to use the redundant on-line version, which complicates somewhat the implementation. Details of this approach are given in [6].

APPENDIX A: SELECTION FUNCTION FOR SQUARE ROOT

We determine here a selection function *dsel* for d_j so that the bounds on the residual $R[j]$ are satisfied. For this, we compute the intervals $[L_k, U_k]$ of $R[j-1]$ so that the value $d_j = k$ ($k = -1, 0, 1$) can be selected and $R[j]$ is inside the allowed interval of bounds defined by (5.4). The expression for $R[j-1]$ is $R[j-1] = 2^{-1}R[j] - z_{j+p}2^{-p-1} + d_jD[j-1] + d_j^2 2^{-j-1}$. The resulting intervals are the following:

for $d_j = 1$,

$$L_1 = -D[j] + 2^{-j-1} + b2^{-p-1} - z_{j+p}2^{-p-1} + D[j-1] + 2^{-j-1}$$
$$= b2^{-p-1} - z_{j+p}2^{-p-1}$$
$$U_1 = D[j] + 2^{-j-1} - a2^{-p-1} - z_{j+p}2^{-p-1} + D[j-1] + 2^{-j-1}$$
$$= 2D[j-1] + 2^{-j+1} - a2^{-p-1} - z_{j+p}2^{-p-1},$$

for $d_j = 0$,

$$L_0 = -D[j] + 2^{-j-1} + b2^{-p-1} - z_{j+p}2^{-p-1}$$
$$= -D[j-1] + 2^{-j-1} + b2^{-p-1} - z_{j+p}2^{-p-1}$$
$$U_0 = D[j-1] + 2^{-j-1} - a2^{-p-1} - z_{j+p}2^{-p-1},$$

for $d_j = -1$,

$$L_{-1} = -D[j] + 2^{-j-1} + b2^{-p-1} - z_{j+p}2^{-p-1} - D[j-1] + 2^{-j-1}$$
$$= -2D[j-1] + 2^{-j+1} + b2^{-p-1} - z_{j+p}2^{-p-1}$$
$$U_{-1} = D[j] + 2^{-j-1} - a2^{-p-1} - z_{j+p}2^{-p-1} - D[j-1] + 2^{-j-1}$$
$$= -a2^{-p-1} - z_{j+p}2^{-p-1}.$$

Since $(-z_{j+p}2^{-p-1})$ appears in all the expressions, we will base the selection on

$$R^*[j-1] = R[j-1] + z_{j+p}2^{-p-1}. \tag{A.1}$$

Consequently,

$$L_1^* = b2^{-p-1}, \qquad U_1^* = 2D[j-1] + 2^{-j+1} - a2^{-p-1}$$
$$L_0^* = -D[j-1] + 2^{-j-1} + b2^{-p-1}, \qquad U_0^* = D[j-1] + 2^{-j-1} - a2^{-p-1}$$
$$L_{-1}^* = -2D[j-1] + 2^{-j+1} + b2^{-p-1}, \qquad U_{-1}^* = -a2^{-p-1}.$$

The containment conditions, $U_1 \leqslant C_1[j-1]$ and $L_{-1} \geqslant C_{-1}[j-1]$, require that $-b \leqslant z_j \leqslant a$ which is satisfied by selecting a and b according to (4.11).

The continuity of the selection intervals and the use of redundant adders require that the interval overlaps satisfy the conditions

$$\Delta_{01} = U_0^* - L_1^* = D[j-1] + 2^{-j-1} - (a+b)2^{-p-1} \geqslant 2^{-t+1}$$

$$\Delta_{\bar{1}0} = U_{-1}^* - L_0^* = D[j-1] - 2^{-j-1} - (a+b)2^{-p-1} \geqslant 2^{-t+1}, \quad \text{(A.2)}$$

where t is the precision of the assimilated remainder $\hat{R}$.

Since $z \geqslant 2^{-2}$ (from the sum of squares), the smallest possible value of $d = 2^{-1}$. Therefore, for $a = 6$ and $b = 0$ (as produced by the sum of squares), we get

$$\Delta_{01}\text{min} = 2^{-1} - 3 \cdot 2^{-p} \geqslant 2^{-t+1} \qquad \Delta_{\bar{1}0}\text{min} = 2^{-1} - 2^{-j-1} - 3 \cdot 2^{-p} \geqslant 2^{-t+1}. \quad \text{(A.3)}$$

Since $d \geqslant \frac{1}{2}$, the first negative digit cannot happen for $j < 3$. Therefore, the positive overlaps imply $p \geqslant 4$ and $t \geqslant 3$.

We now determine the selection function for $p = 4$, $a = 6$, $b = 0$, and $t = 3$. The overlap region for the selection of 0 or 1 is bounded by

$$L_1^*(\text{max}) = b2^{-p-1} = 0$$

$$U_0^*(\text{min}) = D[j-1] + 2^{-j-1} - a2^{-p-1} > 2^{-1} - 6 \cdot 2^{-5} = \tfrac{5}{16}. \quad \text{(A.4)}$$

Similarly, the overlap region for choosing between 0 and -1 is

$$L_0^*(\text{max}) = -D[j-1] + 2^{-j-1} + b2^{-p-1} \leqslant -2^{-1} + 2^{-3-1} + 2^{-4} = -\tfrac{3}{8}$$

$$U_{-1}^*(\text{min}) = -a2^{-p-1} = -\tfrac{3}{16}. \quad \text{(A.5)}$$

Since the error of the estimate is always positive (because of the use of carry-save adders and 2's complement representation) and can be at most $2^{-t+1} = \frac{1}{4}$, we get the selection function

$$d_j = dsel(\hat{R}^*[j-1]) = \begin{cases} 1 & \text{if} \quad \hat{R}^*[j-1] \geqslant 0 \\ 0 & \text{if} \quad \hat{R}^*[j-1] = -\frac{1}{4} \\ -1 & \text{if} \quad \hat{R}^*[j-1] \leqslant -\frac{1}{2}. \end{cases}$$

APPENDIX B: QUOTIENT SELECTION FUNCTION

We determine here the selection function $qsel$ for the quotient digit s_j. This selection function is such that the remainder is inside the required bounds.

First we calculate the actual bounds on $P[j]$ and then the intervals of $W[j]$ for which a selection of $s_j = k$ can be made.

From the recurrence, since $|S[j]| \leqslant 2^{-p}$, we get

$$L_k - 2^{-p} - kD[j] \leqslant P[j] \leqslant U_k + 2^{-p} - kD[j],$$

where $[L_k, U_k]$ is the interval of $2P[j-1]$ for which it is possible to select $s_j = k$.

The containment of the remainder requires that

$$c_{-1} = \frac{L_{-1}}{2} \leqslant P[j] \leqslant \frac{U_1}{2} = c_1 .$$

Consequently,

$$U_1 = 2D[j] - 2^{-p+1}, \qquad L_{-1} = -2D[j] + 2^{-p+1}$$

and

$$c_1 = -c_{-1} = c = D[j] - 2^{-p}. \tag{B.1}$$

The selection is based on $W[j]$ defined by

$$W[j] = 2P[j-1] - S[j-1]d_j = P[j] + D[j]s_j. \tag{B.2}$$

We now determine the selection intervals $[A_k, B_k]$ of $W[j]$ such that $s_j = k$ can be selected. From (B.1) and (B.2) we get

$$A_k = -c + kD_j \leqslant W_j \leqslant c + kD_j = B_k \tag{B.3}$$

and replacing c, $A_k = (k-1)D[j] + 2^{-p}$ and $B_k = (k+1)D[j] - 2^{-p}$.

The value of p is determined to ensure sufficient overlap between the intervals. Since we want to perform the computation of $W[j]$ using carry-save addition, and base the selection on an estimate $\hat{W}$ produced with an assimilation over t fractional bits, the overlap must be at least 2^{-t+1}. Consequently,

$$\Delta(k, k-1) = B_{k-1} - A_k = D[j] - 2^{-p+1} \geqslant 2^{-t+1} \tag{B.4}$$

so that

$$2^{-p} \leqslant \frac{D[j] - 2^{-t+1}}{2} . \tag{B.5}$$

For $D[j] \geqslant \frac{1}{2}$ we get $2^{-p} \leqslant (1 - 2^{-t+2})/4$. A possible solution to this inequality is $p = 3$ and $t = 3$.

Using these values we determine the selection intervals and the selection function:

s_j	$\hat{W}[j]$
-1	$[-2D[j]+2^{-3}, -2^{-3})$
0	$[-D[j]+2^{-3}, D[j]-2^{-3})$
1	$[2^{-3}, 2D[j]-2^{-3})$

To have a selection that is independent of the value of $D[j]$ we look at the values of $D[j]$ that produce the smallest intervals. This occurs in all cases for $D[j] = \frac{1}{2}$. The resulting intervals are

s_j	$\hat{W}[j]$
-1	$[-\frac{7}{8}, -\frac{1}{8})$
0	$[-\frac{3}{8}, +\frac{3}{8})$
1	$[+\frac{1}{8}, +\frac{7}{8})$

Since the use of carry-save addition and 2's complement representation produce $\hat{W} = W - \epsilon$ with $0 \leqslant \epsilon \leqslant 2^{-t+1}$, we get as selection function

$$s_j = qsel(\hat{W}[j]) = \begin{cases} 1 & \text{if} \quad \hat{W}[j] \geqslant \frac{2}{8} \\ 0 & \text{if} \quad -\frac{2}{8} \leqslant \hat{W}[j] \leqslant \frac{1}{8} \\ -1 & \text{if} \quad \hat{W}[j] \leqslant -\frac{3}{8}. \end{cases}$$

APPENDIX C: APPENDING OPERATION FOR SQUARE ROOT ALGORITHM

The operand produced by the selection/appending module APPEND* in the square root scheme is

$$q = -(2d_i D[i-1] + d_i^2 2^{-i}).$$

Consider the three possible values of d_i.

(i) For $d_i = 1$, $q = -(2D[i-1] + 2^{-i})$ which, as a string of bits, is

$$-[D, 0, 1_i, 0, 0, 0 \cdots, 0] \qquad (1_i \text{ to indicate } i\text{th position}).$$

So, the 2's complement is

$$\begin{array}{r} \bar{D}, 1, 0_i, 1, 1, 1, \ldots, 1 \\ 1 \\ \hline \bar{D}, 1, 1_i, 0, 0, 0, \ldots, 0, \end{array}$$

where $\bar{D}$ is the bit complement of D. That is, in this case we concatenate $\bar{D}$ with 11.

(ii) For $d_i = -1$, $q = 2D[i-1] - 2^{-i}$ which as a weighted string can be written as

$$D, 0, -1_i, 0, 0, 0, \ldots, 0 = (D-1), 1, 1_i, 0, 0, 0, \ldots, 0.$$

Since $D^* = D - 1$ because of the conversion algorithm, we obtain

$$D^*, 1, 1_i, 0, 0, 0, \ldots, 0.$$

That is, in this case we concatenate D^* with 11.

(iii) For $d_i = 0$, $q = 0$.

ACKNOWLEDGMENTS

We thank Dr. J. G. Nash of Hughes Research Laboratories, Malibu, for his interest and support, and referees for their comments.

REFERENCES

1. Ahmed, H. M., Delosme, J. M., and Morf, M. Highly concurrent computing structures for matrix arithmetic and signal processing. *IEEE Comput.* **15,** 1 (Jan. 1982), 65–82.
2. Ciminiera, L., Serra, A., and Valenzano, A. Fast and accurate matrix triangularization using an iterative structure. *Proc. 5th IEEE Symposium on Computer Arithmetic,* Ann Arbor, 1981, pp. 215–221.
3. Ercegovac, M. D. An on-line square root algorithm. *Proc. 4th IEEE Symposium on Computer Arithmetic,* Santa Monica, 1978, pp. 183–189.
4. Ercegovac, M. D. On-line arithmetic: An overview. *Proc. SPIE 1984,* Vol. 495, *Real Time Signal Processing VII,* 1984, pp. 86–93.
5. Ercegovac, M. D., and Lang, T. On-the-fly conversion of redundant into conventional representations. *IEEE Trans. Comput.* (July 1987), 895–897.
6. Ercegovac, M. D., and Lang, T. *Redundant and On-Line CORDIC: Application to Matrix Triangularization and SVD.* UCLA Computer Science Department, Tech. Rep. CSD-870046, Sept. 1987.
7. Golub, G. H., and Van Loan, C. F. *Matrix Computations.* The Johns Hopkins Univ. Press, Baltimore, 1983.
8. Irwin, M. J., and Owens, R. M. Digit-pipelined arithmetic as illustrated by the paste-up system: A tutorial. *IEEE Comput.* (Apr. 1987), 61–73.
9. Luk, F. T. Architectures for computing eigenvalues and SVDs. *SPIE Proc. Highly Parallel Signal Processing Architectures,* Vol. 614, 1986, pp. 24–33.
10. Oklobdzija, V. G., and Ercegovac, M. D. An on-line square root algorithm. *IEEE Trans. Comput.* **C-31,** 1 (Jan. 1982), 70–75.
11. Owens, R. M. Compound algorithms for digit online arithmetic. *Proc. 5th Symposium on Computer Arithmetic,* May 1981, pp. 64–71.
12. Trivedi, K. S., and Ercegovac, M. D. On-line algorithms for division and multiplication. *IEEE Trans. Comput.* **C-26,** 7 (July 1977), 667–680.

Reprinted from *IEEE Transactions on Computers*, Volume C-36, Number 7, July 1987, pages 895-897.

On-the-Fly Conversion of Redundant into Conventional Representations

MILOŠ D. ERCEGOVAC AND TOMAS LANG

Abstract—**An algorithm to convert redundant number representations into conventional representations is presented. The algorithm is performed concurrently with the digit-by-digit generation of redundant forms by schemes such as SRT division. It has a step delay roughly equivalent to the delay of a carry-save adder and simple implementation. The conversion scheme is applicable in arithmetic algorithms such as nonrestoring division, square root, and on-line operations in which redundantly represented results are generated in a digit-by-digit manner, from most significant to least significant.**

Index Terms—**Conditional sum addition, conversion, most significant digit first algorithms, on-line algorithms, redundant number representation, square-root, SRT division, 2's complement.**

I. Introduction

We consider an algorithm and implementation to convert a signed integer from a redundant (signed digit) into a conventional range-complement representation. Such a conversion is necessary, for example, in SRT division or square root algorithms that produce this redundant result [1]-[5], or to convert results produced by on-line algorithms into the equivalent conventional forms [6]-[8]. The standard approach for the conversion is to separate the digit vector into two, one formed by the positive digits and the other by the negative ones, and add them in a carry-propagate adder. However, this adds the delay of the adder to the total operation time. To reduce this delay we propose here an algorithm that has the following characteristics:

- it performs the conversion on the fly, as the digits of the result (e.g., of the division or square root) are obtained in a serial fashion from most to least significant,
- it uses two conditional forms of the current result, similar to the conditional sum technique [9], and
- it has a delay which is compatible with one step of a fast division algorithm [4]. This delay is roughly of a carry-save adder.

We begin with a description of the algorithm for a radix-2 representation and then generalize to radix r.

II. Radix-2 Conversion Algorithm and Implementation

We want to convert the redundant digit-vector P (with $p_i \in \{-1, 0, 1\}$) representing the normalized fraction p into the digit-vector Q (with $q_i \in \{0, 1\}$). That is,

$$p = \sum_{i=1}^{m} p_i 2^{-i}, \qquad p_i \in \{-1, 0, 1\}$$

is to be converted into its conventional 2's complement form

$$q = -q_0 + \sum_{i=1}^{m} q_i 2^{-i}, \qquad q_i \in \{0, 1\}.$$

As mentioned before, the transformation is to be performed as the digits of P are produced, from most significant to least significant. A simple algorithm would be to form $q[k]$ such that

$$q[k] = q[k-1] + p_k 2^{-k}$$

which results in $q = q[m]$.

However, this algorithm requires the propagation of a carry when $p_k = -1$. The propagation of this carry is up to the previous bit of Q having value 1. That is, if for example,

$$q_i, q_{i+1}, \cdots, q_{k-2}, q_{k-1} = 1, 0 \cdots 0, 0$$

then after adding -2^{-k} we get

$$q_i, q_{i+1}, \cdots, q_{k-2}, q_{k-1}, q_k = 0, 1, \cdots, 1, 1, 1.$$

The approach we follow is to avoid the propagation of the carry by keeping two conditional forms, one ($A[k-1]$) expecting $p_k = 1$ or $p_k = 0$ and the other ($B[k-1]$) expecting $p_k = -1$, and selecting the correct one when p_k is known.

That is,

$$q[k] = \begin{cases} A[k-1] + 2^{-k} & \text{if } p_k = 1 \\ A[k-1] & \text{if } p_k = 0 \\ B[k-1] + 2^{-k} & \text{if } p_k = -1. \end{cases} \tag{1}$$

To obtain this result it is necessary that

$$A[k-1] = -q_0 + \sum_{i=1}^{k-1} q_i 2^{-i} = q[k-1] \tag{2}$$

and

$$B[k-1] = A[k-1] - 2^{-(k-1)}. \tag{3}$$

The final result is obtained as

$$q = A[m].$$

We now present the recurrence to compute $A[k]$ and $B[k]$ forms. Since p is normalized, the initial conditions are

$$A[1] = \begin{cases} +1/2 \quad (0.1) & \text{if } p > 0 \quad (p_1 = +1) \\ -1/2 \quad (1.1) & \text{if } p < 0 \quad (p_1 = -1) \end{cases}$$

$$B[1] = \begin{cases} +0 \quad (0.0) & \text{if } p > 0 \quad (p_1 = +1) \\ -1 \quad (1.0) & \text{if } p < 0 \quad (p_1 = -1). \end{cases}$$

For $k > 1$

$$A[k+1] = \begin{cases} A[k] + 2^{-(k+1)} & \text{if } p_{k+1} = 1 & (4.1) \\ A[k] & \text{if } p_{k+1} = 0 & (4.2) \\ B[k] + 2^{-(k+1)} & \text{if } p_{k+1} = -1 & (4.3) \end{cases}$$

$$B[k+1] = \begin{cases} A[k] & \text{if } p_{k+1} = 1 & (4.4) \\ B[k] + 2^{-(k+1)} & \text{if } p_{k+1} = 0 & (4.5) \\ B[k] & \text{if } p_{k+1} = -1. & (4.6) \end{cases}$$

Note that none of these requires a propagation of carries.

Proof: By induction. We assume that (2) and (3) are true for k and show that they are satisfied for $k+1$.

Basis: Both (2) and (3) are satisfied for $k = 2$ by the initial conditions.

Induction Step: We consider all possible values of p_{k+1}.

For $p_{k+1} = 1$, we have

$$A[k+1] = A[k] + 2^{-(k+1)} \text{ by (4.1)}$$

$$= q[k] + p_{k+1} 2^{-(k+1)} \qquad \text{by induction hypothesis}$$

$$= q[k+1]$$

Manuscript received September 13, 1985. This work was supported in part by the Office of Naval Research under Contract N00014-85-K-0159.

The authors are with the Department of Computer Science, University of California, Los Angeles, 90024.

IEEE Log Number 8713391.

and

$$B[k+1] = A[k] \text{ by } (4.4)$$
$$= A[k+1] - 2^{-(k+1)} \quad \text{by } (4.1).$$

For $p_{k+1} = 0$ we have

$$A[k+1] = A[k] \text{ by } (4.2)$$
$$= q[k] + p_{k+1}2^{-(k+1)} \quad \text{by induction hypothesis}$$
$$= q[k+1]$$

and

$$B[k+1] = B[k] + 2^{-(k+1)} \quad \text{by } (4.5)$$
$$= A[k] - 2^{-k} + 2^{-(k+1)} \quad \text{by induction hypothesis}$$
$$= A[k+1] - 2^{-(k+1)} \quad \text{by } (4.2).$$

For $p_{k+1} = -1$

$$A[k+1] = B[k] + 2^{-(k+1)} \quad \text{by } (4.3)$$
$$= A[k] - 2^{-k} + 2^{-(k+1)} \quad \text{by induction hypothesis}$$
$$= A[k] - 2^{-(k+1)}$$
$$= q[k] + p_{k+1}2^{-(k+1)} \quad \text{by induction hypothesis}$$
$$= q[k+1]$$

and

$$B[k+1] = B[k] \quad \text{by } (4.6)$$
$$= A[k+1] - 2^{-(k+1)} \quad \text{by } (4.3).$$

Consequently, the recurrence satisfies the conditions of the conversion. □

As an example consider the conversion of $P = 0.1101\bar{1}00\bar{1}1010$ where $\bar{1} = -1$.

k	p_k	$A[k]$	$B[k]$	$q[k]$
0	0			
1	1	0.1	0.0	0.1
2	1	0.11	0.10	0.11
3	0	0.110	0.101	0.110
4	1	0.1101	0.1100	0.1101
5	−1	0.11001	0.11000	0.11001
6	0	0.110010	0.110001	0.110010
7	0	0.1100100	0.1100011	0.1100100
8	−1	0.11000111	0.11000110	0.11000111
9	1	0.110001111	0.110001110	0.110001111
10	0	0.1100011110	0.1100011101	0.1100011110
11	1	0.11000111101	0.11000111100	0.11000111101
12	0	0.110001111010	0.110001111001	0.110001111010

The converted fraction is $q = 0.110001111010$.

Similarly, for a negative case such as $P = 0.\bar{1}0\bar{1}100\bar{1}1$ the conversion is

k	p_k	$A[k]$	$B[k]$	$q[k]$
0	0			
1	−1	1.1	1.0	1.1
2	0	1.10	1.01	1.10
3	−1	1.011	1.010	1.011
4	1	1.0111	1.0110	1.0111
5	0	1.01110	1.01101	1.01110
6	0	1.011100	1.011011	1.011100
7	−1	1.0110111	1.0110110	1.0110111
8	1	1.01101111	1.01101110	1.01101111

The converted fraction is $q = 1.01101111$.

The implementation of the algorithm requires two registers to hold $A[k]$ and $B[k]$, respectively. These registers can be shifted one bit left with insertion in the least significant bit depending on the value of p_k. They also require parallel loading to load $A[k]$ with $B[k]$ and vice versa. This implementation is shown in Fig. 1.

III. RADIX-r CONVERSION

We now generalize the previous algorithm to the radix-r case. That is, we want to convert

$$p = \sum_{i=1}^{m} p_i r^{-i},\ p_i \in \{-a, \cdots, 0, \cdots, a\} \qquad r/2 \le |a| \le r-1$$

into

$$q = -q_0 + \sum_{i=1}^{m} q_i r^{-i}, \qquad q_i \in \{0, \cdots, r-1\}.$$

To keep the characteristics of the algorithm, the expression (1) generalizes to

$$q[k] = \begin{cases} A[k-1] + p_k r^{-k} & \text{if } p_k \ge 0 \\ B[k-1] + (r - |p_k|)r^{-k} & \text{if } p_k < 0. \end{cases}$$

Consequently, to obtain this we need that

$$A[k-1] = q[k-1]$$

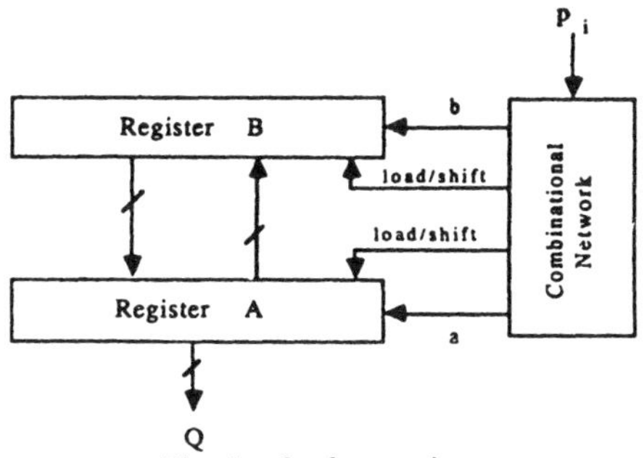

Fig. 1. Implementation.

and

$$B[k-1] = A[k-1] - r^{-(k-1)}.$$

The initial conditions extend to

$$A[1] = \begin{cases} +p_1 r^{-1} \ (0.p_1) & \text{if } p > 0 \\ -|p_1| r^{-1} \ (1.(r - |p_1|)) & \text{if } p < 0 \end{cases}$$

$$B[1] = \begin{cases} +(p_1 - 1)r^{-1} \ (0.(p_1 - 1))^* & \text{if } p > 0 \\ -(|p_1| + 1)r^{-1} \ (1.(r - 1 - |p_1|)) & \text{if } p < 0 \end{cases}$$

*$p_1 > 0$ since p is normalized.

The recurrence presented for the radix-2 case is transformed as follows. For $k > 1$

$$A[k+1] = \begin{cases} A[k] + p_{k+1} r^{-(k+1)} & \text{if } p_{k+1} \ge 0 \\ B[k] + (r - |p_{k+1}|) r^{-(k+1)} & \text{if } p_{k+1} < 0 \end{cases}$$

$$B[k+1] = \begin{cases} A[k] + (p_{k+1} - 1) r^{-(k+1)} & \text{if } p_{k+1} > 0 \\ B[k] + ((r-1) - |p_{k+1}|) r^{-(k+1)} & \text{if } p_{k+1} \le 0. \end{cases}$$

The proof follows the same scheme as that for the radix-2 case.

The implementation is also similar. It requires a one-digit adder to compute $r - |p_{k+1}|$.

IV. Conversion from Radix-4 Redundant to Conventional 2's Complement

As a detailed example, we now apply the conversion algorithm to the radix-4 case, compatible with the radix-4 division scheme [4]. The conversion rules from a radix-4 redundant representation into the conventional binary 2's complement are specified in Table I. The digit appended to bit-vector $A[k]$ is $a = (a_1, a_0)$ with value $2a_1 + a_0 \in \{0, 1, 2, 3\}$. Similarly, $b = (b_1, b_0)$ is appended to bit-vector $B[k]$.

The implementation consists of two left-shift registers (A and B) that contain $A[k]$ and $B[k]$, respectively. The operations on these registers are

$$A \leftarrow \begin{cases} \text{Shift } A \text{ with insert } (a_1, a_0) & \text{if } C_{SHA}=1 \\ \text{Shift } B \text{ with insert } (a_1, a_0) & \text{if } C_{LDA}=1 \end{cases}$$

$$B \leftarrow \begin{cases} \text{Shift } B \text{ with insert } (b_1, b_0) & \text{if } C_{SHB}=1 \\ \text{Shift } A \text{ with insert } (b_1, b_0) & \text{if } C_{LDB}=1. \end{cases}$$

By definition, $C_{LDA} = C'_{SHA}$ and $C_{LDB} = C'_{SHB}$.

For the minimally redundant case $p_i \in \{-2, -1, 0, 1, 2\}$ the radix-4 digit values are represented by the following sign-and-magnitude code:

p_i	s	m_1	m_0
0	0	0	0
1	0	0	1
2	0	1	0
−1	1	0	1
−2	1	1	0

TABLE I
RADIX-4 CONVERSION

p_{k+1}	$A[k+1]$	$B[k+1]$
0	$(A[k], 0, 0)$	$(B[k], 1, 1)$
1	$(A[k], 0, 1)$	$(A[k], 0, 0)$
−1	$(B[k], 1, 1)$	$(B[k], 1, 0)$
2	$(A[k], 1, 0)$	$(A[k], 0, 1)$
−2	$(B[k], 1, 0)$	$(B[k], 0, 1)$
3	$(A[k], 1, 1)$	$(A[k], 1, 0)$
−3	$(B[k], 0, 1)$	$(B[k], 0, 0)$

where s is the sign bit, and m_1 and m_0 are the magnitude bits. Using Table I we determine the following switching expressions for the rightmost radix-4 digits $a = (a_1, a_0)$ and $b = (b_1, b_0)$ where $a, b \in \{0, 1, 2, 3\}$:

$$a_1 = s + m_1$$

$$a_0 = m_0$$

$$b_1 = m'_0 m'_1 + sm'_1$$

$$b_0 = m'_0.$$

The shift and load control signals for the registers are also obtained from Table I:

$$C_{LDA} = s$$

$$C_{SHA} = s'$$

$$C_{LDB} = (s + m'_0 m'_1)' = C'_{SHB}$$

$$C_{SHB} = s + m'_0 m'_1.$$

V. Summary

An algorithm for converting redundant forms into range-complement conventional forms concurrently with digit-by-digit generation is presented. The algorithm has a simple implementation and a delay independent of the working precision, roughly equal to two logic levels plus a register shift/load time. It is applicable in nonrestoring division and square root algorithms, and in producing conventional results in on-line algorithms.

Acknowledgment

We thank Dr. J. G. Nash of Hughes Research Laboratories for his interest and support, and one of the referees for helpful comments.

References

[1] J. E. Robertson, "A new class of digital division methods," *IRE Trans. Electron. Comput.*, vol. EC-7, pp. 218-222, Sept. 1958.
[2] D. E. Atkins, "Higher-radix division using estimates of the divisor and partial remainders," *IEEE Trans. Comput.*, vol. C-17, pp. 925-934, Oct. 1968.
[3] G. Metze, "Minimal square rooting," *IEEE Trans. Electron. Comput.*, vol. EC-14, pp. 181-185, Feb. 1965.
[4] M. D. Ercegovac and T. Lang, "A division algorithm with prediction of quotient digits," in *Proc. 7th IEEE Symp. Comput. Arithmetic*, Urbana, IL, 1985, pp. 51-56.
[5] G. S. Taylor, "Radix 16 SRT dividers with overlapped quotient selection stages," in *Proc. 7th IEEE Symp. Comput. Arithmetic*, Urbana, IL, 1985, pp. 64-71.
[6] M. D. Ercegovac, "A general hardware-oriented method for evaluation of functions and computations in a digital computer," *IEEE Trans. Comput.*, vol. C-26, pp. 667-680, July 1977.
[7] K. S. Trivedi and M. D. Ercegovac, "On-line algorithms for division and multiplication," *IEEE Trans. Comput.*, vol. C-26, pp. 667-680, July 1977.
[8] M. D. Ercegovac, "On-line arithmetic: An overview," in *Proc. SPIE 1984, Vol. 495—Real Time Signal Processing VII*, 1984, pp. 86-93.
[9] J. Sklansky, "Conditional sum addition logic," *IRE Trans. Electron. Comput.*, vol. EC-9, pp. 226-231, June 1960.

Part III:
VLSI ADDER IMPLEMENTATIONS

Editor's Comments on Papers 14 Through 18

14. **Atkins and Ong**
    Time-Component Complexity of Two Approaches to Multioperand Binary Addition

15. **Oklobdzija and Barnes**
    Some Optimal Schemes for ALU Implementation in VLSI Technology

16. **Bayoumi, Jullien and Miller**
    An Area-Time Efficient NMOS Adder

17. **Ngai, Irwin and Rawat**
    Regular, Area-Time Efficient Carry-Lookahead Adders

18. **Johnson, Aylor and Hana**
    Efficient Use of Time and Hardware Redundancy for Concurrent Error Detection in a 32-bit VLSI Adder

Adders are a key arithmetic component of most digital systems. For many years (in fact since the commercial introduction of the 74181/74182 ALU/CLA in the late 1960s) the carry lookahead adder has been the standard for high performance systems. The first few papers in this Chapter compare the standard carry lookahead adder with other options for fast addition. The next group of papers examine the efficient implementation in silicon of the carry lookahead adder. Finally, a detailed design for an error detecting adder completes this part.

In the first paper, Atkins and Ong compare carry save and carry lookahead adders for the summation of many operands. The comparison includes estimates of the hardware complexity (i.e., the gate count) and the time measured in gate delays. The analysis shows that the carry save approach is simpler and faster for most applications.

Oklobdzija and Barnes examine carry skip adders in the next paper. The carry skip adder divides the words to be added into segments (i.e., bit slices of arbitrary size) and detects when a carry entering a segment will be propagated through that segment to the next. This paper analyses the performance for carry skip adders as a function of the size of the carry skip segments. It is shown that a simpler carry skip adder achieves performance that is as good or better than that achieved by carry lookahead adders.

Bayoumi, Jullien, and Miller present the detailed silicon design of a carry lookahead adder for implementation in NMOS. This design considers the delays resulting from capacitive loading which is a function of the interconnect length and the fan out. The paper presents an example design of a four bit adder that achieves an efficient layout and a regular cell placement and interconnection.

The next paper, by Ngai, Irwin, and Rawat, also presents a layout for a carry lookahead adder. The authors consider design regularity, area-time efficiency and the length of interconnections. They present designs of 27-bit and 108-bit two's complement carry lookahead adders and also include design extensions to accommodate sign magnitude operands and to pipeline the adder to give greater throughput.

The final paper, by Johnson, Aylor and Hana, presents the design of an error detecting 32-bit adder. The paper presents a comparison, which shows that recomputation using duplication with comparison is faster and simpler than other approaches to concurrent error detection.

Time-Component Complexity of Two Approaches to Multioperand Binary Addition

DANIEL E. ATKINS, MEMBER, IEEE, AND SHAU-CHI ONG, STUDENT MEMBER, IEEE

Reprinted from *IEEE Transactions on Computers*, Volume C-28, Number 12, December 1979, pages 918-926.

***Abstract*—Component and time complexity measures in terms of number of gates and gate delays, respectively, are derived for two multioperand adder structures: a tree of carry-save adders and a tree of carry-lookahead adders. The parameters of the complexity measures are operand length, number of operands, and maximum fanin to the gates. The analysis involves a constructive derivation of the complexity of carry-lookahead networks including optimization by removal of redundant paths. An alternate measure of component complexity, number of packages with a given number of pinouts, rather than number of gates, is also presented.**

The general conclusion of the paper is that, under the assumptions made, for a large set of design parameters, the carry-save adder structure requires both less time and fewer components than the functionally equivalent carry-lookahead adder structure.

***Index Terms*—Adder tree, binary addition, carry-lookahead adder, carry-save adder, multioperand addition, time-component complexity.**

INTRODUCTION

THIS paper presents the results of a study concerning methods for addition of $P > 2$ numbers, each encoded as a vector of binary digits of length N. Such multioperand addition has been studied most often in the context of reducing a set of partial products to a single result in the implementation of multiplication. More generalized multioperand addition, most notably in the form of inner product calculations is, however, central to numerous scientific applications of digital computers. Although multioperand addition is trivially accomplished by accumulation (iteration in time) in any general purpose machine, demands for very high-speed computation, typified by 2- and 3-D signal processing, prompt implementation of dedicated, hardware-intensive structures for multioperand addition. This study, for example, is motivated in part by requirements for rapid simultaneous addition of up to 100, 16 bit operands in the design of a dedicated processor for real-time reconstruction of 3-D images of the beating heart and breathing lungs [1], [2].

The variety of tradeoffs between iteration in time and hardware, the application of the commutative and transitive properties of addition, and the possibility of restricting carry propagation by using redundant encoding of numbers give rise to numerous published schemes for multioperand addition. Examples of time-intensive sequential schemes are discussed in [3]–[9], and the use of arrays of carry-save adders (deferred carry assimilation structures) are described in [10]–[16]. If we view the rows of the bit vectors to be added as forming a matrix of digits, then the columns of the matrix may each be simultaneously compressed from P bits to a vector of about length $\log P$ bits. Subsequent column compressions will lead to a single binary digit vector representation of the sum of the P operands. These "column compression" ideas are included in [6], [9], [11], [12], [17]–[21]. References [22]–[24] describe schemes which fall in the middle ground with respect to a time-hardware tradeoff.

Our group is considering the time-hardware complexity of multioperand addition across a broad set of finite number representation systems [25], including as a special case, variations in radix. This paper, however, is restricted to the specific topic of a cost and performance comparison of multioperand adders for unsigned, or two's complement binary encoded operands implemented in the following two ways suggested by commercially available MSI/LSI components:

1) As a tree of (nonredundant, carry-propagate) adders which accept two binary digit vectors and produce one;

2) As a tree of carry-save adders which accept three binary digit vectors and produce two.

The analysis will be conducted assuming three variables, namely, number of operands P, length of the operands N (all uniform length), and the maximum fanin of the device gates FANIN.

Informally the problem is as follows: carry-save adders produce a result rapidly since they avoid solving the carry-propagation recurrence at the expense of reduced operand compression (3-to-2 instead of 2-to-1). Adders which assimi-

Manuscript received March 15, 1979; revised June 20, 1979. This work was supported by the National Science Foundation, Division of Mathematical and Computer Science under Grant MCS-77-03310 and by the U.S. Public Health Service Grants HL-04664, RR-00007 from the National Institutes of Health. A preliminary version of this paper was presented at the 4th Symposium on Computer Arithmetic, Santa Monica, CA, Oct. 25–27, 1978.

The authors are with the Program in Computer, Information and Control Engineering, and the Systems Engineering Laboratory of the Department of Electrical and Computer Engineering, University of Michigan, Ann Arbor, MI 48109.

late all carries yield a 2-to-1 operand compression but require auxiliary logic (carry-lookahead logic) in order to solve the carry recurrence. Which is better for multioperand addition? Carry-ripple adders avoid the lookahead logic but, in general, would be so much slower than a carry-save adder that a comparison would be uninteresting.

In the next section we derive cost and performance measures for a tree of 2-to-1, carry-lookahead structures. In subsequent sections, we perform the analogous task for a tree of 3-to-2 carry-save structures, compare the results, and then recompare with the introduction of restrictions on packaging imposed by pin-out limitations. Since for most of the cost and performance functions we have not found a closed form expression, APL [26], [27] versions of the functions are used. These functions are included for completeness and to enable the interested reader to compute time and component complexity for specific design parameters. A detailed understanding of the APL functions is not essential to the flow of the paper. A review of the principles of binary carry-lookahead and carry-save adders may be found in [28], [29].

Cost and Performance of Two-to-One Structures

In this section we first develop a cost and performance function for a carry-lookahead adder which accepts two N-bit operands and produces a $N + 1$-bit result. We then apply these results to the analysis of a tree structure which accepts P, N-bit operands to produce one $N + \lceil \log_2 P \rceil$-bit result. $\lceil x \rceil$ is the ceiling function, i.e., the smallest integer greater than or equal to x.

Two Operand Carry-Lookahead Adder

Consider the block diagram in Fig. 1 which combines two N-bit operands, A and B, to form a sum S using carry-lookahead (CL) techniques to compute the carry vector in the CL box. Assuming that the delay through a logic gate is a uniform time t, then a time t is required to produce carry transmission and generation terms T and G, respectively. Assuming that an inverter also requires time t, then at best $2t$ is needed to produce carries C in the CL block and $3t$ more to develop S. For FANIN > 3 each operand bit requires one gate for T, one for G and eight for S as shown in Fig. 2. For FANIN $= 3$, $4t$ and nine gates are required to form S.

The structure of a CL network which produces 4 carry bits plus a group generate and a group propagate is shown in Fig. 3. The dotted lines show a slight variation which is used to produce a true group carry-out at certain boundary positions in both the balanced and optimized adder trees to be described. The number of positions across which lookahead takes place (i.e., the length of the carry vector produced) is restricted by the fanin of the constituent gates assuming that the operation time does not exceed $2t$. If we further impose the restriction that the true group carry-out must be formed without introducing more delay, then the number of bit positions across which first-level lookahead takes place W is given by

$$W = \text{FANIN} - 1, \tag{1}$$

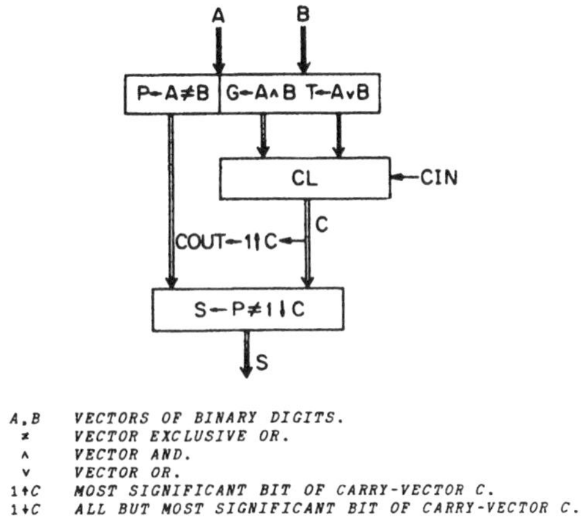

Fig. 1. Block diagram of carry-lookahead adder.

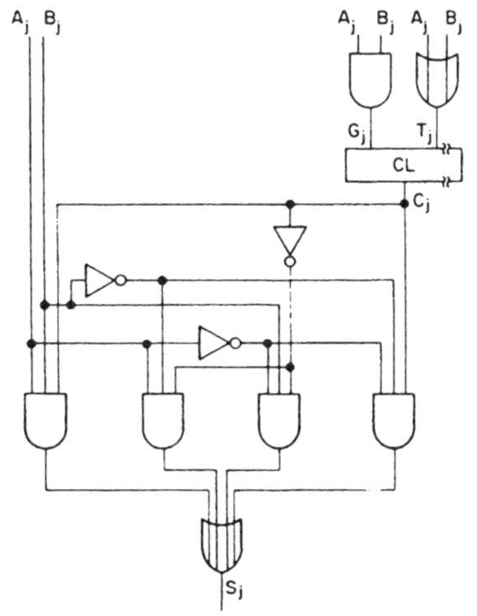

Fig. 2. One position of a carry-lookahead adder.

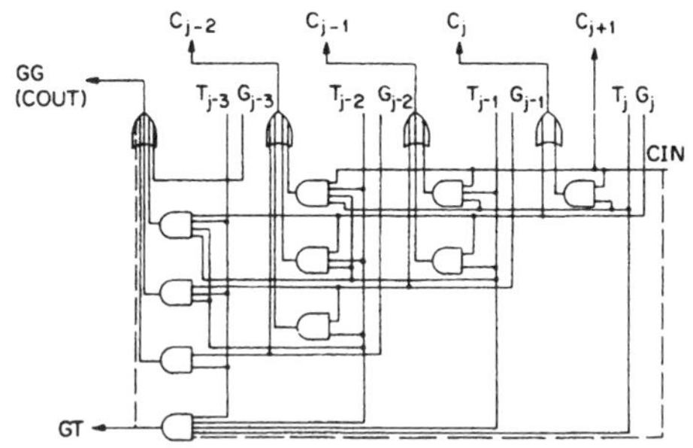

Fig. 3. Four bit carry-lookahead (CL) unit.

where FANIN is the maximum number of inputs to a primitive gate. In the following equations we use W as an implicit function of FANIN.

The number of gates required to implement a CL structure of the type shown in Fig. 3 across W bits, is the sum of W OR gates and a geometric series of AND gates, namely,

$$G[\mathrm{cl}] = \frac{W^2 + 3W}{2}, \tag{2}$$

where W is defined in terms of FANIN in (1).

In general, the operand length N is greater than W and consequently, multilevel carry-lookahead across groups of first level units may be employed. The number of levels of W-wide CL units required to produce the carry vector of length N using a balanced carry-lookahead (bcl) structure is given by

$$L[\text{bcl}] = \lceil \log_W N \rceil. \tag{3}$$

It follows, under these assumptions, that the time required to produce the sum of two operands of N bits using a balanced carry-lookahead adder (bcla) with FANIN > 3 is

$$T[\text{bcla}] = 2(2L[\text{bcl}] + 1)t. \tag{4}$$

The number of CL units is given by $L[\text{bcl}]$ terms of the following

$$U[\text{bcl}] = \left\lceil \frac{N}{W} \right\rceil + \frac{\left\lceil \frac{N}{W} \right\rceil}{W} + \cdots \tag{5}$$

The number of gates required to implement the N-bit, two operands carry-lookahead adder with FANIN > 3 is

$$G[\text{bcla}] = 10N + G[\text{cl}]U[\text{bcl}]. \tag{6}$$

The structure consisting of both the solid and dashed nodes shown in Fig. 4, for example, is the tree for the case $N = 22$, FANIN $= 5$. In this case, 9 CL units of width 4 are required, and the structure produces the final sum in time $14t$. The balanced tree of CL units does not, in general, yield the smallest propagation delay. At some point toward the root of the tree it becomes faster to propagate the carries laterally rather than forward and back down from a higher level CL.

Fig. 4, minus the dashed boxes together with the lateral carry interconnections, yields higher performance with less hardware than the full balanced tree. The pruning of the tree is reasoned as follows:

1) It takes $4t$ more delay and 1 more CL unit to add a level to the tree, but only $2t$ more, without increasing the number of CL units, to form the group carry-out (see Fig. 3). The propagation route from module C to B in Fig. 4 thus saves $2t$ in time and 1 CL in hardware compared to the C-A-B route.

2) It takes the same delay $4t$ for both C-E-D and C-B-D but the former saves 1 CL unit.

3) We need $5t$ in D to form the sum of the most significant bit, either $2t$ in the CL plus $3t$ to form S(sum), or $2t$ for carry-ripple and $3t$ for S. In other words, in some parts of the structure, carry-lookahead had no speed advantage over carry-ripple.

The following steps are a generalization of the method for pruning the balanced tree of CL units to produce the optimized adder:

1) Bypass all nodes which have exactly one successor and one predecessor.

2) Compute k, the number of nodes which are the immediate successor of the root node, exclusive of those nodes

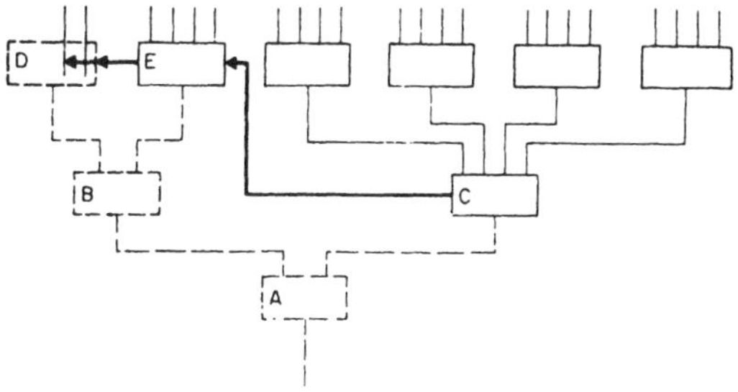

Fig. 4. Tree of CL units for $N = 22$, FANIN $= 5$.

which have become immediate successors due to a node bypass in step 1). If $k > 3$, go to step 3); else disconnect the carries into the root node and connect them serially between the nodes at the next higher level. Delete the root node.

3) Locate the subtree whose root is the leftmost node (closest to the most significant bit), at the next level. Compute k, the number of nodes which are the immediate successor of the root of the subtree, exclusive of those nodes which have become immediate successors due to a node bypass in step 1). If $k > 2$, go to step 4); else disconnect the carries into the root node and connect them serially between the nodes at the next higher level. Delete the root node of the subtree.

4) Repeat step 3) until the highest level (leaf-level) is reached.

Following these principles, the gate delays $T[\text{ocla}]$, the number of levels $L[\text{ocl}]$, the number of CL units $U[\text{ocl}]$ and the number of gates $G[\text{ocla}]$ required by the optimized structure were computed constructively using the APL functions, TOCLA, LOCL, UOCL and GOCLA, respectively, shown in Fig. 5. Figs. 6 and 7 are graphs of the results for $N \leq 64$ and $3 \leq$ FANIN ≤ 9. The number of gate delays in an optimized structure is up to 75 percent (average 14 percent) less than that predicted by (4) for the corresponding balanced structure; and the number of CL units is up to 200 percent (average 27 percent) less than that predicted by (5).

Table I summarizes performance data taken from [30] for adders of length N constructed with the 74S181 ALU and 74S182 carry-lookahead units with $W = 4$. The "Predicted" column lists the number of time units of delay for the N-bit addition using the optimized carry-lookahead adders described above as predicted by the function TOCLA. The last column is the ratio of the given delay in nanoseconds to the largest value of the predicted speed. Note that the manufacturer's data suggest the need for 28 ns to add 17-64 bit operands. The construction used in this paper shows that a uniform delay need not be imposed and that, for example, structures exist which would require only a $10t$ (about 20 ns) delay to add 17-bit operands.

Multioperand Carry-Lookahead Adder

We are now in the position to analyze a multioperand structure realized as a tree of two-operand adders each of which employes multilevel, optimized carry-lookahead techniques as shown in Fig. 8. Note that the operand lengths

```
     ∇ Z←UOCL N;W;L;I;E;NN
[1]    ⍝ THE NUMBER OF CL UNITS FOR TWO N-BIT (N>3)
[2]    ⍝   OPERANDS CLA WITH OPTIMIZED STRUCTURE
[3]    ⍝ FANIN IS DEFINED EXTERNALLY
[4]    W←FANIN-1
[5]    →(W=2)/25
[6]    L←LOCL N
[7]    I←E←0
[8]    NN←N
[9]    Z←N←⌈(N-L+L=1)÷W
[10]   →(0=L←L-1)/0
[11]   E←NN+E-1+W×N
[12]   →(E≥0)/15
[13]   E←0
[14]   →17
[15]   →(E≤2)/17
[16]   E←2
[17]   NN←N
[18]   N←⌈(N+E-2)÷W
[19]   Z←Z+N
[20]   →((E<2)∨I=1)/10
[21]   I←(N=1+W×W)∨(NN≠N×W)∧(N=2+W×W)∨(N=W+2)∨N=2+2×W
[22]   I←((NN+W-1)>W×N)∧I∨(N=W+1)∨N=1+2×W
[23]   Z←Z+I
[24]   →10
[25]   Z←1
[26]   →(N≤4)/0
[27]   Z←2
[28]   →(N≤5)/0
[29]   I←1
[30]   E←4
[31]   NN←N-5
[32]   Z←N-E
[33]   →(0≥NN←NN-2*I)/0
[34]   Z←Z-1
[35]   →(0≥NN←NN-2*I)/0
[36]   Z←Z+1
[37]   →(0≥NN←NN-2*I)/0
[38]   I←I+1
[39]   E←E+1
[40]   →32
     ∇


     ∇ Z←GOCLA N;W;GS
[1]    ⍝ THE NUMBER OF GATES FOR TWO N-BIT (N>3)
[2]    ⍝   OPERANDS CLA WITH OPTIMIZED STRUCTURE
[3]    ⍝ FANIN IS DEFINED EXTERNALLY
[4]    W←FANIN-1
[5]    →(W=2)/8
[6]    GS←10
[7]    →9
[8]    GS←11
[9]    Z←(GS×N)+((W×W+3)÷2)×UOCL N
     ∇

     ∇ Z←TOCLA N;W;K
[1]    ⍝ THE NUMBER OF GATE DELAY FOR TWO N-BIT (N>3)
[2]    ⍝   OPERANDS CLA WITH OPTIMIZED STRUCTURE
[3]    ⍝ FANIN IS DEFINED EXTERNALLY
[4]    W←FANIN-1
[5]    Z←0
[6]    →(0≥N)/0
[7]    Z←4
[8]    →(1≥N)/0
[9]    Z←6
[10]   →(N≤W+1)/0
[11]   K←⌈W⍟1+N×W-1
[12]   Z←4×K-1
[13]   →(N≤(((W*K-1)-1)÷W-1)+W*K-2)/0
[14]   Z←Z+2
     ∇


     ∇ Z←LOCL N;W
[1]    ⍝ THE NUMBER OF CL LEVELS FOR TWO N-BIT (N>3)
[2]    ⍝   OPERANDS CLA WITH OPTIMIZED STRUCTURE
[3]    ⍝ FANIN IS DEFINED EXTERNALLY
[4]    W←FANIN-1
[5]    N←N-3
[6]    Z←0
[7]    N←N-((3×W)-2)×W*Z
[8]    Z←Z+1
[9]    →(N>0)/7
     ∇
```

Fig. 5. Listing of APL functions TOCLA, LOCL, UOCL, and GOCLA.

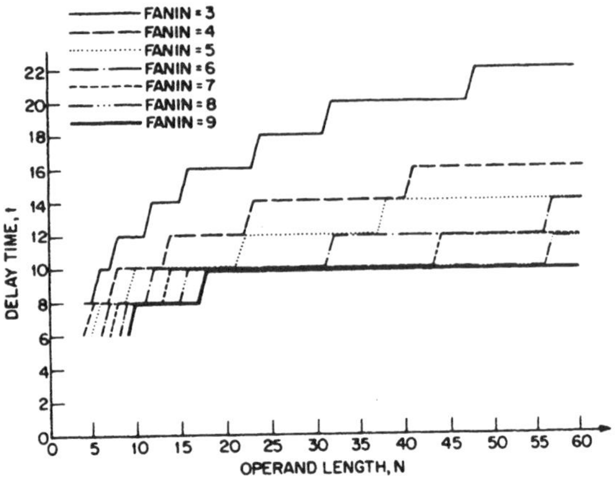

Fig. 6. Operational delay versus operand length and fanin for carry-lookahead adder.

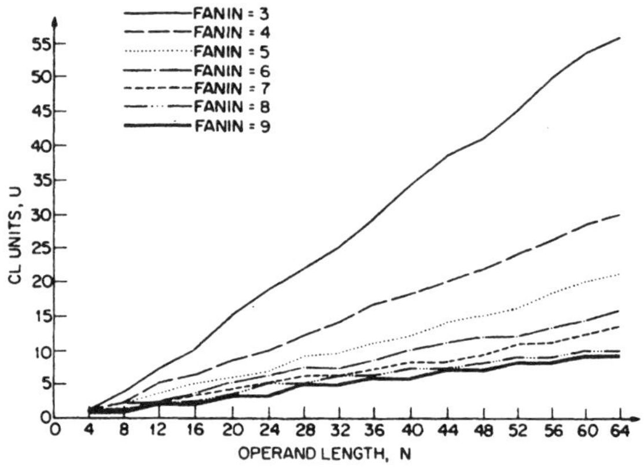

Fig. 7. Required number of CL units versus operand length and fanin for carry-lookahead adder.

TABLE I
COMPARISON OF S181, S182 CARRY-LOOKAHEAD ADDER PERFORMANCE WITH PREDICTED VALUES

N	TI Spec. S181,S182 (ns)	Predicted (t units)	Spec./ Maximum Predicted
1-4	11	4-6	1.83
5-8	18	6-8	2.25
9-16	19	8-10	1.9
17-64	28	10-14	2.0

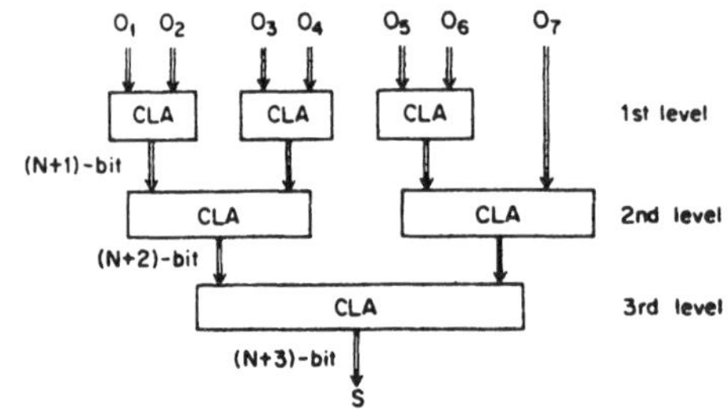

Fig. 8. Example of 2-to-1 tree for $P = 7$.

are $N + i - 1$ bits at the ith level. The number of levels of 2-to-1 adders to combine $P \geq 2$ operands is given by

$$L[21] = \lceil \log_2 P \rceil. \tag{7}$$

Now let $I[21](i)$ and $J[21](i)$ be the number of input operands and the number of carry-lookahead adders (CLA's), respectively, at the ith level. $I[21](i)$ equals the number of operand outputs from level $i - 1$ plus the number of operands which have not been combined at a previous level. It follows that

$$I[21](i) = J[21](i - 1) + I[21](i - 1) - 2J[21](i - 1) \tag{8}$$

with $I[21](1) = P$.

$$J[21](i) = \lfloor I[21](i)/2 \rfloor, \tag{9}$$

where $\lfloor x \rfloor$ is the floor function, i.e., the greatest integer smaller than or equal to x. The total number of CLA units is

$$U[\mathrm{cla}] = P - 1. \tag{10}$$

The total time to compute the N-bit, P operand sum is the total of the delay at each level, namely,

$$T[21] = \sum_{i=1}^{L[21]} T[\mathrm{ocla}](i), \tag{11}$$

where $T[\mathrm{ocla}](i)$ is the time for addition at the ith level using optimized carry-lookahead adders. The number of CL units required is the sum of the number of CLA groups times the number of CL units per CLA for each level.

$$U[\mathrm{cl21}] = \sum_{i=1}^{L[21]} U[\mathrm{ocl}](i)J[21](i), \tag{12}$$

where $U[\mathrm{ocl}](i)$ is the number of CL units to implement a carry-lookahead adder at level i. The total number of gates required for the entire structure is given by

$$G[21] = \sum_{i=1}^{L[21]} G[\mathrm{ocla}](i)J[21](i), \tag{13}$$

where $G[\mathrm{ocla}](i)$ is the number of gates to implement a carry-lookahead adder at level i.

These expressions, $T[21]$, $U[\mathrm{cl21}]$, and $G[21]$ are evaluated using the APL functions TIME21, CLUNIT21, and GATE21 as listed in Fig. 9.

```
    ∇ Z←TIME21 P;I
[1]   ⍝ THE NUMBER OF GATE-DELAYS FOR MULTIPLE OPERAND ADDITION
[2]   ⍝    OF TWO-TO-ONE TREE STRUCTURE
[3]   ⍝ WORDS LENGTH N, FANIN ARE DEFINED EXTERNALLY
[4]   I←(⌈2⍟P)-1
[5]   Z←0
[6]   Z←Z+TOCLA N+I
[7]   →(0≤I←I-1)/6
    ∇

    ∇ Z←CLUNIT21 P;I21;J21
[1]   ⍝ THE NUMBER OF CL UNITS FOR MULTIPLE OPERAND ADDITION
[2]   ⍝    OF TWO-TO-ONE TREE STRUCTURE
[3]   ⍝ WORD LENGTH N, FANIN ARE DEFINED EXTERNALLY
[4]   Z←I21←0
[5]   J21←⌊P÷2
[6]   Z←Z+J21×UOCL N+I21
[7]   P←J21+P-J21×2
[8]   I21←I21+1
[9]   →(P>1)/5
    ∇

    ∇ Z←GATE21 P;I21;J21
[1]   ⍝ THE NUMBER OF GATES FOR MULTIPLE OPERAND ADDITION
[2]   ⍝    OF TWO-TO-ONE TREE STRUCTURE
[3]   ⍝ WORDS LENGTH N, FANIN ARE DEFINED EXTERNALLY
[4]   Z←I21←0
[5]   J21←⌊P÷2
[6]   Z←Z+J21×GOCLA N+I21
[7]   P←J21+P-J21×2
[8]   I21←I21+1
[9]   →(P>1)/5
    ∇
```

Fig. 9. Listing of APL functions TIME21, CLUNIT21, and GATE21.

TABLE II
TRUTH TABLE FOR A CARRY-SAVE ADDER UNIT

A	B	C	PC	PS
0	0	0	0	0
0	0	1	0	1
0	1	0	0	1
0	1	1	1	0
1	0	0	0	1
1	0	1	1	0
1	1	0	1	0
1	1	1	1	1

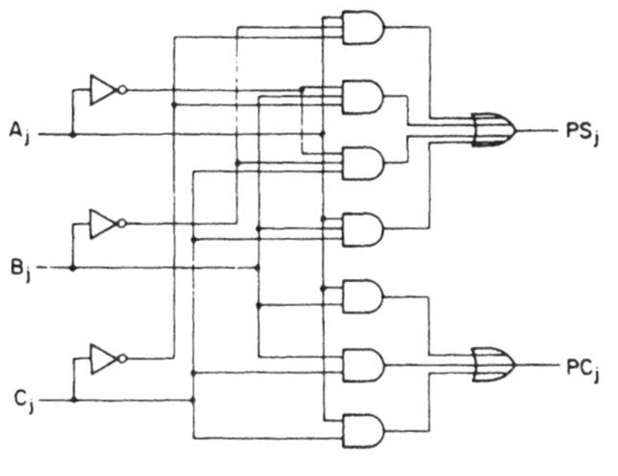

Fig. 10. One position of a carry-save adder.

COST AND PERFORMANCE OF THREE-TO-TWO STRUCTURES

Three-Operand Carry-Save Adder

By "carry-save adder" (CSA) we refer to a logic network which accepts three binary digit vector operands and produces two, the sum of which is equal to the sum of the three. The truth table for one position of a CSA (a CSA unit or full adder) is shown in Table II. This structure is a special case of the carry-save/borrow-save networks studied by Robertson in [31]. Fig. 10 shows a gate-level realization of one position of a CSA for FANIN = 4. If FANIN = 3 then four levels are required, i.e.,

$$T[\mathrm{csa}] = 3t \text{ for FANIN} \geq 4 \tag{14}$$

and the number of gates is given by

$$G[\mathrm{csa}] = 12N \text{ with FANIN} \geq 4. \tag{15}$$

Multioperand Carry-Save Adder

We now consider the 3-to-2 array, sometimes called a Wallace adder tree [10], shown in Fig. 11. Let $I[32](i)$ be the number of input operands to the ith level and let $J[32](i)$ be the number of CSA groups required at the ith level. It follows that

$$I[32](i) = 2J[32](i - 1) + I[32](i - 1) - 3J[32](i - 1) \tag{16}$$

with $I[32](1) = P$.

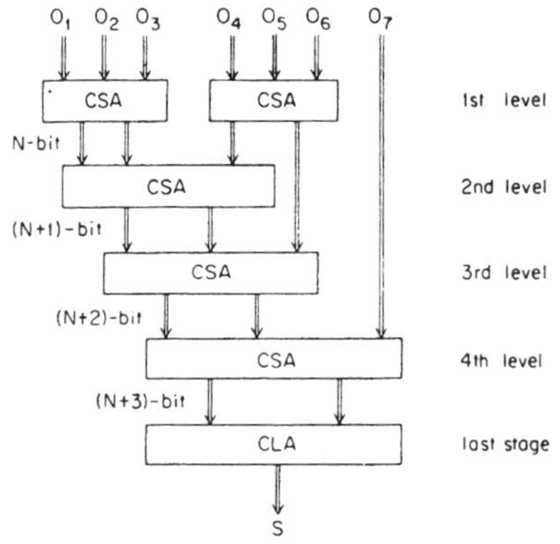

Fig. 11. Example of 3-to-2 tree for $P = 7$.

$$J[32](i) = \lfloor I[32](i)/3 \rfloor. \tag{17}$$

The total number of CSA groups is

$$U[\text{csa}] = P - 2. \tag{18}$$

The number of levels of CSA required, $L[32]$, is computed iteratively as described by the APL function LEVEL32 in Fig. 12. The total time to compute the P operand sum, including carry assimilation as a final step is given by

$$T[32] = T[\text{csa}]L[32] + T[\text{ocla}](L[32] + 1). \tag{19}$$

The number of CSA units required is

$$U[\text{csa32}] = \sum_{i=1}^{L[32]} (N + i - 1)J[32](i). \tag{20}$$

The total number of gates required, including the $N + L[32]$ bit carry-lookahead adder at the output end of the tree, is given by

$$G[32] = G[\text{csa}]U[\text{csa32}] + G[\text{ocla}]. \tag{21}$$

Fig. 12 is a listing of the APL functions which compute $L[32]$, $T[32]$, $U[\text{csa32}]$, and $G[32]$. These functions are denoted LEVEL32, TIME32, CSA32, and GATE32, respectively.

Numerical Evaluation and Comparison of Cost/Performance Functions

Our ultimate concern is with the comparison of two pairs of functions of three variables, namely, $T[21]$, $G[21]$, $T[32]$, and $G[32]$. Our approach has been to evaluate numerical examples and attempt to draw general conclusions. Fig. 13(a) and (b) shows a comparison of the two structures for fixed operand length and fanin. In this case the 3-to-2 structure exhibits greater speed with lower gate count. Fig. 14(a) and (b) compares the structures with fixed values of P and FANIN as N, the operand length, varies. Here the 3-to-2 case is always faster but requires slightly more gates for $N < 12$.

Tables III and IV display some conclusions from trial computations. For FANIN values of 3 through 9, Table III lists the lower bound on N such that the time delay and gate count are smaller for the 3-to-2 structures than the 2-to-1 structures for any P in the range $3 \leq P \leq 40$. For example, if FANIN $= 4$, then for operand lengths of 11 bits or more, a carry-save adder tree is faster and requires fewer gates than a tree of carry-lookahead adders. If $N = 8$ the delay is less but the gate count may not be. If N is less than 4 we cannot make a general statement about the relative cost and performance, but of course these could be computed for specific choices of N, P, and FANIN. Note that for the FANIN range 3–9, if operands are 14 bits or greater in length, then a carry-save adder tree is always faster and requires fewer components than an optimized carry-lookahead adder tree.

```
    ∇ Z←LEVEL32 P;J
[1]   ⍝ THE NUMBER OF LEVELS FOR THREE-TO-TWO TREE STRUCTURE
[2]   Z←0
[3]   →(P≤2)/0
[4]   J←⌊P÷3
[5]   P←(J×2)+P-J×3
[6]   Z←Z+1
[7]   →3
    ∇

    ∇ Z←TIME32 P;I;TCSA
[1]   ⍝ THE NUMBER OF GATE-DELAYS FOR MULTIPLE OPERAND ADDITION
[2]   ⍝    OF THREE-TO-TWO TREE STRUCTURE
[3]   ⍝ WORDS LENGTH N, FANIN ARE DEFINED EXTERNALLY
[4]   I←LEVEL32 P
[5]   →(FANIN<4)/8
[6]   TCSA←3
[7]   →9
[8]   TCSA←4
[9]   Z←(TCSA×I)+TOCLA N+I
    ∇

    ∇ Z←CSA32 P;I32;J32
[1]   ⍝ THE NUMBER OF CARRY SAVE ADDERS FOR MULTIPLE OPERAND
[2]   ⍝    ADDITION OF THREE-TO-TWO TREE STRUCTURE
[3]   Z←I32←0
[4]   →(P≤2)/0
[5]   J32←⌊P÷3
[6]   Z←Z+J32×N+I32
[7]   P←(J32×2)+P-J32×3
[8]   I32←I32+1
[9]   →4
    ∇

    ∇ Z←GATE32 P;GCSA
[1]   ⍝ THE NUMBER OF GATES FOR MULTIPLE OPERAND ADDITION
[2]   ⍝    OF THREE-TO-TWO TREE STRUCTURE
[3]   ⍝ WORDS LENGTH N, FANIN ARE DEFINED EXTERNALLY
[4]   →(FANIN<4)/7
[5]   GCSA←12
[6]   →8
[7]   GCSA←13
[8]   Z←((GCSA×CSA32 P)+GOCLA N+LEVEL32 P)
    ∇
```

Fig. 12. Listing of APL functions LEVEL32, TIME32, CSA32 and GATE32.

Table IV lists the lower bound on P such that the time delay and gate count are smaller for the 3-to-2 structures than the 2-to-1 structures. Again, if P is less than the bound, we cannot make a general statement.

Cost Comparisons Under Packaging Constraints

Our measure of component complexity has so far been in terms of the required number of gates assuming a given limit on fanin. One of our reasons for concentrating on structures based upon carry-lookahead and carry-save adders was the fact that off-the-shelf integrated circuit components are available to implement these devices. At this point we redefine the component complexity measure to be the *number of circuit packages* required under various restrictions on pin-outs.

It requires $3W + 2$ pins for each CL unit ($3W + 3$ pins for the first level CL units which also produce the sum) and 5 pins for each CSA unit and therefore

$$(\text{PIN} - 2) \geq (3W + 2)L \qquad \text{for CL} \tag{22}$$

and

$$(\text{PIN} - 2) \geq 5S \qquad \text{for CSA,} \tag{23}$$

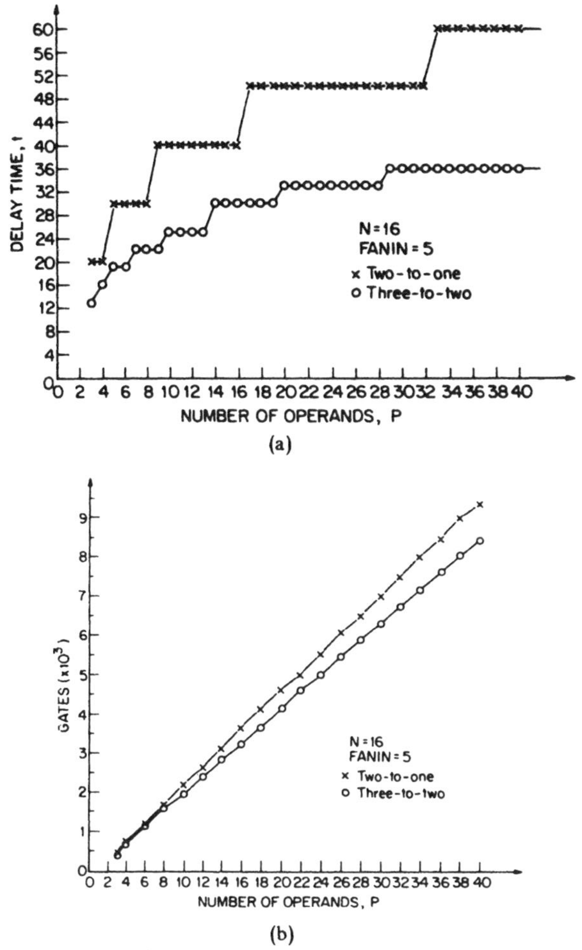

Fig. 13. (a) Operational delay versus number of operands for 2-to-1 and 3-to-2 structures with $N = 16$, FANIN $= 5$. (b) Required number of gates versus number of operands for 2-to-1 and 3-to-2 structures with $N = 16$, FANIN $= 5$.

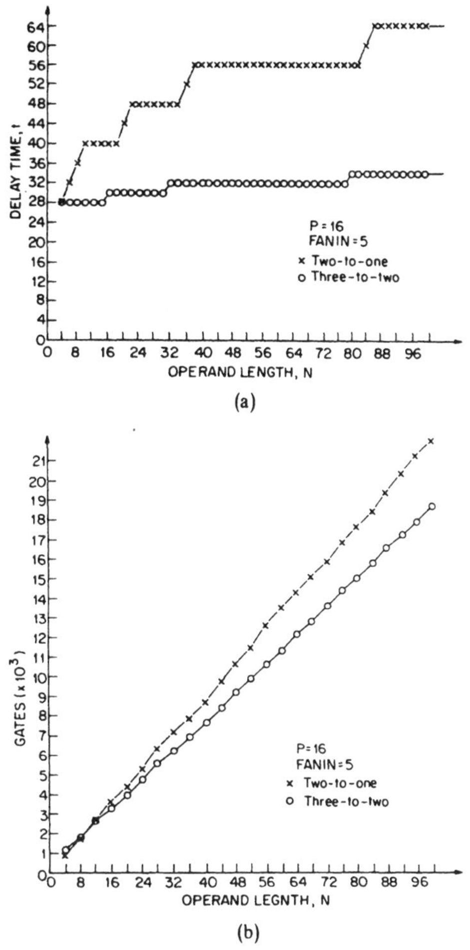

Fig. 14. (a) Operational delay versus operand length for 2-to-1 and 3-to-2 structures with $P = 16$ and FANIN $= 5$. (b) Required number of gates versus operand length for 2-to-1 and 3-to-2 structures with $P = 16$ and FANIN $= 5$.

where L and S are the number of CL and CSA units per package with PIN external connections. Two pins are assumed to be used for power and ground. Fig. 15 is an APL function to calculate the number of packages required for both the CLA and CSA based schemes under restrictions of gate fanin and package pin-outs.

Fig. 16 is a graph of this function for $N = 32$, $W = 4$, and various values of PIN and P. As a practical matter the difference in chip count for the two structures is small; the maximum relative difference is 20 percent; the average relative difference is 7 percent. Recall that we have only modified the component complexity criteria; the CSA structure will always be faster than the CLA structure for $N > 14$.

Fig. 17 summaries results from multiple evaluations of the function CHIP. In general the CLA structures do not appear to offer significant advantage over the CSA structures in terms of package count and never offer a lower cost/performance ratio.

TABLE III
Operand Length such that 3-to-2 Scheme is Faster and/or Requires Fewer Gates than 2-to-1 ($4 \le N \le 64$, $3 \le P \le 40$)

FANIN	TIME-DELAY	GATES
3	N>5	N>14
4	N≥4	N≥11
5	N≥5	N≥14
6	N≥6	N≥12
7	N≥7	N≥8
8	N≥8	N≥9
9	N≥9	N≥10

TABLE IV
Number of Operands such that 3-to-2 Scheme is Faster and/or Requires Fewer Gates than 2-to-1 ($4 \le N \le 64$, $3 \le P \le 40$)

FANIN	TIME-DELAY	GATES
3	P>9	-
4	P≥3	-
5	P≥5	-
6	P≥9	-
7	P≥17	P>17
8	P≥33	P≥8
5	P≥33	P≥9

"-" denotes that the 3-to-2 case may not require less gates.

```
      ∇ Z←CHIP P;W;D1;D2;PP;X;Y;I;J;C21;C32
[1]     ⍝ THE NUMBER OF LSI/MSI CHIPS FOR MULTIPLE OPERAND ADDITION
[2]     ⍝    OF TWO-TO-ONE AND THREE-TO-TWO TREE STRUCTURE
[3]     ⍝ WORD LENGTH N, FANIN, PIN ARE DEFINED EXTERNALLY
[4]     W←FANIN-1
[5]     D1←⌊(PIN-2)÷3+3×W
[6]     →(0=D1)/22
[7]     D2←⌊(PIN-2)÷2+3×W
[8]     PP←P
[9]     X←Y←I←0
[10]    J←⌊PP÷2
[11]    X←X+J×(UOCL N+I)-⌊(N+I)÷W
[12]    Y←Y+J×⌊(N+I)÷W
[13]    I←I+1
[14]    →(1<PP←J+PP-J×2)/10
[15]    C21←(⌈Y÷D1)+⌈X÷D2
[16]    X←N+LEVEL32 P
[17]    Y←UOCL X
[18]    C32←(⌈(CSA32 P)÷⌊(PIN-2)÷5)+(⌈(Y-⌊X÷W)÷D2)+⌈(⌊X÷W)÷D1
[19]    'CHIP21:  ';C21
[20]    'CHIP32:  ';C32
[21]    →0
[22]    'FAN IN IS TOO LARGE'
      ∇
```

Fig. 15. Listing of APL function CHIP.

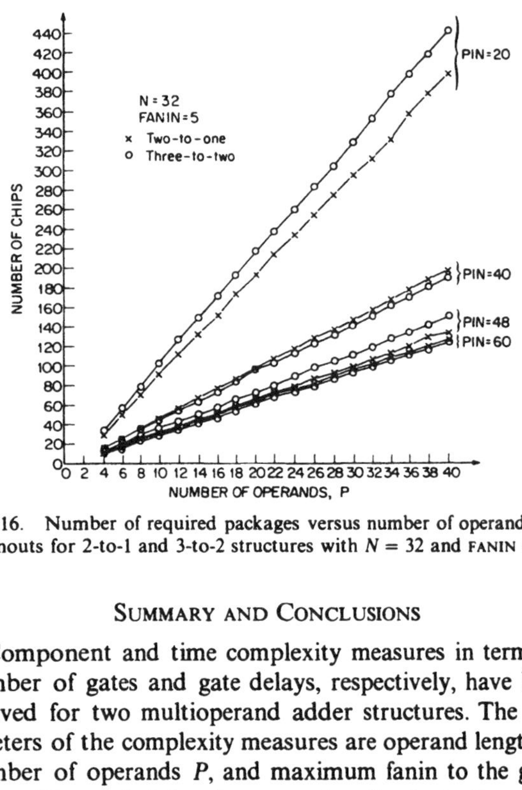

Fig. 16. Number of required packages versus number of operands and pinouts for 2-to-1 and 3-to-2 structures with $N = 32$ and FANIN = 5.

OPERAND LENGTH	FANIN 3	4	5	6	7	8	9	
8	L	L	L	L	X	X	X	
16	L	S	L	L	X	X	X	PIN=20
32	S	S	L	L	X	X	X	
64	S	S	L	L	X	X	X	
8	L	L	L	L	?	L	L	
16	?	L	L	L	S	L	L	PIN=40
32	S	?	S	L	S	S	?	
64	S	S	S	L	S	S	L	
8	L	L	L	?	L	?	L	
15	S	?	L	L	L	S	?	PIN=48
32	S	S	L	L	L	S	S	
64	S	S	L	S	L	S	S	
8	L	L	L	L	L	L	L	
16	?	L	L	L	L	L	L	PIN=60
32	S	S	S	L	S	L	L	
64	S	S	S	L	S	L	L	

L denotes that the carry-Lookahead scheme requires fewer chips.
S denotes that the carry-Save scheme requires fewer chips.
? denotes that the relative costs cross over depending upon P.
X indicates that the gate fanin is limited by number of package connections.

Fig. 17. Comparison of package counts for 3-to-2 versus 2-to-1 schemes under fanin and package connection constraints.

Summary and Conclusions

Component and time complexity measures in terms of number of gates and gate delays, respectively, have been derived for two multioperand adder structures. The parameters of the complexity measures are operand length N, number of operands P, and maximum fanin to the gates FANIN. The analysis involves a constructive derivation of the complexity of carry-lookahead networks including optimization by removal of redundant paths. An alternate measure of component complexity, number of packages with a given number of pinouts, rather than number of gates, is also presented.

Under the assumptions made, the general conclusions are as follows:

1) For FANIN in the range 3–9, operand lengths of 14 bits or greater, and any number of operands P, a carry-save adder tree exhibits less delay and requires fewer gates than a tree of carry-lookahead adders, even with "optimized" carry-lookahead networks.

2) For operand lengths less than 14 bits but greater than 8 bits, the carry-save adder tree exhibits less delay but may require a few more gates.

3) Depending upon the number of pinouts, the carry-save adder tree may require up to 20 percent (average of 7 percent) more packages than the carry-lookahead structure but it will always be faster under the conditions stated in 1.

References

[1] E. E. Swartzlander, Jr., B. K. Gilbert, and I. S. Reed, "Inner product computers," *IEEE Trans. Comput.*, vol. C-27, pp. 21–31, Jan. 1978.

[2] B. K. Gilbert *et al.*, "Ultra high speed transaxial image reconstruction of the heart, lungs, and circulation via numerical approximation methods and optimized processor architecture," *Comput. Biomed. Res.*, vol. 12, pp. 17–30, Feb. 1979.

[3] O. L. MacSorley, "High speed arithmetic in binary computers," *Proc. IRE*, vol. 49, pp. 67–91, Jan. 1961.

[4] H. L. Garner, "Number systems and arithmetic," in *Advances in Computers*, vol. 6, F. L. Alt and M. Rubinoff, Eds. New York: Academic, 1965, pp. 131–194.

[5] A. J. Atrubin, "A one-dimensional real-time iterative multiplier," *IEEE Trans. Electron. Comput.*, vol. EC-14, pp. 394–399, June 1965.

[6] A. Svoboda, "Adder with distributed control," *IEEE Trans. Comput.*, vol. C-19, pp. 749–751, Aug. 1970.

[7] N. G. Kingsbury, "High speed binary multiplier," *Electron. Lett.*, vol. 7, pp. 277–278, May 20, 1971.

[8] R. O. Berg and L. C. Kinney, "Serial adders with overflow correction," *IEEE Trans. Comput.*, vol. C-20, pp. 668–671, June 1971.

[9] D. J. Kuck, *The Structure of Computers and Computations*, vol. I. New York: Wiley, 1978.

[10] C. S. Wallace, "A suggestion for a fast multiplier," *IEEE Trans. Electron. Comput.*, vol. EC-13, pp. 14–17, Feb. 1964.
[11] L. Dadda, "Some schemes for parallel multipliers," *Alta Freq.*, vol. 34, pp. 349–356, May 1965.
[12] L. Dadda and D. Ferrari, "Digital multipliers: A unified approach," *Alta Freq.*, vol. 37, pp. 1079–1086, Nov. 1968.
[13] A. Habibi and P. A. Wintz, "Fast multipliers," *IEEE Trans. Comput.*, vol. C-19, pp. 153–157, Feb. 1970.
[14] J. C. Majithia and R. Kitai, "An iterative array for multiplication of signed binary numbers," *IEEE Trans. Comput.*, vol. C-20, pp. 214–216, Feb. 1971.
[15] S. D. Pezaris, "A 40-ns 17-bit by 17-bit array multiplier," *IEEE Trans. Comput.*, vol. C-20, pp. 442–447, Apr. 1971.
[16] J. A. Gibson and R. W. Gibbard, "Synthesis and comparison of two's complement parallel multipliers," *IEEE Trans. Comput.*, vol. C-24, pp. 1020–1027, Oct. 1975.
[17] N. Kouvaras, D. Lagoyannis, and L. Ponticopoulos, "A digital system of simultaneous addition of several binary numbers," *IEEE Trans. Comput.*, vol. C-17, pp. 992–997, Oct. 1968.
[18] I. T. Ho and T. C. Chen, "Multiple addition by residue threshold functions and their representation by array logic," *IEEE Trans. Comput.*, vol. C-22, pp. 762–767, Aug. 1973.
[19] E. E. Swartzlander, Jr., "Parallel counters," *IEEE Trans. Comput.*, vol. C-22, pp. 1021–1024, Nov. 1973.
[20] L. Dadda, "On parallel digital multipliers," *Alta Freq.*, vol. 45, pp. 574–580, Oct. 1976.
[21] W. J. Stenzel, W. J. Kubitz, and G. H. Garcia, "A compact high-speed parallel multiplication scheme," *IEEE Trans. Comput.*, vol. C-26, pp. 948–957, Oct. 1977.
[22] S. Singh and R. Waxman, "Multiple operand addition and multiplication," *IEEE Trans. Comput.*, vol. C-22, pp. 113–120, Feb. 1973.
[23] E. E. Swartzlander, Jr., "The quasi-serial multiplier," *IEEE Trans. Comput.*, vol. C-22, pp. 317–321, Apr. 1973.
[24] L. Dadda, "Multiple addition of binary serial numbers," in *Proc. 4th Symp. Computer Arithmetic*, IEEE Computer Society Cat. 78CH1412-6C, pp. 140–148, Oct. 1978.
[25] D. E. Atkins, "A suggested approach to computer arithmetic for designers of multi-valued logic processors," in *Proc. 8th Int. Symp. Multi-Valued Logic*, IEEE 78CH1366-4C, May 1978.
[26] L. Gilman and A. Rose, *APL 360, An Interactive Approach.* New York: Wiley, 1970.
[27] S. Pakin, *APL 360 Reference Manual*, 2nd ed. Chicago, IL: SRA, 1972.
[28] F. J. Hill and G. R. Peterson, *Digital Systems: Hardware Organization and Design.* New York: Wiley, 1973, ch. 11.
[29] G. A. Blaauw, *Digital System Implementation.* Englewood Cliffs, NJ: Prentice-Hall, 1976, ch. 2.
[30] Texas Instruments Inc., *The TTL Data Book for Design Engineers*, pp. 7-(271–286), 1976.
[31] J. E. Robertson, "A deterministic procedure for the design of carry-save adders and borrow-save subtracters," Dep. Comput. Sci. Rep. 235, Univ. of Illinois, Urbana, IL, July 1967.

Daniel E. Atkins was born in Jacksonville, FL on April 12, 1943. He received the B.S. degree in electrical engineering from Bucknell University, Lewisburg, PA in 1965, the M.S. degree in electrical engineering, and the Ph.D. degree in computer science from the University of Illinois, Urbana, in 1967 and 1970, respectively.

Between 1963 and 1967 he held positions with the Freas-Rooke Computing Center, Bucknell University, and the U.S. Naval Ordnance Laboratory, Silver Spring, MD. While attending the University of Illinois he participated in design and implementation of the Illinois Pattern Recognition Computer (Illiac III). He conducted research in the area of computer arithmetic and in the use of register-transfer design languages. He has held position as Visiting Assistant Professor at the University of Maryland and Assistant Professor of Computer Science, Bucknell University, Lewisburg, PA. While at Bucknell he directed the design and implementation of an on-line library circulation control system, and participated in the design of an interactive, student information management system. He taught in the areas of data structures, digital system design, and applications programming. He is currently an Associate Professor of Electrical and Computer Engineering at The University of Michigan, Ann Arbor, where he is teaching and conducting research in the area of computer architecture and digital system design, particularly high performance, application-directed structures. He is a member of the Systems Engineering Laboratory and a faculty participant in the graduate program in Computer, Information and Control Engineering (CICE). He is chairman of the Technical Committee of Computer Architecture of the IEEE and is program committee chairman for the 7th International Symposium on Computer Architecture. He has helped organize several symposia on computer arithmetic. He is also a consultant to several firms.

Dr. Atkins is a member of the Association for Computing Machinery, Sigma Xi, and Tau Beta Pi, and was recently awarded a University Distinguished Service Award.

Shau-Chi Ong (S'78) was born in Hong Kong, on March 19, 1952. He received the B.S. degree in electronic engineering in 1974 from the National Chiao Tung University, Taiwan, Republic of China. He is currently working towards a Ph.D. degree in computer, information and control engineering at the University of Michigan, Ann Arbor.

Since 1977 he has been a Research Assistant in the Systems Engineering Laboratory, Department of Electrical and Computer Engineering. His research interests include computer arithmetic and computer architecture. He is helping to organize the 5th Symposium of Computer Arithmetic scheduled for May 1981.

Mr. Ong is a member of Tau Beta Pi and Association for Computing Machinery.

SOME OPTIMAL SCHEMES FOR ALU IMPLEMENTATION IN VLSI TECHNOLOGY

Vojin G. Oklobdzija and Earl R. Barnes

IBM T.J.Watson Research Center
P.O.Box 218
Yorktown Heights, NY 10598

ABSTRACT

An efficient scheme for carry propagation in an ALU implemented in n-MOS technology is presented. An algorithm that determines the optimum division of the carry chain of a parallel adder for various data path sizes is developed. This yields an implementation of a fast ALU which due to its regular structure occupies a modest amount of silicon. The speed of the implementation described is comparable to the carry look-ahead scheme. Our method is based on the optimization of the carry path implemented in n-MOS technology but the results can be applied to other technologies.

1. INTRODUCTION

An efficient implementation of an ALU in VLSI technology depends on many parameters. We consider an *efficient* implementation to be one which is fast, of a small and regular area, and low power. In many VLSI designs achieving the ultimate speed is not always important goal especially if this is achieved by consuming excessive area and power. Therefore Carry-Lookahead (CLA) scheme is not very attractive for VLSI implementations where area, regularity of structure and power are important. Also the determining factor, in case of an ALU, is wheather it is part of a critical path or not.

In this paper we considered Carry-Skip (CSA) scheme [1],[2],[3],[4] because it met our objective of reasonable performance achieved with a relatively small and regular area. This adder is in essence a Carry Lookahead for which the carry-generate portion which consumes a large amount of logic, has been eliminated. As in a Carry Lookahead adder the bits to be added are divided into groups. A circuit is provided for detecting when a carry signal entering a group will ripple through the group. When this condition is detected, the carry is allowed to skip over the group. Carry-Skip Adder (CSA) does not require excessive amount of logic (area) and the "skip" portion of the logic can be added to the existing carry chain of Ripple-Carry Adder (RCA), therefore not disturbing the inherently regular bit-slice implementation of an RCA (Fig.1.).

The power requirements of a CSA are considerably lower than that of a CLA-ALU. In this paper we show how the carry chain in a CSA can be optimized to yield better speed which in the case of a multi-level optimized CSA-ALU can even outperform the speed of a CLA divided into the groups of constant size.

Lehman and Burla [3] studied a design of a CSA and suggested varying the size of the groups. By varying the sizes of the groups one can influence the maximum delay a carry signal can experience in propagating through the adder. Lehman and Burla [3] posed the problem of determining the optimal group sizes for minimizing the maximum delay [9]. They gave a heuristic method for obtaining economical group sizes. However, they did not solve the problem of determining optimal group sizes. For example, for a 48 bit adder they gave the group sizes 4 5 6 7 8 8 7 6 5 4 yielding the maximum delay of 14 [3]. We show that, under the same assumptions an optimal subdivision results in a delay of 12.

They also discussed the problem of achieving even faster addition by allowing carry signals to skip over blocks of groups. The problem of optimizing the carry chain is now complicated by having to choose both the optimal number of blocks and the optimal sizes of groups within blocks. Some rules for choosing economical block and group sizes are given [3]. However, the problem of determining the optimal sizes remains unsolved.

In this paper we consider the problem of designing a carry-skip adder in FET technology and give some optimal solutions. Actually, our solutions are more general in that we generally assume

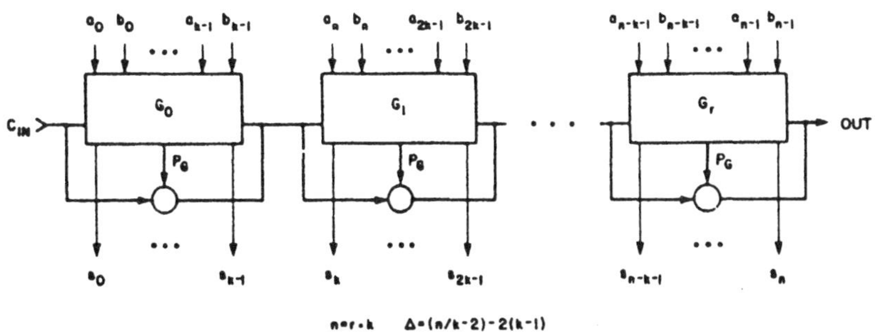

Fig.1. Carry-Skip Adder

that the time required for a carry signal to skip over a group of bits is longer than the time required for the carry to ripple through a single bit. This assumption is relevant for adders designed in n-MOS technology. Lehman and Burla assumed their adder to be designed using discrete components where these two times are equal. Our analysis will include their problem as a special case.

2. DIVIDING THE ADDER INTO GROUPS

Let n denote the number of bits in a carry skip adder and let m denote the number of groups into which the bits are divided. Let $x_1, \ldots, x_m$ denote the sizes of the groups beginning with the most significant bit. Let T denote the time required for a carry signal to skip over a group of bits. To be precise we should write $T = T(x)$ to indicate that T depends on the size x of the group over which the carry is skipped. However, T changes very slowly with x over the range of group sizes that concern us. So we assume that T is constant.
For a given n, the following three-step procedure gives an optimal way of dividing an n bit adder into groups of bits.

Procedure 2:

2(i) Let m be the smallest positive integer such that

$$n \le m + \frac{1}{2}mT + \frac{1}{4}m^2T + (1 - (-1)^m)\frac{T}{8}. \qquad (1)$$

2(ii) Let

$$y_i = \min\{1 + iT,\ 1 + (m + 1 - i)T\}, \quad i = 1, \ldots, m$$

and construct a histogram whose i-th column has height y_i. For example, for $T = 3$ and $n = 48$, we have $m = 7$ and the following histogram.

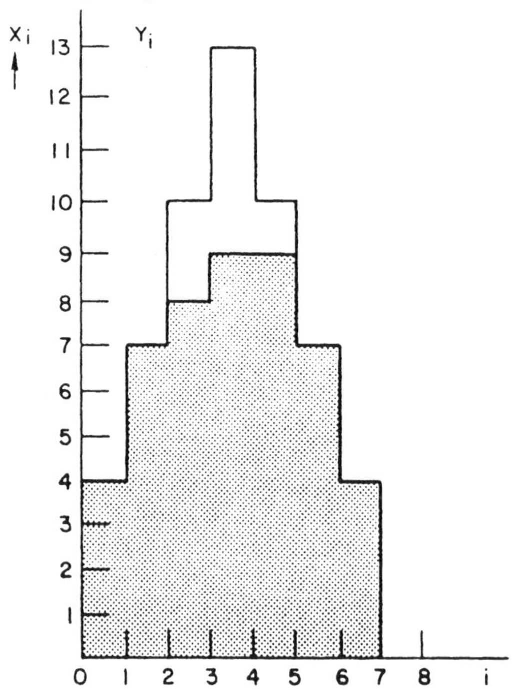

Fig.2. Histogram of segments y_i, x_i

2(iii) It is easily verified that the area of the histogram in (ii) is

$$m + \frac{1}{2}mT + \frac{1}{4}m^2T + (1 - (-1)^m)\frac{T}{8} \ge n$$

so these are at least n unit squares in the histogram. Starting with the first row, shade in n of the squares, row by row. Let x_i denote the number of shaded squares in column i of the histogram, $i = 1, \ldots, m$.

Then $x_1, \ldots, x_m$ is an optimal division of the adder. The maximum delay of a carry signal is mT.

Example 1. For a 48 bit adder we have, from Figure 2. $x_1 = x_7 = 4$, $x_2 = x_6 = 7$, $x_3 = 8$ and $x_4 = x_5 = 9$.

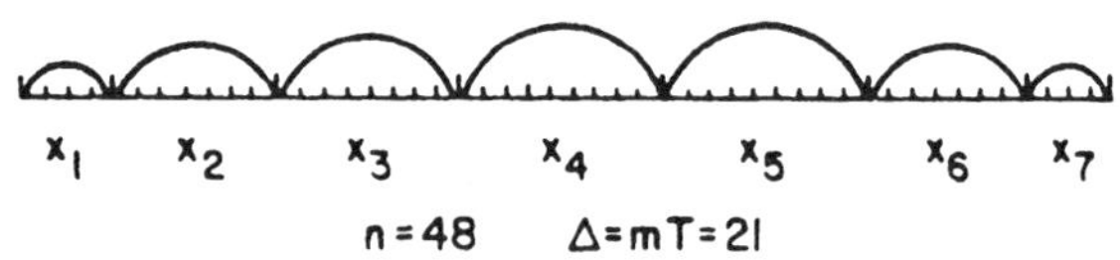

Fig.3. Carry chain of a 48-bit adder: n=48, mT=21

The maximum delay is experienced by a signal generated in the second bit position and terminating in the 47th bit position. The delay is $mT = 21$.

Example 2. Consider a 54 bit adder. From 2(i) we see that again $m = 7$. If we shade 54 squares in Figure 2, we see that

$$x_1 = x_7 = 4,\ x_2 = x_6 = 7,\ x_3 = x_5 = 10 \text{ and } x_4 = 12$$

yields an optimal division of the adder. Again the maximum delay is $mT = 21$.

Example 3. Consider a 64 bit adder. From 2(i) we compute $m = 8$. The corresponding histogram is shown in Figure 4. The optimal group sizes are:

$$x_1 = x_8 = 4,\ x_2 = x_7 = 7,\ x_3 = x_6 = 10,\ x_4 = x_5 = 11.$$

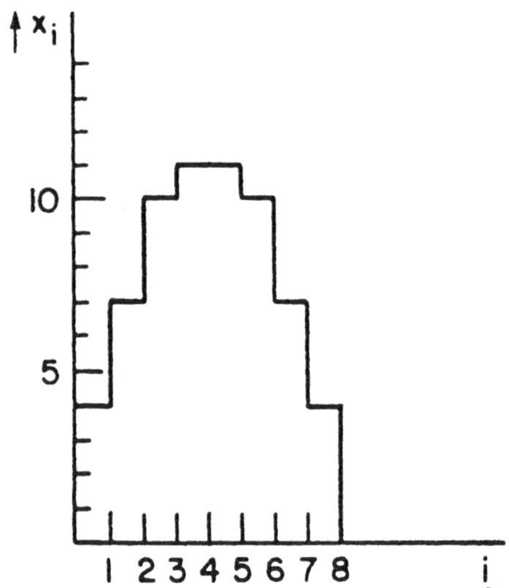

Fig.4. Segment histogram for a 64-bit adder: m=8

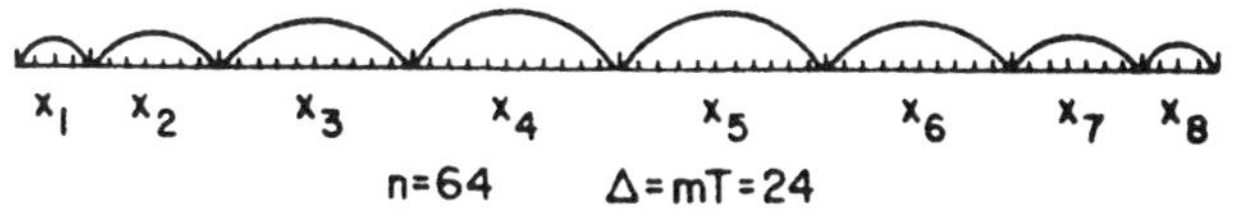

Fig.5. Carry chain of a 64-bit adder: m=8, mT=24

The delay of the longest signal is $mT = 24$.

Proof of Optimality We are going to prove optimality of the division of the carry chain described in 2(i)-(iii). First we need a lemma.

Lemma 1. *When the bits of a carry skip adder are grouped according to the scheme (i)-(iii), the maximum propagation time of a carry signal is* mT.

Proof. The carry generated at the 2nd bit position and terminating at the (n-1) clearly has propagation time mT. We must show that any other signal has propagation time smaller than or equal to mT. Consider a signal originating in the ith group and terminating in the jth, $i < j$. Denote its propagation time by P. Clearly

$$P \le (x_i - 1) + (j - i - 1)T + (x_j - 1).$$

By construction, $x_i \le \min\{1 + iT,\ 1 + (m + 1 - i)T\}$ for each i so

$$\begin{aligned} P \le\ & \min\{1 + iT,\ 1 + (m + 1 - i)T\} \\ & + \min\{1 + jT,\ 1 + (m + 1 - j)T\} + (j - i - 1)T - 2. \end{aligned}$$

There are three cases to consider.

1. First assume

$$\min\{1 + iT,\ 1 + (m + 1 - i)T\} = 1 + iT$$

and

$$\min\{1 + jT,\ 1 + (m + 1 - j)T\} = 1 + jT.$$

In this case $1 + jT \le 1 + (m + 1 - j)T \Rightarrow 2jT - T \le mT$. It follows that

$$\begin{aligned} P &\le 1 + iT + 1 + jT + (j - i - 1)T - 2 \\ &= 2jT - T \le mT. \end{aligned}$$

2. Now assume that

$$\min\{1 + iT,\ 1 + (m + 1 - i)T\} = 1 + iT$$

and

$$\min\{1 + jT,\ 1 + (m + 1 - j)T\} = 1 + (m + 1 - j)T.$$

In this case we have

$$P \le 1 + iT + 1 + (m + 1 - j)T + (j - i - 1)T - 2 = mT.$$

3. Finally, assume that

$$\min\{1 + iT,\ 1 + (m + 1 - i)T\} = 1 + (m + 1 - i)T$$

and

$$\min\{1 + jT,\ 1 + (m + 1 - j)T\} = 1 + (m + 1 - j)T.$$

It follows that

$$\begin{aligned} P &\le 1 + (m + 1 - i)T + 1 + (m + 1 - j)T + (j - i - 1)T - 2 \\ &= 2mT - (2iT - T) \le 2mT - mT = mT. \end{aligned}$$

This completes the proof of the lemma.

Lemma 2. *Let* Δ *denote the maximum delay of a carry signal in a n bit carry skip adder with group sizes chosen optimally.* Then

$$(m - 1)T \le \Delta \le mT.$$

Proof. Since we have exhibited a division of the carry chain into groups in such a way that the maximum delay of a carry signal is mT we clearly have $\Delta \le mT$.

Let $x_1, x_2, \ldots, x_r$ denote the optimal group sizes corresponding to Δ. For the moment assume that $r = 2k$ is even. By considering carries originating in group i and terminating in group $r - i + 1$, $i = 1, \ldots, k$, we deduce the following system of inequalities.

$$\begin{array}{lllll} (x_1 - 1) & + & (r - 2)T & + & (x_r - 1) \le \Delta \\ (x_2 - 1) & + & (r - 4)T & + & (x_{r-1} - 1) \le \Delta \\ \vdots & & & & \vdots \\ (x_k - 1) & + & (r - 2k)T & + & (x_{k+1} - 1) \le \Delta \\ & & rT & & \le \Delta \end{array}$$

If we add these inequalities and use the fact that $\sum_{i=1}^{r} x_i = n$, we obtain the inequality

$$n - 2k + (k + 1)rT - k(k + 1)T \le (k + 1)\Delta$$

which simplifies to

$$\frac{n - 2k}{k + 1} + kT \le \Delta.$$

The left hand side of this inequality assumes its minimum value at

$$k + 1 = \frac{\sqrt{n + 2}}{T}.$$

It follows that

$$\Delta \ge -(T + 2) + \sqrt{4nT + 8T}. \tag{2}$$

Assume now that $r = 2k + 1$ is odd. We then have the system of inequalities

$$\begin{array}{lllll} (x_1 - 1) & + & (r - 2)T & + & (x_r - 1) \le \Delta \\ (x_2 - 1) & + & (r - 4)T & + & (x_{r-1} - 1) \le \Delta \\ \vdots & & & & \vdots \\ (x_k - 1) & + & (r - 2k)T & + & (x_{k+1} - 1) \le \Delta \\ & & (x_{k+1} - 1) & + & kT \le \Delta, \end{array}$$

which implies that

$$\frac{n - 2k - 1}{k + 1} + kT \le \Delta.$$

By minimizing the left hand side of this inequality with respect to k we find that

$$\Delta \ge -(T + 2) + \sqrt{4nT + 4T}. \tag{3}$$

By comparing this with (2) we see that (3) holds in all cases.

We will now produce an upper bound on mT. Since m is the smallest positive integer satisfying (1) we have

$$n > (m - 1) + \frac{1}{2}(m - 1)T + \frac{1}{4}(m - 1)^2 T + (1 - (-1)^{m-1})\frac{T}{8}.$$

Since each size of this inequality is an integer, we can increase the right hand side by 1 to obtain

$$n \geq m - \frac{1}{4}T + \frac{1}{4}m^2T + (1 - (-1)^{m-1})\frac{T}{8}.$$

Solving this inequality for mT gives

$$mT \leq -2 + \sqrt{4nT + 4 + T^2 - (1 - (-1)^{m-1})\frac{T^2}{2}}. \quad (4)$$

Combining this with (2) and (3) gives

$$\Delta \leq (x_i - 1) + sT + (x_j - 1) \\ \leq x_i + x_j - 2 + (m - 3)T$$

So that either $x_i \geq T + 1$ or $x_j \geq T + 1$. This means that we can subdivide one of the groups i,j without increasing Δ. Continuing in this way, we can always increase the number r of group in an optimal division of a carry chain by 1 without increasing Δ if $r < m$. This means that we can arrive at an optimal division of the carry chain into m groups. We must then have $\Delta \geq mT$ which, together with Lemma 2, implies $\Delta = mT$. This completes the proof of the theorem.

$$mT - \Delta \leq \begin{cases} T + \dfrac{T^2 - 8T + 4 - (1 - (-1)^{m-1})\frac{T^2}{2}}{\sqrt{4nT + 8T} + \sqrt{4nT + 4 + T^2 - (1 - (-1)^{m-1})\frac{T^2}{2}}}, & r\text{-even} \\ T + \dfrac{T^2 - 4T + 4 - (1 - (-1)^{m-1})\frac{T^2}{2}}{\sqrt{4nT + 4T} + \sqrt{4nT + 4 + T^2 - (1 - (-1)^{m-1})\frac{T^2}{2}}}, & r\text{-odd} \end{cases} \quad (5)$$

For n sufficiently large we have $mT - \Delta < T + 1$ and since $mT - \Delta$ is an integer, $mT - \Delta \leq T$. This completes the proof of the lemma.

Theorem 1 *The scheme 2(i)-2(iii) given above for dividing the bits of a carry skip adder into groups is optimal for $2 \leq T \leq 7$.*

Proof. Assume the scheme is not optimal and let Δ be the maximum delay corresponding to an optimal division of the bits into groups. Assume there are r groups in the optimal division. Since a carry in signal to the least significant bit group can skip over each group we have $rT \leq \Delta \leq mT$ so $r \leq m$. If $r = m$ then $\Delta = mT$ and the theorem holds by Lemma 1. If $r = m - 1$ m and r have different parities and it follows from (5) that $mT - \Delta < T$ for $2 \leq T \leq 7$ so that $\Delta > (m - 1)T = rT$. This means that a signal which skips over each of the r groups has delay less than the maximum Δ. Similarly, if $r < m - 1$, $\Delta \geq (m - 1)T > rT$ so that a signal which skips over each group has delay $< \Delta$. It follows that a signal with delay Δ must start in a group i, ripple to the end of this group, then skip over $s < r$ groups and either terminate, or ripple through the first few bits of a group $j > i$. Let x_i and x_j denote the lengths of the ith and jth groups respectively. Assume that i is chosen as small as possible and j as large as possible. A signal originating in group i, rippling to the end of this group and then skipping over the next s group has delay

$$\Delta \leq (x_i - 1) + sT \leq (x_i - 1) + (r - 1)T \\ \leq (x_i - 1) + (m - 2)T.$$

Since $\Delta \geq (m - 1)T$ this implies that $x_i \geq T + 1$. Divide group i into two groups such that the group containing the most significant bits has size T. Since the i-th group is the first group in which a signal having maximum delay can originate, this subdivision does not increase the delay of any carry signal of maximum delay. However, it increases the number of groups by 1.

Suppose now that a carry signal originates in a group i, ripples to its end, skips over $s \leq r - 2$ groups and finally ripples through the first few bits of a group j and terminates. We then have

3. DIVIDING GROUPS INTO BLOCKS

It is clear that the maximum delay of a carry signal in a carry skip adder can be further reduced if signals are allowed to skip over blocks of groups. We define a block to be an additional path allowing carry signal to skip directly over groups. In this section we will describe an efficient scheme for dividing the carry chain into blocks of groups. We assume that the time required for a carry signal to skip over a block of groups is T_b. Actually, the time required for a carry to skip over a block T_b is slightly longer than the time T_g required to skip over a group. But for the sake of simplifying the analysis we will assume these two times to be equal i.e. $T_b = T_g$. However, our technique extends to the case where $T_g \neq T_b$.

Let M denote the number of blocks into which the groups of bits are divided. Let Δ denote the maximum delay a carry signal can have in an adder divided into M blocks. Clearly, $\Delta \geq MT$. We will show how to choose the blocks such that $\Delta = MT$. We will also show how to choose M for an adder of length n.

Our blocks are chosen in such a way that the maximum delay of a signal originating and terminating in block i and $M + 1 - i$ is iT.

Consider a signal originating in the first of these blocks and terminating in the second. Such a signal will skip over $M - 2i$ blocks and will accordingly have delay $\leq (iT) + (M - 2i)T + iT = MT$ as desired. It follows from our work in Section 2 that in order for a signal originating and terminating in block i to have delay less or equal iT we must choose the length of the ith and $(M + 1 - i)th$ blocks to be less or equal the number of unit squares in a histogram with base of width i. Thus the maximum length of the ith and $(M + 1 - i)th$ blocks must be

$$i + \frac{1}{2}iT + \frac{1}{4}i^2T + (1 - (-1)^i)\frac{T}{8}, \quad i \leq \lceil \frac{M}{2} \rceil.$$

(Here we use the symbol $\lceil I \rceil$ to denote the smallest integer $\geq I$.) It follows that the maximum length of an adder divided into M blocks must be

$$2\sum_{i=1}^{\lceil \frac{M-1}{2}\rceil}\left\{i+\frac{1}{2}iT+\frac{1}{4}i^2T+(1-(-1)^i)\frac{T}{8}\right\} \quad (3.1)$$
$$+\frac{(1-(-1)^M)}{2}\left\{\lceil\frac{M}{2}\rceil+\frac{1}{2}\lceil\frac{M}{2}\rceil T+\frac{1}{4}\lceil\frac{M}{2}\rceil^2 T + (1-(-1)^{\lceil M/2\rceil})\frac{T}{8}\right\}.$$

Thus for a given adder length n, we choose M to be the smallest positive integer such that the expression (3.1) exceeds or equals n. M is then the number of blocks into which our adder must be divided. The formal statement of our algorithm is as follows.

3(i) Choose M to be the smallest positive integer such that

$$n \le 2\sum_{i=1}^{\lceil \frac{M-1}{2}\rceil}\left\{i+\frac{1}{2}iT+\frac{1}{4}i^2T+(1-(-1)^i)\frac{T}{8}\right\}$$
$$+\frac{(1-(-1)^M)}{2}\left\{\lceil\frac{M}{2}\rceil+\frac{1}{2}\lceil\frac{M}{2}\rceil T+\frac{1}{4}\lceil\frac{M}{2}\rceil^2 T + (1-(-1)^{\lceil M/2\rceil})\frac{T}{8}\right\}.$$

3(ii) Form M blocks labeled 1, 2, ..., M, with blocks i and $M+1-i$ each containing

$$i+\frac{1}{2}iT+\frac{1}{4}i^2T+(1-(-1)^i)\frac{T}{8}$$

bits, $i \le \lceil \frac{M}{2} \rceil$.

This construction is analogous to the construction of the histogram in 2(ii). If necessary, delete bits from the largest blocks in this chain until a total of exactly n bits remain in the M blocks.

3(iii) Treat each of the final blocks in 3(ii) as a complete carry chain and divide it into groups optimally using the algorithms 2(i)-2(iii).

Example 3.1. Consider a 32-bit adder. For $i = 1, 2, 3, \ldots$, and $T = 3$ the numbers

$$i+\frac{1}{2}iT+\frac{1}{4}i^2T+(1-(-1)^i)\frac{T}{8}$$

take on the values 4, 8, 15, 22, 32,... respectively. Since

$$32 \le 2\{4 + 8\} + 15$$

we must have $M = 5$ blocks in step 3(ii). These blocks have sizes 4, 8, 15, 8, 4 respectively. If we delete 7 units from the middle block we obtain block sizes 4, 8, 8, 8, 4 which add up to 32. Dividing each block into groups by the procedure 2(i)-2(iii) we obtain the following chain where each group has size 4.

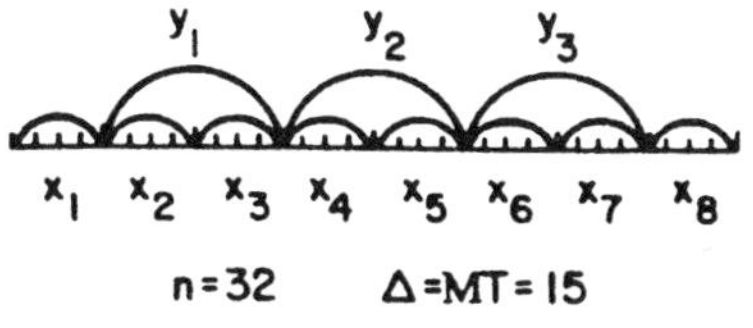

Fig.6. Carry chain of an 32-bit adder: MT=15

The maximum delay of a carry signal is $MT = 15$.

Example 3.2. Consider a 48 bit adder. Again we assume $T = 3$. Since

$$48 \le 2\{4 + 8 + 15\}$$

we must take $M = 6$ corresponding to block sizes 4, 8, 15, 15, 8, 4. The total number of units is 54. So we reduce the size of the two middle blocks by 3 each. This gives block sizes 4, 8, 12, 12, 8, 4 adding up to 48. If we divide each block into groups by the procedure 2(i)-2(iii), each group has size 4. The maximum delay of a carry signal is given by $MT = 18$.

Example 3.3. Consider a 64 bit adder and assume $T = 3$. Since

$$64 \le 2\{4 + 8 + 15\} + 22$$

we take $M = 7$ and start with blocks of sizes 4, 8, 15, 22, 15, 8, 4 respectively. The lengths of these blocks total 76. So we reduce the middle block by 12. The new block sizes are 4, 8, 15, 10, 15, 8, 4. The optimal division of these blocks into groups is given in Figure 7. We could have just

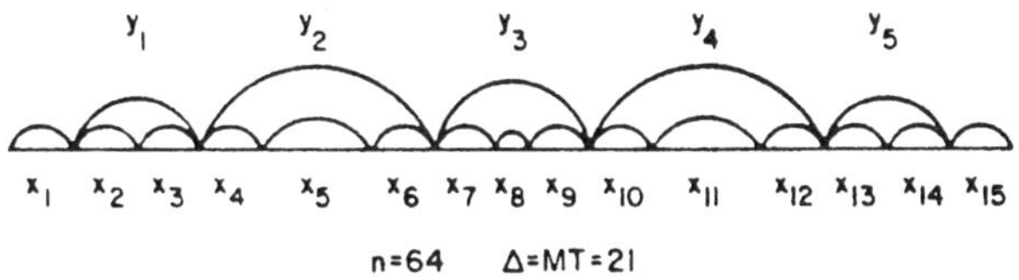

Fig.7. Carry chain of a 64-bit adder: MT = 21

as well reduced the sizes of the three middle blocks obtaining final blocks of sizes 4, 8, 13, 14, 13, 8, 4. The maximum delay of a carry signal would still be $MT = 21$.

4. COMPARISON

In the literature comparison of schemes for ALU implementation is done mainly on the basis of the number of gates, propagation delay per gate and power consumed per gate [5]. However, if a VLSI implementation of a high-speed ALU is considered, these measures are not easily applied. For example, the propagation delay in terms of number of gates is not an adequate measure unless care is taken to implement the function exactly as specified by its logic (gate) representation. This is often not the case since the function is frequently merged into a group of transistors or implemented by using pass-transistors, precharge or other techniques applyed by the circuit designer in order to minimize delay and power.

In general, if the function is implemented in two levels of logic, the delay is not necessarily smaller than the implementation of the same function in three or more logic levels. This is due to the fact that in n-MOS technology the delay is heavily dependent on several factors:

1. *Gate type* : NOR gates are faster than NAND gates.
2. *Fan-in* : for a NAND gate, the delay is directly proportional to the number of inputs, since inside the n-input NAND gate the signal has to propagate through n-transistors. In case of

a NOR gate, delay is not strongly affected by the number of inputs, and therefore the use of NOR gate is preferable.

3. *Fan-out* : the speed of a gate will be different if the fan-out is larger than in the case of small fan-outs.

4. *Wiring* : speed is also dependent on the length of the wires i.e. "wiring capacitance" and the resistance of the long wires.

4.1 Comparison with Carry-Lookahead Scheme

For the purpose of our analysis we consider as a rough measure of delay the number of FET transistors through which the signal needs to propagate in order to reach the destination point. This seems to be satisfactory for measuring delay through a VLSI FET network. In addition we modify the delay equations for the points known to have either a substantial fan-out or considerable capacitive loading due to the long wires carrying signal to the distant points within the VLSI macro block. For example, we assume that the time required for a carry signal to skip over a block of groups is three times the time to ripple through one bit position of the carry-chain. These assumptions were confirmed by simulation performed on a 32-bit ALU which was actually implemented in n-MOS technology.

The Case T = 1.

In order to compare the speed of our carry skip adder with that of the Carry Lookahead Adder (CLA) we take $T = 1$. This modification is made in order to compare our results with the estimates for the CLA adder which are based on the gate delay calculations in which case it is assumed that each gate level introduces a delay $t_G = 1$ regardless of the gate fan-in or fan-out [5]. This assumption works favorably for CLA when calculating its speed. In practice we expect CLA to show worse performance than estimated while the calculation for our CSA schemes should be more accurate.

For our comparison we consider full-CLA with the groups of size $G_s = 4$. CLA adder of size n=32 bits is shown in the Fig.8. with the delays indicated at each signal input and output to or from the block. Delay calculation for CLA of the sizes n = 16, 32, 48, 64 shows delays for the critical path of the carry signal to be Δ = 6, 8, 10, 10 respectively [5].

First we consider our case where the adder is simply divided into groups of bits. The procedure in Section 2 shows that the maximum delay of a carry signal in an n bit adder is m, where m is the smallest integer satisfying

$$n \le m + \frac{1}{2}m + \frac{1}{4}m^2 + (1 - (-1)^m)\frac{1}{8}.$$

The values of a delay m for our method compared with the CLA delays for several values of n are shown in the Table 1.

n	$\Delta(CSA)$	$\Delta(CLA)$
16	6	6
32	9	8
48	12	10
64	14	10

Table 1. Delay comparison with the CLA for the method of Section 2.

Consider now the case where the adder is divided into blocks of groups according to the procedure of Section 3. This algorithm shows that the maximum delay of a carry signal in an n bit carry skip adder is M, the smallest positive integer satisfying

$$n \le 2 \sum_{i=1}^{\lceil \frac{M-1}{2} \rceil} \{i + \frac{1}{2}i + \frac{1}{4}i^2 + (1 - (-1)^i)\frac{1}{8}\}$$
$$+ \frac{(1 - (-1)^M)}{2}\left\{\lceil \frac{M}{2} \rceil + \frac{1}{2}\lceil \frac{M}{2} \rceil + \frac{1}{4}\lceil \frac{M}{2} \rceil^2 + (1 - (-1)^{\lceil M/2 \rceil})\frac{1}{8}\right\}.$$

Recall that $\lceil r \rceil$ is the smallest integer $\ge$ r for any real number r.

Table 2. shows the values of delay M for CSA of Section 3 compared to that of CLA adder corresponding to several values of n.

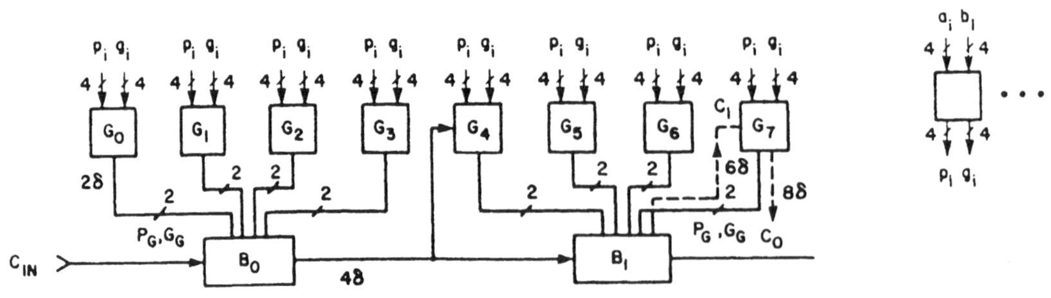

n=32 , k=4 , Δ=8

Fig.8. Carry Signal Delay of a CLA Adder of the size n=32 bits

n	Δ(CSA)	Δ(CLA)
16	5	6
32	7	8
48	9	10
64	10	10

Table 2. Delay comparison between CSA of Section 3. and CLA

In the first case (Table 1.) we notice that CLA is faster than our method of Section 1. resulting in equal delay for n=16 and ending up with 40% advantage for n=64. However, in the second case (Table 2.) our method of Section 3. shows advantage in speed over CLA of approximately 20% for n=16 and no advantage for n=64.

Our scheme exhibits regularity in fan-in and fan-out throughout the entire carry chain and therefore yields simpler and regular implementation. The amount of logic required to implement our scheme should therefore be smaller then that of CLA. Therefore its implementation in VLSI technology should yield even better results.

5. CONCLUSION

Our goal in implementing this scheme was to achieve a regular structure without using an excessive chip area. Comparison of various ALU implementation schemes for area and power as a parameter show that after some point small improvement in speed is achieved by a large investment in area or power [6],[7]. In our approach we argue that this incremental improvement in speed is diminished due to the overhead in the wiring capacitance and device size. Therefore our approach is to maximize the speed not by increasing the power or adding a substantial amount of logic but rather by optimizing on the path of the signal with the critical timing (carry signal) by designing the ALU around the carry path. The method described in Sections 2. and 3. is especially applicable for the floating-point fraction ALU which is usually of large size n.

REFERENCES

[1] T.Kilburn, D.B.G.Edwards and D.Aspinall, *Parallel Addition in Digital Computers: A New Fast "Carry" Circuit*, Proc. IEE, Vol.106, Pt.B. P.464, September 1959.

[2] M.Lehman, *A Comparative Study of Propagation Speed-up Circuits in Binary Arithmetic Units*, Information Processing, 1962, Elsevier-North Holand, Amsterdam, 1963, p.671.

[3] M.Lehman and N.Burla, *Skip Techniques for High-Speed Carry-Propagation in Binary Arithmetic Units*, IRE Transactions on Electronic Computers, December 1961, p.691.

[4] L.P.Morgan and D.B.Jarvis, *Transistor Logic Using Current Switching and Routing Techniques and its Application to a Fast "Carry" Propagation Adder*, Proc. IEE, pt.B, Vol.106, p.467, September 1969.

[5] Kai Hwang , *Computer Arithmetic: Principles Architecture and Design*, John Wiley and Sons, 1979.

[6] S.Ong and D.E.Atkins, *A Comparison of ALU Structures for VLSI Technology*, Proc. of the 6th Symposium on Computer Arithmetic, June 20-22, 1983, Aarhus University, Aarhus, Denmark.

[7] Robert K. Montoye, P.W.Cook, *Automatically Generated Area, Power, and Delay Optimized ALUs*, IEEE ISSCC Digest of Technical Papers, February 23-24, 1983, New York.

[8] A.Bilgory and D.D.Gajski, *Automatic Generation of Cells for Recurrence Structures*, Proc. of 18th Design Automation Conference, Nashville, Tennessee 1981.

[9] V.F.Demyanov, V.N.Malozemov, *Vvedenie v Minimaks*, Izdatelstvo Nauka , Fiziko-Matematicheskoi Literaturi, Moskva 1972.

An area-time efficient NMOS adder *

M.A. Bayoumi, G.A. Jullien and W.C. Miller

Department of Electrical Engineering, University of Windsor, Windsor, Ontario, Canada N9B 3P4

Received 5 July 1983

Abstract. A model of computation for VLSI systems has been developed based on the Mead and Conway approach. This model accommodates the fan-out dependency in NMOS technology. Based on this model, a method for producing area-time efficient carry lookahead adders in NMOS has been developed. This method coordinates between the structural level (cells and interconnections) and the physical layout level (size of individual transistor). The proposed procedure exhibits modularity and regularity. Finally, an example of designing a 4-bit adder is given.

Keywords. Models of computation, fan-out, VLSI, addition, carry lookahead, area-time complexity.

1. Introduction

Adders are considered to be fundamental computational elements in any arithmetic architecture. The major delay in most of those processors is due to the carry propagation in the adder elements. The addition problem has been considered by many authors [1–6]. The tradeoff in those implementations is the time and the number of gates which is proper in MSI and LSI environment. However, fan-in and fan-out problems have not been analyzed.

This paper describes the implementation of a fast adder using NMOS technology. In VLSI implementation, minimizing the chip area and the propagation time delay are the main goals [7,8]. Recently, Brent and Kung [9] have considered the

* The authors would like to acknowledge the Natural Sciences and Engineering Research Council of Canada (NSERC) for their financial support of this research.

North-Holland
INTEGRATION, the VLSI journal 1 (1983) 317–334

layout of an adder based on a regular pattern, that design has considered the interconnection area but not the effect of the various capacitances in the system. In this paper, a model of computation has been developed based on the layout implementation such that it expresses real area and time. It is a more realistic model, since it is a lower level model which considers the capacitive load by wires and interconnections. The model is efficient for any number of bits (not necessarily a power of 2). The layout is a mapping from a functional level to a transistor level rather than mapping from the logical level [10]. This model offers good performance. The layout exhibits regularity and modularity.

Magdy A. Bayoumi is an instructor at the Department of Electrical Engineering, University of Windsor, Windsor, Ontario, Canada. His research interests include digital signal processing architectures, parallel processing and VLSI.

He holds a B.A.Sc. and M.A.Sc. degrees from Cairo University. He obtained an M.A.Sc. from Washington University. He is currently a doctoral candidate.

Mr. Bayoumi teaches courses on microcomputer systems and logic circuits design. He is a member of the IEEE, the ACM and the Special Interest Group in Computer Architecture (SIGARCH).

Graham A. Jullien (M'71) was born in Wolverhampton, England, on June 16, 1943. He received the B.Tech. degree from Loughborough University of Technology, Loughborough, England, in 1965, the M.Sc. degree from the University of Birmingham, Birmingham, England, in 1967 and the Ph.D. degree from Aston University, Birmingham, England in 1969, all in electrical engineering.

From 1961–1966 he worked for English Electric Computers, Kidsgrove, England, first as a Student Apprentice and then as a Data Processing Engineer. From 1967–1969 he was employed as a Research Assistant at Aston University. Since 1969 he has been with the Electrical Engineering Department, the University of Windsor, Windsor, Ontario, Canada, and currently holds the rank of Professor. From 1975 until 1976, he was a Visiting Senior Research Engineer at the Central Research Laboratories of EMI Ltd., Hayes, Middlesex, England. He is currently engaged in research in the areas of digital signal processing and high speed digital hardware. He also teaches courses on electronic circuits, microcomputer systems, and digital signal processing.

William C. Miller (S'57–M'62) was born in Toronto, Ontario, Canada, 1937. He received the B.S.E. degree in Electrical Engineering from the University of Michigan, Ann Arbor in 1960 and the M.A.Sc. and Ph.D. degrees in electrical engineering from the University of Waterloo, Waterloo, Ontario, Canada in 1961 and 1968, respectively.

He joined the Department of Electrical Engineering, the University of Windsor, in 1968 and is presently a Professor and Head of the Department. His research interests are oriented towards digital signal processing and microprocessor applications. He also teaches courses in the area of circuit theory and digital signal processing.

Dr. Miller is a registered Professional Engineer in the Province of Ontario.

The proposed implementation is based on the carry lookahead addition (CLA) algorithm. The CLA algorithm has been considered by many authors [4,5,9,11]. Various descriptions of this algorithm have been given in all of the above references. CLA speeds up the carry propagation process. In the next section, definitions and equations that are necessary for this paper are presented.

2. CLA algorithm

The CLA is comprised of three main functional modules: carry generate/propagate, carry lookahead and summation (see Fig. 1). Let X and Y be two unsigned n-bit binary digits vectors, where

$$X = x_{n-1}x_{n-2}\cdots x_1x_0, \qquad Y = y_{n-1}y_{n-2}\cdots y_1y_0.$$

X and Y are inputs to a carry generate/propagate unit which produces two n-bit output vectors, a carry generate vector G_1 and a carry propagate vector P_1, where

$$G = g_{n-1}g_{n-2}\cdots g_1g_0, \qquad P = p_{n-1}p_{n-2}\cdots p_1p_0,$$

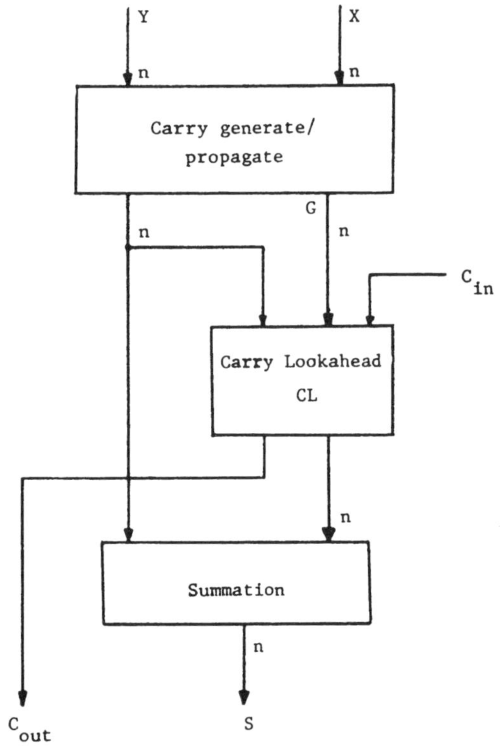

Fig. 1. A functional block diagram of an n-bit CLA.

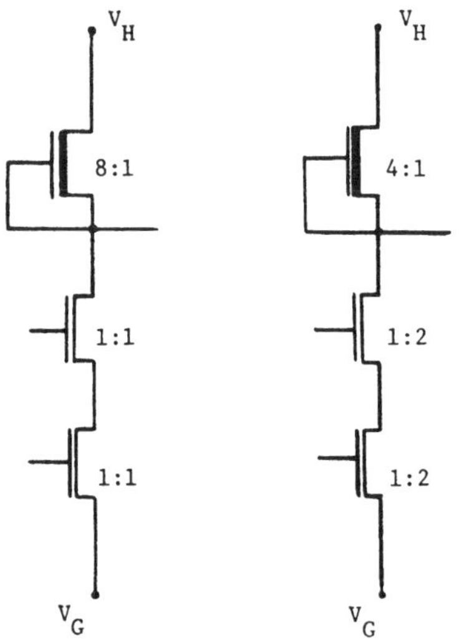

Fig. 2. 2-input NAND gate with $K = 4/1$.

such that

$$g_i = x_i \wedge y_i, \tag{1}$$

$$p_i = x_i * y_i. \tag{2}$$

The sum and carry outputs of the ith stage are denoted by s_i and c_i, respectively, and are defined as follows:

$$s_i = p_i \otimes c_{i-1}, \tag{3}$$

$$c_i = g_i \vee (p_i \wedge c_{i-1}), \tag{4}$$

i.e., if $g_i = 1$, then the ith addition stage produces a carry out. And if $p_i = 1$, then a carry into the ith addition stage propagates through the stage and produces a carry out for the ith stage, p_i and g_i are generated simultaneously.

3. Model of computation

In NMOS technology the fan-out problem plays an important part in determining the performance measures area and time. In the developed model, delays are a continuous function of the ratio between the capacitive load and the capacitance of the driving transistor. In bipolar technologies, delay can be considered constant up to a certain fan-put, limited by the capability of output transistors to provide sufficient output current to drive a set of low impedance transistors while remaining at an acceptable output voltage. While, in NMOS technology, the gate always delivers the required charge to the gates that are being driven. Following [12] a transistor, laid out with gate length L and width W, has capacitance C_g ($C_g \propto L * W$).

The delay for driving a load (composed of other transistors, interconnections and wires) of capacitance γC_g is $(\gamma + 1)\tau$, which expresses the fan-out dependence. An increase in transistor width will reduce delays in driving any load attached to the transistor. To get the same dealy of an inverter for a 2-input NAND gate, the transistor's ratios have to be as shown in Fig. 2. The delay of an inverter pair, shown by the diagram below, is analyzed as follows:

$$\text{for } \_\!\!\_\!\!\lceil\!\!\_\ \ T_r = \tau C_{g_2}/C_{g_1} + K_2 \tau C_{\text{total}}/C_{g_2}, \tag{5}$$

$$\text{for } \_\!\!\rceil\!\!\_\ \ T_f = K_1 \tau C_{g_2}/C_{g_1} + \tau C_{\text{total}}/C_{g_2}, \tag{6}$$

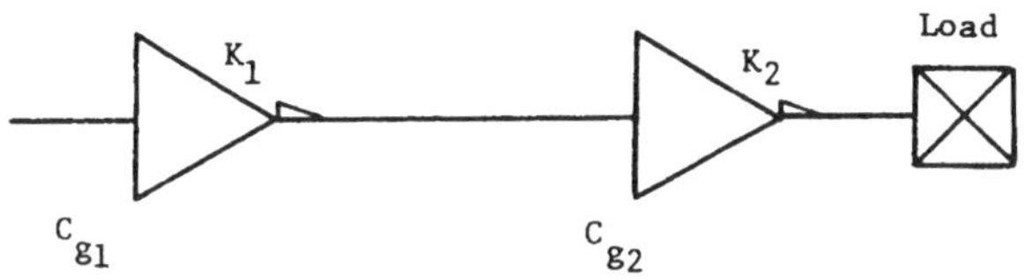

where

τ	=	basic time constant of a minimum transistor,
C_g	=	gate capacitance of the minimum transistor,
C_{total}	=	load capacitance + interconnection capacitance,

Table 1
Evaluation of the use of a logic intermediate specification level

	Fig. 4	Fig. 5(a)	Fig. 5(b)
Depletion mode transistor count	3	2	1
Enhancement mode transistor count	6	5	4
Required area (in unit area)	18	13	8

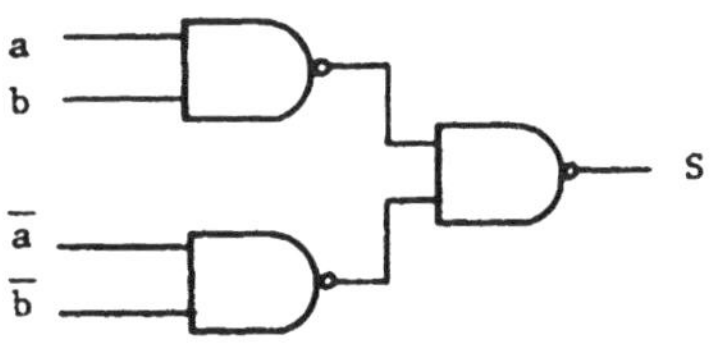

Fig. 3. An exclusive OR gate.

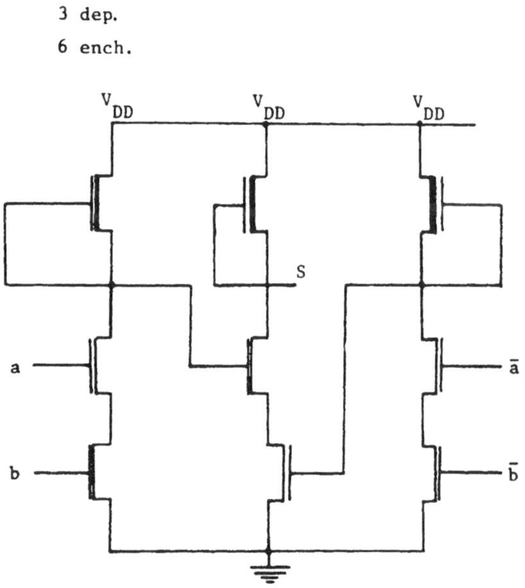

Fig. 4. Transistor diagram obtained via a logic level specification.

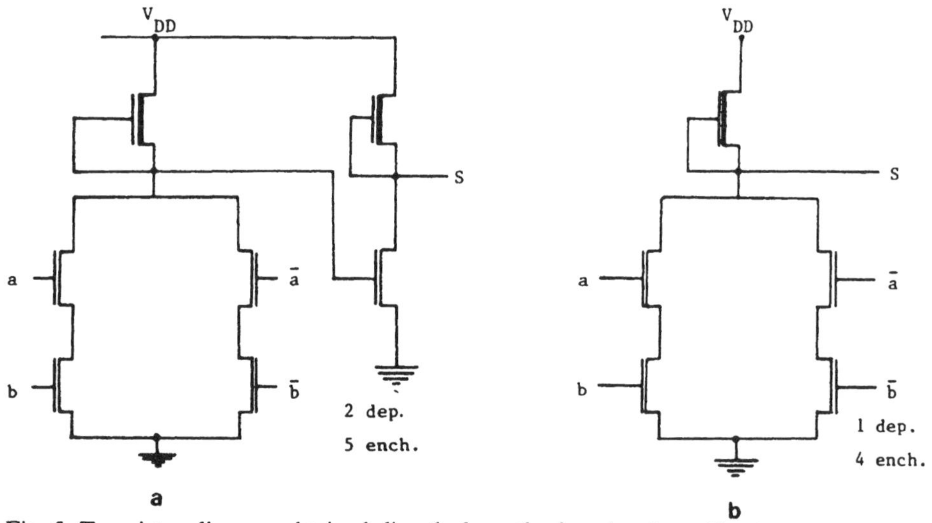

Fig. 5. Transistor diagram obtained directly from the functional specification.

T_r = propagation time delay for a signal going from low to high,
T_f = propagation time delay for a signal going from high to low,
K = Z_{pu}/Z_{pd} where pu = pull-up and pd = pull-down, $Z = L/W$.

An average time delay can be defined, using (5) and (6), as follows:

$$T_{av} = \tau\left[\tfrac{1}{2}(K_1+1)C_{g_2}/C_{g_1} + \tfrac{1}{2}(K_2+1)C_{total}/C_{g_2}\right]. \tag{7}$$

4. The design methodology

Following the hierarchal approach [10], mapping from functional level into logical level and then into layout does not produce efficient area. As an example, consider a simple exclusive OR gate as given in Fig. 3. Mapping from the logic circuit into transistor level produces the circuit in Fig. 4. But, mapping directly from the functional level may produce one of the circuits shown in Fig. 5. Table 1 shows a comparison among those realizations. Assuming that the enhancement mode transistor occupies a unit area and the depletion mode transistor occupies four unit area, the direct realization offers area saving from 27 to 54%.

5. The adder design

In the proposed design, the implementation performance, time and area are compromized to get the most efficient realization. The NMOS constraints have

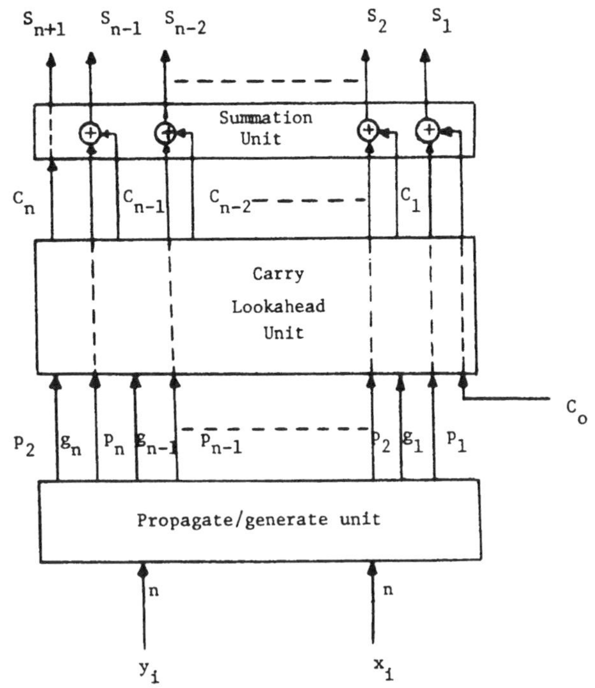

Fig. 6. CLA organization.

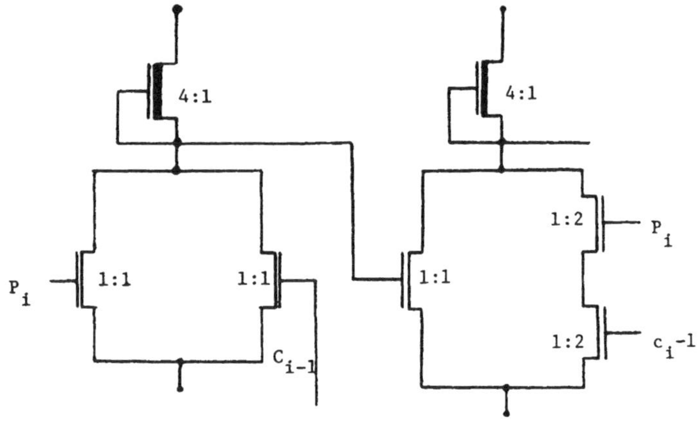

Fig. 7. The basic circuit of an s_i subunit.

been considered such as fan-in, fan-out and interconnections. The unit area used (Δ) is the area of the minimum enhancement mode transistor. The unit time (τ) is the time required to charge the gate capacitance of that transistor. The adder is organized as shown in Fig. 6. It is comprised of three units: summation unit, carry lookahead unit and propagate/generate unit.

5.1. *The summation unit* (S)

This unit is comprised of subunits each to calculate a summation bit. Each subunit has the same area complexity regardless of the bit position. A proposed layout of the basic circuit of a subunit is shown in Fig. 7. The time delay of each subunit is also constant regardless of the bit position. The circuit can be simulated by an inverter pair each of $K = 4/1$, driving two transistors of $1:2$ and $1:1$ ($C_{\text{load}} = 3\ C_g$). The basic area is

$$A_{s_i} = 15\Delta. \tag{8}$$

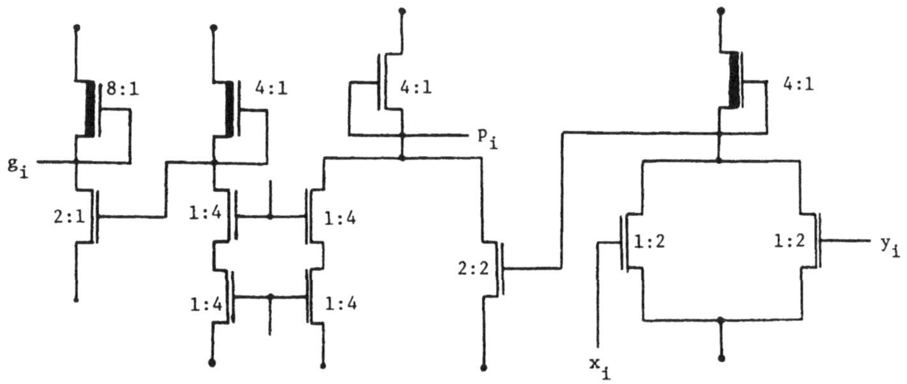

Fig. 8. A basic circuit of a PG subunit.

Applying (7), we have

$$T_S = 2.5 + 2.5(C_{\text{load}} + \beta C_{\text{load}})/C_g\tau = 2.5(4 + 3\beta)\tau, \tag{9}$$

where β is a constant contributed by interconnection and stray capacitance.

5.2. The generate / propagate (PG) unit

This unit comprises of subunits each to calculate p_i and g_i. Each subunit has the same area complexity regardless the bit position. Fig. 8 shows the layout of

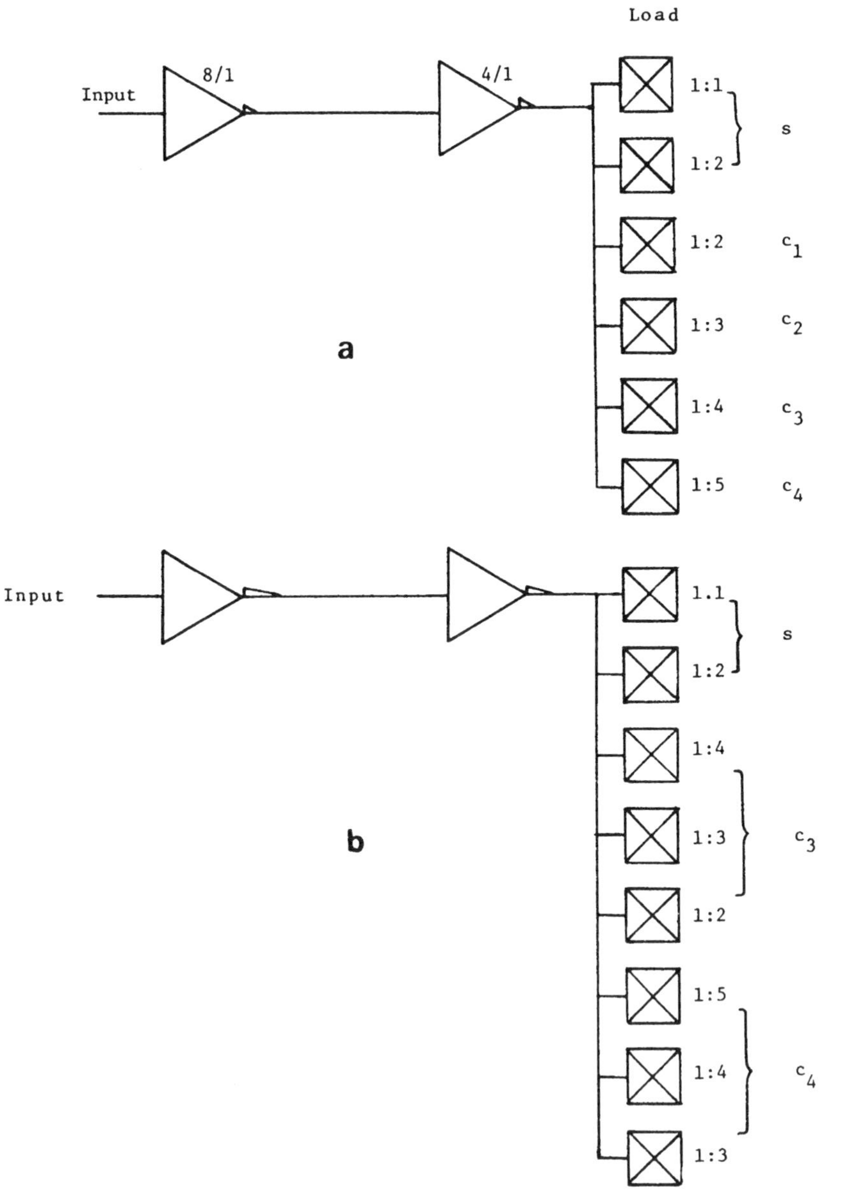

Fig. 9. The load driven by a P section, (a) for p_1, (b) for p_3.

the basic circuit of a subunit. The transistor's ratios are selected to keep the area of all subunits the same. The basic area of each subunit is

$$A_{pg_i} = 44\Delta. \tag{10}$$

To evaluate the time delay, the fan-out plays the main role [12–14]. The propagate section can be simulated by an inverter pair of ratios 8 : 1 and 4 : 1 driving a load of both S and CL units. The load differs according to the subunit position, as follows:

$$C_{L_{p_i}} = \sum_{j=1}^{n-i+1} \sum_{k=1}^{i} (j+k)C_g. \tag{11}$$

Fig. 9 shows the loads for p_1 and p_3 as examples. Applying (7), we have

$$T_{p_i} = 4.5 * \tfrac{4}{2} + 2.5 * \tfrac{1}{4}(C_{\text{load}} + \beta C_{\text{load}})\tau = 9 + 0.625(1+\beta)C_{\text{load}}\tau. \tag{12}$$

The generate section can be simulated as an inverter pair of ratios 8 : 1 and 4 : 1 driving a load of CL unit. The load differs according to the bit position as follows:

$$C_{L_{g_i}} = \sum_{j=1}^{n-i+1} jC_g. \tag{13}$$

Fig. 10 shows the loads for g_1 and g_3 as examples. Applying (7), we have

$$T_{g_i} = 4.5 + 1.25(1+\beta)C_{\text{load}}\tau. \tag{14}$$

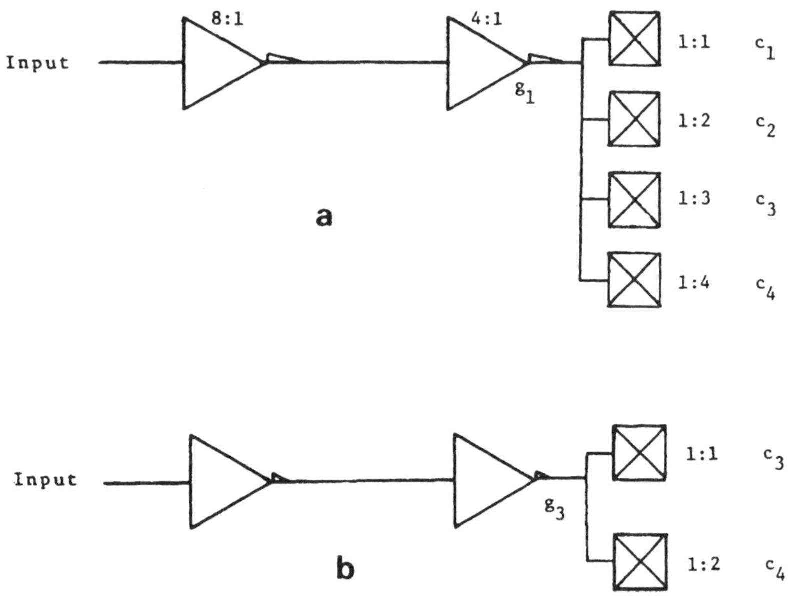

Fig. 10. The load driven by a G section, (a) for g_1, (b) for g_3.

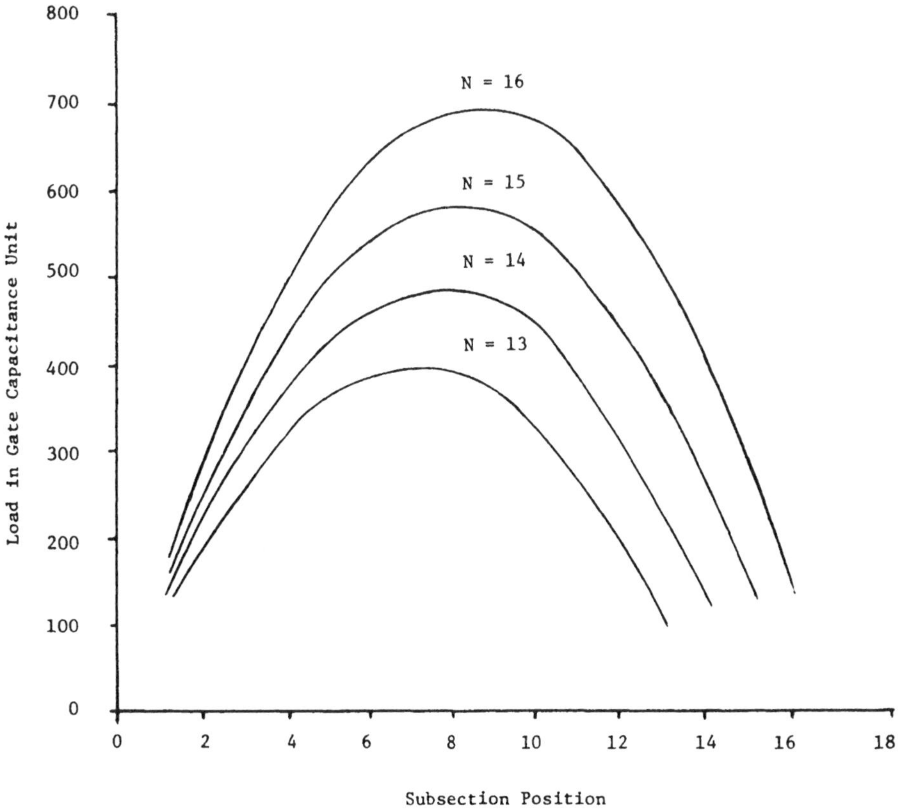

Fig. 11. Load distribution of propagate section.

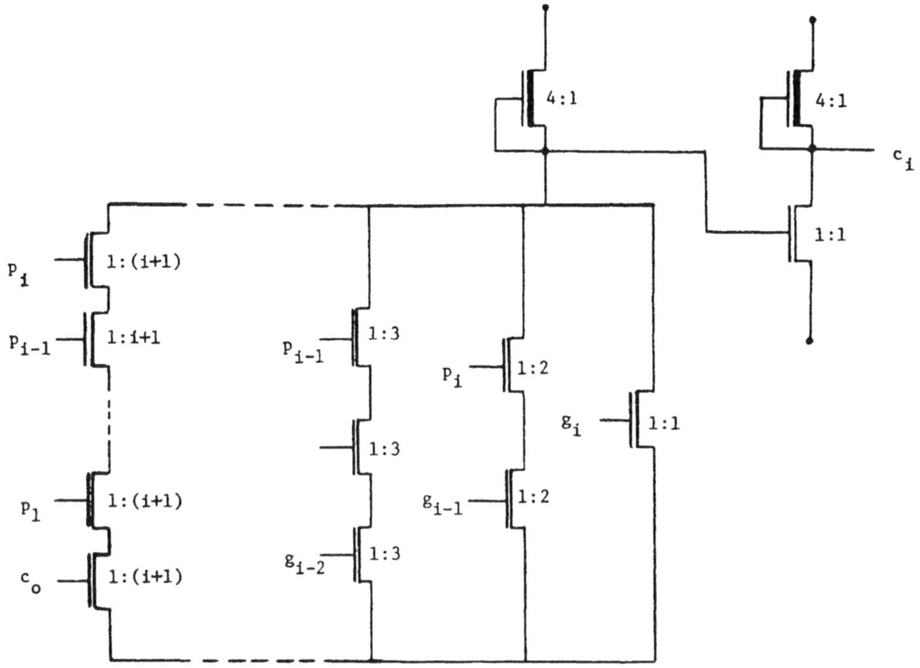

Fig. 12. The basic circuit of subunit i of CL.

The bigger of T_{p_i} and T_{g_i} determines the time delay of the subunit i. Fig. 11 shows the load variation of the propagate section. T_{g_1} is the largest for the generate section.

5.3. The carry lookahead (CL) unit

The g_i's and p_i's are generated in parallel; however, the c_i's must be generated recursively. The CL unit comprises of n subunits, each to calculate a carry bit c_i. The area complexity of each subunit is different according to the bit position. The time delay of each subunit is the same except the last one c_n, where its load is different as c_n is considered to be S_{n+1}. The transistor's ratios are selected to keep the time delay the same for each subsection, and keep a regular layout pattern. The basic circuit layout of a subunit is shown in Fig. 12. The basic area of a

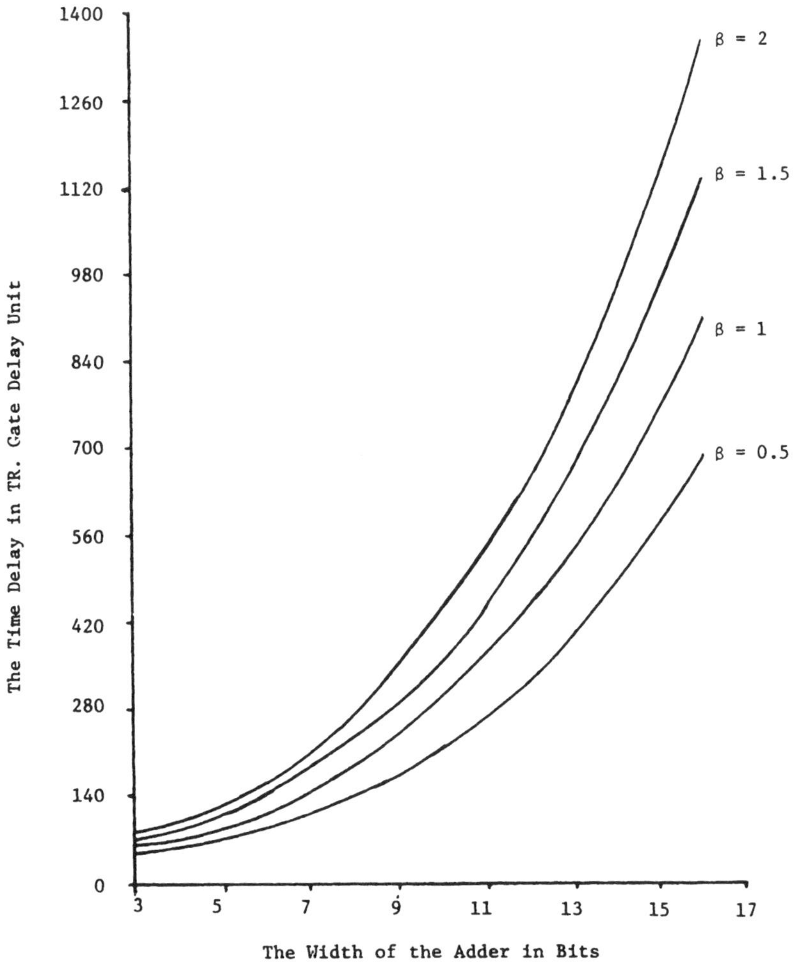

Fig. 13. Time complexity of the adder.

subsection i is as follows:

$$A_{c_i} = \left(9 + \sum_{j=1}^{i+1} j^2\right)\Delta. \tag{15}$$

To evaluate the time-delay of each subunit, it can be simulated as an inverter pair of ratio 4 : 1, driving a load in S unit evaluated by $3C_g$, applying (7):

$$T_{c_i} = 2.5 + 2.5(C_{\text{load}} + \beta C_{\text{load}})/C_g\tau = 2.5(4 + 3\beta)\tau. \tag{16}$$

Table 2 summarizes the space-time complexity required for subunit i.

To determine the time delay of the propagate/generate unit, the largest of T_{p_i} is determined (T_p) and then compared with T_{g_1} (which is the largest of the T_{g_i}). Generally, the time-delay of the adder is given by

$$T = T_S + \max\{T_p, T_{g_i}\} + T_C. \tag{17}$$

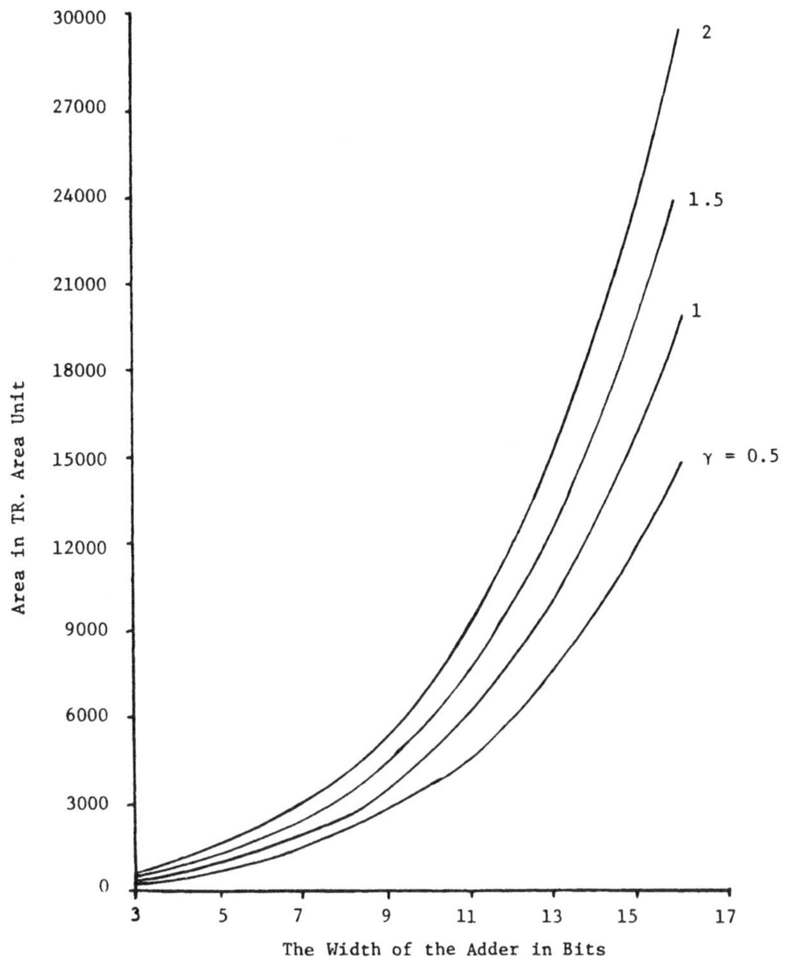

Fig. 14. Area complexity of the adder.

Table 2
Space-time complexity required for the subunits

	Area in unit area Δ	Time-delay in unit time τ
s_i	15	$2.5(4+3\beta)$
p_i	44	$9+0.625(1+\beta)\sum_{j=1}^{n-i+1}\sum_{k=1}^{i}(j+k)$
g_i		$4.5+1.25(1+\beta)\sum_{j=1}^{n-i+1} j$
c_i	$9+\sum_{j=1}^{i+1} j^2$	$2.5(4+3\beta)$

The area of the logical elements required for an n-bit adder is given by

$$A_{\mathrm{L}} = 59n + \sum_{i=1}^{n}\left(9 + \sum_{j=1}^{i+1} j^2\right)\Delta. \tag{18}$$

The total area of the adder A is contributed by both the logical and interconnection area:

$$A = (1+\gamma)A_{\mathrm{L}}\Delta, \tag{19}$$

where γ is a constant contributed by the interconnection area. Figs. 13 and 14 show the performance of the proposed design.

If $c_0 = 0$, the area will be reduced according to the following modifications:

(1) $p_1 \equiv s_1$. (2) $g_1 \equiv c_1$.

In this case the time-delay is the same as in (17), the area required by the logical

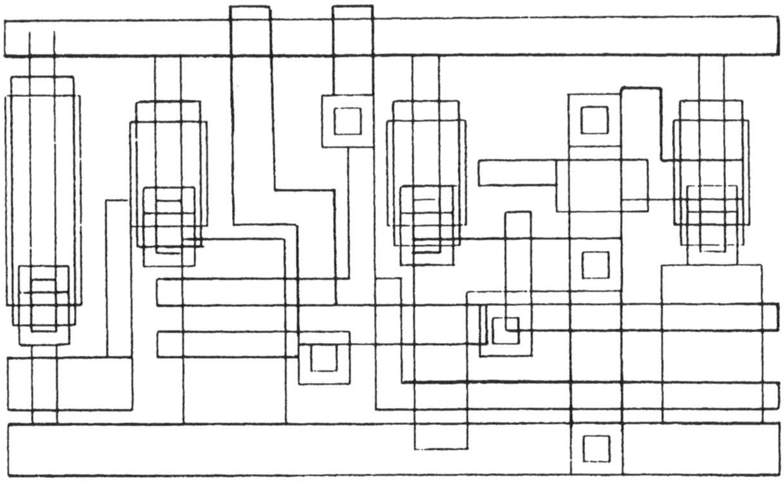

Fig. 15. A layout of PG subunit.

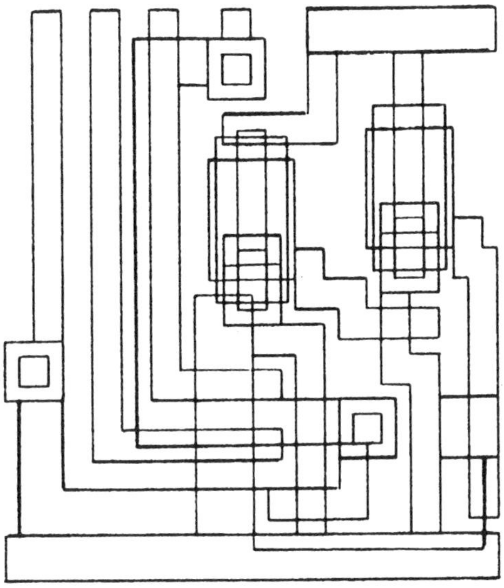

Fig. 16. A layout of C1 subunit.

elements is given by

$$A_L = 59n - 15 + \sum_{i=2}^{n} \left(9 + \sum_{j=1}^{i} j^2 \right) \Delta. \tag{20}$$

6. Example: An implementation of a 4-bit adder

As an example, the layout of a 4-bit adder is discussed based on the explained model of computation and the described adder layout methodology. The basic

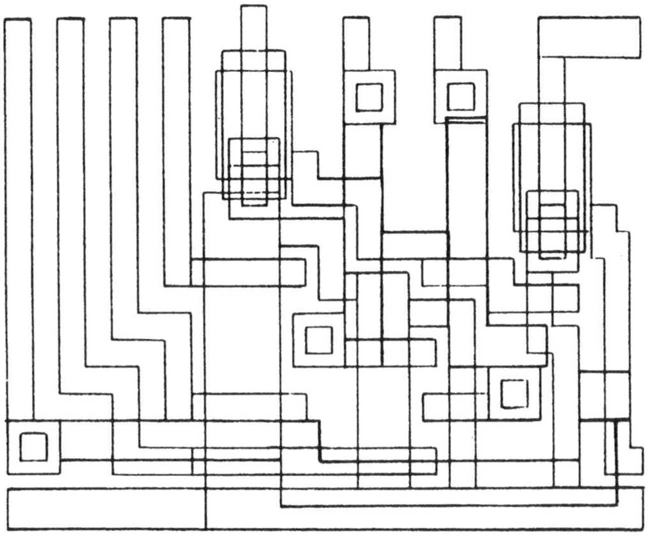

Fig. 17. A layout of C2 subunit.

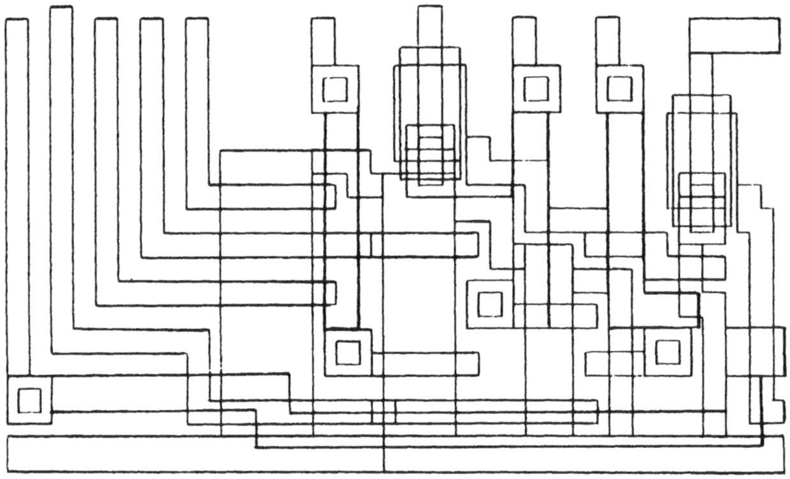

Fig. 18. A layout of C3 subunit.

concepts of the layout are the following:

(1) Keeping regularity in layout, placement and interconnections.

(2) Units are placed such that routing is done in a channel, i.e., channel routing.

(3) The heights of all modules are the same leading to a regular channel.

(4) Routing is performed in two layers, metal for horizontal and polysilicon for vertical.

Fig. 15 shows a layout for a PG subunit in which branches of NOR gates are perpendicular instead of parallel to get more efficient area. Figs. 16–19 show the

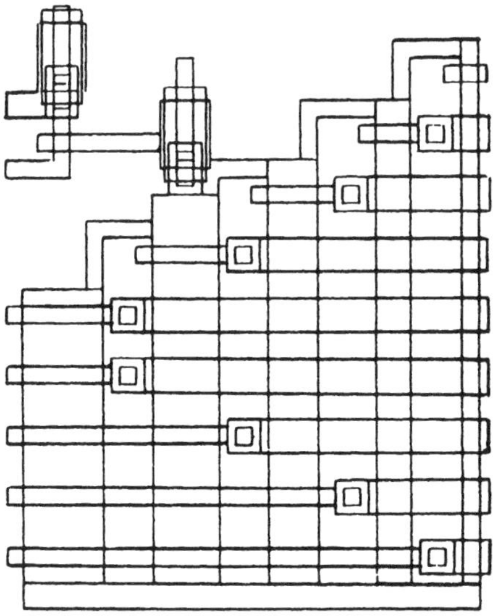

Fig. 19. A layout of C4 subunit.

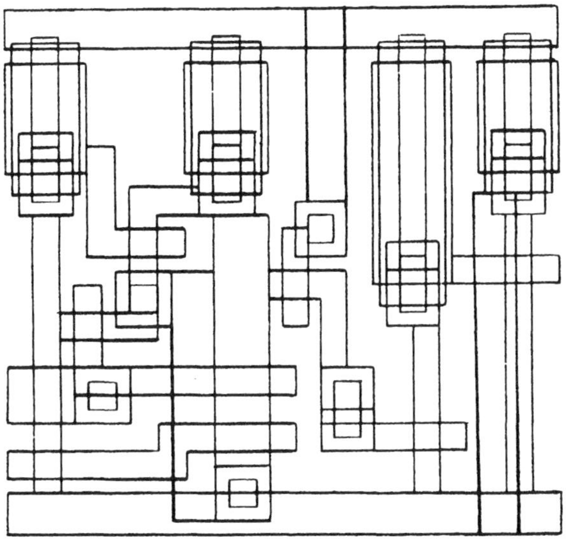

Fig. 20. A layout of an S subunit.

layout of CL subunits c_1–c_4; they have different area but the same height. Fig. 20 shows the layout of an S subunit connected to an F.F. for the output stability, it has the same height as the C subunits. Fig. 21 shows the layout of the whole PG unit where the subunits fit together in an efficient fashion. Fig. 22 shows an assembly of CL unit and S unit interleaved in a beautiful way getting a very efficient area and creating a regular channel for interconnections. Fig. 23 shows a floor-plan for the developed adder.

7. Conclusions

The problem of designing an efficient adder in NMOS technology has been considered. A model has been developed; it is a low level model which reflects the fan-out dependency of the NMOS technology. It is a more realistic model than those developed before. The main philosophy of the followed design methodology is mapping from the functional level into layout level directly which seems very efficient for modest size circuits. The area-time complexity of each subunit of the adder has been analyzed. The given example shows the efficiency of the developed method which offers regularity in layout, placement and interconnection.

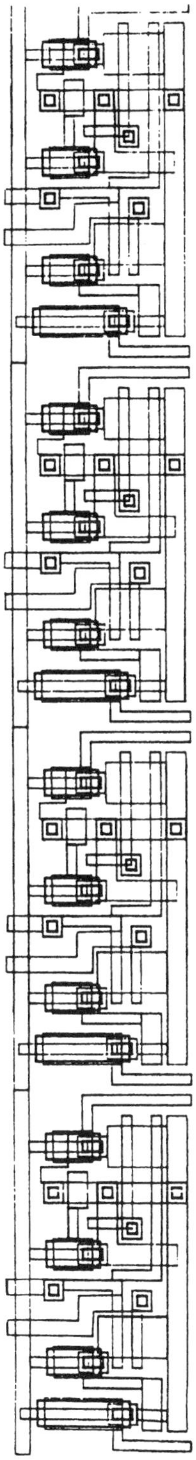

Fig. 21. A layout of the PG unit.

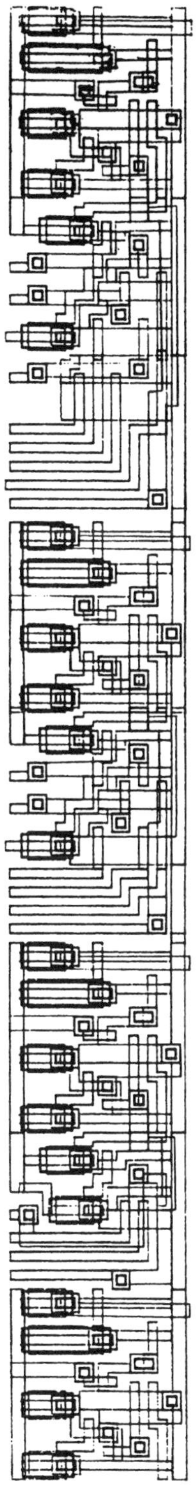

Fig. 22. A layout of CL and S units.

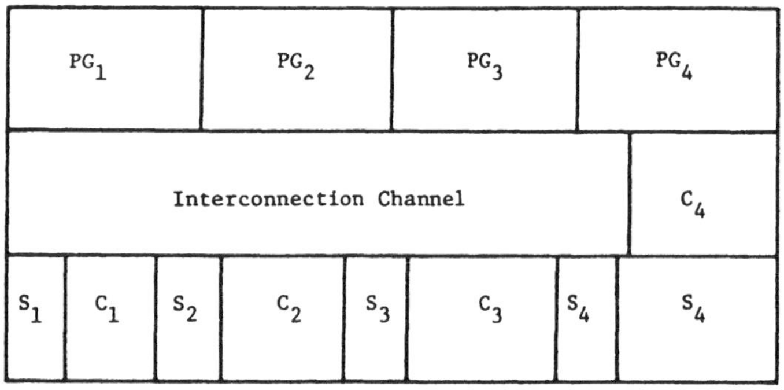

Fig. 23. A floor-plan of the designed adder.

References

[1] Winograd, S., On the time required to perform addition, *J. ACM* **12** (1965) 277–285.

[2] Brent, R.P., On the addition of binary numbers, *IEEE Trans. Comput.* **C-19** (1970) 758–759.

[3] Kuck, D.J., *The Structure of Computers and Computations Vol. 1* (Wiley, New York, 1978).

[4] Sklansky, J., An evaluation of several two summand binary adders, *Trans. Electronic Comput.* **9** (2) (1960) 213–226.

[5] MacSorley, O.L., High-speed arithmetic in binary computers, *IRE Proc.* **49** (1) (1961) 67–91.

[6] Garner, H.L., A survey of some recent contributions to computer arithmetic, *IEEE Trans. Comput.* **C-25** (1976) 1277–1282.

[7] Brent, R.P. and H.T. Kung, The chip complexity of binary arithmetic, *Proc. 12th Ann. ACM Symp. on Theory of Computing* (1980) pp. 190–200.

[8] Waser, S., State-of-the-art in high speed arithmetic integrated circuits, *Comput. Design* **17** (7) (1978) 67–75.

[9] Brent, R.P. and H.T. Kung, A regular layout for parallel adders, *IEEE Trans. Comput.* **31** (3) (1982) 260–264.

[10] Cleemput, W.M. van, Hierarchical design for VLSI problems and advantages, *Proc. Caltech Conf. on VLSI* (1979) pp. 259–274.

[11] Hwang, K., *Computer Arithmetic: Principles, Architecture and Design* (Wiley, New York, 1979).

[12] Mead, C.A. and L.A. Conway, *Introduction to VLSI Systems* (Addison-Wesley, Reading, MA, 1980).

[13] Mohsen, A.M. and C.A. Mead, Delay-time optimization for deriving and sensing of signals on high capacitance paths of VLSI systems, *IEEE Trans. Electronic Devices* **ED-26** (1979) 542–548.

[14] Mead, C.A. and R. Rem, Minimum propagation delay in VLSI, *Proc. Caltech Conference on VLSI* (1981).

Regular, Area–Time Efficient Carry-Lookahead Adders*

TIN-FOOK NGAI

Department of Electrical Engineering, Stanford University, Stanford, California 94305

AND

MARY JANE IRWIN AND SHISHPAL RAWAT

Department of Computer Science, Pennsylvania State University, University Park, Pennsylvania 16802

Received November 15, 1984

For fast binary addition, a carry-lookahead (CLA) design is the obvious choice (S. Ong and D. E. Atkins, *Proc. Sixth Symposium of Computer Arithmetic*, Aarhus, Denmark, June 1983; M. A. Bayoumi, G. A. Jullien, and W. C. Miller, *INTEGRATION* **1** (1983)). However, the direct implementation of a CLA adder in VLSI faces some undesirable limitations. Either the design lacks regularity, thus increasing the design and implementation costs, or the interconnection wires are too long, thus causing area–time inefficiency and limits on the size of addition. R. P. Brent and H. T. Kung (*IEEE Trans. Comput.* **C-31** (Mar. 1982)) solved the regularity problem by reformulating the carry chain computation. They showed that an n-bit addition can be performed in time $O(\log n)$, using area $O(n \log n)$ with maximum interconnection wire length $O(n)$. In this paper, we give an alternative $\log n$ stage design which is nearly optimum with respect to regularity, area–time efficiency, and maximum interconnection wire length.

INTRODUCTION

Our design follows Mead and Conway's λ design rules [4]. We assume two layers of interconnect (metal, silicide, or polysilicon). The only circuits used in the design are NMOS complex gates [5] and transmission gates (pass transistors). All interconnection wires are specified to be of length less than

*This research is supported by ARO Contract DAAG29-83-K-0126. A preliminary version of this paper appeared in "Proceedings of the Seventh Symposium on Computer Arithmetic," Urbana, Ill., June 1985.

L_{max}, which is assumed to be a constant for a particular fabrication technology. The time delay for driving a signal along a wire is assumed to be proportional to the length of the wire. Thus, constant propagation delay can be achieved by having the channel width of the driving transistor proportional to the length of the wire [7, 8]. Finally, the computation is performed in a planar region.

First we introduce our carry-lookahead (CLA) adder, which is based upon traditional block carry-lookahead techniques [6]. Then we derive the NMOS design using a negative logic scheme. For the initial design, the external inputs and outputs are assumed to be available where they are required or generated. Finally, we lift this assumption and consider the practical input/output problem in depth.

The Carry-Lookahead Scheme

Let $a_{n-1}a_{n-2} \cdots a_0$ and $b_{n-1}b_{n-2} \cdots b_0$ be two n-bit binary numbers with a sum of $s_{n-1} \cdots s_0$. The carry-lookahead scheme computes the s_i's by

$$c_{i+1} = g_i + p_i c_i$$
$$s_i = a_i \oplus b_i \oplus c_i \qquad \text{for } i = 0, 1, \ldots, n-1,$$

where

$$g_i = a_i b_i \qquad (\textit{carry generate})$$
$$p_i = a_i + b_i \qquad (\textit{carry propagate})$$

$+$ denotes *logical or*

xy denotes x and y

$\oplus$ denotes *exclusive or*.

It is easy to show that

$$c_{i+m+1} = g_{i+m} + \sum_{l=0}^{m-1}\left(\prod_{j=l+1}^{m} p_{i+j}\right)g_{i+l} + \left(\prod_{j=0}^{m} p_{i+j}\right)c_i.$$

For large m, the above carry computation is difficult to implement due to the practical limitations on fan-in and fan-out. In order to reduce the complexity, it is common practice to group carries into blocks [6]. The block carry scheme is

$$c'_{i+1} = g'_i + p'_i c'_i$$
$$c_{ri} = c'_i$$

$$c_{ri+m+1} = g_{ri+m} + \sum_{l=0}^{m-1}\left(\prod_{j=l+1}^{m} p_{ri+j}\right)g_{ri+l} + \left(\prod_{j=0}^{m} p_{ri+j}\right)c'_i \qquad \text{for } 0 \le m \le r-2,$$

where

$$g'_i = g_{ri+(r-1)} + \sum_{l=0}^{r-2}\left(\prod_{j=l+1}^{r-1} p_{ri+j}\right)g_{ri+l} \qquad \textit{(block carry generate)},$$

$$p'_i = \prod_{j=0}^{r-1} p_{ri+j} \qquad \textit{(block carry propagate)}$$

r is the blocking factor.

The same technique can be applied iteratively to compute the block carries. This scheme is illustrated in Fig. 1 for a general CLA adder.

Because we are using NMOS technology as our model, all complex gates use complementary logic. Thus, it is desirable to modify the above scheme

Primitive unit (P)
- Inputs
  - operands ($a_i s$, $b_i s$)
  - low order carry-in (c_i^1)
- Produces
  - block generate (g_i^1)
  - block propagate (p_i^1)
  - sum s ($s_i s$) with internal lookahead

Block carry unit (BC)
- Inputs
  - block generates ($g_i^k s$)
  - block propagates ($p_i^k s$)
  - block carry-in (c_i^{k+1})
- Produces
  - block generate (g_i^{k+1})
  - block propagate (p_i^{k+1})
  - block carry-outs ($c_i^k s$)

Block generate unit (BG)
- Inputs
  - block generates ($g_i^K s$)
  - block propagates ($p_i^K s$)
  - low order carry-in (c_0)
- Produces
  - block carry-outs ($c_i^K s$)

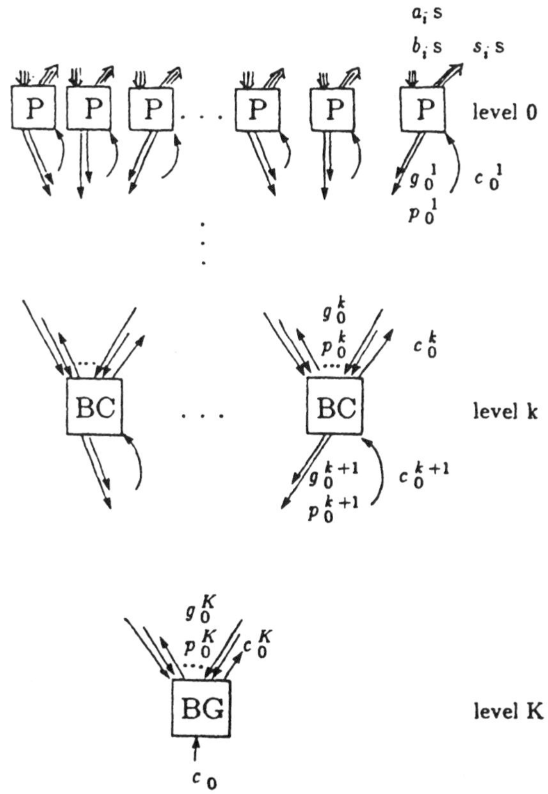

FIG. 1. CLA adder components.

to reflect the use of a negative logic system. Assume that at the kth level of the iteration, the block carry computation is

$$c_{i+1}^k = g_i^k + p_i^k c_i^k.$$

We can reformulate this computation into negative logic as

$$c_{i+1}^k = \neg[G_i^k + P_i^k(\neg c_i^k)],$$

where

$$G_i^k = \neg(g_i^k + p_i^k)$$
$$P_i^k = \neg(g_i^k)$$

$\neg x$ denotes the complement of x (*not* x).

Then for the next level of iteration, we have

$$c_{i+1}^{k+1} = g_i^{k+1} + p_i^{k+1} c_i^{k+1}$$
$$c_{ri}^k = \neg(c_i^{k+1})$$
$$c_{ri+m+1}^k = \neg\left[G_{ri+m}^k + \sum_{l=0}^{m-1}\left(\prod_{j=l+1}^{m} P_{ri+j}^k\right)G_{ri+l}^k + \left(\prod_{j=0}^{m} P_{ri+j}^k\right)c_i^{k+1}\right] \quad \text{for } 0 \le m \le r-2,$$

where

$$g_i^{k+1} = G_{ri+(r-1)}^k + \sum_{l=0}^{r-2}\left(\prod_{j=l+1}^{r-1} P_{ri+j}^k\right)G_{ri+l}^k$$
$$= \neg\left[\prod_{l=0}^{r-1}\left(\sum_{j=l}^{r-1} g_{ri+j}^k + p_{ri+l}^k\right)\right]$$
$$p_i^{k+1} = \prod_{j=0}^{r-1} P_{ri+j}^k = \neg\left(\sum_{j=0}^{r-1} g_{ri+j}^k\right).$$

With this reformulation, all block carries, block carry generates, and block carry propagates can be directly implemented with complex gates.

Adder Design

We found that a blocking factor of three for the first two levels of the iteration and four for all higher levels resulted in a more area efficient VLSI

NMOS design, as is demonstrated in the next section. With this scheme there are three types of basic units to be designed. These are the primitive unit (P), block carry unit (BC), and the block carry generation unit (BG). NMOS layouts of these units are shown in Fig. 2.

Primitive Units (P2 *and* P3). These basic units include four subunits: input/output, block propagate and block generate, carry generation and sum-

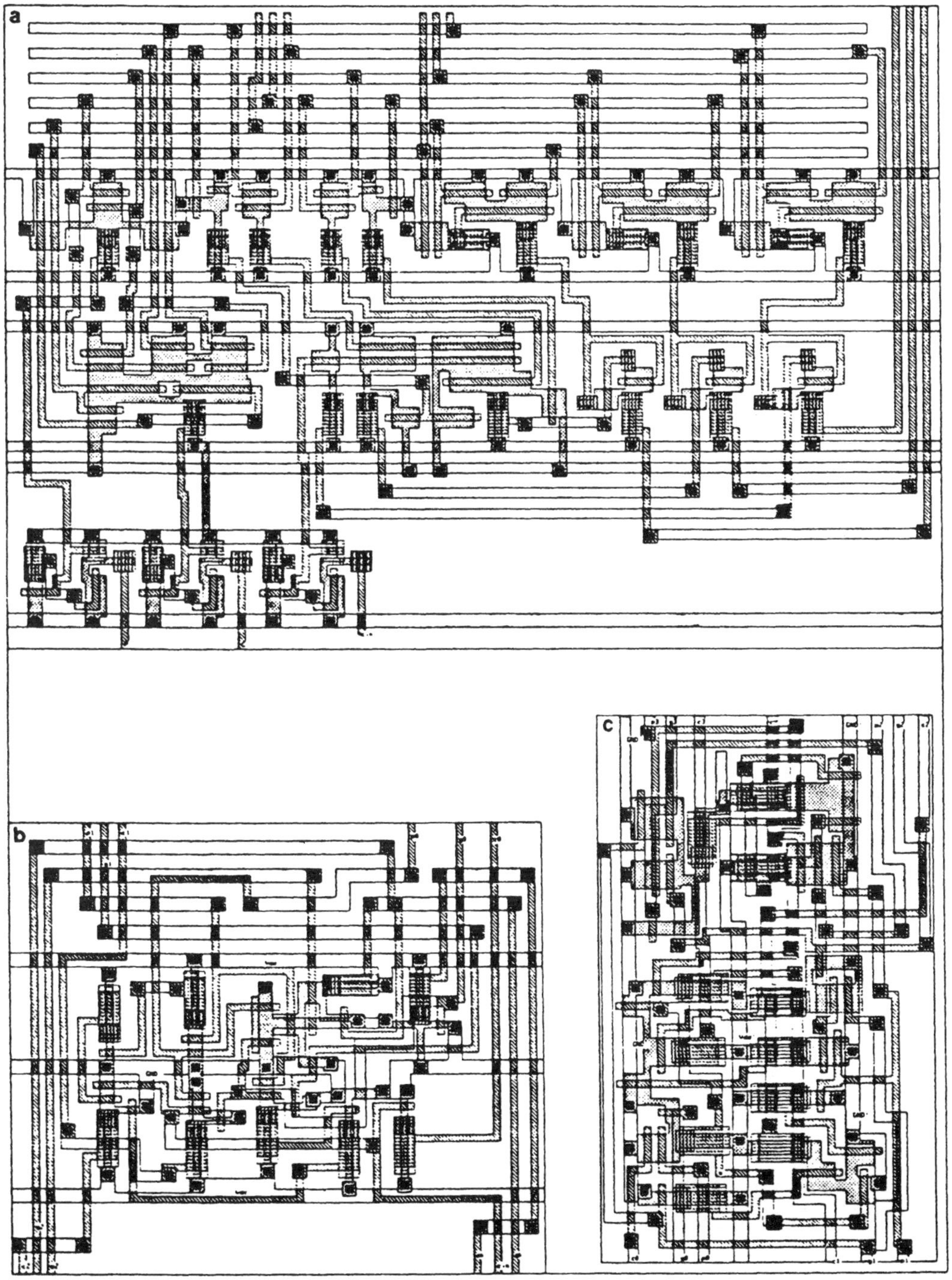

FIG. 2. (a) The primitive unit layout. (b) The BC3 layout. (c) The BG layout.

mation. We postpone the discussion of the input/output subunit. Here we assume that the input/output subunit is of low complexity and uses constant area. In the block propagate and block generate subunit of P3, block propagate p^1 and block generate g^1 are formed directly from the operand bits $a_2a_1a_0$ and $b_2b_1b_0$. (We restrict our attention to the least significant unit for notational simplicity without loss of generality.)

$$\begin{aligned} g^1 &= \neg[(g_2 + p_2)(g_2 + g_1 + p_1)(g_2 + g_1 + g_0 + p_0)] \\ &= \neg[a_2b_2 + (a_2 + b_2)(a_1b_1 + (a_1 + b_1)(a_0 + b_0))] \\ p^1 &= \neg(g_2 + g_1 + g_0) = \neg(a_2b_2 + a_1b_1 + a_0b_0). \end{aligned}$$

The carry generation subunit computes all the carry bits internal to the primitive unit from the block carry c_0^1 which is input to the unit.

$$\begin{aligned} c_0 &= \neg(c_0^1) \\ c_1 &= \neg(G_0 + P_0c_0^1) \\ c_2 &= \neg(G_1 + P_1G_0 + P_1P_0c_0^1)\ \dagger \end{aligned}$$

where

$$\begin{aligned} G_0 &= \neg(g_0 + p_0) = \neg(a_0 + b_0) \\ G_1 &= \neg(g_1 + p_1) = \neg(a_1 + b_1)\ \dagger \\ P_0 &= \neg\, g_0 = \neg(a_0b_0) \\ P_1 &= \neg\, g_1 = \neg(a_1b_1)\ \dagger \end{aligned}$$

($\dagger$ in P3 only).

The summation subunit computes the sum bits from the internal carries and the operand bits.

$$s_i = a_i \oplus b_i \oplus c_i \qquad \text{for } i = 0, 1, 2.$$

The corresponding NMOS complex gate layout for the P3 unit is shown in Fig. 2a.

Block Carry Units (BC3 *and* BC4). These basic units form the block carry generates and block carry propagates from either three (BC3) or four (BC4) inputs which will serve as inputs to the next higher level and also compute the block carries which will serve as inputs to the subsequent lower level. Implementation of the block carry generates and propagates using complex gates is straightforward. For example, in BC4

$$g^{k+1} = \neg\{g_3^k + p_3^k[g_2^k + p_2^k(g_1^k + p_1^k(g_0^k + p_0^k))]\}$$
$$p^{k+1} = \neg(g_3^k + g_2^k + g_1^k + g_0^k).$$

The block carries are computed only when the block carry generated by the next higher level becomes available.

$$c_0^k = \neg(c_0^{k+1})$$
$$c_1^k = \neg(g_0^k + p_0^k c_0^{k+1})$$
$$c_2^k = \neg(G_1 + P_1 c_0^{k+1})$$
$$c_3^k = \neg(G_2 + P_2 c_0^{k+1})\ \dagger$$

where

$$G_1 = g_1^k + p_1^k g_0^k$$
$$G_2 = g_2^k + p_2^k(g_1^k + p_1^k g_0^k)\ \dagger$$
$$P_1 = p_1^k p_0^k$$
$$P_2 = p_2^k p_1^k p_0^k\ \dagger$$
$$(\dagger \text{ in BC4 only}).$$

Figure 2b shows the corresponding NMOS complex gate layout for the BC3 unit.

Block Carry Generation Unit (BG3 *and* BG4). These basic units form the block carries at the highest level of iteration. If K is the total number of iteration levels, then

$$c_0^K = \begin{cases} \neg c_0 & \text{if } K \text{ is odd} \\ c_0 & \text{if } K \text{ is even} \end{cases}$$

and

$$c_1^K = g_0^K + p_0^K c_0^K$$
$$c_2^K = g_1^K + p_1^K(g_0^K + p_0^K c_0^K)$$
$$c_3^K = (g_2^K + p_2^K g_1^K) + (p_2^K p_1^K)(g_0^K + p_0^K c_0^K)\ \dagger$$
$$(\dagger \text{ in BG4 only}).$$

Figure 2c shows the corresponding NMOS complex gate layout for the BG unit.

TABLE I

Unit	Size in λ	Delay in nS
P2	160 × 165	20
P3	265 × 180	45
BC3	150 × 125	25
BC4	195 × 140	30
BG3	95 × 95	21
BG4	155 × 95	25

Layout size and SPICE estimated speeds of each of the basic units are given in Table I assuming a load capacitance of 0.10 pF and maintaining a logic '1' voltage level of at least 4.5 V and a logic '0' voltage level of no more than 0.25 V.

Area and Time Measures

Figure 3 is an NMOS layout for a 27-bit adder using the scheme derived above. This layout has been fit in area $O(n)$ by using a recursive technique for embedding tree-like structures in the plane [10]. As can be seen in the floor-plan schematic (inset in Fig. 3), a 27-bit adder can be packed most densely by using a blocking factor of three at each of the levels. For larger adders, a blocking factor of three for the first two levels and four thereafter leads to the most efficient layout, as seen in Figs. 4a and b. The flexibility that comes from being able to vary the blocking factor at various levels can be used to make optimum use of space. Note that all interconnections between the basic units involve only nearest neighbors with a maximum wire length of $O(\sqrt{n})$.

Brent and Kung's adder [2] can also be embedded in $O(n)$ area with a maximum wire length of $O(\sqrt{n})$, but the lack of flexibility in the blocking factor (only two) results in a much less densely packed layout with considerably more area dedicated to interconnect.

Obviously, any $\log n$ stage adder can add two n-bit numbers in time $O(\log n)$. Another motivation in using negative logic was to speed up the addition even more. At each level, we need to form the carry generate and carry propagate terms and to form the block carries out after receiving the block carry in. While a positive logic approach would require four NMOS gate delays to accomplish these operations, the theoretical lower bound for the negative logic approach should be two gate delays. SPICE simulations have indicated that three gate delays may be a more realistic measure for our complex gates. Table II gives a comparison of the computation time for a serial scheme, Brent and Kung's scheme, and our scheme for various n. T is the average delay of a simple NAND/NOR gate measured at the inversion voltage level. The delay of an exclusive or gate is assumed to be $2T$. Our goal

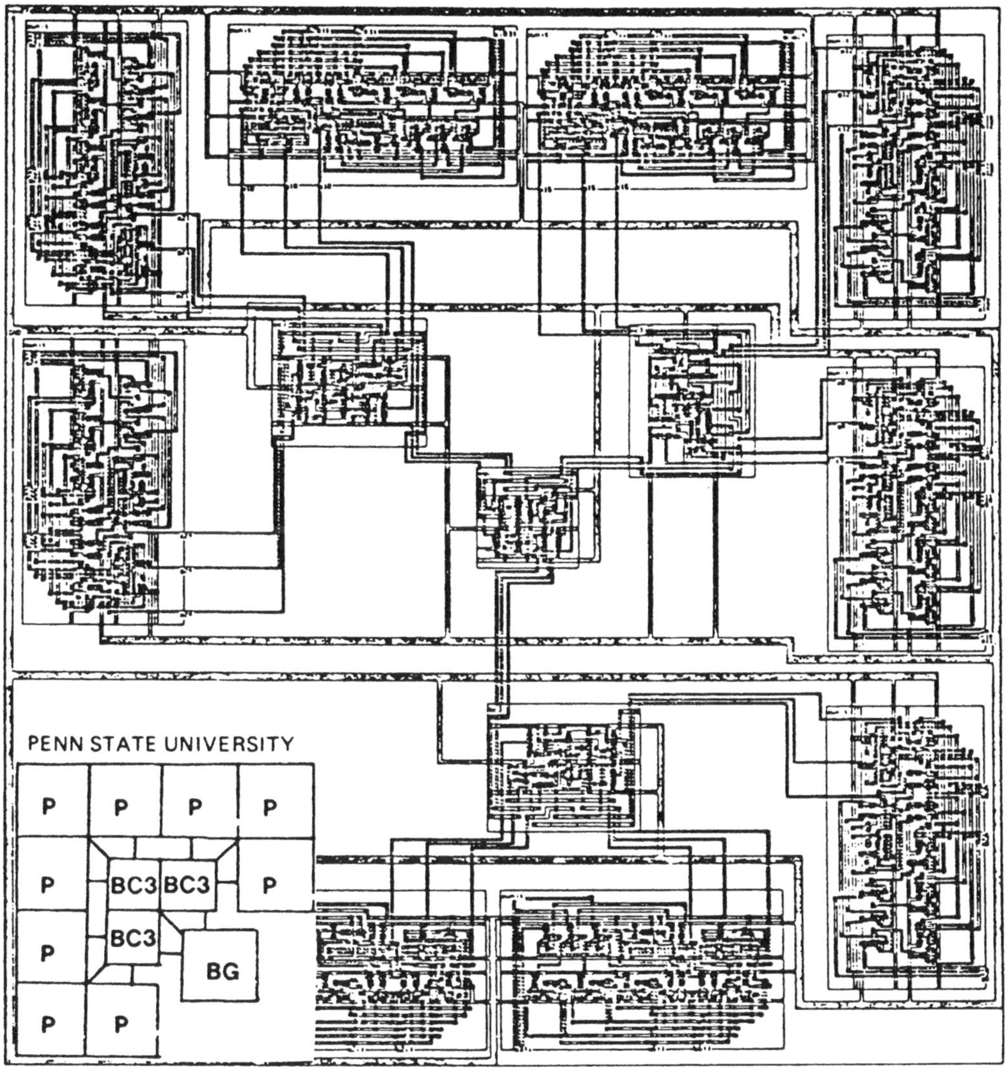

FIG. 3. Floorplan and CIF of a 27-bit CLA adder (9P3 − 3BC3 − BG3, 1046λ × 985λ, 126 nS).

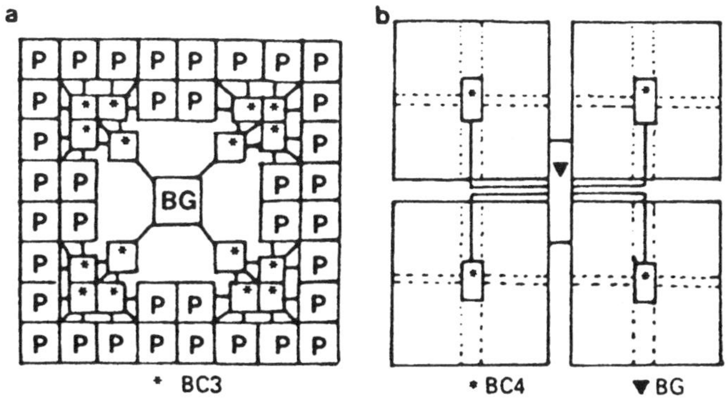

FIG. 4. (a) A 108-bit CLA floorplan. (b) A recursive CLA floorplan.

TABLE II

n	Ripple	B&K (T)	N&I (T)	Component construction
6	14		8	3P2 – BG3
8	18	15		4P2 – BG4
9	20			3P3 – BG3
12	26			4P3 – BG4
16	34	19	12	8P2 – 2BC4 – BG2
18	38			9P2 – 3BC3 – BG3
24	50			12P2 – 4BC3 – BG4
27	56			9P3 – 3BC3 – BG3
32	66	23		16P2 – 4BC4 – BG4
36	74			12P3 – 4BC3 – BG4
48	98			16P3 – 4BC4 – BG4
64	130	27	16	32P2 – 8BC4 – 2BC4 – BG2
72	146			36P2 – 12BC3 – 4BC3 – BG4
108	218			36P3 – 12BC3 – 4BC3 – BG4
128	258			64P2 – 16BC4 – 4BC4 – BG4
192	386			64P3 – 16BC4 – 4BC4 – BG4
256	514	35	20	128P2 – 32BC4 – 8BC4 – 2BC4 – BG2
288	578			144P2 – 48BC3 – 16BC3 – 4BC4 – BG4
432	866			144P3 – 48BC3 – 16BC3 – 4BC4 – BG4

is to maximize the utilization of our basic units maintaining optimal layout throughout the process (assuming the P, BC, and BG are approximately equal in size). For large n, our design uses $2T \times \log n$ as compared to $2T \times 2 \log n$ for Brent and Kung's design.

Pipelining the Adder

An adder suitable for use in a digital filtering scheme was designed using these basic units. Besides the normal add, subtract, minimum, maximum, and pass operations required for the arithmetic cell, we also had to implement an absolute value subtraction operation. Two's complement arithmetic is not particularly well suited for absolute value subtraction, since another add step is required whenever the result happens to be negative. We decided to use the sign magnitude representation for doing the arithmetic operations, which in turn introduces the following extra steps (see Fig. 5).

1. Exnor logic is required prior to addition (converting from sign magnitude to 1's complement).
2. End around carry to perform 1's complement add.
3. Exnor logic is required after addition (converting back to sign magnitude).

This was not a difficult proposition in our case since we were manipulating 12-bit numbers. Taking the overall complexity and speed into consideration

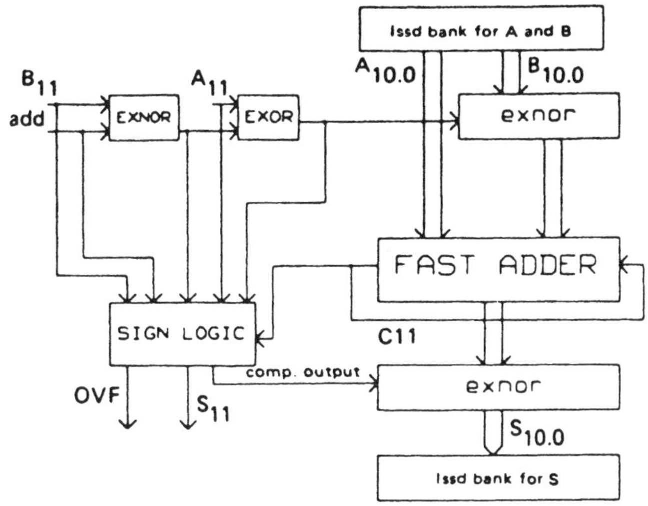

FIG. 5. Sign magnitude adder.

we decided to implement the 12-bit sign magnitude adder. The end around carry introduced only four additional gate delays at the carry generation stage (BG unit).

After adding gating registers and control logic we found the adder to be slower than required by the application (clock speed = 13.4 MHz). The adder was originally conceived as a stage in a digital filter. Since we were more concerned with throughput than latency we decided to further subdivide the adder in two stages. This led us to an innovative construction of a pipelined adder that involved two stages. See Fig. 6.

1. Stage 1 would generate the appropriate block carries and propagate them to Stage 2.
2. Stage 2 would generate the individual carries and generate the individual sum bits. This arrangement allows the pre-add logic correction and post-add logic correction to be divided between the two stages.

The above scheme can be generalized to any n-bit adder, where the first stage of the pipeline generates Block P and G signals, and the following stages further combine these P and G signals to form P and G signals for even bigger blocks (a total of log n stages). The next log n stages would be devoted to carry generation. The last carry generation stage can be combined with the sum generating units. Thus the adder will have 2 log n stages. Such pipelining is very useful in cases where the adder will be under constant use (e.g., digital filters). There is an area penalty involved with the scheme in the absolute sense, since there are so many stages. In the theoretical sense, however, we still have optimal time log n and optimal area n log n. The interconnection is regular and $O(\log n)$ at each stage. The number of wires that need to be transmitted from one stage to another never exceed $3n$ for a blocking factor of 3. For a more detailed discussion please see [9].

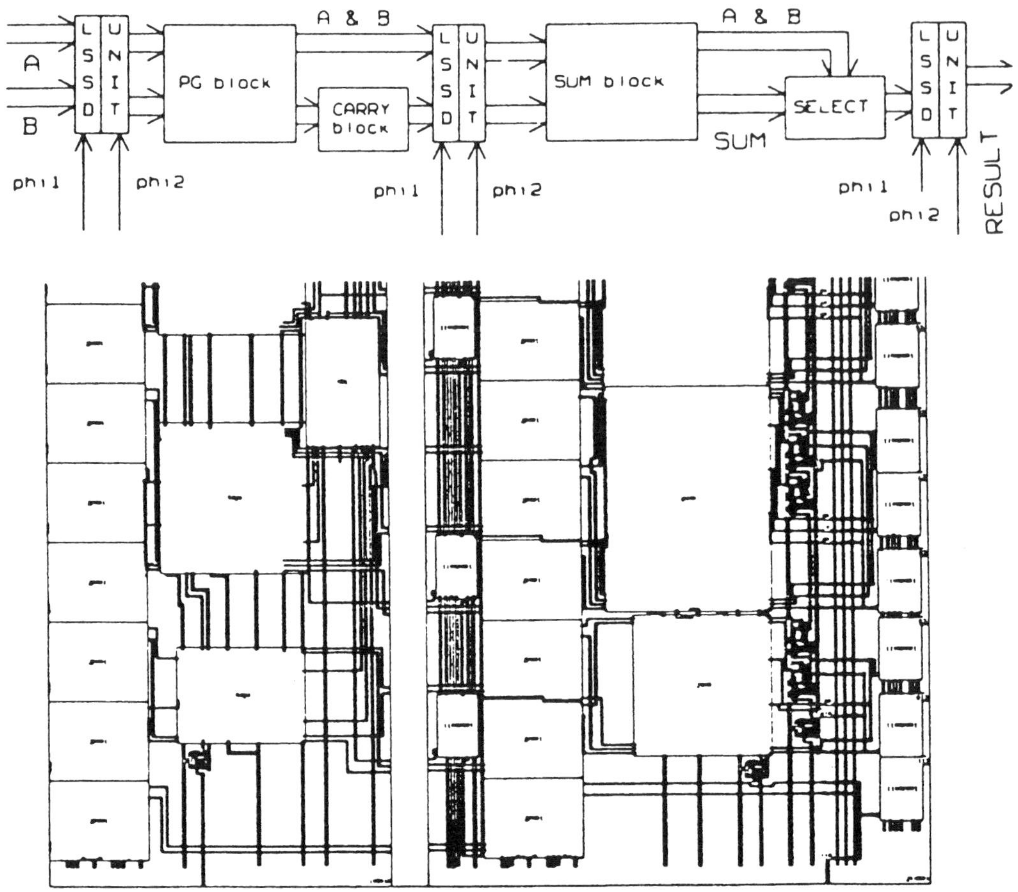

FIG. 6. Pipelined adder: Block diagram and partial CIF plot.

INPUT-OUTPUT CONSIDERATION

In the previous section, our design was based on the assumption that external inputs and outputs are locally available at the primitive units. When more realistic input and output constraints are considered, we face the following well-known dilemma: one can pack n processing elements densely in an area of $O(n)$ with the smallest linear dimension (such as in a circle or in a regular polygon) but if all inputs and outputs are only available on the boundary of the area, how can the $O(n)$ I/O wires be routed across the boundary of length $O(\sqrt{n})$! In our CLA design, there are two obvious possible solutions.

The first solution is to relax the area efficiency requirement; let all I/O wires be routed across the boundary, as shown in Fig. 7a. The area is then of $\Omega(n^2)$. As n increases, the portion of the area used for interconnection increases significantly. Obviously for a very large n, this solution uses the most area for input/output wires and the most time for input and output along the I/O wires. Area and time complexities become $O(n^2)$, and $O(n)$, respectively. Although such asymptotic behavior is undesirable, this solution

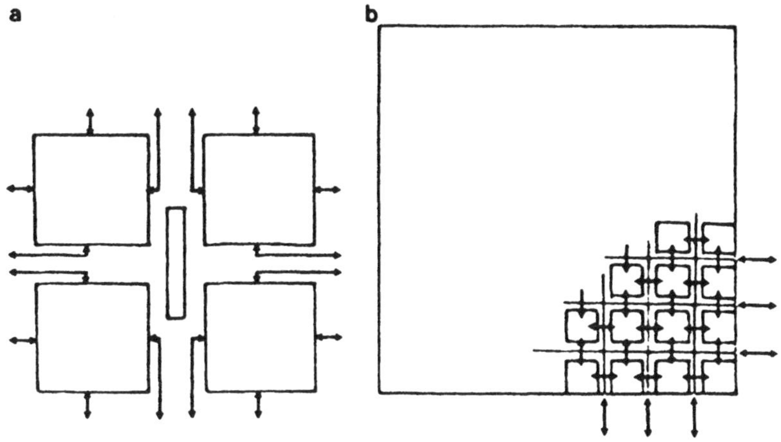

FIG. 7. (a) Boundary I/O. (b) Shared bus I/O. □, 108-bit structure.

may be appealing for not too large n's. It is estimated that for n in the thousands, the total area is only several times that of the previous "generic" adder and the total time (including the input/output time) is less than twice that of the previous generic adder. Furthermore, although the length of the I/O wires is of $O(n)$, the maximum size of addition allowed is still large due to the fact that the proportional constant with n is small.

The other solution relaxes the time efficiency requirement by assuming that inputs and outputs are time multiplexed. In this case the number of I/O wires can be reduced to match the length of the boundary. However, now input and output will take time $\Omega(\sqrt{n})$. Besides the data lines, common address lines and clock and control lines must be provided, as shown in Fig. 7b. In each primitive unit, the input–output subunit has registers and the decoding circuit. With the appropriate address on the bus, data may be loaded into or read out of the registers. Asymptotically, this solution uses area proportional to $n \log^2 n$. But practically for an n as large as 10^5, the addition I/O wire area is still no more than that of the generic adder. This solution will require $c\sqrt{n}$ clocks for each input (output) with the minimum clock period allowed proportional to $\sqrt{n}$. The total I/O time is then proportional to n, which is much longer than the addition time. Therefore, this solution would not be desirable for additions only. By expanding the functionality of the primitive unit, such as performing an accumulation function, the design may be appealing. For example, multiplication can then be done in time proportional to $n \log n$. This outweighs the I/O time and results in a multiplier of time $O(n \log n)$. This is only desirable in the cases where processing time is much longer than the input/output time. Furthermore, the input–output wire reduction also conforms to the limited input/output pin restriction for VLSI chips.

The input–output difficulties discussed above are due to the planar restriction on VLSI processing. Recently, a considerable amount of research effort has been put into developing 3-D VLSI processing and 3-D micro-assembly techniques [11]. Our generic adder would be ideal for these new emerging technologies.

Conclusion

We have shown that using the conventional iterative CLA scheme, nearly optimal CLA adders can be obtained. The layout of the adder is highly regular. Neglecting input–output restriction, the addition of two n-bit numbers can be performed in time $O(\log n)$, using area $O(n)$ with maximum interconnection wire length $O(\sqrt{n})$. When input–output is considered, two different alternative designs are evaluated. Even for practically large n's, it is shown that reasonably good area and time efficiencies can be achieved.

Acknowledgments

This work is supported in part by the Army Research Office under Contract DAAG29-83-K-0126. Thanks go to Poras Balsara for work on the CIF layout plots.

References

1. Ong, S., and Atkins, D. E. A comparison of ALU structures for VLSI technology. *Proc. Sixth Symposium on Computer Arithmetic,* Aarhus, Denmark, June 1983.
2. Brent, R. P., and Kung, H. T. A regular layout for parallel adders. *IEEE Trans. Comput.* **C-31** (Mar. 1982).
3. Bayoumi, M. A., Jullien, G. A., and Miller, W. C. An area–time efficient NMOS adder. *INTEGRATION* (*VLSI J.*) **1** (1983).
4. Mead, C. A., and Conway, L. A. *Introduction to VLSI Systems*. Addison–Wesley, Reading, Mass., 1980.
5. Mavor, J., Jack, M. A., and Denyer, P. B. *Introduction to MOS LSI Design.* Addison–Wesley, Reading, Mass., 1983.
6. Waser, S., and Flynn, M. J. *Introduction to Arithmetic for Digital Systems Designers.* Holt, Rinehart & Winston, New York, 1982.
7. Ngai, T. F. On the capacitive model of computation in VLSI: Its limitation and applicability. Tech. Rep. CS-84-15, Department of Computer Science, Pennsylvania State University, University Park, Pa 16801.
8. Bilardi, G., Pracchi, M., and Preparata, F. P. A critique and an appraisal of VLSI models of computation. In Kung, H. T., *et al.* (Eds.). *VLSI Systems and Computations*. Computer Sci. Press, Rockville, Md., 1981.
9. Rawat, S., Balsara, P., and Irwin, M. J. Pipelined adder. Tech. Rep. CS-85-27, Department of Computer Science, Pennsylvania State University, University Park, Pa 16801.
10. Ullman, J. *Computational Aspects of VLSI*. Computer Sci. Press, Rockville, Md., 1984.
11. Grinberg, J., Nudd, G. R., and Etahcells, R. D. A cellular VLSI architecture. *IEEE Comput.* (Jan. 1984).

Efficient Use of Time and Hardware Redundancy for Concurrent Error Detection in a 32-bit VLSI Adder

BARRY W. JOHNSON, MEMBER, IEEE, JAMES H. AYLOR, SENIOR MEMBER, IEEE, AND HAYTHAM H. HANA, STUDENT MEMBER, IEEE

***Abstract*—This paper presents the design and analysis of a 32-bit adder with concurrent error detection capability. The adder is intended to be used as a building block in the design of more complex circuits and systems using very large scale integration (VLSI). An efficient approach to error detection has been selected through extensive comparisons of several methods that use hardware, time, and hybrid redundancy. Simulation and analysis results are presented to illustrate the adder's timing characteristics, hardware requirements, and error detection capabilities. One unique feature of the analysis is the introduction of error latency as a means of comparing the error detection capabilities of several alternative approaches.**

***Key Words*—building blocks, concurrent error detection, error latency, fault-tolerant computing, hardware redundancy, time redundancy, VLSI.**

I. Introduction

THE ADVENT OF very large scale integration (VLSI) has made the implementation of many approaches to fault tolerance practical because of size, weight, power, and cost reductions. However, the complexity of today's electronic systems has resulted in the incorporation of fault tolerance becoming an overwhelming problem in many applications. System designers are rapidly understanding that fault tolerance, reliability, availability, safety, and similar attributes must be developed as integral parts of the design process if the designs are to be successful. In other words, fault tolerance must be *designed into* a system rather than being *added* at the end of the design process. Consequently, new VLSI-based circuits and design methodologies must be created to support the design of reliable and fault-tolerant electronic systems.

One possible approach to fault-tolerant system design is the use of standard building blocks [1]. The concept is analogous to the standard-cell approach to circuit design. Certain basic functions, such as adders, multipliers, multiplexers, memories, or processors, are designed and placed in a library to be used as building blocks in the construction of larger, more complex systems. Error detection capabilities are designed into the building blocks, and a methodology is defined to allow the building blocks to be interconnected to achieve the functional and fault tolerance requirements of the system. For example, if a processor is a building block and that processor can detect nearly 100 percent of its own errors, an autonomously reconfigurable processing system may be designed using each processor's internal error detection capability to initiate reconfigurations. Clearly, it is easier to design a fault-tolerant system using modules that provide their own error detection than with modules that do not.

The building-block approach to designing fault-tolerant systems provides a versatile medium for the designer. A design library consisting of macros adhering to the design rules specified as a part of the fault-tolerant system design methodology can be created. The availability of the design library and the methodology would facilitate automating the design process for fault-tolerant systems.

This paper addresses the problem of error detection at the building block level. Specifically, a 32-bit adder with concurrent error detection is designed and extensively analyzed. Several approaches for concurrent error detection are considered, and the speed and hardware efficiencies of each are compared. The uniqueness of the design is the efficiency with which the error detection is accomplished. Also, the concept of error latency [2] is introduced and investigated as a means of comparing concurrent error detection approaches.

II. Concurrent Error Detection

Concurrent error detection is the process of detecting and reporting errors while, at the same time, performing the normal operations of the system [3]. Error detection is one of the cornerstones of many fault tolerance techniques. In general, error detection requires some form of redundancy in either hardware, information, time, or software. Certainly the most common form of redundancy is hardware redundancy. For example, a typical hardware redundancy method is duplication with comparison (DWC) [3]. In DWC, as the name implies, the module used is

Manuscript received April 1, 1987; revised June 30, 1987. This work was supported in part by the Virginia Center for Innovative Technology under Contract 5-30960.

The authors are with the Center for Semicustom Integrated Systems, Department of Electrical Engineering, University of Virginia, Charlottesville, VA 22901.

IEEE Log Number 8717252.

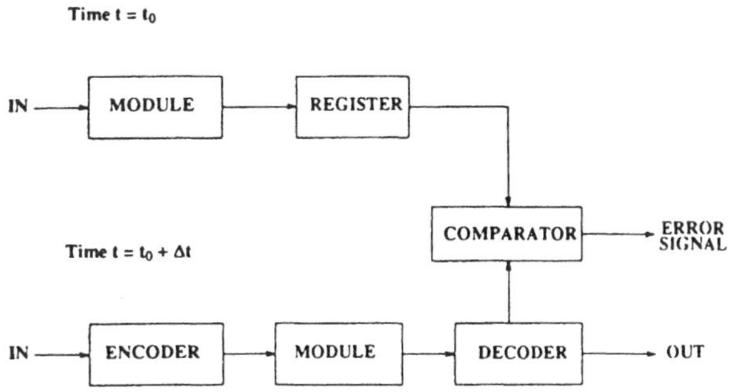

Fig. 1. Time redundancy for permanent fault detection.

duplicated and the results are compared using a comparator. DWC involves a simple and straightforward design that is effective in detecting all single faults that can result in an error. The fundamental problem with hardware redundancy is its impact on physical weight, size, power consumption, and cost. The DWC method, for example, uses at least 100 percent redundancy in the hardware of the unit. Therefore, many applications cannot afford the DWC method.

To overcome some of the difficulties with hardware redundancy, time redundancy has recently received much attention [4]–[7]. Time redundancy attempts to reduce the amount of extra hardware at the expense of using extra time. In many applications, additional time may be much more affordable than extra hardware. For example, the speed at which operations are performed internal to an integrated circuit (IC) may be five to ten times the speed at which information can be transferred across the IC boundaries [8]. Therefore, it may be completely reasonable to perform operations internal to an IC multiple times.

The basic concept of time redundancy is the repetition of computations in ways that allow errors to be detected. To allow time redundancy to be used to detect permanent faults, the repeated computations are performed differently, as illustrated in Fig. 1. During the first computation at time t_0, the operands are used directly, and the results are stored for later comparison. During the second computation at time $t_0 + \Delta t$, the operands are modified, prior to use, in such a way that errors resulting from permanent faults in the module are different in the first calculation than in the second and can be detected when the results are compared. The basic concept of time redundancy is that the same hardware is used multiple times in differing ways such that comparison of the results obtained at the two times will allow error detection.

Various techniques that use time redundancy as the sole means of error detection are presented in the literature. Alternating logic [4], recomputing with shifted operands (RESO) [5], and recomputing with swapped operands (RESWO) [6] are basically the leading ideas in this subject. The difference between these techniques resides in the encoding and decoding functions used. Complementation, arithmetic shifting, and swapping are used with alternating logic, RESO, and RESWO, respectively. Each technique has been shown to possess the ability to detect single errors, assuming that the circuits used are appropriately structured. For example, alternating logic requires that the circuit be self-dual [4], while RESO and RESWO are both developed assuming iterative circuits are used.

Most of the time redundancy techniques, however, require a considerable amount of redundant hardware. The alternating logic technique, for example, may require an increase of 100 percent in hardware to make a function self-dual [4]. Also, RESO and RESWO require the addition of extra circuitry to implement their coding functions. Moreover, the time required to complete the computation is increased due to the repetition of the calculation. In fact, the time required to complete a calculation is at least doubled in RESO, RESWO, and alternating logic.

III. Recomputing Using Duplication with Comparison

In an effort to develop a concurrent error detection technique that possesses the speed of hardware redundancy schemes but utilizes less hardware, a combination of DWC and RESWO was chosen. The method, which is referred to in this paper as recomputing using duplication with comparison (REDWC), is best described as a hybrid redundancy technique. The method with which error detection is accomplished resembles that of DWC. Time redundancy is then used to complete the calculation and obtain the final result. For example, a 32-bit addition operation can be realized by using each of two 16-bit adders twice. During each calculation, the error detection is accomplished by comparing the results from using the two adders. REDWC is similar in many respects to a method described by Toy [9].

To best describe the basis of REDWC, the operation of an n-bit full adder was chosen as an example. The n-bit adder, as well as the operands, is divided into lower and upper halves with each half consisting of $n/2$ bits. [6]. To perform an addition, two calculations are executed. In the *first calculation*, the lower halves of the operands are added in both halves of the adder. In other words, the adder performs the addition of the lower halves of the operands twice but in parallel on the upper and lower halves of the adder, as seen in Fig. 2. The results are then compared and one result is stored to represent the lower half of the final output. The adder hardware is then used a second time to perform a *second calculation*. The second calculation starts by selecting the upper halves of the operands with the output carry of the first computation to be added in both halves of the adder. The results of the second calculation are then compared and are available as the upper half of the final output.

Selection of the appropriate operand half is performed using multiplexers (MUX's). Also, the carries at the boundaries of the adder are handled using one MUX. Fig. 3 shows a block diagram of a 32-bit adder that uses the REDWC technique for error detection.

The checking for errors is performed during each calculation by comparing the outputs of the upper and lower

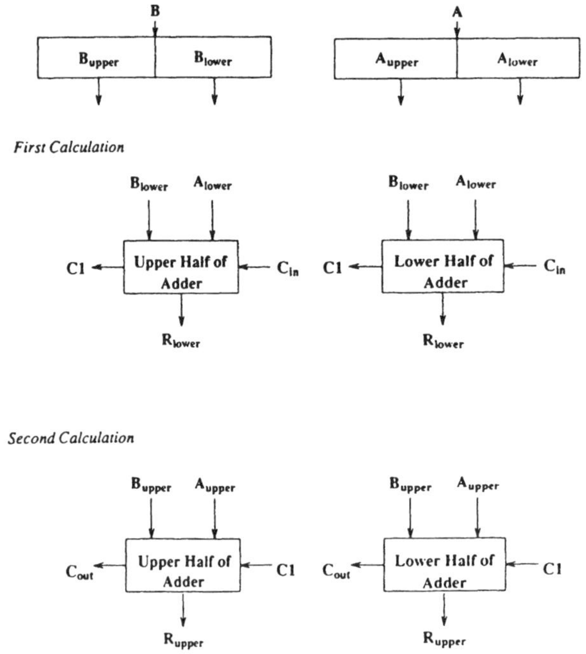

Fig. 2. Addition operation using REDWC.

halves of the adder. Although error detection is accomplished by a simple comparison between the results of the two halves, the same hardware is used twice to obtain the final results. Therefore, the concept of duplication with comparison is employed as the error detection mechanism, and the concept of time redundancy is applied to complete the calculation and obtain the final output.

Since the error detection mechanism of REDWC is identical to that of duplication with comparison, both techniques possess the same ability to detect faults; that is, the ability to detect all single faults confined to one half of the adder in arithmetic and logic operations, as long as both halves do not fail in a similar manner and at the same time. REDWC is also capable of detecting errors in lookahead-carry adders provided that the lookahead-carry operation does not overlap the boundary between the two halves of the adder. In other words, the carry between the two halves must ripple from the lower to the upper half of the adder, but within each half of the adder, lookahead-carry operations may be performed.

IV. A VLSI Adder Using REDWC

A 32-bit adder was designed using REDWC as the sole means of achieving error detection. The adder is designed such that the user provides the input stimuli at time t_0 and reads the result after t_1 nanoseconds have elapsed. Therefore, it is assumed that the input operands are available throughout the calculation. However, since the complete result is available in two halves at two separate times, a register is provided to store the lower half of the result awaiting the upper half to be provided after the second calculation.

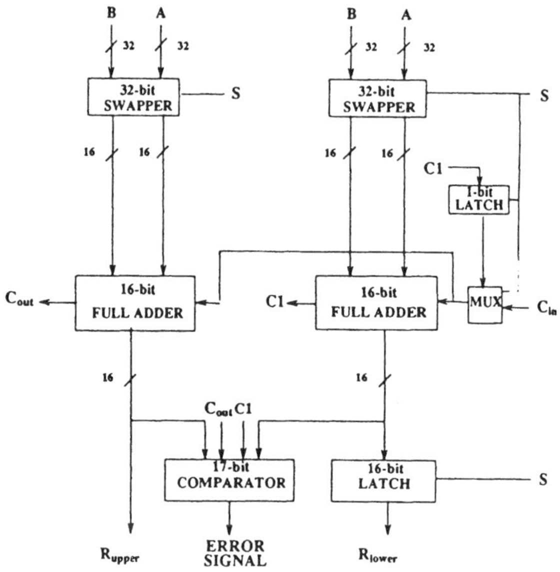

Fig. 3. 32-bit adder using REDWC for error detection.

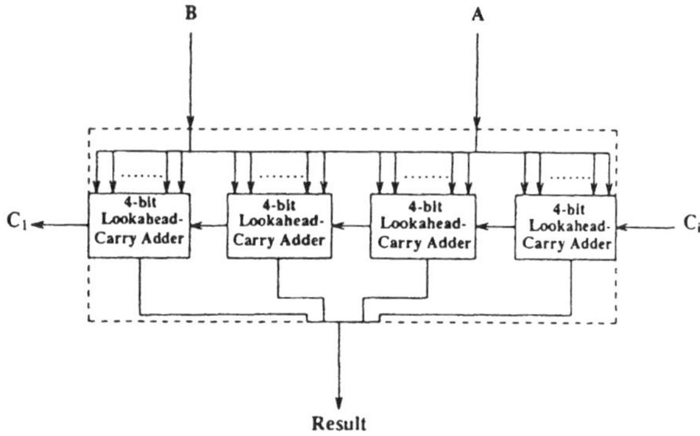

Fig. 4. 16-bit lookahead-carry adder.

The adder, shown in Fig. 3, consists of two "swappers," the adder itself, a MUX, a storage register, and a comparator. Each swapper consists of 32 2-1 MUX's that select the appropriate operand half. A control signal S is provided to manipulate the MUX's and, consequently, control the operation of the adder. When S is low, the lower halves are selected to be added in both halves of the adder; and when S is high, the upper halves are selected. Each half of the adder consists of four 4-bit lookahead-carry adder cells. Each cell possesses a lookahead-carry capability but the carry is rippled between each cell, as shown in Fig. 4. The output carry of the lower half (C_1) is stored in a 1-bit latch to be used as the input carry for the second calculation. The 2-1 MUX is used to appropriately handle the carry at the adder boundaries.

Fig. 5 shows a timing diagram of the sequence of operations that occurs in the adder for one set of input operands. The shaded area represents an ambiguous result (i.e., X) that appears on the output of a module before it assumes a definite value. The ambiguous value appearing in the timing diagram is due to the unavailability of the results in the beginning and during the calculation. The

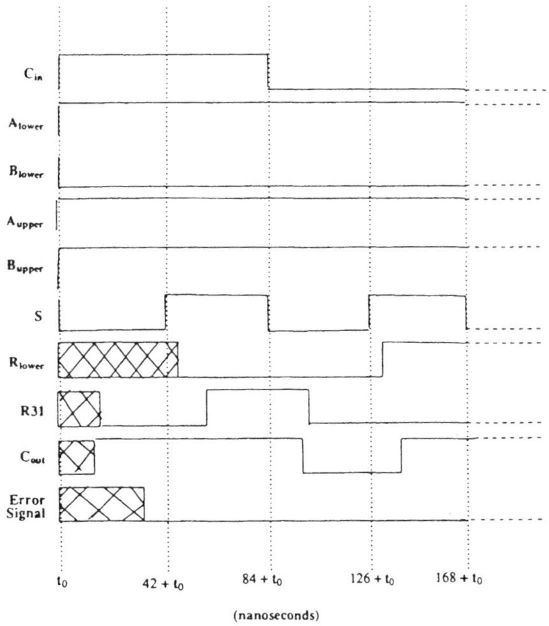

Fig. 5. Timing diagram of REDWC adder.

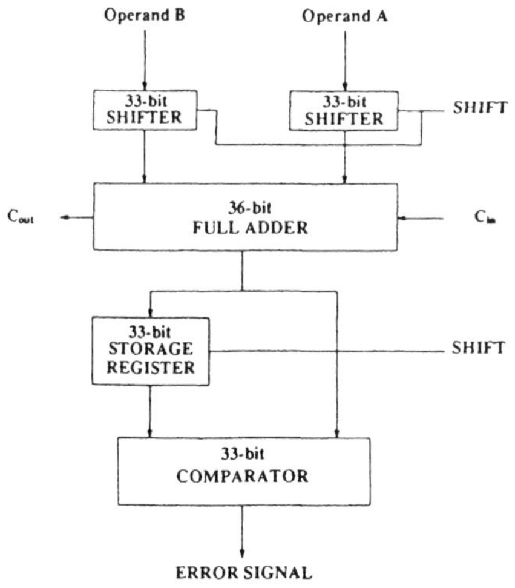

input operands shown in Fig. 5 are assumed to have the same value throughout the calculation; for example, in the timing diagram of Fig. 5, the bits of A_{lower} are all high during the calculations. The signal S controls the calculation and determines the speed with which the adder is driven. The frequency of S was determined according to the critical path of the adder (i.e., the maximum time required to complete one calculation plus obtain the error signal on the comparator's output).

During the period when S is low, the lower half of the result will be available on the output of both adder halves. A 17-bit comparator is used to compare these results and to provide an error signal as an indication of the comparison's outcome. On the transition of S from 0 to 1, several actions occur: (1) the lower half of the result is stored in a 16-bit register, (2) C_1 is stored in the 1-bit latch, (3) the upper halves of the operands are selected (as well as C_1) through the MUX's, and (4) the second calculation is started. When S is high, the lower half of the result will be available as R_{lower} on the output of the storage register and the upper half of the result R_{upper} can be read immediately from the adder output as soon as it becomes available.

To provide a measure of the hardware cost of the REDWC technique, an adder was implemented using the General Electric (GE) 2-μm IGC20000 CMOS gate array technology [10]. The number of gate array cells required for implementation was calculated and compared with that of an RESO adder, a DWC adder, and the nonredundant version of the adder. All the adders are assumed to perform a 32-bit addition and are implemented using the same macro design library [11]. A block diagram of a 32-bit adder using RESO for error detection is shown in Fig. 6. The adder is 36 bits wide because of the extension needed to accommodate the shift operation. The smallest

TABLE I
THE MACROS USED IN DESIGNING THE ADDERS

Macro Name	Macro Symbol	# of Cells	
		Best Case	Worst Case
Two-Input NAND	ND2	2/3	1
Three-Input NAND	ND3	1	1
Four-Input NAND	ND4	1 1/3	2
Four-Input NOR	NR4	1 1/3	2
Two-Input EXCLUSIVE-NOR	XN	1 2/3	2
Inverter	I1X	1/3	1
Inverter	I2X*	2/3	1
Four-bit Adder	F74283	31 2/3	40
One-bit D-Flip Flop	DFR	4	4
Eight-bit D-Flip Flop	F74273	38	40
2-1 MULTIPLEXER	SWAP1†	2 1/3	4
One-bit Latch	LATCH†	4 2/3	5

* I2X has twice the drive capability of I1X.
† Modules constructed using primitive macros.

adder macro available in the design library is 4 bits wide. Therefore, in order to extend the adder by 1 bit to accommodate the shift operation, a 4-bit adder macro has to be used resulting in a 36-bit adder. The shifters, shown in Fig. 6, are implemented using 2-1 MUX's by appropriately connecting the inputs such that the desired arithmetic shifting operation is performed. Also, in order to decode the result of the adder in the second calculation, the adder's output is appropriately routed to the storage register such that the desired bits, with their correct locations, are stored.

V. HARDWARE COMPARISONS

The number of cells required to build the adders and the area, on the chip, needed for these cells depends on the way these cells are placed and routed. Therefore, a *best-* and a *worst*-case cell count were conducted based on the method with which each macro would be utilized to give a lower and an upper bound of the cost, as shown in Table I. For example, the (I1X) inverter requires 1/3 cell to be

TABLE II
BEST- AND WORST-CASE CELL COUNT OF THE 32-BIT ADDERS

Technique	# of Cells		Average # of Cells	% Increase in Average # of Cells
	Best case	Worst case		
Non Redundant	253.33	320	286.67	0
Duplication with Comparison	542.33	685	613.67	114
RESO	665	876	770.5	169
REDWC	446.67	582	514.34	79

TABLE III
TIMING COMPARISONS

Technique	Calculation Time (nanoseconds)	% Increase
Non Redundant	37.5	0
Duplication with Comparison	48.3	28.80
RESO	83.5	122.67
REDWC	52.6	40.27

implemented whereas the EXCLUSIVE-NOR (XNOR) gate requires $1\frac{2}{3}$ cells. Therefore, in the best routing case, the unused 1/3 of the cell from the XNOR can be utilized to contain the inverter, resulting in a cell count of 2. However, in the worst case, an entire cell will be devoted to implementing the inverter and the XNOR will require 2 cells, resulting in a total of 3 cells. Table II shows the best- and the worst-case cell count, as well as the average cell count of each adder. The average number of cells is used as a means of comparison because routing efficiency may vary between candidate designs. For example, the nonredundant design may produce worst-case routing while REDWC yields best-case, or the opposite may occur. As a result, the actual efficiency of REDWC could vary, so an average has been used.

From Table II one can see that the amount of hardware required to implement an adder using the REDWC technique is less than duplication with comparison and RESO. The increase in hardware for the DWC adder comes from the physical duplication of the adder plus the use of the comparator. In the RESO adder, however, the entire output of the adder from the first calculation must be stored, resulting in the use of a 33-bit storage register. Also, the comparison is performed on the full width of the result, creating the need for a 33-bit comparator. The arguments differ for the REDWC adder because only one half of the adder's result must be stored. Also, the comparison is performed on one half of the result, leading to the use of a smaller size register and comparator. Therefore, less extra hardware is needed to implement a REDWC adder than the other techniques, resulting in a significant impact on size, weight, and power consumption.

VI. SPEED COMPARISONS

An equally important criterion for comparison is the resulting system throughput. Simulations of the above designs were conducted using the TEGAS-5[1] simulation system [12], [13]. The simulations were used to provide both functional verification of the REDWC technique and performance comparison. Actual delays were injected in the circuit components so that a timing analysis could be made. Table III summarizes the simulation results for the nonredundant, RESO, DWC, and REDWC adders. The table shows the amount of time (in nanoseconds) required to complete a calculation (including error detection), as well as the percentage increase in calculation time over that of the nonredundant adder. The calculation time represents the maximum time between the application of the input carry to the adder and the availability of the error signal on the output of the comparator. For the REDWC technique, the second error signal marks the end of one complete computation and error checking cycle. The comparators used for the various adders have the same 10.8 ns of propagation delay. The result of the addition is actually available 10.8 ns earlier than the times shown in Table III; however, the error signal is not available until approximately 10.8 ns after the addition is completed.

From Table III, one can see that the penalty, in time, paid by using REDWC for error detection is considerably less than that of RESO and comparable to that of DWC. Although time redundancy is part of the REDWC technique, the extra time required to complete the calculation is only 40.27 percent over that of the nonredundant operation. Therefore, both hardware and time are saved when REDWC is used rather than other time redundancy methods.

The extra time needed to perform the computation in REDWC adder might slow the system down but can provide extra error detection capability over that of DWC. The next section discusses this aspect from a statistical standpoint.

VII. PROBABILISTIC ANALYSIS OF ERROR LATENCY

The fact that the REDWC adder exercises the hardware twice during each calculation leads to the assumption that REDWC will show an erroneous result, in the presence of fault, sooner than a DWC adder. Therefore, to investigate the validity of this assumption, the impact of *error latency* on concurrent error detection is analyzed.

The period of time between the occurrence of a fault and the detection of the error caused by that fault is called error latency, as illustrated in Fig. 7. Error latency is very important in digital circuits because it affects a system's life expectancy [2]. Also, minimizing error latency is important in decreasing the probability of encountering a second fault before the first fault is detected. The probability that the error latency (EL) of a fault is exactly n input patterns is given by [2]

$$P[EL = n] = (1-q)^{n-1}q \tag{1}$$

[1] TEGAS-5 is a registered trademark of the GE Calma Corporation.

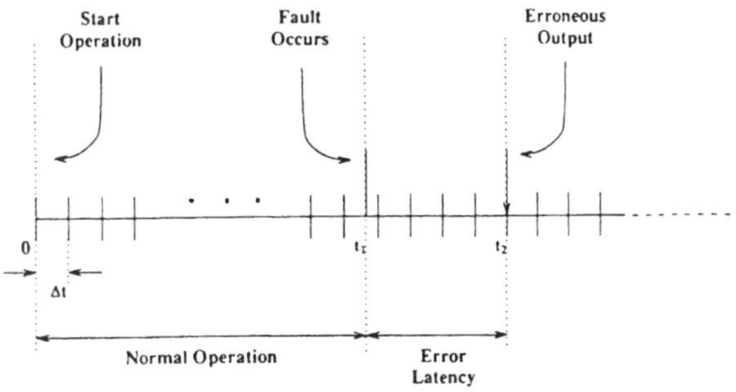

Fig. 7. Error latency.

where q (the error probability) is the conditional probability of an incorrect output upon application of a randomly chosen input vector, given that the fault is active. An active fault is a fault that divides the set of primary input patterns into two subsets; one that will sensitize the fault and another subset that will not. In other words, an active fault is a detectable fault.

The probability distribution of EL is the probability that EL is less than or equal to n input patterns and is given by [2]

$$P[EL \leqslant n] = 1 - (1-q)^n, \qquad n \geqslant 1. \tag{2}$$

In other words, $P[EL \leqslant n]$ is the probability that it will take no more than n input patterns to produce an erroneous output given that the circuit contains a particular fault.

The time required to apply n input patterns to a DWC adder is given by

$$t = n \times (T_{\mathrm{DWC}}) \tag{3}$$

where T_{DWC} is the calculation time in the DWC adder for one input pattern. Therefore,

$$n_{\mathrm{DWC}} = \frac{t}{T_{\mathrm{DWC}}} \tag{4}$$

where n_{DWC} is the number of input patterns that can be applied to the DWC adder in t seconds. Also, the time required to apply n input patterns to the REDWC adder is

$$t = n \times (T_{\mathrm{REDWC}}) \tag{5}$$

or,

$$n_{\mathrm{REDWC}} = \frac{t}{T_{\mathrm{REDWC}}} \tag{6}$$

where n_{REDWC} is the number of input patterns that can be applied to the REDWC adder in t seconds, and T_{REDWC} is the calculation time in the REDWC adder for one input pattern. From Table III we have

$$T_{\mathrm{DWC}} = 48.3 \text{ ns} \tag{7}$$

$$T_{\mathrm{REDWC}} = \frac{52.6}{2} \text{ ns} \tag{8}$$

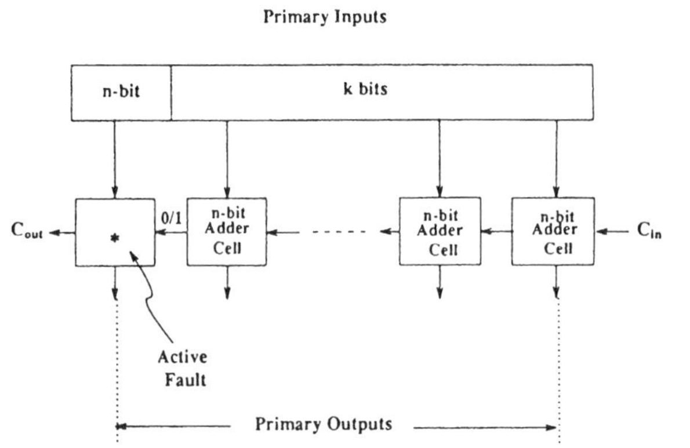

Fig. 8. Fault sensitization in the adder circuit.

because in every computation, the REDWC exercises the adder twice by applying two input patterns (i.e., the two halves of the input operands) whereas the DWC adder uses one input pattern per computation. Hence,

$$n_{\mathrm{DWC}} = \frac{t}{48.3} \tag{9}$$

and

$$n_{\mathrm{REDWC}} = \frac{2t}{52.6}. \tag{10}$$

Therefore,

$$P[EL \leqslant n]_{\mathrm{DWC}} = 1 - (1-q)^{t/48.3} \tag{11}$$

and

$$P[EL \leqslant n]_{\mathrm{REDWC}} = 1 - (1-q)^{2t/52.6}. \tag{12}$$

The error probability (q) is determined by considering the adder structure, as shown in Fig. 8, consisting of a number of n-bit adder cells. Suppose that one, and only one, of the n-bit cells contains a fault, and the fault is detectable. For convenience, Fig. 8 shows the leftmost cell containing the fault; however, the following results are valid regardless of which cell contains the fault. The faulty n-bit cell has a total of $n+1$ input bits (including the carry bits), so there are a total of 2^{n+1} different input combinations that may be applied to that cell. If the detectable fault can be detected by some number m of the 2^{n+1} different input combinations, then the error probability is

$$q = \frac{m}{2^{n+1}}. \tag{13}$$

In other words, the error probability is the probability that an input pattern selected at random will detect the fault.

Any one of the m input patterns that detects the fault will require a specific value to be applied to the carry input of the faulty cell. The carry input can be correctly generated by appropriately selecting one of the 2^{k+1} possible input combinations that may be applied to the previous cells in the circuit. More specifically, we know that each value (i.e., either a 0 or a 1) of the input carry to the faulty

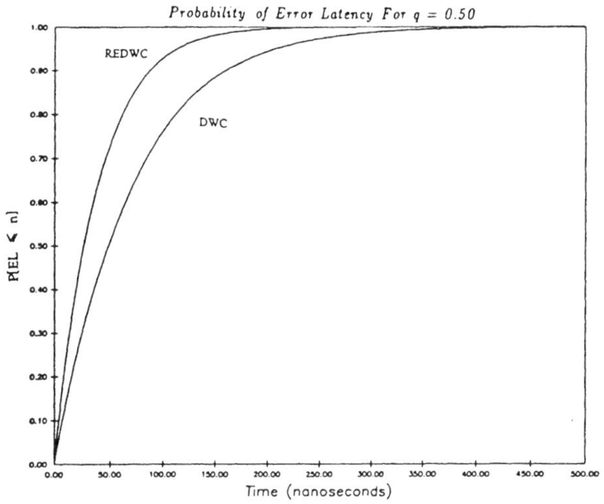

Fig. 9. $P[EL \leqslant n]$ of REDWC and DWC for $q = 0.5$.

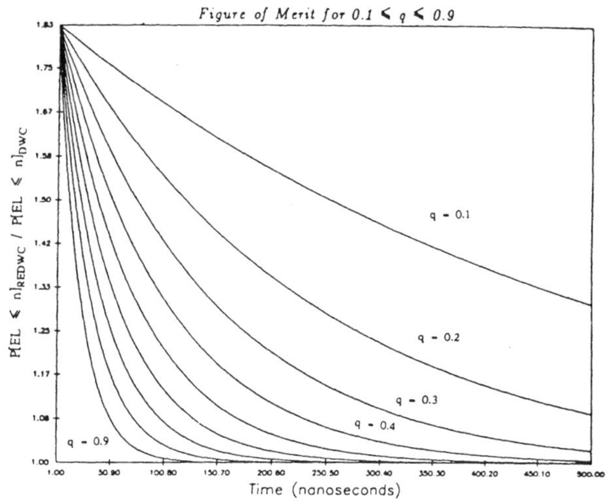

Fig. 10. Figure of merit for different values of q versus time.

cell can be generated by exactly half of the input combinations (namely $2^{k+1}/2$) that can be applied to the previous cells in the adder. Hence, the total number of primary input combinations that can detect the fault is $m \times (2^{k+1}/2)$. Therefore,

$$q = \frac{m \times \left(\dfrac{2^{k+1}}{2} \right)}{2^{n+k+1}} = \frac{m}{2^{n+1}} \tag{14}$$

which implies that the error probability is independent of the number of adder cells. Consequently, the q value for a fault in one cell of the DWC adder is the same as the q value for that same fault in one cell of the REDWC adder. Specifically,

$$q = \frac{m}{2^9} \tag{15}$$

which depends only on the value of m.

Fig. 9 shows a plot of (11) and (12) for $q = 0.5$. At any given time, one can see that $P[EL \leqslant n]_{\text{REDWC}}$ is greater than $P[EL \leqslant n]_{\text{DWC}}$. The REDWC technique has a larger probability of detecting faults than DWC over the same period of time because the number of input patterns that is applied to the REDWC adder is larger than that of the DWC adder. This result indicates that the REDWC adder requires less time, on the average, to detect the error after a fault has occurred than the DWC technique.

Dividing (12) by (11) yields a measure of how the two probabilities vary with respect to time. This measure, which is identified as a *figure of merit*, is the ratio between $P[EL \leqslant n]_{\text{REDWC}}$ and $P[EL \leqslant n]_{\text{DWC}}$, as seen below:

$$\text{figure of merit} = \frac{P[EL \leqslant n]_{\text{REDWC}}}{P[EL \leqslant n]_{\text{DWC}}} = \frac{1-(1-q)^{2t/52.6}}{1-(1-q)^{t/48.3}}. \tag{16}$$

A plot of the above equation as a function of time and q is shown in Fig. 10.

VIII. Conclusions

The design and analysis of an adder with concurrent error detection have been presented The error detection technique employed has been shown to require less hardware than that of a comparable caliber. REDWC also requires less computation time than other time redundancy techniques. Moreover, the REDWC technique possesses the ability to detect faults in a shorter period of time than DWC because of the utilization of time redundancy. Therefore, the error latency of the technique was shown to be less than that of DWC.

Acknowledgment

The authors would like to acknowledge GE-Calma Corporation and GE Semiconductor Business Division for donating the TEGAS-5 and the gate array CAD systems, respectively, for educational purposes.

References

[1] D. A. Rennels, A. Avizienis, and M. Ercegovac, "A study of standard building blocks for the design of fault-tolerant distributed computer systems," in *Proc. 8th Ann. Int. Conf. Fault-Tolerant Comput.*, June 21–23, 1978, pp. 144–149.
[2] J. J. Shedlesky and E. J. McCluskey, "The error latency of a fault in a combinational digital circuit," in *Proc. Int. Symp. Fault-Tolerant Comput.*, 1975, pp. 210–214.
[3] D. Siewiorek and R. Swarz, *The Theory and Practice of Reliable System Design*. Bedford, MA: Digital Press, 1982.
[4] D. Reynolds and G. Metze, "Fault detection capabilities of alternating logic," *IEEE Trans. Comput.*, vol. C-27, pp. 157–162, Dec. 1978.
[5] J. Patel and L. Fung, "Concurrent error detection in ALU's by recomputing with shifted operands," *IEEE Trans. Comput.*, vol. C-31, pp. 417–422, July 1982.
[6] H. H. Hana and B. W. Johnson, "Concurrent error detection in VLSI circuits using time redundancy," in *Proc. IEEE Southeastcon '86 Regional Conf.*, Mar. 23–25, 1986, pp. 208–212.
[7] J. Shedlesky, "Error correction by alternate data retry," *IEEE Trans. Comput.*, vol. C-27, pp. 106–112, Feb. 1978.
[8] D. Patterson and C. Sequin, "Design considerations for single-chip computers of the future," *IEEE Trans. Comput.*, vol. C-29, pp. 108–116, Feb. 1980.
[9] W. N. Toy, "Self-checking arithmetic unit," U.S. Patent Number 4 314 350, Feb. 2, 1982.

[10] J. H. Aylor, E. A. Parrish, and K. L. Pocek, "A semicustom VLSI system design course supported by the General Electric Microelectronics Center," *IEEE Trans. Education*, vol. E-29, no. 2, pp. 85–89, May 1985.

[11] *IGC20000 Gate Array Designer's Reference Manual*, General Electric Semiconductor Business Division, 1985.

[12] *TEGAS Design Language Simulation Reference Manual*, COMSAT General Integrated Systems, Inc. Digital Design Test Division, Austin, TX, 1980.

[13] *Gate Array Design Procedure*, General Electric Semiconductor Business Division, Research Triangle Park, NC, Aug. 2, 1985.

Barry W. Johnson (S'78–M'82) received the B.S., M.E., and Ph.D. degrees in electrical engineering from the University of Virginia, Charlottesville, in 1979, 1980, and 1983, respectively.

Currently he is an Assistant Professor in the Department of Electrical Engineering at the University of Virginia. He is also a member of the Center for Semicustom Integrated Systems, a Technology Development Center of the Virginia Center for Innovative Technology. Prior to joining the University, he was with Harris Corporation in Melbourne, FL, where he participated in the design and analysis of fault-tolerant computer systems for aerospace applications. His research interests include fault-tolerant computing, VLSI architectures, VLSI testing, and microprocessor-based systems.

Dr. Johnson has authored or coauthored approximately 30 articles in the above areas and is presently preparing a text on fault-tolerant computing. He is presently active in the IEEE Computer Society as Chair of the Membership Development Committee, as a member of the Membership and Information Board, and as a member of the Editorial Board of *IEEE Micro*. Dr. Johnson is a member of the IEEE Computer Society, Tau Beta Pi, Eta Kappa Nu, and Sigma Xi.

James H. Aylor (S'71–M'76–SM'82) received the B.S., M.S., and Ph.D. degrees in electrical engineering from the University of Virginia, Charlottesville, VA, in 1968, 1971, and 1977, respectively.

Currently he is an Associate Professor in the Department of Electrical Engineering at the University of Virginia and Director of the Center for Semicustom Integrated Systems. During 1983, he was a Visiting Scientist with the Federal Systems Division of IBM. His research interests are in the areas of design automation of digital systems, VLSI systems, fault tolerance and testing, hardware description languages, and microcomputer applications. He has published approximately 55 technical papers covering the above related areas.

Dr. Aylor is a member of Sigma Xi, Tau Beta Pi, AAAS, and the IEEE Computer Society, where he is currently Vice President for Conferences and Tutorials. He has also served as a member of the Governing Board and the *IEEE Micro* Editorial Board, Chairman of the Computer Society Press, and founder of the Technical Committee on Computing and the Handicapped within the Computer Society.

Haytham H. Hana (S'87) is a digital designer in the digital module compiler design group at Seattle Silicon Corporation. He joined Seattle Silicon in 1987 after receiving the M.S. degree in electrical engineering from the University of Virginia and the B.S. degree in physics from Virginia Commonwealth University in 1987 and 1985, respectively. His interests are in fault-tolerant computing, VLSI, and computer design.

Mr. Hana is a member of the IEEE Computer Society and Phi Kappa Phi.

Part IV:
VLSI MULTIPLIER IMPLEMENTATIONS

Editor's Comments on Papers 19 Through 24

It has often been suggested that the development of the single chip VLSI multiplier is directly responsible for the growth of interest in the field of digital signal processing. This part examines implementations of VLSI multipliers.

The first paper, by McIver, Miller, and O'Shaughnessy, describes the design of the first single chip parallel multiplier. A revised design, the TRW MPY-16, was first offered for commercial sale in 1976. The commercial product differs from the developmental part described here in directly outputing the most significant half of the product (not multiplexed as shown on Figure 3), implementing a two's complement algorithm, and achieving speeds of 160 nsec (upon introduction in 1976). Current CMOS parts achieve delays of 50 nsec, while bipolar parts achieve delays of less than 10 nsec. This is a very regular design, which is important in minimizing the design cost, although the maximum carry path consists of 2n stages (for an n by n multiplier). Use of non-thresholding logic produces a delay proportional to $\sqrt{n}$ which is comparable to the delay of (less regular) fast multipliers.

In the second paper, Iwano and Steiglitz examine the design of full adder cells for use in implementing array multipliers. They optimize the transistor sizes of the adders for a saturating logic version of the array multiplier of the previous paper.

In the next paper, Cappello and Steiglitz describe the implementation of a Dadda style multiplier with unary to binary converters (i.e., parallel counters). The concept developed contemporaneously by Dadda and Wallace consists of a reduction of the n by n bit product matrix to a two row matrix in log n steps followed by summation of the two row matrix in a carry lookahead adder. The Dadda approach is of significant theoretical importance since, as shown here, it achieves optimal speed for both pipelined and non-pipelined implementations. It has not been widely implemented in VLSI due to its physical irregularity.

Vuillemin presents a fast multiplication algorithm in the next paper. His approach sums the partial product matrix with a tree of carry save adders. This is shown to require O(Log(n)) delays to compute the product of two n-bit numbers.

In the next paper, Harata, *et al.* present the design of an array multiplier that internally uses signed digit numbers (as described by Avizienis in Paper 6). The paper briefly describes signed digit arithmetic, describes conversion between two's complement and signed digit numbers, shows the resulting implementation of a 16 by 16 multiplier in a 230 mil by 250 mil chip using 2.7 μm NMOS technology, and shows that this design compares favorably with other designs.

The final paper of this Part, by Oowaki, *et al.*, shows the performance attainable with advanced (0.6 μm twin-tub CMOS) technology. The net result is a chip with an active area of 40 mils by 80 mils that attains a multiplication time of 7.4 nsec and dissipates only 400 mW when clocked at a 10 MHz rate.

A Monolithic 16 x 16 Digital Multiplier

George W. McIver, Ralph W. Miller and Timothy G. O'Shaughnessy

TRW Systems Group

Redondo Beach, Cal.

THE DESIGN, FABRICATION and testing of a 16 x 16 multiplier fabricated on a single monolithic chip will be described.

The chip was designed for use in a small, fast, multipurpose processor being built for avionics applications. The multiplier is set up for a two's complement number system to be compatible with that processor. The chip also features tri-state buffers for the output signals and holding registers for the input signals. This feature not only allows the external leads in the system to be time shared, but also allows the chip signal leads to be time shared, reducing the total number of leads to 32 data lines, 4 control signals, plus several leads for power supply and ground.

Figure 1 shows a schematic for the most commonly repeated cell. One important difference in the design of this circuit from most previous circuits can be determined by studying this schematic; the logic levels are not restored at the output of each block. The reasons for this is that risetimes do not add linearly but approximately as root sum of squares. Thus, if one stage has τ risetime to some circuit threshold, N stages between thresholds have $N^{1/2}\tau$ delay rather than $N\tau$ as one would get with thresholding at each stage. This makes an important difference in the delay times for large asynchronous circuits. The reason one can afford to do this for an LSI chip is that the on-chip environment is much less subject to switching transients, ground noise, etc.

This technique of using many stages of logic before restoration of logic levels is particularly appropriate for a *double-railed* logic system, where a signal and its complement are separately synthesized and both are sent to the next logic elements; eliminating intermediate inverters which would have threshold effects leading to linearly accumulating delay times.

Figure 2 is a photograph of a completed chip, showing that portion of the circuit which corresponds to Figure 1. Corresponding circuit elements occupy about the same realtive positions in both Figure 1 and Figure 2. Figure 3 is a block layout showing how the individual cells are assembled to form the complete circuit. The circuit is exceedingly repetitive; this was used to good advantage, as mask generation was accomplished via layout on automated layout equipment*. The layout of *cell* A was optimized for minimum dimensions, and the other cells were made to fit.

This circuit was fabricated by the triple diffusion process[1] starting with homogenous P-type wafers of 3 Ω-cm resistivity. In this process, the collector, base and emitter are formed by successive diffusion steps. This process is actually very old in the history of integrated circuits, but was abandoned in favor of buried layer epitaxial construction for most applications. There were two major reasons: (1)–the lightly doped collector diffusion was very difficult to control; (2)–the lower r_c' of the buried layer. This second objection is still present, but is not an important item in the design of emitter follower logic circuits.

With the present technology it is easy to control the collector doping sufficiently well to make LSI circuits. Using ion implantation for the collector deposition, the collector sheet can be held to ±5 per cent with relative ease. Most resistors are made using the collector under a base diffusion, however, and the accuracy of these resistors is only ±20 per cent because of diffusion variability, Figure 4 is a cross section of triple diffused devices, showing a coalesced NPN, PNP and load resistor.

Counting each emitter or resistor as a device, the device count for this chip is about 16,700. The chip size is 301 x 279 mils, giving a figure about 5 mil^2/device. This device density is somewhat higher than for standard epitaxial construction due to the fact that all devices are self-isolated and may be closely packed.

Yield of functional devices at wafer probe test has been excellent, averaging about 3 good die per wafer – out of only 19 possible. This good yield is due to the combination of the triple diffusion process – which gives very few defects, albeit poorer transistors than epitaxial construction – with emitter follower logic, which does not require exceptional device performance.

The wafer probe test is performed in the following way. A read-only memory is programmed to have a sequence of input data in one block and the corresponding correct answers in another block; the input words are fed to the chip and the output of the chip compared bit-by-bit to the correct answer. The results of the comparison are displayed on an oscilloscope to give the operator instant visibility as to which bits are in error and in which word position. The display is a plot with X-position corresponding to input word number and Y-position corresponding to output bit number. The Z-axis is modulated in such a way that one gets a spot at each position where a bit did not compare. In this way systematic errors can be detected and the mask set corrected.

Tests have been performed on packaged units. The results of these tests indicate a fixed delay of about 200 ns for all data outputs, plus a delay proportional to $N^{1/2}$ where N is the number of logic delay stages in the matrix associated with the output. This last term is about 130 ns for 31 stages (the maximum) giving a total multiply time of 330 ns.

---

*Applicon 762

[1] Buie, J. L., and Breuer, D. R., "A Large-Scale Integrated Correlator," *The Journal of Solid-State Circuits*", p. 357-363; Oct., 1972.

Operation of the Circuit

To load data into the holding registers, the tri-state buffers for that register are disabled by applying 2.5 V to the READ terminal, the data lines are energized, and the appropriate LOAD terminal is toggled high. After about 330 ns, the data may be taken from the multiplier by enabling the READ commands. Data may be loaded into both registers simultaneously if desired, and output of the most significant half product may be taken at the same time as the least significant half product. Alternatively, the load may be accomplished one word at a time and either the most significant half or least significant half product may be read out first.

Figure 5 shows a set of typical waveforms for this circuit.

Acknowledgments

The authors would like to express their appreciation to K. Morishige for an excellent layout, and to G. Graves for assistance in the design and testing of these circuits.

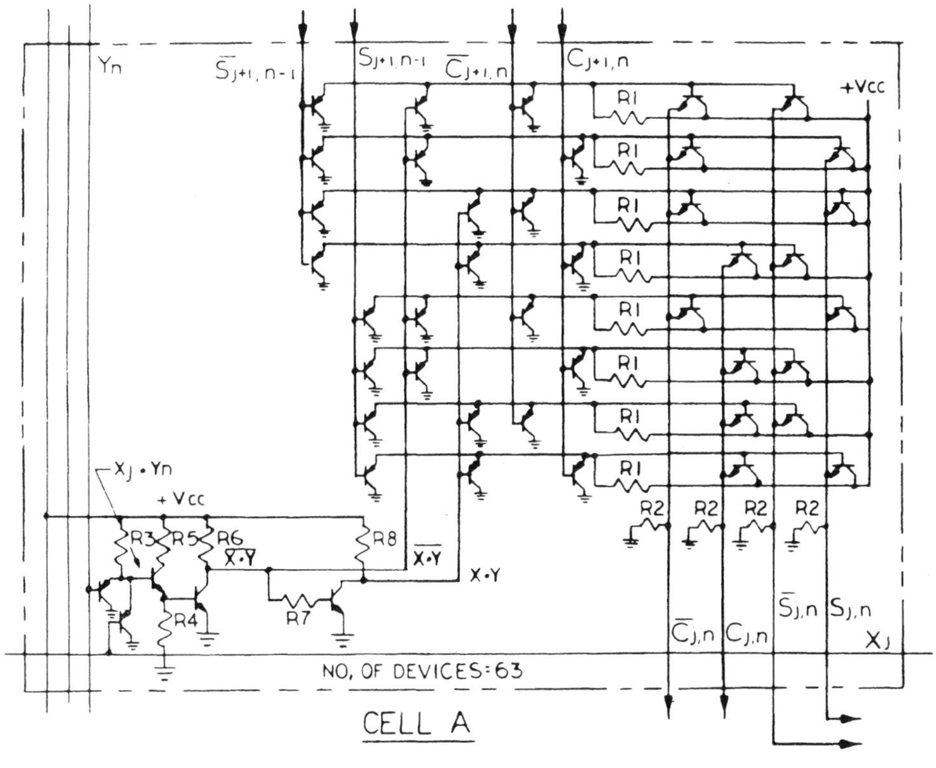

FIGURE 1—Schematic diagram for most commonly repeated cell.

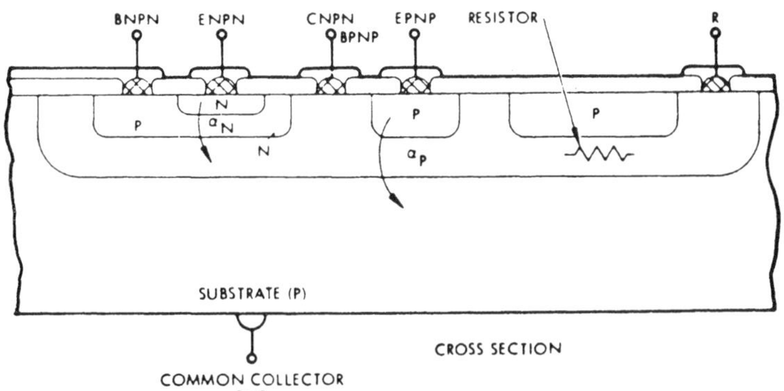

FIGURE 4—Cross-section of NPN and resistor. This figure illustrates how these devices may be coalesced to save space.

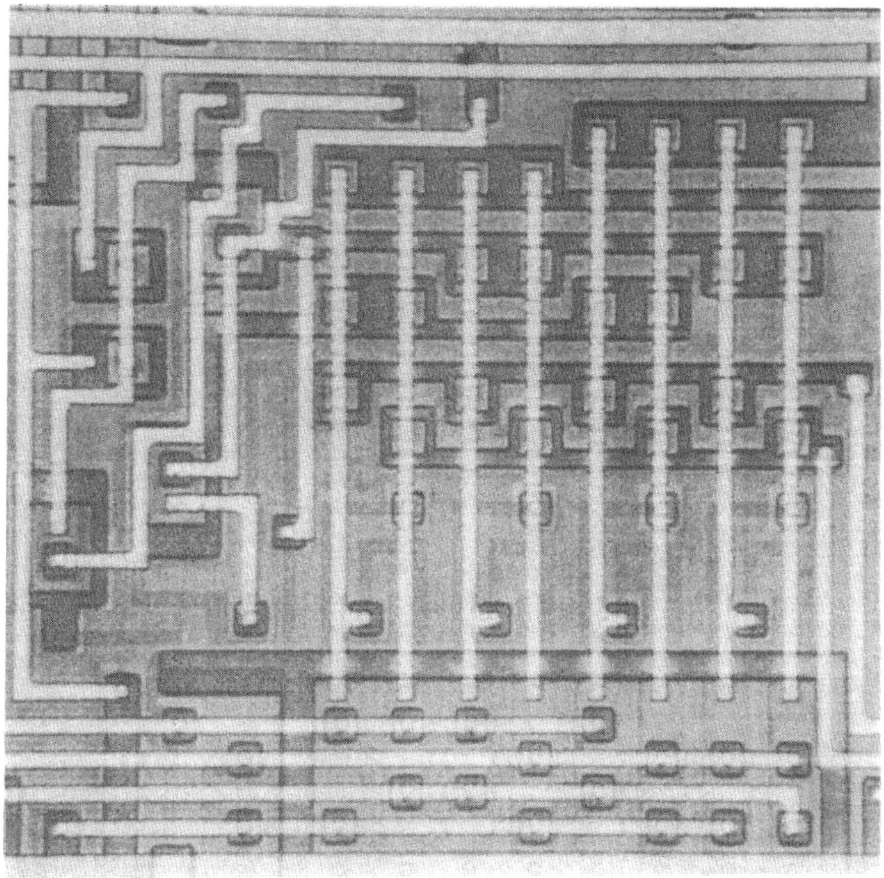

FIGURE 2—Photograph of most commonly repeated cell on a completed chip.

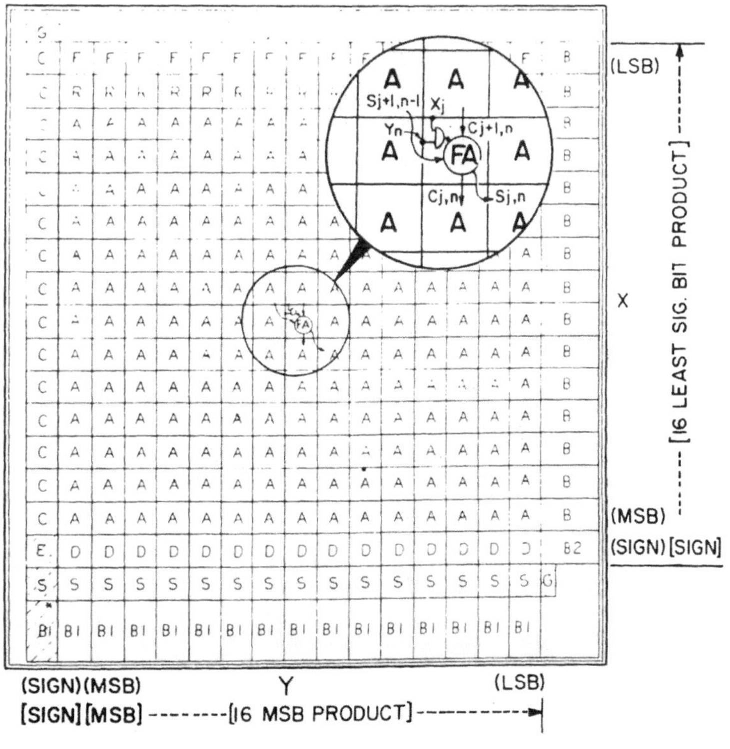

[Left]

FIGURE 3—Block layout showing how individual cells are assembled to make a multiplier.

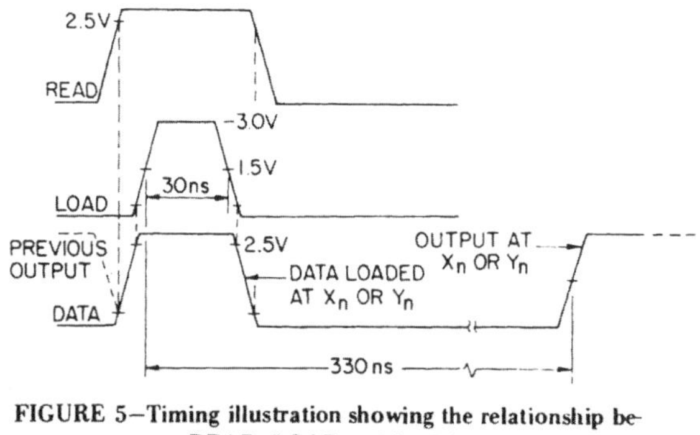

FIGURE 5—Timing illustration showing the relationship between READ, LOAD and DATA signals.

Optimization of One-Bit Full Adders Embedded in Regular Structures

KAZUO IWANO AND KENNETH STEIGLITZ, FELLOW, IEEE

Abstract—We study the problem of optimizing the transistor sizes in the one-bit nMOS full adder either isolated or embedded in a regular array. A local optimization method that we call the *critical-path optimization method* is developed. In this method, two parameters at a time are changed along the critical path until a locally optimal choice of transistor sizes is found. The critical-path optimization method uses the Berkeley VLSI tools and the hierarchical layout language ALLENDE developed at Princeton. First, we optimize the isolated one-bit full adder implemented in three ways: as a PLA, data selector, and with random logic. The details of the critical-path optimization method and power-time tradeoff curves are illustrated here. Second, we optimize the one-bit full adder embedded in a simple array multiplier. The entire 3 × 3, 4 × 4, 8 × 8, and 10 × 10 multipliers are optimized and their local optima are compared. Because the optimization of the entire circuit becomes less practical when the circuit becomes larger, we develop a method that makes use of circuit regularity. We prove that some small array of one-bit full adders, called the *canonical configuration*, has the same local optima as the $n \times n$ multiplier for large n, with the criterion of minimizing the delay time T. Hence, we can greatly reduce the computation load by optimizing this canonical configuration instead of optimizing the entire circuit. Experimental results confirm the effectiveness of this approach.

I. Introduction

REGULAR arrays of cells are used often in custom chips for digital signal processing. Such regular arrays lead to designs that are easy to lay out efficiently and have high throughput. For example, bit-parallel and bit-serial multipliers can be constructed from one- and two-dimensional arrays of one-bit full adders, as well as a wide variety of pipelined FIR and IIR filters (see, for example, [1]–[5], [9], [12], [16], [25]). This paper is aimed at the problem of optimizing such large arrays. The technology used throughout is 4 μ nMOS, but the general approach described is applicable to other technologies as well.

We will develop what we call the *critical-path optimization method*. This is a heuristic method for finding a locally optimal choice of transistor sizes by using systematic variation of the parameters along the critical path. The optimization loop uses the Berkeley tools [15] CRYSTAL (for timing) [19], POWEST (for estimating the power consumption), and the constraint-based high-level layout language ALLENDE, developed at Princeton [11], [13]. We illustrate the critical-path optimization method and resulting power-time tradeoff curves by optimizing an isolated one-bit full adder implemented in three topologies, namely, the PLA, data selector, and with random logic.

Investigating the optimization of the one-bit full adder embedded in an array multiplier, we will next study how to take advantage of a circuit's regularity to reduce the optimization workload. Here we develop the *canonical configuration method*, which is shown to be practical for optimizing large regular structures. We will analyze the critical paths of the $n \times n$ multiplier using a finite automaton and extract its *canonical configuration*. Then we will prove that the optimization of this *canonical configuration* provides, in the limit of large n, the same local optima as the $n \times n$ multiplier.

The problem of optimizing the transistor sizes has been studied by several authors. ANDY, developed by Trimberger [17], sizes transistors in a symbolic description of a chip to match the load the transistors are driving, then performs power optimization off the critical path. Glasser and Hoyte [6] developed what they called *macromodels* of VLSI circuits and optimize the transistor sizes in a critical path. However, their macromodel is sometimes inaccurate and leads them to errors of as much as 70 percent when compared to the SPICE [15] circuit simulation. Matson [14] improved their macromodels to be more accurate and computationally faster, and used them for nonlinear optimization of transistor sizing. Other related work is reported by Strojwas, Nassif, and Director [18], and Jouppi [23]. Our efforts will concentrate on taking advantage of circuit regularity to make practical the optimization of large arrays.

II. Critical-Path Optimization Method

The design of VLSI chips often involves the difficult task of effecting tradeoffs among three important measures, that is, the delay time T; the peak or average power dissipation P_{max} or P_{ave}, and the area A. There are many circuit choices which can be used for controlling these tradeoffs. For example, we can control the choice of an appropriate topology, use precharging, superbuffers, insert or delete logic stages to control the appropriate fanout factor, etc. However, we will concentrate in this paper on the choice of pulldown diffusion widths, because the

Manuscript received July 30, 1985; revised March 21, 1986. This work was supported in part by the National Science Foundation under Grant ECS-8414674, by the U.S. Army Research–Durham Contract DAAG29-85-K-0191, DARPA Contract N00014-82-K-0549, by the Office of Naval Research under Grant N00014-83-K-0275, and by IBM–Japan.

The authors are with the Department of Computer Science, Princeton University, Princeton, NJ 08544.

IEEE Log Number 8609627.

problem of sizing transistors is important but very tedious work for chip designers, and lends itself well to efficient solution by automated methods.

The constraint-based high-level layout language ALLENDE [13] enables us to parameterize a circuit; that is, ALLENDE accepts a circuit parameter vector π and produces the layout $C(\pi)$. Since the circuit performance (such as the delay time T, the power dissipation P, and the area A, etc.) is determined by the circuit, the vector (P, T, A) can be expressed as a function of C. Since the circuit C is parameterized as $C(\pi)$, the vector (P, T, A) can finally be expressed as a function $g(\pi)$.

In general, therefore, our optimization can be formalized as follows:

$$\min_{\pi} f(P, T, A) = f(g(\pi))$$

subject to constraints on P, T, and/or A. Here $f(\cdot)$ is the cost function to be optimized, and π is a circuit parameter vector $\pi = (d_1, d_2, \cdots, d_n)$. Since we optimize the transistor sizes of pulldowns, we treat each pullup/pulldown pair as a *node*. Typically, each node represents an inverter, NAND, or NOR gate. Each layout is characterized by the parameter vector $\pi = (d_1, d_2, \cdots, d_n)$, which means that the pulldown diffusion width of node i is $d_i\lambda$. We also use the vector $\kappa = (k_1, k_2, \cdots, k_n)$ to mean that the pullup-to-pulldown ratio of node i is k_i [22]. The vector κ is fixed for each circuit.

The choice of the cost function $f(\cdot)$ and constraints depends on the design issues. For example, in one application the clock period may be fixed at a known value T_0, and it would therefore be senseless to make the cell faster. On the other hand, peak power may be a real constraint because of heat dissipation limitations. At the same time, it may be important to keep the area small so as to fit as many cells on one chip as possible. We might, therefore, try to minimize some measure of the peak power and area (the product, $P_{max}T$, for example), while enforcing the constraint $T \leq T_0$. In other applications, speed may be critical, and it may be important to minimize T while observing constraints on P and A, and so on. In general, we would like to have enough information about the tradeoffs among the measures P, T, and A to make intelligent design decisions. As we will see, the P–T tradeoff is often of most interest, since the area is often a less sensitive function of design parameters (at least for fixed topology).

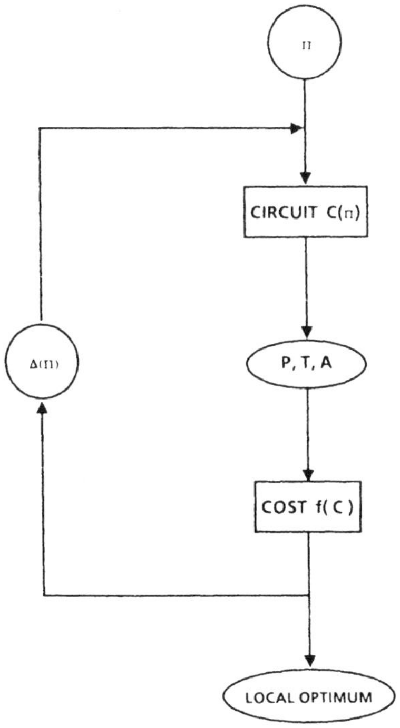

Fig. 1. Critical-path optimization.

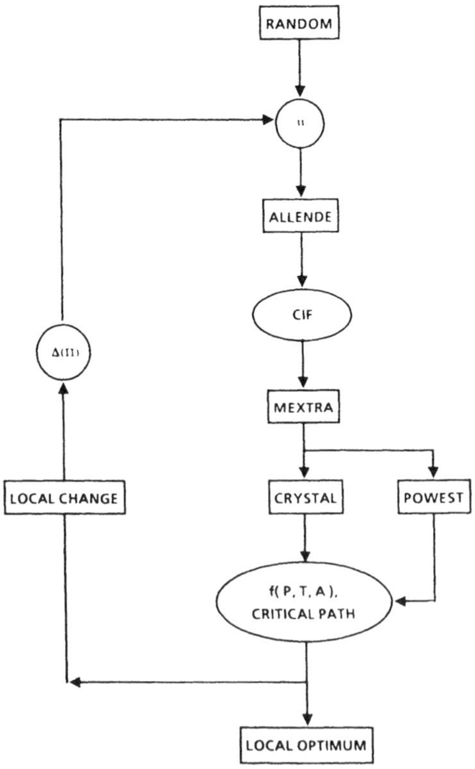

Fig. 2. Detailed flowchart of the critical-path optimization method.

III. Implementation of the Critical-Path Optimization

We use a heuristic optimization method based on a critical path, and we will call our optimization method a *critical-path optimization method*. The general concept of the method is shown in Fig. 1. A circuit $C(\pi)$ is generated based on a circuit parameter π. The cost function $f(C)$ of the circuit $C(\pi)$ is computed next. Then a desired variation $\delta(\pi)$ of the parameter vector π is computed, based on the critical path. Finally, $\delta(\pi)$ is added to π and the new parameter $(\pi + \delta\pi)$ replaces π if its cost is better. This is repeated until a local optimum is found. Fig. 2 shows a detailed flowchart of our implementation, which

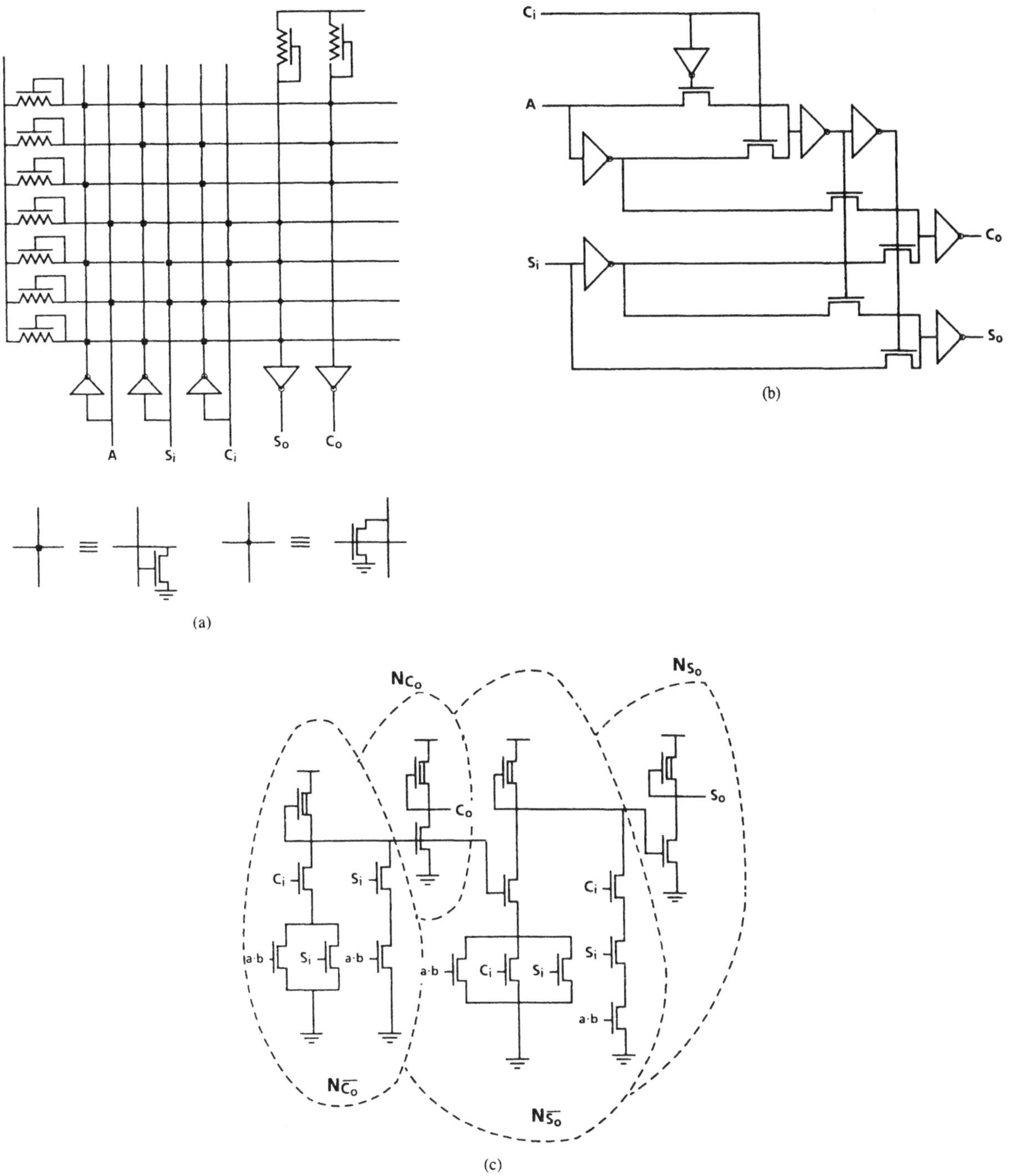

Fig. 3. (a) Circuit diagram of the PLA. (b) Circuit diagram of the data selector. (c) Circuit diagram of the random logic circuit.

uses the tools ALLENDE, MEXTRA, CRYSTAL, POWEST. A short description of each follows below.

1) ALLENDE: This procedural constraint-based VLSI layout language produces an integrated circuit layout in Caltech Intermediate Form (CIF) corresponding to the specified circuit parameter π [14].

2) MEXTRA: MEXTRA reads CIF and extracts the nodes to create a circuit description for further analysis [15].

3) CRYSTAL: CRYSTAL is used for finding the critical path and the delay time of the circuit [15], [19].

4) POWEST: POWEST is used for finding the average and maximum power consumption of the circuit [15].

The basic approach we take will be to search for local improvements from random initial designs. The search strategy will be to consider all single or double changes of the current parameter vector π along the critical path. The idea is that the critical path indicates which parameters are most important to performance at any given point in the analysis. The "k-change" method is described below in general, with simple and double change corresponding to $k = 1$ and 2.

Given a current parameter vector $\pi = (d_1, d_2, \cdots, d_n)$ and the critical path nodes which are on a critical path, say, $cpn = (d_{i_1}, d_{i_2}, \cdots, d_{i_m})$, the "$k$-change" method picks k nodes from cpn, say, $d_{j_1}, d_{j_2}, \cdots$, and d_{j_k}, then changes each of d_{j_l}, $1 \le l \le k$ by one unit and keeps the others the same. For example, "2-change" produces $\binom{m}{2} \times 2^2 = 2m(m-1)$ sets of parameters. Then each parameter is analyzed in a fixed order. When the first cost improvement is met, the current parameter is picked as the parameter for the next iteration. That is, the first improvement found is adopted.

IV. Full-Adder Circuit Implementations

We used the one-bit full-adder circuit as an example for experiments because it is relatively simple, but is a basic arithmetic logic circuit and has a wide variety of uses. The one-bit fill-adder circuit can be implemented in many ways. We chose three kinds of circuits: the PLA, data selector, and random logic. All implemented circuits have been verified by ESIM [15] or SIMULATE [13].

1) PLA: Fig. 3(a) shows the full-adder circuit diagram implemented by a programmable logic array (PLA). The π and κ of the PLA are as follows:

$$\pi = (d_{\text{and}_1}, \cdots, d_{\text{and}_7}, d_{\text{or}_1}, d_{\text{or}_2}, d_{\text{in}_1}, d_{\text{in}_2}, d_{\text{in}_3}, d_{\text{out}_1}, d_{\text{out}_2})$$

$$k = (4, 4, 4, 4, 4, 4, 4, 4, 4, 4, 4, 4, 4, 4).$$

2) Data Selector: Fig. 3(b) shows the full-adder circuit diagram of a data selector implementation [20]. The following truth table is used:

C_i	S_i	S_o	C_o
0	0	A	C_i (or S_i)
0	1	$\overline{A}$	A
1	0	$\overline{A}$	A
1	1	A	C_i (or S_i)

This circuit selects inputs (A, $\overline{A}$, or C_i) instead of calculating S_0 and C_0. Here C_i is the input carry signal, C_0 is the output carry signal, and S_0 is the output sum signal. A and S_i denote the two other inputs. This layout has the following 7 parameters.

$$\pi = (d_A, d_{S_i}, d_{C_i}, d_1, d_2, d_{C_o}, d_{S_o})$$

$$\kappa = (4, 4, 4, 8, 4, 8, 8).$$

3) Random Logic: Fig. 3(c) shows the circuit diagram of the random logic implementation [21]. This layout has the following 4 parameters: one node for computing $\overline{carry}$, one for $\overline{sum}$, one for *carry*, and one for *sum*.

$$\pi = (d_{\overline{C_o}}, d_{\overline{S_o}}, d_{C_o}, d_{S_o})$$

$$\kappa = (8, 12, 4, 4).$$

The following logical equations describe the circuit:

TABLE I
Performance Comparison (One-Bit Full Adder)

Type	A	P_{ave}	P_{max}	T	APT	PT
PLA	21.2	7.2	10.1	9.7	2074	980
	21.3	7.5	10.6	9.7	2185	1026
Data Selector	8.2	3.3	5.2	14.7	628	760
	8.4	3.5	5.6	14.7	699	830
Random Logic	9.3	1.7	2.4	7.2	161	173
	9.3	1.8	2.6	7.3	179	197

$$C_o = A \cdot (S_i + C_i) + C_i \cdot S_i,$$

$$S_o = \overline{C_o} \cdot (A + S_i + C_i) + C_i \cdot S_i \cdot A.$$

V. Computational Results of the Full-Adder Circuits Optimization

Table I shows a comparison of the performance of our implementations. Each row represents one point locally optimal with respect to T. The units of A, P_{ave}, P_{max}, T, APT, and PT are $10^3 \cdot \lambda^2$, (mW), (mW), ns, $(10^{-12} \cdot \lambda^2 \cdot W \cdot ns)$, and $(10^{-8} \cdot W \cdot ns)$, respectively, in all tables.

Fig. 4 shows P_{max} versus T curves of the one-bit full adder for different topologies when minimizing T. These P_{max} versus T tradeoff curves are obtained as follows. During a critical-path optimization process, every time a current parameter π shows an improvement in cost (in this case, T), we plot the associated values of $P_{max}(\pi)$ and $T(\pi)$. We then obtain a trajectory from each initial random starting point to a locally optimal point. By drawing an envelope of these points on many trajectories, we finally obtain an approximate P_{max} versus T tradeoff curve.

As shown in Fig. 4, the P_{max} versus T tradeoff curve of the random logic circuit is below that of the data selector circuit and that of the PLA.

Table II shows a normalized performance comparison of the best locally optimal point for each layout, minimizing T. The random logic seems to be the best choice in all respects except A. The product $P_{max}T$ of the random logic is about 1/4.4 that of the data selector, while it is about 1/5.7 that of the PLA.

Table III is the performance comparison table between the T-locally optimal circuit and the circuit designed using the minimum sizes (2λ). Our optimization shows a good improvement of the delay time (improvement from 55 to 73 percent) in any implementation. Matson optimized the same random logic circuit and obtained a delay time of 8.0 ns in [14], while our locally optimal circuit has 7.2 ns, providing an independent check of the effectiveness of our optimization method.

VI. Full Optimization of $n \times n$ Multipliers

In this section, we take up the problem of optimizing the one-bit full adder when it is embedded in a regular array, using the array multiplier illustrated in Fig. 5(a) for 4 bits.

Complete layouts of the 3×3, 4×4, 8×8, and 10×10 multipliers were optimized using the critical-path optimization method. Fig. 6 shows the possible tradeoff

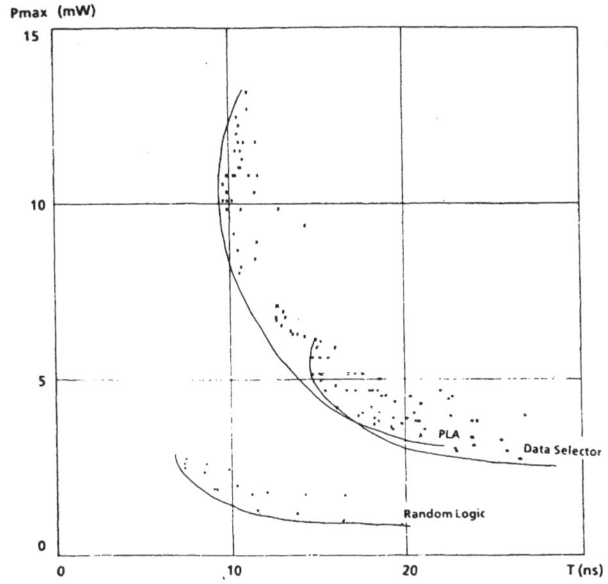

Fig. 4. The P-T tradeoff curves of one-bit full-adder circuits.

TABLE II
NORMALIZED PERFORMANCE COMPARISON (ONE-BIT FULL ADDER)

Type	A	P_{ave}	P_{max}	T	APT	PT
PLA	228	424	421	135	1288	566
Data Selector	88	194	217	204	390	439
Random Logic	100	100	100	100	100	100

TABLE III
PERFORMANCE IMPROVEMENT RATIO

Type	Cost	PLA	Data Selector	Random Logic
min size (ns)	T	21.7	53.2	26.9
T-opt (ns)	T	9.7	14.7	7.2
improvement	T	−55.2 percent	−72.3 percent	−73.2 percent

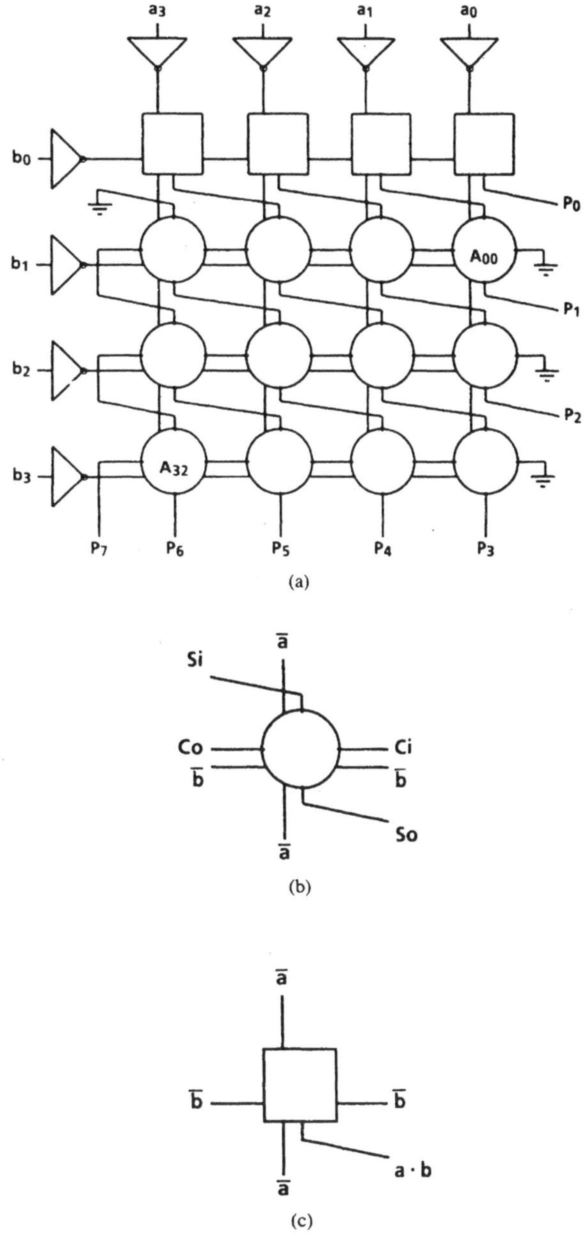

Fig. 5. (a) 4 × 4 multipler. (b) One-bit full-adder cells. (c) AND cell.

of power against delay time obtained in the same way as Fig. 4.

Table IV gives the results of starting from 11 different initial parameter vectors. They yield only four distinct local optima, corresponding to the parameters $\pi_1 = (4, 16, 8, 8, 8)$, $\pi_2 = (8, 16, 8, 8, 8)$, $\pi_3 = (12, 12, 8, 8, 8)$, and $\pi_4 = (12, 16, 8, 8, 8)$. Note that in Table IV, ** indicates that the associated π is not a local optimum.

As we can see in Table V, the running time of this optimization method increases quickly with the size of the array, growing approximately as the number of basic cells in the circuit, or n^2 for an $n \times n$ multiplier. This means that optimizing the entire circuit at once is a very costly operation, practical for a relatively small circuit, but not for large circuits. We will see later how to take advantage of the circuit regularity to reduce the computational workload.

In the next section we will define a class of circuits consisting of rectangular arrays of one-bit full adders. In succeeding sections, we will analyze the critical paths of their circuits, and show that they can be constructed from information obtained by optimization of a small "representative" circuit, called a *canonical configuration*. We will then prove that this optimization also yields the same locally optimal one-bit full adders as direct optimization of large circuits. Finally, we give experimental results which confirm this fact, and show the utility of this approach.

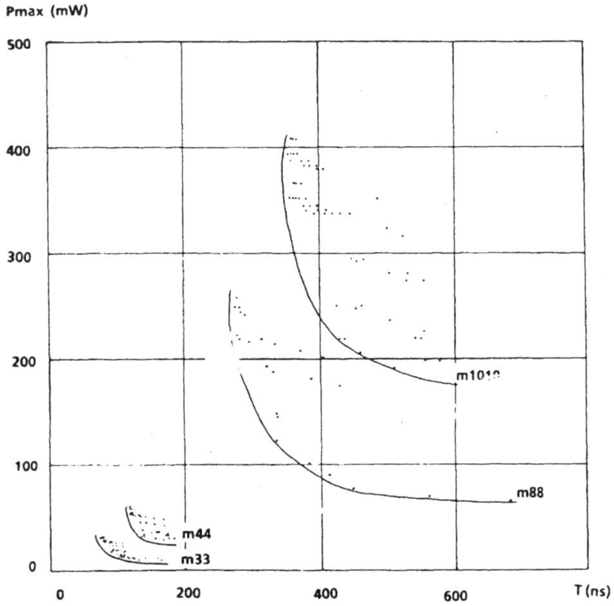

Fig. 6. The P–T tradeoff curves of multipliers.

TABLE IV
LOCAL OPTIMA (** MEANS THAT THIS IS NOT A LOCAL OPTIMUM)

Type	Cost	π_1	π_2	π_3	π_4
3 × 3	T	78.9	78.6	74.1	80.7(**)
	P_{max}	30.1	32.9	32.9	32.9
	P_{ave}	18.4	20.5	20.5	20.5
	A	82.7	84.0	82.4	87.9
4 × 4	T	118.9	118.1	110.2	120.3(**)
	P_{max}	54.6	60.2	60.2	60.2
	P_{ave}	33.4	37.6	37.6	37.6
	A	152.3	154.7	151.5	162.0
8 × 8	T	279.6	277.1	377.9(**)	283.5
	P_{max}	224.7	251.0	224.6	251.0
	P_{ave}	138.4	158.1	138.3	158.1
	A	636.2	646.6	632.9	678.8
10 × 10	T	359.2	355.9	487.2(**)	364.1
	P_{max}	353.0	395.3	352.9	395.3
	P_{ave}	217.7	249.4	217.6	249.4
	A	1001.7	1018.1	996.5	1069.3

TABLE V
OPTIMIZATION COMPUTING COST

	$m33$	$m44$	$m88$	$m1010$
Number (basic cells)	6	12	56	90
CPU (one point)	2 min	3 min	14 min	22 min
Average number of (iterations)	5.0	5.1	5.0	6.0
Average number (points searched)	61	84	61	100

VII. A CLASS OF CIRCUITS

We discuss a class of circuits consisting of one-bit full adders which have four nodes as shown in Fig. 3(c). They are $N_{\overline{C_o}}$, N_{C_o}, $N_{\overline{S_o}}$, and N_{S_o}, which yield the $\overline{carry}$, *carry*, $\overline{sum}$, and *sum* signals, respectively. Note that there are inputs C_i and S_i to the nodes $N_{\overline{C_o}}$ and $N_{\overline{S_o}}$. The relations among nodes are shown in Fig. 7. Next we define a class of circuits as follows.

Definition: $C_{FA} \equiv \{C|$ the circuit C has the following three properties.$\}$

1) The circuit C is an $m \times n$ subarray of the two-dimensional infinite array of identical one-bit full adders shown in Fig. 8, for some m and n.

2) The one-bit full adder cell used in the array is the random logic circuit shown in Fig. 3(c). Note that the input $A = a \cdot b$ is created from two other inputs $\bar{a}$ and $\bar{b}$ by a NOR circuit. The one-bit full adder is characterized by the circuit parameter $\pi = (d_{a \cdot b}, d_{\overline{C_o}}, d_{\overline{S_o}}, d_{C_o}, d_{S_o})$. Since we use identical one-bit full adders in the entire array as mentioned above, we regard the circuit parameter π of the one-bit full adder as the circuit parameter of the entire circuit C.

3) The interconnection scheme of the one-bit full adders is shown in Fig. 9.

In Fig. 8 each cell is a one-bit full adder with coordinates (x, y). We measure x to the left and y down, and $A_{x,y}$ designates a cell located in position (x, y). Each cell has two inputs (S_i: the sum input and C_i: the carry input) and two outputs (S_o: the sum output and C_o: the carry output) as shown in Fig. 9. Precisely speaking, $A_{x,y}$ has two more inputs $\bar{a}$ and $\bar{b}$, as shown in Fig. 5(b), but we do not include these inputs in our model because they will not appear in any critical path. In other words, we assume that two inputs $\bar{a}$ and $\bar{b}$ are available to any cell when a critical path reaches that cell. As shown in Fig. 9, the carry output C_o of $A_{x,y}$ is propagated to the carry input of $A_{x+1,y}$, while the sum output S_o of $A_{x,y}$ is propagated to the sum input S_i of $A_{x-1,y+1}$.

Here we analyze an $n \times n$ multiplier which has n columns and $(n - 1)$ rows of full adders. Hence, the $n \times n$ multiplier corresponds to the rectangle bounded by the corner cells $A_{0,0}$, $A_{n-1,0}$, $A_{n-1,n-2}$, and $A_{0,n-2}$ in Fig. 8.

VIII. DEFINITION OF CRITICAL PATHS

In this section we define a critical path between two cells in a circuit $C \in C_{FA}$ and analyze the behavior of signals on a critical path.

Suppose that a path α exits the one-bit full adder cell $A_{x,y}$ from either the carry output C_o or the sum output S_o with high or low signal. Hence, in order to identify the state of the path α at the exit of the cell $A_{x,y}$, we can use the representation (A, a) where A is either *carry* or *sum*, and a is either *high* or *low*. Let $(C, 0)$ denote the state in which a path exits from the carry output with a low signal, and let $(C, 1)$, $(S, 0)$, and $(S, 1)$ be defined analogously. Then the behavior of the path α can be represented by a sequence of states. For example, the expression $(C, 0) \rightarrow (C, 0) \rightarrow (S, 1) \rightarrow (S, 0)$ represents the path which exits $A_{0,0}$ with the low carry signal, exits $A_{1,0}$ with the low carry signal, exits $A_{2,0}$ with the high sum signal, and finally exits $A_{1,1}$ with the low sum signal as shown in Fig. 8. We can thus represent the behavior of the path by the state transition diagram D as shown in Fig. 10(a). Note

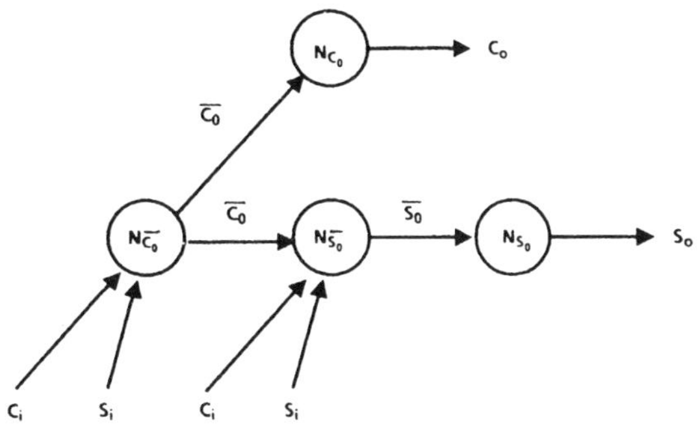

Fig. 7. The nodes in the one-bit full adder.

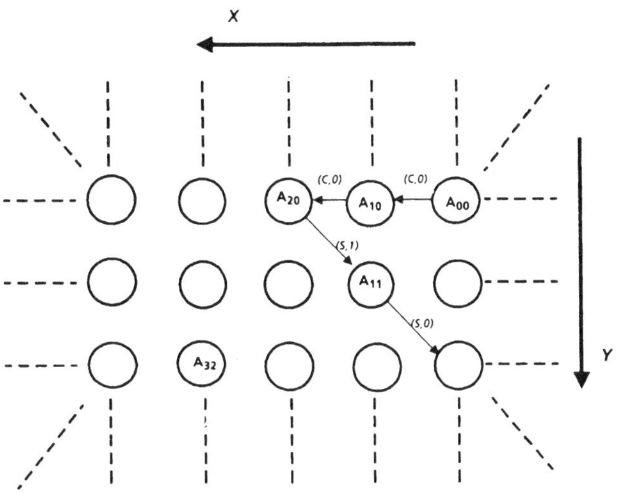

Fig. 8. Two-dimensional array of one-bit full adders.

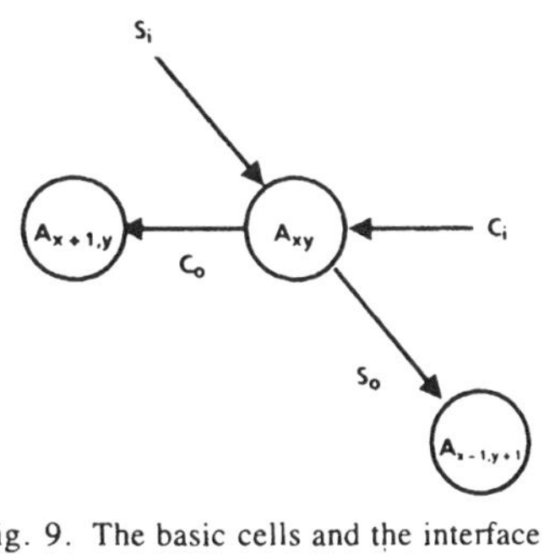

Fig. 9. The basic cells and the interface rule.

that in this diagram, the states C, D, S, and T are used instead of $(C, 0)$, $(C, 1)$, $(D, 0)$, and $(D, 1)$, respectively.

Lemma 1: The state transition diagram D in Fig. 10(a) correctly describes the behavior of the signal along any path in the array of Fig. 8.

Proof: In Fig. 7, each time a path passes through a node, the signal on that path changes from high (low) to low (high). For example, suppose we have a path α which exits from the carry output C_o with a high signal. The path α must pass through the nodes $N_{\overline{C_o}}$ and N_{C_o}. Since a signal changes from low (high) to high (low) when a path passes through a node, the $\overline{C_o}$ signal is low, and the input on the

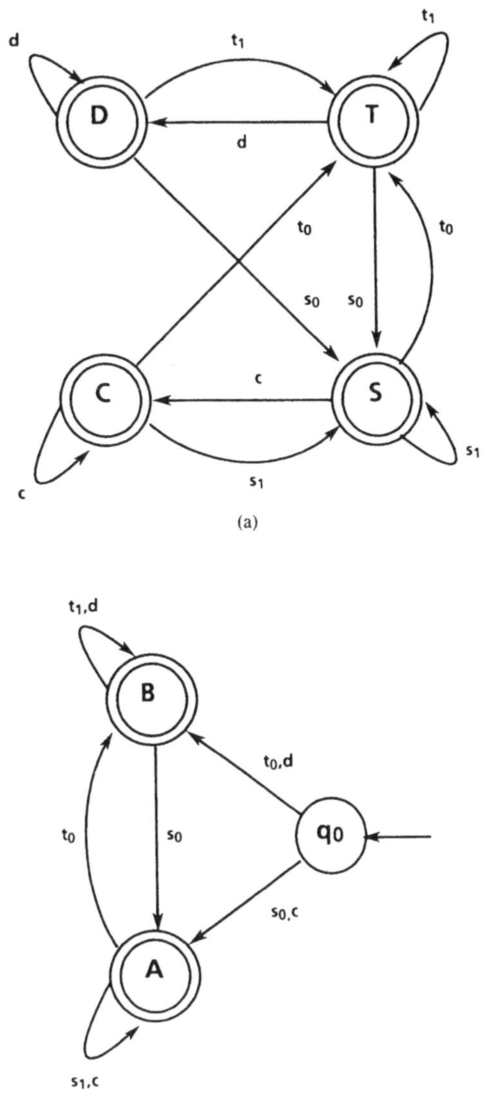

Fig. 10. (a) State transition diagram along a critical path. (b) Minimized state transition diagram.

path into the node $N_{\overline{C_o}}$ is high. Hence, the state $(C, 1)$ can be reached either from the state $(C, 1)$ or $(S, 1)$.

In the same way we can determine other state transitions. Note that a path to the sum output S_o results from either $N_{\overline{C_o}} \rightarrow N_{\overline{S_o}} \rightarrow N_{S_o}$ or $N_{\overline{S_o}} \rightarrow N_{S_o}$. □

From the above discussion, we can define a finite automaton M that represents the state transitions along a path.

Definition: The finite automaton M is defined as follows:

$$M = (Q, \Sigma, \delta, q_0, F),$$

where Q is the set of states, Σ is the alphabet, δ is the state transition function, q_0 is the initial state, and F is the set of final states.

$$Q = \{q_0, C, D, S, T\}$$

$$\Sigma = \{c, d, s_0, s_1, t_0, t_1\},$$

and

$$F = Q,$$

where the states C, D, S, and T represent the states $(C, 0)$, $(C, 1)$, $(S, 0)$, and $(S, 1)$, respectively. The symbol $c(d)$ indicates the transition from the node $N_{\overline{C_o}}$ with a low (high) input to the node N_{C_o} with a high (low) output, while the symbol $s_0(t_0)$ indicates the transition from the node $N_{\overline{C_o}}$ with a high (low) input to the node N_{S_o} with a low (high) output. And the symbol $s_1(t_1)$ indicates the transition from the node $N_{\overline{S_o}}$ with a low (high) input to the node N_{S_o} with a low (high) output. The transition function δ is shown in Fig. 10(a) where q_0 and transitions from q_0 such as $\delta(q_0, c) = C$, $\delta(q_0, d) = D$, $\delta(q_0, s_0) = S$, and $\delta(q_0, t_0) = T$ are not shown.

Since we use the fixed one-bit full adder shown in Fig. 7, there is a fixed delay time associated with each state transition when the circuit parameter π is fixed. Hence, we can define a delay-time function w as follows.

Definition: Given a circuit parameter π and a finite automaton $M = (Q, \Sigma, \delta, q_0, F)$, defined above, the *delay-time function* w_π is defined as follows: w_π: $Q \times Q \rightarrow R$ such that $w_\pi(q_1, q_2)$ is the delay time for calculating the corresponding output signal and propagating this signal when the transition (q_1, q_2) is in δ. When the transition (q_1, q_2) is not defined in δ, $w_\pi(q_1, q_2)$ is not defined. We use w instead of w_π when a circuit parameter π is fixed in the discussion. Since from the definition of the symbols, any transition with the same symbol has the same delay time, we also use the symbol to mean its delay time. Note that $s_0 > s_1$ and $t_0 > t_1$, since the transition from the node $N_{\overline{S_o}}$ to the node N_{S_o} is a part of the transition from the node $N_{\overline{C_o}}$ to the node N_{S_o} through $N_{\overline{S_o}}$.

Let $L(M)$ be the language accepted by the finite automaton M. The language $L(M)$ corresponds to the set of all possible paths in our two-dimensional array. Let $L_{mn} = \{\alpha \in L(M) \mid |\alpha|_c + |\alpha|_d = m, |\alpha|_{s0} + |\alpha|_{s_1} + |\alpha|_{t0} + |\alpha|_{t_1} = n\}$ where $|\alpha|_a$ indicates the number of times the symbol "a" in the string α. Thus, L_{mn} corresponds to the set of paths from $A_{0,0}$ to $A_{m-n-1,m-1}$, or in other words, L_{mn} is the set of paths that have m carry stages and n sum stages.

We use the notation $C_{FA,m,n}$ for representing the set of circuits $C \in C_{FA}$ whose critical paths are in L_{mn}. We define the delay-time function w on a string in $L(M)$ as follows. Let $\alpha \in L(M)$ and $\alpha = a_1 a_2 \cdots a_n$ where $a_i \in \{c, d, s_0, s_1, t_0, t_1\}$ for $1 \leq i \leq n$. Then $w(\alpha) \equiv \Sigma_{i=1}^{n} w(a_i)$.

We can now define a *critical path* in our terms.

Definition: We call $\alpha \in L_{mn} \subseteq L(M)$ a *critical path* when $w(\alpha) \geq w(\beta)$ for all $\beta \in L_{mn}$. Define CPN as the set of all critical paths of all subcircuits in C_{FA} and $\text{CPN}_{mn} = \text{CPN} \cap L_{mn}$. We say that two paths are *equivalent* when their delay times are equal. When $w(\alpha) < w(\beta)$ ($w(\alpha) > w(\beta)$, $w(\alpha) = w(\beta)$) for two paths α and β, we denote that by $\alpha <_w \beta$ ($\alpha >_w \beta$, $\alpha =_w \beta$, respectively).

Our problem of finding a critical path in the $m \times n$ multiplier then corresponds to the problem of finding a critical path from $A_{0,0}$ to another cell $A_{m-1,n-2}$, as shown in Section VII.

Lemma 2: Every critical path of the $n \times n$ multiplier has $(2n - 3)$ carry calculation stages and $(n - 1)$ sum calculation stages.

Proof: Every time a sum signal appears in a path, the y coordinate increases by 1, while the x coordinate decreases by 1. Every time a carry signal appears in a path, the x coordinate increases by 1, while the y coordinate remains the same. Hence, a path from the cell $A_{0,0}$ with a_s sum stages and a_c carry stages reaches the cell $A_{a_c - a_s, a_s}$. In the $n \times n$ multiplier, a critical path starting from the cell $A_{0,0}$ finally comes out of the cell $A_{n-1,n-2}$ with the sum signal. Hence, $n - 1 = a_c - a_s$ and $n - 2 = a_s$. Thus, $a_c = n - 1 + a_s = 2n - 3$. Hence, we proved that every critical path α of the $n \times n$ multiplier has $(2n - 3)$ carry calculation stages, and $(n - 2)$ sum calculation stages, plus another sum calculation in the cell $A_{n-1,n-2}$. □

IX. Analysis of Critical Paths

In Section VIII, we saw that the behavior of the signals on the path can be described by the weighted state transition diagram M. In this section, we investigate the problem of finding the critical paths effectively, given the weighted state transition diagram.

Since the longest path between two nodes can be computed given the delay time of all paths between two nodes, we have the following theorem.

Theorem 1: Given a fixed parameter π and the delay-time function w_π: $Q \times Q \rightarrow R$, we can effectively construct the set CPN.

We next describe a more practical way of constructing a critical path in L_{mn}. We use $R(a_1, a_2, \cdots, a_k)$ to represent the pair (i, a_i) where $a_i = \max(a_1, a_2, \cdots, a_k)$; that is, R tells us which argument is maximum.

Theorem 2: Given a circuit parameter π, knowledge of

$$R(c, d),\ R(s_0, t_0),\quad R(s_0 + t_0, s_0 + s_1, t_0 + t_1),$$

and

$$R(s_0 + t_0, 2s_1, 2t_1)$$

is both necessary and sufficient to construct a critical path in L_{mn} for $m \geq 0$ and $n \geq 0$.

We need the following lemmas to prove theorem 2. The first lemma shows that we can simplify our finite automaton.

Lemma 3: Let L be the set of strings accepted by the finite automaton $M = (Q, \Sigma, \delta, q_0, F)$ defined in Section VIII and shown in Fig. 10(a). Let L_1 be the set of strings accepted by the finite automaton $M_1 = (Q_1, \Sigma_1, \delta_1, q_0, F_1)$ shown in Fig. 10(b), where $Q_1 = \{q_0, A, B\}$. $F_1 = \{A, B\}$, and δ_1 is shown in Fig. 10(b). Then $L(M) = L(M_1)$.

Proof: Use the state minimization algorithm in [24]. □

From now on, we will concentrate our attention on the reduced state automaton M_1. Next we will characterize the strings accepted by M_1.

Lemma 4: Let E be the regular expression of the strings accepted by M_1. Then

$$E = a_0 a_1^* (t_0 b_1^* s_0)^* a_1^* (\epsilon + t_0)$$
$$+ b_0 b_1^* (s_0 a_1^* t_0)^* b_1^* (\epsilon + s_0)$$

where $a_i = s_i + c$, $b_i = t_i + d$ for $i = 0, 1$, and ϵ is the empty input.

Proof: Let $r_A (r_B)$ be the regular expressions of strings starting from and ending at state $A(B)$. Then $r_A = a_1^* (t_0 b_1^* s_0)^* a_1^*$ and $r_B = b_1^* (s_0 a_1^* t_0)^* b_1^*$. Let $E_A (E_B)$ be the regular expressions of strings accepted at state $A (B)$. Then $E_A = a_0 r_A + b_0 r_B s_0$ and $E_B = b_0 r_B + a_0 r_A t_0$. Therefore, $E = E_A + E_B$ is the desired regular expression. □

Definition: For two regular expressions E_1 and E_2, we define $E_1 \sim E_2$ iff for any $e_1 \in E_1 (e_2 \in E_2)$, there exists $e_2 \in E_2 (e_1 \in E_1)$ such that $e_1 =_w e_2$, that is, $T(e_1) = T(e_2)$. We also use $e_1 \sim e_2$ when $e_1 =_w e_2$.

Lemma 5: $E \sim (s_0 t_0 + c t_0 + d s_0 + s_0 + c + d) s_1^* c^* (s_0 + t_0)^* t_1^* d^*$.

Proof: Note that for $a, b \in \Sigma$, $(a + b)^* \sim a^* b^*$, $ab \sim ba$. Hence, $a_1^* = (s_1 + c)^* \sim s_1^* c^*$, $b_1^* = (t_1 + d)^* \sim t_1^* d^*$, $(t_0 b_1^* s_0)^* \sim (s_0 t_0)^* b_1^*$, and $(s_0 a_1^* t_0)^* \sim (s_0 t_0)^* a_1^*$. From lemma 4, we can obtain desired result. □

Let $y_3 = \max(s_0 + t_0, s_0 + s_1, t_0 + t_1)$ and $y_4 = \max(s_0 + t_0, 2s_1, 2t_1)$. Then define z_1, z_2, z_3, and z_4 as follows:

$$z_1 = \max(c, d), \quad z_2 = \max(s_0, t_0),$$

$$z_3 = \begin{cases} s_0 t_0 & \text{if } y_3 = s_0 + t_0 \\ s_0 s_1 & \text{if } y_3 = s_0 + s_1, \\ t_0 t_1 & \text{if } y_3 = t_0 + t_1 \end{cases}$$

$$z_4 = \begin{cases} s_0 t_0 & \text{if } y_4 = s_0 + t_0 \\ s_1^2 & \text{if } y_4 = 2s_1 \\ t_1^2 & \text{if } y_4 = 2t_1. \end{cases}$$

Lemma 6: Let α be a critical path in $L_{m,n}$. Then

$$\alpha \le_w \begin{cases} z_1^m z_2 z_4^k & \text{if } n = 2k + 1 \\ z_1^m z_3 z_4^{k-1} & \text{if } n = 2k. \end{cases}$$

Proof: Let β be a path in $L_{m,n}$. From lemma 5, $\beta \sim c^{x_1} d^{x_2} \gamma$ where $x_1 + x_2 = m$ and $\gamma \in (s_0 + t_0 + s_1 + t_1)^n$. Clearly, $\beta \le_w z_1^m \gamma = \max(c, d)^m \gamma$. Hence, we only have to think about a critical path in $L_{0,n}$. Let β be a path in $L_{0,n}$. From lemma 5, $\beta \sim (s_0 t_0 + s_0 + t_0) s_1^{x_1} t_1^{x_2} (s_0 t_0)^{x_3}$.

We take the following two cases: 1) $n = 2k + 1$, and 2) $n = 2k$.

1) $n = 2k + 1$: $\beta \sim \gamma s_1^{x_1} t_1^{x_2} (s_0 t_0)^{x_3}$ where $\gamma \in (s_0 t_0 + s_0 + t_0)$.

a) If $\gamma = s_0 t_0$, then either x_1 or x_2 is odd. Without loss of generality, we assume that $x_1 = 2k_1 + 1$ and $x_2 = 2k_2$. Since $s_1 <_w s_0$, $s_1^2 \le_w z_4$, $t_1^2 \le_w z_4$, $s_0 t_0 \le_w z_4$, and $s_0 \le_w z_2$, we have

$$\beta \sim s_0 t_0 s_1 s_1^{2k_1} t_1^{2k_2} (s_0 t_0)^{x_3} <_w z_4 s_0 z_4^{k_1} z_4^{k_2} z_4^{x_3} \le_w z_2 z_4^k.$$

b) If $\gamma = s_0$ or t_0, then $\beta \sim \gamma s_1^{x_1} t_1^{x_2} (s_0 t_0)^{x_3}$ and $x_1 \equiv x_2 \pmod 2$. Note that $\gamma \le_w z_2$. If $x_1 = 2k_1 + 1$ and $x_2 = 2k_2 + 1$, then

$$\beta \sim \gamma (s_1 t_1) s_1^{2k_1} t_1^{2k_2} (s_0 t_0)^{x_3} \le_w z_2 (s_0 t_0) z_4^{k_1} z_4^{k_2} z_4^{x_3} \le_w z_2 z_4^k.$$

If $x_1 = 2k_1$ and $x_2 = 2k_2$, then

$$\beta \sim \gamma s_1^{2k_1} t_1^{2k_2} (s_0 t_0)^{x_3} \le_w z_2 z_4^k.$$

Hence, $\beta \le z_2 z_4^k$.

2) $n = 2k$: $\beta \sim \gamma s_1^{x_1} t_1^{x_2} (s_0 t_0)^{x_3}$ where $\gamma \in (s_0 t_0 + s_0 s_1 + t_0 t_1 + s_0 t_1 + s_1 t_0)$. Since $n = 2 + x_1 + x_2 + 2x_3$ is even, we know $x_1 \equiv x_2 \pmod 2$. Since $s_0 t_1 \le_w s_0 t_0$ and $s_1 t_0 \le_w s_0 t_0$, we have $\beta \le_w z_3 s_1^{x_1} t_1^{x_2} (s_0 t_0)^{x_3}$. In the same way as in b) above, $s_1^{x_1} t_1^{x_2} (s_0 t_0)^{x_3} \le_w z_4^{k-1}$. Thus, $\beta \le_w z_3 z_4^{k-1}$ □

Definition: A string $\beta \in \Sigma^*$ is said to be *constructible* when there exists a path $\alpha \in L(M_1)$ such that $\alpha \sim \beta$.

Lemma 7: For any integers $m \ge 0$ and $k \ge 1$, the strings $z_1^m z_2 z_4^k$ and $z_1^m z_3 z_4^{k-1}$ are constructible, and therefore, the upper bounds in lemma 6 are attained.

Proof: We will prove this for $n = 2k + 1$. The proof for n even is similar. Without loss of generality, we assume that $z_2 = s_0$. Now we consider the constructibility of $z_2 z_4^k$. We will find a string $\alpha_{0,n} \in L_{0,n}$ such that $\alpha_{0,n} \sim z_2 z_4^k$. We have the following three cases for z_4.

1) $z_4 = s_0 t_0$: Let $\alpha_{0,n} = s_0 (t_0 s_0)^k \in L_{0,n}$. Then $\alpha_{0,n} \in L(M_1)$ and $\alpha_{0,n} \sim z_2 z_4^k$.

2) $z_4 = s_1^2$: Let $\alpha_{0,n} = s_0 s_1^{2k} \in L_{0,n}$. Then $\alpha_{0,n} \in L(M_1)$ and $\alpha_{0,n} \sim z_2 z_4^k$.

3) $z_4 = t_1^2$: Since $s_0 + t_0 \le 2t_1 < t_0 + t_1$, we have $s_0 < t_1 < t_0$. However, we assumed that $z_2 = \max(s_0, t_0) = s_0$. Thus, this case does not happen.

From 1), 2), and 3) we see that the string $z_2 z_4^k$ is constructible.

Now we prove that $z_1^m z_2 z_4^k$ is constructible. Let $y_1 = s_0 c^m$ if $z_1 = c$, or let $y_1 = d^m s_0$ if $z_1 = d$. Let $\alpha_{0,n} = s_0 y_2$. Let $\alpha = y_1 y_2$. Here we know that $\alpha \in L_{m,n}$ and $\alpha \sim z_1^m z_2 z_4^k$. Thus, we proved that $z_1^m z_2 z_4^k$ is constructible. □

The following example shows how to construct a critical path β in $L_{10,5}$ from knowledge of R functions, given a fixed parameter $\pi = (12, 16, 12, 8, 8)$. By computation we obtain a critical path $\alpha = c c t_0 s_0$ in $L_{2,2}$, $w_c = 10.3$ ns, $w_{t_0} = 18.7$ ns, and $w_{s_0} = 15.2$ ns. Since $w(c c t_0 s_0) \ge w(d d s_0 t_0)$, we have $\max(c, d) = c$. Since $c c t_0 s_0 \ge_w c c t_0 t_1$, we have $s_0 + t_0 \ge t_0 + t_1 > 2t_1$. Since $\max(s_0, t_0) = t_0$, we know $s_0 + t_0 \ge 2s_0 > 2s_1$ and $t_0 \ge s_0 > s_1$. Thus, $\max(s_0 + t_0, 2s_1, 2t_1) = s_0 + t_0$ and $\max(s_0 + t_0, s_0 + s_1, t_0 + t_1) = s_0 + t_0$. Then from lemma 7, we can construct a critical path $\beta = c^{10} (t_0 s_0)^2 t_0 \in L_{10,5}$ and calculate the delay time $w(\beta) = 189.5$ ns. In fact, actual computation shows that a critical path of $L_{10,5}$ is $\gamma = c^7 (t_0 s_0) c^2 (t_0 s_0) c t_0$ and its delay time $w(\gamma)$ is 189.3 ns. Note that the critical paths γ and β are equivalent in the sense that $w(\beta) = w(\gamma) = 10c + 3t_0 + 2s_0$.

Lemma 8: Given a critical path α in $L_{1,3}$, we know the values of the R functions in theorem 2.

Proof: Let α be a critical path in $L_{1,3}$. $\alpha \sim \beta\gamma$ where $\beta \in (c + d)$ and $\gamma \in (s_0 + s_1 + t_0 + t_1)^3$. Clearly, $\beta = \max(c, d) = z_1$. Since $s_0 s_1 t_0 <_w s_0 t_0 s_0$ and $s_0 t_0 t_1 <_w t_0 s_0 t_0$, we have $\gamma \sim (s_0 s_1^2 + s_0 t_0 s_0 + t_0 s_0 t_0 + t_0 t_1^2)$. Suppose $\gamma \sim s_0 s_1^2$. Then we can find the values of the R functions in theorem 2 as follows. Since $s_0 s_1 s_1 \geq_w s_0 t_0 s_0$, we have $s_0 + t_0 \leq 2s_1 < s_1 + s_0$. Thus, $t_0 < s_1 < s_0$. Hence, $\max(s_0, t_0) = s_0$, $\max(s_0 + t_0, s_0 + s_1, t_0 + t_1) = s_0 + s_1$ and $\max(s_0 + t_0, 2s_1, 2t_1) = 2s_1$. We can also compute the R functions in the other cases as above. □

Now we prove theorem 2.

Proof of Theorem 2: Suppose we know the values of the R functions in theorem 2. From lemma 7, there is a path β in $L_{m,n}$ such that

$$\beta \sim \begin{cases} z_1^m z_2 z_4^k & \text{if } n = 2k + 1 \\ z_1^m z_3 z_4^{k-1} & \text{if } n = 2k. \end{cases}$$

From lemma 6, the delay time of the path β is worse than any path in $L_{m,n}$. This means that the path β itself is a critical path. Conversely, we can find the desired R functions from a critical path in $L_{1,3}$ from lemma 8. □

From theorem 2 and lemma 8, we have found a new way to implement our critical-path optimization method, as shown in Fig. 11. By analyzing the small configuration (the circuit in $C_{FA,1,3}$), we can avoid analyzing the entire circuit. This means that our optimization workload will be reduced significantly when the circuit is large.

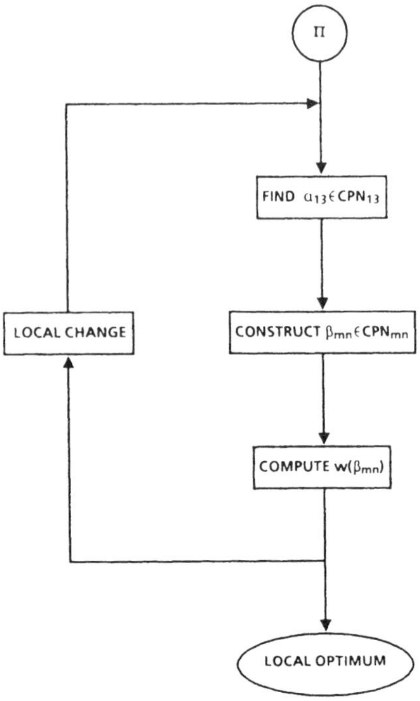

Fig. 11. New implementation of the critical-path optimization method.

X. A Canonical Configuration for the $n \times n$ Multiplier

Although we found an effective way of computing a critical path of L_{mn} and its delay time, we still have a major question left. That is, does there exist an effective way to find a locally optimal parameter for the large $n \times n$ multiplier? In this section we prove that the answer to the question is yes and, furthermore, we will show a stronger result by introducing the idea of the *canonical configuration*.

Definition: A circuit CC $\in C_{FA}$ is called the *canonical configuration* of the $n \times n$ multiplier iff the optimization of CC yields the same parameters as the optimization of the $n \times n$ multiplier, for all sufficiently large n.

We now consider the following problem.

Problem 1: Is there a canonical configuration CC of the $n \times n$ multiplier? If a CC exists, what is it? How can it be found?

Let $T_\pi(n)$ denote the delay time of the critical path of the $n \times n$ multiplier given a circuit parameter π. We use $T(n)$ instead of $T_\pi(n)$ when π is fixed in the discussion. We can now give an explicit formula $T_\pi(n)$ for the $n \times n$ multiplier.

Theorem 3: Given a circuit parameter π, then $T_\pi(n)$ can be represented as follows:

$$T_\pi(n) = kx_1 + x_2, \tag{1}$$

where $k = \lfloor n/2 \rfloor - 1$,

$$x_1 = 4\max(c, d) + \max(t_0 + s_0, 2s_1, 2t_1)$$

$$x_2 = \begin{cases} \max(c, d) + \max(t_0, s_0) & \text{if } n = 2k \\ 3\max(c, d) + \max(t_0 + s_0, s_0 + s_1, t_0 + t_1) & \text{if } n = 2k + 1. \end{cases}$$

Proof: Let z_1, z_2, z_3, and z_4 be as defined in Section IX. From lemma 2, a critical path α of the $n \times n$ multiplier is in $L_{2n-3,n-1}$. And from lemma 6, we know that

$$\alpha \sim \begin{cases} z_1^{2n-3} z_2 z_4^k & \text{if } n - 1 = 2k + 1 \\ z_1^{2n-3} z_3 z_4^{k-1} & \text{if } n - 1 = 2k \end{cases} \qquad n = 2k + 2 \text{ or } n = 2k + 1,$$

$$\alpha \sim \begin{cases} z_1^{4k-3} z_2 z_4^{k-1} & \text{if } n = 2k \\ z_1^{4k-1} z_3 z_4^{k-1} & \text{if } n = 2k + 1 \end{cases}$$

thus,

$$\alpha \sim \begin{cases} (z_1^4 z_4)^{k-1}(z_1 z_2) & \text{if } n = 2k \\ (z_1^4 z_4)^{k-1}(z_1^3 z_3) & \text{if } n = 2k + 1. \end{cases}$$

Note that $k = \lfloor n/2 \rfloor - 1$.

Corollary 1: Given a circuit parameter π, then $T_\pi(n)$ is asymptotically proportional to n.

Proof: This is clear from (1). □

In fact, from Table IV in Section V, we can obtain empirically the formula $T_{\pi_1} = 40(n - 1)$ when $\pi_1 = (4, 16, 8, 8)$.

Corollary 2: The circuit $C_1 \in C_{FA,4,2}$ shown in Fig. 12 is a canonical configuration of the $n \times n$ multiplier. The critical path of C_1 has four carry stages and two sum stages.

Proof: From theorem 3, we have $T(n) = \alpha_{4,2} \lfloor n/2 \rfloor + \beta$, where $\alpha_{4,2}$ is the delay time of four carry stages and β is the constant delay time. Hence, an optimal parameter of $L_{4,2}$ is, asymptotically for large n, also an optimal parameter L_{mn}. □

From lemma 2 in Section VII, there are $(2n - 3)$ carry stages and $(n - 1)$ sum stages in any critical path of the $n \times n$ multiplier. Thus, we might expect that the circuit $C_0 \in C_{FA,2,1}$, whose critical path has two carry stages and one sum stage, is a canonical configuration of the $n \times n$ multiplier. However, as we saw in this section, the circuit $C_1 \in C_{FA,4,2}$ is a canonical configuration, but the circuit $C_0 \in C_{FA,2,1}$ is not a canonical configuration. The reason is that optimization of the circuit C_0 cannot determine $\max(s_0 + t_0, 2s_1, 2t_1)$.

XI. The Optimization of a Subcircuit

In this section, we optimize the two circuits C_0 and C_1 discussed in the previous section, verifying that the circuit C_1 works well as a canonical configuration, but the circuit C_0 does not. The circuit C_1 is indicated by a solid line in Fig. 12, while the circuit C_0 is indicated by a dotted line. The critical path from cell $A_{0,0}$ to cell $A_{2,2}$ is analyzed for C_1 in Fig. 12, while the critical path from cell $A_{0,0}$ to cell $A_{1,1}$ is analyzed for C_0.

Table VI shows which parameters are locally optimal for $m33$, $m44$, $m88$, $m1010$, C_0, and C_1. The symbol × indicates a local optimum. The same set of 11 random initial parameters were used for each circuit, and only these 5 distinct local optima were obtained.

The most important result is that every local optimum of $m88$ and $m1010$ is also a local optimum of C_1. This is not true for C_0, nor is it true for the smaller circuits $m33$ and $m44$. Thus, we can say that the circuit C_1 is indeed appropriate as a representative subcircuit of the $n \times n$ multiplier. In this way, corollary 2 in Section IX is confirmed very well by numerical experiments.

XII. Conclusions

We have described a general approach for sizing the transistors in a cell that is embedded in a regular array, using local search along the critical path. The simplest, most regular array multiplier structure was used as an example, with delay time (not throughput) as a criterion. No attempt was made to incorporate intermediate clocking, precharging, or superbuffers.

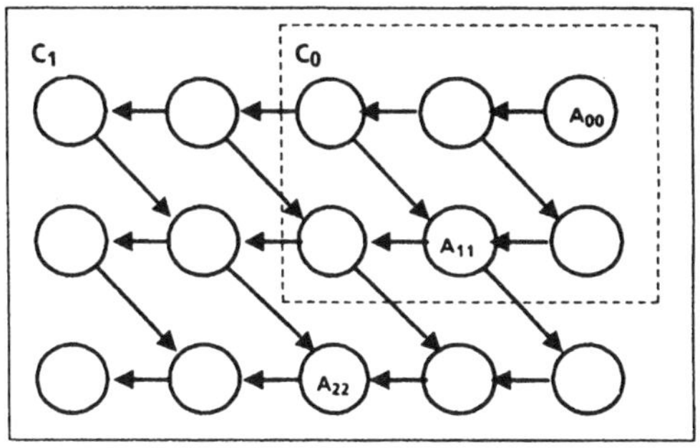

Fig. 12. The circuits C_0 and C_1.

TABLE VI
Locally Optimal Parameters for Each Circuit

	π_1	π_2	π_3	π_4	π_5
C_0	×				×
C_1	×	×		×	
$m1010$	×	×		×	
$m88$	×	×		×	
$m44$	×	×	×		
$m33$	×	×	×		

We quickly encountered the problem that the running time of the optimization increases rapidly when we increase the size of the multiplier. We therefore tried to make our optimization method more practical by making use of the circuit's regularity and developed what we call the *canonical configuration method*. In this method we locally optimize the small circuit C_1 instead of the entire circuit. We showed how to extract this canonical configuration C_1 for the $n \times n$ multiplier, and gave experimental results that illustrate the savings offered by this method.

The following problems are the subject of future investigation. Do canonical configurations exist in more general regular arrays? If the canonical configuration exists for some regular array, how can we extract it?

References

[1] P. R. Cappello and K. Steiglitz, "Digital signal processing applications of systolic algorithms," in *CMU Conference on VLSI Systems and Computations*, H. T. Kung, B. Sproull, and G. Steel, Eds. Rockville, MD: Computer Science Press, 1981.
[2] ——, "Completely pipelined architectures for digital signal processing," *IEEE Trans. Acoust., Speech, Signal Processing*, vol. ASSP-31, pp. 1016–1022, Aug. 1983.
[3] ——, "A note on 'free accumulation' in VLSI filter architectures," *IEEE Trans. Circuits Syst.*, in press.
[4] C. Caraiscos and B. Liu, "Bit serial VLSI implementations of FIR and IIR digital filters," in *Proc. IEEE Int. Symp. Circuits Syst.*, May 1983.
[5] P. B. Denyer and D. J. Myers, "Carry-save arrays for VLSI signal processing," in *VLSI 81: Very Large Scale Integration*, J. P. Gray, Ed. London, England: Academic, 1983. (Also, in *Proc. Conf. Very Large Scale Integration*, Univ. Edinburgh, U.K., Aug. 18–21, 1981.)
[6] L. A. Glasser and L. P. J. Hoyte, "Delay and power optimization in VLSI circuits," in *Proc. IEEE 21st Design Automat. Conf.*, 1984, pp. 529–535.
[7] K. Iwano and K. Steiglitz, "Some experiments in VLSI leaf-cell optimization," presented at the 1984 IEEE Workshop on VLSI Signal Processing, Univ. Southern Calif.. Nov. 12–14, 1984, pp. 387–395.
[8] ——, "Time-power-area tradeoffs for the nMOS VLSI full-adder" in

1985 Proc. Int. Conf. Acoust., Speech, Signal Processing, Tampa FL, Mar. 1985. pp. 1453-1456.
[9] H. T. Kung, L. M. Ruane, and D. W. L. Yen, "A two-level pipelined systolic array for convolutions," *CMU Conference on VLSI Systems and Computations*, H. T. Kung, B. Sproull, and G. Steele, Eds. Rockville, MD: Computer Science Press, 1981.
[10] E. Lawler, *Combinatorial Optimization*. New York: Holt, Reinhart, and Winston, 1976.
[11] R. J. Lipton, S. C. North, R. Sedgewick, J. Valdes, and G. Vijayan, "VLSI layout as programming," *ACM Trans. Program. Languages Syst.*, July 1983.
[12] R. F. Lyon, "A bit-serial VLSI architecture methodology for signal processing," in *VLSI 81: Very Large Scale Integration*, J. P. Gray, Ed. London, England: Academic, 1981. (Also, in *Proc. Int. Conf. Very Large Scale Integration*, Univ. Edinburgh, U.K., Aug. 18-21, 1981.)
[13] J. Mata, "ALLENDE: A procedural language for the hierarchical specification of VLSI layouts," in *Proc. IEEE 22nd Design Automat. Conf.*, 1985.
[14] M. D. Matson, "Macromodeling and optimization of digital MOS VLSI circuits," Ph.D. dissertation, Dept. EECS, M.I.T., Cambridge, MA, Jan. 1985.
[15] R. N. Mayo, J. K. Ousterhout, and W. S. Scott, "1983 VLSI tools," Comput. Sci. Div. (EECS), Univ. Calif., Berkeley, CA, Rep. UCB/CSD 83/115, Mar. 1983.
[16] J. V. McCanny, J. G. McWhirter, J. B. G. Roberts, D. J. Day, and T. L. Thorp, "Bit level systolic arrays," in *Proc. 15th Asilomar Conf. Circuits, Syst., Comput.*, Nov. 1981.
[17] S. Trimberger, "Automated performance optimization of custom integrated circuits," in *VLSI 83: VLSI Design of Digital Systems*, F. Anceau and E. J. Aas, Eds. Amsterdam, The Netherlands: North-Holland, 1983. (Also, in *Proc. IFIP Int. Conf. Very Large Scale Integration*, Trondheim, Norway, Aug. 1983.)
[18] A. J. Strojwas, S. R. Nassif, and S. W. Director, "Optimal design of VLSI minicells using a statistical process simulator," in *Proc. IEEE Int. Conf. Circuit Syst.*, 1983, pp. 202-205.
[19] J. K. Ousterhout, "Switch-level delay models for digital MOS VLSI," in *Proc. IEEE 21st Design Automat. Conf.*, 1984, pp. 542-547.
[20] D. J. Myers, "Multipliers for LSI and VLSI signal processing applications," Masters thesis, Edinburgh Univ., Edinburgh, England, Sept. 1981.
[21] J. Allen, "VLSI architectures for signal processing," in *VLSI Architecture*, B. Randell and P. C. Treleaven, Eds. Englewood Cliffs, NJ: Prentice-Hall, 1983, pp. 242-254.
[22] C. Mead and L. Conway, *Introduction to VLSI Systems*. Reading, MA: Addison-Wesley, 1980.
[23] N. Jouppi, "Timing analysis for nMOS VLSI," in *Proc. IEEE 20th Design Automat. Conf.*, June 1983, pp. 411-418.
[24] J. E. Hopcroft and J. P. Ullman, *Introduction to Automata Theory, Languages and Computation*. Reading, MA: Addison-Wesley, 1979.
[25] C. W. Wu, P. R. Cappello, and M. Saboff, "An FIR filter tissue," presented at the 1985 Proc. 19th Asilomar Conference on Circuit, Systems, and Computers, Pacific Grove, CA, Nov. 1985.

Kazuo Iwano received the B.S. degree in mathematics from the University of Tokyo, Tokyo, Japan, in 1975, and the M.S.E. and the M.A. degrees in electrical engineering and computer science from Princeton University, Princeton, NJ, in 1983 and 1984, respectively.

He is currently a Ph.D. student at Princeton University in the Department of Computer Science. When not in graduate school, he has been working for IBM-Japan, Ltd. since 1975. His current interests include the VLSI design, graph theory, and the design and analysis of algorithms.

Kenneth Steiglitz (S'57-M'64-SM'79-F'81) was born in Weehawken, NJ, on January 30, 1939. He received the B.E.E., M.E.E., and Eng.Sc.D. degrees from New York University, New York, NY, in 1959, 1960, and 1963, respectively.

Since September 1963 he has been at Princeton University, Princeton, NJ, where he is now a Professor of Computer Science, teaching and conducting research on VLSI design and implementation of signal processing, optimization algorithms, and the foundations of computing. He is the author of *Introduction to Discrete Systems* (New York: Wiley, 1974), and co-author, with C. H. Papadimitriou, of *Combinatorial Optimization: Algorithms and Complexity* (Englewood Cliffs, NJ: Prentice-Hall, 1982).

Dr. Steiglitz is a member of the VLSI Committee of the IEEE ASSP Society, is serving his second term as member of the Administrative Committee, and has also served on the Digital Signal Processing Committee, and as Awards Chairman of that Society. He is an Associate Editor of the journal *Networks*, and is a former Associate Editor of the *Journal of the Association for Computing Machinery*. He is a member of Eta Kappa Nu, Tau Beta Pi, and Sigma Xi, and he received the Technical Achievement Award of the ASSP Society in 1981, and the IEEE Centennial Medal in 1984.

A VLSI Layout for a Pipelined Dadda Multiplier

PETER R. CAPPELLO and KENNETH STEIGLITZ
Princeton University

Parallel counters (unary-to-binary converters) are the principal component of a Dadda multiplier. We specify a design first for a pipelined parallel counter, and then for a complete multiplier. As a result of its structural regularity, the layout is suitable for use in a VLSI implementation.

We analyze the complexity of the resulting design using a VLSI model of computation, showing that it is optimal with respect to both its period and latency. In this sense the design compares favorably with other recent VLSI multiplier designs.

Categories and Subject Descriptors: B.2.1 [**Arithmetic and Logic Structures**]: Design Styles—*parallel*; *pipeline*; C.5.4 [**Computer System Implementation**]: VLSI Systems

General Terms: Algorithms, Design, Performance

Additional Key Words and Phrases: multiplier, VLSI, layout, complexity

1. INTRODUCTION

There has been considerable interest in applying VLSI technology to the task of integer multiplication [21]. Lower bounds for this problem have been presented on the area [24], area-time product [1, 4], and data rate [24]. Upper bounds have also appeared for the area-time product (e.g., [4, 13, 17, 18]).

These bounds are asymptotic, and before we go further we should point out that it is important to consider carefully the meaning of such bounds in our context, and especially the effect of using either the Fast Fourier Transform (FFT) or the Discrete Fourier Transform (DFT) as a subalgorithm for multiplication. Designs that do use the DFT or FFT can have good asymptotic properties, but are generally practical only for very long word lengths. Upper bounds obtained this way are viewed as being mainly of theoretical interest. The design

This work was supported in part by the National Science Foundation under Grant ECS-7916292, and in part by the U.S. Army Research Office, Durham, North Carolina, under Grant DAAG29-79-C-0024.

Authors' addresses: P. R. Cappello, Dept. of Computer Science, University of California, Santa Barbara, CA 93106; K. Steiglitz, Dept. of Electrical Engineering and Computer Science, Princeton University, Princeton, NJ 08544.

This paper was originally submitted for publication in the *Communications of the ACM*. The CACM department editor responsible for the paper was Duncan H. Lawrie. Both the authors and editor kindly agreed to publish the paper in the *ACM Transactions on Computer Systems*.

proposed in this paper does not use the FFT or DFT, and appears to be quite practical in the sense that the constants in the asymptotic upper bounds are relatively small.

The following remark appears in [4]: "We do not know if there is any practical design having $AT^{2\alpha} = o(n^{1+2\alpha})$ for $\alpha \in [0, 1]$." Brent and Kung go on to present a design and show that their lower bound is tight to within log factors, but they deem their design to be primarily of theoretical interest because it multiplies via the DFT. Preparata and Vuillemin [18] give an AT^2 optimal ($O(n^2)$) multiplier using the cube-connected cycles topology; it uses the FFT. Preparata [17] improves this theoretical result by achieving the AT^2 lower bound with a simpler topology: the two-dimensional mesh. This design is interesting because such a mesh can be embedded in the plane without long edges, but the design uses the DFT. It *may* be practical for sufficiently large integers. Jackson, Kaiser, and McDonald [11] and Lyon [14], on the other hand, describe a practical multiplier, but its area is $O(n)$ and its time is $O(n)$ so that its $AT^{2\alpha} = O(n^{1+2\alpha}) \neq o(n^{1+2\alpha})$.

Luk, in direct response to the remark of Brent and Kung for $\alpha = 1$, presents a multiplier layout with $T = O(\log^2 n)$ [13]. He states: "The AT^2 measure of this multiplier layout is nearly optimal, being $O(n^2 \log^4 n)$; so it answers the question posed by Brent-Kung that the existence of a practical multiplier having $AT^2 = o(n^3)$ measure." The multiplier design presented by Luk is recursive, performing an n-bit multiply via four $n/2$-bit multiplies and two additions. Its AT^2 complexity is $O(n^2 \log^6 n)$. His design employs the shuffle-exchange network. Reference was made to an alternate design where an n-bit multiply would be obtained by recursively performing three $n/2$-bit multiplies (see [2]). Luk stated that this design had an AT^2 complexity of $O(n^2 \log^4 n)$. It was not presented, however, because the layout is less regular.

The purpose of this paper is to present and analyze a VLSI layout for a multiplier that also has $AT^2 = o(n^3)$. The model we use here is the one used by [4, 13, 17, 18], the *synchronous* model of VLSI [3, 4, 23]. The design we present has time and period that are asymptotically optimal ($T = O(\log n)$ and $P = O(1)$); its AT^2 measure is $O(n^2 \log^3 n)$. Because its period is $O(1)$, it is potentially useful in applications that require both fast response and high throughput. A VLSI measure that favors fast response and high throughput without ignoring area is AP^2T^2. Under this measure, a lower bound for n-bit multiplication is $\Omega(n^2 \log^2 n)$, and the scheme presented has complexity $O(n^2 \log^3 n)$. Moreover, the architecture falls under the rubric of a Dadda scheme [8]. Such architectures are already used in some high-speed computers, attesting to their practicality in demanding situations. The layouts given for the parallel counter and overall multiplier are structurally regular, and so are suitable for a VLSI implementation. Table I summarizes the practicality and asymptotics of the designs mentioned above.

The remainder of the paper is organized as follows. Section 2 defines some terms used in the paper. Section 3 presents a design for a parallel counter (which we call a unary-to-binary converter). In Section 4 we review the Dadda scheme for K-ary addition. Then we present a Dadda design for K-ary addition that employs the unary-to-binary design presented in the previous section. In Section 5 we present a layout for a multiplier using the K-ary adder of Section 4. This is followed by some historical and concluding remarks.

Table I. Summmary of Multiplier Designs

Design	Note	Period	Time	Area	AP^2T^2
Lower bound (Ω)	—	1	$\log n$	n	$n^2\log^2 n$
[4]	DFT	$n^{1/2}\log n$	$n^{1/2}\log n$	$n \log n$	$n^3\log^5 n$
[18]	FFT	$\log^2 n$	$\log^2 n$	$(n/\log n)^2$	$n^2\log^4 n$
[17]	DFT	$n^{1/2}$	$n^{1/2}$	n	n^3
[14]	—	n	n	n	n^5
[15]	—	1	n	n^2	n^4
[13]	—	1	$\log^2 n$	$n^2\log^2 n$	$n^2\log^6 n$
This paper	—	1	$\log n$	$n^2\log n$	$n^2\log^3 n$

Note: Designs that multiply via a DFT computation are generally regarded as being practical only for large values of n.

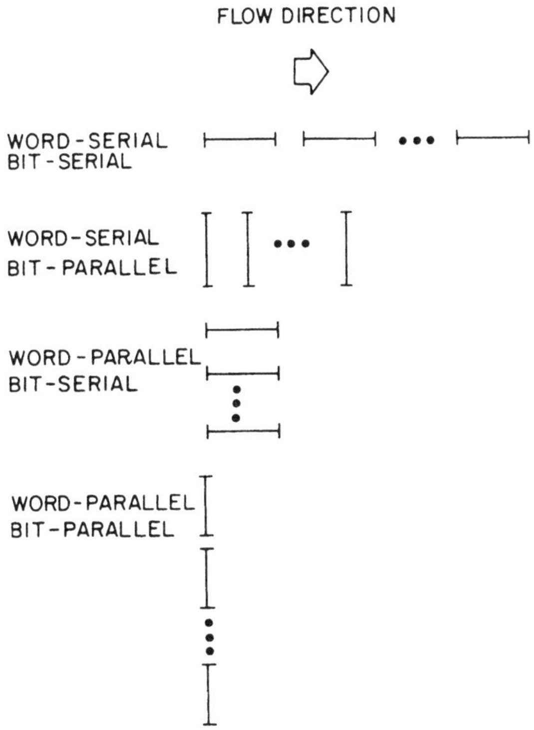

Fig. 1. Use of the following terms: word-serial, word-parallel, bit-serial, and bit-parallel.

2. TERMS

In this section we introduce some terminology. The terms word-serial, word-parallel, bit-serial, and bit-parallel will be used in the obvious ways, as shown in Figure 1.

We deal with classes of functions (and circuits) that are parameterized by a vector, π. For example, when we consider addition of K operands of word size B, the parameter vector is $\pi = (K, B)$. Asymptotic complexity will be measured with respect to a parameter vector, π, throughout this paper.

We want to make clear important distinctions among three time measures of interest:

Definition. The *functional latency* (or time [24]) of a circuit is the amount of time separating the appearance of the first input bit on some port from the appearance of the last output bit on some port, for one computation of the function, f, denoted T_f. This corresponds to the usual use of the term *speed of operation.* A "100-nanosecond" multiplier, for example, means that 100 nanoseconds elapses between the appearance of the first input bit of the multiplicands and the last output bit of the product. The term does not take pipelining into account.

In order to define a circuit's *cycle time* we appeal to a finite state machine model of computation [10]. A circuit C can be defined as follows:

$$C = (Q, \Sigma, \Delta, \delta, q_0)$$

where

Q is a set of states,
Σ is an input alphabet,
Δ is an output alphabet,
q_0 is distinguished as the machine's initial state, and
δ is a (state transition) function that maps a state and an input to a new state and an output, $\delta : Q \times \Sigma \rightarrow Q \times \Delta$.

Definition. The *cycle time* of a circuit is T_δ (the latency of the circuit with respect to its transition function, δ).

In some circuit architectures, the cycle time depends on the *size* of the function being implemented. For example, an n-bit array multiplier [19] may have cycle time, $\tau = n\tau_c + n\tau_s$, where τ_c and τ_s are the carry and sum bit times of a 1-bit full adder. Such a circuit architecture has a cycle time of $O(n)$.

Definition. The *functional period* of a circuit is the number of cycles separating corresponding bits of successive inputs (outputs) of function f, denoted P_f. Period is the reciprocal of throughput rate. This term does take pipelining into account. Note that $P_\delta = T_\delta$; δ is an indivisible function with respect to these time measures.

Definition. A circuit is *completely pipelined* with respect to function f when $P_f = 1$ and its cycle time is $O(1)$ (see Proposition 2 of [6]).

Note. We consider in this paper only completely pipelined circuits ($P_f = O(1)$ *seconds*).

For example, in our B-bit (parallel) multiplier, $\Sigma = \{0, 1\}^{2B}$ and $\Delta = \{0, 1\}^{2B}$. In our K-ary B-bit (parallel) adder, $\Sigma = \{0, 1\}^{KB}$ and $\Delta = \{0, 1\}^{B+\log K}$. (See [15] for a variant of the array multiplier that is pipelined). Thus in these architectures, any combinational logic that can be performed in one cycle has depth $O(1)$. The chip's actual bandwidth is then within a constant factor of its maximum bandwidth as determined by the I/O delay of the technology.

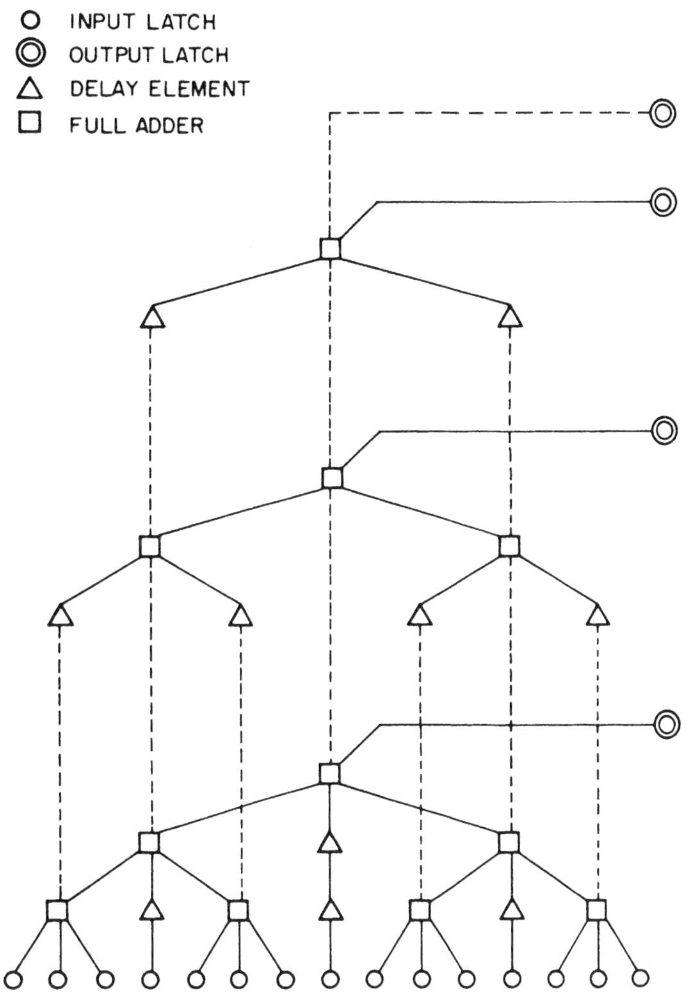

Fig. 2. A unary-to-binary conversion (UBC) structure for $B = 15$.

3. PARALLEL COUNTING

Parallel counting lies at the heart of all the algorithms presented in this paper. In this section we present a parallel counting scheme. First, we state the problem, which we refer to as *unary-to-binary conversion* (UBC).

Problem: Unary-to-binary conversion (UBC)
Input: B bits
Output: $\lfloor \log B \rfloor + 1$ bits—the binary representation of the number of 1 values among the input bits

We now give the UBC design.[1] Figure 2 illustrates a structure for UBC. Each square (□) in the figure represents a latched 1-bit full adder (FA). The triangles

[1] We assume $B = 2^K - 1$ for some integer K without loss of generality as far as asymptotic complexity is concerned.

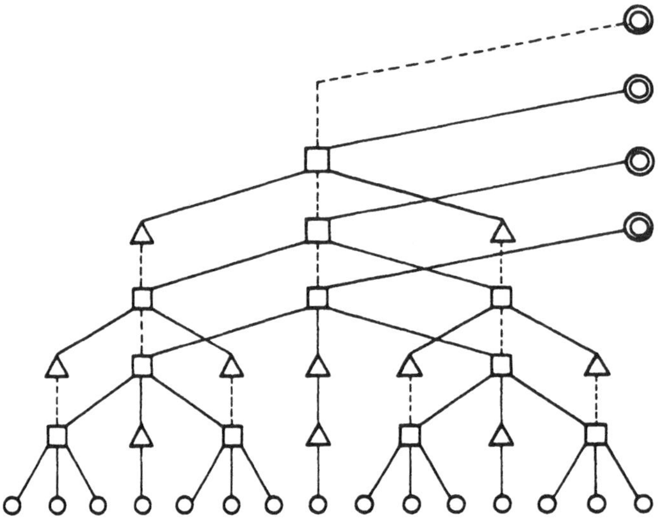

Fig. 3. A more compact UBC structure for $B = 15$.

(Δ) are delay elements. The computation proceeds as follows. Input enters the leaves of the *bottom* tree. The *dashed* output line transmits the carry bit (as depicted in the figure). The *solid* output line transmits the sum bit. Thus, the sum bit transmitted from the root node of the *bottom* tree is the low-order bit of the answer. The dashed lines emanating from the bottom tree, the carry bit lines, enter a second, smaller tree. The sum bit transmitted from that tree's root node is the next-to-low-order bit of the answer, and so on. The number of such trees needed to perform the entire UBC is log n. If the design was such that one tree finished its computation before its successor tree began, then the time complexity

$$T(n) = \sum_{i=\log((n+1)/2)}^{1} i = O(\log^2 n).$$

If the structure had the layout of Figure 2, then the area complexity

$$A(n) = n \times \left[\sum_{i=\log((n+1)/2)}^{1} i \right] = O(n \log^2 n).$$

The trees nest into each other, however, as depicted in Figure 3, and so do the computations. We do not have to finish computing in one tree before starting in the next. Thus,

$$T_{\text{UBC}}(n) = \log((n+1)/2) + \log((n+1)/4) = O(\log n)$$

$$A_{\text{UBC}}(n) = n[\log((n+1)/2) + \log((n+1)/4)] = O(n \log n).$$

The structure, in fact, has an $A(n) = O(n)$ layout since it has an $O(\log n)$-separator (see [12]). Such a layout would be useless though; it would not place the inputs on the layout boundary. Figure 4 illustrates the UBC design with one set of input bits. Each successively higher level in the structure gets values at successively later cycles (as indicated by the cycle labels of Figure 4). This

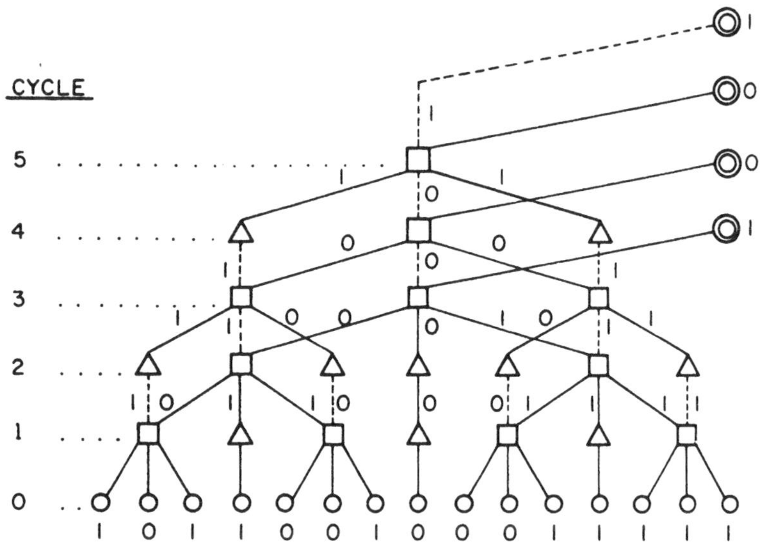

Fig. 4. An example computation on a UBC-network for $B = 15$.

{/* The algorithm starts with a pile of the K numbers to be added. */
Do sequentially from low order columns to high order columns:
{
Add the digits in the column;
Post the low order digit of this sum in the sum digit for this column;
Post the remaining (carry) digits in the appropriate columns;
/* to be included in the summation of those columns */
}
}

Fig. 5. Elementary school addition algorithm.

illustrates, as well, an important aspect of the design: it is pipelined, so a new set of inputs can be accepted every cycle.

The design has complexity $AP^2T^2_{\mathrm{UBC}}(n) = O(n \log^3 n)$. Using the techniques of [24] we argue that $AP^2_{\mathrm{UBC}}(n) = \Omega(n)$. $T_{\mathrm{UBC}}(n) = \Omega(\log n)$ due to bounded fan-in. So $AP^2T^2_{\mathrm{UBC}}(n) = \Omega(n \log^2 n)$.

In sum, the design is within a log factor of asymptotic optimality and has a simple, regular structure.

4. *K*-ARY ADDITION

In this section we present a K-ary adder (for adding K B-bit numbers) that is based on UBC. We begin by reviewing the K-ary addition algorithm taught in elementary school, given in Figure 5. The algorithm presented is a parallelized version of this.

A parallelized version of this algorithm (reported in [8]) proceeds roughly as follows.

First we do in *parallel* the following (for all columns):

1. Add the "digits" in the column.
2. Distribute the $\log K$ "digits" to the appropriate columns.

		COLUMN						
		6	5	4	3	2	1	0
5	≡	0	0	0	0	1	0	1
-10	≡	1	1	1	0	1	1	0
1	≡	0	0	0	0	0	0	1
-13	≡	1	1	1	0	0	1	1
12	≡	0	0	0	1	1	0	0
30	≡	0	0	1	1	1	1	0

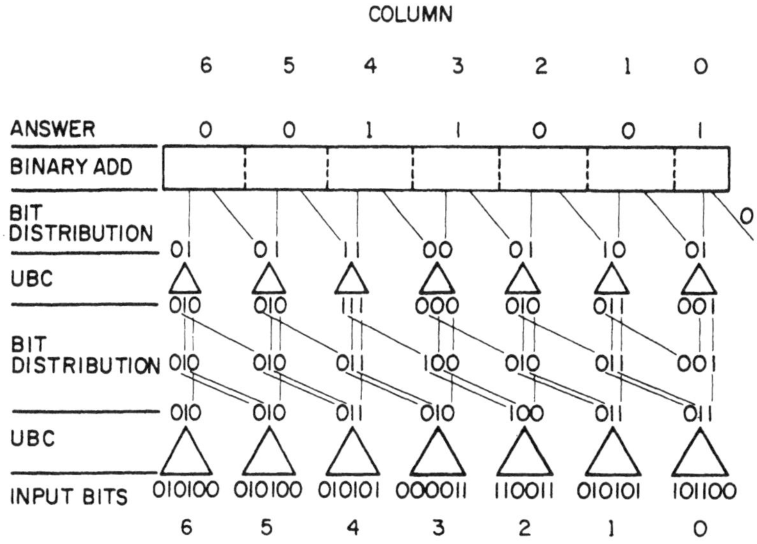

Fig. 6. K-ary addition for $K = 6$, $B = 7$, sum $= 25_{10} \equiv 011001_2$.

```
{
repeat until columns contain 2 bits
    do for all columns
        {
        perform UBC on the column;
        Distribute the sum bits;   /* create new columns */
        }
sum ← carry-lookahead binary-addition;
}
```

Fig. 7. A parallel K-ary addition algorithm.

These "digits" now constitute new columns to be added. We have reduced our K-ary addition problem to a log K-ary addition problem. The process is repeated until a binary addition problem remains. We use a carry-lookahead binary adder for the final addition.

We propose that a VLSI implementation of this algorithm employ the UBC-network design of the previous section, and the adder of Brent and Kung [5]. Figure 6 shows the structure and an example. Figure 7 is a restatement of the algorithm that assumes a binary number representation.

We now analyze the period, latency, and area complexity of this design using the *synchronous* model of VLSI computation [3, 4].

Since the UBCs and binary adder [5] are completely pipelined, the entire K-ary B-bit adder is completely pipelined: $P_{K\text{-add}} = 1$ *cycle*.

The latency of this algorithm is the sum of the time to do the following.

1. Repeat block, which is the sum of the time for doing UBCs and bit distribution ("column" recreation).
2. B-bit parallel binary addition.

The asymptotic latency is

$$
\begin{aligned}
T_{\text{add}}(K, B) &= O\left(\sum_{i=1}^{\beta(K)} \underbrace{\log \cdots \log}_{i} K\right) && \text{(UBC)} \\
&\quad + \sum_{i=1}^{\beta(K)} O(1) && \text{(Bit distribution)} \\
&\quad + O(\log B) && \text{(Binary addition)} \\
&= O(\log KB)
\end{aligned}
$$

where $\beta(n)$ is a function that grows slower than $\log n$, but is not a constant. See Appendix.

The area complexity is taken as the product of the width and length of the structure:

$$
\begin{aligned}
A_{\text{add}}(K, B) &= O(KB) && \text{(Base)} \\
&\quad \times \left[O\left(\sum_{i=1}^{\beta(K)} \underbrace{\log \cdots \log}_{i} K\right) \right. && \text{(UBC)} \\
&\quad + O\left(\sum_{i=1}^{\beta(K)} \underbrace{\log \cdots \log}_{i} K\right) && \text{(Bit distribution)} \\
&\quad \left. + O(\log B) \right] && \text{(Binary addition)} \\
&= O(KB \log KB)
\end{aligned}
$$

The design has complexity $AP^2T^2_{\text{add}}(K, B) = O(KB \log^3 KB)$. Using the techniques of [24], we argue that $AP^2_{\text{add}}(K, B) = \Omega(KB)$. $T_{\text{add}}(K, B) = \Omega(\log KB)$ due to bounded fan-in. So $AP^2T^2_{\text{add}}(K, B) = \Omega(KB \log^2 KB)$.

Again, the design is within a log factor of asymptotic optimality and has a simple, regular structure.

5. BINARY MULTIPLICATION

In this section we present a binary multiplier. The algorithm, based on K-ary addition, is suggested by the following definition of binary multiplication.[2]

$$
x \times y = \sum_{b=0}^{B-1} 2^b \times x_b \times y \tag{1}
$$

[2] We work with positive numbers without loss of generality, assuming two's complement arithmetic. The structures described here require a small modification to compute a product with $2B$ correct bits.

$y_3 \times x_3$	$y_2 \times x_3$	$y_1 \times x_3$	$y_0 \times x_3$			
	$y_3 \times x_2$	$y_2 \times x_2$	$y_1 \times x_2$	$y_0 \times x_2$		
		$y_3 \times x_1$	$y_2 \times x_1$	$y_1 \times x_1$	$y_0 \times x_1$	
			$y_3 \times x_0$	$y_2 \times x_0$	$y_1 \times x_0$	$y_0 \times x_0$

Fig. 8. The summands of eq. (1).

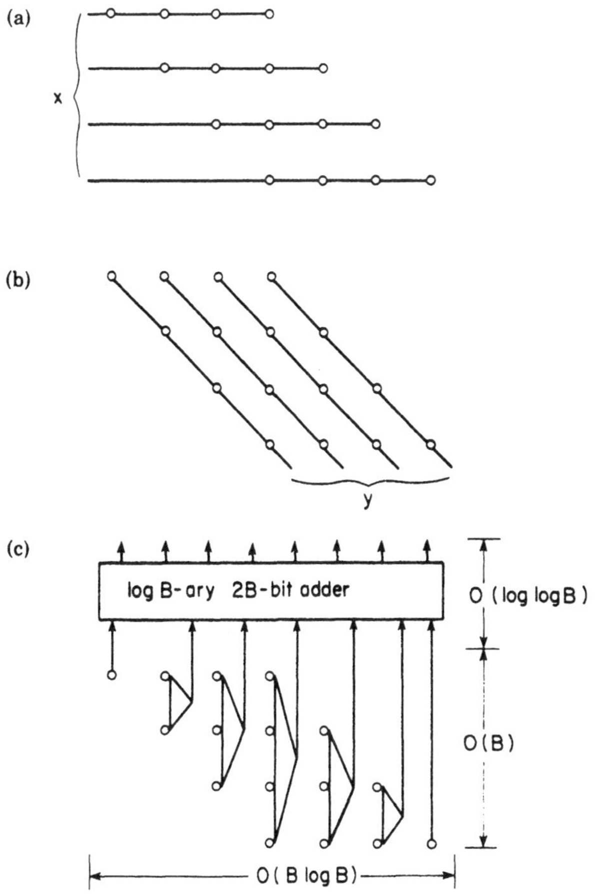

Fig. 9. The multiplication structure is the superposition of the above structures: (a) a structure for distributing x values ($B = 4$); (b) a structure for distributing y values ($B = 4$); (c) a structure for performing the B-ary addition ($B = 4$).

Figure 8 depicts the summands of eq. 1 for $B = 4$. The multiplication algorithm is simple:

1. Distribute x and y values.
2. B-ary add the resulting pile of numbers.

Figure 9(a, b, c) illustrates the structures for distributing x values, y values, and B-ary addition. The multiplication structure is the superposition of these structures.

We now analyze the VLSI complexity of this UBC-network Dadda multiplier with respect to its period, latency, and area. Since the input distribution and AND operation can be completely pipelined into the B-ary B-bit adder which is also completely pipelined, the entire multiplier is completely pipelined: $P_{\text{mult}} = 1$ *cycle*.

The latency of this algorithm is the sum of the time to do the following.

1. Distribute x and y values.
2. AND(x_i, y_j) for every i, j.
3. B-ary add the result.

Thus the asymptotic latency, dominated by the UBC-network latency, is

$$T_{\text{mult}}(B) = O(1) + O(1) + O(\log B) = O(\log B)$$

The asymptotic area complexity, also dominated by the UBC-network areas, is

$$A_{\text{mult}}(B) = O(B^2 \log B)$$

as indicated by Figure 9c.

The design has complexity $AP^2T^2_{\text{mult}}(B) = O(B^2 \log^3 B)$. Regarding VLSI lower bounds we note the following. Vuillemin has shown [24] that:

1. Integer multiplication is a transitive function, and
2. Any circuit computing a transitive function at data rate D must have wire area $A_w \geq a_w \times D^2$, for some technology-dependent constant a_w.

Since the period $P = B/D$ where B is the number of input bits and D is the rate at which they are read in, we have that for transitive functions

$$AP^2(B) = \Omega(B^2).$$

$T_{\text{mult}}(B) = \Omega(\log B)$ due to bounded fan-in. So $AP^2T^2_{\text{mult}}(B) = \Omega(B^2 \log^2 B)$.

As in the UBC and K-ary adder the design is within a log factor of asymptotic optimality and has a simple, regular structure.

6. HISTORICAL COMMENT

We would like to place this VLSI design in some historical perspective. *Gelosia* multiplication methods are those that

1. Construct a two-dimensional array of componentwise products, and
2. Take the B-ary sum of these component products.

The gelosia method (grating method) for multiplication was apparently quite popular during the fifteenth and sixteenth centuries. The method, illustrated in Figure 10, was "probably first developed in India, for it appears in a commentary on the *Lilavati* of Bhaskara (1114 to ca. 1185) and in other Hindu works" [9].

A *Dadda* multiplication scheme is a gelosia multiplication scheme where the "diagonal" K-ary addition is performed in parallel, as presented in Section 4. That is, the K-ary addition problem is reduced to a log k-ary addition problem,

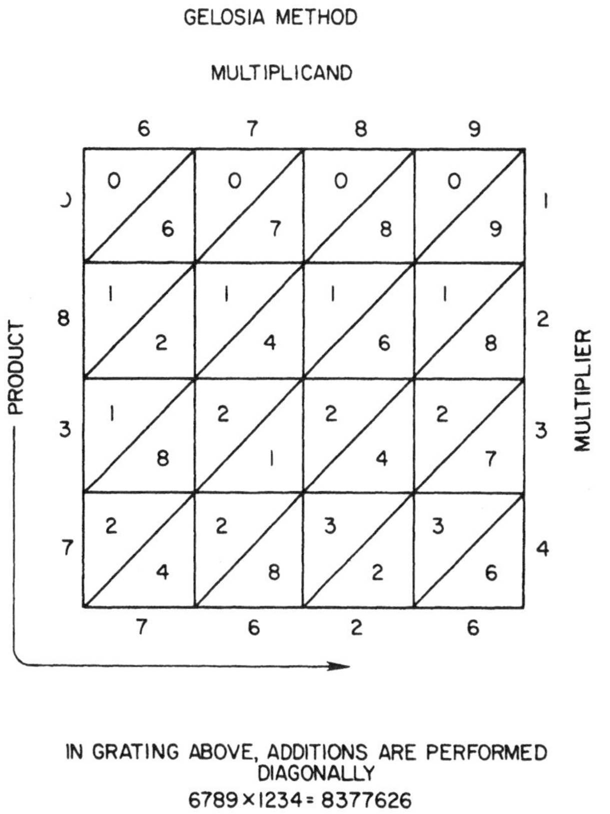

Fig. 10. Gelosia method for multiplication (after [9]). (In the grating illustrated above, additions are performed diagonally.

and that is reduced to a loglog k-ary addition problem, and so on, until a binary addition problem remains. In order to achieve asymptotic optimality with respect to period and latency, this approach requires performing the following:

1. Unary-to-binary conversions asymptotically optimally, and
2. Final binary addition asymptotically optimally.

The adder in [5] does the latter, the UBC-network of Section 3 does the former. And, what is important for VLSI implementation, they do them with simple, regular, compact layouts.

7. DISCUSSION

VLSI techniques are such that the time has now or will soon come when the performance characteristics of a multiplier can be matched to its intended application. For applications where neither fast response nor high throughput are crucial, the multiplier in [11, 14] may be a good match; it has an asymptotically minimal area and a simple, regular structure. For applications where high throughput is crucial but fast response is not, the design in [15] provides an

$AP^2(B)$ asymptotically optimal multiplier. This simple hex-connected array of latched 1-bit full adders is also practical.

We believe the UBC-network Dadda design presented here is a practical design for applications that require both short latency and high throughput. It is *flexible* as well, in that

1. Latency can be minimized by eliminating all latches and delay elements, yielding a purely combinational multiplier of circuit depth (hence cycle time) $O(\log B)$,
2. Throughput can be maximized by latching the UBC-networks at every level within the UBC-network, as presented in this paper, and
3. Some trade-off between latency and throughput is possible by latching every k level within a UBC-network (for $1 \leq k \leq O(\log B)$).

As long as the UBC-networks are latched every O(1) levels, the design is asymptotically optimal in both period and latency.

Finally, it may be that:

1. B is large.
2. The technology is such that propagation delay over a wire of length l is asymptotically greater than $\Omega(l^{1/2}/\log l)$, a situation not modeled properly by the synchronous VLSI model (e.g., [7, 20]).
3. High throughput is not as important as short latency.

For these applications, the design given in [17] merits consideration due to its two-dimensional mesh topology and $T(B) = O(B^{1/2})$ latency.

Digital signal processing is one area which stands to gain a great deal from VLSI implementation of arithmetic, both with small period (for throughput) and small latency (for fast adaptation, for example). Many such applications involve K-ary addition and can benefit from the Dadda approach. One important example is the *second-order section* of an Infinite Impulse Response (IIR) filter. This computation can be defined as follows [16]:

$$y_k \leftarrow a_0 x_k + a_1 x_{k-1} + a_2 x_{k-2} - b_1 y_{k-1} - b_2 y_{k-2}$$

The Dadda-like design for this "sum of products" involves using:

1. The gelosia method to create bit arrays for the five products, and
2. A UBC-network Dadda scheme to sum them *directly*, computing y_k (which Swartzlander calls *merged arithmetic* [22]).

Figure 11 illustrates the use of a UBC-network Dadda scheme in the design of a second-order section. The top part of that figure shows the *distribution* of the x and y bits, and the bottom part shows the *collection* of the results using UBC networks in a way analogous to Fig. 9c. Note that this structure can *not* be completely pipelined, because we need y_k to compute y_{k+1}. The latency of O(log B) is asymptotically as small as possible, so one cannot build a second-order section with higher asymptotic throughput. Using strictly combinational

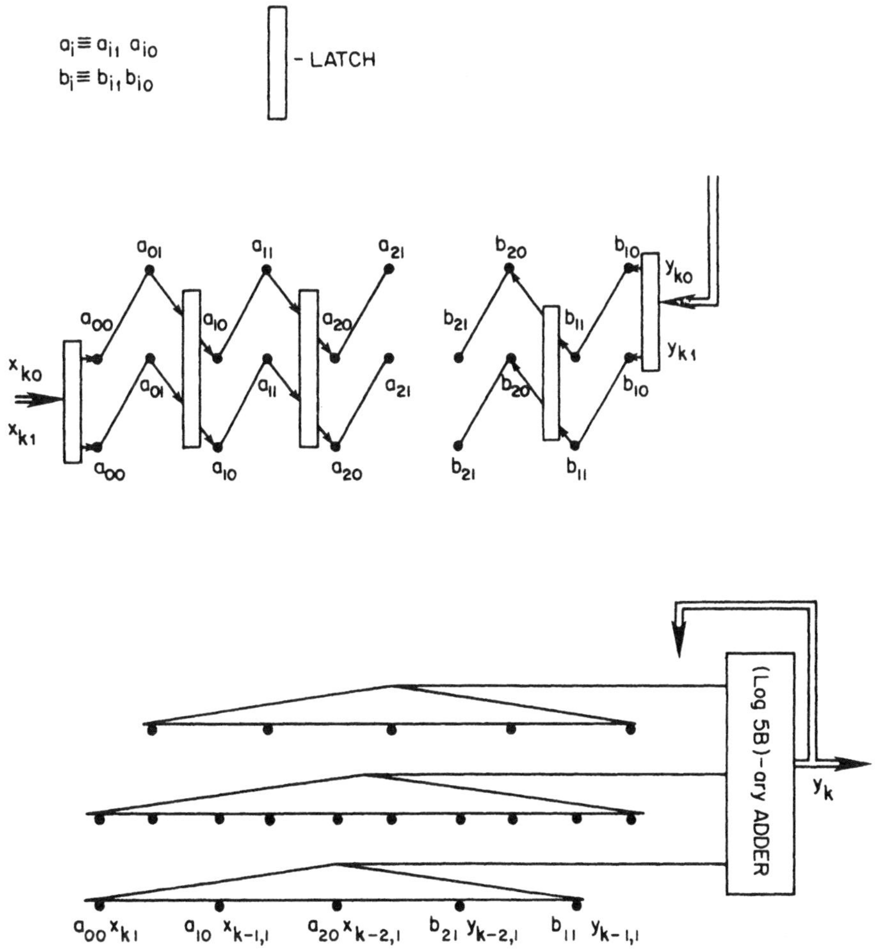

Fig. 11. The superposition of these structures is a UBC-network Dadda scheme for a second-order section ($B = 2$).

logic (no latches and no delays) would simplify the control and reduce latency. The cycle time would then be $O(\log B)$ seconds, as would the latency.

Figures 12 and 13 show similar designs for convolution and fixed-matrix vector product. In these problems there is no feedback, and the structures are completely pipelined. Table II summarizes the complexity of several functions designed this way.

Two points are worthy of emphasis here. First, the latency of these designs is in all cases asymptotically optimal. It is on the order of the *logarithm* of the input size. Second, these designs are *completely pipelined* (with the exception of the second-order section). That is, the designs are such that a new set of input data can be accepted every cycle and that the cycle time is *independent* of the parameters of the function (e.g., B-bits in the case of multiplication) being implemented.

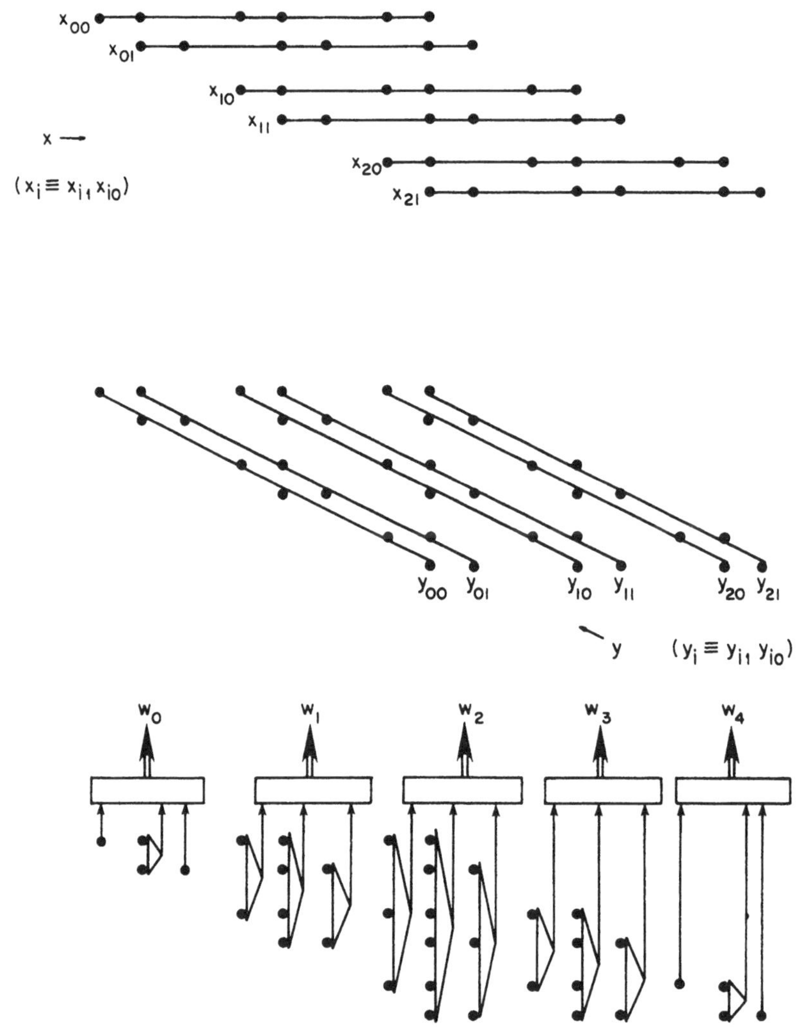

Fig. 12. The superposition of these structures is a UBC-network Dadda scheme for convolution ($K = 3$, $B = 2$).

8. CONCLUSIONS

We have presented a design for completely pipelined multiplication, and we have analyzed its period, latency, and area complexities using a VLSI model of computation. Such a multiplier is useful for applications where both short latency and high throughput are very important; its complexity is within one log factor of asymptotic optimality with respect to an appropriate measure: AP^2T^2. Its latency and period are asymptotically optimal. The cell types are simple, being no more complex than 1-bit full adders, and the layout has a regular structure. Consequently, we feel that the design is practical.

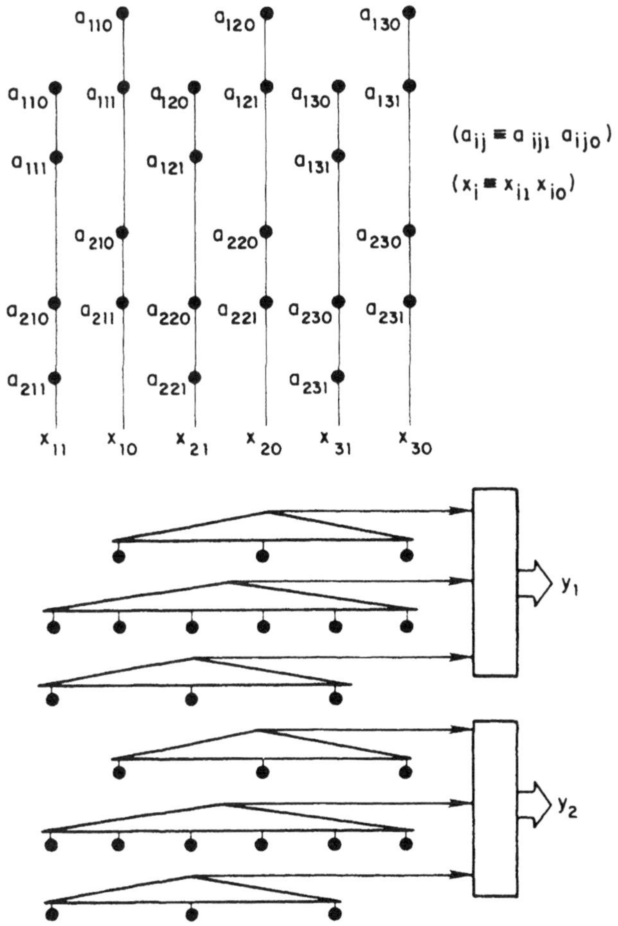

Fig. 13. The superposition of these structures is a UBC-network Dadda scheme for fixed-matrix product (the matrix is 2×3, $B = 2$).

Table II. Asymptotic Growth Using UBC-Network Dadda Scheme Design

	Area Bound	AP^2T^2 Bounds	
Function	Upper area	Lower	Upper
Integer product	$B^2 \log B$	$B^2 \log^2 B$	$B^2 \log^3 B$
Fixed-matrix vector product	$K^2B^2 \log KB$	$(KB)^2 \log^2 KB$	$(KB)^2 \log^3 KB$
Convolution	$K^2B^2 \log KB$	$(KB)^2 \log^2 KB$	$(KB)^2 \log^3 KB$
Matrix product	$K^4B^2 \log KB$	$K^2(KB)^2 \log^2 KB$	$K^2(KB)^2 \log^3 KB$

Table III. The function $\beta(n)$

n	$\beta(n)$
2	0
3	1
4–7	2
8–127	3
$128\text{-}2^{128}-1$	4

APPENDIX

$\beta(n)$ is defined as follows.

$$\beta(n) = \begin{cases} 0 & \text{if} \quad n = 2 \\ 1 + \beta(\lfloor \log n \rfloor) & \text{if} \quad n > 2 \end{cases}$$

It is essentially the same as $G(n)$ [2]:

$$G(n) = \begin{cases} 0 & \text{if} \quad n = 1 \\ 1 + G(\lceil \log n \rceil) & \text{if} \quad n > 1 \end{cases}$$

For practical purposes, $\beta(n)$ is less than 5, as can be seen from Table III.

ACKNOWLEDGMENT

We thank Peter Reusens for his many helpful suggestions on an earlier version of this paper. They improved the result considerably. We also benefited from the constructive criticism of the referees.

REFERENCES

1. ABELSON, H., AND ANDREAE, P. Information transfer and area-time tradeoffs for VLSI multiplication. *Commun. ACM 23*, 1 (Jan. 1980), 20–23.
2. AHO, A. V., HOPCROFT, J. E., AND ULLMAN, J. D. *The Design and Analysis of Computer Algorithms.* Addison-Wesley, Reading, Mass., 1974.
3. BILARDI, G., PRACCHI, M., AND PREPARATA, F. P. A critique and an appraisal of VLSI models of computation. In *VLSI Systems and Computations*, edited by H. T. Kung, Bob Sproull, and Guy Steele, Computer Science Press, Rockville, Md., 1981.
4. BRENT, R. P., AND KUNG, H. T. The area-time complexity of binary multiplication. *J. ACM 28*, 3 (July 1981), 521–534.
5. BRENT, R. P., AND KUNG, H. T. The chip complexity of binary arithmetic. In *Proc. 12th Annual ACM Symp. Theory of Computing*, (Los Angeles, Cal., April 28–30, 1980), ACM, New York, pp. 190–200.
6. CAPPELLO, P. R., AND STEIGLITZ, K. Completely pipelined architectures for digital signal processing. Tech. Rep. 303, Dept. of Electrical Engineering and Computer Science, Princeton Univ., Princeton, N.J., June 1982.
7. CHAZELLE, B., AND MONIER, L. A model of computation for VLSI with related complexity results. In *Proc. 13th Annual Symp. on Theory of Computing*, (Milwaukee, Wisconsin, May 11–13, 1981), ACM, New York, pp. 318–325.
8. DADDA, L. Some schemes for parallel multipliers. *Alta Frequenza 34*, (May 1965), 349–356. Reprinted in *Computer Design Development*, E. E. Swartzlander, Jr., Ed., Hayden Book, Rochelle Park, N.J., 1976.
9. EVES, H. W. *Mathematical Circles Revisited.* Prindle, Weber & Schmidt, Boston, Mass., 1971.
10. HOPCROFT, J. E., AND ULLMAN, J. D. *Introduction to Automata Theory, Languages, and Computation.* Addison-Wesley, Reading, Mass., 1979.

11. JACKSCN, L. B., KAISER, S. F., AND MCDONALD, H. S. An approach to the implementation of digital filters. *IEEE Trans. Audio and Electro-Acoustic, AU-16* (Sept. 1968), 413–421.
12. LEISERSON, C. E. Area-efficient graph layouts (for VLSI). In *Proc. IEEE 21st Annual Symp. Foundations of Computer Science.* Syracuse, N.Y., 1980.
13. LUK, W. K. A regular layout for parallel multiplier of $O(\log^2 n)$ time. In *VLSI Systems and Computations*, H. T. Kung, Bob Sproull, and Guy Steele, Eds., Computer Science Press, Rockville, Md., 1981.
14. LYON, R. F. Two's complement pipeline multipliers. *IEEE Trans. Comm., COM-24*, 4 (April 1976), 418–425.
15. MCCANNY, J. V., MCWHIRTER, J. G., ROBERTS, J. B. G., DAY, D. J., AND THORP, T. L. Bit level systolic arrays. In *Proc. 15th Asilomar Conf. on Circuits, Systems, & Computers*, Nov. 1981.
16. OPPENHEIM, A. V., AND SCHAFER, R. W. *Digital Signal Processing.* Prentice-Hall, Englewood Cliffs, N.J., 1975.
17. PREPARATA, F. P. A mesh-connected area-time optimal VLSI integer multiplier. In *VLSI Systems and Computations*, H. T. Kung, Bob Sproull, and Guy Steele, Eds., Computer Science Press, Rockville, Md., 1981.
18. PREPARATA, F. P., AND VUILLEMIN, J. E. The cube-connected-cycles: A versatile network for parallel computation. *Commun. ACM 24*, 5 (May 1981), 300–309.
19. RABINER, L. R., AND GOLD, B. *Theory and Application of Digital Signal Processing*, Prentice-Hall, Englewood Cliffs, N.J., 1975.
20. RAMACHANDRAN, V. On driving many long wires in a VLSI layout. Tech. Rep. 300, Department of Electrical Engineering and Computer Science, Princeton, Univ., Princeton, N.J., April 1982.
21. REUSENS, P., KU, W., AND MAO, Y. Fixed-point high-speed multipliers in VLSI. In *VLSI Systems and Computations*, H. T. Kung, Bob Sproull, and Guy Steele, Eds., Computer Science Press, Rockville, Md., 1981.
22. SWARTZLANDER, E. E., JR., Merged arithmetic for signal processing. In *Proc. IEEE 4th Symp. Computer Arithmetic*, Santa Monica, Calif., 1978.
23. THOMPSON, C. D. Area-time complexity for VLSI. In *Proc. 11th Annual Symp. Theory of Computing*, (Atlanta, Georgia, April 30–May 2, 1979), ACM, New York, pp. 81–88.
24. VUILLEMIN, J. A combinatorial limit to the computing power of VLSI circuits. In *Proc. IEEE 21st Annual Symposium of Foundations of Computer Science.* Syracuse, N.Y., 1980.

Received October 1981; revised June 1982; accepted November 1982

A very fast multiplication algorithm for VLSI implementation

Jean Vuillemin

INRIA, Domaine de Voluceau - Rocquencourt, B.P. 105, 78153 Le Chesnay Cedex, France

Received 28 December 1982

Abstract. We present a simple recursive algorithm for multiplying two binary N-bit numbers in parallel O(log N) time. The simplicity of the design allows for a regular layout. The area requirement of this algorithm is comparable with that of much slower designs classically used in monolithic multipliers and in signal processing chips, hence the construction has definite practical impact.

Keywords. Recursive multiplication, O(log N) time, regular layout.

1. Introduction

While O(log N) time algorithms for computing the product of two N-bits binary numbers have been known for close to twenty years [1,2], they have not much been used in modern integrated circuits. Because of their complex nature, such algorithms have generally been discarded by designers of integrated multipliers.

Designers have chosen instead to implement a comparatively slow [3,4,5] O(N) time *Seq-Mult* algorithm, which trades theoretical speed for regularity of design and silicon area.

The O(log N) time multiplication algorithm *Par-Mult* proposed here is simple enough to admit a regular layout whose area requirement is only marginally bigger than that of *Seq-Mult*. Its recursive definition makes it ideally suited for

North-Holland Publishing Company
INTEGRATION, the VLSI journal 1 (1983) 39–52

macro-generating regular mask descriptions from a high level algorithmic specification [6]. We thus propose *Par-Mult* as a practical alternative to current VLSI multiplier design, whenever speed is the dominant design criterion.

Indeed, we estimate that *Par-Mult* computes 16 (resp. 32) bit-product 2 (resp. 3) times faster than *Seq-Mult*; yet it only uses 30% (resp. 40%) more area.

A prototype circuit and complete analysis of *Par-Mult* is performed by Luk [6].

2. Description of the algorithm

We describe the algorithms *Par-Mult* and *Seq-Mult* in this section, postponing layout considerations until Section 3.

2.1. Binary notation and statement of the problem

Let $A = \langle a[n-1], \ldots, a[0] \rangle$ be an n-bit binary sequence. Such a sequence denotes the natural number $\mathrm{val}\langle A \rangle = \mathrm{SUM}\langle 0 =< i < n\colon a[i] * (2 ** i) \rangle$. Function *val*, which maps binary sequences $\{0, 1\} ** n$ into natural numbers $\{a\colon 0 <= a < (2 ** n)\}$ defines the *semantics* of binary notation.

A binary multiplier is an algorithm for transforming two binary sequences A and B into $P = \mathrm{MULT}(A, B)$, the binary representation of the product $\mathrm{val}\langle A \rangle * \mathrm{val}\langle B \rangle$. If A and B are respectively m- and n-bit sequences, P is an $(m+n)$-bit sequence. For the algorithm to be correct, sequence P must satisfy:

$$\begin{aligned} \mathrm{val}\langle P \rangle &= \mathrm{SUM}\langle k >= 0\colon p[k] * (2 ** k) \rangle \\ &= \mathrm{SUM}\langle i, j >= 0\colon a[i] * b[j] * (2 ** (i + j)) \rangle \\ &= \mathrm{val}\langle A \rangle * \mathrm{val}\langle B \rangle. \end{aligned} \tag{1}$$

Multipliers realized as integrated circuits encode bit values in the physical world as states of a bi-stable electronic device. For example, in MOS technologies bits are represented by the presence or absence of electrical charges on localised capacitive wires. Inputs and outputs to such circuits must appear in this representation.

To implement a multiplication algorithm on silicon we must decompose its description to the point where it is entirely made of atomic actions, whose functionality is exactly matched by an electronic realisation (gate) in the technology. Designing such algorithms is conceptually not different from producing machine code from high-level descriptions.

We attempt here to describe the silicon-structure of the proposed multiplier by successively refining a high-level description of the algorithm in a top down (and ultimately error-free) manner.

2.2. Primitive building blocs for multiplication

If we represent 1 by *true* and 0 by *false*, the product $a * b$ of bits a and b is implemented as the logical *and*(a, b), an atomic gate of the technology.

Using m copies of the primitive *and*, we can construct the operator *bit-product*, computing the product of an m-bit binary number A by bit b:

$$\textbf{bit-product}(A, b) = \langle \textbf{and}(a[m-1], b), \ldots, \textbf{and}(a[0], b)\rangle$$
$$\text{val}\langle \textbf{bit-product}(A, b)\rangle = \text{val}\langle A\rangle * \text{val}\langle b\rangle$$
$$= \textbf{if } b = 0 \textbf{ then } 0 \textbf{ else } A. \quad (2)$$

In hardware implementations a parallel broadcast of bit b on a common bus (wire) ensures that *bit-product* introduces a (small) delay, independent of m, the number of bits in A.

Another useful primitive is *shift*(i, A), which adds i zeroes to the right (least significant bit) of sequence A:

$$\textbf{shift}(i, A) = \langle a[m-1], \ldots, a[0], 0, \ldots, 0\rangle$$
$$\text{val}\langle \textbf{shift}(i, A)\rangle = \text{val}[A] * (2 * * i). \quad (3)$$

In parallel multipliers, the *shift* operator is 'hard-wired' so as to introduce zero delay.

We shall fix later (see Section 2.4) the structure of the adders from which we assemble our multiplier. It will suffice for the time being to consider addition as a primitive operator *add* satisfying:

$$\text{val}\langle \textbf{add}(A, B)\rangle = \text{val}\langle A\rangle + \text{val}\langle B\rangle. \quad (4)$$

2.3. *Algorithms Seq-Mult and Par-Mult*

Formula (1), defining multiplication, can be rewritten as

$$\text{val}\langle P\rangle = \text{val}\langle A\rangle * \text{val}\langle B\rangle = \text{SUM}\langle j >= 0\text{: val}\langle A\rangle * b[j] * (2 * * j)\rangle. \quad (5)$$

It follows that we can compute $P = \text{MULT}(A, B)$ by initially forming all summands $\text{val}\langle A\rangle * b[j] * (2 * * j)$, and adding them all.

The classical algorithm *Seq-Mult* essentially sums up the operands in $n - 1$ sequential stages.

We propose for *Par-Mult* to sum up the operands in a parallel tree, 2 by 2, then 4 by 4, ..., in $\log(n)$ parallel stages.

The following recursive definitions of both algorithms show their similarities and differences:

```
(Seq-Mult)
  SEQ-MULT(A, B)::=
    if 0 =< val(B) =< 1
      then bit-product(a, b[0])
      else add(shift(k, SEQ-MULT(A, B1)), SEQ-MULT(A, B0))
```

where $k = 1$,

$$B = \langle b[n-1], \ldots, b[0]\rangle, \qquad B1 = \langle b[n-1], \ldots, b[1]\rangle, \qquad B0 = \langle b[0]\rangle.$$

```
(Par-Mult)
  PAR-MULT(A, B)::=
    if 0 =< val⟨B⟩ =< 1
      then bit-product(A, b[0])
      else add(shift(k, PAR-MULT(A, B1)), PAR-MULT(A, B0))
```

where $k = \frac{1}{2}n + 1$,

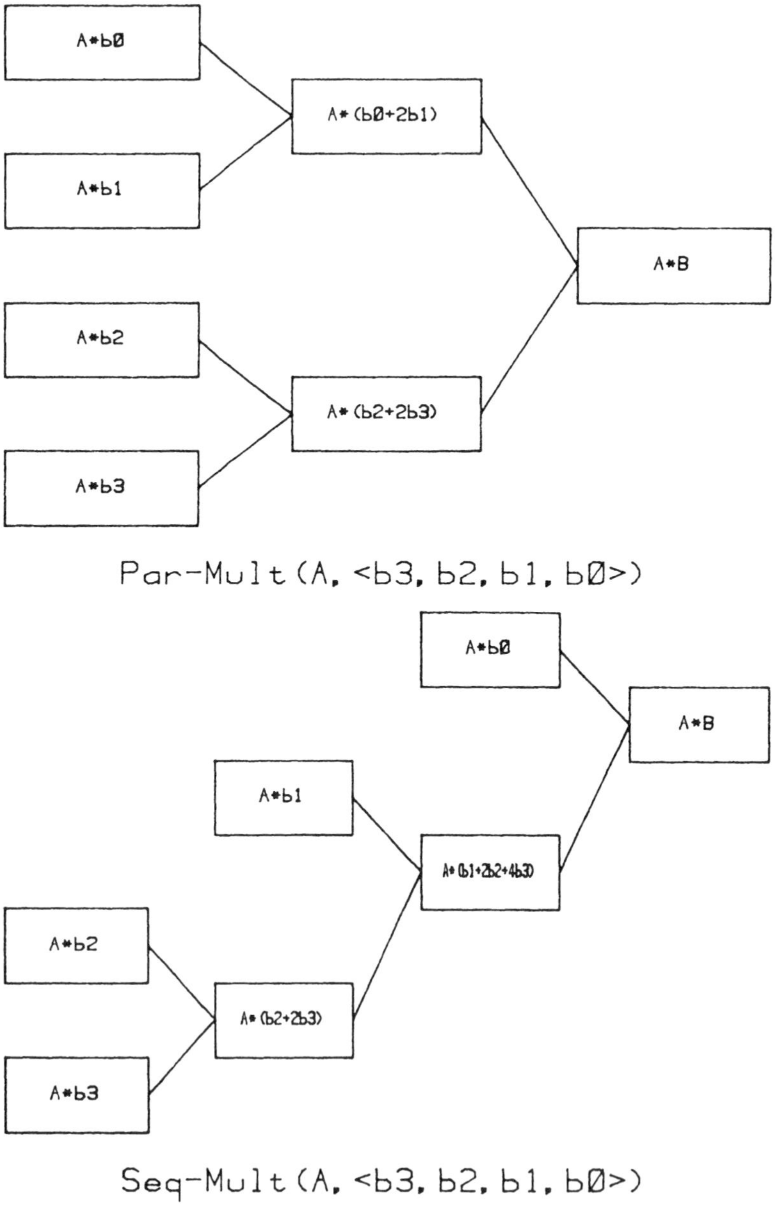

Fig. 1. Addition trees for *Seq-Mult* and *Par-Mult*.

$$B = \langle b[n-1], \ldots, b[0] \rangle, \qquad B1 = \langle b[n-1], \ldots, b[k] \rangle,$$
$$B0 = \langle b[k-1], \ldots, b[0] \rangle.$$

The correctness of both algorithms, as expressed by (1) and (5), follows from (2), (3) and (4), together with

$$\text{val}\langle B \rangle = \text{val}\langle B1 \rangle * (2 * * k), \qquad (6)$$

this last expression holding true for both algorithms ($k = 1$ or $k = \frac{1}{2}n + 1$).

Both algorithms perform exactly $n - 1$ additions in the course of multiplying an m-bit sequence A by the n bits of B. Because of parallelism, these additions are completed within log(n) (base 2 logarithm) stages in *Par-Mult*, while *Seq-Mult* requires $n - 1$ such stages.

Each algorithm can be depicted by a tree of successive additions as in Fig. 1.

2.4. Using carry save representation for intermediate products

A *carry save number M* is a pair $[R, S]$ of binary sequences R and S. The value of M is the sum $\text{val}\langle M \rangle = \text{val}\langle R \rangle + \text{val}\langle S \rangle$ of the values if its components R and S. A convenient representation of carry save numbers by sequences is $M = \langle m[n-1], \ldots, m[0] \rangle$, where digit i is $m[i] = r[i] + s[i]$, with value in 0, 1, 2 as shown in Fig. 2.

The addition *add-csb*($[R, S]$, T) of a carry save number $M = [R, S]$ with a binary integer T, yielding a carry save result $[R', S']$ can be performed in a bit-wise manner:

$$\textbf{add-csb}([R, S], T) ::= [R', S'] \qquad (7)$$

where

$$S' = \langle \text{fa} - \text{sum}(r[n-1], s[n-1], t[n-1]), \ldots, \text{fa} - \text{sum}(r[0], s[0], t[0]) \rangle$$

and

$$R' = \text{shift}(1, \langle \text{fa-carry}(r[n-1], s[n-1], t[n-1]), \ldots,$$
$$\text{fa-carry}(r[0], s[0], t[0]) \rangle).$$

```
Binary representation:
<100101011>

Carry save representations:
299=[<110101111>, <010101000>]
   =<120210111>
   =<121010111>
```

Fig. 2. Carry save representation.

The bit-wise *full adder* is defined by

$$s' = \textbf{fa-sum}(r, s, t) ::= \{(r + s + t) \textbf{ mod } 2\} \tag{8}$$

and

$$r' = \textbf{fa-carry}(r, s, t) ::= \{(r + s + t) > 1\},$$

so that $(2r' + s') = (r + s + t)$.

The sum *add-cscs*($[R, S], [T, U]$) of two carry save numbers $M = [R, S]$ and $N = [T, U]$ is computed in two steps:

$$\textbf{add-cscs}([R, S], [T, U]) ::= \textbf{add-csb}(\textbf{add-csb}([R, S], T), U). \tag{9}$$

It follows directly from (7), (8) and (9) that

$$val\{\textbf{add-cscs}(M, N)\} = val\{M\} + val\{N\}. \tag{10}$$

The delay Tadd introduced by the parallel computation of *add-cscs* is independent of the length of the operands, namely

$$Tadd - cscs = 2 * Tadd\text{-}csb = 2 * Tfa \tag{11}$$

where Tfa is the delay introduced by one stage of full adder.

2.5. *Par-Mult and Seq-Mult revisited*

Assuming the operands A and B of multiplication to be initially presented in binary form, the first level of addition in the trees of Fig. 1 is used to obtain the intermediate sums $[A * b[2j], shift(1, A * b[2j + 1])]$ in carry save form. There is no delay involved in the initial conversion *form-cs* which is a mere convention.

In the case of *Seq-Mult,* all subsequence additions combine a carry save and a binary number, thus we use *add-csb.*

For *Par-Mult*, all intermediate results after the first stage are in carry save form, and they are combined two by two using *add-cscs.*

In both cases, the final addition stage yields a carry save representation of the product $P = A * B$. The two binary sequences forming this carry save product must then be added in order to produce the final result in true binary form. For this purpose we use a fast carry-look-ahead adder *cla-add,* such as described for example by Vuillemin and Guibas [7].

Computation of the numerical product 23 * 13 with both algorithms is shown in Fig. 3.

2.6. *Timing analysis*

If m and n are the number of bits of A and B respectively, the time required by *Seq-Mult* and *Par-Mult* for computing the $(n + m)$ bits of product $P = A * B$ is given by

$$Tseq\text{-}mult(m, n) = T0 + (n - 2) * Tfa + Tcla - add(n + m), \tag{12}$$

$$Tpar\text{-}mult(m, n) = T0 + 2(\log(n) - 1) * Tfa + Tcla - add(n + m). \tag{13}$$

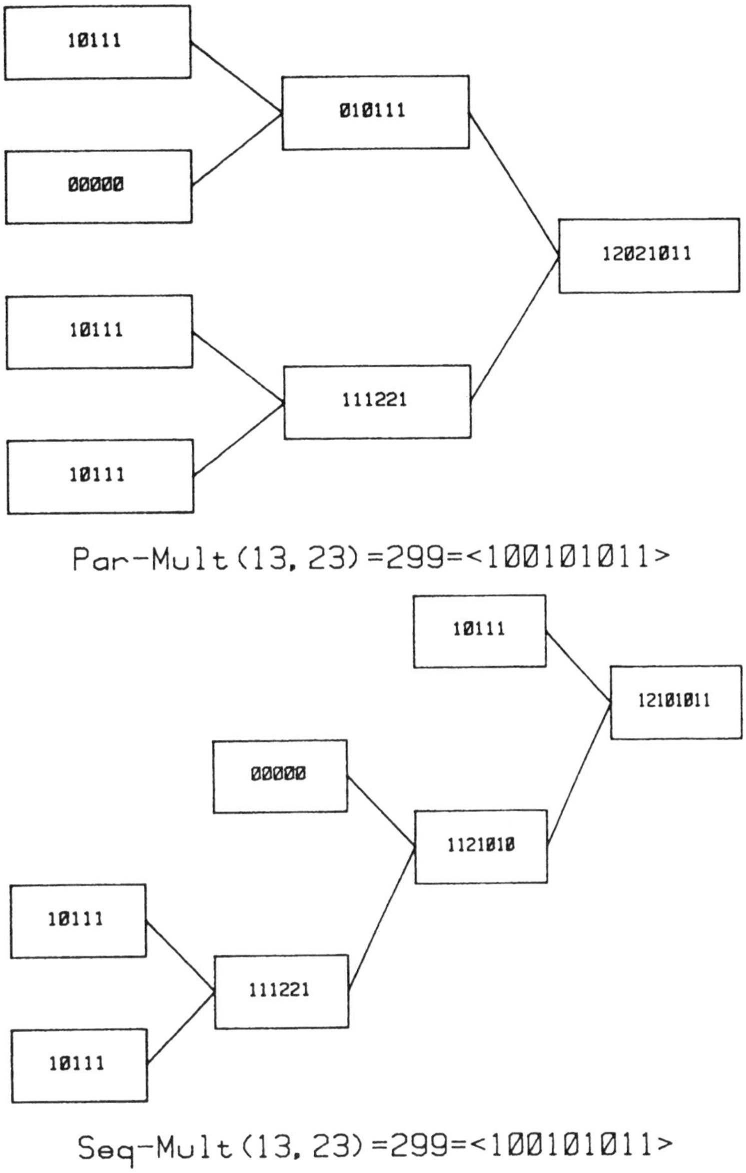

Fig. 3. Computation of 13 * 23 with *Seq-Add* and *Par-Add*.

Here, $T0$ represents the time required for input distribution and bit-wise products. It is shown by Vuillemin and Guibas [7] that integrated carry look ahead adders can be designed with logarithmic delay $\text{Tcla} - \text{add}(p) = T1 * \log(p)$. It follows that *Par-Mult* also has a logarithmic computing time, and it proves *faster* than *Seq-Mult* for *all* values of n and m.

To be fair, we must point out that *Seq-Mult* is almost invariably implemented in conjunction with Booth's recoding of operand B, as described for example in [4]. Such a recoding divides the number of summands by two, improving the

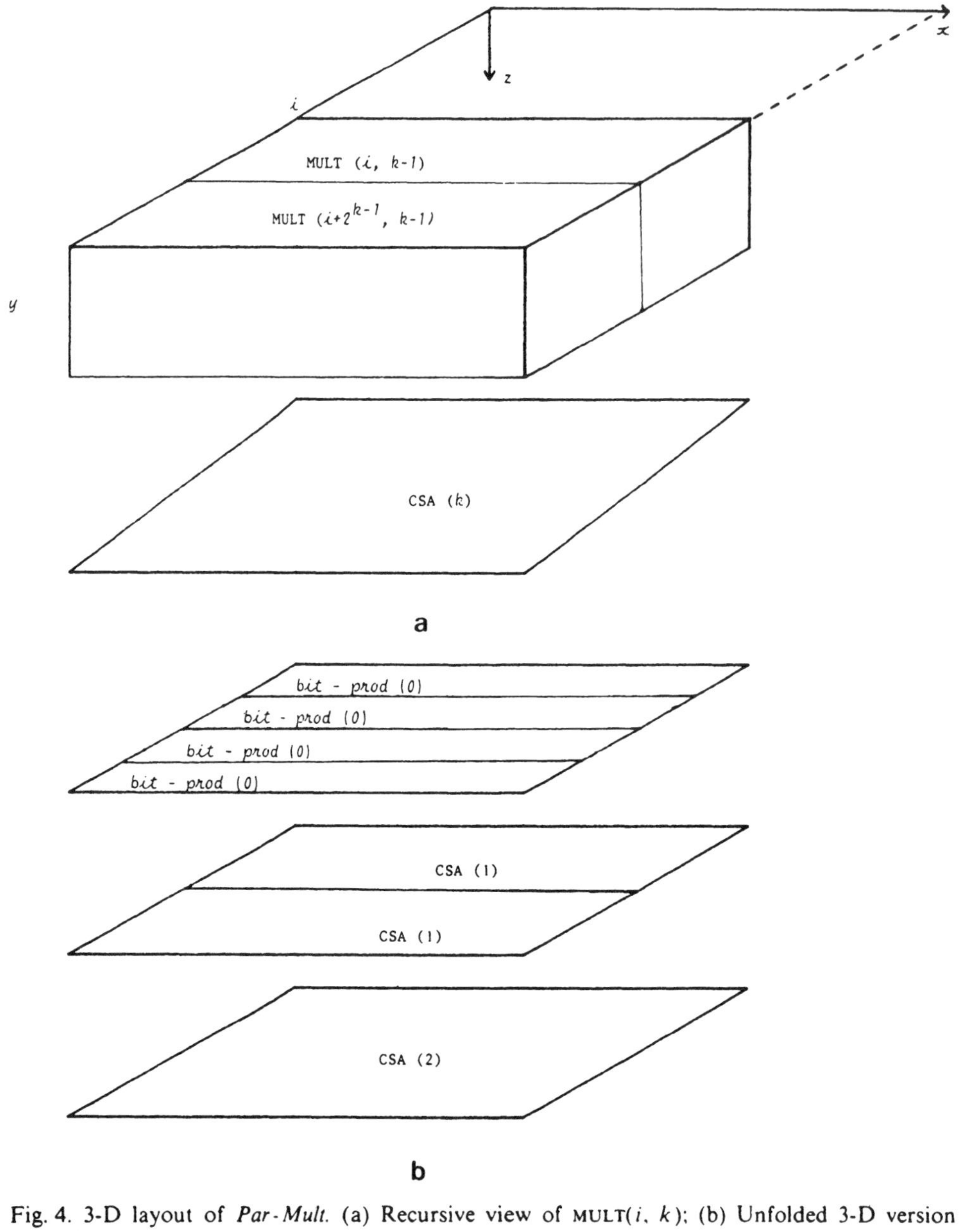

Fig. 4. 3-D layout of *Par-Mult.* (a) Recursive view of MULT(i, k); (b) Unfolded 3-D version of MULT(0, 2).

A pictorial view of the 3-D multiplier is given in Fig. 4.

3.1.1. Layout of bit-product

The primitive operation bit-product-layout(i) generates the planar layout of a rectangular circuit, formed by a row of $(n+m)$ identical cells. Each cell has a horizontal (x-axis) input $b[i]$, a diagonal (x&y-axis) input $a[j]$, and computes the

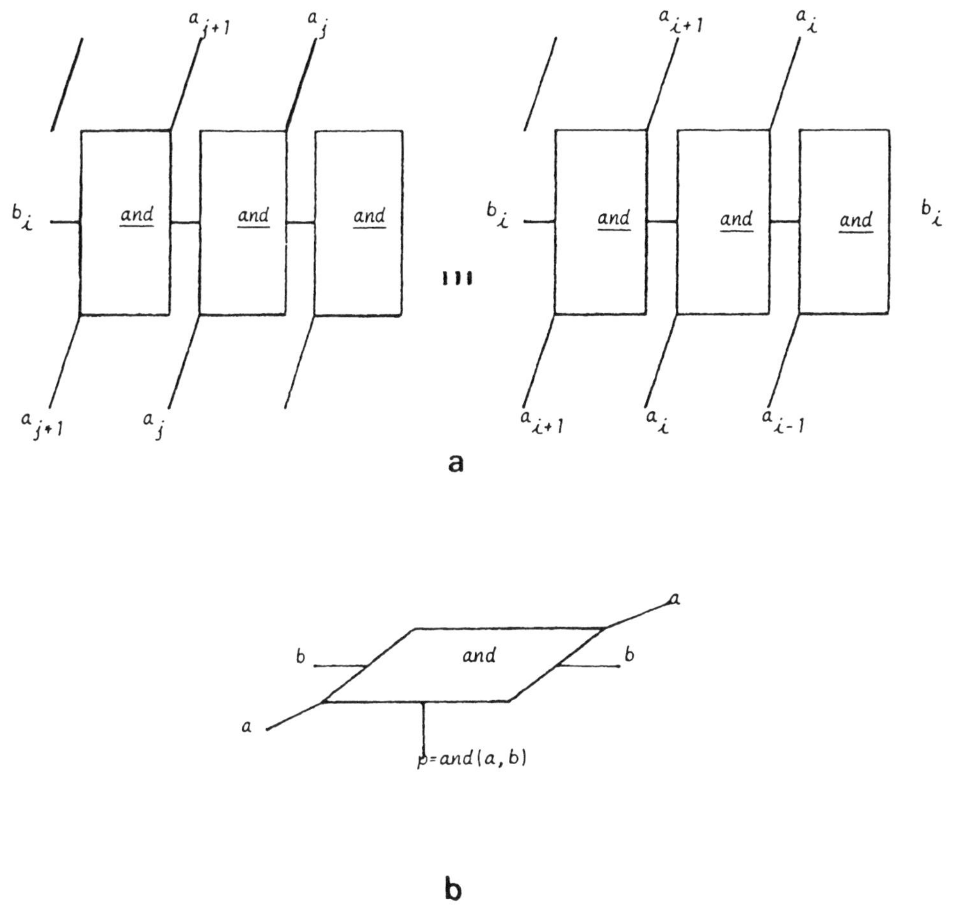

Fig. 5. Bit-product layout. (a) Result of *bit-product-layout*(i); (b) A 3-D view of the *and* cell.

product *and*($b[i]$, $a[j]$) along the third (z-axis) dimension as in Fig. 5. The index j of $a[j]$ runs from $n+m-i-1$ to $j=-i$ with the convention $a[j]=0$ whenever j is outside the range $0 \leqslant j < m$.

Although Fig. 5 uses diagonal connections, it is easy enough to replace these by connections which are all parallel to one of the coordinate axis, x, y or z.

3.1.2. Carry save adder layout

The operator add-cscs-layout(k) generates the layout of a circuit for adding two carry-save numbers, entirely made of interconnected full-adders *fa*. There are two such *fa*'s in each bit-slice. The first *fa* receives its three operands $s[i]$, $t[i]$, $u[i]$ from level $(k-1)$, producing two outputs $q[i+1]$, $p[i]$ such that $s[i]+t[i]+u[i]=2*q[i+1]+p[i]$. The second *fa* receives operand $r[i]$ from level $(k-1)$, operands $p[i]$ and $q[i]$ from level k. It delivers its outputs $n[i+1]$ and $m[i]$ at level $(k+1)$ as shown in Fig. 6.

As indicated in Section 2.5 the adder add-cscs-layout(1) at level 1 is not

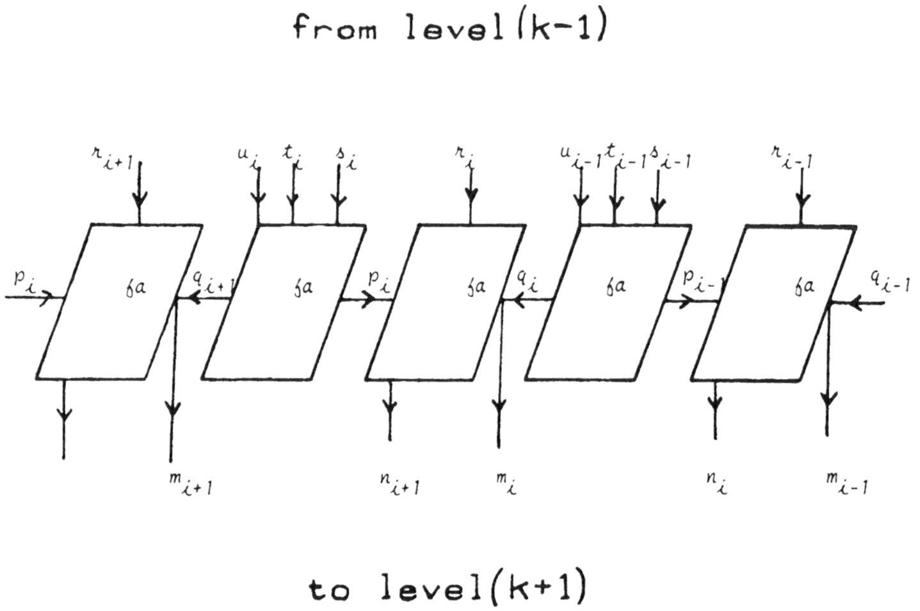

Fig. 6. Layout of add-cscs.

performing any operation; it merely combines two binary numbers originating at level 0, into a pair, representing the carry-save input to level 2. Thus, level 1 comprises only *routing* of signals.

3.2. Mapping 3-D into 2-D layout

In order to imbed our layout in a planar technology, we use the following rules:

Rule 1. We keep the row structure, by alternating bit-products and additions, according to Fig. 7, obtained by the recursive definition

```
par-mult-2D-layout(i, k)::=
  if k = 0 then bit-product-layout(i)
        else JY(par-mult-2D-layout(i, k − 1),
              add-cscs-layout(k),
              par-mult-2D-layout(i + 2 * * (k − 1), k − 1)).     (15)
```

Rule 2. We keep the bit-slice structure of columns by allocating $k = \log(n)$ vertical channels within each column. Channel j is dedicated to the layout of level j.

Rule 3. (x) Horizontal (x-axis) wires are mapped into horizontal wires within the same row.

(y) Vertical (y-axis) wires are mapped into vertical wires with the corresponding bit-slice channel.

(z) Third dimension (z-axis) wires between levels i and $i+1$ run horizontally between the ith and the $(i+1)$st channel of the corresponding row and column.

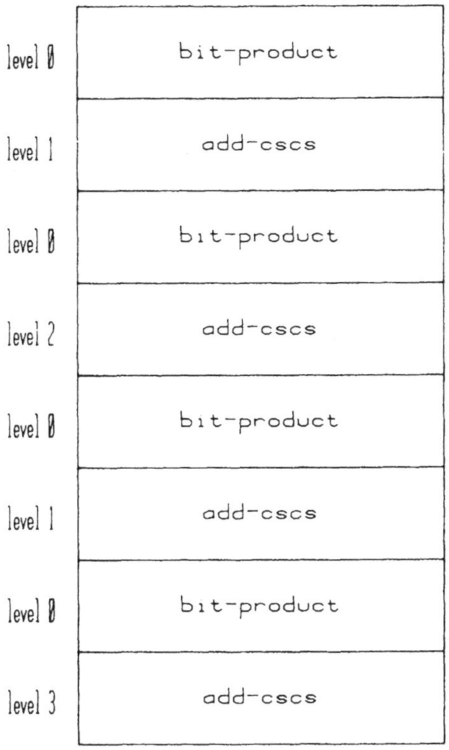

Fig. 7. Row-layout of *Par-Mult*.

Fig. 8 shows the channel structure of a bit-slice and the corresponding routing strategy.

Such a layout keeps the tree structure adopted in *Par-Mult* for summing up partial product apparent in each bit-slice, as in Fig. 8. The area penalty incurred over *Seq-Mult* thus only arises because of the O(log(n)) extra wires required by the tree structure connexion. For n up to 32 this area is less than half of that devoted to the full-adder logic layout. Hence, *Par-Mult* can be layed out within 150% of the area of *Seq-Mult*.

3.3. On truly large time O(log(n)) multipliers

When n gets very large ($n > 64$), the tree structure of Fig. 8 no longer performs addition in time O(log(n)): although the signal only traverses log(n) gates, gates no longer possess unit delay, because their parasitic capacitance increases with the length of output wires. As a consequence, gates become slower as they get closer to the root of the tree. Indeed, wire lengths double with each level up in the tree.

A radical solution to this problem, as presented in [7] or [8], uses the layout of Fig. 9, which has signal amplification built into the tree structure. Tree nodes at level $h + 1$ have twice the speed and size of tree nodes at level h. This means that transistor gates at level $h + 1$ are twice longer than their homologue at level h.

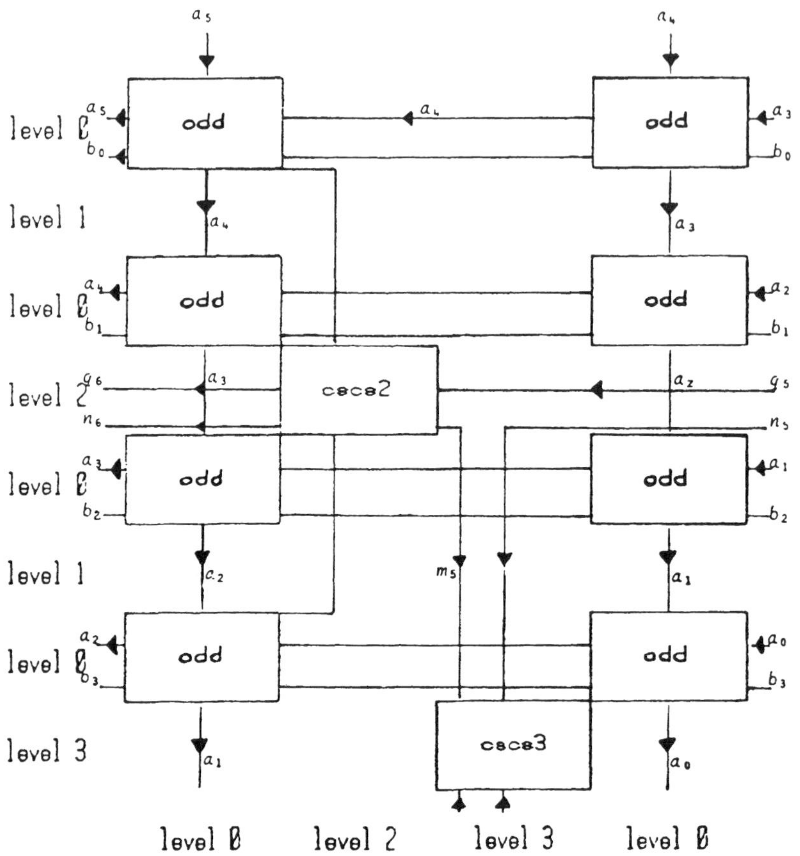

Fig. 8. Column layout of *Par-Mult*.

They are therefore able to drive output capacitance twice bigger within the same time constant.

Using this technique allows to design tree structures in which all gates have the

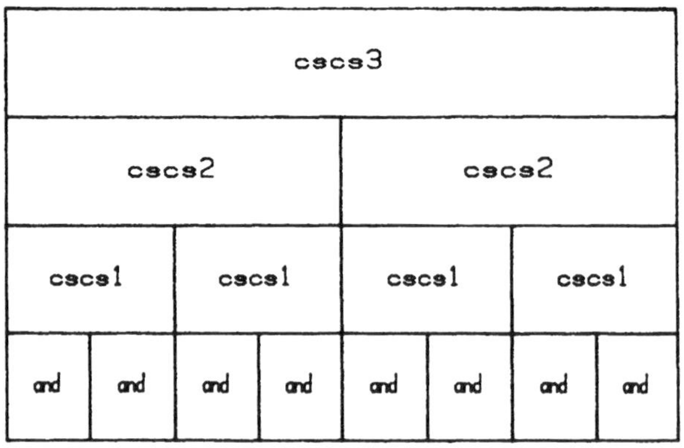

Fig. 9. Bit-slice layout of Fig. 8 with signal amplification.

same timing characteristic, regardless of their height in the tree. Consequently, it is possible to design time O(log(n)) multipliers in MOS technologies for arbitrary large values of n. This requires however *two* levels of interconnect, since connections of Fig. 8 still have to overlap the gates of Fig. 9.

4. Conclusion

We feel that the algorithm *Par-Mult* and its silicon layout are interesting from at least two points of view:

(a) It provides a mathematically elegant and easy to program technique for multiplying two N-bit binary integers in time O(log N).

(b) It provides an attractive practical alternative to *Seq-Mult* in designing fast integrated multipliers, for 8 bits and over. This holds true for a large class of technologies, including n-MOS, c-MOS and bipolar.

References

[1] Wallace, C.S., A suggestion for parallel multipliers, *IEEE Trans. Electron. Comput.* **EC-13** (1964) 14–17.

[2] Dadda, L., Some schemes for parallel multipliers, *Alta Frequenza* **34** (1965) 349–356.

[3] TRW, MPY-LSI multipliers: AJ 8×8, 12×12 and 16×16, LSI Products, TRW, Redondo Beach, CA, 1977.

[4] Matsumoto, R.T., The design of a 16×16 multiplier, LAMBDA No. 1 (1980) 15–21.

[5] Cand, M., P. Le Scan and A. Rosset, A single chip digital signal processor, *Proc. IEEE* **429** (1982) 356–359.

[6] Luk, W.K., Silicon compilation of a fast parallel multiplier, Thèse de 3eme Cycle, Université de Paris, 1983.

[7] Vuillemin, J. and L. Guibas, On fast binary addition in MOS technologies, *Proc. IEEE* **429** (1982) 147–150.

[8] Mead, C. and M. Rem, Minimum propagation delays in VLSI, *Proc. 2nd Caltech Conf. on VLSI*, 1981.

A High-Speed Multiplier Using a Redundant Binary Adder Tree

YOSHIHISA HARATA, YOSHIO NAKAMURA, HIROSHI NAGASE, MITSUHARU TAKIGAWA, ASSOCIATE MEMBER, IEEE, AND NAOFUMI TAKAGI, MEMBER, IEEE

Abstract—This paper describes a 16-bit × 16-bit multiplier for 2 two's complement binary numbers based on a new algorithm. This multiplier has been fabricated on an LSI chip using a standard n-E/D MOS process technology with a 2.7-μm design rule. This multiplier is characterized by use of a binary tree of redundant binary adders. In the new algorithm, n-bit multiplication is performed in a time proportional to $\log_2 n$ and the physical design of the multiplier is constructed of a regular cellular array. This new algorithm has been proposed by Takagi *et al.* The 16-bit × 16-bit multiplier chip size is 5.8 × 6.3 mm² using the new layout for a binary adder tree. The chip contains about 10 600 transistors, and the longest logic path includes 46 gates. The multiplication time was measured as 120 ns. We estimate that a 32-bit × 32-bit multiplication time is about 140 ns.

I. Introduction

MULTIPLICATION is one of the basic operations in digital signal processing systems providing filtering, correlation, and range measurement. A high-speed multiplier is required for real-time signal processing systems. Various high-speed multipliers have been proposed and realized with a one-chip LSI (e.g., [1]–[9]).

An array multiplier and a multiplier using a Wallace tree [10] are well-known for their high-speed multiplication. The former has a good repeatability of unit cells and is favorable for VLSI implementation. Therefore many multipliers have been fabricated using this method (e.g., [1]–[6]). However, the n-bit multiplication time is proportional to n, and this method requires a long computation time for larger n's. On the other hand, in the latter, the n-bit multiplication time is proportional to $\log_2 n$. Thus the multiplication time increases slowly for larger n's. However, the physical design is rather difficult because of the complex interconnection in an LSI chip. So there are only a few multipliers based on this method [7], [8].

In this paper, a 16-bit × 16-bit multiplier LSI based on a new algorithm using the redundant binary representation is presented. This multiplier is fabricated using a standard n-E/D MOS process technology. The new algorithm has been proposed by Takagi *et al.* [11]–[13]. In this new algorithm, the multiplication time is proportional to the logarithm of the word length of operands. Therefore the computation time is almost the same as that of a multiplier using a Wallace tree and much shorter than that of a conventional array multiplier for large n's. In addition, the physical design is as easy as in array multipliers due to its good repeatability.

The multiplier using the new algorithm offers the advantages of high-speed multiplication and easy physical design. On the other hand, the number of signal lines in this multiplier is about twice as many as that of conventional multipliers. Therefore we paid special attention to signal wiring during the physical design of the multiplier.

Manuscript received December 2, 1985; revised July 7, 1986.

Y. Harata, Y. Nakamura, H. Nagase, and M. Takigawa are with the Toyota Central Research and Development Laboratories, Inc., Nagakute, Aichi 480-11, Japan.

N. Takagi is with the Department of Information Science, Faculty of Engineering, Kyoto University, Kyoto 606, Japan.

IEEE Log Number 8611325.

II. Algorithm of the Multiplier

A. The Multiplication Time Based on the New Algorithm

This multiplier internally uses a redundant binary representation which is one of the signed digit representations proposed by Avizienis [14]. In the n-bit multiplier based on the new algorithm as shown in Fig. 1, n partial products are added up pairwise in a $\log_2 n$ level binary adder tree.

Addition of two n-digit redundant binary numbers is performed in parallel in a constant time. The summation of the partial products is performed in a time proportional to $\log_2 n$. As the sum is represented by a redundant binary number, conversion of the number into a two's complement binary number is required. This conversion is carried out by a carry lookahead adder with a time proportional to $\log_2 n$. As a result, n-bit multiplication is performed in a time proportional to $\log_2 n$.

B. Addition in Redundant Binary Representation

In a binary number system, addition of two numbers requires a computation time at least proportional to the logarithm of the word length of the operands, because of carry propagation. However, in a redundant binary number system, since carry propagation can be eliminated, addition of two numbers can be performed in a constant time independent of the word length. This addition is one of the features of our multiplier.

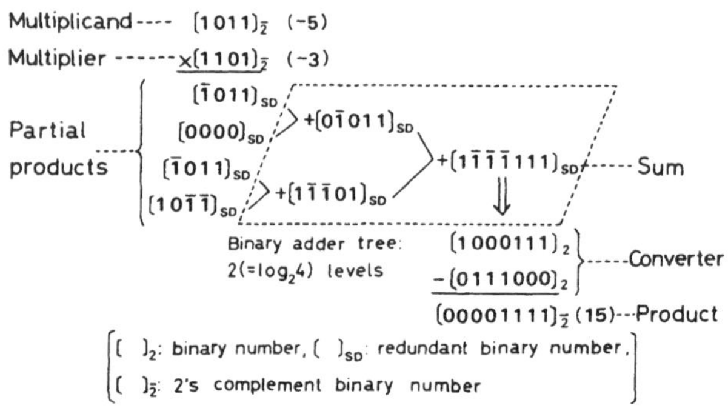

Fig. 1. Example of multiplication by use of a redundant binary adder tree.

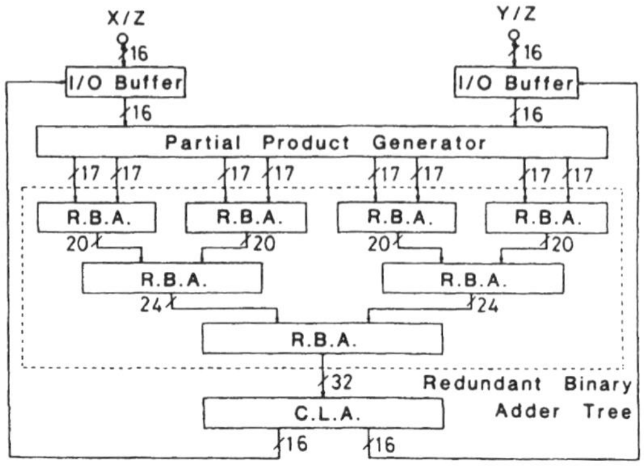

Fig. 3. Block diagram of the 16-bit×16-bit multiplier.

TABLE I
COMPUTATION RULES FOR THE FIRST STEP IN ADDITION WITHOUT CARRY PROPAGATION

TYPE No.	Combination	Next-lower-order position	Carry C_i	Sum S_i
1	$\{1,1\}$	———	1	0
2	$\{1,0\}$	Neither are negative	1	$\bar{1}$
		Otherwise	0	1
3	$\{1,\bar{1}\}$	———	0	0
4	$\{0,0\}$	———	0	0
5	$\{0,\bar{1}\}$	Neither are negative	0	$\bar{1}$
		Otherwise	$\bar{1}$	1
6	$\{\bar{1},\bar{1}\}$	———	$\bar{1}$	0

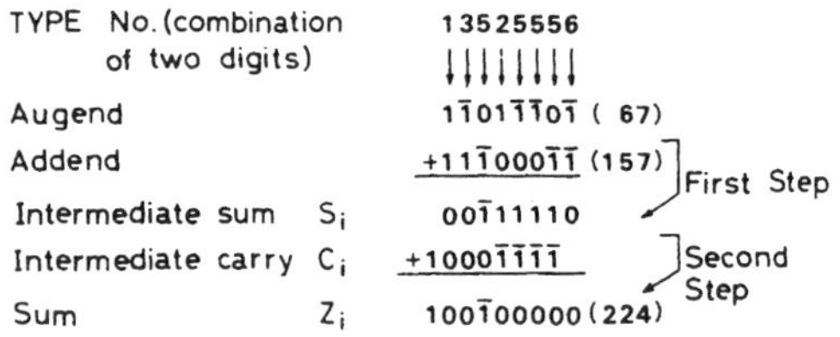

Fig. 2. Example of addition without carry propagation.

The redundant binary representation used in our multiplier has a digit set $\{0,1,\bar{1}\}$, where $\bar{1}$ denotes -1. Addition of two n-digit redundant binary numbers is performed in a constant time utilizing the redundancy. This addition is performed in the following two steps.

In the first step, two digits in the same figure positions of an augend and an addend are added. There are six types of combinations of two digits in addition as shown in Table I. The results of addition are two signals: intermediate carry digit C_i and intermediate sum digit S_i defined as listed in this table.

The second step is to obtain the sum digit Z_i at each position by adding S_i and C_{i-1} from the next lower order position. After the first step, there is no carry generation at any position in the second step.

Fig. 2 shows an example of addition in accordance with the above rule. Therefore addition of two n-digit redundant binary numbers is performed in a constant time.

C. Conversion Between a Redundant Binary Number and a Two's Complement Binary Number

Because two's complement numbers are used in the external representation of a multiplier, the conversion between a redundant binary number and a two's complement binary number must be done. This conversion can be done easily as follows. The conversion of a two's complement binary number into a redundant binary number is performed by changing the most significant digit 1 to $\bar{1}$. As an n-digit redundant binary number

$$A\left(=\sum_{i=0}^{n-1} a_i * 2^i, \qquad a_i \in \{0,1,\bar{1}\}\right)$$

is equal to

$$A^+\left(=\sum_{a_i=1} a_i * 2^i\right) - A^-\left(=\sum_{a_i=\bar{1}} (-a_i) * 2^i\right)$$

the conversion of a redundant binary number into a two's complement binary number is performed by addition A^+ and two's complement of A^-.

For example, we explain the conversion of $[\bar{1}\ 0\ 1\ \bar{1}]_{SD}$ into $[1\ \ 1\ \ 0\ \ 0\ \ 1]_{\bar{2}}$ as follows:

$$\begin{aligned}[\bar{1}\ \ 0\ \ 1\ \ \bar{1}]_{SD} &= [0\ \ 0\ \ 1\ \ 0]_2 - [1\ \ 0\ \ 0\ \ 1]_2 \\ &= [0\ \ 0\ \ 0\ \ 1\ \ 0]_{\bar{2}} + [1\ \ 0\ \ 1\ \ 1\ \ 1]_{\bar{2}} \\ &= [1\ \ 1\ \ 0\ \ 0\ \ 1]_{\bar{2}}.\end{aligned}$$

III. CONSTRUCTION OF THE 16-BIT×16-BIT MULTIPLIER

A. Block Diagram

The 16-bit×16-bit multiplier of 2 two's complement binary numbers was realized on an LSI chip. The block diagram of the multiplier is shown in Fig. 3. Each of the input/output buffer circuits contains a latch circuit. The partial product generator is based on a modified Booth's algorithm, and the eight partial products are converted into redundant binary numbers in this generator circuit. This generator is designed to act dynamically to reduce the area of the circuit. A three-level partial product adder tree is constructed by connecting seven redundant binary adders as shown in Fig. 3. There is no sign term in this multiplier to correct sign bits of the partial products in two's complement binary number, which is required in other methods. A redundant binary adder is constructed by repetitive arrangement of redundant binary adder cells.

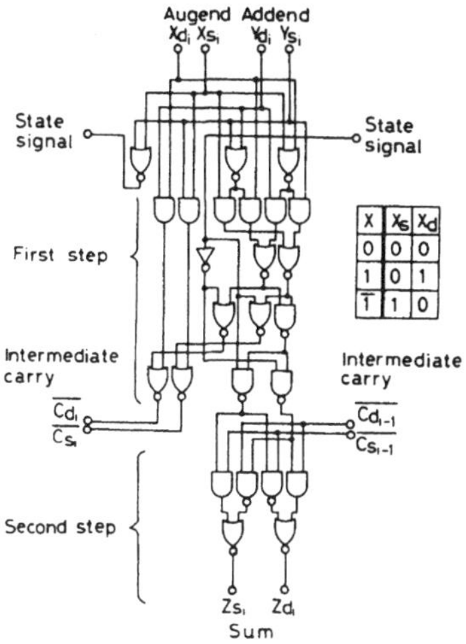

Fig. 4. Redundant binary adder cell.

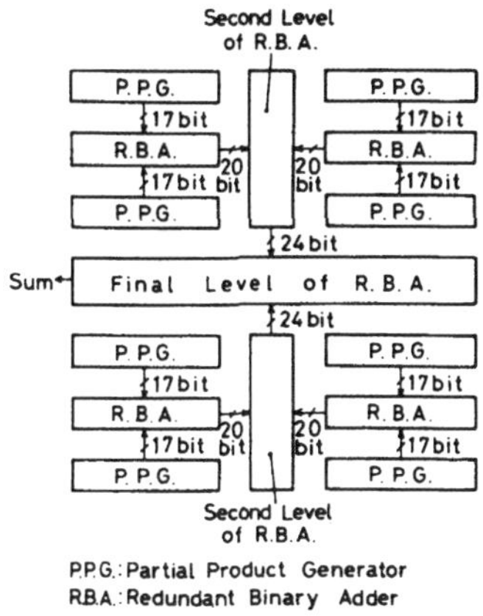

Fig. 5. Layout by H-binary tree.

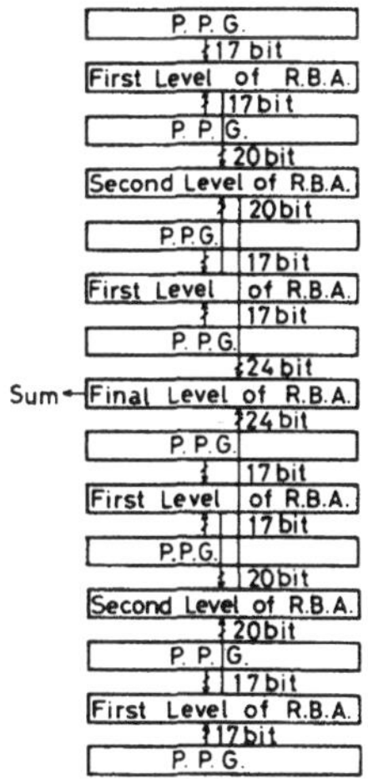

Fig. 6. Layout by L-binary tree.

Each adder at the first, second, and final levels consists of 18, 20, and 24 cells, respectively.

The arrangement of these cells has a good regularity. The redundant binary adder cell is shown in Fig. 4. In this cell, a set of two digits $\{X_s, X_d\}$ is necessary to express one redundant binary digit $\{X\}$, and $\{0,0\}$, $\{0,1\}$, and $\{1,0\}$ are assigned to $\{0\}$, $\{1\}$, and $\{\bar{1}\}$, respectively. A 32-bit carry lookahead adder designed by Unger's method [15] is used to convert a redundant binary number into a two's complement one. This carry lookahead adder has a tree structure. The computation time is proportional to $\log_2 n$.

B. Chip Layout

The following three points were important in the layout of the chip:

a) the layout is designed so that the signal flows on the chip are as simple as possible;
b) the layout is easily achieved by making good use of repeatability; and
c) the layout has good extensibility for larger n's.

Especially, a) is a key point in deciding chip size, because the number of signal lines in our multiplier is about twice that of conventional multipliers.

In a multiplier, an addition circuit summing up partial products occupies the major part of the area. Therefore the physical design of the addition circuit is most important. The addition circuit is constructed with a redundant binary adder tree in our multiplier. Several layouts (H-, L-, V- and 3D-binary trees) for a binary tree structure in a multiplier were already discussed by Luk and Vuillemin [16]. The paper concluded that the L-binary tree was the best layout. However, their multiplier was not based on Booth's algorithm and did not use a redundant binary representation. Then we reinvestigated three layouts (H-, L- and V-binary trees) for physical design of a redundant binary adder tree using a 2.7-μm rule and single-layer metal wiring.

1. H-Binary Tree (Fig. 5): In the case of a multiplier, of which the node of a binary tree is an adder, the layout of the node becomes slender. In H-binary tree layout, signal flows go along the long side of the node. Furthermore, there are many interconnections between nodes in a redundant binary adder tree. Then, a wiring area occupies a very large area on an LSI chip. H-binary tree layout is not suitable for a redundant binary adder tree.

2. L-Binary Tree (Fig. 6): A generator circuit of a partial product based on Booth's algorithm needs more signals and gates than a normal generator circuit. Therefore it is not so good to arrange partial product generators inside the adder tree. Furthermore, signal flows take forward and backward moves in one direction. Then, many signal lines are piled up in the center of the layout especially for a redundant binary adder tree.

3. V-Binary Tree (Fig. 7): Since the signal flows go forward in one direction in this layout, a small wiring area is achieved compared with the other layouts. This layout seems to be the best for a redundant binary adder tree. However, the layout becomes long and slender, if we use usual layout techniques. The result of estimation comes to 7.5×2.5 mm^2.

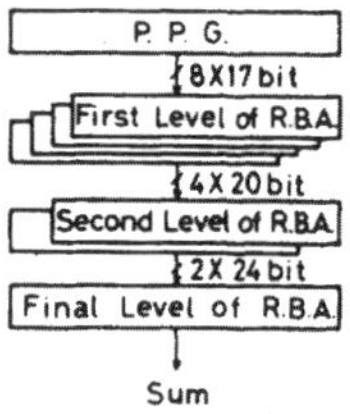

Fig. 7. Layout by V-binary tree.

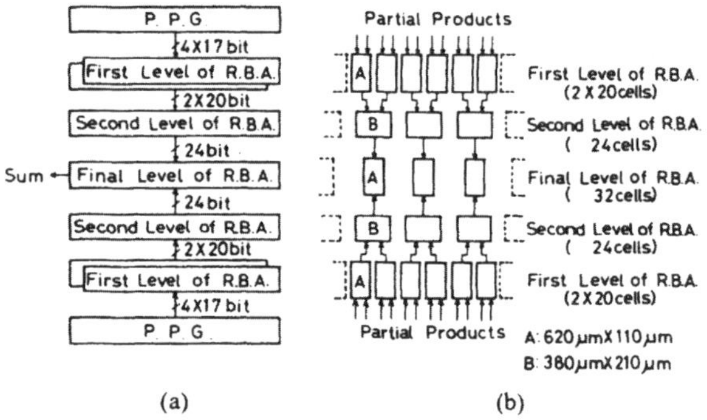

Fig. 8. (a), (b) Layout for a redundant binary adder tree.

Fig. 9. Microphotograph of the 16-bit×16-bit multiplier.

From the result of the above consideration, a new layout of an improved V-binary tree has been adopted as shown in Fig. 8(a). The signal flows on the chip go forward in one direction. Redundant binary adders are constructed by using the same adder cells in two physical patterns. One type is for the first and the final level of redundant binary adders. Its area on physical pattern is 620×110 μm^2. The other type is for the second level of redundant binary adders. Its area is 380×210 μm^2. These redundant binary adder cells connect as shown in Fig. 8(b). This redundant binary adder tree occupies an area of about 4×4 mm^2 on the chip.

Fig. 9 shows the microphotograph of the fabricated 16-bit×16-bit multiplier LSI chip. This LSI chip was fabricated using a standard n-E/D MOS process technology with a 2.7-μm design rule and single-layer metal wiring. The chip size of the multiplier is 5.8×6.3 mm^2, and the number of transistors is about 10 600. The redundant binary adder tree and the metal wiring area for interconnections between the redundant binary adder tree and the conversion circuit are located in the middle part of the LSI chip. The area is about 40 percent of the total chip area. A redundant binary number to two's complement number converter is arranged on the left side of the adder tree area and the partial product generator circuits are arranged on the other three sides. The area of the two circuits is about 25 percent to the whole space of the chip. The remaining 35 percent is occupied by the input and output buffer, metal wiring to other blocks, and 39 bonding pads.

In order to pack the multiplier chip in a dual-in-line package with 40 pins, 32 bidirectional buffers are used for input data and output data. This multiplier has an input enable pin and an output enable pin to control data input/output.

IV. Results and Discussion

A. Computation Time and Logic Path

The multiplication time of the fabricated 16-bit×16-bit multiplier is 120 ns, which is nearly the expected value. This multiplication time includes the signal propagation time of the input/output buffers and the intermediate buffers between circuit blocks.

The longest logic path from input to output includes 46 gates, while the intrinsic logic path includes 29 gates. The intrinsic logic path means the logical gate depth which is indispensable to accomplish multiplication function. The detail of the 29 gates on the intrinsic logic path is as follows: three gates are in a partial product generator circuit, 18 gates are in a redundant binary adder tree, and eight gates are in a conversion circuit from a redundant binary number to a two's complement one. In this design, the maximum fan-in of the gate circuits is restricted to two except the recode circuit in the partial product generator circuit using the modified Booth's algorithm.

A conversion circuit occupies eight of 29 gates on the intrinsic logic path. Therefore if we assume a redundant binary system in which all arithmetic operations are executed in redundant binary representation, a product is obtained in 21 gates. As there are several reports for arithmetic operations using redundant binary representation [17]–[20] besides multiplication and addition, it is possible to use this multiplier as a key element of such a redundant binary system.

B. Study of a 32-bit×32-bit Multiplier

Multiplication time of the n-bit multiplier using the new algorithm is proportional to $\log_2 n$. Therefore the multiplication time increases slowly for larger n. As n increases from 16 to 32, the longest logic path increases by only eight gates (the intrinsic logic path increases by only seven

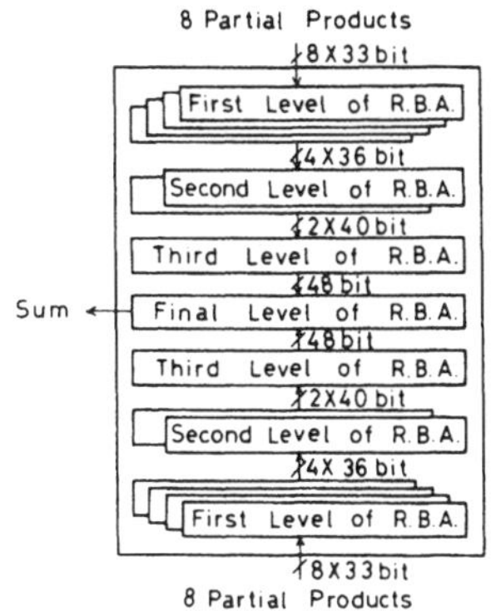

Fig. 10. Layout of a redundant binary adder tree for a 32-bit × 32-bit multiplier.

gates). Assuming that an average gate delay of a 32-bit × 32-bit multiplier is equal to that of the realized 16-bit × 16-bit multiplier, a 32-bit × 32-bit multiplication time is estimated at about 140 ns. The computation time of the 32-bit × 32-bit multiplier that is estimated increases by only 17 percent to that of the 16-bit × 16-bit multiplier and the number of transistors is estimated at about 35 000.

A chip layout of the redundant binary adder tree for the 32-bit × 32-bit multiplier is shown in Fig. 10. The chip size of the 32-bit × 32-bit multiplier based on Fig. 10 will be about 10×11 mm^2 using a 2.7-μm design rule. We estimate the chip size and the computation time for the 32-bit × 32-bit multiplier using a 1.5-μm design rule and double-layer metal wiring. A redundant binary adder cell area on physical pattern is about 35×400 μm^2. According to Fig. 10, a redundant binary adder tree area is about 4.6×3.0 mm^2. Then, a part of the third level and final level of redundant binary adders leaves a space. It is possible to put a converter from redundant binary to two's complement numbers into that space. Partial product generator circuits and recode circuits for Booth's algorithm are arranged on the sides of a redundant binary adder tree. A chip size within 5.0×5.0 mm^2 is expected.

The total logic path of the 32-bit × 32-bit multiplier increases only eight gates to that of the 16-bit × 16-bit multiplier. Based on the scaling principle, delay time is reduced to a κth for scaling factor κ [21]. Using double-layer metal wiring, the resistance and the capacitance of the wiring is also reduced significantly. Especially, the number of wirings in the multiplier utilized redundant binary representation is twice that of conventional multipliers. Therefore the reduction of the resistance and capacitance is effective in achieving high-speed computation. Then the computation time of the 32-bit × 32-bit multiplier using a 1.5-μm design rule and double metal wiring will be expected to be less than 50 ns. For example, it is reported that the computation time of a 16-bit × 16-bit multiplier using a 1.5-μm design rule is less than 40 ns [7]. This multiplier utilizes a Wallace tree and the total logic path is nearly equal to our 16-bit × 16-bit multiplier.

TABLE II
LOGIC PATH AND NUMBER OF TRANSISTORS OF THREE TYPES OF MULTIPLIERS

	Logic path		Number of transistors	
	16-bit	32-bit	16-bit	32-bit
Our multiplier	29	36	9,593	35,000
Array multiplier	93*	187*	10,700	43,300
Multiplier using a wallace tree	25*	32*	7,700	31,200

* Full adder : 3, Half adder : 2

C. Comparison with Other Methods

The features of the multiplier using the redundant binary numbers as compared with other methods are the next three points.

a) The n-bit multiplication time is proportional to $\log_2 n$ and the physical layout has a good repeatability.

b) The multiplier using the redundant binary numbers does not require any optional sign bits in adding up partial products.

c) For partial products with minus signs of Booth's algorithm, a process of adding ONE that is indispensable in general to generate two's complement of multiplicand is not required because the redundant binary representation has the digit of -1.

We compare our multiplier having such features with other high-speed multipliers. Table II shows a comparison of our multiplier, an array multiplier, and a multiplier using a Wallace tree for 16-bit and 32-bit multipliers for the intrinsic logic path and the number of transistors of NMOS circuits. For an array multiplier, the Booth's algorithm and a carry lookahead adder are not used. For a multiplier using a Wallace tree, the Booth's algorithm and a carry lookahead adder are used. If the computation time is proportional to the intrinsic logic path, the multiplication time of our multiplier is similar to that of a multiplier using a Wallace tree and is three times shorter for 16 bit and four times shorter for 32 bit than an array multiplier. The number of transistors of our multiplier is fewer than that of an array multiplier and is more than that of a multiplier using a Wallace tree. A multiplier using a Wallace tree is the best scheme regarding the speed and the complexity. However, the physical design is considerably difficult. Since our multiplier has a good repeatability similar to an array multiplier, our multiplier may be superior to a conventional one for VLSI implementation.

V. CONCLUSION

A 16-bit × 16-bit multiplier based on the new algorithm has been designed for LSI implementation using a standard n-E/D MOS process technology with a 2.7-μm de-

TABLE III
CHARACTERISTICS OF THE FABRICATED MULTIPLIER

Process technology	n-E/D MOS process
Gate length	2.7 μm
Transistor configuration	10,600 Transistors
Chip size	5.8 mm X 6.3 mm
Intrinsic logic path	29 gates
Longest logic path	46 gates
Multiplication time	120 nsec

sign rule. The principal feature of the new algorithm is the use of a redundant binary adder tree for addition of partial products. Therefore in this new multiplier the n-bit multiplication time is proportional to $\log_2 n$ and the physical layout has a good repeatability. In comparison with other high-speed multipliers, the intrinsic logic path is similar to that of a multiplier using a Wallace tree and is three to four times shorter than that of an array multiplier. The pattern layout of this multiplier has a good repeatability similar to an array multiplier.

Three layouts (H-, L- and V-binary trees) were examined for physical design. Because of its compact arrangement, a new layout of improved V-binary tree was adopted. As a result, the area of the redundant binary adder tree is about 4×4 mm^2.

The longest logic path from input to output of our multiplier includes 46 gates. The multiplication time of our multiplier is 120 ns, which is nearly the expected value. A multiplication time of a 32-bit$\times$32-bit multiplier based on the new algorithm is estimated at about 140 ns according to our experience with the actual 16-bit$\times$16-bit multiplier. The characteristics of the fabricated multiplier are summarized in Table III.

REFERENCES

[1] TRW, "MPY-LSI multipliers: AJ 8×8, 12×12 and 16×16," LSI Projects, TRW, Redondo Beach, CA, Mar. 1978.

[2] J. M. Anderson, B. L. Troutman, and R. A. Allen, "A CMOS LSI 16×16 multiplier/multiplier-accumulator," presented at the Int. Solid-State Circuits Conf., Feb. 1982, THAM 10.3.

[3] C. P. Lerouge, P. Girard, and J. S. Colardelle, "A fast 16 bit NMOS parallel multiplier," in *Proc. 9th EUROPEAN Solid-State Circuits Conf.*, Sept. 1983, pp. 29–32.

[4] F. P. J. M. Welten and J. Lohstroh, "A 25/50 MHz dual-mode parallel multiplier/accumulator," presented at the Int. Solid-State Circuits Conf., Feb. 1984, WPM 8.2.

[5] M. Uya, K. Kaneko, and J. Yasui, "A CMOS floating point multiplier," presented at the Int. Solid-State Circuits Conf., Feb. 1984, WPM 8.4.

[6] J. Iwamura, K. Suganuma, M. Kimura, and S. Taguchi, "A CMOS/SOS multiplier," presented at the Int. Solid-State Circuits Conf., Feb. 1984, WPM 8.5.

[7] N. C. Wittmer, J. A. Michejda, J. W. Gannett, P. F. Bechtold, G. W. Taylor, N. Latshitz, D. C. Dennis, and R. J. Bayruns, "A NMOS LSI 16×16 multiplier," presented at the Int. Solid-State Circuits Conf., Feb. 1983, WAM 2.5.

[8] Y. Kaji, N. Sugiyama, Y. Kitamura, S. Ohya, and M. Kikuchi, "A 45 ns 16×16 CMOS multiplier," presented at the Int. Solid-State Circuits Conf., Feb. 1984, WPM 8.1.

[9] W. K. Luk and J. E. Vuillemin, "Recursive implementation of optimal time VLSI integer multipliers," in *Proc. VLSI*, F. Anceau and E. J. Aas, Eds. Amsterdam: Elsevier, 1983, pp. 155–168.

[10] C. S. Wallace, "A suggestion for a fast multiplier," *IEEE Trans. Electron. Comput.*, vol. EC-13, no. 1, pp. 14–17, Feb. 1964.

[11] N. Takagi, H. Yasuura, and S. Yajima, "A VLSI-oriented O(log n) stage high-speed multiplier using a redundant binary addition tree," Tech. Group on Automata and Languages, IECE Japan, Rep. AL82-31, 1982 (in Japanese).

[12] N. Takagi, H. Yasuura, and S. Yajima, "A VLSI-oriented high-speed multiplier using a redundant binary addition tree," *Trans. IECE Japan*, vol. J66-D, no. 6, pp. 683–690, June 1982 (in Japanese).

[13] N. Takagi, H. Yasuura, and S. Yajima, "A high-speed VLSI multiplier using a binary tree of redundant binary adders," Yajima Lab., Kyoto Univ., Kyoto, Japan, Res. Rep. ER83-03, Dec. 1983.

[14] A. Avizienis, "Signed-digit number representations for fast parallel arithmetic," *IRE Trans. Electron. Comput.*, vol. EC-10, no. 9, pp. 389–400, Sept. 1961.

[15] S. Unger, "Tree realization of iterative circuits," *IEEE Trans. Computers*, vol. C-26, no. 4, pp. 365–383, Apr. 1977.

[16] J. E. Vuillemin, "A very fast multiplication algorithm for VLSI implementation," INRIA, 1983.

[17] N. Takagi, H. Yasuura, and S. Yajima, "A VLSI-oriented high-speed divider using redundant binary representation," *Trans. IECE Japan*, vol. J67-D, no. 4, pp. 450–457, Apr. 1984 (in Japanese).

[18] N. Takagi, H. Yasuura, and S. Yajima, "Hardware algorithms for division and square rooting internally using redundant binary representation" Research Inst. Mathematical Sciences, Kyoto Univ., Kyoto, Japan, no. 494, pp. 223–235, June 1983.

[19] N. Takagi and S. Yajima "Hardware algorithms for logarithmic and exponential functions using redundant binary representation," Tech. Group on Automata and Languages, IECE Japan, Rep. AL83-70, Feb. 1984 (in Japanese).

[20] N. Takagi, T. Asada, and S. Yajima, "A hardware algorithm for trigonometric functions using redundant binary representation," Tech. Group on Automata and Languages, IECE Japan, Rep. AL84-59, Jan. 1985 (in Japanese).

[21] R. H. Dennard, F. H. Gaensslen, H. Yu, V. L. Rideout, E. Bassous, and A. Le-Blanc, "Design of ion implanted MOS FET's with very small physical dimension," *IEEE J. Solid-State Circuits*, vol. SC-9, p. 9256, 1974.

Yoshihisa Harata was born in Nagoya, Japan, in 1949. He received the B.E. degree from the Nagoya Institute of Technology, Nagoya, Japan, in 1973, and the M.E. degree from the Tokyo Institute of Technology, Tokyo, Japan, in 1975.

He joined Toyota Central Research and Development Laboratories, Inc., Nagoya, Japan, in 1975 and was initially engaged in development and application of sensors. Since 1979 he has been engaged in logical design for LSI.

Mr. Harata is a member of the Institute of Electronics and Communication Engineers of Japan and the Society of Instrument and Control Engineers. He received the outstanding paper award of the IEEE International Conference on Computer Design 1984.

Yoshio Nakamura was born in Ni-i-gata Prefecture, Japan, in 1948. He received the B.S. degree from the University of Electro-communications in 1971 and the M.S. degree in electronics engineering from the Tokyo Institute of Technology, Tokyo, Japan, in 1973.

He joined Oki Electric Industry Co., Ltd., Tokyo, Japan, in 1973, where he worked on bipolar LSI design. In 1980 he joined Toyota Central Research and Development Laboratories, Inc., Nagoya, Japan, where he has been engaged in LSI design and the development of power devices.

Mr. Nakamura is a member of the Institute of Electronics and Communication Engineers of Japan. He received the outstanding paper award of the IEEE International Conference on Computer Design 1984.

Hiroshi Nagase was born in Nagoya, Japan, in 1944. He received the B.E. and M.E. degrees in measurement and control engineering from the Nagoya Institute of Technology, Nagoya, Japan, in 1967 and 1969, respectively.

In 1969 he joined Toyota Central Research and Development Laboratories, Inc., Nagoya, Japan, where he worked on development and application of semiconductor sensors. He has worked on IC design since 1979 and is now Group Leader of the No. 342 Research Group in the Division of Semiconductor and IC Technology.

Mr. Nagase is a member of the Institute of Electronics and Communication Engineers of Japan and the Society of Instrument and Control Engineers. He received the outstanding paper award of the IEEE International Conference on Computer Design 1984.

Mitsuharu Takigawa (A'85) was born in Hokkaido, Japan, in 1936. He received the S.B. degree from the Tokyo Electrical Engineering University, Tokyo, Japan, in 1962. He received the Dr. Eng. degree from Osaka University, Osaka, Japan, in 1975.

From 1962 to 1978 he was with the Faculty of Science and Technology at Sophia University, Tokyo, Japan, where his research interests were in semiconductor physics, crystal growth, and IC technology. In 1978 he joined the Division of Semiconductor Physics and IC Technology at Toyota Central Research and Development Laboratories, Inc., Nagoya, Japan. He is now Manager of the No. 34 Laboratory.

Dr. Takigawa is a member of the Japan Society of Applied Physics. He received the outstanding paper award of the IEEE International Conference on Computer Design 1984.

Naofumi Takagi (S'82–M'84) was born in Osaka, Japan, in 1959. He received the B.E. and M.E. degrees in information science from Kyoto University, Kyoto, Japan, in 1981 and 1983, respectively.

He is an Instructor at the Department of Information Science, Faculty of Engineering, Kyoto University. His current interests include design and analysis of hardware algorithms, computer arithmetic, computer architecture, and CAD/DA for VLSI logic design.

Mr. Takagi received the IEEE Centennial Medal and the outstanding paper award of the IEEE International Conference on Computer Design 1984.

A Sub-10-ns 16×16 Multiplier Using 0.6-μm CMOS Technology

YUKIHITO OOWAKI, KENJI NUMATA, KENJI TSUCHIYA, KAZUSHI TSUDA, HIROSHI TAKATO, NAOKO TAKENOUCHI, AKIHIRO NITAYAMA, TAKAYUKI KOBAYASHI, MASAHIKO CHIBA, SHIGEYOSHI WATANABE, KAZUNORI OHUCHI, AND AKIMICHI HOJO

Abstract —**A 16×16-bit parallel multiplier fabricated in a 0.6-μm CMOS technology is described. The chip uses a modified array scheme incorporated with a Booth's algorithm to reduce the number of adding stages of partial products. The combination of scaled 0.6-μm CMOS technology and advanced arithmetic architecture achieves a multiplication time of 7.4 ns while dissipating only 400 mW. This multiplication time is shorter than other MOS high-speed multipliers previously reported and is comparable to those for advanced bipolar and GaAs multipliers.**

I. Introduction

HIGH-SPEED multiplication is indispensable to digital signal processing. Digital communication and image processing require many operations in time periods of 50 ns or less. Multiplication time is a critical design parameter for these digital signal processing operations, and therefore demands for higher speed multiplication are growing strongly. So far very-high-speed multipliers, operating at less than 10 ns for 16×16-bit multiplications, have been obtained only by high-speed current-mode logic (CML) or emitter-coupled logic (ECL) bipolar products or GaAs products [1]–[4]. However, the high power consumption for bipolar devices, 2–4 W per device, and the rather low manufacturability of GaAs devices tend to limit utilization of a very-high-speed multiplier as a macro for a larger LSI or VLSI chip.

Scaled CMOS is projected as the most promising device technology due to its high speed and low power dissipation along with its ease of manufacturability. Using advanced CMOS technology and advanced arithmetic architecture, it has become possible to obtain a sub-10-ns multiplier with very low power consumption. This paper describes the architecture, circuits, and performance of a 16×16 multiplier fabricated with 0.6-μm CMOS technology [5].

II. Chip Design

Multiplication arithmetic is usually realized as additions of partial products. The multiplication time is proportional to the number of additions, since the generation and propagation delay times for sum and carry signals dominate the operation time. Therefore, one of the most important design constraints is to reduce the number of additions in order to increase the arithmetic speed. There are two ways to meet the above requirement. One is to reduce the number of partial products to be added, and the other is to reduce the number of sum and adding stages. The Booth's method may be used for the former approach and a Wallace tree and modified array scheme may be used for the latter approach [6]–[8]. A Wallace tree incorporated with a Booth's implementation has been prevailing in high-speed multipliers [3], [9], [10]. However, the Wallace tree suffers from poor structure regularity. Regularity is necessary not only to enhance the design productivity and multiplier bit-width extendability but also to enhance the performance, by allowing interconnection wiring which is generally less complicated and shorter.

The multiplier performs 16×16-bit multiplication in signed two's complement. A functional block diagram of the 16×16-bit multiplier is shown in Fig. 1. The chip consists of input registers, a Booth's encoder, a multiplier array, a carry propagation stage, and an output register. Input operands X and Y are clocked into two 16-bit input registers. The output of the Y register is encoded into eight sets of five Booth's factors. Booth's factors and the output of the X register are fed into the multiplier array and processed in an asynchronous manner. The output of the array is fed into the final carry propagation stage and the final product is latched into an output register.

This multiplier utilizes a further modified array implementing a Booth's algorithm. Fig. 2(a) shows a basic cell which consists of a Booth's partial-product generator and a full adder. Solid lines show sum signals and dotted lines show carry signals. Fig. 2(b) shows a fraction of the further modified array. The array consists of odd rows and even rows where sum and carry signals generated in an odd row are fed to the next odd row and signals generated in an even row are concurrently transferred to the next even row. Therefore, the array has two signal streams in a column in parallel. In the following stage, the sum of odd rows and the sum of even rows are added together. Using this modified array implementing a Booth's algorithm, the maximum number of adder stages in a column is reduced

Manuscript received March 27, 1987; revised June 4, 1987.
The authors are with the VLSI Research Center, Toshiba Corporation, Komukai, 1, Saiwai-ku, Kawasaki 210, Japan.
IEEE Log Number 8716099.

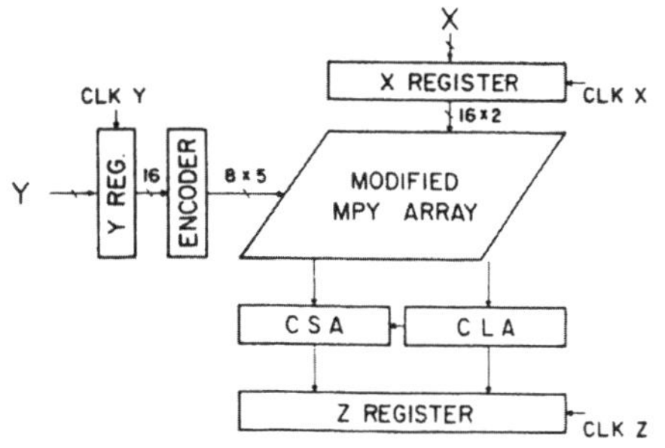

Fig. 1. A functional block diagram of the 16×16-bit multiplier.

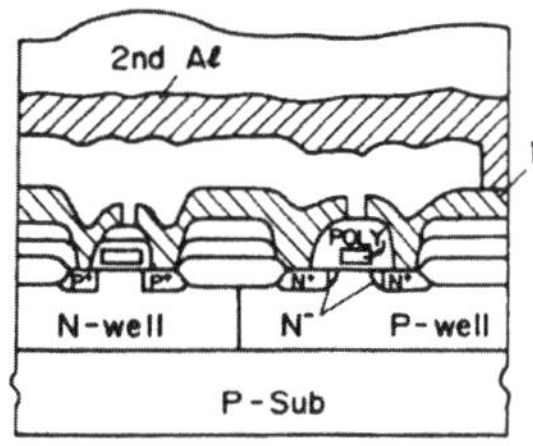

(a)

Well	Twin-Tub
Tox	10nm
Gate Length	NMOS 0.6μm
	PMOS 0.8μm
Contact hole	1.0μm x 1.3μm
1st Aℓ Pitch	2.4μm
Through hole	1.3μm x 1.3μm
2nd Aℓ Pitch	2.7μm

(b)

Fig. 4. (a) A device cross section, and (b) related device parameters.

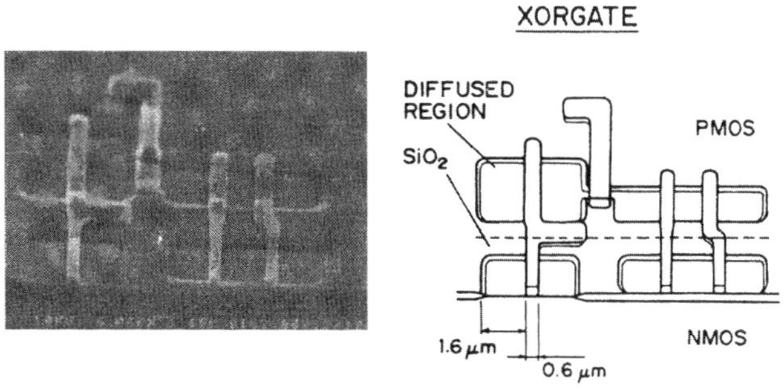

Fig. 5. An SEM micrograph of the multiplier.

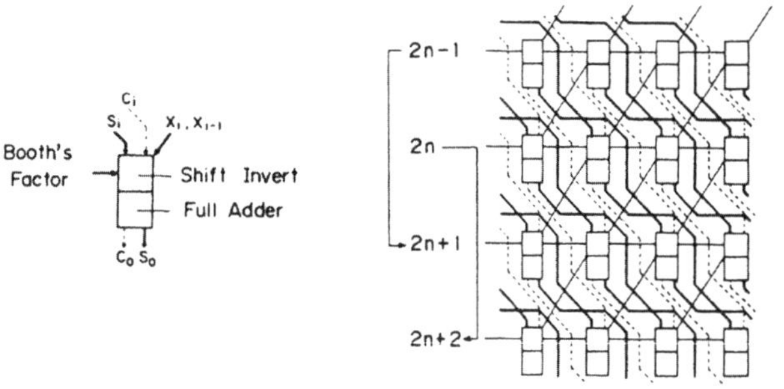

(a) (b)

Fig. 2. (a) A basic full-adder cell with a partial-product generator. (b) A modified multiplier array implementing a Booth's algorithm.

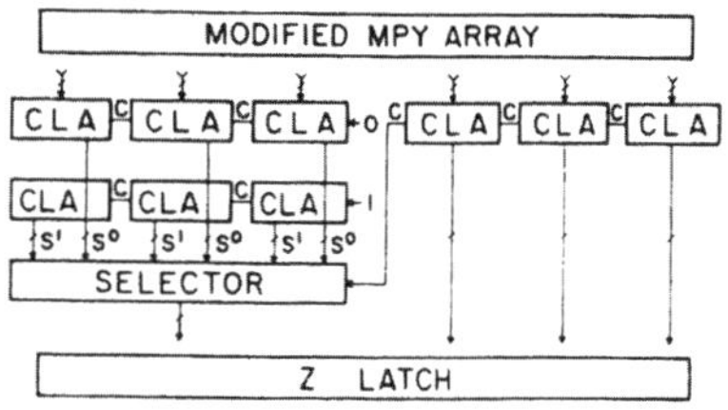

Fig. 3. A block diagram of the final carry propagation stage.

to five, while conventional carry-save arrays require 15 stages and carry-save arrays incorporating Booth's algorithm require seven stages. Moreover, this array achieves a highly regular cell configuration and affords good design productivity and a short design time.

Fig. 3 shows a block diagram of the final carry propagation stage. Besides reduction in the number of adder stages in the array, carry lookahead adders (CLA's) are fully utilized to speed up the carry propagations in the final stage. Furthermore, duplicate CLA's are employed on the most critical paths of the product. Duplicate CLA's perform carry propagations in parallel assuming the carry input from the lower CLA is 0 or 1. Then the final product is selected by 2-to-1 data selectors. These parallel processing methods for increasing the arithmetic speed require more logic gates compared with simple architecture, but the low-power and high-density characteristics of CMOS circuits allow these parallel processing methods to be used with minimal design penalty.

III. Chip Fabrication

The multiplier was fabricated with 0.6-μm twin-tub CMOS technology. A schematic device cross section and related device parameters used for the multiplier are shown in Fig. 4. P-type 10-Ω·cm (100) orientation wafers were used as the starting material. After implanting the tubs, a 1200°C 200-min drive-in was performed, producing an N-well diffusion depth of 4 μm.

The polysilicon gate is formed with N^+-poly over 10-nm-thick gate oxide. Minimum polysilicon gate lengths are 0.6 μm for NMOS and 0.8 μm for PMOS, with a lightly doped drain (LDD) structure used for NMOS. From two-dimensional process simulations, effective channel lengths are estimated as 0.4 μm for both NMOS and PMOS devices.

A two-level aluminum metallization technology was employed for interconnections in order to reduce wiring resistance and to minimize propagation delays. The minimum interconnection pitches are 2.4 μm for the first aluminum layer and 2.7 μm for the second aluminum layer. The minimum contact hole size to polysilicon and diffusion is 1.0×1.3 μm and the minimum throughhole size between metal layers is 1.3 μm^2. Fig. 5 shows an SEM micrograph of an XOR gate in the multiplier. This photograph was taken after the polysilicon gate had been defined.

The characteristics for typical NMOS and PMOS devices are shown in Fig. 6(a) and (b), respectively. It can be seen that there are minimal short-channel effects and that the saturation drain currents in both devices are quite large. A 10-μm-wide 0.6-μm-long NMOS device with a threshold voltage of 0.5 V produces a saturation drain

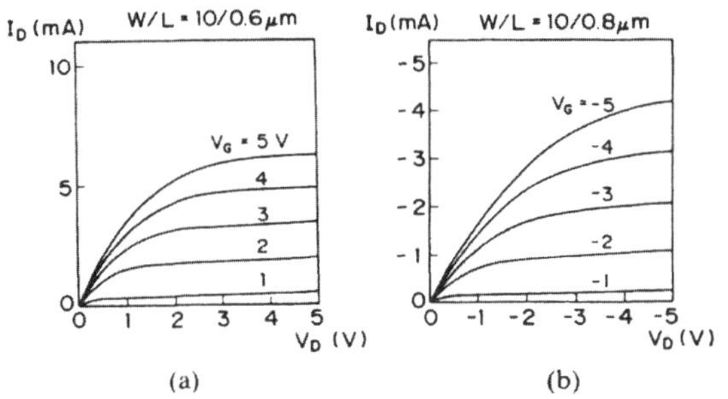

Fig. 6. (a) $I_d - V_d$ curve for a 0.6-μm NMOS device. $V_{sub} = 0$ V. (b) $I_d - V_d$ curve for a 0.8-μm PMOS device. $V_{well} = 0$ V.

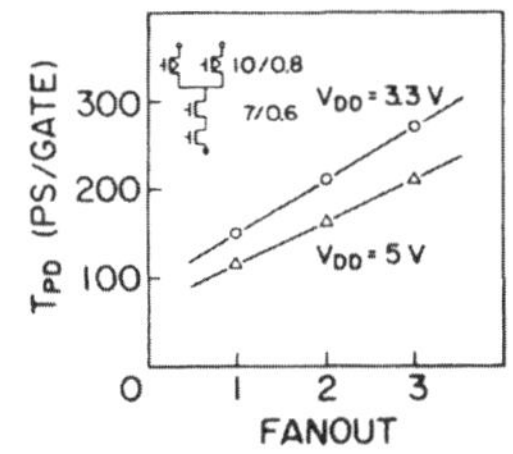

Fig. 7. Performance for a two-input NAND gate.

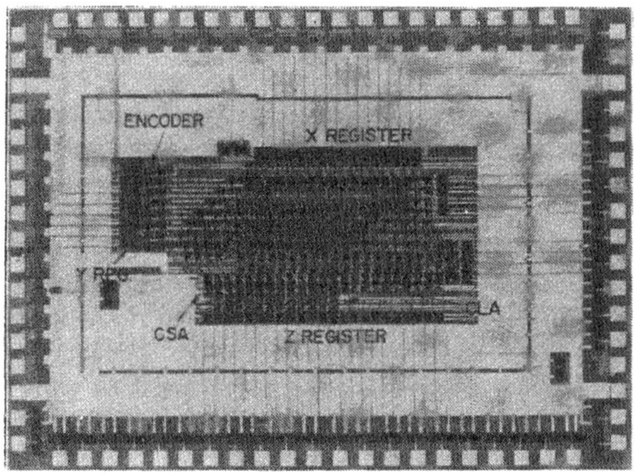

Fig. 8. A chip photograph.

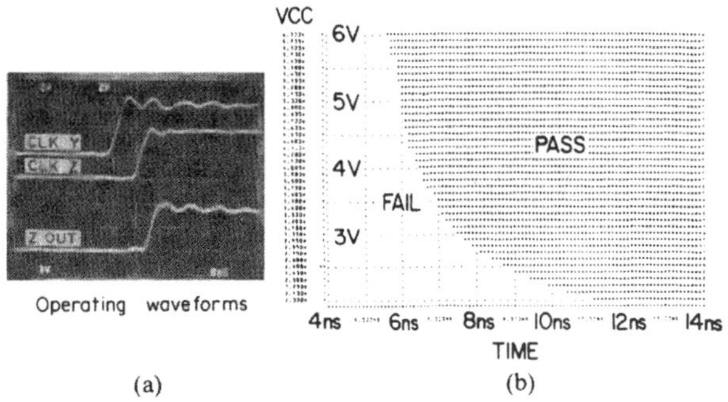

Fig. 9. (a) Typical operating waveforms. (b) A schmoo plot for the multiplier.

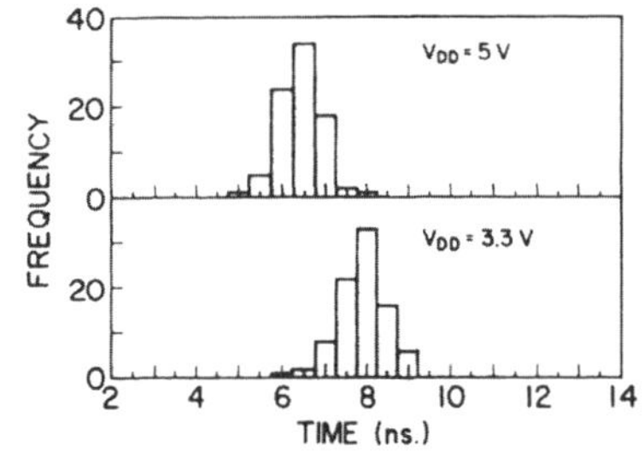

Fig. 10. Speed distribution for the multiplier.

current of 3.4 mA at a 3-V supply, while a 0.8-μm-long PMOS device with a threshold voltage of −0.6 V produces a saturation drain current of 1.8 mA using the same device width and supply.

Due to these excellent NMOS and PMOS device electrical characteristics, very high switching performance has been obtained as demonstrated by the delay per gate for a two-input NAND gate plotted versus fan-out conditions in Fig. 7. The obtained delays are 150 and 115 ps/gate for power supply voltages of 3.3 and 5.0, respectively, under unloaded conditions. Fan-out dependences are 60 and 48 ps per fan-out for supply voltages of 3.3 and 5.0, respectively. The obtained gate delay is comparable to those for advanced ECL, CML, and GaAs gates [4], [11], [12].

Fig. 8 shows a fabricated chip which is 2.4 × 3.2 mm. However, this size is defined by 76 bondpads, as actually 11 000 FET's are integrated in 1.0 × 2.0 mm.

IV. Chip Performance

Fabricated multiplier chips were mounted on ceramic pin-grid-array packages and tested using an Advantest-3341 automatic tester. A sequence of more than 1200 test vectors, including a series of worst-case patterns such as 8001 × FFFF − 1FFF × FBB7 − 0000 × FFC0, were used for the measurement. Fig. 9(a) shows typical operating waveforms. The multiplication time is the delay from the leading edge of clock X and/or clock Y, which latches two 16-bit inputs as a multiplicand or multiplier, to the leading edge of clock Z, which places the arithmetic result into the output register. Fig. 9(b) shows the typical schmoo plot of supply voltage versus multiplication time. On the right of the schmoo is the pass region and on the left is the fail region. The multiplier shows wide operational margins of power supply voltage up to 6 V. Fig. 10 shows the speed distribution for the multiplier. The functional yield of the chip was as high as 95 percent.

At a 3.3-V supply, the typical multiplication time was about 7.4 ns while the fastest one achieved was 6.0 ns, Since the critical path of the multiplier has 38 gate stages, the average loaded propagation delay per gate was 195 ps.

At a 5-V supply, the typical multiplication time was around 6 ns while the fastest one achieved was 5 ns. While the multiplication time is very short, further reliability studies are needed for actual 5-V use [13].

Typical power dissipation was 400 mW at 3.3-V 10-MHz operation, but most of the power, 310 mW, was consumed by 31 output buffers driving 75 pF each. The multiplier array and peripheral circuits consumed only 90 mW.

TABLE I
SALIENT FEATURES OF THE MULTIPLIER

Technology	Twin Tub CMOS
Gate Length	NMOS 0.6um
	PMOS 0.8um
Chip Size	2.4 X 3.2mm^2
Transistor Count	11000
Multiplication Time	7.4ns
Power Supply	3.3V
Power Dissipation	400mW

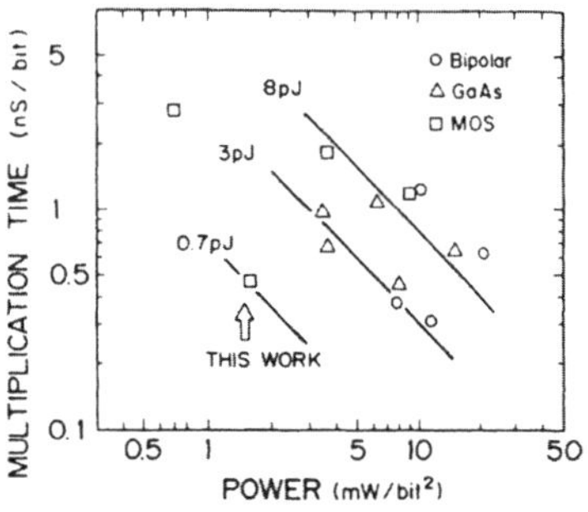

Fig. 11. A comparison of high-speed multipliers.

The key features of the chip are shown in Table I, while Fig. 11 compares the chip's performance to high-speed multipliers previously reported. In this figure, normalized multiplication time is plotted versus normalized power dissipation and the sloped lines represent the power–delay product. The typical multiplication time for this chip was 0.46 ns/bit and power dissipation was 1.5 mW/bit^2. Thus, the multiplier achieved ECL comparable speed at much lower power consumption. Consequently the obtained power–delay product for this chip is smaller than those for high-speed multipliers demonstrated to date [1]–[4], [8]–[10], [14], [15].

V. CONCLUSION

A very high-speed and low-power 16×16-bit multiplier was successfully designed and fabricated using 0.6-μm CMOS technology. The combination of scaled 0.6-μm CMOS technology and a modified array incorporating a Booth's encoder has achieved a typical multiplication time of 7.4 ns and a typical power dissipation of 400 mW at 3.3-V 10-MHz operation. The average loaded gate delay is 195 ps at a 3.3-V supply, and only 90 mW of the chip power is dissipated in the array while the remaining 310 mW is dissipated in the loaded output buffers. The power–delay product of the chip is the smallest to date while the functional yield for the chips has been as high as 95 percent.

Scaled CMOS has been demonstrated to be the most promising device technology for very-high-speed LSI's and VLSI's, not only due to its low power dissipation and high density, but also due to its very-high-speed characteristics and high manufacturability.

ACKNOWLEDGMENT

The authors would like to thank Dr. H. Iizuka for his continuous encouragement. They are indebted to J. Iwamura, M. Kimura, and S. Tanaka for their helpful discussions, and to S. Yokogawa and C. Takubo for their technical support.

REFERENCES

[1] T. Sakai, S. Konaka, Y. Yamamoto, and M. Suzuki, "Prospect of SST technology for high speed LSI," in *IEDM Dig. Tech. Papers*, Dec. 1985, pp. 18–21.
[2] B. E. Miller and R. E. Owen, "A sub 10 ns low power bipolar 16×16 bit multiplier," in *Proc. CICC*, May 1986, pp. 97–99.
[3] H. Shimizu, Y. Nakayama, S. Yokogawa, H. Kawata, and H. Nishi, "GaAs 16×16 bit parallel multiplier using CLA adder," in *1985 Nat. Conf. Rec. Semiconductor Devices and Materials, IECE Japan*, 1985, p. 176.
[4] S. Horiguchi, M. Suzuki, H. Ichino, S. Konaka, and T. Sakaki, "An 80 ps 2500-gate bipolar macrocell array," in *ISSCC Dig. Tech. Papers*, 1985, pp. 198–199.
[5] Y. Oowaki, K. Numata, K. Tsuchiya, K. Tsuda, A. Nitayama, and S. Watanabe, "A 7.4 ns CMOS 16×16 multiplier," in *ISSCC Dig. Tech. Papers*, 1987, pp. 52–53.
[6] S. Waser and M. Flynn, *Introduction to Arithmetic for Digital Systems Designers*. New York: Holt, Rinehart and Winston, 1982.
[7] C. S. Wallace, "A suggestion for a fast multiplier," *IEEE Trans. Electron. Comput.*, vol. EC-13, no. 1, pp. 14–17, 1964.
[8] J. Iwamura, K. Suganuma, M. Kimura, and S. Taguchi, "A CMOS/SOS multiplier," in *ISSCC Dig. Tech. Papers*, Feb. 1984, pp. 92–93.
[9] Y. Kaji, N. Sugiyama, Y. Kitamura, S. Ohya, and M. Kiuchi "A 45 ns 16×16 CMOS multiplier," in *ISSCC Dig. Tech. Papers*, 1983, pp. 32–33.
[10] A. G. Gamal, D. Gluss, P-H Ang, J. Greene, and J. Reyneri, "A CMOS 32b Wallace tree multiplier-accumulator," in *ISSCC Dig. Tech. Papers*, 1986, pp. 194–195.
[11] M. Tatsuki, S. Kato, M. Okabe, H. Yakushiji, and Y. Kuramitsu, "An ECL 5000-gate gate array with 190-ps gate delay," *IEEE J. Solid-State Circuits*, vol. SC-21, pp. 234–239, Apr. 1986.
[12] K. Gonoi, I. Honbori, M. Wada, K. Togashi, and Y. Kato "A GaAs 8×8-bit multiplier/accumulator using JFET DCFL," *IEEE J. Solid-State Circuits*, vol. SC-21, pp. 523–529, Aug. 1986.
[13] A. Nitayama, N. Takenouchi, T. Hamamoto, and Y. Oowaki, "Hot-carrier-induced degradation phenomena in asymmetric structured submicron MOSFETs," in *Proc. 172nd Meeting Electro Chemical Soc.*
[14] J. Y. Lee, H. L. Garvin, C. W. Slayman, and R. P. Mento, "A 8×8b parallel multiplier in submicron technology," in *ISSCC Dig. Tech. Papers*, 1985, pp. 84–85.
[15] Y. Nakayama, H. Suyama, H. Shimizu, N. Yokoyama, A. Shibatomi, and H. Ishikawa, "A GaAs 16×16 b parallel multiplier using self-alignment technology," in *ISSCC Dig. Tech. Papers*, Feb. 1983, pp. 48–49.

Yukihito Oowaki was born in Kagoshima, Japan, on December 19, 1958. He received the B.E. and M.E. degrees in applied physics from the University of Tokyo, Tokyo, Japan, in 1981 and 1983, respectively.

In 1983 he joined the Toshiba Research and Development Center, Toshiba Corporation, Kawasaki, Japan, where he worked on the development of an NMOS 1-Mbit DRAM. Since 1985 he has been engaged in the research and development of submicrometer CMOS DRAM's at the VLSI Research Center, Toshiba Corporation.

Mr. Oowaki is a member of the Physical Society of Japan.

Kenji Numata was born in Fukushima Prefecture, Japan, on January 9, 1958. He received the B.E. and M.E. degrees in electrical and electronic engineering from Toyohashi University of Technology, Achi, Japan, in 1980 and 1982, respectively.

In 1982 he joined the Research and Development Center of Toshiba Corporation, Saiwai-ku, Kawasaki, Japan, where he has been working on the research and development of MOS memories at the Toshiba VLSI Research Center.

Mr. Numata is a member of the Institute of Electronics, Information and Communication Engineers of Japan and the Japan Society of Applied Physics.

Kenji Tsuchiya was born in Tokyo, Japan, in 1959. He received the B.E. and M.E. degrees in electrical engineering from Meiji University, Kawasaki, Japan, in 1982 and 1984, respectively.

In 1984 he joined the Toshiba VLSI Research Center, Kawasaki, Japan, where he has been engaged in the research and development of submicrometer devices and process technologies.

Kazushi Tsuda was born in Fukuoka, Japan, on December 7, 1958. He received the B.E. and M.E. degrees in electrical engineering from Chuo University, Tokyo, Japan, in 1982 and 1984, respectively.

In 1984 he joined the VLSI Research Center, Toshiba Corporation, Kawasaki, Japan, where he has been engaged in the research and development of submicrometer device technologies.

Mr. Tsuda is a member of the Japan Society of Applied Physics.

Hiroshi Takato was born in Fukuoka, Japan, on August 9, 1959. He received the B.S. degree in physics from Shizuoka University in 1983 and the M.E. degree in electromagnetic energy engineering from Osaka University in 1985.

In 1985 he joined the Toshiba VLSI Research Center, Kawasaki, Japan. He has been engaged in the development of VLSI devices and fabrication processes.

Mr. Takato is a member of the Physical Society of Japan.

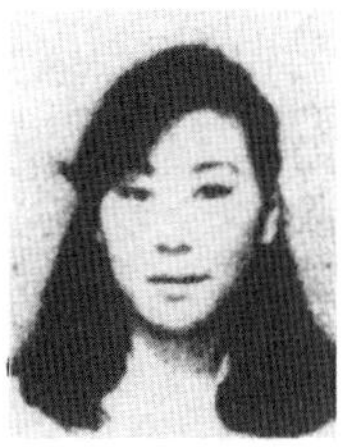

Naoko Takenouchi was born in Tokyo, Japan, on September 7, 1962. She received the B.S. degree in physics from Sophia University, Tokyo, Japan, in 1985.

In 1985 she joined the VLSI Research Center, Toshiba Corporation, Kawasaki, Japan, where she has been engaged in the research and development of submicrometer CMOS devices. Her research interests are in device physics and reliability problems.

Akihiro Nitayama was born in Tokyo, Japan, on December 27, 1955. He received the B.E. degree in electrical engineering in 1978, and the M.E. and Ph.D. degrees in electronics engineering in 1980 and 1983, respectively, from Tokyo University, Tokyo, Japan.

In 1983 he joined the Toshiba Research and Development Center, Toshiba Corporation, Kawasaki, Japan. Since 1984 he has been engaged in the development of megabit DRAM's at the VLSI Research Center.

Dr. Nitayama is a member of the Japan Society of Applied Physics.

Takayuki Kobayashi was born in Hyogo, Japan, on March 5, 1960. He received the B.E. and M.E. degrees, both in electronic engineering, from Waseda University, Tokyo, Japan, in 1983 and 1985, respectively.

In 1985 he joined the Toshiba Research and Development Center, Toshiba Corporation, Kawasaki, Japan. Since 1985 he has been engaged in the development of megabit DRAM's.

Mr. Kobayashi is a member of the Institute of Electronics, Information and Communication Engineers of Japan.

Masahiko Chiba was born in Aomori, Japan, on February 6, 1961. He received the B.E. degree in mechanical engineering from Keio University in 1985.

In 1985 he joined the VLSI Research Center, Toshiba Corporation, Kawasaki, Japan. Since 1985 he has been engaged in the development of dynamic memories.

Mr. Chiba is a member of the Institute of Electronics, Information and Communication Engineers of Japan.

Shigeyoshi Watanabe was born in Mie, Japan, on March 19, 1955. He received the B.E. degree in instrumental engineering from Keio University in 1977 and the M.S. degree in applied physics from Tokyo Institute of Technology, Tokyo, Japan, in 1979.

In 1979 he joined the Semiconductor Device Engineering Laboratory of Toshiba Corporation, Kawasaki, Japan, where he was engaged in the design of floating-gate-type memory. In 1982 he was engaged in the design of dynamic memory in Toshiba's Memory Devices Engineering Department. In 1984 he joined the VLSI Research Center where he is currently working on megabit dynamic memories.

Mr. Watanabe is a member of the Physical Society of Japan.

Kazunori Ohuchi was born in Tochigi, Japan, in June 1947. He received the B.S. degree in electronic engineering from Tohoku University, Sendai, Japan, in 1970.

In 1970 he joined the Toshiba Research and Development Center, Toshiba Corporation, Kawasaki, Japan. Since then he has been working on the circuit and device design of MOS memories, including EPROM, static RAM, and dynamic RAM. His current interests are in the advanced device structures and circuits for 4-Mbit dynamic RAM's and beyond.

Mr. Ohuchi is a member of the Institute of Electronics and Communication Engineers of Japan.

Akimichi Hojo received the B.E., M.E., and Ph.D. degrees in applied physics from the University of Tokyo, Tokyo, Japan, in 1965, 1967, and 1970, respectively.

In 1970 he joined the Toshiba Research and Development Center, Kawasaki, Japan. He worked on the research and development of GaAs liquid phase epitaxial growth, GaAs Hall devices, and GaAs IC's. He is currently responsible for the development of advanced CMOS DRAM and other memory devices.

Dr. Hojo is a member of the Institute of Electronics and Communication Engineers of Japan and the Physical Society of Japan.

Part V:
FLOATING-POINT VLSI CHIPS

Editor's Comments on Papers 25 Through 30

In contrast to the previous two Parts, which have described the design and implementation of fixed point adders and multipliers, this Chapter focuses on floating-point implementations. In fixed point arithmetic, multiplication is more difficult (i.e., requiring either more time or more area or both!) then addition. In floating point arithmetic, the situation is reversed and addition is more difficult than multiplication.

In the first paper, Wolrich, McLellan, Harada, Montanaro and Yodlowski describe a floating point coprocessor for a 16/32 bit microprocessor. The chip performs 32 and 64 bit floating-point add, subtract, multiply, and divide. Implemented with 3 μm NMOS, the typical operation times range from 1.1 μsec for 32 or 64 bit addition/subtraction to 4.7 μsec for 64 bit division. A unique aspect of the design is the use of data dependent timing, where the normal clock cycle time is less than the time required for a maximum length carry propagation. Carry completion sensing is used to cause the clock to be delayed if a long carry propagation is required.

Next, Ware, McAllister, Carlson, Sun and Vlach describe three floating-point processors. These are an adder/subtracter, a multiplier, and a divider. They are implemented in 4 μm CMOS silicon-on-sapphire technology. Carry lookahead addition was selected to implement the internal adders as it is three times faster than a ripple carry adder, but only about 25% more complex. A conditional sum adder would be another factor of two faster, but the complexity would double. The three chips are capable of performing one million–point operations per second.

In the third paper, Uya, Kaneko, and Yasui describe a floating-point multiplier that multiplies 32 bit numbers in less than 80 nsec. The chip, which conforms to the IEEE standard 32 bit format was designed with 2 μm n-well CMOS technology. The paper describes the functional design, provides design details of the exponent adder, and the mantissa multiplier (realized with a modified Booth algorithm that is implemented using carry save adders), and concludes with a discussion of the testing procedure and results.

Next, Takeda, Ishino, Ito, Kasai and Nakashima describe a single chip processor for 80 bit floating-point operands. This 400 mil square chip, fabricated in 1.2 μm n-well CMOS technology, implements add, subtract, and multiply directly (in 360–420 nsec, 540–600 nsec, and 180 nsec, respectively), and other operations such as division, square rooting, exponentiation, etc. are performed by executing sequences of addition, subtraction, and multiplication operations. The paper describes the functional design, the structure of the chip, the computer aided design system (including an interesting view of the tradeoff between chip size and design effort), and concludes with a description of the performance of the chip.

In the fifth paper, Fandrianto and Woo compare several commercial floating-point chips. They consider two classes of chips: general purpose floating-point instructions (with rather slow performance) and building block floating-point processors which execute only a few instructions (at high speed). The paper concludes with a brief description of 32 and 64 bit adder and multiplier.

The chapter concludes with a paper by Bose, Pei, Taylor and Patterson on a floating-point multiplier/divider designed as part of a floating-point processor. The authors selected an iterative multiplier/divider as part of their effort to produce a balanced design where the multiplier/divider occupies 40% of the chip area (roughly the same as the percentage of floating-point operations that are multiplies or divides). Multiplication and division algorithms are compared on the basis of chip area and speed for 2 μm and 1 μm CMOS.

A High Performance Floating Point Coprocessor

GIL WOLRICH, EDWARD McLELLAN, LARRY HARADA, JAMES MONTANARO, AND ROBERT A. J. YODLOWSKI

Reprinted from *IEEE Journal of Solid-State Circuits*, Volume SC-19, Number 5, October 1984, pages 690-696.

***Abstract*—A 34000 transistor single-chip floating point coprocessor fabricated in 3 μm double metal NMOS technology is described. The fraction data path, including a shifter and 60 bit carry propagate ALU, is cycled in 100 ns for all operations requiring less than 19 bits of consecutive carry. A versatile carry length detection scheme, which requires minimal additional logic, is used to extend the microcycle for the small percentage of operations in which a *long* carry exists. Three bit per cycle multiplication and one and one half bit per cycle division algorithms were used to achieve excellent overall performance.**

Introduction

THIS paper describes a single-chip floating point accelerator (FPA) for the J-11, 16/32 bit microprocessor chip set. Fabricated in a 3 μm drawn double-metal NMOS process, the FPA implements a floating point instruction set of 46 instructions. The FPA supports four data types: single and double precision floating point and 16 and 32 bit integer (Table I).

The FPA measures 7.6×6.3 mm, contains 34000 transistors, and dissipates 2 W. It is packaged in a 40 pin DIP. A photomicrograph is shown in Fig. 1. The chip contains six externally addressable floating point data registers, a floating exception code register, and a floating point status and mode control register. The principal functional blocks in the FPA are the EU (execution unit), consisting of fraction, exponent, and sign datapaths, and the BIU (bus interface unit).

Microarchitecture

The fraction processor is a 60 bit wide data path (Fig. 2) consisting of 7 single-ported RAM registers, 8 ROM constants, an ALU, an 8-position argument shifter (left 2 to right 5), 2 operand registers, a Q register (multipler/quotient/output), a 4-position Q shifter (left 2, left 1, no shift, or right 3), and an input register. The fraction shifters provide sufficient range for 3 bit/cycle multiplication, 1.5 bit/cycle variable shift division, 5 bit/cycle alignment, and 3 bit/cycle normalization algorithms.

The EU sequencer contains a two-bank folded PLA with 13 inputs, 160 total AND terms, and 36 outputs. The exponent processor is a 10 bit data path used to process the exponent and sign of floating point operands in parallel with the fraction. The BIU controls all I/O for the FPA and contains a second-state sequencer allowing operation independent of the EU.

Manuscript received March 22, 1984; revised May 16, 1984.
The authors are with the Digital Equipment Corporation, Hudson, MA 01749.

TABLE I
FLOATING POINT DATA TYPES

F = Floating ⇒	1 bit sign,	8 bit exponent,	24 bit fraction
D = double ⇒	1 bit sign,	8 bit exponent	56 bit fraction
G (VAX only) ⇒	1 bit sign,	11 bit exponent,	53 bit fraction
32 bit integers,	16 bit integers		

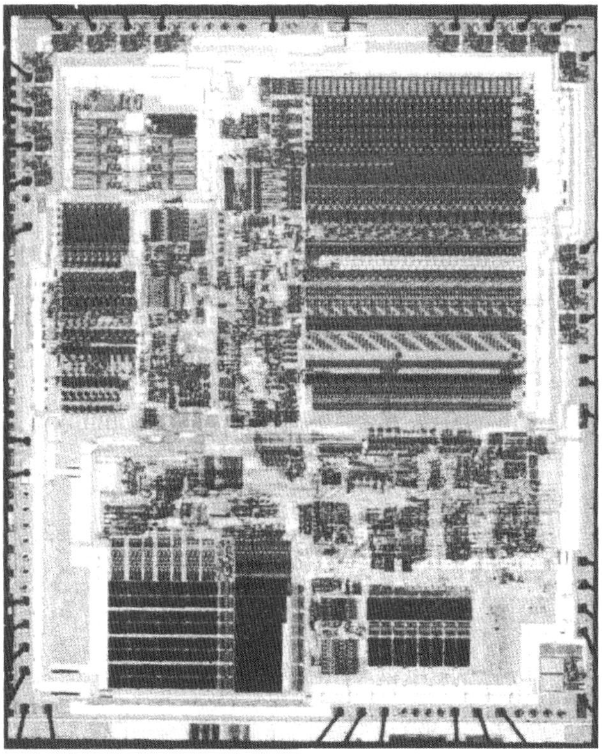

Fig. 1. Photomicrograph of FPA.

Carry Length Detection

Hardware floating point processing is characterized by very wide data paths. A great deal of hardware in the form of carry lookahead schemes or carry save adders are often used in computing systems to reduce the penalty associated

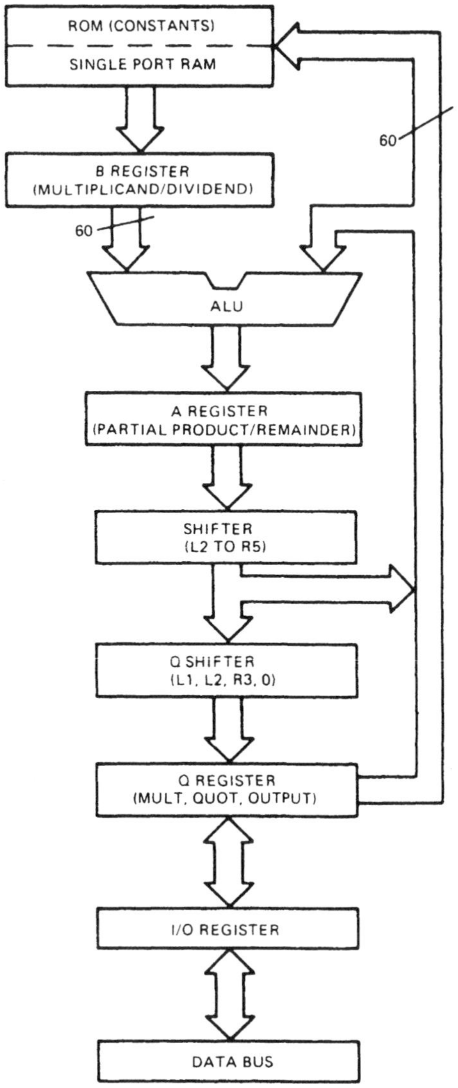

Fig. 2. Block diagram of fraction data path.

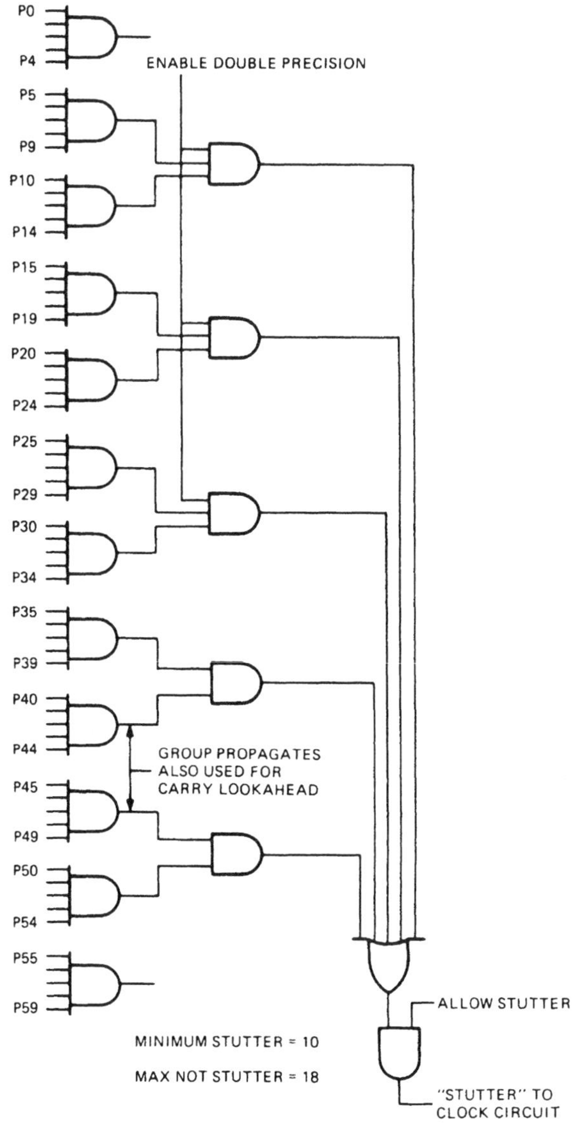

Fig. 3. FPA 60 bit ALU "stutter circuit."

with long carry propagation delays. The FPA, by comparison, achieves a fast (100 ns) microcycle time, including a 60 bit ALU operation using a new technique well suited to VLSI applications, as well as designs using standard parts.

A simple carry length detection scheme is used to produce a *stutter* signal that stretches the final phase of the EU clock if a *long* carry propagation path exists. The method takes advantage of the fact that most ALU operations have a largest maximum carry length which is much less than the width of the ALU. By detecting a *long* carry and providing additional time for the ALU to complete for a small percentage of operations, a data path can be run at a fast rate for most (> 95 percent) ALU cycles.

Fig. 3 shows the stutter circuit for the 60 bit wide ALU used in the J-11 floating point accelerator chip. In Fig. 3 the propagates produced in bit positions 54 to 5 of the ALU's PG (propagate generate) logic are gated with a minimum of logic to produce a detection signal *stutter* for all operations which have a carry length of 19 or greater. Two levels of AND gating are used because the first level of gating is already present for the minimal 5 bit group carry lookahead logic. The ANDing of two group propagate signals indicates whether or not all of the propagates in that group of 10 bits are asserted.

Single precision processing uses only the upper half of the fraction data path. In order to avoid unnecessary stutter cycles caused by data in the lower half of the data path, an additional enable signal is included in the detection gates covering bits 34 to 5. The *stutter* signal may set for as few as 10 consecutive propagates, but might not set for as many as 18. A propagate is a necessary but not sufficient condition to imply an actual carry. For this reason, an *allow stutter* signal is used to gate the stutter signal to the clock circuitry. The *allow stutter* signal is not set for ALU operations in which all bit positions will produce a generate. Unnecessary stutter cycles are therefore prevented for ALU operations in which carry propagation is not a factor (i.e., $A \leftarrow A$ generates a carry at all bit positions).

The optimal width of the AND function used to produce the group propagate signals will depend on the technology in which the ALU is to be implemented. Fig. 4 shows an example of the stutter circuit where the width of the group propagate is 4 bits. If the individual propagate bits are not available, as in standard part ALU slices, then the ANDing of the group propagates can be used directly. The stutter circuit is very inexpensive to implement, due to the fact that the group propagates are already required for any ALU with lookahead carry.

Propagate terms become valid after a fixed combinational delay allowing enough time to control the length of a system clock phase. The current design uses only one stutter signal to force a 100 ns clock phase extension sufficient for the longest possible carry. A second stutter signal detecting wider carry lengths could also be used to more finely tune the clock phase extension (Fig. 5).

The difference between the maximum carry length which always stutters and the minimum carry length which might stutter can be reduced by adding additional sets of AND functions which produce overlapped groupings of consecutive propagates (Fig. 6).

For the circuit described in Fig. 3 the probability of requiring a stutter on random data is

$$P = ((w - m)/n)(2**(w - n))/2**w$$

where

w = width of the ALU

m = total bits not included in any detection gate

n = width of the detection gates

(this equation is an upper bound since data which sets more than 1 detection gate are counted more than once)

double precision $w = 60,\ m = 10,\ n = 10$

$P = 5/1024$

single precision $w = 30,\ m = 10,\ n = 10$

$P = 2/1024.$

The 60 bit ALU/shifter cycle of the J-11 FPA can be completed in 100 ns for all operations in which the maximum consecutive carry is less than 19 bits.

Algorithms

The FPA executes four data path assisted microinstructions which serve as the basis for executing the multiplication, division, alignment, and normalization algorithms.

The FPA uses a fixed 3 bit shift algorithm to perform multiplication. The algorithm requires the generation of multiples 0, 2, 4, 6, and 8 times the multiplicand. The multiples are added or subtracted to the partial product and the result is shifted to account for the three multiplier

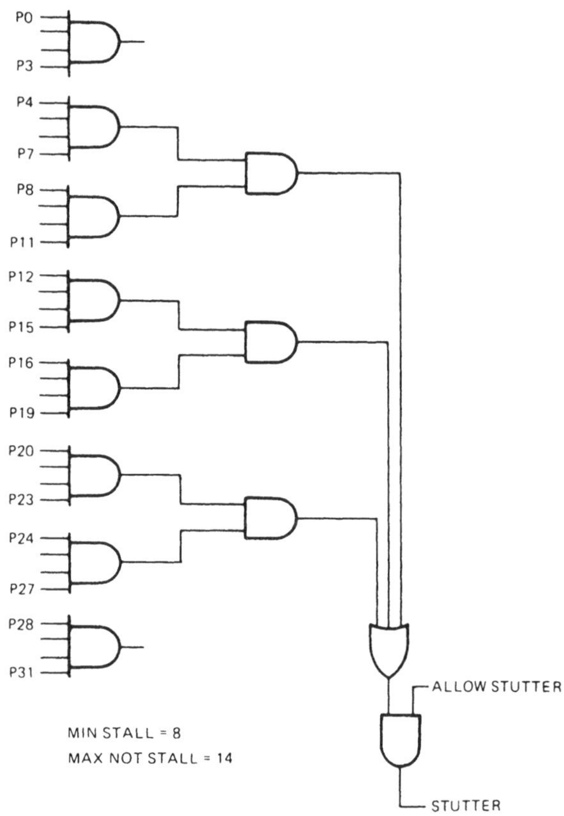

Fig. 4. Stutter detect for 32 bit ALU with 4 bit lookahead groupings.

bits retired. The required shift and ALU function for each cycle is determined by examining the multiplier bits from the LSB's of the Q register (bits 38:35 for F, bits 5:2 for D). Since the main data path has only one shifter, the total shift required for each cycle combines the shift necessary to align the binary point for the present multiple, and the post shift required to complete the 3 bit retirement from the previous cycle. The previous group of multiplier bits are held in a delay register which is initially cleared allowing the algorithm to begin without any shifts being owed from a previous cycle.

Prior to the start of the multiplication, 3/4 times the multiplicand is calculated and placed in the scratch register for use in generating the multiple of 6. If the multiples 2, 4, or 8 are required, the normal multiplicand register is accessed and the partial product is appropriately shifted. When a multiple of 6 times the multiplicand is required, the contents of the scratch register are used instead of the multiplicand and added to the partial product shifted right 3 times. A special microinstruction is used to establish an initial partial product of either zero if the LSB of the multiplier is zero, or minus one times the multiplicand if the multiplier LSB is a one. Table II details the single shifter 3 bit multiplication algorithm implemented in the FPA.

The FPA uses a normalizing nonrestoring division algorithm which produces a quotient at a rate of 1.5 bits/cycle. If the partial remainder will be normalized for

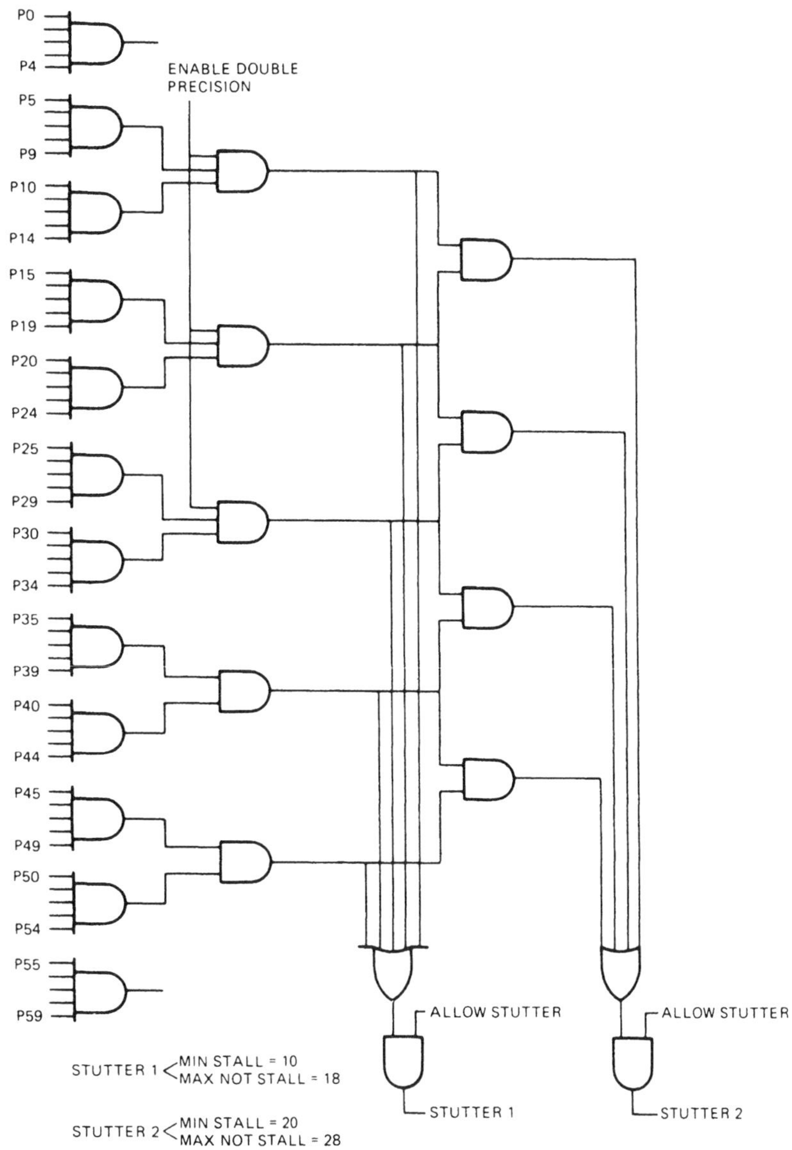

Fig. 5. Stutter circuit with second tier.

division ($-1 \leqslant R < -1/2$, $1/2 \leqslant R < 1$) by a left shift of one, then one new quotient bit is determined; when the partial remainder requires more than a single left shift to become normalized, two quotient bits are determined. Table III describes the next shift, ALU operation, and quotient bits derived as a function of the MSB's of the partial remainder. If the 4 MSB's of the partial remainder equal all ones or all zeros then a left shift of 2 will not normalize the present partial remainder and the next cycle ALU operation is $A \leftarrow A$. This insures that the next partial remainder will remain in the range $-1/2 \leqslant R < 1/2$. If the present partial remainder can be normalized by a left shift of one or two, the next ALU operation adds or subtracts the divisor depending on the sign of the remainder in order to drive it toward zero. The quotient bits are inserted at the guard bit and LSB positions of the Q register (bits 35:34 for F, bits 3:2 for D) and the Q register shifts either 1 or 2 bits left as required. When $Q57 = 1$, a Q shift of left one is forced and only one more quotient bit is accepted. When $Q58 = 1$, the division is completed and the normalized quotient is in the Q register. The normalized quotient can be in the range $1/2 \leqslant Q < 1$ or $1 \leqslant Q < 2$, depending on the ratio of the initial dividend and divisor. If the initial subtraction of the divisor from the dividend is positive, then $1 \leqslant$ quotient < 2 and the final exponent is incremented.

The important feature of the FPA alignment and normalization algorithms is that although the main shifter has limited range, the shift probability data for floating point addition and subtraction (Table IV) show this range to be all that is required for most operations. The FPA performs 78 percent (up to 5 bits of exponent difference) of the

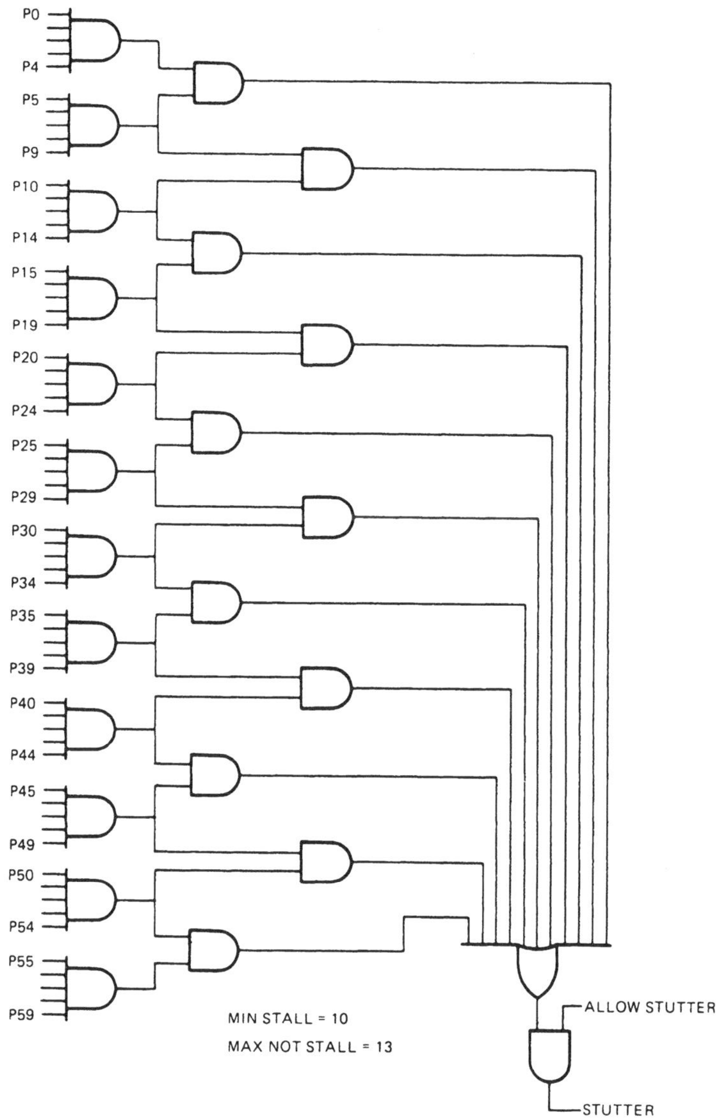

Fig. 6. Stutter circuit with overlapped detection gates.

TABLE II
FPA 3 BIT/CYCLE MULTIPLICATION ALGORITHM

ALU/REG	Present Multiplier Group	Alignment Shift for Present	Previous Multiplier Group	Shift owed from Previous
$A \leftarrow A$	0000	3	0000	0
$A \leftarrow A+B;\ B = AC$	0001	1	0001	2
$A \leftarrow A+B;\ B = AC$	0010	1	0010	2
$A \leftarrow A+B;\ B = AC$	0011	2	0011	1
$A \leftarrow A+B;\ B = AC$	0100	2	0100	1
$A \leftarrow A+B;\ B = SCR$	0101	3	0101	0
$A \leftarrow A+B;\ B = SCR$	0110	3	0110	0
$A \leftarrow A+B;\ B = AC$	0111	3	0111	0
$A \leftarrow A-B;\ B = AC$	1000	3	1000	0
$A \leftarrow A-B;\ B = SCR$	1001	3	1001	0
$A \leftarrow A-B;\ B = SCR$	1010	3	1010	0
$A \leftarrow A-B;\ B = AC$	1011	2	1011	1
$A \leftarrow A-B;\ B = AC$	1100	2	1100	1
$A \leftarrow A-B;\ B = AC$	1101	1	1101	2
$A \leftarrow A-B;\ B = AC$	1110	1	1110	2
$A \leftarrow A$	1111	3	1111	0

$AC \leftarrow$ Multiplicand
$SCR \leftarrow 3/4$ Multiplicand

TABLE III
FPA 1.5 BIT/CYCLE DIVISION ALGORITHM

						Quotient Bit(s) Formed If Partial Remainder Is Derived by			
				Next	Next	ADD/SUB		Shift Left 2	
$F59$	$F58$	$F57$	$F56$	Shift	ALU	$Q3$	$Q2$	$Q3$	$Q2$
0	0	0	0	SHL2	$A \leftarrow A$	1	0	0	0
0	0	0	1	SHL2	SUB	1	0	0	0
0	0	1	0	SHL1	SUB		1		0
0	0	1	1	SHL1	SUB		1		0
1	1	0	0	SHL1	ADD		0		1
1	1	0	1	SHL1	ADD		0		1
1	1	1	0	SHL2	ADD	0	1	1	1
1	1	1	1	SHL2	$A \leftarrow A$	0	1	1	1

$F59$ = Fraction MSB
$Q3$ = Quotient Register LSB
$Q2$ = Quotient Register Guard Bit

Note: Quotient bit(s) are inserted at bit positions $Q35$ and $Q34$ for single precision operation instead of $Q3$ and $Q2$.

TABLE IV
WEIGHTED DATA ON ALIGNMENT AND NORMALIZATION

Shift	Alignments (percent)	FPA Align Cycles	Normalizations (percent)	FPA Normalize Cycles
Result 0			0.82	1
Overflow			17.43	1
0	26.28	1	60.25	1
1	13.29	1	8.01	1
2	8.77	1	3.17	1
3	6.53	1	1.54	3
4	7.77	1	1.33	3
5	8.07	1	1.03	3
6	5.13	2	0.63	4
7	3.78	2	1.02	4
8	1.10	2	0.93	4
9	1.23	2	0.56	5
10	1.84	2	0.66	5
11	1.35	3	0.16	5
12	1.54	3	0.22	6
13	0.81	3	0.31	6
14	0.48	3	0.16	6
15	0.58	3	0.03	7
16	0.29	4	0.06	7
17	0.31	4	0.09	7
18	0.50	4	0.17	8
19	0.32	4	0.15	8
20	0.26	4	0.08	8
21	0.40	5	0.07	9
22	0.30	5	0.07	9
23	0.24	5	0.19	9
24	0.25	5	0.49	10
25	0.26	5	0.09	10
26	0.16	6	0.28	10
27–53	0.86			
54–255	7.30			

*Reference D. W. Sweeney.

aligns and 90 percent of the normalizations with a single shift cycle. The average number of shift cycles is 1.5 for alignment and 1.2 for normalization. Alignment proceeds at a 5 bit/cycle rate if the exponent difference is between 6 and the length of the data type. Only one cycle is needed for normalizations requiring shifts in the range of right one to left two. If more than 2 bits of left shifting is required, a no shift cycle occurs to examine additional fraction bits. Normalization then proceeds at 3 bits/cycle. The left shift function of the fraction ALU is combined with the 2 bit left shift capability of the fraction shifter to accomplish the 3 bit shift.

TABLE V
J-11 FPA TYPICAL REGISTER-TO-REGISTER EXECUTION TIMES

ADDF/SUBF	1.1 μs	F	1 bit sign
MULF	1.6 μs	floating =	8 bit exponent
DIVF	2.7 μs		24 bit fraction
ADDD/SUBD	1.1 μs	D	1 bit sign
MULD	2.8 μs	double =	8 bit exponent
DIVD	4.7 μs		56 bit fraction

PERFORMANCE

The combination of a 100 ns internal cycle and optimized arithmetic algorithms provides excellent performance. Table V shows typical execution times for register-to-register operations.

The J-11 FPA interfaces as a true coprocessor. The BIU inputs all instruction stream data and decodes instructions in parallel with the base processor. Support microcode in the CPU initiates all I/O cycles required by the FPA. As a coprocessor, floating point instruction execution can occur simultaneously with integer code. This overlap can effectively be used to reduce the execution time of mixed code by interleaving floating point and nonfloating point instructions.

A second and more frequent type of instruction overlap which provides a substantial performance gain in floating point intensive code is also achieved by the FPA. The BIU supports the overlap of operand data loading for the next floating point instruction, while the EU completes the processing of the current instruction. The FPA asserts a *stall* signal to prevent the CPU from initiating more than one new floating point instruction while the EU is still busy. Only the portion of FPA execution time, if any, that the CPU is stalled actually effects system performance.

Two additional floating point processor chips have been developed by extending the J-11 FPA design to the VAX architecture. The G floating point format and the extended multiply and integerize instructions are supported by increasing the fraction and exponent data paths to 67 and 13 bits, respectively. The three designs achieve similar performance. The carry length detection method and the 3 bit multiplication are especially beneficial in applications requiring wide data paths, as evidenced by the performance of these three floating point processor chips.

References

[1] D. W. Sweeney, "An analysis of floating-point addition," *IBM Syst. J.*, vol. 4, pp. 31–42, 1965.

[2] O. L. MacSorley, "High-speed arithmetic in binary computers," *Proc. IRE*, vol. 49, pp. 67–91, 1961.

[3] E. Swartzlander, Ed., *Computer Arithmetic*. New York: Dowden, Hutchinson, and Ross, 1980.

Gil Wolrich received the B.S. degree in electrical engineering from Rensselaer Polytechnic Institute, Troy, NY, in 1971, and the M.S. in electrical engineering from Northeastern University, Boston, MA, in 1978.

He joined the Digital Equipment Corporation, Hudson, MA, in April 1979. He is currently a Principal Engineer with the Semiconductor Engineering Group in Hudson, MA.

Edward McLellan received the B.S. degree in computer and systems engineering from Rensselaer Polytechnic Institute, Troy, NY, in 1980.

He is currently a Senior Engineer with the Semiconductor Engineering Group of the Digital Equipment Corporation, Hudson, MA.

Larry Harada received the B.S. degree in electrical engineering in 1980, and the M.E. degree in 1981, both from Cornell University, Ithaca, NY.

He joined Digital Equipment Corporation, Hudson, MA in July 1981. He is currently working for the Digital LSI Manufacturing Group in Hudson.

James Montanaro received the B.S. and M.S. degrees in electrical engineering from Massachusetts Institute of Technology, Cambridge, MA, in 1980.

Prior to joining Digital Equipment Corporation, Hudson, MA, in 1982, he was with Integrated Circuit Systems, Incorporated, Westboro, MA.

Robert A. J. Yodlowski received the B.S. degree in engineering physics from Cornell University, Ithaca, NY, in 1968, and the M.S. degree in electrical engineering from Syracuse University, Syracuse, NY, in 1970.

He has been employed with the Digital Equipment Corporation, Hudson, MA, since 1977. He is currently a Principal Engineer with the Semiconductor Engineering Group in Hudson, MA. His interests include MOS circuit and logic design.

64 Bit Monolithic Floating Point Processors

FREDERICK A. WARE, MEMBER, IEEE, WILLIAM H. McALLISTER, JOHN R. CARLSON, DAN K. SUN, AND RICHARD J. VLACH

Reprinted from *IEEE Journal of Solid-State Circuits*, Volume SC-17, Number 5, October 1982, pages 898-907.

Abstract—This paper describes a set of three processor chips capable of performing 32 and 64 bit floating point add/subtract, multiply, and divide operations. The chips can perform over one million scalar floating point operations per second, and over four million vector operations per second. The set is implemented in a four micron CMOS-on-sapphire process. Each chip has between 30 000 and 60 000 devices, and is about 250 mils on a side. Although asynchronous data paths are used within the chips, their interface to external system buses is synchronous with a maximum data bandwidth of over 70 Mbytes/s. The set has been designed for use in Hewlett-Packard computer and instrument systems.

I. Introduction

HEWLETT-PACKARD has developed a set of three floating point processor chips (an adder/subtracter, a multiplier, and a divider) that are to be utilized by a variety of computer and instrument systems within the company. This paper provides a description of the design and operation of the individual chips.

The principle design objective was maximum performance of 32 and 64 bit floating point scalar (single element) operations. This was accomplished primarily with the three chip partitioning which allows each fundamentally different operation to be optimized with its own data path. Also, combinational circuitry is provided for minimum execution delay of each operation.

A second objective was to maximize the ease of use of the chip set for the system designer. Static logic is used throughout the design, and all internal registers are edge-triggered with a single clock. All inputs and outputs are TTL compatible, and outputs are tristated. Power consumption is low because the chip set is implemented with static CMOS circuitry.

The third design objective was maximum vector performance. The add/subtract and multiply chips each contain a pipeline storage register that allows the combinational logic delay to be overlapped with loading and unloading operations. Also, it is possible to parallel several chips of one type and interleave their operation, thereby improving the vector (multiple element) bandwidth.

II. Chip Operation

As shown in the block diagrams of Figs. 3(a), 8(a), and 9(a), the chips have similar interface and control circuitry, but the data path logic is unique to each. Each chip has three 16 bit unidirectional data buses (two input and one output) that have maximum bandwidths of over 12 MHz. The 32 and 64 bit operands and results must be time-multiplexed for loading and unloading. The floating and fixed point data formats used are those of the HP1000 minicomputer.

Operands and the function opcode are loaded into a chip on two or four clock edges. The data propagate through combinational data path logic, and may be directly unloaded in another two or four clock cycle. Results may alternately be passed through a pipeline register allowing two operations to be overlapped. Unloading this register requires two or four cycles as before.

The divide chip contains a single clocked feedback path within its data path logic, and thus is not purely combinational. Its internal delay is about two to four times the 400 ns-600 ns delay required for add/subtract and multiply. Unlike the add/subtract and multiply chips, it has no pipeline register since the internal delay is far greater than load/unload delays.

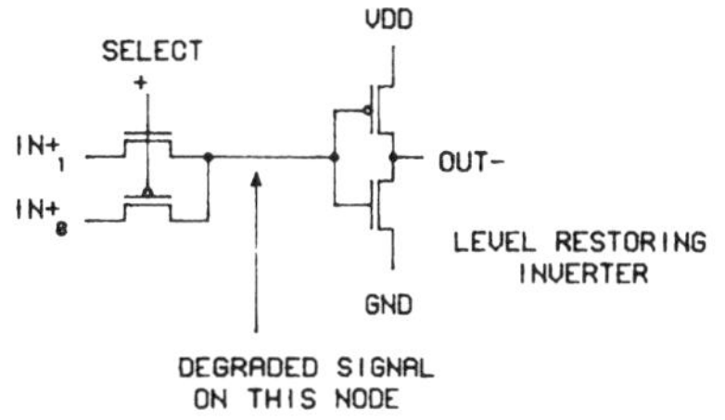

Fig. 1. Two-to-one multiplexer implemented with single-sided pass devices (p devices have bubbles, n devices do not). Half the time the devices pass a degraded signal which is restored by an inverter. Disadvantages of this technique are loss of noise margin and performance and an increase in dc power dissipation in the inverter. The principle advantage of this technique is its high functional density.

III. Circuit Design Techniques

The CMOS-on-sapphire process used for the chip set has four micron feature sizes and metal gate devices. The supply voltage is relatively large (12 V) compared to the threshold voltages (1.5-2.0 V). Also, there is less performance degradation due to the body effect with SOS devices than with bulk CMOS devices. This allows circuit designers limited use of devices operating as source-follower drivers, i.e., p devices passing a low level or n devices passing a high level.

A two-to-one multiplexer can be implemented with two single-sided pass devices, and with a level restoring inverter as in Fig. 1. This requires about half the area of an implementation utilizing double-sided pass devices [Fig. 1(b)]. The performance of the single-sided circuit is less than that of the other implementation, but it is still acceptably fast, particu-

Manuscript received February 23, 1982; revised May 5, 1982.
The authors are with the Computer Systems Division, Hewlett-Packard, Inc., Palo Alto, CA 94304.

$$C_{i+1} = P_i C_i + \overline{P_i} B_i$$

$$C_{i+2} = P_{i+1} P_i C_i + P_{i+1} \overline{P_i} B_i + \overline{P_{i+1}} B_{i+1}$$

$$C_{i+3} = P_{i+2} P_{i+1} P_i C_i + P_{i+2} P_{i+1} \overline{P_i} B_i + P_{i+2} \overline{P_{i+1}} B_{i+1} + \overline{P_{i+2}} B_{i+1}$$

$$G_{i+4} = P_{i+3} P_{i+2} P_{i+1} P_i C_i + P_{i+3} P_{i+2} P_{i+1} \overline{P_i} B_i + P_{i+3} P_{i+2} \overline{P_{i+1}} B_{i+1} + P_{i+3} \overline{P_{i+2}} B_{i+2} + \overline{P_{i+3}} B_{i+3}$$

$$PP_i = P_{i+3} P_{i+2} P_{i+1} P_i$$

$$C_{i+4} = PP_i C_i + \overline{PP}_i G_{i+4}$$

(a)

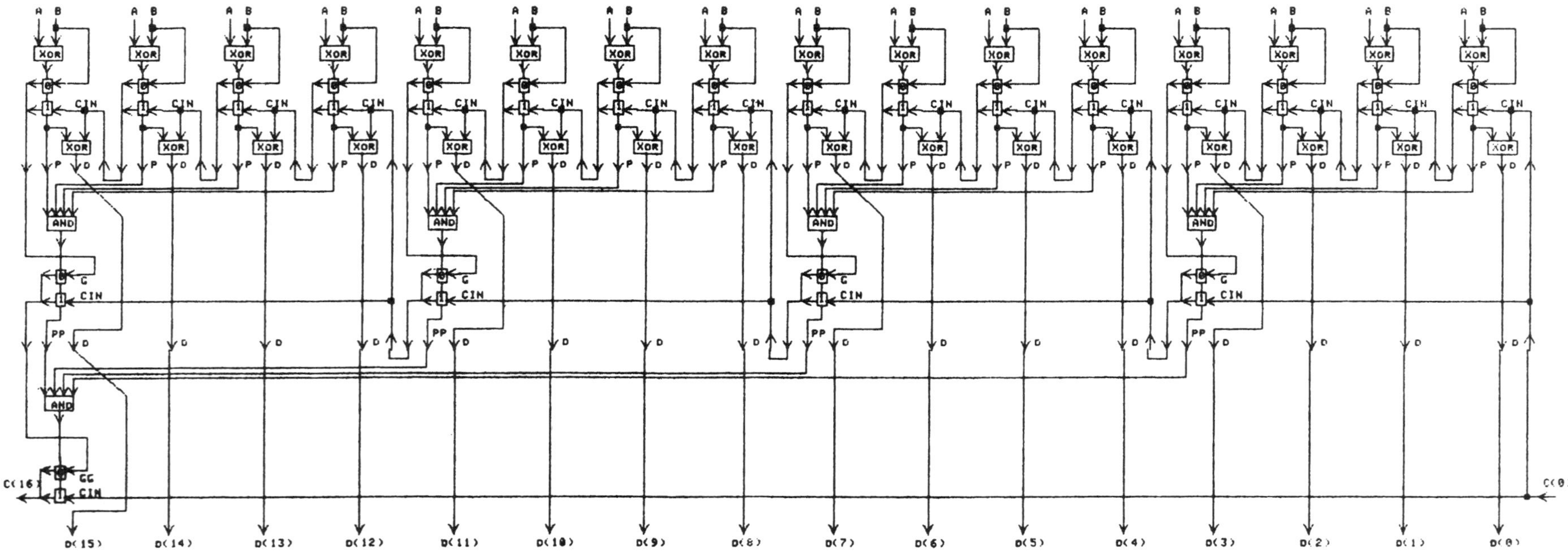

(b)

Fig. 2. (a) Boolean equations for a four bit adder slice with indexes i through $i + 3$. The Ai, Bi, and Pi terms are, as before, for each of the four bits. The $Ci + 3$, $Ci + 2$, and $Ci + 1$ terms are expressed as functions of any of the Bi and Pi and of only the first carry signal Ci. The carry-out of the four bit slice $Ci + 4$ is then generated from two intermediate terms $Ci + 4$ and PPi. The equations may be efficiently implemented with two-to-one multiplexers and Exclusive-OR gates as in Fig. 2(b). (b) This is a 16 bit slice of a 64 bit carry propagate adder with three levels of look-ahead logic. Each four bit slice is equivalent to the equations of Fig. 2(a). Ai and Bi are the bits of two 64 bit numbers to be added. Pi is the Exclusive-OR of Ai and Bi, and is used to control a two-to-one multiplexer to form the carry out $Ci + 1$. $Ci + 1$ is either given the value of Ci (PASS) or Bi (KILL/GENERATE) if Pi is one or zero, respectively. Pi is also Exclusive-ORED with Ci to form the sum bit Di. The ripple carry chain is broken at four and sixteen bit intervals. If all four or sixteen bits have a propagate signal Pi equal to one, the carry-in to that slice may skip over it. The above logic is implemented with twenty transistors/bit–sixteen for the full adder and four for the look-ahead circuity.

larly when compared to equivalent implementations with bulk rather than SOS devices. The density advantages of the single-sided circuit are far more important for the processors than the additional 20 percent or 30 percent overall propagation delay that must be tolerated.

IV. Carry Propagate Adder

The most common digital function required for the chip set was that of fixed point addition with carry propagation. This operation may be implemented in a variety of ways depending upon the exact area-performance tradeoff required. The 64 bit precision of the chip set meant that silicon area for carry propagation should be minimized as far as was practical.

An efficient carry look-ahead addition scheme which requires only 20 transistors/bit is shown in Fig. 2(b) along with the Boolean equations in Fig. 2(a). It has an equivalent delay of 15 to 20 full adder delays, giving three to four times the performance of a ripple adder for about 25 percent additional silicon area (a ripple adder requires 16 transistors/bit).

It is possible to design 64 bit adders with up to five or six times the ripple adder performance by utilizing conditional sum techniques [1]. However, the device count per bit is two to three times that of the ripple adder. Such an approach was not judged to be a good area-performance tradeoff.

V. Floating Point Add/Subtract Chip

A block diagram and microphotograph of the floating point add/subtract chip are shown in Fig. 3(a) and (b). There are about 30 000 devices on a die that is 230 × 255 mils in size. Most of the chip area is devoted to the fraction data path in the center and right. The major hardware structures in this path are operand registers, right shifter, adder and priority encoder, left shifter, and incrementer and output register. The shifters make the distinctive slanting patterns.

The exponent data path is in the upper left and control logic in the lower left. Operand input pads are along the top and right sides, result pads are along the bottom, and control pads are along the left side.

When the two operands are first loaded, their exponents are compared and the larger one passed on. The fraction with the smaller exponent is right shifted by N bits, where N is the difference of the two exponents. The right shift logic consists of six two-to-one multiplexers/bit, and the wiring to perform power-of-two shifts.

Next, the addition or subtraction of fractions is performed exactly as with fixed point operands, except that at least three extra bits of precision must be included to provide accuracy of one half of the LSB (least significant bit). When the two fractions have the same sign, a fraction overflow may occur. If this happens, the fraction is right shifted once to put it back into the range, and the exponent is incremented.

If the operand fractions have different signs, there may be cancellation of leading significant bits, yielding an unnormalized result. Normalization is performed with a priority encoder and an N bit left shifter. The priority encode logic uses a look-ahead circuit (similar to that of the carry propagate adder) to detect the position of the first significant bit from the left. This position is encoded into a six bit value that is subtracted from the exponent, and is also used as the shift amount by the left shifter.

The extra bits of precision that were carried are now used to perform rounding of the normalized result fraction. An incrementer (with a look ahead unit identical to that of the carry propagate adder) conditionally adds one LSB to the fraction. If the fraction overflows, the exponent is incremented and the fraction right shifted one place.

The final step is checking the exponent for overflow or underflow. If either occurs the appropriate constants are generated for the exponent and fraction of the result, and the status bits are set.

VI. Multiplication Techniques

Fig. 4 illustrates the application of signed digit encoding to an 8 × 8 two's complement multiplier. The technique allows the number of partial products to be reduced by a factor of two, and requires only a small amount of shifting and multiplexing logic.

The Boolean truth table for a commonly used encoding scheme (modified Booth encoding) is shown in Fig. 5. The original Booth encoding technique [2] is able to skip over arbitrarily long strings of ones or strings of zeros; the modified technique more rigidly examines only pairs of multiplier bits and converts them into the set of five signed digits [3]-[5].

An alternate encoding scheme not previously reported is possible. It uses the reduced signed digit set +2, +1, 0, -1 by modifying line five of the truth table to give a multiplier of *XP*2 and a signed-digit-carry-out of zero. The disadvantage of this is that the signed-digit-carry-out is now a function of the signed-digit-carry-in, unlike the original scheme. It might be thought that this would impact the multiplier performance since the signed-digit-carries must ripple down the multiplier bit pairs. However, since the partial products must ripple down rows of carry save adders in the array as they accumulate, there is in fact no performance penalty. The advantage of the technique is that no *XM2(J)* version of the multiplicand needs to be generated in the multiply array, allowing an efficient four-to-one multiplexer to perform the encoding operation in each array cell.

A complete 8 × 8 recoded multiplier is shown in Fig. 6(a) along with a summary of Boolean equations for its logic blocks in Fig. 6(b) [the equations work for either encoding technique]. The 8 bit operands are A(7-0) and B(7-0). The inputs *TCA* and *TCB* indicate whether the A and B operands are two's complement (=1) or unsigned (=0). The 16 bit result of the multiplication is D(15-0), and the *F*ADD and *H*ADD blocks are conventional full adders and half adders.

The schematic for the critically important array cell for the multiplier is shown in Fig. 7(a). A microphotograph of the array is in Fig. 7(b). The metal interconnect layer runs horizontally and is used for power, ground, and the encoded B operand lines. The silicon island layer runs vertically and is used for the A operand lines. There is significant series resistance in these vertical silicon lines, but because they are

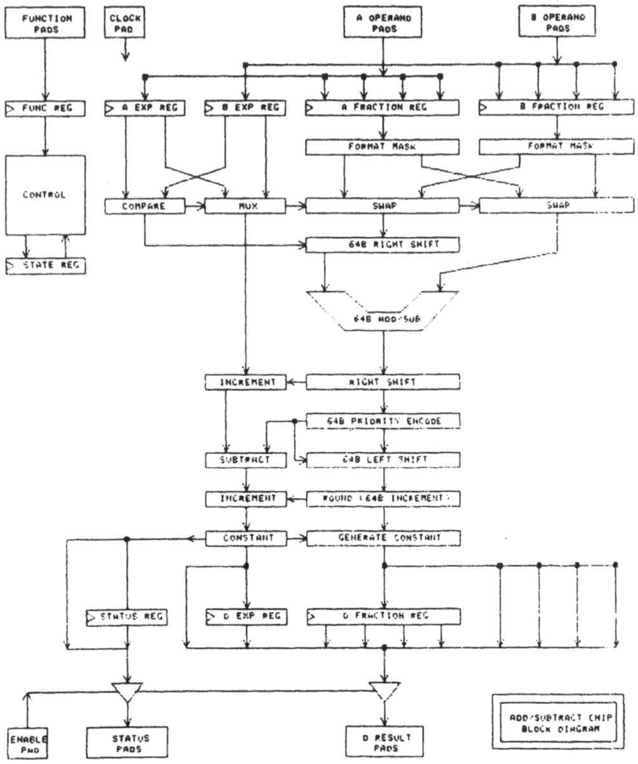

(a)

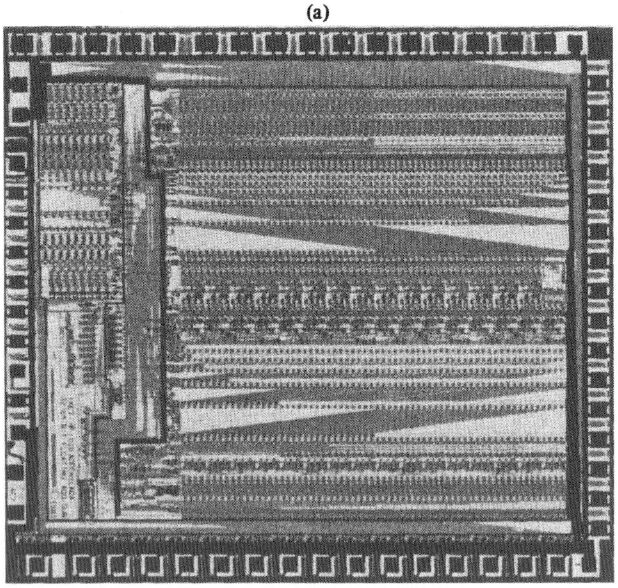

(b)

Fig. 3. (a) Floating point add/subtract chip block diagram. (b) Floating point add/subtract chip microphotograph.

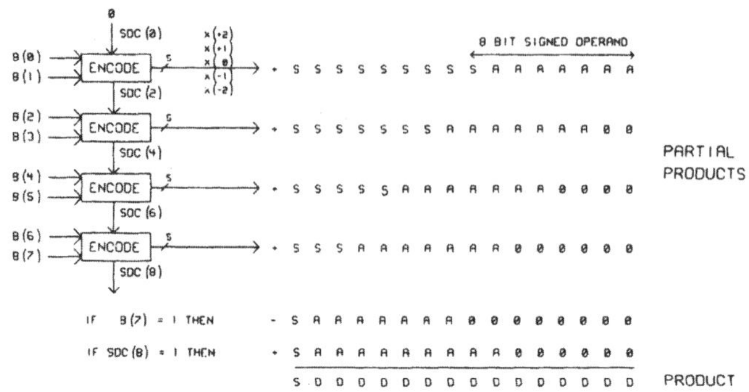

Fig. 4. Two's complement 8 × 8 multiplication with encoding. Pairs of multiplier bits are encoded into the signed digit set +2, +1, 0, −1, −2. A signed-digit-carry is passed between pairs of B multiplier bits to allow the bit pair "11" to be represented (it is represented by a −1 signed digit and a carry-out of one). The Boolean table for the encoding logic is given in Fig. 5. The signed digit from the encoding process multiplies the left-shifted, sign-extended A operand and accumulates it. If the B operand is negative ($B(7) = 1$), then the indicated correction term must be subtracted. If $SDC(8) = 1$, then the same term must be added. These two corrections are performed by a single addition, a single subtraction, or by no action.

B(2J+1)	B(2J)	SDC(2J)	MULTIPLIER	SDC(2J+2)
0	0	0	X0	0
0	0	1	XP1	0
0	1	0	XP1	0
0	1	1	XP2	0
1	0	0	XM2	1
1	0	1	XM1	1
1	1	0	XM1	1
1	1	1	X0	1

Fig. 5. Boolean table for the two bit encoding scheme utilizing the full set of signed digits: +2, +1, 0, −1, −2. The signed-digit multiplier ($XP2$, $XP1$, $X0$, $XM1$, $XM2$) and signed-digit-carry-out are functions of the signed-digit-carry-in and the two B operand bits. The two operand bits have relative weights of $x2$ and $x1$, and the carry-in has a weight of $x1$. The weight of the signed digits +2, +1, 0, −1, −2 is $x1$, and the weight of the carry-out is $x4$. In each of the eight lines of the table, the sum of the three weighted input terms $B(2J+1)$, $B(2J)$, and $SDC(2J)$ equals the sum of the two weighted outputs. The index J runs from 0 through 3 for an 8 × 8 multiplier.

driven from the top of the chip and because the rows of full adder cells ripple from top to bottom, the full RC time penalty is not paid.

Note that a four-to-one multiplexer is used to select the $XP2$, $XP1$, $X0$, or $XM1$ version of the A operand. Because single sided transfer gates are used (along with a level restoring inverter), only six devices and two metal control lines are required. This compact circuit could not be used if the conventional encoding scheme with five signed digits were utilized.

VII. Floating Point Multiply Chip

Floating point multiplication is relatively simple compared to floating point add and subtract. Referring to Fig. 8(a) and (b), it is seen that most of the silicon is used to perform integer multiplication of the fraction parts (center and right). The circuitry at the lower left of the chip contains the exponent data path, which occupies a very small portion of the chip compared to the exponent path of the floating point adder. The multiplier is roughly 245 × 280 mils, and contains about 60 000 devices.

When the operands are first loaded, the two exponents are added, and the fractions are masked to the proper precision. One operand is encoded as previously described, and conventional integer multiplication is then performed. The integer product (in carry propagate form) is normalized with a zero or one bit left shift, and then rounded to the final precision. The result exponent is incremented if there was a zero bit shift or if the fraction overflowed after rounding. The exponent is checked for exponent overflow or underflow and if either oc-

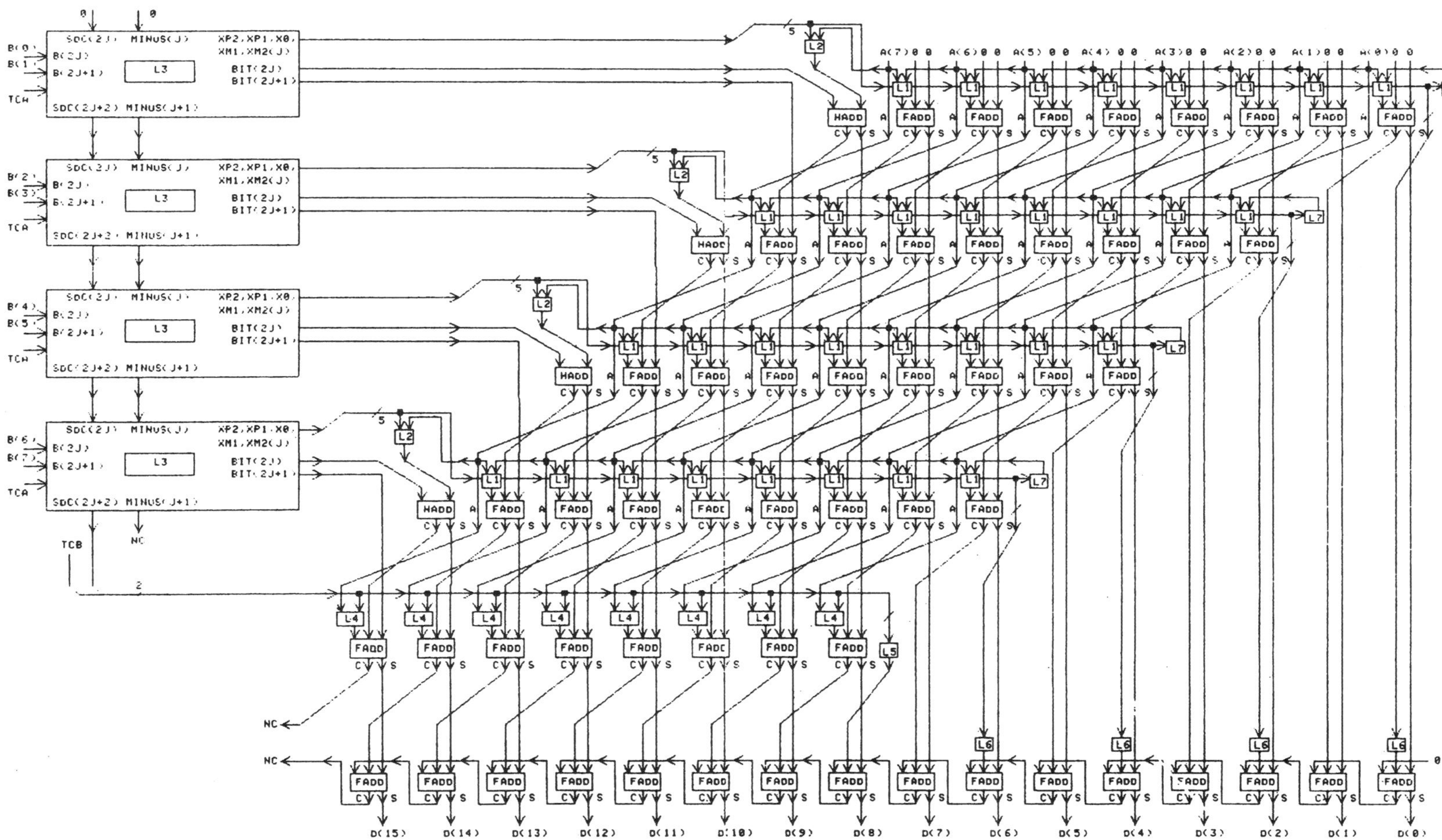

(a)

Fig. 6. (a) Complete 8 × 8 encoded multiplier. The J index runs from 0 through 3 and is used for the (vertical) multiplier B(7-0). The K index runs from 0 through 7 and is used for the (horizontal) multiplicand A(7-0). The inputs *TCA* and *TCB* indicate whether each operand is two's complemeht (=1) or unsigned (=0). The output result is D(15-0). Logic block $L3$ accepts pairs of multiplier bits $B(2J)$ and $B(2J+1)$ and a signed-digit-carry-in $SCD(2J)$, and produces signed-digit-carry-out SCD $(2J+2)$ and asserts one of the five signed-digit multiplier lines $XP2$, $XP1$, $X0$, $XM1$, or $XM2$. This logic block also uses the minus(J) signal to generate two signals bit($2J$) and bit($2J+1$) that perform in effective sign extension of each partial product to 16 bits. Logic blocks $L1, L2, L6$, and $L7$ perform the shifting and multiplexing operations that multiply the multiplicand A(7-0) by one of the signed digits +2, +1, 0, −1, −2. The FADD and HADD blocks are conventional full adders and half adders. Logic blocks $L4$ and $L5$ perform sign correction for the B operand, and correct for a signed-digit-carry-out of the last $L3$ block [$SDC(8)$]. The final carry propagation of the product D(15-0) is shown using ripple adders, although a real circuit would use a look-ahead technique for the upper 8 bits.

```
L1 LOGIC
    L1(K,J)      = [XP2(J) * A(K-1)] + [XP1(J) * A(K)]
                     + [XM1(J) * Ā(K)] + [XM2(J) * Ā(K-1)]
L2 LOGIC
    L2(2J)       = [XP2(J) * A(7)] + [XM2(J) * Ā(7)]
L3 LOGIC
    NEGA         = TCA * A(7)
    P            = [XM2(J) * NEGA‾] + [XP2(J) * NEGA]
    Q            = [XM1(J) * NEGA‾] + [XP1(J) * NEGA]
    BIT(2J+1)    = [MINUS(J) XOR P] + [MINUS‾(J) * Q]
    BIT(2J)      = MINUS(J) XOR Q
    MINUS(J+1)   = MINUS(J) + P + Q
L4,L5,L6,L7 LOGIC
    NEGB         = TCB * B 7
    L4(K)        = [NEGB XOR SDC(8)] * [NEGB XOR A(K)]
    L5           = NEGB * SDC(8)‾
    L6(J)        = XM1(J) + XM2(J)
    L7(J)        = A(-1) = 0
```

(b)

Fig. 6. *(Continued.)* (b) Summary of Boolean equations for the logic blocks of the 8 × 8 multiplier of Fig. 6(a). The index J runs from 0 through 3, and the index K runs from 0 through 7. The variables $NEGA$, $NEGP$, P, and Q are used within a single logic block, and therefore do not appear in Fig. 6(a).

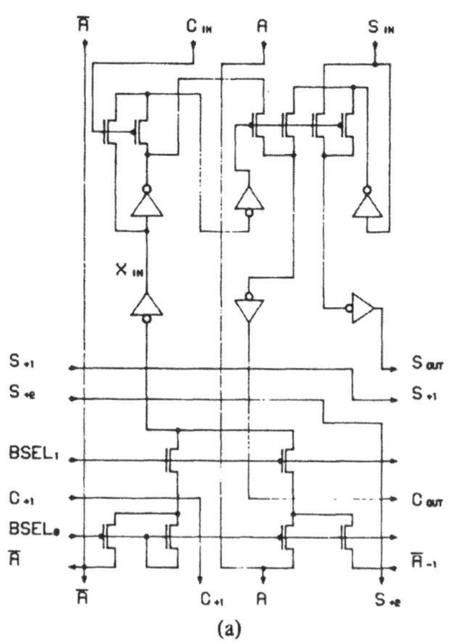

(a)

(b)

Fig. 7. (a) Schematic of 2 × 1 multiply cell. This shows the 24 NMOS and PMOS devices used to implement the $L1$ logic block and FADD block of Fig. 6(b) (p devices have a bubble on the gate). Two select lines are encoded from each pair of multiplier bits, and indicate whether +2, +1, 0, or −1 is to multiply the operand A to form a partial product. The true and complement of the appropriate A bit pass vertically through the cell and are used by the four-to-one multiplexer at the bottom. A +2 multiplier implies using $A(-1)$, +1 implies $A(0)$, 0 implies zero, and −1 implies complemented $A(0)$. The multiplexer output X_{in} is accumulated in the full adder with C_{in} and S_{in} to produce S_{out} and C_{out}. The S_{out} (sum-out) line is passed two cells to the right before being used by the next row, and the C_{out} (carry-out) line is passed one cell. (b) Microphotograph of 2 × 1 multiply cell. The schematic of Fig. 7(a) is quite similar topologically to this photo. Metal lines are dark and run horizontally, and island silicon is lighter and runs vertically. The cell is outlined by a heavy black line. Power and ground run through the center of the cell in metal, powering the six inverters. Three two-to-one multiplexers are above the inverters, and one four-to-one multiplexer is below. The cell is 88 × 144 μm (11 island pitches by 16 metal pitches).

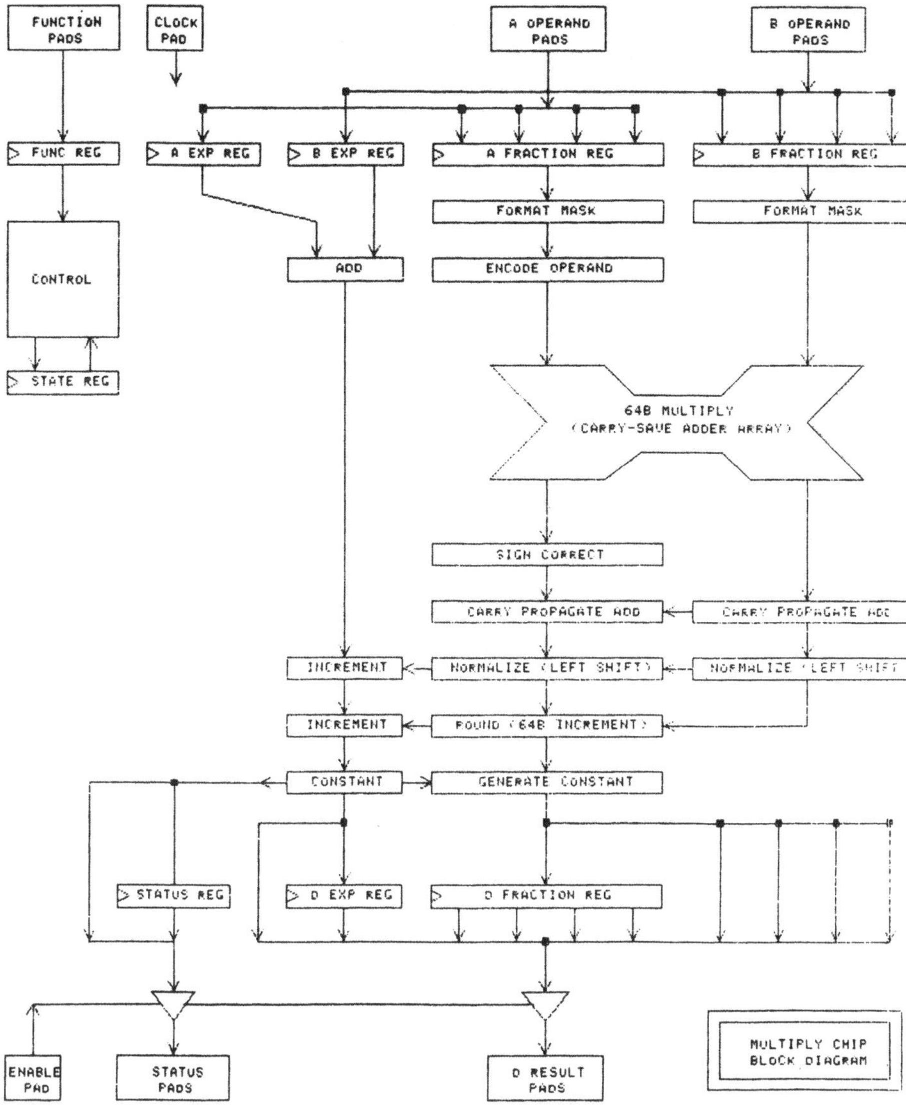

(a)

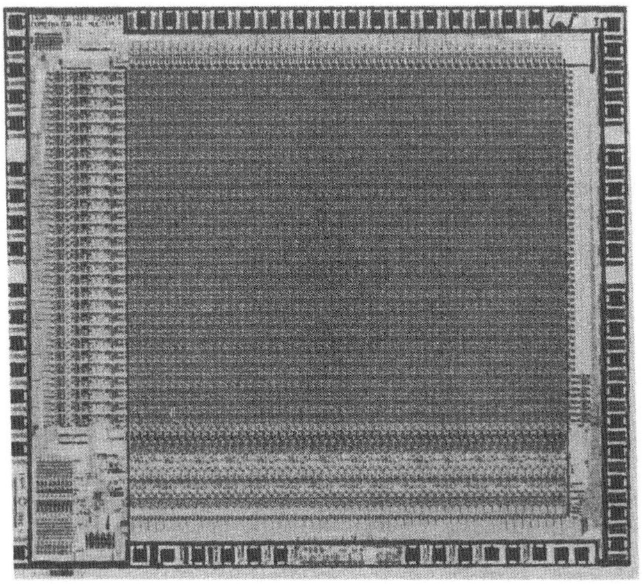

(b)

Fig. 8(a) Floating point multiply chip block diagram. (b) Floating point multiply chip microphotograph.

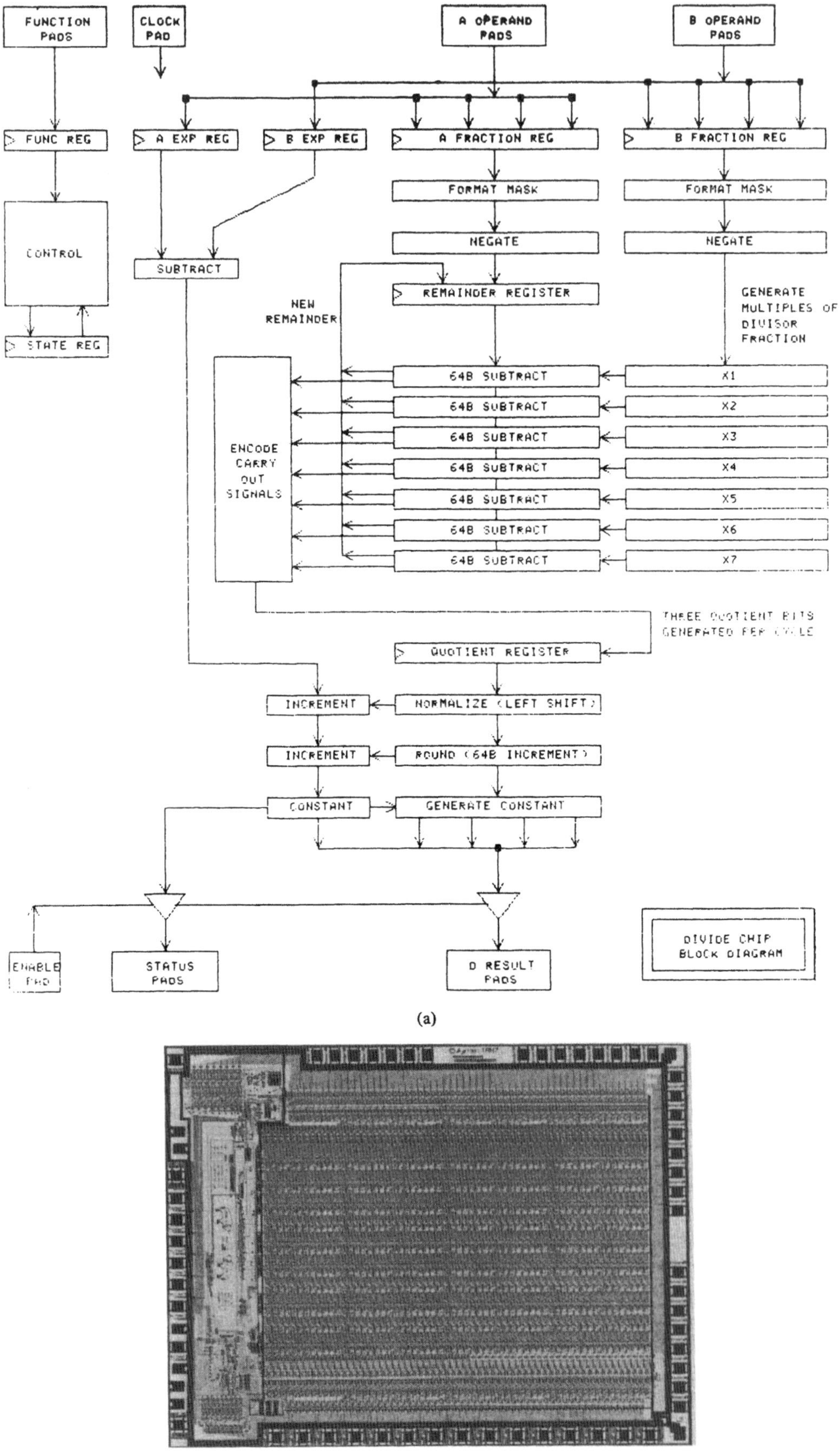

(a)

(b)

Fig. 9. (a) Floating point divide chip block diagram. (b) Floating point divide chip microphotograph.

curs the appropriate constants are forced and the status outputs set.

VIII. Floating Point Divide Chip

Floating point division, like multiplication, is not much harder than its fixed point counterpart. In Fig. 9(a) and (b) it is seen that the divide chip has a small exponent path in the upper and lower left. Most of the silicon is devoted to the fraction data path in the center and right of the chip. The chip is about 210 × 290 mils in size, and contains about 35 000 devices.

When the operands are first loaded, the two exponents are subtracted and the fractions masked to the proper precision. Conditional two's complement negation is performed so the dividend fraction is positive and the divisor fraction negative. The first seven integral multiples of the divisor are generated combinationally with three carry propagate adders and four left shifters. An extra load cycle is required at this point to transfer the dividend fraction into the remainder register.

The following sequence of events is repeated anywhere from nine to twenty times depending upon the fraction precision. The remainder fraction is passed to an array of seven carry propagate adders which add it to the seven negative divisor multiples to produce seven new remainders. The smallest positive remainder is passed back to the remainder register, and its encoded three bit position is shifted into the quotient register.

This technique is inherently faster than the method of using a full combinational array of 64 64-bit carry propagate adders [6]-[8]. The first technique generates three bits in the time to perform one carry propagation, to drive horizontal enable lines, and to drive the remainder into and out of the array. The combinational technique generates one bit in the time to perform one carry propagation and drive horizontal enable lines. Technique one is thus about two to three times faster.

When the quotient is completely loaded, it is normalized with a zero or one bit left shift, and then rounded to its final precision. The result exponent is incremented if there was a zero bit shift or if the fraction overflowed after rounding. The exponent is checked for exponent overflow or underflow and if either occurs the appropriate constants are forced and the status outputs set.

IX. Summary

This chip set will provide a powerful (up to one million floating point operations per second), yet inexpensive floating point execution unit for Hewlett-Packard minicomputer systems. In addition, multiple chip sets may be interleaved to implement vector processing units (four MFLOPS per chip set) in technically oriented minicomputers [9], [10].

The chips may also be used as building blocks to construct processing systems for specific applications. Such applications include signal and image processing, graphics processing, and the numerical solution of linear systems. Virtually any structured algorithm can benefit from the chip set in the implementation of high performance floating point systems.

Acknowledgment

The authors wish to acknowledge the efforts of T. Shillingburg, J. Pelayo, H. Shishido, V. Wilson, L. Andres, T. Corsick, S. Boschma, B. Ames, and D. Zuras.

References

[1] J. Sklansky, "Conditional sum addition logic," *IRE Trans. Electron. Comput.*, vol. EC-9, pp. 226-231, June 1960.
[2] A. D. Booth, "A signed binary multiplication technique," *Quart. J. Mech. Appl. Math.*, vol. 4, pp. 236-240, 1951.
[3] S. Waser, "High speed monolithic multipliers," *Computer*, pp. 19-29, Oct. 1978.
[4] C. S. Wallace, "A suggestion for a fast multiplier," *IEEE Trans. Electron. Comput.*, pp. 14-17, Feb. 1964.
[5] A. Habibi and P. A. Wintz, "Fast multipliers," *IEEE Trans. Comput.*, pp. 153-157, Feb. 1970.
[6] D. P. Agrawal, "High speed arithmetic arrays," *IEEE Trans. Comput.*, vol. C-28, pp. 215-224, Mar. 1979.
[7] I. Flores, *The Logic of Computer Arithmetic*. Englewood Cliffs, NJ: Prentice-Hall, 1963.
[8] K. Hwang, *Computer Arithmetic*. New York: Wiley, 1979.
[9] F. Ware, "64b monolithic floating point processors," in *ISSCC, Conf. Dig.*, vol. 25, Feb. 1982, pp. 24-25.
[10] F. Ware and W. McAllister, "CMOS floating point chip set," *Electron.*, vol. 55, no. 3, pp. 149-153, Feb. 1982.

Frederick A. Ware (M'77) was born in Omaha, NE, on April 14, 1952. He received the B.S. degree in applied physics from the California Institute of Technology, Pasadena, in 1974, and the M.S. degree in electrical engineering from the University of California, Berkeley, in 1975.

He joined Hewlett-Packard's Data Systems Division in 1975, where he worked on custom integrated circuits for the HP300 central processing unit. In 1978 he began the development of CMOS/SOS floating point processor circuits. He is currently managing the design of a second set of floating point processors that will be compatible with the new IEEE standard.

William H. McAllister received the B.S.E.E. degree from the University of California, Santa Barbara, in 1974, and the M.S.E.E. from Stanford University, Stanford, CA, in 1976.

After graduating he joined TRW, Inc., in Redondo Beach, CA, where he worked on the design and analysis of LSI for special-purpose signal processors. In 1980 he joined Hewlett-Packard, where he is currently working on custom VLSI chip design for arithmetic processors. His special interest is in the area of VLSI design methodologies.

John R. Carlson received the B.S. degree in mathematics from California State University, Hayward, in 1973, the M.S. in computer science from the University of California, Davis, in 1975, and he is currently studying for the M.S. in electrical engineering at Stanford University, Stanford, CA.

He joined Hewlett-Packard in 1975, working on microcode. He has worked on the development of a microcode compiler and is currently working on NMOS floating point chips.

Dan K. Sun was born in China. He received the B.S. degree in computer science from the University of San Francisco, San Francisco, CA, in 1969.

Since joining Hewlett-Packard in 1977, he has been participating in SOS-CMOS process VLSI device testing.

Richard J. Vlach, a native of Ohio and graduate of CEI, moved to California in 1962. He worked for Lockheed Missile and Space Corporation helping in the development of a thin film process. He then worked in the areas of bipolar mask design, industrial engineering, and mask development at Stewart Warner Microcircuits. He later moved to National Semiconductor to design masks in bipolar, linear, and CMOS technologies, also making contributions in automated assembly and teaching mask design. He joined Hewlett-Packard in 1980 to work on the floating point processor family.

A CMOS Floating Point Multiplier

MASARU UYA, KATSUYUKI KANEKO, AND JURO YASUI

Reprinted from *IEEE Journal of Solid-State Circuits*, Volume SC-19, Number 5, October 1984, pages 697-702.

***Abstract*—This paper describes a 32-bit CMOS floating point multiplier. The chip can perform 32-bit floating point multiplication (based on the proposed IEEE Standard format) and 24-bit fixed point multiplication (two's complement format) in less than 78.7 and 71.1 ns, respectively, and the typical power dissipation is 195 mW at 10 million operations per second.**

High-speed multiplication techniques—a modified Booth's algorithm, a carry save adder scheme, a high-speed CMOS full adder, and a modified carry select adder—are used to achieve the above high performance.

The chip is designed for compatibility with 16-bit microcomputer systems, and is fabricated in 2 μm n-well CMOS technology; it contains about 23000 transistors of 5.75×5.67 mm² in size.

I. Introduction

RECENTLY, the need for high-speed and high-precision computation has been increasing in applications for image processing, computer graphics, robotics, model simulation, CAD/CAM, measurement, and so on. Floating point computation is most suitable for these applications because it keeps the precision of operation high in spite of the wide-dynamic range. For microcomputer systems, not only high-speed but also small-sized floating point devices are in great demand.

The authors have developed a high-speed, low-power 32-bit floating point multiplier employing 2 μm design rule n-well CMOS technology. The data format is based on the proposed IEEE P754 Standard, which is suitable for microcomputer systems.

First, the configuration of the chip is described. Then, several techniques employed to achieve high-speed operation are discussed. The last part of the paper describes the performance and characteristics as well as considerations relating to chip operation.

II. Chip Configuration

According to the single precision format of the proposed IEEE P754 Standard, the value of a multiplier X (normalized number) is described as

$$X=(-1)^{Sx}\cdot 2^{Ex-127}\cdot(1.Fx)$$

where Sx, Ex, and Fx are the 1-bit sign, the 8-bit exponent biased by 127, and the 23-bit fraction, respectively $(1.Fx)$ is the 24-bit mantissa consisting of an implied leading "1" and Fx. So, the range of the mantissa is $1 \leqslant (1.Fx) < 2$.

Manuscript received March 29, 1984; revised May 22, 1984.
The authors are with the Matsushita Electric Industrial Co., Ltd., Central Research Laboratory, Osaka 570, Japan.

Also, the value of a multiplicand Y has the expression

$$Y=(-1)^{Sy}\cdot 2^{Ey-127}\cdot(1.Fy).$$

Then, the product $P = X\cdot Y$ comes to

$$P=(-1)^{Sx\oplus Sy}\cdot 2^{(Ex+Ey-127+a)-127}\cdot[(1.Fx)\cdot(1.Fy)]\cdot 2^{-a}$$

where "a" results from

$$a=0: \quad 0 \leqslant (1.Fx)\cdot(1.Fy) < 2$$
$$a=1: \quad 2 \leqslant (1.Fx)\cdot(1.Fy) < 4.$$

The chip is designed to fit 16-bit microcomputer systems. The 32-bit multiplier/multiplicand and product are loaded and taken out by 16 bits on two clocks each. An error signal is output for use as an interrupt to the host microcomputer when a multiplication finishes abnormally, and then overflow/underflow flags can be read if necessary. Aside from 32-bit floating point multiplication, fixed point multiplication can be provided in 24-bit two's complement data format.

A block diagram of the floating point multiplier is shown in Fig. 1. The chip consists of input/mode registers, an exponent adder, a mantissa multiplier, and output/status registers. Multiplier X, multiplicand Y, and mode control signals (FIX, RND) are clocked into 16-bit input registers (XH, XL), (YH, YL), and a mode register (FIX, RND), respectively, through a bidirectional 16-bit bus ($D15$-$D0$). The FIX is operation mode control ($FIX=0$: 32-bit floating point operation, $FIX=1$: 24-bit fixed point operation). The RND is the rounding mode control [$RND=0$: round toward zero, $RND=1$: round to the nearest value (add $\frac{1}{2}$ LSB to the result)]. Selection signals ($X/\overline{Y}$, $H/\overline{L}$, $S/\overline{D}$) and a clock CLK are used for time multiplexing the bus. $\overline{OE}$ is the output enable signal for the bus. ERR is an error signal indicating the occurrence of overflow or underflow of multiplication. Exponent adder block executes the operation of the sign and exponent in floating point multiplication mode. Besides the 8-bit exponent addition, this block detects invalid input data and predetects overflow/underflow. The final overflow/underflow is determined in the overflow/underflow detector ($OVF/UDF\ DET$).

Mantissa multiplier is a 24×24-bit asynchronous parallel multiplier which can perform both signed 2's complement and unsigned magnitude multiplication with/without rounding. The 24-bit result taken out of it is sent to the normalizer, where the above result (1-bit sign, 8-bit expo-

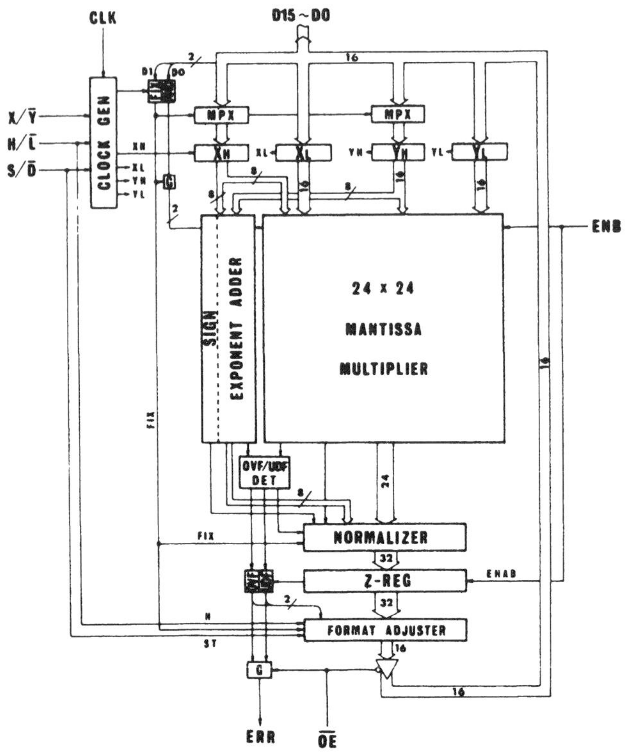

Fig. 1. Block diagram of the floating point multiplier.

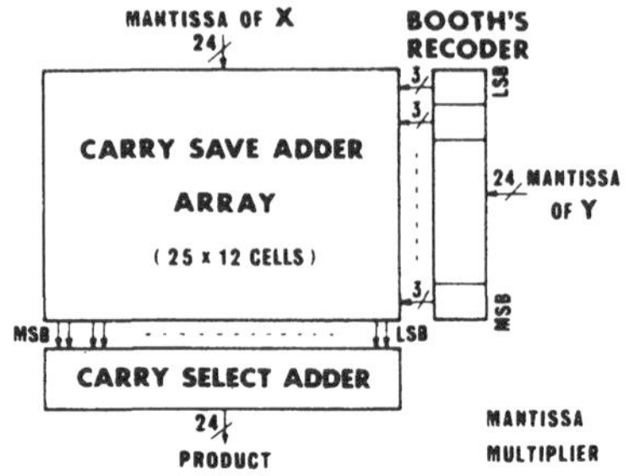

Fig. 2. Block diagram of the mantissa multiplier.

nent, and 24-bit mantissa) is formed into IEEE standard format normalized numbers and are sent to the output register Z. The normalizer forces the result to the true zero ($Ez = Fz = 0$) when underflow occurs. The 32-bit result and the status are put on the bus 16 bits at a time.

ENB is the enable signal for the multiplier and is provided to measure the multiply time. The execution of multiplication is triggered by the rising edge of the *ENB*, and the output/status registers are latched by the falling edge of the *ENB*. Thus, the pulse width of the *ENB* is the multiply time.

As mentioned above, this chip is designed to configure easily with 16-bit microcomputer systems.

III. Exponent Adder Design

As the result of mantissa multiplication is in the range $1 \leqslant (1.Fz) < 4$, the exponent Ez must be incremented by 1 when $2 \leqslant (1.Fz) < 4$. This normalization (increment or not) of the exponent cannot be done until mantissa multiplication is completed. That is, this operation is on the critical path of the floating point multiplication.

High-speed operation in the exponent adder part is achieved by selecting one out of two previously prepared values: $Ex + Ey - 127$ and $Ex + Ey - 127 + 1$. The exponent adder consists of an 8-bit adder performing the addition of Ex, Ey, and -127, an incrementer incrementing the output of the 8-bit adder, and an 8-bit 2-to-1 data selector. The output of the incrementer is the exponent of product Z in the case of a product where the mantissa multiplier is equal or greater than 2 (normalizing is required). As soon as the multiplication of the mantissa multiplier is completed, the data selector selects out a true exponent Ez between the output of the 8-bit adder and that of the incrementer. Thus, the chip can perform the normalization in only 5 gate delays.

IV. Mantissa Multiplier Design

A block diagram of the mantissa multiplier is shown in Fig. 2. The mantissa multiplier performs the 24×24-bit multiplication of two two's complement binary numbers (in fixed point mode) or two unsigned magnitude binary numbers consisting of 23-bit fraction and implicit leading "1" (in floating point mode). High-speed operation is achieved by the following techniques.

A modified Booth's algorithm reduces the number of partial products to half. The 24-bit mantissa of multiplicand Y is recoded into 12 sets of 3-bit codes by twelve Booth's recoders, so that the 24×24 multiplication becomes 25×12 recoded multiplication, and the number of partial products is reduced from 24 to 12. The critical paths of the Booth's recoder and decoder (a partial product generator) are 3 and 3 gate stages, respectively.

A carry save adder (CSA) scheme is employed. The CSA array in Fig. 2 is made up of 25 columns by 12 rows, and performs the high-speed addition of the 12 partial products. The longest path of propagation in the CSA array is 12 CSA's. Each terminal in the CSA array is interconnected so that the propagation delay time on the critical path should be the shortest. The final carry propagation for the lower 24 bits of the product is packed into the array by using twelve high-speed 2-bit carry generators.

A high-speed CMOS full adder is used in the CSA array. Fig. 3 shows the circuit diagram of the CMOS full adder. This full adder consists of 40 transistors and generates the sum and carry output in only 3 and 2 gate delays including the output buffer, respectively. Thus, the full adder can perform addition at about twice the speed of conventional CMOS full adders. By using the full adder, the critical path of the array made up of 12 CSA's is reduced to 36 gate stages.

The physical design of the full adder cell used in the CSA array is one of the most important aspects because its performance greatly influences the operating speed, total power dissipation, and chip size. Full CMOS construction

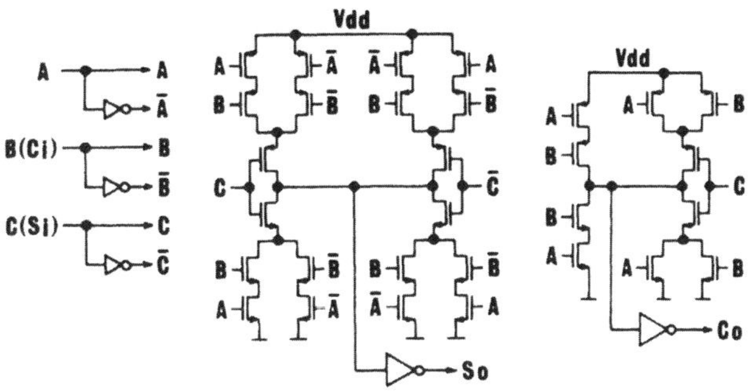

Fig. 3. Circuit diagram of the CMOS full adder.

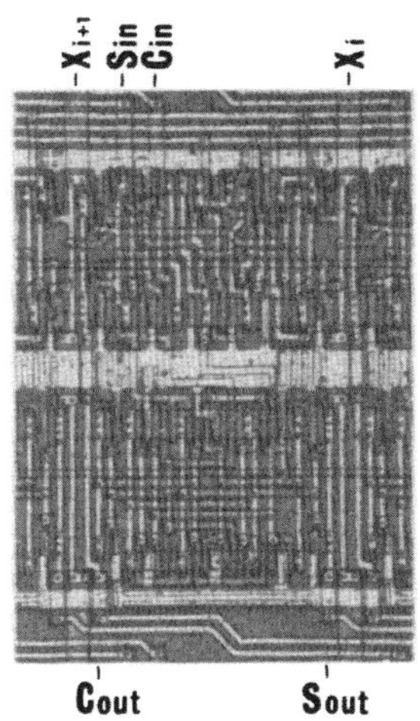

Fig. 4. Photograph of the array cell in the carry save adder array. The cell contains the CMOS full adder and the partial product generator consisting of a 2-1 multiplexer and an Exclusive-OR gate. The cell size is $139 \times 275\ \mu m^2$ (repeatable).

is employed to realize the low-power dissipation. Wiring resistances and parasitic capacitances are minimized by optimum pattern designing. As a result, the array cell of the CSA array is $139 \times 275\ \mu m^2$ in size, including the full adder, the partial product generator (a 2-1 multiplexer and an Exclusive-OR gate), and the interconnection area. A photograph of the array cell is shown in Fig. 4. This array cell can perform the addition in 3.7 ns, typically, as mentioned later.

A modified carry select adder speeds the 24-bit addition in the final adder stage, and is shown in Fig. 5. This adder is made up of five block adders and three block-carry generators (two OR-NAND gates and an AND-OR gate). Each n-bit block adder has two n-bit ripple adders (one and the other provide addition assuming that the carry input is "1" and "0," respectively), and an n-bit 2-to-1 multiplexer.

When the partial product addition in the CSA array is completed, all of the ripple adders start operating simultaneously. The carry output of the n-bit ripple adder is generated after n gate delays. In our circuit design, we made the carry propagation time of the block-carry generator equal to that of a "one-bit" ripple adder (1 gate delay). By designing an $n+1$-bit block adder next to the n-bit block adder, two carry outputs of the ripple adders in the $n+1$-bit block adder are generated at the same time as an output of the block-carry generator in the n-bit block adder. For example, in Fig. 5, C_{19}^1 and C_{19}^0, $\overline{C_{13}^1}$, and $\overline{C_{13}^0}$ are generated at the same time as C_{13} and $\overline{C_8}$, respectively. That is, the higher the order of a block adder, the larger the number of bits of it can be made. For this reason, 4-bit, 5-bit, and 6-bit ripple adders are used for the second, third, and fourth block adders, respectively, so that every data path becomes critical. For instance, $A_0 - C_4 - \overline{C_8} - C_{13} - \overline{C_{19}} - S_{23}$, $B_8 - C_{13} - \overline{C_{19}} - S_{23}$, and $A_{13} - \overline{C_{19}} - S_{23}$ are all critical.

As mentioned above, optimizing the distribution of bits to each block adder makes it possible to reduce the number of block adders and minimize the critical path. This adder reduces the 24-bit addition time to 10 gate delays including the propagation delay time (3 gate delays) of the 2-1 multiplexer.

Besides these high-speed techniques, careful attention was paid to the arrival sequence of input signals in interconnecting gates.

Consequently, the propagation delay on the critical path of the mantissa multiplier is reduced to 52 gate delays.

V. Characteristics

A photograph of the chip is shown in Fig. 6. The 24×24-bit mantissa multiplier has an effective area of $3.8 \times 3.8\ mm^2$. The twelve Booth's recoders are at the right side of the chip. The exponent adder and other control logic are to the left of that.

The chip was fabricated using a conventional 2 μm n-well CMOS process with single level poly-Si and Al. The gate length is 2.0 μm for the n-channel, and 2.25 μm for the p-channel transistor. The gate oxide thickness is 400 Å. The threshold voltage is 0.6 V for the n-channel, and -0.6 V for the p-channel transistor.

Fig. 7 shows the multiply time as the sum of individually measured propagation delay times on the critical path.

The addition time of the CMOS full adder measured 3.7 ns under normal operating conditions ($V_{DD} = 5$ V, $T = 25°C$). This value coincided well with a simulated value of 3.66 ns. Therefore, the critical path of the carry save adder array required 44.1 ns using the CMOS full adders.

A typical addition time for the modified carry select adder measured 18 ns.

As a result, we obtained a typical value of 71.1 ns as the 24-bit two's complement fixed multiply time, and a value of 78.7 ns as the 32-bit floating point multiply time.

We can measure the multiply time including the propagation delay time of an output register and two gates by using the enable control *ENB*. The propagation delay time is equivalent to about 4 gate delays, namely, about 9 ns. In the 24-bit two's complement fixed point multiplication, 80 ns is obtained as the measured value by using *ENB* in the typical condition. The characteristics of multiply time versus supply voltage at room temperature ($T = 25°C$) are shown in Fig. 8.

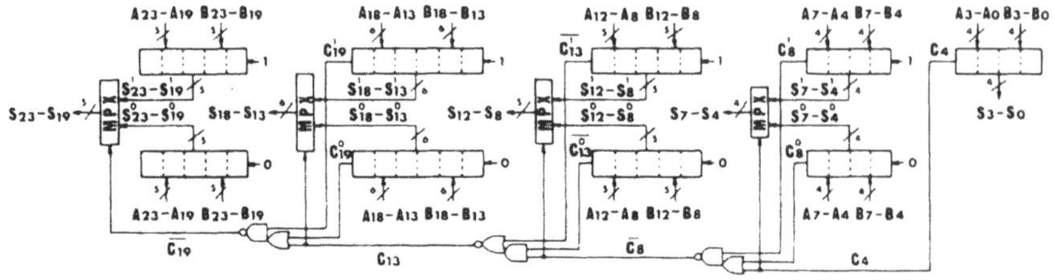

Fig. 5. Circuit diagram of the modified carry select adder. The adder can perform 24-bit addition in the final addition stage in only 10 gate delays. The typical addition time measured 18 ns.

Fig. 6. Photograph of the multiplier chip. The chip size is 5.75×5.67 mm^2.

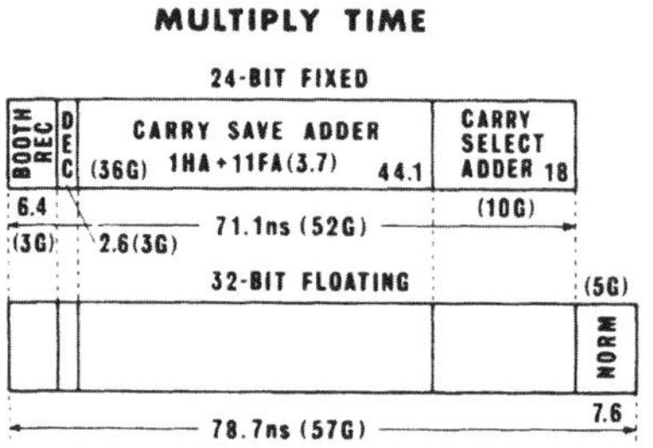

Fig. 7. Multiply time as the sum of individually measured propagation delay times on the critical path.

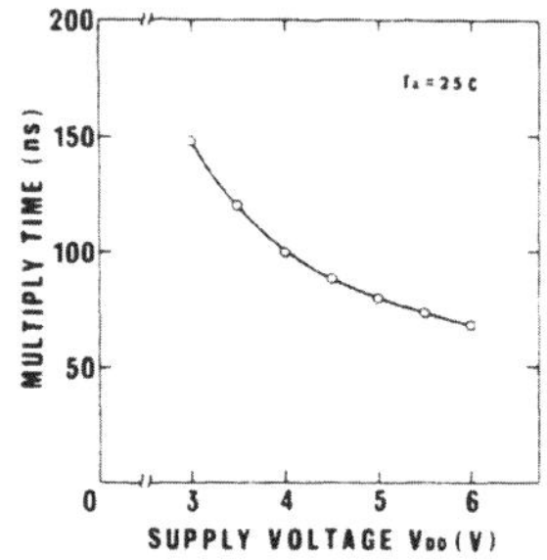

Fig. 8. Characteristics of multiply time versus supply voltage at room temperature ($T = 25$°C).

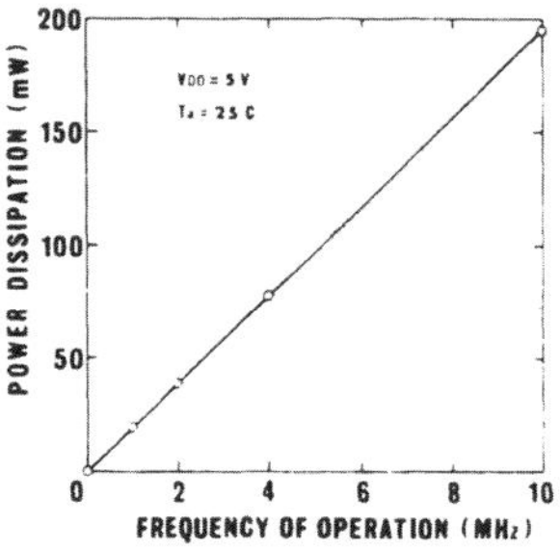

Fig. 9. Characteristics of power dissipation versus frequency of operation in the typical condition ($V_{DD} = 5$ V, $T = 25$°C).

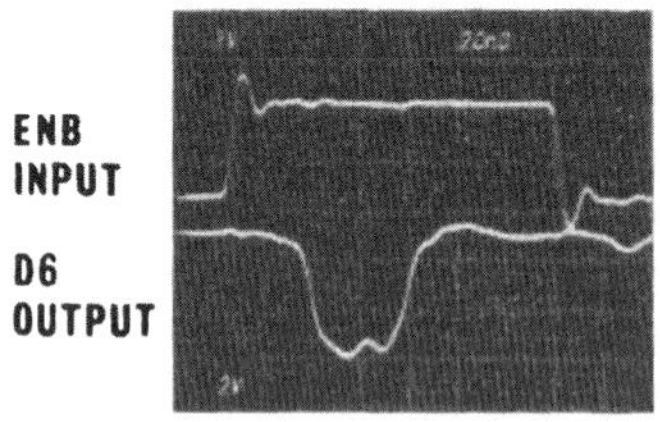

Fig. 10. Oscillograph of the waveforms of ENB input and $D6$ (MSB of a fraction Fz) output. An interval time from a falling edge to a rising edge of the $D6$ output waveform represents the multiply time.

The characteristics of power dissipation versus frequency of operation at $V_{DD} = 5$ V and $T = 25$°C are shown in Fig. 9, and represent the typical power dissipation of 195 mW at 10 MHz (10 million operations per second).

Waveforms of ENB input and $D6$ (MSB of a fraction Fz) output are shown in Fig. 10. This oscillograph is an example of 32-bit floating point multiplication of the case that the multiplier $X = 44400001H$ ($Sx = 0$, $Ex = 88H$, $Fx = 400001H$), the multiplicand $Y = 83FFFFFFH$ ($Sy = 1$, $Ey = 07H$, $Fy = 7FFFFFH$), $RND = 0$, and the product $Z = C8400000H$ ($Sz = 1$, $Ez = 90H$, $Fz = 400000H$).

When the ENB input is low, an input of the output register (this is an output of the mantissa multiplier) is low. As the ENB input goes high, the "low" signal propagates from the input of the output register to the $D6$ output through the output register, the format adjuster, an internal output bus, and an output driver. At the same time, the mantissa multiplier starts the multiplication. After the operation, the MSB of the fraction of the product Z goes high. This "high" signal propagates to the $D6$ output as well as the "low" signal. Therefore, the time interval from a falling edge to a rising edge of the $D6$ output waveform in Fig. 10 represents the multiply time including the propagation delay time of the output register and two gates. In this case, we can read a value of about 40 ns out of the Fig. 10 oscillograph.

```
**********  00000000
**********  00FFFF80
30.00NS     00000000001100000000001000110100
31.00NS     00000000001101010000001000111000
32.00NS     00000000000101010000001000111100
33.00NS     00000000000100010010101010111100
34.00NS     00000000100100010010111010111101
35.00NS     00000000100100010111111010111101
36.00NS     00000000100100011111111010001101
37.00NS     00000000000100000010100010001101
38.00NS     00000000000100000010100010001101
39.00NS     00000000000100000000000000001111
40.00NS     00000000000100000000000000001111
41.00NS     00000000000100000000000000001111
42.00NS     00000000000100000000000000001111
43.00NS     00000000000000000000000000001111
44.00NS     00000000000000000000000000001111
45.00NS     00000000000000010000000000011111
46.00NS     00000000000000010000000000011111
47.00NS     00000000000000010000000000011111
48.00NS     00000000000000010000000111111111
49.00NS     00000000000000010000000111111111
50.00NS     00000000000000110000001111111111
51.00NS     00000000000001110000011111111111
52.00NS     00000000000001111100001111111111
53.00NS     00000000000001111100001111111111
54.00NS     00000000000001111100001111111111
55.00NS     00000000000000000000001111111110
56.00NS     00000000000000000000001111111110
57.00NS     00000000000000000000001111111100
58.00NS     00000000000000000000001111111100
59.00NS     00000000000000000000001111111000
60.00NS     00000000000000000000101111111000
61.00NS     00000000000000000000111111110000
62.00NS     00000000000000000001111111100000
63.00NS     00000000000000000001111111100000
64.00NS     00000000000000000011111111100000
65.00NS     00000000000000000011111111100000
66.00NS     00000000000000000011111000000000
67.00NS     00000000000011111111111000000000
68.00NS     00000000000011111111111000000000
69.00NS     00000000111111111100000000000000
70.00NS     00000000111100000000000000000000
71.00NS     00000000000000000000000000000000
72.00NS     00000000000000000000000000000000
73.00NS     00000000000000000000000000000000
74.00NS     00000000000000000000000000000000
75.00NS     00000000000000000000000000000000
76.00NS     00000000000000000000000000000000
77.00NS     00000000000000000000000000000000
78.00NS     00000000000000000000000000000000
79.00NS     00000000000000000000000000000000
80.00NS     00000000000000000000000000000000
Enable Time Bit-31   Bit-23                 Bit-0
```

Fig. 11. 32-bit output patterns versus enable times.

Fig. 11 shows 32-bit output patterns versus enable times. This is measured and printed out by an LSI tester. The enable time signifies that the time during the enable signal *ENB* is high, namely, an allotted time for multiplication. The 32-bit output patterns are the data latched into the output register *Z* by the falling edge of the *ENB* signal, that is, the halfway result of the multiplication for the enable time. This is the case of $00000000_H \times 00FFFF80_H = 00000000_H$ in 24-bit two's complement fixed point mode. The higher 8 bits (bit-31–bit-24) are always "0"'s in the fixed point format. "1" and "0" in the output patterns represent the wrong and right answer, respectively.

Fig. 11 shows that the first right answer of bit-0 (LSB) is obtained at 55 ns, and that of bit-23 (MSB of 24-bit two's complement result) at 71 ns. The first, second, third, fourth, and fifth block adder of the modified carry select adder put the first right answers at 62, 66, 69, 70, and 71 ns, respectively. As a result, the modified carry select adder stage takes 16 ns (71–55 ns), and the multiply time (including the propagation delay time of the output register and two gates) is 71 ns.

TABLE I
SUMMARY OF THE CHIP FEATURES

PROCESS	2 µm N-well CMOS
GATE LENGTH	2.0 µm (N-ch) 2.25 µm (P-ch)
TRANSISTOR COUNT	23,000 Tr
CHIP SIZE	5.75 x 5.67 mm²
PACKAGE	28-pin DIP
INTERFACE	16-bit BUS
LOGIC LEVEL	TTL Compatible
SUPPLY VOLTAGE	5 V
MULTIPLY TIME	71.1 ns (24x24 FIXED) 78.7 ns (32x32 FLOATING)
POWER DISSIPATION	195 mW (10 MFLOPS)

VI. SUMMARY

A high-speed, low-power CMOS floating point multiplier chip has been developed. The chip can perform 32-bit floating point and 24-bit two's complement fixed multiplication in typically less than 78.7 and 71.1 ns, respectively. The above high performance is obtained by high-speed multiplication techniques including the modified Booth's algorithm, a carry save adder scheme, a high-speed CMOS full adder, and a modified carry select adder. The typical power dissipation is 195 mW at 10 million operations per second.

The chip is designed for 16-bit bus compatibility and can be used as a floating point accelerator for microcomputer systems. Furthermore, the design techniques used are applicable to on-chip multipliers and other arithmetic units.

The features of the chip are summarized in Table I.

REFERENCES

[1] A. D. Booth, "A signed binary multiplication technique," *Quart. J. Mech. Appl. Math.*, vol. 4, part 2, pp. 236–240, 1951.
[2] O. J. Bedrij, "Carry select adders," *IRE Trans. Electron. Comput.*, vol. EC-11, no. 3, pp. 340–346, June 1962.
[3] K. Hwang, *Computer Arithmetic.* New York: Wiley, 1979.
[4] D. Stevenson, "A proposed standard for binary floating point arithmetic," *IEEE Trans. Comput.*, vol. C-14, no. 3, pp. 51–62, Mar. 1981.
[5] F. Ware, W. McAllister, J. Carlson, D. Sun, and R. Vlach, "64 bit monolithic floating point processors," *IEEE J. Solid-State Circuits*, vol. SC-17, no. 5, pp. 898–907, Oct. 1982.
[6] B. Woo, L. Lin, and R. Owen, "ALU, multiplier chips zip through IEEE floating-point operations," *Electronics*, vol. 56, no. 10, pp. 121–126, May 1983.
[7] M. Uya, K. Kaneko, and J. Yasui, "A CMOS floating point multiplier," in *ISSCC Conf. Dig.*, Feb. 1984, vol. 27, pp. 90–91.

Masaru Uya was born in Shimane, Japan, on January 15, 1950. He received the B.S. degree in electrical engineering from Osaka University, Osaka, Japan, in 1972.

In 1972 he joined the Matsushita Electric Industrial Co., Ltd., Osaka, Japan. From 1972 to 1980 he was engaged in the research and development of an electronic music synthesizer in the Wireless Research Laboratory at Matsushita. Since 1980 he has been involved in the research of high-speed CMOS circuit design, and the development of high-speed CMOS LSI circuits such as 16-bit CMOS microprocessor at Matsushita's Central Research Laboratory. His research interests include computer arithmetic, computer architecture, high-speed numerical computation techniques, and LSI circuit design.

Katsuyuki Kaneko was born in Nagano, Japan, on July 21, 1955. He received the B.S. and M.S. degrees in communication engineering from Tohoku University, Sendai, Japan, in 1978 and 1980, respectively.

He joined the Matsushita Electric Industrial Co., Ltd., Osaka, in 1980. He has been working on the design of MOS LSI circuits.

Mr. Kenko is a member of the Institute of Electronics and Communication Engineers of Japan.

Juro Yasui received the B.S. and M.S. degrees in solid-state engineering from Osaka University, Osaka, Japan, in 1969 and 1971, respectively.

In 1974 he joined the Semiconductor Research Laboratory, Matsushita Electric Industrial Co., Ltd., Moriguchi, Osaka, Japan, where he has been working on the development of integrated circuit processing.

A Single-Chip 80-bit Floating Point Processor

KAZUMITSU TAKEDA, MEMBER, IEEE, FUMIAKI ISHINO, YOSHITAKA ITO, RYOTA KASAI, MEMBER, IEEE, AND TAKAYOSHI NAKASHIMA, MEMBER, IEEE

Reprinted from *IEEE Journal of Solid-State Circuits*, Volume SC-20, Number 5, October 1985, pages 986-992.

***Abstract*—A single-chip 80-bit floating point VLSI processor, capable of performing 5.6 million floating point operations per second, has been realized using 1.2-μm n-well CMOS technology. The processor handles 80-bit double-extended floating point data conforming to IEEE standard 754. The chip has 128 microinstructions which are stored in an on-chip ROM. By programming microinstruction sequences in an external control storage, not only basic arithmetic operations but also special arithmetic functions can be performed. A composite design method supported by a hierarchical design automation system was employed to quickly lay out 50 K gates including a 64- × 64-bit multiplier and 15 kbits of memory on a chip with a die size of 10 × 10 mm². Only 11 man-months as required for the design effort.**

I. Introduction

IN RECENT YEARS, various kinds of chips have been developed to enhance the floating point calculations for image processing, computer graphics, model simulations, CAD/CAM, etc. The chips can be grouped into two categories: microcomputer coprocessors and floating point calculation chip sets which include multiplier, divider, adder/subtracter, etc. Some of the former group handle 32/64/80-bit floating point data as well as integer and BCD data [1], [2]. They support various kinds of functions such as add, subtract, multiply, divide, and square root, but they perform at less than 1 million floating point operations per second. Higher performance has been reported in chip sets belonging to the latter group [3]–[5]. However, they require a larger number of chips to construct a floating point system.

The authors sought to develop a single-chip 80-bit floating point processor of higher performance. Objectives and key points of the VLSI design were as follows:

1) High Performance: High-speed (5.6 MFLOPS) and high precision (80 bit) were accomplished by employing 64-bit arithmetic blocks and three 80-bit buses which enable parallel operations on exponent and mantissa parts. Also, 1.2-μm n-well CMOS technology was used. It has a potential speed as high as 1 ns per gate.

2) Easy and Wide Application: A microprogram control method was adopted and a 32-word dual-port register file was used. This register file is useful in constructing various kinds of algorithms.

3) Single Chip and Short Design Time: A composite design method supported by a hierarchical design automation system and 1.2-μm CMOS technology with double metal wiring layers was applied.

In the following sections, chip structure, fabrication and performance, and an example of chip operation as well as the layout design method will be described.

Manuscript received March 20, 1985; revised June 5, 1985.
The authors are with NTT Atsugi ECL,3-1 Morinosata Wakamiya, Atsugishi, Kanagawa 243-01, Japan.

II. Chip Function Design

The floating point data conforming to IEEE standard 754 consists of a 1-bit sign 15-bit exponent biased 16383, and a 64-bit mantissa including a 1-bit integer part as shown in Fig. 1. Floating point multiplication and addition/subtraction are the main functions of this chip and they were designed to be performed as described below. Other complicated functions such as logarithmic and trigonometric functions can be executed by combining their constituent basic operations.

A. Floating Point Multiplication

Floating point multiplication sequences are shown in Fig. 2. In the first and second cycles, the exponent part of the two operands are added and then bias correction is performed. Mantissa multiplication is performed in parallel with these operations during two clock cycles. This is because the critical path of the multiplier is expected to be nearly twice as long as that of the other functional blocks. In the third cycle, the mantissa part is normalized and the exponent part is incremented according to the normalization.

The data loading and unloading were designed to overlap with the normalization cycle and the exponent addition cycles, respectively. In this way, multiplication results can be obtained every three clock cycles.

B. Floating Point Addition/Subtraction

Floating point addition and subtraction sequences are a little more complicated than those of floating point multi-

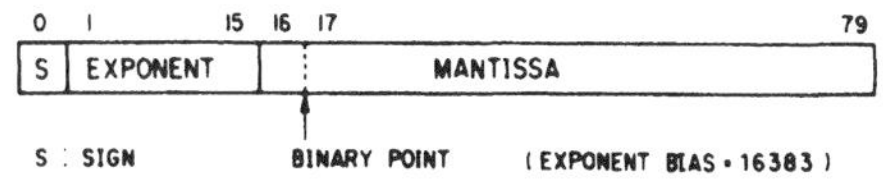

Fig. 1. An 80-bit floating point data format.

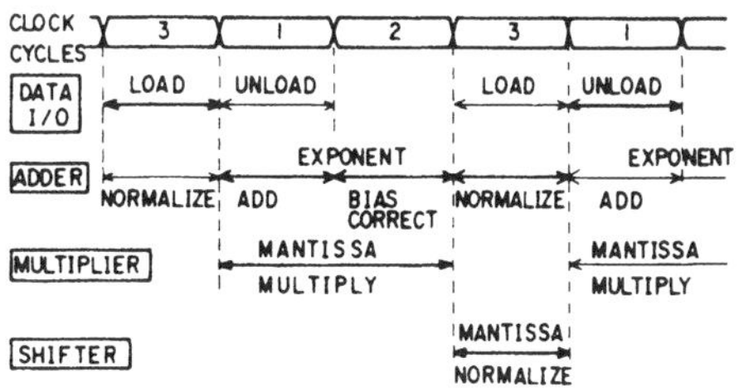

Fig. 2. Floating point multiplication sequence.

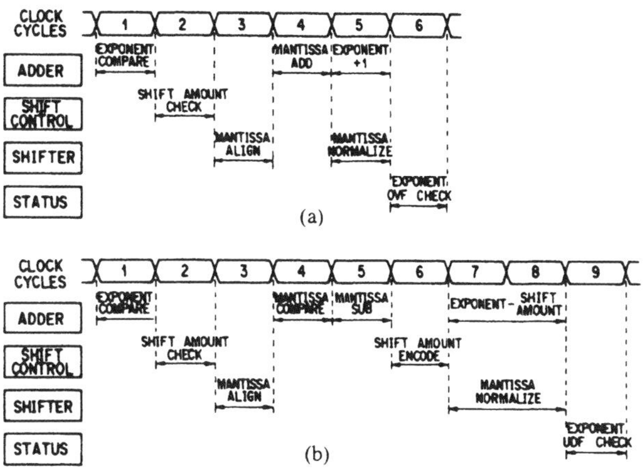

Fig. 3. Floating point addition/subtraction sequence. (a) Addition. (b) Subtraction. When the difference between the two exponents is larger than 32 in floating point add/subtract, the shift operations for mantissa alignment is performed twice during two clock cycles.

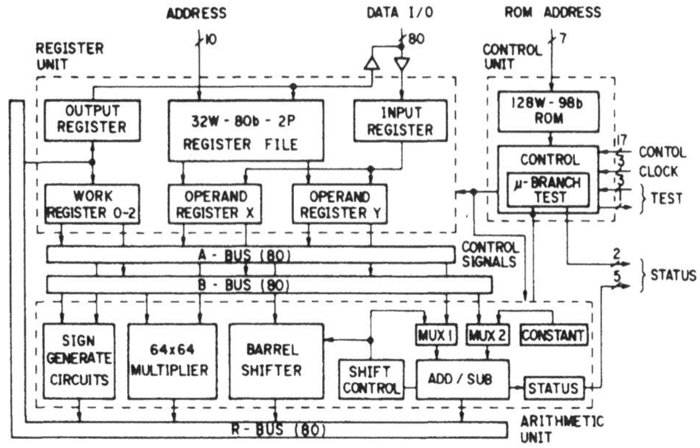

Fig. 4. Block diagram.

plication. They take 6–7 and 9–10 clock cycles, respectively, as shown in Fig. 3. First, the two exponents are compared and then the mantissa with a smaller exponent is right shifted by N by the shift control circuits and the result is left shifted by N bits (this operation is called mantissa alignment), where N is the difference between the two exponents. When N is larger than 32, the SHIFT operation is designed to be performed twice during the two clock cycles.

Next, mantissa addition is performed. This is done when the two operands have the same sign and the instruction indicates floating point add, or when the two operands have different signs and the instruction indicates floating point subtract. If mantissa overflow occurs, the result is right shifted by 1 bit and the exponent is incremented.

When the two operands have the same sign and the instruction indicates floating point subtract, or when the two operands have different signs and the instruction indicates floating point add, two mantissas are compared and then the smaller one is subtracted from the larger one. The most significant non-zero-bit position is detected and encoded into N by the shift control circuits and the result is left shifted by N bits. In parallel with this SHIFT operation, N is subtracted from the exponent.

Finally, exponent overflow/underflow is checked. If either occurs, the status bits are set.

III. Chip Structure

In order to realize the aforementioned chip operations, the chip structure shown in Fig. 4 is implemented. The chip consists of an arithmetic unit, a register unit, and a control unit. Each of them will be presented in detail.

A. Arithmetic Unit

Exponent adding/subtracting and the mantissa multiplying in the floating point multiplication must be executed in parallel. Exponent adding/subtracting/incrementing and the mantissa shifting in the mantissa alignment/normalization must also be parallel. Therefore, three 64-bit execution blocks (multiplier, shifter, adder/subtracter) with three 80-bit buses are accomodated as the main execution blocks of an arithmetic unit. These buses consist of 1-bit sign 15-bit exponent and 64-bit mantissa parts.

The multiplier performs 64×64 multiplication with a modified Booth's algorithm. Two data inputs are from the mantissa part on A-bus and B-bus and the 64-bit multiplied result is output to the mantissa part of R-bus. The shifter performs 0–32 bit right or left shift on the mantissa part of the A-bus and also outputs the result to the mantissa part of the R-bus. The shifter is controlled by a shift control circuit with shift amount data which are calculated by the adder/subtracter for mantissa alignment, or encoded from the data of the most significant non-zero-bit position by the shift control circuit itself for normalization. The shift amount for mantissa alignment is limited to 64 bits when the difference between the two exponents is larger than 64. The adder/subtracter has six functions: $A+B/A-B/B-A/A/B/0$, where A and B denote the two data inputs of the adder/subtracter. For the adder/subtracter to handle the exponent and mantissa data, two multiplexers (MUX1, MUX2) are implemented to select appropriate data. The MUX1 selects one out of exponent and mantissa data in the A-bus and the shift amount from the shift control circuit according to the normalization. The

MUX2 selects one out of exponent and mantissa data in the *B*-bus and a constant for exponent bias correction. The 15-bit EXPONENT operation result by the adder/subtracter is output to the exponent part of the *R*-bus, and the 64-bit MANTISSA operation result is output to the mantissa part of the *R*-bus.

Thanks to this arithmetic unit structure, parallel operations in floating point multiplication and addition/subtraction can be realized.

B. Register Unit

A 32-word×80-bit dual-port register file was adopted to reduce data transfers between the chip and external circuits by using it as a scratch pad memory. This dual-port structure enables a parallel READ operation of the two 80-bit data. The register file has two 5-bit address lines, one of which is also used for write address. This unit has seven other registers; two for operand, three for work, one for input and one for output register.

Simultaneously with the arithmetic operations, data are loaded into the input register and unloaded from the output register through 80-bit bidirectional data I/O terminals. The outputs of the input register and the register file are connected to the operand registers, and the operand and work registers are connected to the arithmetic unit through the *A*- and *B*-buses. The results of the execution blocks are transferred to the output register and then unloaded from the chip or stored in the register file. The *R*-bus is also connected to the work registers to use an intermediate result for the next operation. In order to avoid the malfunctions which may occur because of the racing of data and the clock, three phase clocks (Clock-$A/B/C$) were used and master-slave circuits were used for the work registers. Clock-*A* was used for the master registers of the work registers and for the output register, Clock-*B* for the slave registers of the work registers and for the input register, and Clock-*C* for the operand registers.

C. Control Unit

A horizontal microprogram control method including the next microinstruction addresses in the control storage itself, was adopted for logic design simplicity. The on-chip control unit consists of 128-word×98-bit ROM and other control circuits such as microinstruction branch test circuits. The on-chip ROM is programmed for basic operations such as floating point ADD/SUB/MULTIPLY/DIVIDE, fixed point ADD/SUB/MULTIPLY, and LOAD/UNLOAD operations. A control storage for the next micro-instruction addresses is excluded from this chip to enhance control flexibility. The control method of this chip is classified into the following two categories.

The basic control structure of this chip is shown in Fig. 5(a). The next address field read out from the external control storage with 128 word×7 bit is modified by the microbranch test results and used for addressing both the on-chip ROM and the external control storage. Then, the next microinstruction is read out. In this way, microinstruction sequences are carried out.

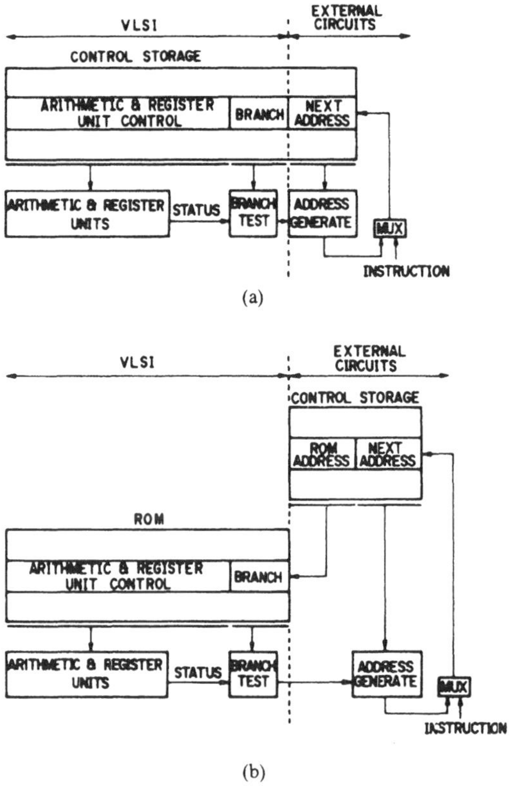

Fig. 5. Control structure of the processor. (a) Basic control structure. (b) Extended control structure.

Another control structure is shown in Fig. 5(b), where the external control storage consists of two address fields: on-chip ROM address and the next microinstruction addresses fields. In this structure, the on-chip ROM address may be considered a 7-bit microinstruction operation code. Various kinds of functions such as logarithmic and trigonometric functions can be realized by programming their sequences on the external control storage.

IV. Layout Design

To obtain a compact chip in a short time, a tradeoff between chip size and design effort was considered acceptable provided that the composite design method [6] supported by a hierarchical design automation system is used.

A. Composite Design Method

The chip architecture needs about 50 000 logic gates and 15 000 bits for memories and registers. This corresponds to about 150 000 transistors. In order to pack the transistors into a single chip with an acceptable chip size and to design the chip within a short time, a composite design method consisting of a modular cell method and a standard cell method was employed. This design method is supported by a hierarchical design automation system.

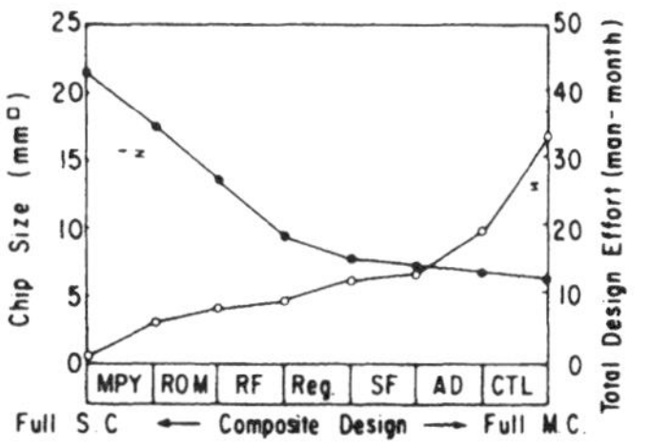

Fig. 6. Tradeoff between chip size and design effort. MPY: multiplier. ROM: control ROM. RF: register file. SF: shifter block. AD: adder block. CTL: control block. Reg.: operand, work and I/0 registers.

The modular cell method refers to a design approach which can construct a certain logic network by simply arranging a few kinds of cells. This method has the advantage of high gate density, since each cell can be constructed of sophisticated circuit configurations and densely laid out by a manual approach. However, this design method requires a long design time. Accordingly, the modular cell method is suitable for highly repetitive networks such as multipliers and registers.

The standard cell method has quite a short design turnaround time thanks to cell standardization and automatic layout program support. However, it results in a relatively low integration density. Therefore, this method is suitable for irregular or random networks.

B. Tradeoff Between Chip Size and Design Effort

The tradeoff between chip size and design effort when the composite design method using a 1.2-μm CMOS technology with double wiring metals is employed is shown in Fig. 6. The horizontal axis denotes the coverage by manual layout design in a total chip design. The block names are arranged in an ascending order of suitability for the modular cell method. Chip size was estimated by an inter-block wiring area estimation and a block size estimation, using the ratio of device area to wiring area which was empirically obtained. Design effort was roughly estimated from layout productivity; five devices/man-day for handcrafted modular cells, ten devices/man-day for standard cell preparation, and a half month for an automatic layout design including power-line wirings. It can be seen in Fig. 6 that with the increase in the use of the manual approach, chip size decreases to 6×6 mm^2, whereas design effort increases to 30 man-months. Conversely, it also shows that with the increase in the use of the standard cell approach, the design effort decreases to 1.5 man-months, but chip size increases to 20×20 mm^2. It is clear that a tradeoff between chip size and design effort must be considered to obtain a compact chip in an acceptable design time.

Considering the chip design objectives, it was decided to design the multiplier, ROM, register file and other registers using manually designed modular cell blocks and the rest with the standard cell method. It was expected that the chip size would be 8×8 mm^2 excluding the peripherals and that the design effort would be about ten man-months.

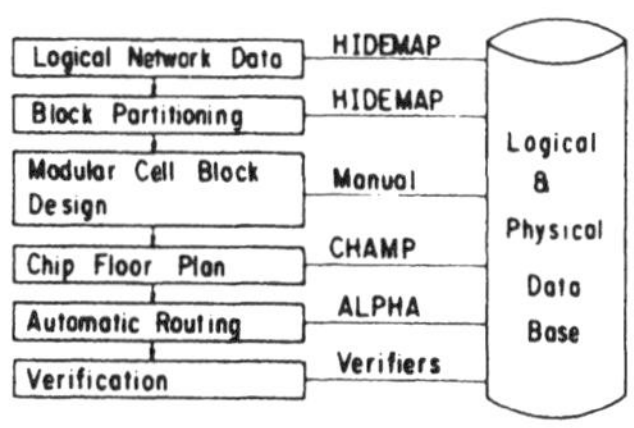

Fig. 7. Design procedure.

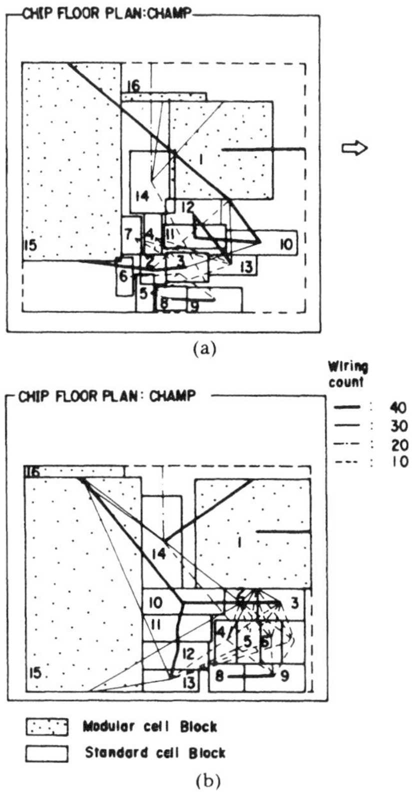

Fig. 8. Interactive chip floor planning. (a) Initial block placement. (b) Final block placement.

C. Layout Design Using a DA System

The chip was hierarchically designed (chip, block, cell) using a DA system. The design procedure and the CAD programs used in each stage of the design are shown in Fig. 7. The original logic file was partitioned into 16 blocks. Three of them were manually designed as modular cell blocks and the others were designed by the standard cell method according to the forementioned tradeoff.

After block partitioning, chip floor planning was performed with the aid of a Chip Floor-Plan Automation Program (CHAMP) [7]. CHAMP generated the initial block placement referring to inter-block network data and the size of standard cells and modular cell blocks as shown in Fig. 8(a). Then, each block was interractively rearranged considering the inter-block wiring, and the aspect of standard cell blocks was adjusted to minimize open spaces. As a result, the final chip floor plan shown in Fig. 8(b) was obtained.

Fig. 9. Chip photomicrograph. MPY: multiplier. ROM: control ROM. RF: register file. SF: shifter block. AD: adder block. CTL: control circuits. RG: operand, work, and I/O registers.

TABLE I
SUMMARY OF LAYOUT RESULTS

Block Name	Circuit Structure	Devices	Area(mm²)
Multiplier	Modular Cells	67K	18.5
ROM	Modular Cells	14K	0.9
Register	Modular Cells	41K	9.2
Adder, etc.	Standard Cells	28K	34.7
Total		150K	63.3

Design Effort 11 man-months

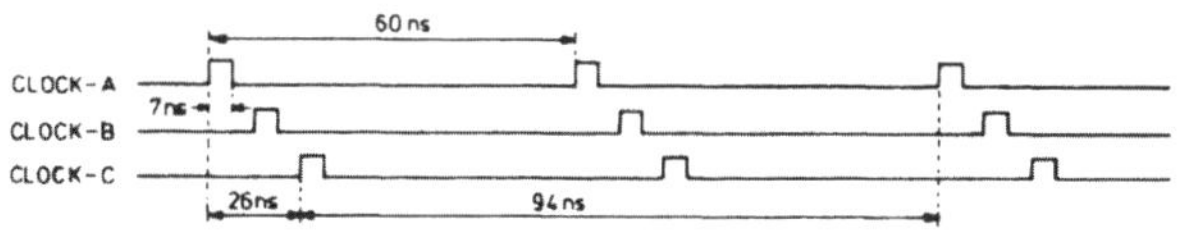

Fig. 10. The mantissa multiplication time which is assigned two clock cycles was 94 ns. Therefore, the minimum clock period was set to 60 ns taking into account the clock pulsewidth and clock skews.

Further layout design, i.e., the placement and wiring for standard cell blocks and peripheral connections between inner circuits and I/O circuits were done fully automatically using *A* Layout Program with Hierarchical Approach (ALPHA) [8]. Power and ground bus lines were semiautomatically laid on four sides of the inner circuits and at appropriate positions determined by calculations of voltage drops and electromigration constraints due to excess current flow. Power and ground bus lines for I/O circuits were routed independently of those for inner circuits to prevent the circuit noise generated in I/O circuits from entering into inner circuits. Finally, the layout results were verified through design rule, connectivity, and timing checks.

D. Layout Results

A chip photomicrograph is shown in Fig. 9. The multiplier labeled MPY contains 66.7 K transistors in a space of 3.15×5.87 mm^2. The standard cell blocks labeled CTL and SF·AD consist of about 28 K transistors and occupy 34.7 mm^2 including the inter-block wiring area. The modular cell blocks have an integration density from 4.5 to 19 times higher than that of standard cell blocks. The total transistor count is nearly 150 K and the chip size measures 10 mm $\times$ 10 mm. The total design effort was about 11 man-months using the DA system including one man-week for chip floor planning and automatic placement and wiring. The layout results are summarized in Table I.

V. CHIP FABRICATION AND PERFORMANCE

A chip was fabricated using silicon gate n-well CMOS technology with double metal wiring layers. Gate lengths of p-channel and n-channel transistors are 1.5 and 1.2 μm, respectively. Effective channel length is from 0.8 to 0.9 μm for both transistors. Wiring pitches of the first and second level metal are 3.4 and 4.2 μm, respectively.

The ring oscillator speed is 180 ps at 5 V and 25°C. The typical propagation delay time of two-input NAND standard cell gate with a gate width-to-length ratio of 18, is 1 ns when there are three fan-outs and 2-mm wiring loads. The basic features of the technology are summarized in Table II.

TABLE II
BASIC FEATURES OF TECHNOLOGY

Process	Si Gate N-Well CMOS
Gate length	1.5 μm (P-ch) 1.2 μm (N-ch)
Gate oxide thickness	270 Å
Threshold voltage	-0.9V (P-ch) 1.0 V (N-ch)
Wiring pitch	3.4 μm (1st Al) 4.2 μm (2nd Al)
Gate delay	1ns (2-input-NAND)

TABLE III
SUMMARY OF CHIP FEATURES

Clock Cycle Time	60ns
Multiplication Time	180ns
Addition Time	360-420ns
Subtraction Time	540-600ns
Supply Voltage	5V
Power Dissipation	1 W
Package	144 PIP

Chip Performance

The critical path and minimum clock cycle were investigated through the following operations. First, the two operands were loaded into the operand registers. Then, mantissa multiplication was performed and the result was latched into the output register. Finally, the output register was unloaded and compared with the projected data.

It was confirmed that the 64×64 multiplication time which was assigned two clock cycles was 94 ns. This consists of the propagation delay time between the operand registers controlled by Clock-C and the multiplier, multiplication time and the propagation delay time between the multiplier and the output register controlled by Clock-A. This propagation delay time was more than twice the other operation time which was assigned one clock cycle such as a shift and a 64-bit add/sub. Therefore, the minimum clock period was set to 60 ns taking into account the clock pulse-width and clock skews as shown in Fig. 10. This clock cycle was the value at 5 V and 25°C. The power dissipation was about 1 W at a clock cycle time of 60 ns and 5-V supply voltage.

The chip is housed in a 144-pin plug-in package. Sixteen pins are used for the power supply and ground to avoid the large current noise generated by current flowing through inductive package wiring when 80 off-chip buffers switch simultaneously. The chip features are summarized in Table III.

VI. SUMMARY

This chip provides high-performance operations on floating point data. For example, an 80-bit floating point multiplication can typically be executed within 180 ns, which corresponds to about 5.6 million floating point operations per second. This high-speed operation was achieved by implementing powerful 64-bit arithmetic blocks connected to various registers through three 80-bit buses and by employing 1.2-μm CMOS technology. The chip has 128 microinstructions. By programming microinstruction sequences in the external control storage, not only basic arithmetic operations but also special arithmetic functions can be easily performed. During the 11 man-months design effort, 150 000 transistors were packed into a 10- × 10-mm² chip using a composite design method supported by the quick turnaround hierarchical DA system. The chip features show that a high-speed floating point execution processor can be easily constructed on a small board using this VLSI processor.

ACKNOWLEDGMENT

The authors would like to thank H. Mukai, T. Sudo, and S. Horiguchi for their direction and encouragement. The authors are also grateful to K. Ueda, T. Adachi, H. Kitazawa, and M. Nagatani for providing layout programs.

REFERENCES

[1] R. Nave and J. Palmer, "A numeric data processor," in *ISSCC Dig. Tech. Pap.* Feb. 1980, pp. 108–109.

[2] A. Komal, K. Goskel, P. W. Diodato, J. A. Fields, U. V. Gumaste, C. K. Kung, K. Lin, M. E. Lega, P. M. Maurer, T. K. Ng, Y. T. Oh, and M. E. Thierbach, "An IEEE standard floating point chip," in *ISSCC Dig. Tech. Pap.* Feb. 1985, pp. 18–19.

[3] F. A. Ware, W. H. McAllister, J. R. Carlson, D. K. Sun, and R. J. Vlach, "64 bit monolithic floating point processors," *IEEE J. Solid-State Circuits*, vol. SC-17, Oct. 1982.

[4] F. Ware, L. Lin, R. Wong, B. Woo, and C. Hansen, "Fast 64bit chip set gangs up for double-precision floating-point work," *Electron.*, pp. 99–103, July 1984.

[5] T. Ushiki, M. Suzuki, and T. Tsuchiya, " A floating point chip," in *IECE Nat. Conv. Rec.*, Mar. 1983, pp. 2–257.

[6] R. Kasai, K. Fukami, K. Tansho, H. Kitazawa, and S. Horiguchi, "An integrated modular and standard cell VLSI design approach," *IEEE J. Solid-State Circuits*, vol. SC-19, pp. 407–417, Feb. 1985.

[7] K. Ueda, H. Kitazawa, and I. Harada, "CHAMP: Chip floor plan for hierarchical VLSI layout design," *IEEE Trans. Computer-Aided Des.*, vol. 4, pp. 12–22, Jan. 1985.

[8] T. Adachi, H. Kitazawa, M. Nagaztani, and T. Sudo, "Hierachical top-down layout design method for VLSI chip," in *Proc. 19th Des. Automation Conf.* (Las Vegas, NV), 1982, pp. 785–791.

Kazumitsu Takeda (M'85) was born in Akita, Japan, on November 19, 1951. He received the B.S. degree in electronic engineering from Tohoku University, Sendai, Japan, in 1974.

Since joining the Electrical Communication Laboratory (ECL), Nippon Telegraph and Telephone (NTT) Public Corporation in 1974, he has been engaged in the development of custom integrated circuits for processors. He is now with the Atsugi ECL, NTT, Kanagawa, Japan.

Mr. Takeda is a member of the Institute of Electronics and Communication Engineers of Japan.

Fumiaki Ishino received the B.E. degree from Iwate University, Iwate, Japan, in 1975 and M.E. degree from Hokkaido University, Hokkaido, Japan, in 1977.

Currently, he is a Staff Engineer in switching service systems section, NTT Musashino ECL, Japan. He is presently doing development research on switching system using the Communication Satellite. Since joining the Musashino Laboratory, in 1977, he has been engaged in the development research on digital data switching systems, especially subscriber systems. His work has also been concerned with packetized voice communication system and VLSI processor.

Mr. Ishino is a member of the Institute of Electronics and Communication Engineers of Japan.

Ryota Kasai (M'85) was born in Tokushima, Japan, on July 28, 1948. He received the B.S. and M.S. degrees in electrical engineering from Osaka University, Osaka, Japan, in 1972 and 1974, respectively.

He joined the Musashino Electrical Communication Laboratory, Nippon Telegraph and Telephone Public Corporation (NTT), Tokyo, in 1974, and has been engaged in the research of FET device modeling and the design of CMOS VLSI's. He is now with the Atsugi ECL, NTT, Kanagawa, Japan.

Mr. Kasai is a member of the Institute of Electronics and Communication Engineers of Japan.

Yoshitaka Ito was born in Tokyo, Japan, on January 29, 1944. He received the B.S. degree from Musashi Institute of Technology, Tokyo, Japan, in 1967.

He joined the Musashino Electrical Communication Laboratory, Nippon Telegraph and Telephone Public Corporation (NTT), Tokyo, Japan, in 1967, and has been engaged in the research and development of central processor for Electronic Switching System, and the design of CMOS VLSI processor. He is now with the Atsugi ECL, NTT, Kanagawa, Japan.

Mr. Ito is a member of the Institute of Electronics and Communication Engineers of Japan.

Takayoshi Nakashima (M'85) was born in Japan, on June 16, 1943. He received his B.S. and M.S. degrees in electrical engineering from Hiroshima University, Hiroshima, Japan, in 1966 and 1968, respectively.

In 1968, he joined the Electrical Communication Laboratory, Nippon Telegraph and Telephone Public Corporation, where he has been engaged in the research and development of logic LSI's and LSI-CAD systems. He is currently interested in VLSI architectures for special purpose uses and CAD system.

VLSI FLOATING–POINT PROCESSORS

Jan Fandrianto – B. Y. Woo

WEITEK CORPORATION
501 Mercury Drive
Sunnyvale, California 94086

ABSTRACT

The advance of VLSI technology has been the enabling factor in the appearance of VLSI circuits handling floating-point arithmetics. These circuits have found their way into many number-crunching applications such as telecommunications, seismic energy exploration, radar, medical imaging, graphics and simulation. Because of the different requirements for different applications, some processors have a rich repertoire of functions but rather low performance, while some processors aim at having the highest throughput for the most frequent operations such as multiply and add. This paper will review the architecture, the technology, and the design techniques for the current VLSI floating-point processors; and it will also report a high performance chipset (*) implementing complete basic arithmetic functions.

I. Introduction

The motivation for using floating-point data format to handle arithmetic operations is due to its large dynamic range and high precision. However, since the hardware cost of implementing full floating-point arithmetic has been prohibitive, many have been using integer or block floating-point arithmetic instead. As a result, extra care needs to be exercised to take care of overflow, underflow and roundoffs. Block floating-point format has the advantage that the dynamic range can be adjusted to the biggest number. However, small numbers would have large noise to signal ratio. There are two driving forces that are changing the picture. First is the appearance of standards, among them the IEEE Binary Floating Point Arithmetic Standard (Task P754) [1] is the dominant one. This will make the software portable. Second is the advances of VLSI technology. As the level of integration rises, it becomes possible to implement the hardware support for all or most of the floating-point arithmetic operations in one or two chips.

It is the intent of this paper to present an overview of the current VLSI floating-point processors, in terms of their architecture, their interface with the host CPU and their functionality. Basically we can categorize them into two main groups: general-purpose and building block processors. General-purpose processors usually have a rich repertoire of functions. Their hardware consists of a set of basic arithmetic elements such as an adder, on-chip storage and control store. They rely on the built-in microcode routines to execute the arithmetic operations. Versatility and moderate performance are their features. On the other hand, the building-blocks optimize the performance on the key arithmetic functions such as add and multiply. They have dedicated hardware such as multiplier array to offer the utmost speed required. Depending on the applications, the building blocks can be configured accordingly to maximize the performance and throughput.

Table 1 lists some of the key application areas for the VLSI floating-point processors.

II. General-purpose Floating-point Processors

Processors falling in this category are Intel 8087/80287, National 32081, Motorola 68881, Zilog 8070, AMD 9511A/9512 and Fairchild F9450.

A. Intel 8087/80287 [2]

1. Architecture and Interface

8087 is designed to couple tightly with the host CPU, which in this case is 8086. In order to maximize the system throughput, the coprocessor duplicates the instruction queue and operand queue of the host. When one of the ESCAPE instructions is encountered, the host calculates the address and fetches the operand, and the coprocessor captures the address and operand and starts the computation and releases the bus. In doing so no extra bus cycle is needed for handshaking.

There are eight 80-bit registers that serve as a standard register set or as a stack. Associated with each of the registers, there is a two-bit Tag Word to indicate whether its content is valid, zero, special (Not-a-Number, infinity, etc.) or empty. On chip there are a 68-bit wide ALU and a 63-bit wide barrel shifter controlled by microcodes. Modified Cordic algorithm is used to calculate transcendental functions.

2. Functionality and Data Formats

The processor conforms to the IEEE standards and handles seven data types:16-bit, 32-bit and 64-bit two's complement integers, 32-bit, 64-bit and 80-bit floating-point numbers and 80-bit signed packed BCD. Arithmetic functions include add, subtract, multiply,

(*) Weitek WTL1164/1165

divide, square root, transcendental functions (exponential, logarithmic and trigonometric functions) and binary to/from decimal conversion. On-chip recovery from overflow and underflow is provided.

B. Motorola 68881 [3]

1. Architecture and Interface

The processor serves as an extension of the host CPU by providing eight 80-bit data registers, a status register, a control register and an instruction register. These registers are memory-mapped into the CPU address space.

The interface between the processor and host CPU is asynchronous. If the host CPU has a direct hardware support for the floating-point processor, such as 68020, it will recognize the 68881 instructions and perform the protocol to exchange machine instructions, operands and results on a 32-bit data bus, which can also be configured as an 8-bit or a 16-bit bus. Otherwise a F-line trap handler is invoked and software routine is used to complete the interface. In any case, the floating-point processor operations start with fetching one of the 68881 instructions (F-line instructions). The host CPU interprets the operation word in the instruction and write the command word to the 68881 command register. 68881 decodes the command word and provides a response to the host in the form of primitives. There are 18 primitives divided into five groups: processor/coprocessor synchronizations, instruction stream manipulation, exception handling, general operand transfer and register transfer. The host will then decide whether no actions is needed or some additional service must be performed, such as calculating the effective address, fetching variable-length operand or handling exception flags.

2. Functionality and Data Formats

68881 implements the IEEE floating-point standards. It handles seven data types: 8-bit (byte), 16-bit (word), 32-bit (long integer) two's complement integers, 32-bit, 64-bit and 80-bit floating-point number and 96-bit packed BCD.

Arithmetic functions include add, subtract, multiply, divide, square root, modulo remainder, IEEE remainder and all transcendental functions. In addition, operations such as compare, absolute value, exponent extract, mantissa extract, scaling and test can also be performed.

Data manipulating instructions include move to and from 68881, branch, save and restore. The last two are for virtual memory support so that the current state can be saved and restored in case of page fault.

C. National Semiconductor 32081 [4]

1. Architecture and Interface

32081 is designed to be a slave to the CPU. CPU is responsible for the fetching of the opcode and operands and sending them to the processor. The communication between the CPU and 32081 is via a two-bit protocol, a slave processor control line and a 16-bit bidirectional data bus.

There are three major functional blocks: the control unit, the execution unit and the interface and storage unit. The micro-instruction includes part of the next instruction's address and part of the condition code. The execution unit has separate hardware for exponent, fraction and sign computation. The interface and storage unit takes care of the control and data interface with external world. There are eight 32-bit data registers, which can be paired together to become four 64-bit registers. In addition, there is a 32-bit Floating Status Register, storing the rounding, underflow, inexact and trap type informations.

2. Functionality and Data Formats

The instruction set includes basic operations such as add, subtract, multiply, divide, compare, negate and absolute value. Also it can perform data type conversion, floating-point to and from integer, single precision to and from double precision. Five data types can be handled: IEEE single and double precision floating-point formats, 8-bit, 16-bit and 32-bit two's complement integers.

D. Others

1. AMD 9511/9512 [5]

These chips have a stack-oriented organization, with only one of the floating-point operands can be supplied from the stack at one time. 9511 handles a special 32-bit floating-point format which is different from the IEEE format, whereas 9512 uses the IEEE single and double precision formats. Data transfer between 9511/9512 and the host is via programmed I/O or DMA. End of execution is signalled as an interrupt to the CPU.

2. Fairchild F9450 [6]

This microprocessor fully implements the military instruction set architecture (MIL-STD-1750A ISA). Data types include bit, byte, 16-bit, 32-bit integer, single-precision(32-bit) and extended-precision (48-bit) floating-point numbers.

3. Zilog Z8070 [7]

Z8070 interprets the opcode fields in the extended instructions and carries out the proper operations. Internally it is structured as two separate processors: the interface processor and the data processor. The former fetches and aligns instructions and data, schedules operations using an instruction queue. The latter carries out the arithmetic operations, which include add, subtract, multiply, divide, square root, remainder, negate, compare and absolute.

Table 2 summarizes the important features of various general-purpose floating-point processors.

III. Building-block Floating-point Processors

The building-block approach is used for the most speed-demanding applications. In the VLSI implementation, the hardware support includes a high-speed multiplier array, a fast carry-look-ahead adder, a priority encoder and barrel shifters (both left and right shifters). As the technology advances, this type of processors starts to emerge, first among them being the Weitek WTL1032/1033 chip-set.

A. Weitek WTL1032/1033 [8]

1. Architecture and Interface

The chip set consists of a multiplier (WTL1032) and an adder (WTL1033). To increase the throughput, the array is partitioned into a three-stage pipe. At a clock rate of 20MHz, a maximum throughput of 10 Megaflops per chip can be achieved. A mode bit can be set to make the pipeline registers transparent in order to reduce the latency for scalar computations. There are three 16-bit data busses, two for the input and one for the output. The I/O bandwidth completely matches with the throughput of the array. The circuits can be configured in such a way that 100% of the pipe speed can be achieved in the applications such as FFT butterfly [9] and a 4 X 4 matrix multiplication for graphics computations.

2. Functionality and Data Formats

This chip-set supports the IEEE single-precision binary floating-point standard. In addition to the basic arithmetic operations such as multiply, add and subtract, the adder can also perform floating-point to integer conversion and vice versa and absolute add. Data types include 32-bit floating-point and 24-bit two's complement integer. To handle gradual underflow, the chip-set also supports operations on wrapped numbers, a normalized floating-point number with the exponent wrapped around to a negative value.

B. TRW TDC1022/1042 [10]

1. Architecture and Interface

TDC 1022 and TDC1042 are a 22-bit floating-point adder and multiplier respectively. Both chips implement a two-stage pipe in order to double the throughput. A feedback path is provided on chip so that continuous accumulation or multiplication can be performed. Both chips have a 22-bit output port. TDC1042 has two 22-bit input ports, whereas TDC1022 has only one.

2. Functionality and Data Formats

Aside from add, subtract and multiply functions, the chips can also perform zeroing, saturating, division by two, normalizing and denormalizing. The floating-point format has a 6-bit two's complement exponent and a 16-bit two's complement mantissa with the binary point to the right of the most significant bit.

C. Weitek WTL1064/1065 [11]

1. Architecture and Interface

The 64-bit floating-point chip-set consists of a multiplier (WTL1064) and an adder (WTL1065). In order to reduce chip area, an iterative multiplier array is used. For single-precision multiply, data has to pass through the array twice, and for double-precision, four times. A two-stage pipe is used for the multiplier and four-stage for the adder. Each of these pipes can be made transparent via a mode bit. At 16 MHz clock, the multiplier delivers 4 Megaflops performance for single precision, 2 Megaflops for double precision, and the adder yields 8 Megaflops for either precision operations.

Normally, the data loading and unloading are via two 32-bit input ports and one 32-bit output port. However, the input and output can be configured via the mode control to be a 64-bit input, single 32-bit port or two 16-bit input ports, and the output, either 32-bit or 16-bit port. User can select whether the most or least significant of operands should be loaded or unloaded first in the case of double precision operations.

2. Functionality and Data Formats

In addition to handling IEEE single and double precision data formats, the chip set also supports DEC F(32-bit) and D(64-bit)formats. Conversion from either of these formats to 32-bit two's complement integer and vice versa can also be performed. Functionality is similar to WTL1032/1033.

C. AMD 29325 [12]

1. Architecture and Interface

This design integrates both adder and multiplier on the same chip. It has a three-port architecture, two 32-bit input ports and one 32-bit output port. It is designed for one cycle latency for all operations. On chip there is a feedback path so that chained operations can be achieved without the results going off chip. The user can have the option of using just one 32-bit input port or have the port split into two 16-bit slices, with the operands loaded at the rising and falling edges of the clock.

2. Functionality and Data Formats.

Besides performing add, subtract and multiply, it can also perform 2-B (for an iterative divide operation where B is one of the input operands), and conversion from floating-point to 32-bit two's complement integer. The processor supports both IEEE single precision and DEC F(32-bit) data formats.

D. Analog Devices ADSP3210/3220 [13]

1. Architecture and Interface

A floating-point multiplier (ADSP3210) and a floating-point adder (ADSP3220) are the components of this chip-set. ADSP3210 has a 32 X 32 parallel multiplier array, implementing Booth and Wallace Tree algorithms. The data path is divided into a two-stage pipe; the first stage performs the mantissa multiply and the second does format adjust, rounding and exception detection. The pipe delay can be 100ns. Thus for single precision multiply, it can deliver 10 Megaflops performance. For double precision, however, iterative multiplication has to be done because of the size of the multiplier array, the performance is thus reduced to 2 Megaflops. Similarly, there is a two-stage pipe on the ADSP3220. The data I/O for both chips is via a 32-bit input bus and a 32-bit output bus. 32-bit operands can be loaded or unloaded every 50ns. Although the pipe delay is only 100ns, for double precision add, the adder can only deliver 5 Megaflops performance because the speed is I/O limited.

2. Functionality and Data Formats

In addition to the arithmetic operations for single and double precision data in IEEE formats, 32-bit two's complement integer arithmetics (multiply and add) and logic functions are supported. It also handles gradual underflow by wrapping denormalized numbers.

Table 3 summarizes the key features of the building-block VLSI processors.

IV. General Purpose Building-block Floating-point Processors

Weitek WTL1164/1165 [14]

The chip-set is designed to provide a high-speed 32-bit and 64-bit floating-point processing capability for general purpose computer systems at supermicrocomputer and superminicomputer levels. The chip-set design goals are to achieve short computational latency for increased scalar throughput, small die sizes for low cost processing and packaging, and an expanded set of operations to meet the requirements for the general purpose application.
The chip-set provides all four basic floating-point operations: add, subtract, multiply, as well as divide, conforming to the IEEE standard [1] with all exception handling including NaN, infinity, overflow and underflow.
Compare operation is also provided, producing comparison result required by the IEEE standard. Other functionality includes 32-bit two's complement numbers conversion to either format of the floating-point number and vice versa.

In order to achieve a superfast arithmetic processing, the chip-set uses dedicated circuit array to perform the required functions. The array flowthrough time for add, subtract, compare and conversions operations is 150ns for both single and double precisions, while multiplication takes 150ns or 250ns for single or double precision respectively. Division operation is implemented to fully conform to the IEEE standard, since it is important in a general purpose computer application to meet the accuracy and precision specified by the IEEE standard. Other division algorithm which uses a look-up table to obtain approximation to the reciprocal is often slower and in some cases will not meet the IEEE standard for accuracy at all.

The chip-set has several mode registers that select optional characteristics which are not often changed. Such mode registers include the IEEE rounding informations, a fast mode operation which substitutes zero for a denormalized number, and ability to control the array latency as a function of the number of clock cycles.

Figure 1 shows a simplified block diagram and the microphotograph of the floating-point multiplier core. The exponent block performs the addition of the exponents of the two operands, and subtract the appropriate bias from it. The exponent block also generates condition for overflow and underflow. The fraction block consists of an array of Carry Save Adders (CSA), and a 53 bit Carry Look-ahead Adder (CLA). To perform the array multiplication, a modified-Booth algorithm is employed to reduce the number of partial products required to be added. For a single precision multiplication, the array is used once and the CLA will provide the final fraction result. For double precision, the array is used twice, and again the CLA will provide the final result.

Figure 2 shows the block diagram and the microphotograph of the adder/divider chip. For add operation, the exponents of both operands are compared and the difference is used to right shift the fraction of the operand with the larger exponent. Now that the exponent is aligned, both fractions are added or subtracted depending on the signs of the operands. A priority encoder hardware is used to detect how many bit position a post-normalization is needed to be performed. The fraction is then left-shifted and the left shift amount is subtracted from the exponent. Detection of overflow and underflow is then performed on the exponent and the rounding is also performed on the fraction at the same time. The datapath is used once for single and double precision add, and conversion among three different data formats (single, double floating-point formats, and 32 bit 2's complement integer). A 8 by 8 'octal' barrel shifter technique is employed to perform up to 63-bit shift. This technique significantly improves the speed and area over a conventional binary shifter. An octal barrel shifter shifts 0 to 7 bit position per stage by the use of 8 to 1 multiplexer gates. Consequently, only 2 instead of 6 MUX stages are required to perform up to 63 bit shift. The gate delay can be roughly compared as log base 8 of 64 versus log base 2 of 64 for octal versus binary shifter. A fast carry look ahead technique (4 bit at a time) is also implemented on the incrementer and priority encoder hardware.

An iterative one-bit at a time restoring divide algorithm is employed to perform single and double precision divisions. On each cycle the divisor is subtracted from the partial remainder, and one quotient bit is generated. It is a '1' if the subtraction result is positive and a '0' otherwise. A 56-bit shift register chain is implemented above the priority encoder hardware. The shift register collects all quotient bits generated on every cycle. 25 and 54 cycles are needed to perform single and double precision division respectively. On the last cycle the priority encoder determines the left shift amount required to normalize a single precision quotient result, since the shift reqister is partially filled. The exponent block, meanwhile, is subtracting the exponent of the divisor from the exponent of the dividend, and an appropriate bias is added accordingly.

Every cycle, an operand may be loaded and unloaded through 32 bit I/O data pins. Internally, a precharged 32-bit bus serves to connect the input and output latches to the I/O pads. To save area, all data bits are stored in half registers (latches) instead of full master-slave registers. As a result of a shared internal I/O bus, a chain of multiplication or addition operations may be performed without having the result to go off-chip first, thereby increasing the scalar performance throughput. Since both chips implement 64 bit wide array datapath, operation involving operands with differing formats can also be handled directly without speed penalty to yield a double precision result (a mixed precision operation).

Figure 3 shows a design of an Extended Floating Point Unit (EFPU) [15] using WTL 1164/1165 as the central arithmetic elements. It is designed as a memory mapped peripheral to work in MC68020 based systems. A peak performance of 2 Megaflops can be expected.

IV. Conclusion

In this paper we have given an overview of the state-of-the-art VLSI implementations for handling floating-point arithmetics. Figure 4 shows where the general purpose building block FP processors stand among other VLSI processors. The IEEE binary floating point standard(P754), although more complex to implement than other standards, is being adopted almost universally because of it is rigorously defined and it offers greater precision and larger dynamic range. Because of the requirements for different applications, two main groups of floating-point processors are developed: general purpose and building blocks. As VLSI technology develops further, we will see that the general-purpose microprocessor will integrate the hardware to handle floating-point arithmetics on the same chip, resulting in the disappearance of the former group of processors. However, the building blocks will still be needed for the most speed-demanding applications. They will have further enhanced functionality and performance and greater I/O capability.

References:

[1]. A Proposed Standard For Binary Floating Point Arithmetic Draft 10.0 of IEEE Task P754; December 2, 1982

[2]. Nave, R. and Palmer, J., " A Numeric Data Processor", ISSCC Digest of Technical Papers, P.108-109; February 1980.

[3]. Huntsman, C. and Cawthron, D.," The MC68881 Floating-point Coprocessor", IEEE Micro, p.44-54, December 1983.

[4]. National Semiconductor, NS16081 Floating Point Unit Data Sheet; May 1983.

[5]. Advanced Micro Devices Am9511A/9512 Floating Processor Manual 1981.

[6]. Mor S., Hingarh H., Vora M., Wilnai D. and Longo T., "A 16b Microprocessor for Realtime Applications", ISSCC Digest of Technical Papers, p.28-29; February 1983.

[7]. Sandesara S., "Zilog's Z8070 Floating-point Processor"

[8]. Woo B., Lin L. and Owen R. E., "ALU, Multiplier Chips Zip Through IEEE Floating-point Operation", Electronics, p.121-126, May 19, 1983.

[9]. Woo B., Lin L. and Ware F., " A High-Speed 32 Bit IEEE Floating-point Chip Set for Digital Signal Processing" ICASSP paper 16.6, March 19-21, 1984.

[10]. Eldon J. and Winter G.E., "Floating-point Chips Carve Out FFT Systems", Electronic Design, p.133-138, August 4, 1983.

[11]. Ware F., Lin L., Wong R., Woo B. and Hansen C., "Fast 64-bit Chip Set Gangs Up for Double-Precision Floating-point Work", Electronics, July 12, 1984.

[12]. Quong D. and Periman R., "Single-chip Accelerators Speed Floating-point and Binary Computations" Electronic Design, p.246-262, November 15, 1984.

[13]. Analog Devices Word Slice Product Family Data Sheet, September, 1984.

[14]. Weitek WTL1164/1165 Preliminary Data Sheet.

[15]. White, G., "Solutions For a High Performance Floating-Point Coprocessor Design", Internal Memo, Feb 14,1985

TABLE 1

KEY APPLICATION FOR FLOATING POINT ARITHMETICS

Computer Aided Engineering
- Simulation
- Modelling

Scientific Computation
- Quantum Physics
- Quantum Chemistry
- EM Boundary Value Problems

Graphics

Seismic Energy Exploration

Medical Imaging

Telecommunications

Radar/Sonar Processing

Weather Analysis

Spectrum Analysis

Noise Cancellation

Digital Filtering

Non-destructive Testing

TABLE 2

Features of General–purpose Floating–Point Processors

1. Intel 8087/80287

Technology	3 micron HMOS		
Arithmetic functions	+,−,x,/,rem,sqrt,transcendental		
Data format/type	IEEE 32,64,80b FP, 16,32,64b int, 18 BCD		
Clock Rate	5 megahertz		
Input/Output	16b bidirectional bus		
Pin Count	40 pins		
Availability	now		
Performance	ADD	MPY	DIV
single prc	17us	19us	39us
double prc	17us	27us	39us

2. Motorola 68881

Technology	2.25 micron HCMOS		
Arithmetic functions	+,−,x,/,rem,mod,sqrt,transcendental		
Data format/type	IEEE 32,64,80b FP, 8,16,32b int, 96b BCD		
Clock Rate	16.67 megahertz		
Input/Output	configurable to 8,16,32b data bus		
Pin Count	64 pin DIP / 68 pin LCC		
Availability	sampling		
Performance	ADD	MPY	DIV
single prc	2.8us	2.8us	3.4us
double prc	2.8us	3.7us	5.6us

3. National 32081

Technology	3 micron XMOS		
Arithmetic functions	+,−,x,/,conversion to integer		
Data format/type	IEEE 32,64b FP, 8,16,32b integer		
Clock Rate	10 megahertz		
Input/Output	16 bit bidirectional bus		
Pin Count	24 pin DIP		
Availability	now		
Performance	ADD	MPY	DIV
single prc	7.4us	4.8us	8.9us
double prc	7.4us	6.2us	12us

4. Fairchild 9450

Technology	I3L–II, bipolar		
Arithmetic functions	+,−,x,/,compare, int convertion		
Data format/type	MIL–STD–1750A ISA 32,48b FP, 8,16,32b integer		
Clock Rate	20 megahertz		
Input/Output	16 bit bidirectional		
Pin Count	64 pin DIP		
Availability	sampling		
Performance	ADD	MPY	DIV
single prc	4.5us	5.6us	9.8us
(48b) double prc	5.75us	12.4us	21.2us

5. Zilog 8070

Technology	2 micron NMOS		
Arithmetic functions	+,−,x,/,remainder,sqrt,compare		
Data format/type	IEEE 32,64,80b FP, 16,32b int, BCD		
Clock Rate	10 megahertz		
Input/Output	16 bit bidirectional		
Pin Count	40 pin DIP		
Availability	no		
Performance	ADD	MPY	DIV
single prc	1.8us	2.8us	2.9us
double prc	1.8us	4.2us	4.3us

6. AMD 9512

Technology	NMOS		
Arithmetic functions	+,−,x,/		
Data format/type	IEEE 32,64b FP		
Clock Rate	3.125 megahertz		
Input/Output	8 bit bidirectional		
Pin Count	24 pin DIP		
Availability	now		
Performance	ADD	MPY	DIV
single prc	163.8us	81.3us	84.5us
double prc	992us	595.2us	1638.4us

TABLE 3

Features of Floating–Point Building–Block

1. Weitek 1032/1033

Technology	3 micron HMOS	
Data format/type	IEEE 32 bit FP, 24 bit integer	
Clock Rate	20 megahertz	
Input/Output	two 16bit input, one 16 bit output	
I/O bandwidth	120 megabytes/s	
Pin Count	64 pin DIP, 68 pin LCC	
Availability	now	
Performance	ADD	MPY
(32b) throughput	10 MFlops	10 MFlops
(32b) latency	500 ns	500 ns

2. TRW TDC1022/1042

Technology	1 micron bipolar triple diffusion	
Data format/type	22 bit FP	
Clock Rate	10 megahertz	
Input/Output	one 22b in(1022), two 22b in(1042), one 22b out	
I/O bandwidth	80 megabytes/s (1022/1042)	
Pin Count	64 pin DIP (1022), 84 pin LCC (1042)	
Availability	now	
Performance	ADD	MPY
(22b) throughput	10MFlops	10 MFlops
(22b) latency	200ns	200ns

Features of Building-block Floating-Point

(continued)

3. Weitek 1064/1065

Technology	2 micron NMOS		
Data format/type	IEEE/DEC 32,64b FP, 32b integer		
Clock Rate	16.67 megahertz		
Input/Output	two 32b input, one 32b output		
I/O bandwidth	200 megabytes/s		
Pin Count	144 pin PGA		
Availability	now		
Performance		ADD	MPY
(32b)	throughput	8 MFlops	4 MFlops
	latency	720ns	600ns
(64b)	throughput	8 MFlops	2 MFlops
	latency	720ns	840ns

4. AMD 29325

Technology	2 micron IMOX bipolar		
Data format/type	IEEE 32b FP, 32b integer, DEC 32b FP		
Clock Rate	10 megahertz		
Input/Output	two 32b input, one 32 bit output		
I/O bandwidth	120 megabytes/s		
Pin Count	144 pin PGA		
Availability	sampling		
Performance		ADD	MPY
(32b)	throughput	10MFlops	10 MFlops
	latency	100ns	100ns

5.Analog Devices ADSP3210/3220

Technology	2 micron CMOS		
Data format/type	IEEE 32,64b FP, 32b integer		
Clock Rate	10 megahertz		
Input/Output	one 32 bit input, one 32 bit output		
I/O bandwidth	160 megabytes/s		
Pin Count	100 pin PGA		
Availability	no		
Performance		ADD	MPY
(32b)	throughput	10 MFlops	10 MFlops
	latency	350ns	350ns
(64b)	throughput	5 MFlops	2 MFlops
	latency	500ns	600ns

6. Weitek 1164/1165

Technology	2.5 micron NMOS			
Data format/type	IEEE 32,64b FP, 32b integer			
Clock Rate	20 megahertz			
Input/Output	one 32 bit multiplexed input/output			
I/O bandwidth	80 megabytes/s			
Pin Count	64 pin DIP, 68 pin LCC			
Availability	now			
Performance		ADD	MPY	DIV
(32b)	throughput	4.4 MFlops	4.4 MFlops	-
	latency	150ns	150ns	1.25us
(64b)	throughput	2.5 MFlops	2.5 MFlops	-
	latency	150ns	250ns	2.7us

FIGURE 1

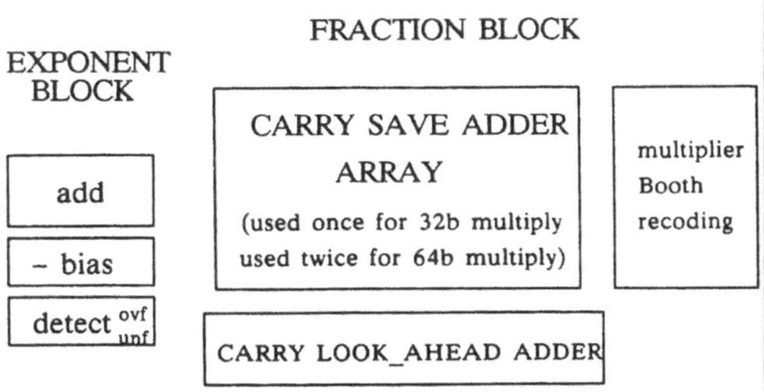

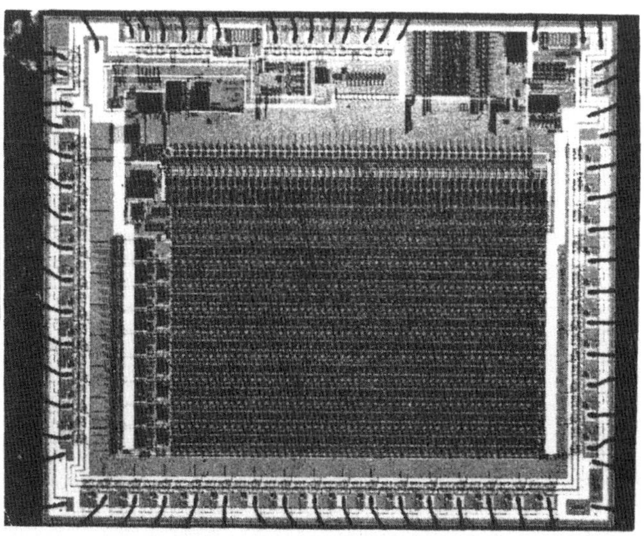

FIGURE 2

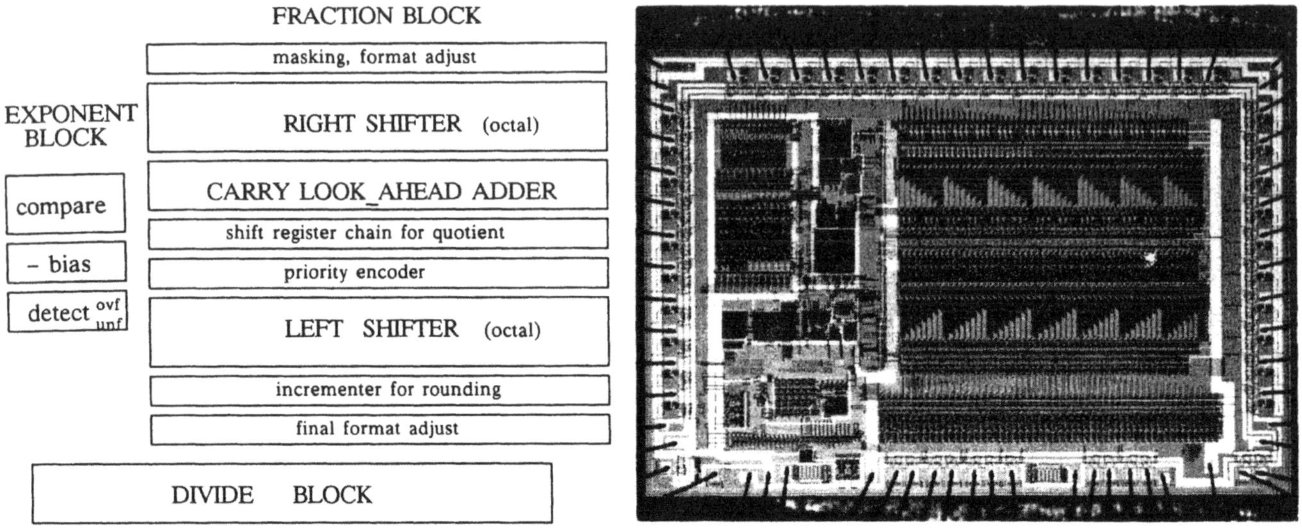

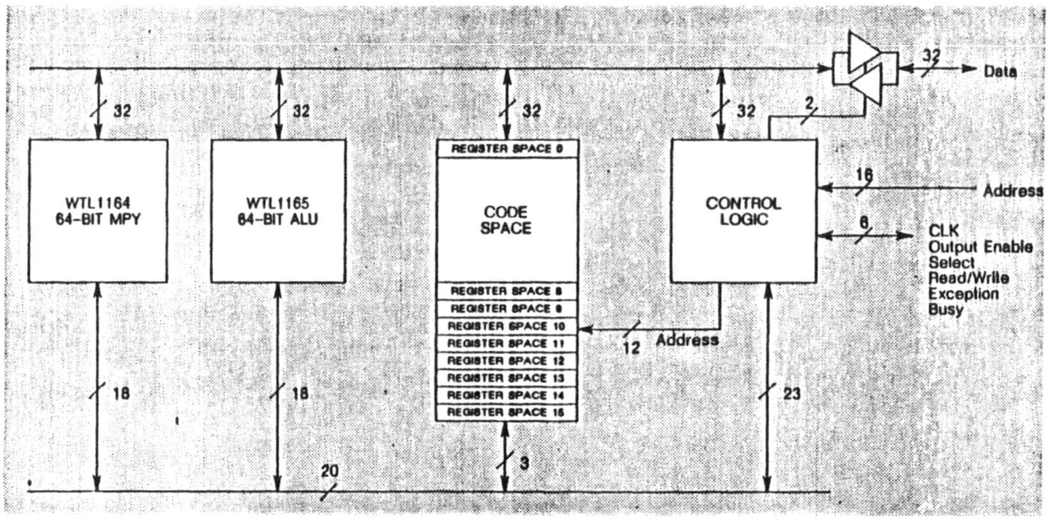

Figure 3 Block diagram of the Extended Floating-Point Unit (EFPU)

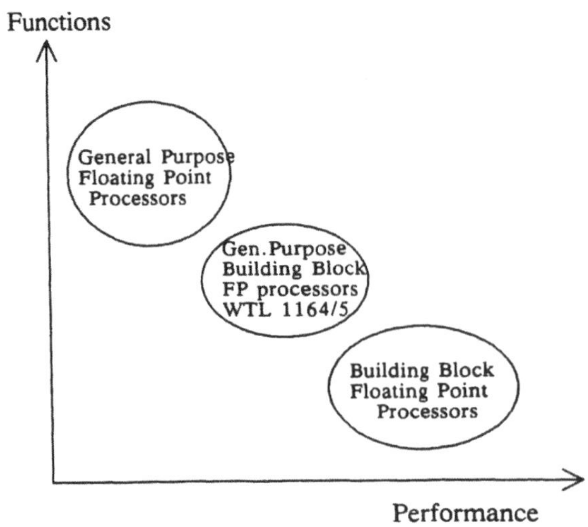

FIGURE 4

Fast Multiply and Divide for a VLSI Floating-Point Unit

B. K. Bose, L. Pei, G. S. Taylor, D. A. Patterson

Department of Electrical Engineering and Computer Sciences
University of California
Berkeley, CA 94720

ABSTRACT

This paper presents the design of a fast and area-efficient multiply-divide unit used in building a VLSI floating-point processor (FPU), conforming to the IEEE standard 754. Details of the algorithms, implementation techniques and design tradeoffs are presented. The multiplier and divider are implemented in 2 micron CMOS technology with two layers of metal, and occupy 23 square mm (23% of the entire FPU). We expect to perform extended-precision multiplication and division in 1.1 and 2.8 microseconds, respectively.

1. Introduction

High speed computation is essential for a large class of problems, like computer modeling and simulation, CAD/CAM, computer graphics, image processing, and robotics, where integer arithmetic lacks the range and precision of most of these real-world needs. Traditionally, floating point arithmetic has been slow in software and expensive in hardware. VLSI technology is now making it possible to have fast, inexpensive floating point arithmetic.

In keeping with increasing CPU performance, floating point computation needs to be fast. A balanced implementation should take into account the relative frequency of different floating point operations and implementation constraints, that in turn affect the choice of algorithms, the micro-architecture, the clocking methodology and the design style.

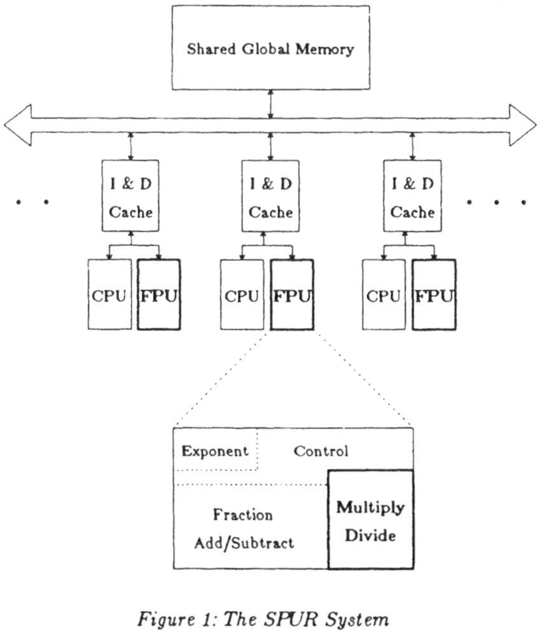

Figure 1: The SPUR System

In order to investigate these design tradeoffs, a single-chip floating-point processor has being designed. It is a tightly coupled coprocessor in the SPUR shared-memory multi-processor system [Hill86], providing instruction set enhancement for fast floating point computation (Figure 1).

The VLSI chip conforms to the IEEE Standard 754, providing the basic arithmetic functions for single, double and extended precisions, IEEE-style rounding, and detection of special operands and exceptions. It is designed in 2 micron, 2-layer-metal CMOS technology. This paper will focus on the multiply/divide section of this chip, which has a target speed of 1.1 microseconds for extended-precision multiply and 2.8 microseconds for extended-precision divide.

Several floating point units are commercially available, and some have recently been described in the literature [Nave80], [Shahan84], [Wolrich84], [Gavrielov86], [Troutman86]. We compare execution times between different implementations, to get an idea of the relative performance of our algorithms and implementation. We do not include multiple-chip designs in this comparison, since the design tradeoffs are then quite different. Table 1 summarizes the results for register-to-register operations.

Table 1: Performance of Multiply and Divide in 1-chip VLSI FPUs

Implementation	Significand	Cycle Time (ns)	Multiply (μs)	Divide (μs)
Intel 8087	64 bits	100	13.80	19.80
National 32081	53 bits	100	6.20	11.30
Motorola 68881	64 bits	60	3.12	5.04
Western Elec. 32106	64 bits	56	2.80	16.80
Digital Equipment*	56 bits	100	2.80	4.70
Fairchild Clipper*	53 bits	30	2.07	5.46
SPUR	64 bits	140	1.12	2.80

*These numbers are typical, while the others are worst-case.

2. Choice of Algorithms

Recent papers [Gamal86], [Uya84] show that with current 2 micron technology, combinational 32 bit multipliers require about 40,000 transistors, and take up about 30 square mm. Full 64-bit array multipliers are still difficult to integrate onto a single chip. Even if we could build a multiplier that computed in a single cycle, it would be difficult to find an adder that was proportionately faster. Again, divide times must improve with multiply; otherwise, algorithm designers will be tempted to avoid divide. Addition and multiplication in VLSI arithmetic accelerators have received a lot of attention lately, but not much work has been reported to build fast division schemes in VLSI. Higher radix non-restoring division is becoming feasible in VLSI, and here again, operation frequencies can serve as a guideline for choosing the most appropriate algorithm and for budgeting hardware. This is especially critical in VLSI, where propagation delay between modules within a chip and inter-chip delay are so different, directly affecting what goes onto a chip. Thus, chip area and I/O pins need to be allocated very carefully. Again, since there is a linear relationship between delay and fanout in MOS technology, putting more functionality on chip can adversely affect cycle time, since control signals have to now drive larger capacitance.

There have been several studies of various programs and benchmarks that are intensive in floating-point computation. We summarize results from Berkeley [Leung86] with those of Knuth[1971] and Gibson[1970] in Table 2. We see that add/subtract operations occur from 1.5 to 2.5 times more frequently than multiply operations, which in turn are 2 to 3 times as frequent as divide operations. The Lattice Filter seems to be an exception in that divisions occur much less often than in the others, and additions occur less frequently than multiplications. This table indicates chip resource allocation for a balanced design, where the proportion of hardware for add : multiply : divide should be close to the ratio of operation frequency. For example, a large chip area invested in an array multiplier may not be cost-effective without a proportionately fast divider.

Table 2: Relative Frequencies of Floating Point Operations

Source	Add,Subtract	Multiply	Divide
Knuth	2.30	1.00	0.38
Gibson Mix	1.80	1.00	0.39
Whetstone	1.80	1.00	0.50
Lattice Filter	0.75	1.00	0.08
SPICE	1.45	1.00	0.35

Operation frequencies are normalized to Multiply.

Algorithms can have quite different area and time costs depending on their implementation technology, whether it is Schottky TTL, ECL MSI, ECL gate arrays or MOS VLSI. Estimates of area and delay depending on gate count ignore such realities as fan-in, fan-out, interconnect and chip crossings. In VLSI, datapath pitch is usually determined by interconnect requirements, like the number of data busses that need to traverse it. Thus, the size of a variety of circuits is the same in one direction, while varying in the other. Naturally, some circuits will be much more densely packed than others, and so merely counting the number of gates in a circuit block can give a misleading idea of the area it requires. In Table 3, we present the relative areas of some basic circuit blocks, together with their relative delays, for driving identical loads.

To illustrate this technology dependence, let us see whether it is advisable to build Booth recoding into an iterative multiplier in the two technologies. Without recoding, we require eight rows of adders to reduce one multiplier byte into its partial 'sum' and 'carry' vectors; and with recoding, we require four 4:1 multiplexors and

Table 3: Circuit Area-Time Relationships in Bipolar LSI and CMOS VLSI

Circuit	Technology			
	Macrocell ECL LSI		Custom CMOS VLSI	
	Area	Time	Area	Time
2:1 Multiplexor	1	1.0	1	1
4:1 Multiplexor	4	2.0	2	1
2-in OR,NAND	1	0.8	1	2
2 2-in Exclusive-OR	1	1.2	2	2
D Flip-Flop	2	1.2	2	3
Full Adder	4	1.2	8	3
8-bit barrel-shift	7	3.0	7	1

Area and time of circuit blocks are normalized to a 2:1 multiplexor in each technology. In ECL, for example, a full adder is the same size as a 4:1 multiplexor, whereas in CMOS, a full adder is four times the size of a 4:1 multiplexor.

four adders. Note that some CSA rows can evaluate in parallel, resulting in 5 and 3 effective adder delays in the two cases. From Table 3, we see that in Bipolar LSI, the areas of the two schemes are the same, and the scheme with Booth recoding is 7% faster. In CMOS, the area with Booth recoding is 37% less than without recoding, and is also 33% faster. Clearly, Booth recoding is preferable in CMOS, whereas it is a toss-up in Bipolar LSI.

Using process yield curves, we estimated that the maximum size of our chip can be 10mm × 10mm. Accounting for periphery, this leaves about 9mm on a side for circuits. Allowing 25% area for control and routing, this leaves 60 square mm. for the exponent and fraction datapaths. Using the relative frequencies in Table 2, multiply and divide together accounts for 40% of the operations, and so are allocated close to that percentage of the datapath, i.e., 24 square mm. Given this area constraint, we moved away from purely combinational algorithms to iterative ones.

3. Implementation Considerations

The custom chips in the SPUR system are being designed in CMOS technology. Complimentary MOS provides a variety of desirable characteristics for implementation of a system of this complexity. CMOS technology, with its characteristic scalability, provides a good speed-power product, making it attractive at high levels of integration. CMOS has high noise immunity, high circuit density, provides a wide choice of design styles, and is compatible with the CAD tools at our disposal.

The maximum width of the fraction datapath is 73 bits (for example, the partial product vectors for multiply). Delay in control signals that run the entire length of the datapath has a large impact in the delay in and between modules in the datapath. If there is any appreciable resistance in these control lines, the RC delay can become a significant fraction of module delay, leading to slower computation rates and large clock non-overlap times to protect against clock skew. We chose a process with two available layers of metal, with control and data signals running orthogonally for the most part in the two metal layers, thus virtually eliminating any resistive delay. To minimize clock skew, we scale the drivers of the control lines to match the capacitances they have to drive, so that control delay is held between very tight tolerances for the entire width of the datapath.

To allow for a mix of static and dynamic design styles, we chose a four-phase clocking scheme, used in the other chips in the SPUR system as well. With an on-chip register file with an access time of one phase, this allows two accesses, one for read and one for write per cycle, with the two intermediate phases going to precharge the dynamic busses. The cycle time is limited by the register file read and write time. The current technology allows transistors with minimum channel length of 2 microns, with a minimum size inverter discharging 1pf capacitor in one phase. The present clocking scheme is shown in Figure 2.

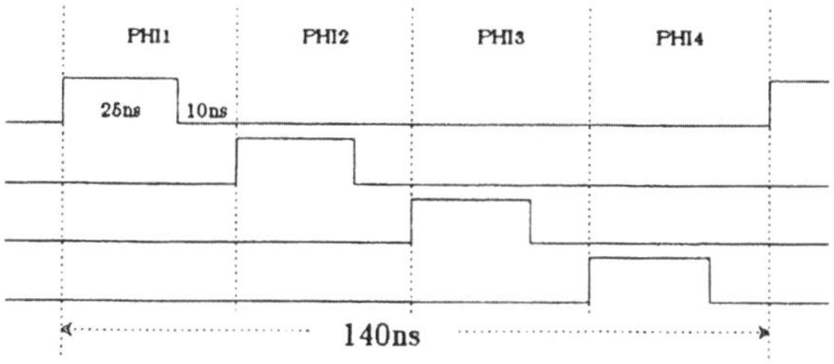

Figure 2: SPUR Clocking Scheme

3.1. Pipelining the Inner Loop

A key factor in making our design area-efficient is that the multiply and divide schemes have much in common, enabling significant sharing of hardware. Figure 3 shows the entire datapath for multiply and divide. A key factor in making our design achieve the speeds it did is the fact that the inner loop, the heart of our iterative algorithms, is completely pipelined, with different stages of the pipeline carefully adjusted for balanced delay. Modules that are density-critical are designed using dynamic circuits, and modules that have rigid timing constraints, and where precharge times cannot be overlapped with evaluation times, are designed using static circuits. The hardware blocks that are not shared are the multipler Booth recoder and the divider quotient selector and accumulator, and account for 8% of the entire multiply/divide unit.

The following sections describe the algorithmic and implementation considerations for this multiply/divide datapath.

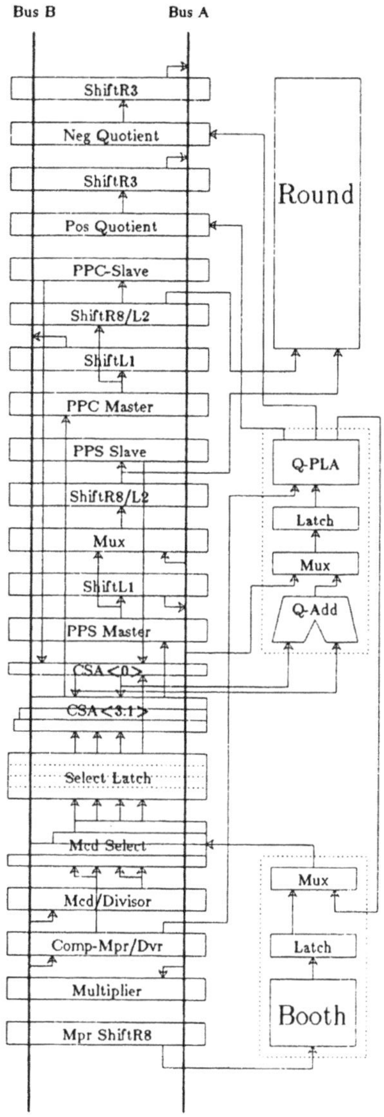

Figure 3: The Multiply-Divide Datapath

The datapath has the following sections, from bottom to top: input latches and multiplier shifter; multiplicand/divisor select; carry-save-adder tree; partial sum and partial carry formation; quotient accumulation.

4. The Multiplier

Since it is not feasible to build a 64 × 64 array multiplier as part of a single-chip FPU with currently available technology, we considered several iterative schemes. A 64 × 32 array requires 2 iterations to compute the full product, but takes up about twice as much area as a 32 × 32 array, which requires 4 iterations. Even the area of a 32 × 32 array just for the multiplier exceeds our area budget for both multiply and divide.

4.1. The Algorithm

Our multiplier is implemented in nine iterative steps. Each iteration implements a 64 bit by 8 bit multiplication. In each iteration, four overlapped triplets of multiplier bits (9 bits) are decoded by a modified Booth recoder. Four multiplicand multiples of magnitude +2MCD, +1MCD, -1MCD and -2MCD are needed per iteration, along with 0. The relative cost of a multiplexor compared to that of an adder makes Booth recoding feasible in CMOS, exchanging six rows of carry-save-adders with four rows of multiplexors and four carry-save-adders in the datapath. Also, separating the recoding from the carry-save-addition allows us to evaluate them in separate time-slots in our pipelined implementation, thus avoiding six CSA rows in the critical path.

The four overlapped triplets of multiplier pairs generate the four multiples of the multiplicand. They are added to the partial 'sum' and 'carry' terms of the previous iteration, using an array of four carry-save-adders (CSA). Note that the four multiples of the MCD are shifted left 2 bits with respect to each other, depending on

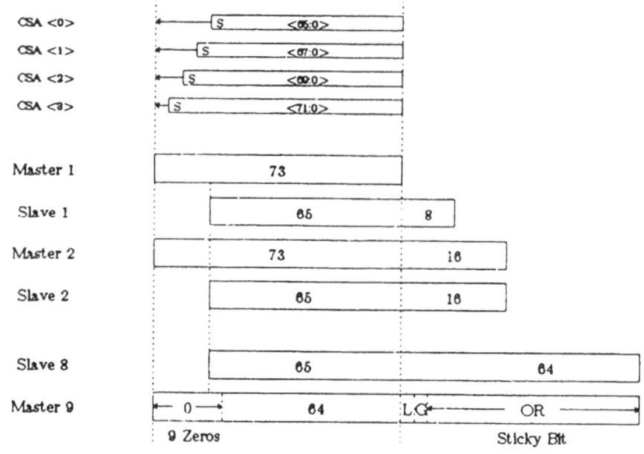

Figure 4: Forming the Product

The width of the CSA, multiplexors and PPS/PPC are 73 bits. After right-shifting 8 bits, the sum of the two partial products is 65 bits (64 bit magnitude and 1 sign bit).

the significance of each multiplier triplet. The partial 'sum' and 'carry' are shifted left 8 bits and 7 bits respectively, when looping them back to be the new inputs of the CSA for the next iteration. Since there are negative as well as positive operands, both in two's complement form, the multiplexers and CSA must be fully sign-extended to the left (MSB) side.

A carry-look-ahead adder is necessary at the start of the multiply operation, to produce the complement of the multiplicand. It is also necessary at the end to form the final result by adding the partial product vectors. Since the fraction unit has such an adder already, we share this module instead of duplicating it in the multiply/divide unit. This increases the setup and completion times by one cycle, but reduces the area of the unit by 14%.

4.2. The Multiply Inner Loop

The CSA tree, containing four rows of adders, is the critical path for the multiplier inner loop. To reduce this delay, two rows of the CSA tree, CSA<1,0>, evaluate in parallel, reducing the net delay to 3 CSA stages. This makes the interconnect less regular, but provides a 25% speed improvement in the CSA stage. For the divider, only one row of the CSA tree is necessary, and so we can use the isolated CSA row to advantage. Figure 5 shows the organization of the CSA tree.

In the multiplier inner loop, multiplicand selection is overlapped with shifting and rounding the partial product vectors. During each cycle, two phases are for signals controlled by the master, and the other two phases are for the evaluation of signals controlled by the slave. Blocks that are controlled by the master are the Booth recoder and the CSA tree, and blocks controlled by the slave are the multiplexor set and the shifter between master and slave partial product latches. Figure 6 shows the multiplier pipeline.

We have shown how the different parts of the multiply inner loop are pipelined. Several of the modules were designed using dynamic circuits to meet area, interconnect and timing constraints. Table 4 shows the design

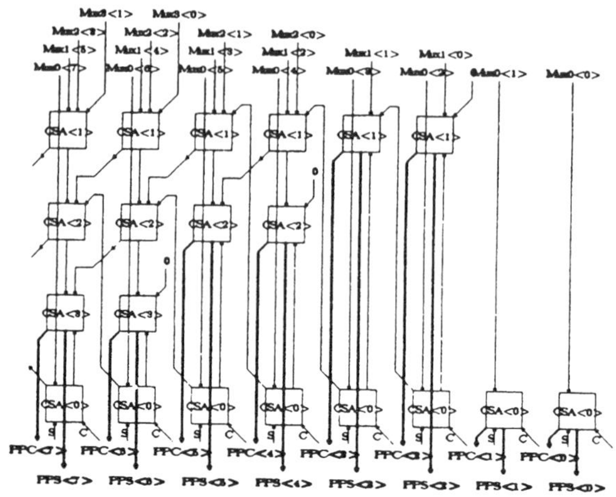

Figure 5: Least Significant Bits of the CSA Tree

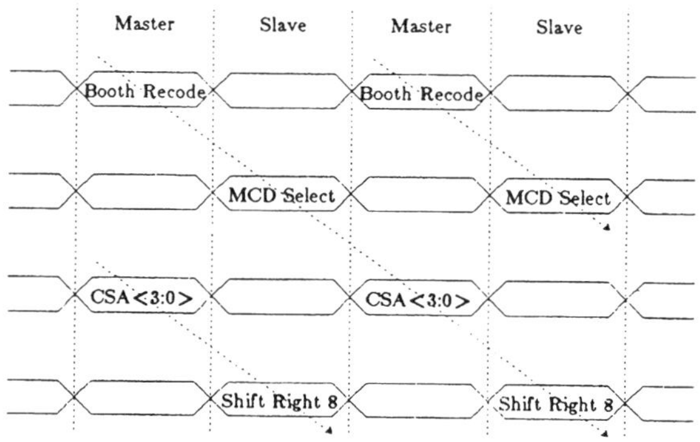

Figure 6: The Multiplier Pipeline

style used for building the modules in the multiply/divide inner loop, together with their area and time relationships, normalized to the Booth recode block.

The four rows of carry-save-adders take the longest time among the pipeline components. The dynamic CSA design was 16% smaller than its static counterpart, and was still able to meet the timing requirements. If we went to a more aggressive clocking scheme, it would be feasible to build the four multiplexors and the CSA tree static, and modify the pipeline to have only two stages. One stage could perform Booth recoding and MCD selection, while the other could do CSA evaluation and the right shift. Not only could the disparity between the stages become smaller, but we could also utilize two out of the four clock non-overlap times.

Table 4: Area-Time Relationships of Multiply Function Modules

Function	Design Style	Area	Time
Booth Recode	Dynamic	1.0	1.0
MCD Select	Dynamic	2.6	0.9
CSA Tree	Dynamic	9.9	4.3
Partial Sum & Carry	Static	8.7	0.5

Area and time are normalized to the Booth recode block. The CSA, for example, is almost ten times larger and more than four times slower than the Booth Recode block.

4.3. Rounding

The ability to perform unbiased rounding with error less than half a unit in the last place requires three extra bits, called the Guard, Round and Sticky bits. The Guard and Round bits are used if the intermediate result of a division is between .5 and 1 and hence requires a one-bit normalizing left shift. The Sticky bit is equal to zero only if all subsequent bits in a result of infinite precision are zero and is used to identify the half-way case for unbiased rounding.

During a multiply operation, the vectors containing the partial products are shifted right eight bits before being returned for the next iteration. These two eight-bit quantities are added together, and ORed to form the 'partial sticky' bit. This is fed back to the rounding adder for the next iteration. At the end of the final rounding addition, bit<0> of the result is the most significant bit of the rounding adder result, followed by the Guard bit. The Round bit is zero, since the result must be in a range between 1 and 4. The OR of the remaining bits of the rounding adder provides the Sticky bit, as shown in Figure 7.

Since each iteration takes half a cycle in this pipelined scheme, partial product evaluation takes four and a half cycles. It takes almost that long, again, for evaluating the MCD complement, the final carry-propagate addition and the rounding, making the total multiply latency 8 cycles. Some of this can be saved by duplicating the fraction ALU in the multiply/divide unit, and also duplicating the rounding PLA that generates the least significant bit. Currently, both of these are shared with the fraction unit; if these were duplicated, the potential time saving can be 20%.

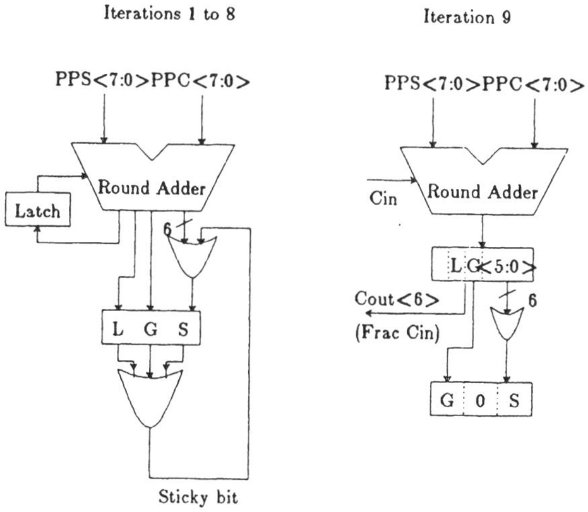

Figure 7: Multiply Rounding

5. The Divider

Restoring divide is the least expensive approach for radix-two division, but the cost increases exponentially with the number of bits generated per iteration. The same is true of parallel-serial schemes. Multiplicative inverse schemes produce incorrectly rounded quotients and inexact remainders, and later fix-ups can be expensive and time-consuming. SRT division [Robertson65], [Atkins67] focuses attention on quotient digit selection, and the remainder iteration does not require back-tracking. Higher radix SRT division schemes are likely to provide significant gains in area and speed, as better ways are found to provide compact quotient-selection logic, and concurrency between different portions of the algorithm (like partial remainder formation and quotient selection) is exploited.

5.1. The Algorithm

Our algorithm is based on radix-four, non-restoring division, using estimates of the divisor and partial remainder. The radix-four quotient digits are expressed using redundant representations of -2, -1, 0 +1 and +2, and the partial remainder is irredundant. This redundancy in the quotient digits permits less precision in comparing the divisor and partial remainder to select a quotient digit. The required precision required in inspecting the partial remainder and the divisor can be determined using P/D plots. It can be shown that six bits of partial remainder and four bits of divisor are needed to determine the next quotient digit [Freiman61],[Atkins68].

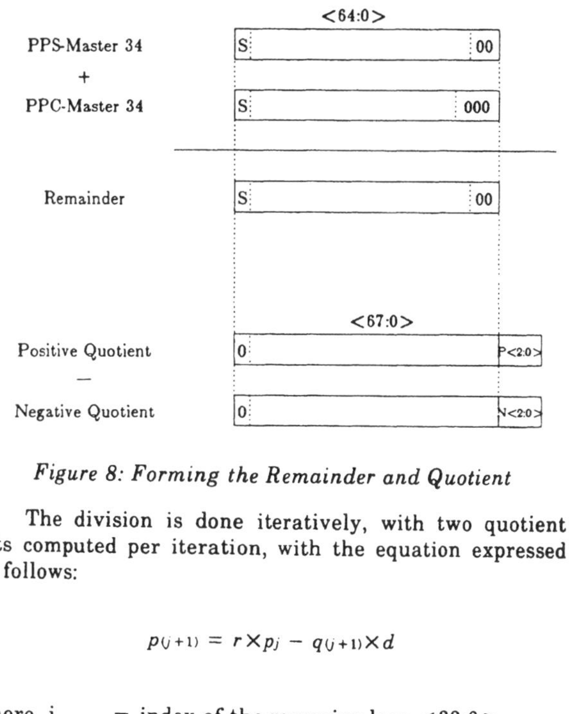

Figure 8: Forming the Remainder and Quotient

The division is done iteratively, with two quotient bits computed per iteration, with the equation expressed as follows:

$$p_{(j+1)} = r \times p_j - q_{(j+1)} \times d$$

where j = index of the recursive loop <33:0>
p_j = partial remainder after the jth loop
p_0 = dividend
$q_{(j+1)}$ = quotient digit after the jth loop
d = divisor
r = radix <4>

The hardware loop for generating the next remainder and the next quotient estimate contains an eight-bit carry-lookahead-adder, which generates six bits of truncated partial remainder. Together with these six bits, four bits of the truncated divisor are sent to the quotient selection logic, which in turn generates three bits, representing one of the five possible values of the quotient digit. Depending on the sign of the quotient digit, it is channeled into one of two registers, one holding positive and the other holding negative quotient estimates. These registers are shifted left two bits per iteration. The quotient selection logic is decoded to control a multiplexor, that decides what multiple of the divisor to use for the next iteration.

5.2. The Divide Inner Loop

For the divider, the partial product latches are used to hold the partial remainders, with one modification. Multiplexors in front of these latches let us load the master with the dividend at the very beginning of a divide. The PPC latch gets loaded with zero. The partial remainders are shifted left by two bits after every iteration.

Since the divider uses only one row of the CSA tree, one of its rows is separated out from the rest of them, to enable this fast and direct path for the divider. Partial remainder evaluation and quotient estimation are done in parallel. In the remainder evaluation, the CSA and left shift operations are split into master and slave phases. For the quotient evaluation, the 8-bit estimation adder is evaluated at the master time, while the quotient selection PLA and the divisor selection is done at the slave time. The divider pipeline is shown in Figure 9.

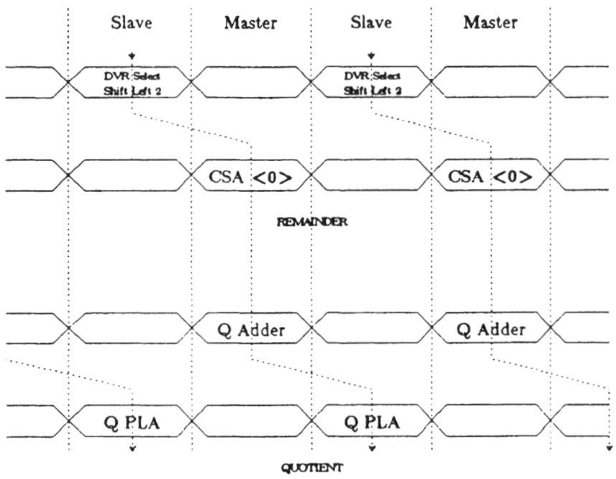

Figure 9: The Divider Pipeline

There are several reasons for our efficient divider. First, we use radix four, allowing the evaluation of two quotient bits per iteration. Second, non-restoring divide allows us to look ahead for quotient selection, keeping exact quotient determination until later, without backtracking at every iteration. Third, a small degree of redundancy in the quotient digit representation keeps us from having to generate more costly multiples of the divisor, albeit requiring an increase in quotient selection logic. Lastly, the concurrency between partial remainder formation and quotient selection [Atkins74] significantly increases algorithm efficiency. Table 5 shows the relative area-time cost for different sections of the divide pipeline, normalized to the divisor select block.

5.3. Quotient Selection

An integral part of the divide scheme is the logic for quotient selection. To keep the quotient selection logic from becoming the critical path, the quotient adder and PLA were split into two phases. The adder can now be

Table 5: Area-Time Relationships of Divide Function Modules

Function	Design Style	Area	Time
Divisor Select	Dynamic	1.0	1.0
CSA Row	Dynamic	4.1	3.2
Quotient Adder	Static	0.4	3.5
Quotient PLA	Static	0.5	4.8
Quotient Accumulate	Static	6.4	0.8

Area and time are normalized to the divisor select block. The quotient PLA, for example, is half as large and almost 5 times slower than the divisor selector.

dynamic, since it can be precharged when the PLA evaluates, thereby saving area. The PLA that generates the two bits of quotient per iteration, gets its inputs at the beginning of every slave phase and has to evaluate the quotient bits by the end of that phase. To achieve this strict timing requirement, we went with a pseudo-static PLA design. At the expense of a little DC power, output evaluation can be done in a single phase time. There are ten inputs to the PLA (6 bits from the quotient adder, and 4 from the divisor), and three outputs (2 quotient bits and the sign bit). Two optimizations were done to reduce the size of the PLA. Firstly, the quotient sign bit is the same as the sign of the partial dividend, and hence does not have to be a PLA output. Secondly, with optimal encoding of the PLA outputs, the number of product terms were reduced significantly. Given the small number of outputs, it was possible to exercise the PLA minimization tools [Rudell86] to find what encoding of outputs resulted in the smallest number of minterms (*). Table 6 summarizes the results.

Quotient selection takes the longest time for radix 4 division; remainder evaluation takes only 40% of the time of quotient selection. For radix 16 division, there is no extra cost in remainder evaluation, but quotient selection area increases by about 6 times. In the critical path, we now require several quotient adders that work in parallel, with the final result muxed out to the quotient PLA. The extra time cost is about 30%.

Table 6: Effect of Output Encoding on the Quotient PLA

Output	Twelve Unique Output Encodings											
Q=0	0	0	0	1	1	1	1	1	1	3	3	3
Q=1	1	1	3	0	0	2	2	3	3	0	1	1
Q=2	2	3	1	2	3	0	3	0	2	1	2	0
P-Terms	43	26	26	26	26	29	30	27	27	26	44	25*

5.4. Rounding

The divider must provide three rounding bits along with a 65-bit result. Since two quotient bits are generated at every iteration, 34 iterations are necessary to generate the partial remainder and quotient vectors. After adding the two partial remainder vectors, the sign of the remainder is returned to the multiply-divide unit. The OR of the rest of the bits provides the 'partial sticky' bit. Using a 3-bit rounding subtractor, which uses the complement of the sign of the remainder as carry, bits <2:0> of the quotient vectors are subtracted. The least significant bit is ORed with the 'partial sticky' bit to form the final Sticky bit, while the other two bits of result provide the Guard and Round bits. The carry out of the subtractor goes out to the fraction adder, as shown in Figure 10. It would be possible to eliminate most of this logic if we had a 68-bit ALU. But since we share the ALU with the fraction unit, which has only a 65-bit ALU, we have to retain this logic.

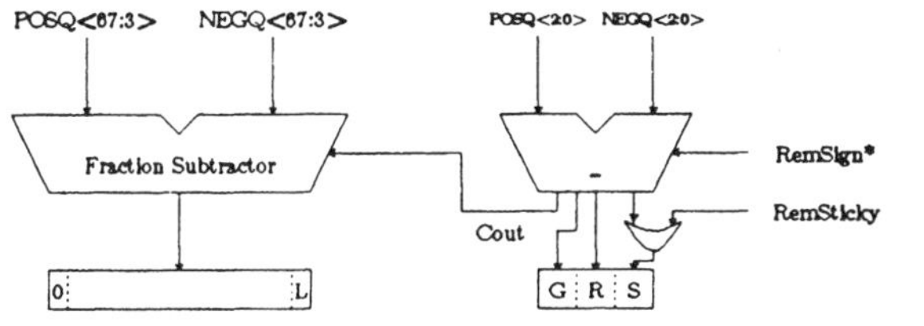

Figure 10: Divide Rounding

6. Scaling Implications

Further enhancement in performance will come from two inter-related causes. As technology continues to scale to smaller and smaller geometries, individual transistors and logic gates will improve in performance. Scaling will also enable more logic to be put on a single chip, allowing for further algorithmic enhancements. Reducing the feature size by α causes area to decrease by the square of α and the speed to increase by almost the same factor. Table 7 indicates the effects of halving the smallest geometry on chip, from 2 microns to 1 micron, on algorithms and their area and time.

Table 7: Technology Impact on Algorithm Area-Time Cost

Algorithm	Technology			
	2 micron		1 micron	
	Area	Time	Area	Time
Multiply (64 × 8, iterative)	1.0	1.00	0.25	0.25
Multiply (64 × 32, array)	1.9	0.65	0.48	0.16
Multiply (64 × 64, array)	3.8	0.35	0.95	0.09
Divide (SRT,r=4)	1.0	1.00	0.25	0.25
Divide (SRT,r=16)	1.5	0.70	0.38	0.18
Divide (Prescale, r=256)	3.5	0.40	0.88	0.10

Area and time are normalized to those for our implementation in 2 micron CMOS. For example, our multiply scheme will be 25% its present size when built in 1 micron CMOS and will be 4 times as fast. Again, a radix 16 divider will be 50% larger than radix 4 in 2 micron CMOS, but 38% its present size in 1 micron CMOS.

We can expect significant speed enhancements by matching technology to algorithm, while retaining a balanced design. Large combinational multipliers will soon be feasible as components in an FPU. Rounding hardware will need to be integrated into the multiply unit, to keep it from becoming the bottleneck.

Dividers, being inherently sequential in nature, are harder to speed up. The size of escalating cost (in area and time) in quotient selection logic will probably limit the use of non-restoring divide to radix 16 [Taylor85]. Partial remainder evaluation will be virtually unaffected, but the quotient selection logic will increase from radix 4 by about 6 times. Prescaling schemes [Ercegovac85] are worth exploring, for generating more quotient digits per iteration. Eight and even sixteen bits per iteration seem feasible, provided initial setup, final remainder adjustment and data flow can be handled efficiently. The estimate in Table 7 takes into account the fact that we need two 64 × 8 multipliers for the prescaling, and the datapath width increases by eight bits; on the other hand, quotient selection logic is greatly simplified.

7. Conclusion

VLSI provides certain unique opportunities and constraints in the design and implementation of high-performance arithmetic. Chip area and delay closely interact to affect all levels of design, from available functionality and choice of algorithms to clocking strategies and circuit design styles. In particular, we study the design tradeoffs offered by CMOS technology, the dominant VLSI technology in the 1980's. In an attempt to explore tradeoffs in this complex and inter-dependent design space, we present the design of a fast and compact multiply/divide unit implemented for a single-chip FPU.

8. Acknowledgements

Prof. Kahan has helped us understand the intricacies of the IEEE Standard and encouraged us all along; Prof. Hodges has always been available to advise us on circuit issues. Glenn Adams wrote portions of the FPU simulator, which Corinna Lee has completed and tested. Albert Wang and Timothy Hu have helped in layout and circuit simulation. Principal funding for the project is by DARPA, under contract N00039-85-C-0269.

9. References

[Atkins67]- D. E. Atkins, 'The Theory and Implementation of SRT Division,' Report No. 230, Dept. of Computer Science, Unviersity of Illinois, June, 1967.

[Atkins68]- D. E. Atkins, 'Higher-Radix Division Using Estimates of the Divisor and Partial Remainders,' IEEE Trans. Computers, Vol. C-17, No. 10, October 1968, pp. 925-934.

[Atkins74]- D. E. Atkins & U. Kalaycioglu, 'Concurrency in Generalized Radix Non-Restoring Division,' Proc. Twelfth Allerton Conference on Circuit and Switching Theory, October 1974, pp. 628-640.

[Ercegovac85] - M. Ercegovac & T. Lang, 'A Division Algorithm with Prediction of Quotient Digits,' Proc. Seventh IEEE Symposium on Computer Arithmetic, June 1985, pp. 51-56.

[Freiman61] - C. V. Freiman, 'Statistical Analysis of Certain Binary Division Algorithms,' Proc.IRE, Vol. 49, January 1961, pp. 91-103.

[Gamal86]- A. E. Gamal et al, 'A CMOS 32b Wallace Tree Multiplier-Accumulator,' Proc. ISSCC, February 1986, pp. 194-195.

[Gavrielov86] - M. Gavrielov & L. Epstein, 'The NS32081 Floating-Point Unit,' IEEE Micro, April 1986, pp. 6-12.

[Gibson70] - J. C. Gibson, 'The Gibson Mix,' IBM Systems Development Div., Tech. Report, June 1970.

[Hill86] - M. D. Hill et al, 'SPUR - A Multiprocessor Workstation,' to appear in IEEE Computer, November 1986.

[Knuth71]- D. E. Knuth, 'An Empirical Study of FORTRAN Programs,' Software: Practice and Experience, Vol. 1, No.2, April 1971, pp. 105-133.

[Leung86] - B. Leung & Y. M. Lin, ' Statistics on Floating Point Arithmetic,' CS 252 Report, May 1986.

[Nave80] - R. Nave & J. Palmer, 'A Numeric Data Processor,' Proc. Intl. Solid-State Circuits Conference, February 1980, pp.108-109.

[Robertson65] - J. E. Robertson, 'Methods of Selection of Quotient Digits during Digital Division,' File No. 663, University of Illinois, Urbana, June 1965.

[Rudell86]- R. Rudell, 'ESPRESSO,' 1986 VLSI Tools, Report No. UCB/CSD 86/272.

[Shahan84] - V. Shahan, 'The MC68881: The IEEE Floating Point Standard Reduced to One VLSI Chip,' Proc. IEEE Computer Conference, March 1984, pp. 172-176.

[Taylor85] - G. S. Taylor, 'Radix 16 SRT Division Methods With Overlapped Quotient Selection Stages,' Proc. Seventh IEEE Symposium on Computer Arithmetic, June 1985, pp. 64-71.

[Troutman86] - W. W. Troutman et al, 'Design of a Standard Floating-Point Chip,' IEEE J. of Solid-State Circuits, Vol. SC-21, No.3, June 1986, pp. 396-399.

[Uya84] - M. Uya et al, ' A CMOS Floating Point Multiplier,' IEEE J. Solid-State Circuits, Vol. SC-19, No.5, October 1984, pp. 697-702.

[Wolrich84] - G. Wolrich et al, 'A High Performance Floating Point Coprocessor,' IEEE J. of Solid-State Circuits, Vol. SC-19, No.5, October 1984, pp. 690 - 696.

Part VI: NUMBER REPRESENTATION

Editor's Comments on Papers 31 Through 37

This part explores system level issues associated with number representation. These papers describe a variety of "exotic" number systems and examine various aspects of floating-point arithmetic.

Paper 31 by Garner introduced the residue number system. This system represents data by their residues (i.e., their smallest positive remainders) with respect to a set of relatively prime moduli. Addition and multiplication are performed individually on the residues of each modulus. These operations can be performed in parallel with simple hardware, resulting in high speed and low complexity. Comparison of two residue numbers to determine which is larger and division are a real problem in the residue system, so it is best suited to specialized applications. More information on residue arithmetic is in the collection edited by Soderstrand, Jenkins, Jullien and Taylor (1986).

In the next paper, Matula provides a framework for analyzing the conversion of "real" numbers into floating point numbers and the conversion of numbers from one base to another. This paper defines floating-point number systems, presents conversion procedures, and then considers compound conversions (where a number may be converted several times during a computation).

In paper 33, Cohen, Hull, and Hamacher describe CADAC, a "user friendly" variable precision decimal floating-point arithmetic unit, including the number representation, system interfaces, instruction set, hardware implementation, realization of extended precision, rounding, and zero counting.

Next, Matula and Kornerup introduce the fixedslash and floating-slash number systems. These systems, which represent numbers as fractions, are shown to have the attractive computational property that all computations on single word operands produce results that can be represented exactly in double precision. Both the fixed-slash and floating-slash number systems (roughly analogous to conventional fixed point and floating-point number systems) are described and their characteristics are derived.

In the fifth paper, Kornerup and Matula present results for Lexicographic Continued Fraction (LCF) number systems. The paper defines the LCF representation, formulates the binary mediant operation, and uses the binary mediant to derive the gap sizes of finite LCF systems

In the next paper, Matsui and Iri present a floating-point number system that avoids the traditional problems of overflow and underflow. In this system the normal integer exponent of a floating-point number is replaced with a floating-point number, which greatly expands the range of the number system. This paper describes the system, its implementation, the error, and concludes with several numerical examples.

In the final paper, Olver presents the level index (li) and symmetric level index (sli) number systems. These systems are similar to the approach presented by Matsui and Iri in paper six. The paper describes the li and sli systems, proves that they are closed for addition, subtraction, multiplication, and division, and addresses precision, and implementation issues.

References

M. A. Soderstrand, w. K. Jenkins, G. A. Jullien, and F. J. Taylor, *Residue Number System Arithmetic: Modern Applications in Digital Signal Processing*, New York: IEEE Press, 1986.

The Residue Number System*

HARVEY L. GARNER†

Summary—**A novel number system called the residue number system is developed from the linear congruence viewpoint. The residue number system is of particular interest because the arithmetic operations of addition, subtraction and multiplication may be executed in the same period of time without the need for carry. The main difficulties of the residue code pertain to the determination of the relative magnitude of two residue representations, and to the division process. A discussion of the arithmetic operations and the conversion process required to convert from a residue code to a weighted code is given. It is concluded that in its present state the residue code is probably not suitable for general purpose computation but is suitable for a special class of control problems. Further research in both components and arithmetic is required if a residue code suitable for general purpose computation is to be obtained.**

Introduction

IN THIS PAPER we develop and investigate the properties of a novel system called the residue code or residue number system. The residue number system is of particular interest because the arithmetic operations of addition and multiplication may be executed in the same time as required for an addition operation. The main difficulty of the residue code relative to arithmetic operations is the determination of the relative magnitude of two numbers expressed in the residue code. The residue code is probably of little utility for general purpose computation, but the code has many characteristics which recommend its use for special purpose computations. The residue code is most easily developed in terms of linear congruences. A brief discussion of the pertinent properties of congruences is presented in the next section.

Congruences

The congruence relationship is expressed as

$$A \equiv \alpha \text{ Mod } b \tag{1}$$

which is read, A is congruent to α Modulo b. The congruence states that the equation

$$A = \alpha + bt \tag{2}$$

is valid for some value of t, where A, α, b and t are integers. α is called the residue and b the base or modulus of the number A.

* Manuscript received by the PGEC, January 20, 1959; revised manuscript received March 31, 1959. An earlier version of this paper was presented at the 1959 WJCC, San Francisco, Calif.

† Dept. Electrical Engineering, University of Michigan, Ann Arbor, Mich.

As examples of congruences consider

$$\begin{aligned} 10 &\equiv 7 \text{ Mod } 3 \\ 10 &\equiv 4 \text{ Mod } 3 \\ 10 &\equiv 1 \text{ Mod } 3. \end{aligned} \tag{3}$$

In these examples the integers 7, 4, and 1 form a residue class of 10 Mod 3. Of particular importance is the least positive residue of the class which in this example is one. The least positive residue is that residue for which $0 \leq \alpha < b$.

Consider the following set of congruences. Given

$$\begin{aligned} A_1 &\equiv \alpha_1 \text{ Mod } b \\ &\vdots \\ A_n &\equiv \alpha_n \text{ Mod } b. \end{aligned} \tag{4}$$

Then:

1) Congruences with respect to the same modulus may be added and the result is a valid congruence.

$$\sum_{i=1}^{n} A_i \equiv \left(\sum_{i=1}^{n} \alpha_i \right) \text{ Mod } b. \tag{5}$$

It follows that terms may be transferred from one side of a congruence to the other by a change of sign and also that congruences may be subtracted and the result is a valid congruence.

2) Congruences with respect to the same modulus may be multiplied and the result is a valid congruence.

$$\prod_{i=1}^{n} A_i \equiv \left(\prod_{i=1}^{n} \alpha_i \right) \text{ Mod } b. \tag{6}$$

It follows that both sides of the congruence may be raised to the same power or multiplied by a constant and the result is a valid congruence.

3) Congruences are transitive. If $A \equiv B$ and $B \equiv C$, then $A \equiv C$.

4) A valid congruence relationship is obtained if the number, the residue and the modulus are divided by a common factor.

5) A valid congruence relationship is obtained if the number and the residue are divided by some common factor relatively prime to the modulus.

The material of this section has presented briefly, without proof, the pertinent concepts of congruences. Additional material on the subject may be found in any standard text on number theory.[1]

[1] G. H. Hardy and E. M. Wright, "An Introduction to the Theory of Numbers," Oxford University Press, London, Eng.; 1956.

Development of the Residue Code

A residue code associated with a particular natural number is formed from the least positive residues of the particular number with respect to different bases. The first requirement for an efficient residue number system is that the bases of the difference digits of the representation must be relatively prime. If a pair of bases are not relatively prime, the effect is the introduction of redundancy. The following example will illustrate this fact. Contrast the residues associated with bases of magnitude 2 and 6 against the residues associated with bases of magnitude 3 and 4. In the first case, the bases are not relatively prime while in the second case the bases are relatively prime. The residues associated with the bases of magnitude 2 and 6 are unique for only 6 states while the residues associated with the bases of magnitude 3 and 4 provide a unique residue representation for 12 states. This is further clarified by Table I.

TABLE I

Redundancy of a Nonrelatively-Primed Base Representation

Number	Least Postive Residue			
	Mod 2	Mod 6	Mod 3	Mod 4
0	0	0	0	0
1	1	1	1	1
2	0	2	2	2
3	1	3	0	3
4	0	4	1	0
5	1	5	2	1
6	0	0	0	2
7	1	1	1	3
8	0	2	2	0
9	1	3	0	1
10	0	4	1	2
11	1	5	2	3
12	0	0	0	0
13	1	1	1	1
14	0	2	2	2

An example of a residue number system is presented in Table II. The number system shown in Table II uses the prime bases 2, 3, 5 and 7. The number system therefore contains 210 states. The 210 states may correspond to the positive integers 0 to 209. Table II shows the residue number representation corresponding to the positive integers 0 to 29. Additional integers of the number system may be found by congruence operations. Let a, b, c and d be the digits associated with the bases 2, 3, 5 and 7, respectively. The following congruences define a, b, c and d for the residue representation of the number N:

$$N \equiv a \text{ Mod } 2$$
$$N \equiv b \text{ Mod } 3$$
$$N \equiv c \text{ Mod } 5$$
$$N \equiv d \text{ Mod } 7. \tag{7}$$

The residue number system is readily extended to include more states. For example, if a base 11 is added to the representation, it is then possible to represent 2310 states. Table III shows the product and sum of the first nine consecutive primes greater than or equal to 2. The product of the primes indicates the number of states of the number system, while the sum of the primes is a measure of the size of the representation in terms of digits. Table III also includes the number of bits required to represent each prime base in the binary number system.

TABLE II

Natural Numbers and Corresponding Residue Numbers

N.N.	2357	N.N.	2357	N.N.	2357
0	0000	10	0103	20	0206
1	1111	11	1214	21	1010
2	0222	12	0025	22	0121
3	1033	13	1136	23	1232
4	0144	14	0240	24	0043
5	1205	15	1001	25	1104
6	0016	16	0112	26	0215
7	1120	17	1223	27	1026
8	0231	18	0034	28	0130
9	1042	19	1145	29	1241

TABLE III

Number of States and Digits Associated with a Residue Representation

i	p_i	$\sum_{i=1}^{n} p_i$	$\prod_{i=1}^{n} p_i$	p_i bits	$\sum p_i$ bits
1	2	2	2	1	1
2	3	5	6	2	3
3	5	10	30	3	6
4	7	17	210	3	9
5	11	28	2,310	4	13
6	13	41	30,030	4	17
7	17	58	510,510	5	22
8	19	77	9,699,690	5	27
9	23	100	223,092,670	5	32

Residue Addition and Multiplication

The residue number representation consists of several digits and is assumed to be in one to one correspondence with some positive integers of the real number system. The digits of the residue representation are the least positive residues of these real positive integers with respect to the different moduli which form the bases of the residue representation. It follows as a direct consequence of the structure of the residue number system and the properties of linear congruences that operations of addition and multiplication are valid in the residue number system subject to one proviso. The proviso is that the residue system must possess a number of states sufficient to represent the generated sum or product. If the residue number system does not have a sufficient number of states to represent the sums and the products generated by particular finite set of real integers then the residue system will overflow and more than one sum or product of the real number system may correspond to one residue representation. For a residue number with a sufficient number of states an isomorphic relation exists with respect to the operations of addition and multiplication in the residue system and a finite system of real positive integers.

Each digit of the residue number system is obtained with respect to a different base or modulus. It follows, therefore, that the rules of arithmetic associated with each digit will be different. For example, the addition and multiplication of the digits associated with moduli 2 and 3 follow rules specified in Table IV. No carry tables are necessary since the residue number system does not have a carry mechanism. Addition of two residue representations is effected by the modulo addition of corresponding digits of the two representations. Corresponding digits must have the same base or modulus. Modulo addition of digits which have different bases is not defined. Multiplication in the residue system is effected by obtaining the modulo product of corresponding digits. The operations of addition and multiplication of two residue numbers are indicated by the following notation:

TABLE IV

MOD 2 AND MOD 3 SUMS AND PRODUCTS

sum Mod 2

$\oplus$	0	1
0	0	1
1	1	0

product Mod 2

$\odot$	0	1
0	0	0
1	0	1

sum Mod 3

$\oplus$	0	1	2
0	0	1	2
1	1	2	0
2	2	0	1

product Mod 3

$\odot$	0	1	2
0	0	0	0
1	0	1	2
2	0	2	1

$$S = A \oplus B$$
$$p = A \odot B. \tag{8}$$

Consider a residue number representation with bases 2, 3, 5 and 7. We assume an isomorphic relation between the residue number system and the real positive numbers 0 to 209. An isomorphic relation then exists for the operations of multiplication and addition only if the product or sum is less than 210. The following examples employing residue numbers illustrate the addition and multiplication operations and the presence of an isomorphism or the lack of isomorphism in the case of overflow. Residue numbers will be distinguished by the use of parentheses.

$$29 + 27 = S = 56$$
$$29 \leftrightarrow (1\ 2\ 4\ 1)$$
$$27 \leftrightarrow (1\ 0\ 2\ 6)$$
$$56 \leftrightarrow (0\ 2\ 1\ 0)$$

$$\oplus \begin{array}{c}(1\ 2\ 4\ 1)\\ \underline{(1\ 0\ 2\ 6)}\\ (0\ 2\ 1\ 0)\end{array}. \tag{9}$$

The following operations are considered in performing the addition of the two residue representations.

$$1 + 1 \equiv 0 \text{ Mod } 2$$
$$2 + 0 \equiv 2 \text{ Mod } 3$$
$$4 + 2 \equiv 1 \text{ Mod } 5$$
$$1 + 6 \equiv 0 \text{ Mod } 7. \tag{10}$$

Consider the addition of two numbers which produce a sum greater than 209.

$$S = 100 + 200$$

$$\oplus \begin{array}{c}(0\ 1\ 0\ 2)\\ \underline{(0\ 2\ 0\ 4)}\\ (0\ 0\ 0\ 6)\end{array}. \tag{11}$$

The residue representation (0 0 0 6) corresponds to the real positive number 90. In this particular example, the sum has overflowed the residue representation. The resulting sum is the correct sum modulo 210.

$$300 \equiv 90 \text{ Modulo } 210.$$

Finite real number systems and residue number systems have the same overflow characteristics. The sum which remains after the overflow is the correct sum with respect to a modulus numerically equal to the number of states in the finite number system.

The following is presented as an example of the process of residue multiplication.

$$p = 10 \times 17 = 170$$
$$10 \leftrightarrow (0\ 1\ 0\ 3)$$
$$17 \leftrightarrow (1\ 2\ 2\ 3)$$
$$170 \leftrightarrow (0\ 2\ 0\ 2)$$

$$\odot \begin{array}{c}(0\ 1\ 0\ 3)\\ \underline{(1\ 2\ 2\ 3)}\\ (0\ 2\ 0\ 2)\end{array}. \tag{12}$$

The process of multiplication involved consideration of the following relations for each digit.

$$1 \times 0 \equiv 0 \text{ Mod } 2$$
$$1 \times 2 \equiv 2 \text{ Mod } 3$$
$$0 \times 2 \equiv 0 \text{ Mod } 5$$
$$3 \times 3 \equiv 2 \text{ Mod } 7. \tag{13}$$

An overflow resulting from a multiplication is no different than the overflow resulting from an addition. Consider the product obtained from the residue multiplication of the numbers 10 and 100. The result in the modulo 210 number system is 160 since

$$1000 \equiv 160 \text{ mod } 210. \tag{14}$$

Subtraction and the Representation of Negative Numbers

The process of subtraction is obtainable in the residue number system by employing a complement representation consisting of the additive inverses of the positive residue representation. The additive inverse always exists, since each of the elements of the residue representation is an element of a field. There is no basic problem associated with the subtraction operation. There is, however, a problem associated with the representation of negative numbers. In particular, some mechanism must be included in the number system which will permit the representation of positive and negative numbers. This problem is discussed in this and the following section.

The additive inverse of a residue number is defined by the following:

$$a \oplus a' = 0. \tag{15}$$

The formula may be considered to apply to a digit of the residue system or equally well to the whole residue representation. Consider the following examples with reference to the modulo 210 residue number system.

$$a = (1\ 2\ 4\ 1) \tag{16}$$

then

$$a' = (1\ 1\ 1\ 6)$$

since

$$\oplus \begin{array}{l} (1\ 2\ 4\ 1) \\ \underline{(1\ 1\ 1\ 6)} \\ (0\ 0\ 0\ 0) \end{array}.$$

The following examples have been chosen to illustrate the subtraction process and to some extent the difficulties associated with the sign of the difference.

$$D = A \ominus B = A \oplus B'. \tag{17}$$

We consider first the case where the magnitude of A is greater than B.

Let

$$A = 200 \qquad B = 100.$$

In residue representation

$$B' = (0\ 2\ 0\ 5) \tag{18}$$

and

$$\oplus \begin{array}{l} (0\ 2\ 0\ 4) \\ \underline{(0\ 2\ 0\ 5)} \\ (0\ 1\ 0\ 2) \end{array}.$$

The residue representation of the difference corresponds to positive 100 in the real number domain. We consider next the difference

$$D = B \ominus A = A' \oplus B \tag{19}$$

where the magnitude of A is greater than B. Then

$$A' = (0\ 1\ 0\ 3) \tag{20}$$

and

$$\oplus \begin{array}{l} (0\ 1\ 0\ 3) \\ \underline{(0\ 1\ 0\ 2)} \\ (0\ 2\ 0\ 5) \end{array}.$$

The difference (0 2 0 5) is the additive inverse of (0 1 0 2). Unless additional information is supplied, the correct interpretation of the representation (0 2 0 5) is in doubt. (0 2 0 5) may correspond to either +110 or −100.

The difficulties associated with whether a residue representation corresponds to a positive or negative integer can be partially removed by the division of the residue number range into two parts. This is exactly the scheme that is employed to obtain a machine representation of positive and negative natural numbers. For the system of natural numbers two different machine representations of the negative numbers may be obtained and are commonly designated the radix complement representation of negative numbers and the diminished radix complement representation of negative numbers.

The complement representation for a residue code is defined in terms of the additive inverse. Thus, the representation of negative A is A' where $A \oplus A' = 0$, and the range of A is restricted to approximately one-half of the total possible range of the residue representation. This can be illustrated by consideration of a specific residue code. This residue representation, employing bases of magnitude 2, 3, 5, and 7, is divided into two parts. The residue representations corresponding to the natural numbers 0 to 104 are considered positive. The residue representations corresponding to the natural numbers 105 to 209 are considered inverse representations and associated with the negative integers from −1 to −105.

The range of this particular number system is from -105 to $+104$. The arithmetic rules pertaining to sign and overflow conventions for this particular number system are the same rules normally associated with radix complement arithmetic.

The complement representation does eliminate in principle any ambiguity concerning the sign of the result of an arithmetic operation. However, there is a practical difficulty. The determination of the sign associated with a particular residue representation requires the establishment of the magnitude of the representation relative to the magnitude which separates the positive and negative representations. The determination of relative magnitude for a residue representation is discussed in the next section. It will be shown that the determination of relative magnitude is not a simple problem.

Conversion from a Residue Code to a Normal Number Representation

It is frequently desirable to determine the natural number associated with a particular residue representation. The need for this conversion occurs frequently when investigating the properties of the residue system. The residue representation is constructed in such a manner that magnitude is not readily obtainable. The presence of digit weights in the normal polynomial type number representation greatly facilitates the determination of magnitude. However, it is possible to assign a weight to each digit of the residue representation in such a manner that the modulo m sum of the digit-weight products is the real natural number in a consistently weighted representation. m is the product of all the bases employed in the residue representation. The conversion technique is known as "The Chinese Remainder Theorem." The material which follows describes the conversion technique but omits the proof. A simple and straightforward proof is found in Dickson.[2] The proof does not refer specifically to residue number systems, but rather to a system of linear congruences. If so regarded, a system of congruences defines a component of a residue number system.

Consider a residue number system with bases $m_1 \cdots m_t$. The corresponding digits are labeled $a_1 \cdots a_t$. The following equations define the conversion process.

$$a_1 A_1 \frac{m}{m_1} + \cdots + a_t A_t \frac{m}{m_t} \equiv S \text{ Mod } m \tag{21}$$

where

$$A_i \frac{m}{m_i} \equiv 1 \text{ Mod } m_i$$

and

$$m = \prod_{j=1}^{t} m_j.$$

The conversion formula for a particular residue number system is now obtained.

$$m_1 = 2 \qquad m_2 = 3 \qquad m_3 = 5 \qquad m_4 = 7$$

$$\begin{aligned}
105\, A_1 &\equiv 1 \text{ Mod } 2 \qquad \text{so } A_1 = 1 \\
70\, A_2 &\equiv 1 \text{ Mod } 3 \qquad \text{so } A_2 = 1 \\
42\, A_3 &\equiv 1 \text{ Mod } 5 \\
2\, A_3 &\equiv 1 \text{ Mod } 5 \qquad \text{so } A_3 = 3 \\
30\, A_4 &\equiv 1 \text{ Mod } 7 \\
2\, A_4 &\equiv 1 \text{ Mod } 7 \qquad \text{so } A_4 = 4
\end{aligned}$$

$$105\, a_1 + 70\, a_2 + 126 a_3 + 120\, a_4 \equiv S \text{ Mod } 210. \tag{22}$$

The conversion formula is now used to determine the natural number corresponding to the residue representation (1 2 0 4).

$$\begin{aligned}
105\,(1) + 70\,(2) + 126\,(0) + 120\,(4) &= 725 \\
725 &\equiv S \text{ Mod } 210 \\
S &= 95.
\end{aligned} \tag{23}$$

The above conversion formula may be obtained in a much more direct manner. The residue number representation may be regarded as a vector with the base vectors (1000), (0100), (0010) and (0001). The residue representation expressed in terms of the base vectors is:

$$(a_1a_2a_3a_4) = a_1(1000) \oplus a_2(0100) \oplus a_3(0010) \oplus a_4(0001). \tag{24}$$

The magnitude of the base vectors is

$$\begin{aligned}
(1000) &\leftrightarrow 105 \\
(0100) &\leftrightarrow 70 \\
(0010) &\leftrightarrow 126 \\
(0001) &\leftrightarrow 120.
\end{aligned} \tag{25}$$

Other conversion techniques exist. In particular it is possible by means of a deductive process to determine the magnitude of a particular residue representation. This requires both a knowledge of the nature of the residue system and the natural number representation associated with at least one residue representation.

Due to the deductive nature of the process, it is more suitable for human computation than for machine computation. The process is explained using the residue number of the previous example (1 2 0 4). The knowledge of the residue representation for unity which is (1 1 1 1) is assumed. Consider the effect of changing the second digit from one to two. The change adds the product $m_1m_3m_4 = 70$ to the number since 70 is congruent 1, modulo 3. The resulting residue representation (1 2 1 1) corresponds to 71. The effect of changing the third digit is to change the magnitude by some multiple of the product $m_1m_2m_4 = 42$. The correct change in magnitude is $42x$ where $42x \equiv 4$ Mod 5. So $42x = 84$ and the

[2] L. E. Dickson, "Modern Elementary Theory of Numbers," University of Chicago Press, Chicago, Ill., p. 16: 1939.

residue representation (1 2 0 1) corresponds to 155. The fourth digit is modified by the addition of a three. The effect of this change is determined by $30x \equiv 3$ Mod 7. The magnitude change is 150. The sum of 150 and 155 modulo 210 yields the correct result 95, in correspondence with (1 2 0 4).

Sign determination for the residue code is dependent on the determination of a greater than or less than relationship. A possible method might involve the conversion techniques described previously. Such a scheme would involve the standard comparison techniques associated with the determination of the relative magnitude of two numbers represented in a weighted representation. An alternate conversion procedure yields a conversion from the residue code to a nonconsistently based polymonial number representation by means of residue arithmetic. Consider a residue code consisting of t digits. The t digits of the residue code are associated with t congruence relationships as follows:

$$S \equiv a_i \text{ Mod } m_i \qquad 1 \leq i \leq t \tag{26}$$

S is the magnitude of the number expressed in normal representation. It is also possible to express the number S as

$$S = a_i + A_i m_i. \tag{27}$$

A_i is the integer part of the quotient of S divided by m_i. In regard to a greater or less than relationship, the determination of A_i divides the range of the residue representation into m/m_i parts. We proceed to calculate A_i from the set of t equations given above. Let

$$S = a_t + A_t m_t \qquad A_t < \frac{m}{m_t}\cdot \tag{28}$$

This equation is then used to replace S in the remaining $t-1$ equations, yielding $t-1$ equations of the form

$$A_t m_t \equiv (a_i + a_t') \text{ Mod } m_i \quad 1 \leq i \leq t-1 \tag{29}$$

or

$$A_t \equiv (a_i + a_t')/m_t^{i} \text{ Mod } m_i$$

$$A_t \equiv d_t^{i} \text{ Mod } m_i$$

where $/m_t^i$ is the multiplicative inverse of m_t with respect to base m_i. The multiplicative inverse is defined as[3]

$$x_t/x_t^{i} \equiv 1 \text{ Mod } m_i \tag{30}$$

d_t^i is the least positive residue of $(a_i+a_t')/m_t$ with respect to base i. a_t' is the additive inverse of a_t. Let A_t be expressed as

$$A_t = d_t^{t-1} + A_{t-1} m_{t-1} \qquad A_{t-1} < \frac{m}{m_t m_{t-1}}\cdot \tag{31}$$

[3] The existence of the multiplicative inverse requires that x_t and m_i be relatively prime.

If this expression is substituted for A_t a set of $t-2$ equations remain. The equations are of the form

$$A_{t-1} \equiv [d_t^{i} + (d_t^{t-1})']/m_{t-1} \text{ Mod } m_i \quad 1 \leq i \leq t-2$$

$$A_{t-1} \equiv d_{t-1}^{i} \text{ Mod } m_i. \tag{32}$$

The system of equations shown below is generated by repetition of the above substitution process until no equations remain.

$$\begin{aligned} S &= a_t + A_t m_t \qquad\qquad (33)\\ A_t &= d_t^{t-1} + A_{t-1} m_{t-1} \\ A_{t-1} &= d_{t-1}^{t-2} + A_{t-2} m_{t-2} \\ &\vdots \\ A_3 &= d_3^{2} + A_2 m_2 \\ A_2 &\equiv d_2^{1} \text{ Mod } m_1. \qquad (34) \end{aligned}$$

The equations are combined to yield:

$$S = a_t + m_t \{ d_t^{t-1} + m_{t-1}[d_{t-1}^{t-2} + m_{t-2} (d_{t-2}^{t-3} + \cdots \tag{35}$$

$$= a_t + m_t d_t^{t-1} + m_t m_{t-1} d_{t-1}^{t-2}$$

$$+ m_t m_{t-1} m_{t-2} d_{t-2}^{t-3} + \cdots + \frac{m}{m_1} d_2^{1} \tag{36}$$

where

$$A_t < m_t$$

$$d_{t-n}^{t-n-1} < m_{t-n-1}.$$

Therefore, S is never equal to or greater than m and d_2^1 divides the range into m_1 parts, d_3^2 divides each of the m_1 parts into m_2 parts, d_4^3 divides each of the m_2 parts into m_3 parts, etc.

The formulas which define the conversion process may be applied recursively to obtain a formula for a greater number of digits. The process has been extended to five variables and the results are shown in Fig. 1.

A somewhat simpler interpretation of the conversion process is obtained if (36) is considered from the vector viewpoint. Let the terms 1, m_t, $m_t m_{t-1}$, $\cdots$, m/m_1 be regarded as base vectors. The residue representation of the base vectors is readily obtainable and the following characteristics with respect to zero and nonzero digits of the base vectors are observed

$$\begin{aligned} &1 \leftrightarrow (xxx \cdots xxx) \\ &m_t \leftrightarrow (xxx \cdots xx0) \\ &m_t m_{t-1} \rightarrow (xxx \cdots x00) \quad x \text{ indicates a nonzero digit} \\ &\vdots \qquad\qquad\qquad\qquad\quad 0 \text{ indicates a zero digit} \\ &\frac{m}{m_1} \leftrightarrow (x00 \cdots 000). \end{aligned} \tag{37}$$

The following equality exists between the vector representation of the residue code and the vector representation of the nonconsistently based code.

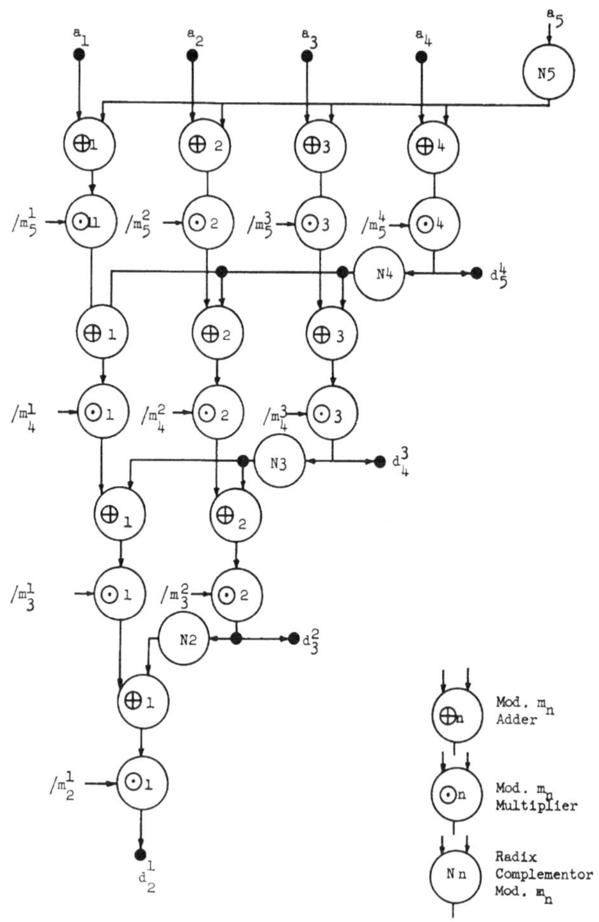

Fig. 1—Logic required to convert from the residue code to a weighted code.

$$a_1(101 \cdots 000) \oplus a_2(010 \cdots 000) \oplus \cdots$$
$$\oplus\ a_{t-1}(000 \cdots 010) \oplus a_t(000 \cdots 001)$$
$$= a_t(111 \cdots 111) \oplus d_t^{t-1}(1xx \cdots xx0) \oplus \cdots$$
$$\oplus\ d_3^2(1x0 \cdots 000) \oplus d_2^1(100 \cdots 000). \tag{38}$$

Equating the coefficients of each coordinate in (38) obtains the following set of equations relating the residue digits and the digit of the nonconsistently based number system.

$$a_t \equiv \alpha_t \bmod m_t$$
$$xd_t^{t-1} + a_t \equiv \alpha_{t-1} \bmod m_{t-1}$$
$$\vdots$$
$$xd_3^2 + \cdots + xd_t^{t-1} + a_t \equiv \alpha_2 \bmod m_2$$
$$d_2^1 + d_3^2 + \cdots + d_t^{t-1} + a_t \equiv \alpha_1 \bmod m_1. \tag{39}$$

The conversion is executed by first subtracting $a_t(111 \cdots 111)$; d_t^{t-1} is then a function of only one residue digit and may be determined. This is followed by the subtraction of $d_t^{t-1}(1xx \cdots xx0)$ etc. Consider the conversion of the residue representation (1204) with bases 2, 3, 5 and 7.

The weights and the corresponding residue representation of the nonconsistently based code are:

$$1 \leftrightarrow (1111)$$
$$m_4 = 7 \leftrightarrow (1120)$$
$$m_4m_3 = 35 \leftrightarrow (1200)$$
$$m_4m_3m_2 = 105 \leftrightarrow (1000). \tag{40}$$

The actual conversion between the residue code and the nonconsistently based code is accomplished by the following operation:

$$\begin{array}{cll} & (120④) & \text{1st digit is a 4 since } 4\cdot 1 \equiv 4 \bmod 7 \\ \oplus & \underline{(0213)} & \text{Subtract off } 4(1111) = (0144) \\ & 11①0 & \text{2nd digit is a 3 since } 3\cdot 2 \equiv 1 \bmod 5 \\ \oplus & \underline{104} & \text{Subtract off } 3(1120) = (1010) \\ & 0①00 & \text{3rd digit is a 2 since } 2\cdot 2 \equiv 1 \bmod 3 \\ \oplus & \underline{02} & \text{Subtract off } 2(1200) = (0100) \\ & 00 & \text{4th digit is 0.} \end{array}$$

Thus

$$S = 105\cdot 0 + 35\cdot 2 + 7\cdot 3 + 1\cdot 4$$
$$= 95. \tag{41}$$

Admittedly, the process required to obtain a magnitude leaves much to be desired though the conversion process is no more complicated than the standard change of base conversion process used for consistently based number systems. One presumed advantage of the residue number system was the absence of a carry process. The greater or less than process is essentially sequential and is in many ways similar to the carry process of ordinary arithmetic. The ultimate usefulness of the residue code for general purpose computation appears very much dependent on the development of simple techniques for the determination of the relative magnitude of two residue code digits.

Division

The division process for residue codes is complicated by two factors. The first is the absence of a multiplicative inverse for the zero element. The second difficulty is the fact that residue division and the normal division process are in one to one correspondence only when the resulting quotient is an integer value. We shall consider first the problem of residue division of the elements of a single field and shall consider later the elements of several fields considered as a residue code. The division process represented in equation form as

$$\frac{a}{b} = q \tag{42}$$

implies the following equation:

$$a = bq. \tag{43}$$

The difference between normal arithmetic and residue arithmetic is that in residue arithmetic the product bq

need not necessarily be equal to a, only the congruence of a and bq is required.

$$bq \equiv a \text{ Mod } m_n. \tag{44}$$

Multiplication by the multiplicative inverse of b designated $/b$ obtains

$$q \equiv a/b \text{ Mod } m_n. \tag{45}$$

The correct interpretation of q in the above equation is that the number a is obtained by forming the modulo sum consisting of b, q representations. The sum is carried out in a closed and finite modulo number system of base m_n. Thus, q corresponds to the quotient only when the quotient has an integer value. Examples may be obtained from the consideration of a modulo 5 number system

$$\frac{4}{2} = q \tag{46}$$

$$2q \equiv 4 \text{ Mod } 5$$

$$q \equiv 2 \text{ Mod } 5$$

$$\frac{4}{3} = q \tag{47}$$

$$3q \equiv 4 \text{ Mod } 5$$

$$q \equiv 3 \text{ Mod } 5$$

$$\text{note } 3 \times 3 \equiv 4 \text{ Mod } 5$$

$$\frac{3}{4} = q \tag{48}$$

$$4q \equiv 3 \text{ Mod } 5$$

$$q \equiv 2 \text{ Mod } 5.$$

In the above examples, q corresponds to the quotient only in the first example.

The residue code representation of a number consists of many digits, $A = (a_1, a_2, \cdots, a_n)$. Each digit of the representation is associated with a different prime base. The number system is a modulo m system where

$$m = \prod_{i=1}^{n} m_i.$$

The division of two numbers in residue code may be expressed by a system of congruences. The solution $Q = (q_1, q_2, \cdots, q_n)$ must satisfy all the congruence relationships of the system. A zero digit in the divisor $B = (b_1, b_2, \cdots, b_n)$ means that B and m are not relatively prime hence the multiplicative inverse of B doesn't exist.

$$QB \not\equiv A \text{ Mod } m. \tag{49}$$

For the special case in which $b_i = 0$ and $a_i = 0$, a valid congruence relationship of the form

$$\frac{QB}{m_i} \equiv \frac{A}{m_i} \text{ Mod } \frac{m}{m_i} \tag{50}$$

is obtainable.

The process of residue division has certain interesting properties and quite possibly has applications in respect to special problems. Unfortunately, the residue division process is not a substitute for normal division. It appears that the only way in which division can be effected in the residue code is by the utilization of techniques similar to those employed for division in a consistently weighted number system. The division process then requires trial and error subtraction or addition and the greater than or less than relationship. The division algorithm could also include trial multiplication since in the residue system addition and multiplication require the same period of time.

Conclusions

The material of this paper forms a preliminary investigation of the applicability of residue number systems to the arithmetic operations of digital computers. The residue system has been found attractive in terms of the operations of multiplication and addition. It is possible to realize practical logical circuitry to yield the product in the same operation time as for the sum since the product is not obtained by the usual procedure of repetitive addition. The main disadvantages of the residue number system are associated with the necessity of determining absolute magnitude. Thus, the division process, the detection of an overflow and the determination of the correct sign of a subtraction operation are processes which at this stage of the investigation seem to involve considerable complexity. Nevertheless, there are certainly many special purpose applications well-suited to the residue code. In particular, there exists a class of control problems characterized by the absence of the need for division, the existence of a well-defined range for the variables and also by the fact that the sign of the variables is known. For the problems of this class, the use of the residue code should result in a reduction of the over-all computation period and give a computer with a higher bandwidth than obtainable with the conventional number system.

The ultimate usefulness of the residue code will probably be determined largely by the success of the circuit designer in perfecting circuitry ideally suited for residue code operations.

Acknowledgment

The material of this paper is essentially Chapter 5 of the author's Ph.D. dissertation.[4] At the time of the completion of the dissertation, the author was unaware of the work of Valach[5] and Svoboda[6,7] in Czechoslovakia, references of which were obtained from recent visitors to the Soviet Union. The author wishes to take this opportunity to acknowledge same.

[4] H. L. Garner, "Error Checking and the Structure of Binary Addition," Ph.D., dissertation, University of Michigan, Ann Arbor, Mich., pp. 105–140; 1958.

[5] M. Valach, "Vznik kodu a ciselne soustavy zbytkovych trid," *Stroje Na Zpracovani Informaci*, Sbornik III; 1955.

[6] A. Svoboda and M. Valach, "Operatorove Obvody," *Stroje Na Zpracovany Informaci*, Sbornik III; 1955.

[7] A. Svoboda, "Rational Numerical System of Residual Classes," *Stroje Na Zpracovani Informaci*, Sbornik V; 1957.

A Formalization of Floating-Point Numeric Base Conversion

DAVID W. MATULA

Reprinted from *IEEE Transactions on Computers*, Volume C-19, Number 8, August 1970, pages 681-692.

Abstract—The process of converting arbitrary real numbers into a floating-point format is formalized as a mapping of the reals into a specified subset of real numbers. The structure of this subset, the set of n significant digit base β floating-point numbers, is analyzed and properties of conversion mappings are determined. For a restricted conversion mapping of the n significant digit base β numbers to the m significant-digit base δ numbers, the one-to-one, onto, and order-preserving properties of the mapping are summarized. Multiple conversions consisting of a composition of individual conversion mappings are investigated and some results of the invariant points of such compound conversions are presented. The hardware and software implications of these results with regard to establishing goals and standards for floating-point formats and conversion procedures are considered.

Index Terms—Accumulated error of successive conversions, base conversion, "equivalent digits" formula, floating-point numbers, mixed-base computational environments, round-off error, significant digits.

I. Introduction and Summary

THE necessity of base conversion of numeric data during some stages of computation on a digital computer is a de facto component of practical numeric computation that must be recognized in any complete analysis of digital computation. On the hardware level, the trend towards establishment of computer networks with possibly differently based machines and, on the software level, the mixed-base flexibility inherent in the PL/I language specifications, both suggest that internal data of certain jobs may be necessarily subjected to multiple conversions before job termination. Thus, references to a purportedly constant floating-point datum occurring at different points during program execution might encounter altered datum values. Both hardware and software designers must recognize this problem, and each can benefit from the fundamental principles obtained by considering base conversion as a mathematical transformation. In this report we shall follow the notation of our previous work [1]–[4], integrating the mainstream of results from those articles with a general formal development of base conversion, and providing new results particularly in the area of multiple conversions.

A fundamental analysis of base conversion is concerned first with determination of the theoretical limitations inherent in any implementation of base conversion. The actual algorithmic mechanics of any theoretically realizable conversion procedure is then a secondary (albeit nontrivial [5]) problem which will not concern us in this paper. The formalization we introduce provides the vehicle for studying the properties and recognizing the inherent anomalies of the conversion process, which must necessarily then guide the performance specifications for floating-point representations and base conversions at both hardware and software levels.

Converting integer and fixed-point data to an "equivalent" differently based number system is generally achieved by utilizing essentially $\log_\delta \beta$ times as many digits in the new base δ as were present for representing numbers in the old base β system. This simplified notion of equivalence does not extend to the conversion of floating-point systems. Actually, conversion between floating-point number systems introduces subtle difficulties peculiar to the structure of these systems so that no such convenient formula for equating the "numbers of significant digits" is even meaningful. Thus, our formalization of floating-point base conversion is preceded in Section II by a careful analysis of floating-point number systems.

Following our previous work [1]–[4], a system of floating-point numbers of n significant digits to the base β is characterized as a *significance space*, S_β^n, and in Theorem 1 the number of elements in S_δ^m relative to the number of elements in S_β^n within a specified interval is shown to converge to $((\delta-1)\delta^{m-1}/((\beta-1)\beta^{n-1}))\log_\delta \beta$ as the interval grows to include the whole real line. This relative density of "total membership" of two significance spaces provides only a gross comparison of the two number systems, for actually there is considerable local variation in the relative density. The *gap function* [1] $\Gamma_\beta^n(x)$ is defined as the relative difference between nearest neighbors of S_β^n at x. A comparison of the graphs of Γ_β^n and Γ_δ^m provides more insight into the comparability of two differently based floating-point systems than any simplified "equivalent digit" formula.

Having characterized floating-point number systems and the gap function, the conversion of the real numbers into S_β^n both by rounding and by truncation procedures are then formalized in Section III. The order-preserving properties of these conversion mappings are detailed, and the base conversion theorem [2] is stated, which gives the necessary

Manuscript received December 12, 1969; revised January 4, 1970, and March 2, 1970. This research was partially supported by the Advanced Research Projects Agency of the Department of Defense under Contract SD-302, and by the National Science Foundation under Contract GJ-446.

The author is with the Department of Applied Mathematics and Computer Science, Washington University, St. Louis, Mo. 63130.

and sufficient conditions for a conversion mapping from S_β^n to S_δ^m to be 1) one-to-one and 2) onto.

The important problems associated with multiple conversions of a datum are analyzed in Section IV. The composition of repeated rounding and/or truncation conversions is termed a *compound conversion*, and the associated *invariant points* (i.e., the points mapped into themselves) of such a compound conversion mapping are analyzed. For a compound truncation conversion, it is shown that the only invariant points of the mapping are the numbers common to all of the significance spaces involved. Thus, considerable importance must be attached to the intersection of significance spaces, and in Theorems 5 and 6 these intersections are shown to exhibit a crucial dependence on the commensurability of the bases. Specifically, significance spaces with commensurable bases will always jointly contain a common significance space. For example, the six-digit hexadecimal numbers and the eight-digit octal numbers both contain all 21-bit binary numbers. On the other hand, the members common to significance spaces with incommensurable bases (e.g., binary and decimal) will be finite in number and, for cases of computational interest, these members will typically all fall in an interval much smaller than the interval range provided by current exponent ranges on digital computers.

As a consequence of the limited membership of points common to two incommensurable significance spaces, the multiple back-and-forth conversion of a "constant datum" between two incommensurable significance spaces by truncation conversion can accumulate error so as to invalidate even the leading digit of the value of this "constant datum." Practically, the process of updating a binary coded decimal (BCD) tape on a binary machine might well subject stored data which is never updated to multiple binary-decimal conversions, so BCD tape updating is very sensitive to such anomalies of compound conversions. Fortunately, we can show that under very general conditions, iterated rounding conversion of a datum between two significance spaces quickly generates a stable pair of values, each of which is a reasonable approximation of the initial datum. Under the stronger conditions given in the in-and-out conversion theorem [3], rounding conversion through an intermediate significance space can be guaranteed to regenerate the initial datum of the original significance space.

In the presence of more than two incommensurable bases, the possibility of cyclic conversions' accumulating error in a datum is shown to exist even under rounding conversion. Our final result resolves the problem of controlling the overall growth of accumulated conversion error within a mixed-base computational environment by a process which standardizes a datum's value in each of the significance spaces involved.

II. Floating-Point Number Systems

A formalization of floating-point number systems must start with a characterization of the set of floating-point numbers, preferably divorced from the cumbersome digit sequence representational notation. In previous articles [1]–[4] this set has been termed a significance space.

Definition: For the integers $\beta \geq 2$, called the *base*, and $n \geq 1$, called the *significance* (or *precision*), let the *significance space* S_β^n be the following set of real numbers:

$$S_\beta^n = \{b \mid b = k\beta^j \text{ for some integers } k, j \text{ where } |k| < \beta^n\}.$$

For clarity we shall utilize the Greek letters α, β, and δ to denote bases; the English letters a, b, c, and d will denote elements of a significance space, i.e., the so called "floating-point" numbers; the letters, i, j, k, l, m, n, p, and q will denote integers; and x, y, and z will denote arbitrary real numbers.

For an element $b = k\beta^j \in S_\beta^n$ the actual floating-point representation of b can be visualized as having the fixed-point integer portion k represented by a sign and n or less digits to the assumed base β, with the exponent portion then represented by the integer j. The floating-point representation just described will not in general be unique, with digit sequence realizations of b corresponding to both "normalized" and "unnormalized" forms possible in some cases. Considerations related to the nonunique determination of k, j in $b = k\beta^j$ will be treated where necessary; however, our main concern is with membership of b in the set of real numbers S_β^n for which the form of representing b is irrelevant. Note that the significance space S_β^n differs from an actual floating-point number system in that there is no bound on the exponent portion j of the members $b = k\beta^j \in S_\beta^n$. Thus S_β^n is actually an infinite set. Since we shall not concern ourselves with underflow and overflow problems, the significance space S_β^n is a perfectly acceptable model of a floating-point number system for our purposes.

It is easy to visualize the change in the set S_β^n caused by varying significance, since increasing n to $n+1$ maintains all members of S_β^n and adds $\beta - 1$ new members uniformly spaced between every neighboring pair of members of S_β^n. The dependence of the membership of S_β^n on the base β is far more subtle. In practice it has been convenient to identify a nondecimal floating-point number system with the "appropriate" decimal based system; however, we now show that such a purported equivalence glosses over certain inherent anomalies between differently based floating-point number systems.

A gross comparison of two differently based significance spaces can be obtained by determining the relative number of members of each space over a comparable range. For example, the 3 (significant)-bit binary numbers between unity and one thousand are 1, $1.01_2 = 1.25$, $1.10_2 = 1.5$, $1.11_2 = 1.75$, $10.0_2 = 2$, $10.1_2 = 2.5$, $11.0_2 = 3$, $11.1_2 = 3.5$, $100_2 = 4$, $101_2 = 5$, $\cdots$, $1100000000_2 = 768$, $1110000000_2 = 896$; and the 1 (significant)-digit decimal numbers over the same range are 1, 2, 3, 4, 5, 6, 7, 8, 9, 10, 20, 30, 40, 50, 60, 70, 80, 90, 100, 200, 300, 400, 500, 600, 700, 800, 900, 1000. In Fig. 1 these members are indicated by tickmarks on the real line plotted on a logarithmic scale so that the log periodic nature of the spacing between floating-point numbers is evident. The ratio of the numbers of members of these systems over this interval is $40/28 = 1.43$. Now this ratio of

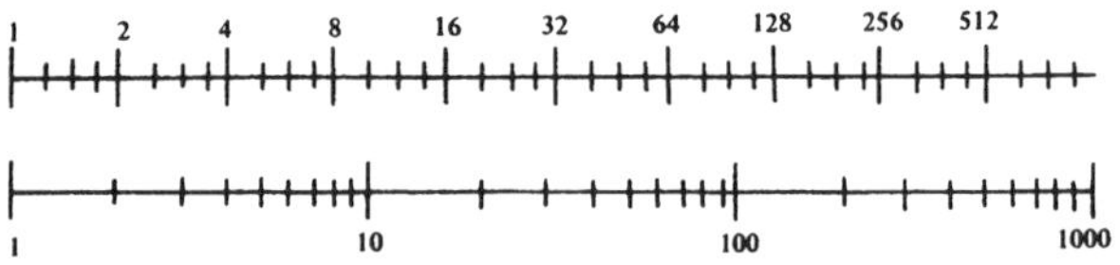

Fig. 1. Tickmarks plotted on a log scale showing (a) the 40 3 (significant)-bit binary numbers and (b) the 28 1 (significant)-digit decimal numbers over the range $[1, 10^3]$.

membership density will vary with the choice of interval; however, a reasonable overall comparison of any two floating-point systems can be calculated by determining the limit of such a ratio as the interval grows to include the whole real line.

Theorem 1: Let S_β^n and S_δ^m be any two significance spaces. Then letting $|S|$ denote the number of members of the set S,

$$\lim_{M\to\infty}\left|\frac{\left\{d \mid d\in S_\delta^m,\ \frac{1}{M}\le |d|\le M\right\}}{\left\{b \mid b\in S_\beta^n,\ \frac{1}{M}\le |b|\le M\right\}}\right| = \frac{(\delta-1)\delta^{m-1}}{(\beta-1)\beta^{n-1}}\log_\delta\beta. \quad (1)$$

Proof: Let $[x]$ denote the greatest integer in x. Then the closed interval $[1/M, M]$ may be divided into $2[\log_\beta M]$ disjoint half-open half-closed intervals of the form $[\beta^j, \beta^{j+1})$ and two subintervals of such intervals. Each interval $[\beta^j, \beta^{j+1})$ contains $(\beta-1)\beta^{n-1}$ distinct members of S_β^n. Noting that $-b\in S_\beta^n \Leftrightarrow b\in S_\beta^n$ we have for $M\ge 1$,

$$\begin{aligned}&\left|\left\{b \mid b\in S_\beta^n,\ \frac{1}{M}\le |b|\le M\right\}\right|\\ &= 2\left|\left\{b \mid b\in S_\beta^n,\ \frac{1}{M}\le b\le M\right\}\right|\\ &= 2(2\lfloor\log_\beta M\rfloor + \varepsilon)(\beta-1)\beta^{n-1} \quad \text{where } 0\le\varepsilon\le 2\\ &= 2(2\log_\beta M + \varepsilon')(\beta-1)\beta^{n-1} \quad \text{where } |\varepsilon'|\le 2,\end{aligned}$$

the latter resulting from removal of the greatest integer brackets. Finally,

$$\begin{aligned}&\lim_{M\to\infty}\left|\frac{\left\{d \mid d\in S_\delta^m,\ \frac{1}{M}\le |d|\le M\right\}}{\left\{b \mid b\in S_\beta^n,\ \frac{1}{M}\le |b|\le M\right\}}\right|\\ &= \lim_{M\to\infty}\frac{2(2\log_\delta M+\varepsilon_2)(\delta-1)\delta^{m-1}}{2(2\log_\beta M+\varepsilon_1)(\beta-1)\beta^{n-1}} \quad \text{where } |\varepsilon_1|, |\varepsilon_2|\le 2\\ &= \frac{(\delta-1)\delta^{m-1}}{(\beta-1)\beta^{n-1}}\log_\delta\beta.\end{aligned}$$

In the folklore on conversion there is an often quoted notion that a decimal digit is equivalent to $\log_2 10 = 3.32\cdots$ bits. Thus a three-digit decimal system should be only slightly less numerous than a ten-bit binary system. However, if we let $|S_\delta^m/S_\beta^n|$ symbolically denote the limiting ratio given in (1), then $|S_{10}^3/S_2^{10}| = 0.529\cdots$, so that there are actually only 53 percent as many real numbers representable with three significant decimal digits as there are real numbers representable with ten significant bits. Furthermore, the ratio of the number of members of S_2^3 to S_{10}^1 over the range shown in Fig. 1 was not atypical since $|S_2^3/S_{10}^1| = 1.476\cdots$, attesting to the fact that there are about 50 percent *more* 3-significant-bit binary numbers than 1-significant-digit decimal numbers, and providing a clear contradiction to the digit = 3.32-bits rule. This anomaly prevails even with more digits, since for large integral m and n chosen such that $10^m/2^n$ approaches unity,

$$\lim_{\substack{m\to\infty\\ 10^m/2^n\to 1}} |S_{10}^m/S_2^n| = \frac{9}{5}\log_{10}2 = 0.5418\cdots$$

The results just obtained are not inexplicable on intuitive grounds. The relation "digit $=\log_2 10$ bits" comes from an entropy argument where all different states (values) of the system (digit sequence) are distinguishable. Despite the applicability of this notion to integer and fixed-point number systems, floating-point systems have redundant representations for some numbers, generally resolved by normalization, so that the β^n patterns of digits possible for k for each j in $k\beta^j$ do not yield β^n new different numbers for each j. Furthermore, the collection of normalized numbers corresponding to a fixed power of the base are spread over an interval whose length depends on the base, and both of these conditions are reflected in the final form of (1).

Now one approach for determining a floating-point "equivalent digit formula" is to equate the right-hand side of (1) to unity and solve for m in terms of n, β, and δ. Thus if S_δ^m is said to be *more dense* than S_β^n when $|S_\delta^m/S_\beta^n| > 1$, then from (1), the following occurs.

Corollary 1: S_δ^m is more dense than S_β^n if and only if

$$m > n\log_\delta\beta + \log_\delta\left(\frac{\delta(\beta-1)}{\beta(\delta-1)}\right) - \log_\delta\log_\delta\beta. \quad (2)$$

Attributing meaning to a nonintegral number of digits, one may propose that

$$m = n\log_\delta\beta + \log_\delta[\delta(\beta-1)/\beta(\delta-1)] - \log_\delta\log_\delta\beta \quad (3)$$

is the "equivalent digit formula for floating-point number systems." For binary-decimal conversion we then would have

$$\#\,\text{bits} = 3.32\cdots\times(\#\,\text{decimal digits}) - 0.884\cdots. \quad (4)$$

The variability of spacing of floating-point numbers of a given S_β^n is such that the simplified formula (3) does not really provide an adequate comparison of differently based floating-point systems and more attention to local magnitude dependent variability must be considered. In studying the internal structure of S_β^n, note that every one of its nonzero members will have both a next largest and next smallest neighbor in S_β^n.

Definition: The *successor* b' of $b\in S_\beta^n$, $b\ne 0$, is given by

$$b' = \min\{d \mid d > b,\ d\in S_\beta^n\}, \quad (5)$$

and since distinct members of S_β^n have distinct successors, b may then be referred to as the (unique) *predecessor* of b' in S_β^n.

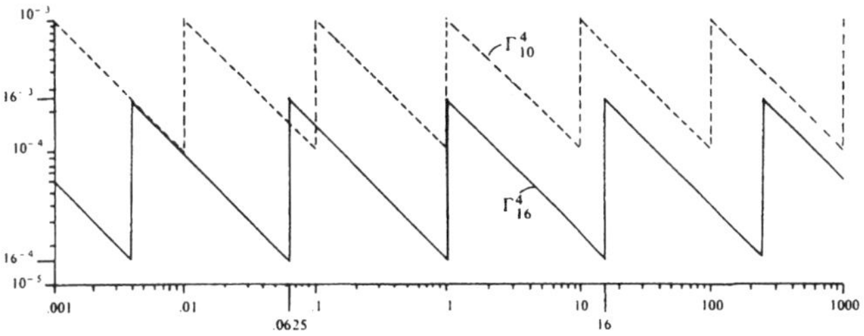

Fig. 2. The gap functions $\Gamma_{10}^4(x)$ and $\Gamma_{16}^4(x)$ for $0.001 \le x \le 1000$.

Now the absolute difference, $b'-b$, will grow with b in S_β^n, however, the relative difference is bounded [1].

Definition: The *gap* $\Gamma_\beta^n(x)$ in S_β^n at $x \neq 0$ is given by

$$\Gamma_\beta^n(x) = \begin{cases} \dfrac{\min\{b|b>x,\, b\in S_\beta^n\} - \max\{b|b\le x,\, b\in S_\beta^n\}}{x} & \text{for } x>0 \\ \Gamma_\beta^n(-x) & \text{for } x<0. \end{cases} \quad (6)$$

Specifically then, $\Gamma_\beta^n(a)=(a'-a)/a$ for $0<a\in S_\beta^n$. From the structure of floating-point number systems it is evident that $\Gamma_\beta^n(\beta x)=\Gamma_\beta^n(x)$, so Γ_β^n will experience a log periodic behavior. For $1\le x<\beta$, the numerator of (6) will have the constant value β^{1-n}, so that on a log–log scale the gap function appears as a saw tooth function. In Fig. 2, sections of the gap functions Γ_{10}^4 and Γ_{16}^4 are illustrated. Note that the variation in the magnitude of the gap function is greater for larger bases.

From Theorem 1 we can calculate that $|S_{16}^4/S_{10}^4|=5.66\cdots$, so that overall there are more than five times as many four-significant-digit hexadecimal numbers as four-significant-digit decimal numbers. Yet from the gap functions in Fig. 2, it is apparent that over the interval (0.0625, 0.1000) there are more four-significant-digit decimal numbers than four-significant-digit hexadecimal numbers since $\Gamma_{10}^4(x)<\Gamma_{16}^4(x)$ over that interval.

From the log periodic behavior of Γ_β^n the following bounds are immediate.

Theorem 2: The function Γ_β^n attains both a minimum and a maximum value over the nonzero members of S_β^n given by

$$\begin{aligned} \min\{\Gamma_\beta^n(b)|b\in S_\beta^n,\, b\neq 0\} &= 1/(\beta^n-1) \\ \max\{\Gamma_\beta^n(b)|b\in S_\beta^n,\, b\neq 0\} &= 1/\beta^{n-1}, \end{aligned} \quad (7)$$

and over the nonzero reals the bounds on $\Gamma_\beta^n(x)$ are given by

$$\begin{aligned} \inf\{\Gamma_\beta^n(x)|x\neq 0\} &= 1/\beta^n \\ \max\{\Gamma_\beta^n(x)|x\neq 0\} &= 1/\beta^{n-1}. \end{aligned} \quad (8)$$

Thus the gap function presents a more complete picture of the structure of a floating-point number system than any "equivalent digit" notion, and simplified formulas such as (3) and (4) must be used with extreme caution in any comparison of differently based floating-point number systems.

III. Conversion Mappings

A conversion procedure determines a specific value in S_β^n for each real number x. Thus, a conversion process may be characterized as either a function or a mapping. Formalization of conversion as a mapping appears preferable since it is often useful to invert the question and refer to the set of points which map into a given element of S_β^n, and this notion is then readily available by considering the inverse mapping. Certainly any conversion mapping of the reals to S_β^n should be the identity on S_β^n and in addition it is desirable for this mapping to satisfy certain order-preserving properties.

Definition: A mapping M of a set R' of real numbers into the reals is

1) *weakly order-preserving* (*isotone* [6], *monotone*) *on* R' if $x<y \Rightarrow M(x)\le M(y)$ for all $x, y\in R'$
2) *strongly order-preserving on* R' if $x<y\Rightarrow M(x)<M(y)$ for all $x, y\in R'$.

Furthermore, the mapping is said to be *weakly* (*strongly*) *order-preserving* if it is weakly (strongly) order-preserving on the reals.

No conversion mapping of the reals into S_β^n can be strongly order-preserving; however, we can expect that a conversion process should at least be weakly order-preserving in addition to being the identity on S_β^n. These latter two conditions do assure that the inverse image of any $b\in S_\beta^n$ under a conversion mapping is an interval containing b. Formally we shall limit our discussion to the rounding and truncation (sometimes called chopping) conversion procedures usually encountered in computerized numeric processing.

Definition: The *truncation conversion mapping* T_β^n and the *rounding conversion mapping* R_β^n of the real numbers into S_β^n are defined for all integers $\beta\ge 2$, $n\ge 1$ as follows.

Truncation Conversion:

$$T_\beta^n(x) = \begin{cases} \max\{b|b\le x,\, b\in S_\beta^n\} & \text{for } x\ge 0 \\ \min\{b|b\ge x,\, b\in S_\beta^n\} & \text{for } x<0. \end{cases} \quad (9)$$

Rounding Conversion:

$$R_\beta^n(x) = \begin{cases} \min\left\{b\middle|\dfrac{b+b'}{2} > x,\, b\in S_\beta^n\right\} & \text{for } x>0 \\ \min\left\{b\middle|\dfrac{b+b'}{2} \ge x,\, b\in S_\beta^n\right\} & \text{for } x<0 \\ 0 & \text{for } x=0. \end{cases} \quad (10)$$

The effects of these conversion mappings in the neighborhood of a power of the base are shown in Fig. 3. Note that the distinctions in the definitions of the mappings of positive and negative values are required to achieve the desired *sign complementary* relations:

$$T_\beta^n(-x) = -T_\beta^n(x), \qquad R_\beta^n(-x) = -R_\beta^n(x). \quad (11)$$

Note that $R_\beta^n((b+b')/2)=b'$ for all positive $b\in S_\beta^n$, so that we have not imposed the additional symmetric rounding condition for midpoints dependent on the "parity" of b that some researchers prefer [7]. Although this refinement could be added without materially affecting our results, we prefer the definition above.

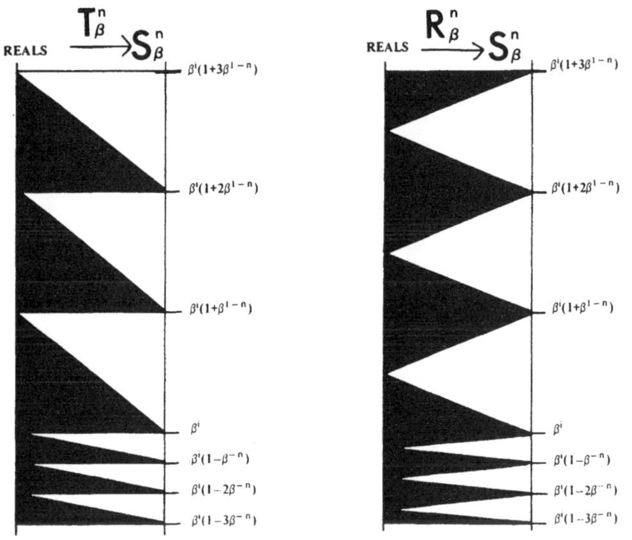

Fig. 3. Conversion of the real numbers to the n (significant) digit base β numbers in the neighborhood of a power of the base by truncation conversion T_β^n and rounding conversion R_β^n.

It is evident that T_β^n and R_β^n both are weakly order-preserving mappings which are identities on S_β^n as desired. In practical numeric computation we often need to convert data already expressed in floating-point form to a differently based floating-point form. Thus we are interested in the properties of the restricted mappings $R_\beta^n|S_\delta^m \to S_\beta^n$ and $T_\beta^n|S_\delta^m \to S_\beta^n$.

From consideration of the algorithmic mechanics of conversion, there is evidently a considerable difference between binary–hexadecimal conversion and binary–decimal conversion. Since special relationships between bases do affect the properties of restricted conversion mappings, there is need to characterize this important "commensurability" relationship between bases.

Definition: Let $\beta \geq 2$ be a root-free integer, i.e., β has no integral ith root for any i. Then the numbers β, β^2, β^3, $\cdots$ form a *commensurable family* of bases termed the *β-family* of bases. Two or more bases belonging to the same commensurable family of bases are *commensurable bases.* Two bases which do not belong to a common commensurable family are termed *incommensurable bases.* Furthermore, two or more significance spaces will be termed *commensurable* when their bases are commensurable.

Thus two, eight, and sixteen are commensurable bases, whereas base ten is incommensurable with any member of the binary family of bases. The root-free condition on β in this definition simply assures that each base is in precisely one family.

A useful equivalent characterization of commensurable and incommensurable bases avoiding explicit mention of the respective families to which the bases belong is provided in the following.

Lemma 1: The bases β and δ are commensurable if and only if $\beta^i = \delta^j$ for some nonzero integers i, j.

Corollary 2: The bases β and δ are commensurable if and only if $\log_\delta \beta$ is rational.

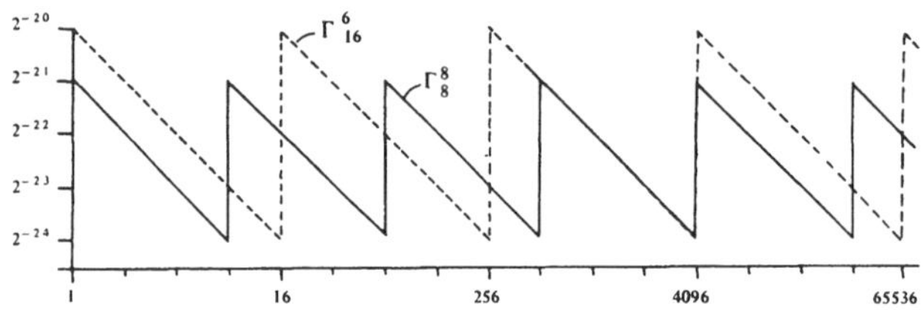

Fig. 4. Sections of the gap functions Γ_{16}^6 for the 6 (significant)-digit hexadecimal numbers and Γ_8^8, for the 8 (significant)-digit octal numbers. Hexadecimal and octal are commensurable bases and share a common period of 2^{12} on the log scale. Note that the six-digit hexadecimal numbers and the eight-digit octal numbers are not equivalent floating-point systems.

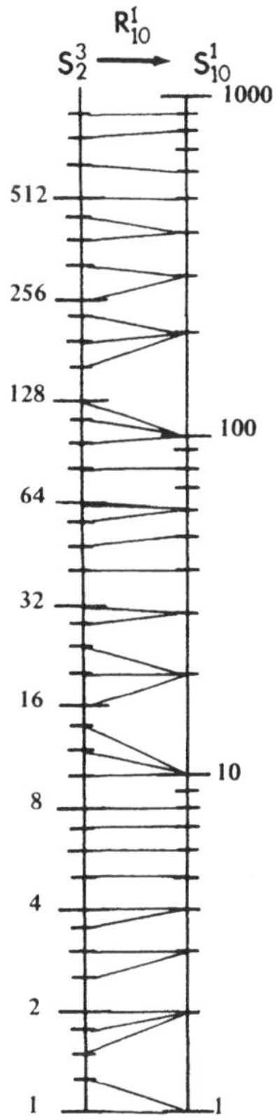

Fig. 5. Conversion by rounding of the 3 (significant)-bit binary numbers to the 1 (significant)-digit decimal numbers over the range [1, 1000], indicating that $R_{10}^1|S_2^3$ is neither one-to-one nor onto.

Thus, the gap functions Γ_β^n and Γ_δ^m plotted on a log–log scale will share a common period when β and δ are commensurable, as in Fig. 4, and otherwise will not (see Fig. 2).

It is readily verified that the conversion by rounding or truncation from a fixed-point number system with n β-ary digits to the right of the radix point to another fixed-point system with m δ-ary digits to the right of the radix point will be one-to-one if $m \geq n \log_\delta \beta$ and onto if $m \leq n \log_\delta \beta$. Thus a conversion between fixed-point number systems must be either one-to-one or onto; however, a conversion between floating-point number systems need be neither one-to-one nor onto, as seen in Fig. 5. The necessary and sufficient conditions for rounding and truncation conversions between incommensurable significance spaces to be 1) one-to-one mappings and 2) onto mappings have been determined [2] in the base conversion theorem.

Theorem 3 (*Base Conversion Theorem*): For incommensurable bases β and δ, the truncation (rounding) conversion mapping of S_β^n to S_δ^m, i.e., $T_\delta^m|S_\beta^n \to S_\delta^m$ ($R_\delta^m|S_\beta^n \to S_\delta^m$)

$$\begin{aligned} &1) \text{ is one-to-one if and only if } \delta^{m-1} \geq \beta^n - 1 \\ &2) \text{ is onto if and only if } \beta^{n-1} \geq \delta^m - 1. \end{aligned} \tag{12}$$

The details of the proof of this theorem are given in [2]; however, an intuitive understanding for the result can be gleamed from comparing the gap functions (see Fig. 2). Certainly if Γ_δ^m is uniformly less than Γ_β^n, then the mapping of S_β^n to S_δ^m should be one-to-one. Conversely, if the maximum of Γ_δ^m falls above the minimum (restricted to S_β^n) of Γ_β^n, then from a theorem of Kronecker [8] (that the integer multiples of an irrational number mod 1 are dense in the unit interval) it can be shown that for some $b \in S_\beta^n$, $\Gamma_\delta^m(b) > \Gamma_\beta^n(b)$.

In summary, the essential effect of formulae (12) regarding the conversion of the initial system S_β^n to the target system S_δ^m is that to assure one-to-one conversion a digit must be sacrificed in the target system and to assure onto conversion a digit must be sacrificed in the initial system.

The conditions for one-to-one conversion guarantee another desirable property of the conversion mapping, since it is readily shown that a weakly order-preserving mapping on R' is strongly order-preserving on R' if and only if the mapping is one-to-one from R' to the reals.

Corollary 3: For β and δ incommensurable, T_δ^m and R_δ^m are each strongly order-preserving on S_β^n if and only if $\delta^{m-1} \geq \beta^n - 1$.

It is instructive to consider the implications of the base conversion theorem for decimal-to-binary conversion. It follows that the mapping by rounding or truncation conversion of a decimal-based significance space to a binary-based significance space will be one-to-one (and strongly order-preserving) if and only if

$$\#\text{bits} > 3.32\cdots \times (\#\text{digits}) + 1 \tag{13}$$

and onto if and only if

$$\#\text{bits} \leq 3.32\cdots \times (\#\text{digits}) - 3.32\cdots. \tag{14}$$

Relating these inequalities (13), (14) to the membership density formula (1), we conclude that decimal-to-binary floating-point conversion will be 1) one-to-one and strongly order-preserving if and only if the decimal system has less than 0.9 $(\log_{10} 2) = 0.270$ times the membership of the binary system, and 2) onto if and only if the decimal system has more than 18 $(\log_{10} 2) = 5.418$ times the membership of the binary system. Thus conversions between floating-point systems of nearly equal density will be neither one-to-one nor onto.

The properties of decimal–binary conversion are succinctly presented in Fig. 6. The lattice point n, m corresponds to the binary system S_2^n and the decimal system S_{10}^m. Lattice points falling to the left of the line $n = (\log_2 10)\, m + 1 = 3.32m + 1$ correspond to decimal-to-binary conversions which are one-to-one and strongly order-preserving, as well as binary-to-decimal conversions which are onto. Lattice points falling to the right of the line $n = 3.32m - 3.32$ correspond to decimal-to-binary conversions which are onto and to binary-to-decimal conversions which are one-to-one and strongly order-preserving. Lattice points falling between $n = 3.32m + 1$ and $n = 3.32m - 3.32$ correspond to decimal–binary conversions which have none of these three properties. The equal density line $n = 3.32m - 0.88$ separates the lattice points so that those to the left correspond to binary systems which are more dense than the decimal systems, and lattice points to the right correspond to the decimal system being more dense. An increase of one unit on the n axis increases $|S_2^n/S_{10}^m|$ by a factor of 2, and a one unit in-

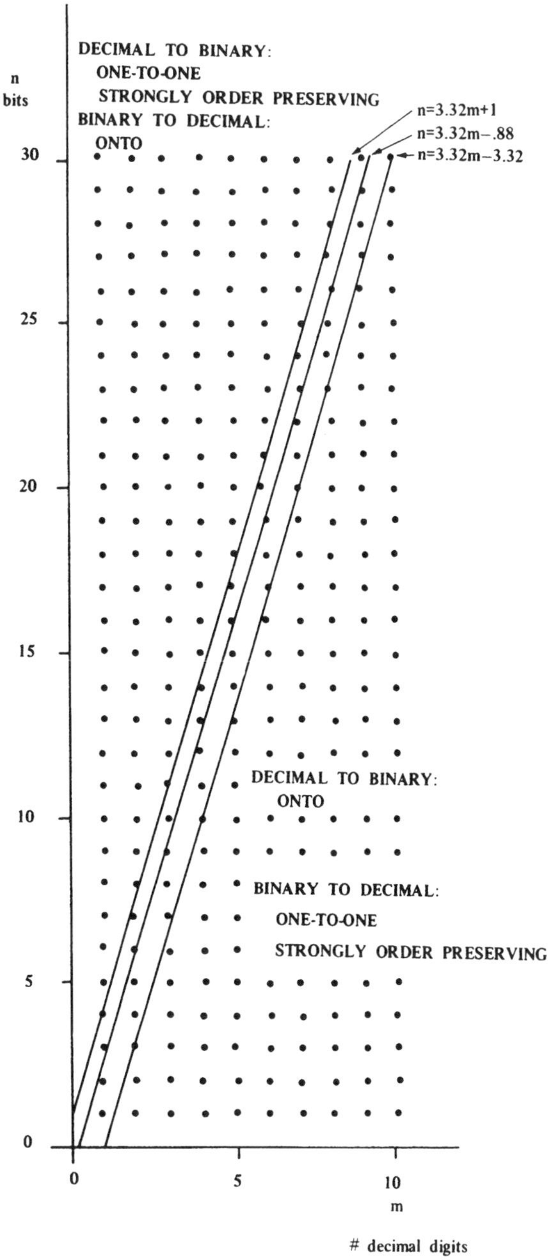

Fig. 6. A summary of the properties of floating-point decimal–binary base conversion.

crease on the m axis decreases this density ratio by a factor of 10, so $|S_2^n/S_{10}^m|$ may be easily estimated by determining the distance of the lattice point n, m from the equal density line.

IV. Compound Conversions

The preceding section dealt only with the properties of a single conversion mapping. We have indicated previously that computing environments such as PL/I language programming, multicomputer networks and BCD tape updating on a binary machine present situations where data may be subjected to multiple base conversions during overall job execution. Considerable care must be provided in such mixed-base computing environments to avoid excessive accumulation of error in purportedly "constant" data. To critically analyze this problem, the notion of compound conversion is formally introduced.

Definition: For all $n \geq 1$, $\beta \geq 2$,

1) T_β^n and R_β^n are *1-fold compound conversions through* S_β^n
2) for Q a k-fold compound conversion through $S_{\beta_1}^{n_1}$, $S_{\beta_2}^{n_2}, \cdots, S_{\beta_k}^{n_k}$, $T_\beta^n Q$ and $R_\beta^n Q$ are *$(k+1)$-fold compound conversions through* $S_{\beta_1}^{n_1}, S_{\beta_2}^{n_2}, \cdots, S_{\beta_k}^{n_k}, S_\beta^n$.

Furthermore, Q is a *compound truncation (rounding) conversion* if all the individual conversions composing Q are truncation (rounding) conversions.

Thus $R_\beta^n R_\delta^m$ is a 2-fold compound rounding conversion through S_δ^m, S_β^n, and $R_\delta^m T_\alpha^i R_\delta^m R_\beta^n$ is a 4-fold compound conversion through S_β^n, S_δ^m, S_α^i, S_δ^m (see Fig. 7).

A compound conversion is a composition of mappings, and many properties of individual mappings readily carry over to compositions of such mappings. Thus the following important properties of compound conversions are immediate.

Lemma 2: If Q is a compound conversion, then $Q(-x) = -Q(x)$ for all x.

Lemma 3: Compound conversions are weakly order-preserving.

An evident property of truncation conversion is that the magnitude of $T_\beta^n(x)$ is never larger than the magnitude of x. Thus truncation conversion performs a contraction of the reals towards zero [see Fig. 3].

Definition: The function M : Reals→Reals is a *contraction* if $M(x)$ has the same sign as x and $|M(x)| \leq |x|$ for all x.

Clearly a composition of contractions is a contraction.

Lemma 4: Compound truncation conversions are contractions.

The mappings R_β^n and T_β^n are identities on their image space S_β^n; however, $T_2^4 T_{10}^2(1.8) = 1.75$ and $T_2^4 T_{10}^2(1.75) = 1.625$, so that members of the image space of a compound conversion Q are not necessarily carried into themselves by the mapping Q. For the compound conversion Q illustrated in Fig. 7, note that q_1, q_5, and q_6 are mapped into themselves by Q, whereas q_2 and q_3 are points of the image space of Q which are not mapped into themselves.

In general, the points mapped into themselves by a compound conversion may be difficult to determine, but they are nonetheless important, since the weakly order-preserv-

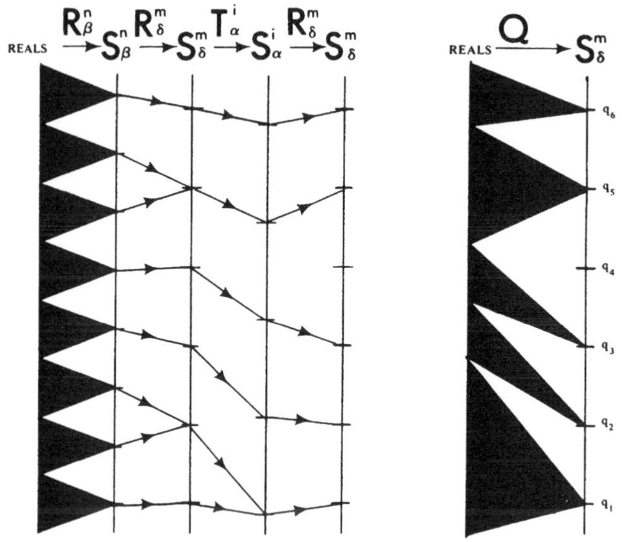

Fig. 7. The composition of a compound conversion from successive conversions. (a) The sequence of conversions R_β^n, R_δ^m, T_α^i, R_δ^m. (b) The 4-fold compound conversion $Q = R_\delta^m T_\alpha^i R_\delta^m R_\beta^n$.

ing property of compound conversions assures us that no point in an interval between two such invariant points of a compound conversion can be mapped outside that interval by that compound conversion.

Formally points x such that $M(x) = x$ are generally referred to [6] as fixed points of the mapping M. However, to avoid confusion with the computational notion of fixed-point numbers, we shall refer to fixed points of mappings as invariant points.

Definition: Let M be a mapping of the reals to the reals. Then the *invariant set of* M, denoted $f(M)$, is given by

$$f(M) = \{x | M(x) = x\}, \tag{15}$$

and for $x \in f(M)$ we say that x is an *invariant point* of the mapping M.

Our first result characterizing the invariant sets of compound conversions states that any element common to all significance spaces through which a compound conversion passes is an invariant point of that compound conversion.

Lemma 5: If Q is a k-fold compound conversion through

$$S_{\beta_1}^{n_1}, S_{\beta_2}^{n_2}, \cdots, S_{\beta_k}^{n_k}, \quad \text{then} \quad \bigcap_{i=1}^{k} S_{\beta_i}^{n_i} \subset f(Q).$$

Proof: For any $x \in \cap S_{\beta_i}^{n_i}$, each of the conversions composing Q maps x into itself since $R_{\beta_i}^{n_i}$ and $T_{\beta_i}^{n_i}$ are identities on $S_{\beta_i}^{n_i}$ for $i = 1, \cdots, k$. Thus $Q(x) = x$, and $x \in f(Q)$.

Furthermore, it is now shown that if Q is a compound truncation conversion, then the points in the intersection of the $S_{\beta_i}^{n_i}$ are the only invariant points of the mapping.

Theorem 4: If Q is a k-fold compound truncation conversion through $S_{\beta_1}^{n_1}, S_{\beta_2}^{n_2}, \cdots, S_{\beta_k}^{n_k}$, then

$$f(Q) = \bigcap_{i=1}^{k} S_{\beta_i}^{n_i}.$$

Proof: For $Q = T^{n_k}_{\beta_k} T^{n_{k-1}}_{\beta_{k-1}} \cdots T^{n_1}_{\beta_1}$, by Lemma 5,

$$\bigcap_{i=1}^{k} S^{n_i}_{\beta_i} \subset f(Q).$$

Now assume

$$x \notin \bigcap_{i=1}^{k} S^{n_i}_{\beta_i},$$

and let us show $Q(x) \neq x$. From sign complementarity of Q (Lemma 2), $x > 0$ may be assumed. Let j be some index such that $x \notin S^{n_j}_{\beta_j}$. Compound truncation conversions are contractions and truncation conversion is weakly order-preserving so that

$$x > T^{n_j}_{\beta_j}(x) \geq T^{n_j}_{\beta_j}(T^{n_{j-1}}_{\beta_{j-1}} \cdots T^{n_1}_{\beta_1}(x)) \geq Q(x).$$

Thus $x \neq Q(x)$, proving the theorem.

Lemma 5 and Theorem 4 show the importance of the intersection of significance spaces with regard to determining the invariant points of compound conversion mappings. In turn, the character of the intersection of significance spaces depends in large part on the commensurability or incommensurability of the significance spaces involved. Specifically, the base sixteen is in the binary family, and every hexadecimal number may be easily converted to binary by writing each hexadecimal digit as the appropriate four bits. For $b \in S^6_{16}$, the 24-bit binary representation of the six hexadecimal digits of b may have up to three leading zeros, or some 22-bit binary numbers may not be representable in S^6_{16}, e.g. $(1+2^{-21}) \notin S^6_{16}$; however, every 21-bit binary number will be contained in S^6_{16}. In generalizing this notion it is evident that if $\beta = \delta^p$, then each base β digit may be represented by p base δ digits. Thus an n-digit base β integer yields an np-digit base δ representation of which no more than $p-1$ leading digits may be zero. Setting $m = (n-1)p + 1$, we then have $S^m_\delta \subset S^n_\beta$, $S^{m+1}_\delta \not\subset S^n_\beta$, and the following theorem is derived.

Theorem 5: Let $\beta_1, \beta_2, \cdots, \beta_k$ be commensurable bases of the δ family. Then for $m = 1 + \min_i \{(n_i - 1)\log_\delta \beta_i\}$,

$$S^m_\delta \subset \bigcap_{i=1}^{k} S^{n_i}_{\beta_i} \quad \text{and} \quad S^{m+1}_\delta \not\subset \bigcap_{i=1}^{k} S^{n_i}_{\beta_i}. \tag{16}$$

Actually Theorem 5 can be sharpened if we also consider the different intervals $[\beta^j, \beta^{j+1}]$. Reasoning as above, it can be shown that by letting $p_i = \log_\delta \beta_i$, and setting $m = 1 + \min_i \{p_i(n_i - 1) + (j \bmod p_i)\}$, then we have

$$S^m_\delta \cap [\beta^j, \beta^{j+1}] = \bigcap_{i=1}^{k} S^{n_i}_{\beta_i} \cap [\beta^j, \beta^{j+1}]. \tag{17}$$

Furthermore, if q is the minimizing value of i in the above definition of m, then also $S^m_\delta = S^{n_q}_{\beta_q}$ over the interval $[\beta^j, \beta^{j+1}]$.

These observations along with Theorem 4 yield important properties of certain compound conversions. If Q is a compound truncation conversion through commensurable significance spaces, then $Q(x) \in f(Q)$ for all real x. Hence $QQ = Q$, and if Q^* is any compound truncation conversion through the same commensurable significance spaces as Q but in any order, then $Q^* = Q$. Furthermore, when $Q = T^{n_1}_{\beta_1} \cdots T^{n_k}_{\beta_k}$ with $\beta_1, \cdots, \beta_k$ in the same commensurable family, then $|x - Q(x)|/x < \max_i \Gamma^{n_i}_{\beta_i}(x)$, so accumulated conversion error is effectively controlled.

Now it will be shown that the intersection of incommensurable significance spaces does not contain any common significance space, for such an intersection has only a *finite number* of elements.

Theorem 6: If β and δ are incommensurable, then $S^n_\beta \cap S^m_\delta$ has no more than $2(\beta^n - 1)(\delta^m - 1) + 1$ members.

Proof: Let $\mathscr{P}$ be the following set of ordered pairs of integers, $\mathscr{P} = \{(k, k^*) \mid |k| < \beta^n, |k^*| < \delta^m \text{ and } kk^* \geq 1 \text{ or } k = k^* = 0\}$. Define a mapping $P: S^n_\beta \cap S^m_\delta \to \mathscr{P}$ as follows. $P(0) = (0, 0)$. Suppose $b \in S^n_\beta \cap S^m_\delta$, $b \neq 0$. Then $b = k\beta^j$ where k and j may be chosen uniquely such that $|k| < \beta^n$ and β does not divide k. Similarly, there are unique k^*, j^* such that $b = k^* \delta^{j^*}$ where $|k^*| < \delta^m$ and δ does not divide k^*. In this case $P(b) = (k, k^*)$. Thus P yields the normalized (right-shifted) integer portions of the representations of any element common to S^n_β and S^m_δ. It is now shown that P is a one-to-one mapping of $S^n_\beta \cap S^m_\delta$ into $\mathscr{P}$.

Clearly only zero is mapped into (0, 0), so let $a, b \in S^n_\beta \cap S^m_\delta$ be any two nonzero elements where $P(a) = P(b)$. Then there must exist $k, k^* \neq 0$, and j, j^*, i, i^* such that $a = k\beta^j = k^*\delta^{j^*}$, and $b = k\beta^i = k^*\delta^{i^*}$ where β does not divide k and δ does not divide k^*. Then $k/k^* = \delta^{j^*}/\beta^j = \delta^{i^*}/\beta^i$, so that $\delta^{j^* - i^*} = \beta^{j-i}$. Then $j^* = i^*$ and $j = i$ since β and δ are incommensurable, so $a = b$ and P is one-to-one. Now $S^n_\beta \cap S^m_\delta$ can have no more members than $\mathscr{P}$ which has $2(\beta^n - 1)(\delta^m - 1) + 1$ elements, thus proving the theorem.

Utilizing the fundamental theorem of arithmetic (unique prime decomposition), much more can be said about the members of $S^n_\beta \cap S^m_\delta$ for particular β and δ. If β and δ are relatively prime, then all members of $S^n_\beta \cap S^m_\delta$ are integers, with the largest such integer strictly less than $\beta^n \delta^m$. If the greatest common divisor of the incommensurable β and δ is a prime number (as in the binary–decimal case), then the smallest positive element of $S^n_\beta \cap S^m_\delta$ can be shown to be a negative power of that prime. This smallest positive number exactly representable in both systems can be considerably larger than the underflow bound on a typical computer, for example $2^{-10} = 0.0009765625$ is the smallest positive member of $S^{24}_2 \cap S^7_{10}$.

For a compound truncation conversion Q through $S^{n_j}_{\beta_j}$, $j = 1, \cdots, k$, we have shown that $Q(x) = x$ if and only if x is in the intersection of all significance spaces. Furthermore, if at least one pair of the β_j are not commensurable, then the intersection is finite and $Q(x) \neq x$ for any positive x lower than the minimum positive element of the intersection. For such an x the compound truncation conversion Q would have $Q(x) > QQ(x) > QQQ(x) > \cdots > Q^{(i)}(x) > \cdots$, and in fact $\lim_{i \to \infty} Q^{(i)}(x) = 0$ since zero is the only finite accumulation point of $S^{n_j}_{\beta_j}$. Thus for all i, $Q^i \neq Q^{i-1}$ for any compound truncation conversion Q involving at least two incommensurable bases, in sharp contrast to a compound truncation conversion through commensurable significance spaces where the iterated conversions immediately converged. Practically speaking, the successive updatings of a

BCD tape [4] on a binary machine could cause some of the "constant" floating-point data to be iteratively converted back-and-forth with each updating. If truncation conversion were adhered to as a standard, some of this data could drift lower in value (see Fig. 8), losing all accuracy, with no error indication provided by the system. Thus truncation conversion should be avoided in mixed-base computation unless all bases are in the same commensurable family.

With rounding conversion, the error accumulation upon successive conversions of a datum among incommensurable as well as commensurable bases is much better controlled. For example, it has been shown [9], [3] that with suitably high significance in the intermediate space S_δ^m, the 2-fold compound conversions $R_\beta^n R_\delta^m$ and $R_\beta^n T_\delta^m$ can both reduce to the identity on S_β^n.

Theorem 7 (In-and-Out Conversion Theorem): For β and δ incommensurable,

1) $R_\beta^n R_\delta^m$ is the identity on $S_\beta^n \Leftrightarrow \delta^{m-1} > \beta^n$
2) $R_\beta^n T_\delta^m$ is the identity on $S_\beta^n \Leftrightarrow \delta^{m-1} \geq 2\beta^n - 1$.

Thus from Theorem 7 we see that for certain compound conversions the invariant set will be the whole image space.

Corollary 4: For β and δ incommensurable,

$$\begin{aligned} &1)\ f(R_\beta^n R_\delta^m) = S_\beta^n \Leftrightarrow \delta^{m-1} > \beta^n \\ &2)\ f(R_\beta^n T_\delta^m) = S_\beta^n \Leftrightarrow \delta^{m-1} \geq 2\beta^n - 1. \end{aligned} \tag{18}$$

The in-and-out conversion theorem [3] has been stated for the case of incommensurable bases; however, incorporating the techniques of Theorem 5, a more general result encompassing both commensurable and incommensurable bases can be conveniently derived for the 2-fold compound rounding conversion case.

Corollary 5: For $\beta, \delta \geq 2$, let γ be the greatest common root of β and δ when β and δ are commensurable, and let $\gamma = 1$ otherwise. Then

$$R_\beta^n R_\delta^m \text{ is the identity on } S_\beta^n \Leftrightarrow \gamma\delta^{m-1} \geq \beta^n. \tag{19}$$

Proof: When β and δ are incommensurable, $\gamma = 1$ and $\delta^{m-1} \neq \beta^n$, so (19) follows from Theorem 7. Otherwise when β and δ are commensurable with greatest common root γ, then $\delta = \gamma^i$, $\beta = \gamma^j$ where i and j are relatively prime. Hence $S_\beta^n \subset S_\gamma^{jn}$ and $S_\gamma^{(m-1)i+1} \subset S_\delta^m$. Now $\gamma\delta^{m-1} \geq \beta^n$ implies that $(m-1)i + 1 \geq jn$, so $S_\beta^n \subset S_\delta^m$ and $R_\beta^n R_\delta^m$ is clearly the identity on S_β^n. Alternatively, assuming $\gamma\delta^{m-1} < \beta^n$ means that $(m-1)i + 1 < jn$. Now since i and j are relatively prime, there exists a k such that $k = j - 1 \bmod j$, $k = 0 \bmod i$. But then $S_\beta^n = S_\gamma^{jn}$ over the interval $[\gamma^k, \gamma^{k+1}]$ and also $S_\delta^m = S_\gamma^{(m-1)i+1}$ over the same interval, so that $(m-1)i + 1 < jn$ means S_δ^m is contained in but not equal to S_β^n over the interval $[\gamma^k, \gamma^{k+1}]$. Thus $R_\beta^n R_\delta^m$ cannot be the identity on S_β^n, completing the corollary.

When the condition $\gamma\delta^{m-1} \geq \beta^n$ is not obtained, then clearly $S_\beta^n \cap S_\delta^m \subset f(R_\beta^n R_\delta^m) \subsetneq S_\beta^n$, and it is of interest to give some alternative properties sufficient to characterize an invariant point of the compound conversion $R_\beta^n R_\delta^m$.

In general, if the image space of the compound conversion

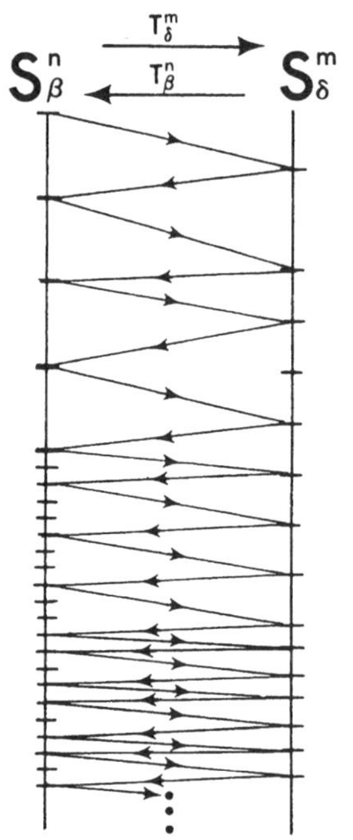

Fig. 8. The possible drift in value of a "constant datum" under iterated truncation conversion between incommensurable significance spaces.

Q were exactly equal to $f(Q)$, then $QQ = Q$, and a desirable situation controlling accumulated error is obtained. If the k-fold compound conversion Q ends with a truncation conversion, we have shown that QQ may not equal Q. Even if the final conversion of such a Q is a rounding conversion, the image space of Q can contain some points which are not invariant points of Q. For example, let $Q = R_3^2 R_2^4$. Since $2^{4-1} = 8 = 3^2 - 1$, by Theorem 3 the mapping $R_3^2 | S_2^4 \to S_3^2$ is onto, so Q covers all of S_3^2. However, from Corollary 4, $f(R_3^2 R_2^4) \neq S_3^2$.

Some detailed criteria for determining invariant points of $R_\beta^n R_\delta^m$ will now be considered.

Lemma 6: Assume that $b \in S_\beta^n$, $b \neq \pm\beta^i$ for any i, and for some $d \in S_\delta^m$, $R_\beta^n(d) = b$. Then $R_\beta^n R_\delta^m(b) = b$, i.e., $b \in f(R_\beta^n R_\delta^m)$.

Proof: Let β and δ be commensurable of the γ family. Let $b \in S_\beta^n$, $b \neq \pm\beta^i$ for any i and assume $\gamma^j \leq |b| < \gamma^{j+1}$. Over the intervals $[\gamma^j, \gamma^{j+1}]$ and $[-\gamma^j, -\gamma^{j+1}]$ either $S_\beta^n \subset S_\delta^m$ or $S_\delta^m \subset S_\beta^n$, and in either case with $d \in S_\delta^m$, $R_\beta^n(d) = b$ implies that $b \in S_\delta^m \cap S_\beta^n$. Therefore $b \in f(R_\beta^n R_\delta^m)$.

Alternatively, let us assume that β and δ are incommensurable. Let $0 < b' \in S_\beta^n$ have predecessor b and successor b'', and assume that $b' \neq \beta^i$ for any i. Then $b'' - b' = b' - b$. We assume there exists a $d \in S_\delta^m$ such that $R_\beta^n(d) = b'$, so

$$|d - b'| \leq (b' - b)/2. \tag{20}$$

From the definition of rounding (10), it is evident that $|x - R_\delta^m(x)| = \min_{a \in S_\delta^m} \{|x - a|\}$. Therefore with $x = b'$

$$|b' - R_\delta^m(b')| \leq |b' - d|, \tag{21}$$

and again with $x = R_\delta^m(b')$

$$|R_\delta^m(b') - R_\beta^n R_\delta^m(b')| \leq |R_\delta^m(b') - b'|. \tag{22}$$

and finally combining (20)–(22)

$$|b' - R_\beta^n R_\delta^m(b')| \leq b' - b. \tag{23}$$

If (23) were an equality, then $d \neq R_\delta^m(b')$ and (20)–(22) would all be equalities. Therefore $|d - R_\delta^m(b')| = b' - b$, and furthermore there could be no members of S_δ^m between d and $R_\delta^m(b')$. Also then

$$b' = (d + R_\delta^m(b'))/2 \tag{24}$$

so that $R_\delta^m(b')$ would be the successor of d in S_δ^m. Now any element and its successor in S_δ^m must differ by an amount δ^j for some j. Similarly, $b' - b = \beta^i$ for some i, so that with equality in (23), $\delta^j = \beta^i$. But then $i = j = 0$ since δ and β have been assumed incommensurable. Now $b' - b = \beta^0 = 1$ only if b and b' are consecutive integers, and similarly d and $d' = R_\delta^m(b')$ would be consecutive integers, contradicting (24). Therefore (23) must be a strict inequality, so that $R_\beta^n R_\delta^m(b')$ must equal b'. The case for negative elements of S_β^n follows from sign complementarity, and, of course, $R_\beta^n R_\delta^m(0) = 0$, completing the lemma.

As a consequence of this lemma, it is now shown that a comparison of the gap functions Γ_β^n and Γ_δ^m will suffice to determine many of the invariant points of $R_\beta^n R_\delta^m$.

Corollary 6: Assume $b \in S_\beta^n$, $b \neq \pm\beta^i$ for any i, and $\Gamma_\beta^n(b) \geq \Gamma_\delta^m(b)$. Then $b \in f(R_\beta^n R_\delta^m)$.

Proof: If $0 < b' \in S_\beta^n$ with $b' \neq \beta^i$ for any i, then $b' - b = b'' - b'$, and the interval mapping into b' under R_β^n is the half-open half-closed interval $[(b'-b)/2, (b''-b')/2]$. Suppose no member of S_δ^m falls in this interval. Then

$$\max\{d | d < b', d \in S_\delta^m\} \leq (b' - b)/2$$
$$\min\{d | d \geq b', d \in S_\delta^m\} > (b'' - b')/2$$

so that evaluating Γ_δ^m at the real number b' (see (6))

$$\Gamma_\delta^m(b') > (b'' - b')/b' = \Gamma_\beta^n(b').$$

Thus if $\Gamma_\beta^n(b') \geq \Gamma_\delta^m(b')$, then some member of S_δ^m must map into b' under R_β^n, and by Lemma 6 $b' \in f(R_\beta^n R_\delta^m)$. The result for negative $b \in S_\beta^n$, $b \neq -\beta^i$, then follows from sign complementarity.

Thus it is readily possible to find sequences of invariant points of $R_\beta^n R_\delta^m$ from a gap function comparison.

Corollary 7: For $\beta^i < \delta^j$, assume that $\Gamma_\beta^n \geq \Gamma_\delta^m$ over the open interval $(\beta^i, \min\{\beta^{i+1}, \delta^j\})$. Then

$$(\beta^i, \min\{\beta^{i+1}, \delta^j\}) \cap S_\beta^n \subset f(R_\beta^n R_\delta^m).$$

Referring back to Fig. 2, it is evident from Corollary 7 that all members of S_{16}^4 greater than 0.0625 and less than 0.1 are invariant points of $R_{16}^4 R_{10}^4$. Generally not all members of S_β^n from an interval I where $\Gamma_\beta^n < \Gamma_\delta^m$ will be members of $f(R_\beta^n R_\delta^m)$, but certainly any point of $R_\beta^n(S_\delta^m)$ other than a power of β must be an invariant point of $R_\beta^n R_\delta^m$. Since R_β^n restricted to $S_\delta^m \cap I$ is one-to-one to S_β^n, we may surmise that the number of members of $f(R_\beta^n R_\delta^m)$ in a neighborhood of x is comparable to the lesser of the numbers of members of S_β^n and S_δ^m in that neighborhood. Hence although the members of $f(R_\beta^n R_\delta^m)$ are more erratically spaced, the relative difference between neighboring points of $f(R_\beta^n R_\delta^m)$ will still be bounded with a bound perhaps larger but not by an order of magnitude than the largest value attained by Γ_β^n and Γ_δ^m.

The preceding lemma and its corollaries do not quite provide the full story about $f(R_\beta^n R_\delta^m)$, since the integral powers of β and δ represent break points in their respective gap functions, and by the preceding theory they must be treated separately. We have already pointed out that the image space of Q need not be identical to $f(Q)$ even for $Q = R_\beta^n R_\delta^m$, so the question of whether the iterates of Q, namely Q, QQ, $QQQ, \cdots, Q^{(k)}, \cdots$, will converge is still unanswered for $Q = R_\beta^n R_\delta^m$. This question is a very practical one since we would like to know if iterated rounding conversion between binary and decimal based systems will allow indefinite drift in the value of a "constant" as did truncation conversion, or if a stable pair of values must be achieved after a fixed number of rounding conversions back and forth. The following theorem is a surprisingly general and reassuring answer to this question.

Theorem 8 (Iterated Conversion Theorem): Let $Q = R_\delta^m R_\beta^n$ where $m, n \geq 2$. Then $QQQ = QQ$. Furthermore, this result is best possible in the sense that there exists $\beta, \delta, n, m \geq 2$ such that $R_\beta^n QQ \neq R_\beta^n Q$.

Proof: For any real x, $R_\delta^m R_\beta^n R_\delta^m R_\beta^n(x) = R_\delta^m R_\beta^n(x)$ unless $R_\delta^m R_\beta^n(x) = \pm\delta^i$ for some i by Lemma 6. Similarly, with $Q = R_\delta^m R_\beta^n$, $QQQ(x) = QQ(x)$ unless $QQ(x) = \pm\delta^j$ with $j \neq i$. Assuming $QQQ(x) \neq QQ(x)$, the definition of rounding conversion assures that

$$|R_\delta^m R_\beta^n(x) - R_\beta^n(x)| \geq |R_\beta^n R_\delta^m R_\beta^n(x) - R_\delta^m R_\beta^n(x)|$$
$$\geq |R_\delta^m R_\beta^n R_\delta^m R_\beta^n(x) - R_\beta^n R_\delta^m R_\beta^n(x)|,$$

so that with $|Q(x)| = \delta^i$, $|QQ(x)| = \delta^j \neq \delta^i$,

$$|QQ(x) - Q(x)| \leq 2|R_\delta^m R_\beta^n(x) - R_\beta^n(x)| < \delta^i(1 + \delta^{1-m}) - \delta^i$$

and also

$$|QQ(x) - Q(x)| = |\delta^i - \delta^j| = \delta^i|1 - \delta^{j-i}|,$$

and since $m \geq 2$ by assumption in the theorem,

$$|1 - \delta^{j-i}| < \delta^{1-m} \leq 1/\delta \leq 1/2.$$

Now $|1 - \delta^{j-i}|$ has a positive integer value for $j > i$, and for $j < i$, the smallest value of $|1 - \delta^{j-i}|$ is $1 - 1/\delta \geq 1/2$, and this is a contradiction. Hence $QQQ(x) = QQ(x)$ for all x.

The remainder of the theorem demonstrating that $R_\beta^n QQ \neq R_\beta^n Q$ for some $n, m, \beta, \delta \geq 2$ is shown in the example of Fig. 9.

Computer hardware involving the bases 2, 8, and 16 of the binary family and the base 10 is in popular usage. Our results so far discuss accumulated error under compound conversion 1) within a commensurable family and 2) between two incommensurable significance spaces. The modification of our results to cover compound conversion between more than two significance spaces, all from just two different commensurable families, is straightforward, so that our theory does cover the current situations where one might expect to encounter mixed base computation.

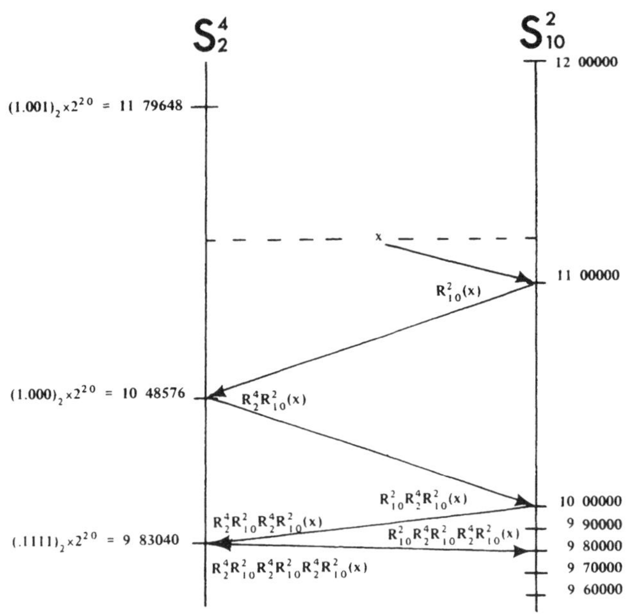

Fig. 9. Iterated conversions of $x = 1\ 120\ 000$ by the compound conversion $R_2^4R_{10}^2$, showing $R_{10}^2R_2^4R_{10}^2R_2^4R_{10}^2 \neq R_{10}^2R_2^4R_{10}^2$.

If a suitable three-state device were perfected for use in computer hardware, a ternary based floating-point system would undoubtedly be implemented and the possibility of mixed base computation among the bases 2, 3, and 10 would ensue. Our final thoughts will then be concerned with compound conversion through a collection of incommensurable significance spaces.

For the significance spaces S_{11}^5, S_2^{14}, and S_5^7, we have $1\ 960\ 563 = (11\ 1A000)_{11} \in S_{11}^5$, $1\ 960\ 576 = (1\ 11011\ 11010\ 10100\ 00000)_2 \in S_2^{14}$, and $1\ 960\ 500 = (10002\ 14000)_5 \in S_5^7$, and the absolute difference between an element and its successor in this neighborhood is 121, 128, and 125, respectively (see Fig. 10). Thus the iterates of the compound rounding conversion $Q = R_5^7R_2^{14}R_{11}^5$ in this neighborhood will be different for a number of cycles. As seen in Fig. 10, $Q^{(7)}(1\ 960\ 563) = 1\ 961\ 375$, and $Q^{(k)}(1\ 960\ 563) = 1\ 961\ 500$ for $k \geq 8$, hence $Q^{(k)} \neq Q^{(k-1)}$ for at least all $k \leq 8$. Examples such as this one show that utilizing rounding conversion is not enough to successfully restrict the drift in value of a constant datum under compound conversion. From the generalized result on in-and-out conversion given in Corollary 5, a resolution to the problem of controlling overall error growth in the presence of more than two incommensurable bases by intermediate reconversions to a standard significance space is feasible.

Specifically, let $S_{\beta_i}^{n_i}$, $1 \leq i \leq k$ be a collection of significance spaces representing the different floating-point data formats of a mixed-base computational environment. Suppose we introduce an intermediate space S_δ^m with the significance m small enough such that $R_\delta^m R_{\beta_i}^{n_i}$ is the identity on S_δ^m for all i. Then let all data introduced into the mixed-base computational environment first be converted by rounding to S_δ^m, and let subsequent conversions from $S_{\beta_i}^{n_i}$ to $S_{\beta_j}^{n_j}$ be preceded by a reconversion to S_δ^m, (i.e., $R_{\beta_j}^{n_j}R_\delta^m|S_{\beta_i}^{n_i} \to S_{\beta_j}^{n_j}$). Note that the

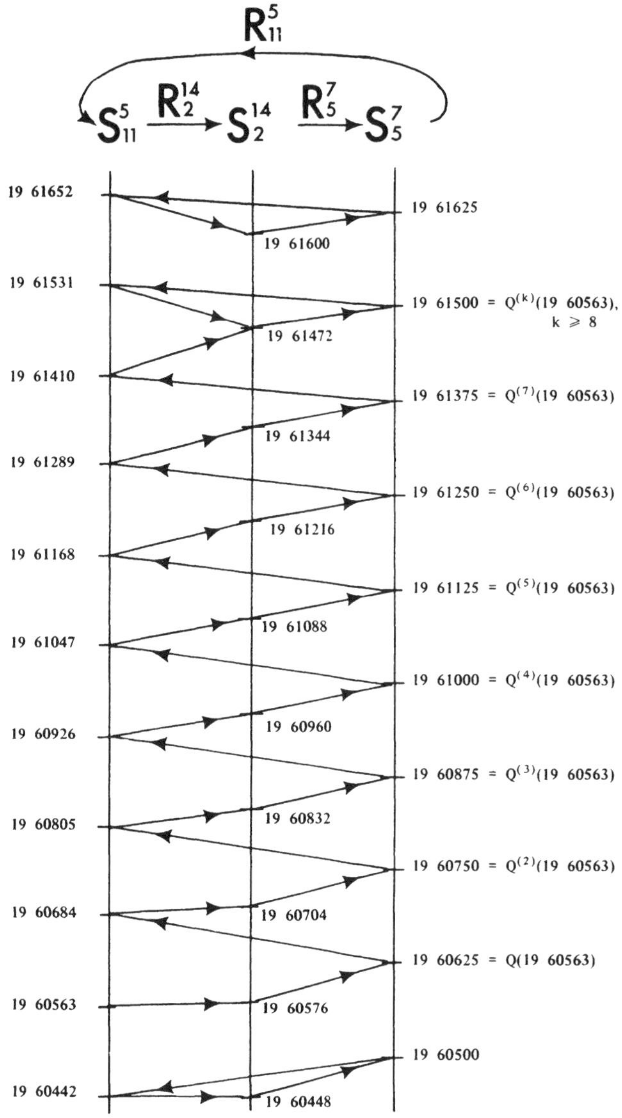

Fig. 10. Iterates of the compound rounding conversion $Q = R_{11}^5R_2^{14}R_5^7$ showing the drift in value of a "constant datum" under successive conversions.

conversion from $S_{\beta_i}^{n_i}$ to S_δ^m always regenerates the same value, $R_\delta^m(x)$, in S_δ^m for an initial datum x, since $R_\delta^m R_{\beta_i}^{n_i}$ is the identity on S_δ^m. Thus the value of the initial datum x whenever encountered in $S_{\beta_i}^{n_i}$, even after numerous intermediate conversions, is given by $R_{\beta_i}^{n_i}R_\delta^m(x)$, and this standardization of a constant's value with regard to each $S_{\beta_i}^{n_i}$ provides a highly desirable property for mixed-base computation.

Now the range of possible values achievable in $S_{\beta_i}^{n_i}$ is $R_{\beta_i}^{n_i}(S_\delta^m)$, and it is of course desirable to have as large an m as possible so that the conversion through S_δ^m does not introduce too much error. Yet it should be kept in mind that if m is chosen so large that one of the $R_\delta^m R_{\beta_i}^{n_i}$ is not the identity on S_δ^m, then conversion error may accumulate and generate a greater overall error than if a smaller m (meaning less initial accuracy) were chosen. These observations can be formalized as an additional corollary to Theorem 7 and Corollary 5.

Corollary 8: For $S_{\beta_i}^{n_i}$, $1 \le i \le k$, let γ_i be the greatest common root of δ and β_i when they are commensurable, and let γ_i be unity otherwise. Let

$$m \le \min_i \{(n_i - 1)\log_\delta \beta_i + \log_\delta \gamma_i\} \tag{25}$$

and let $Q_i = R_{\beta_i}^{n_i} R_\delta^m$ for $1 \le i \le k$. If $Q = Q_j Q'$ for any $1 \le j \le k$ where Q' is composed from the mappings Q_i, $1 \le i \le k$, then $Q = Q_j$.

The fact that m must be bounded from above in (25) in order to guarantee control of accumulated error and avoid situations such as that exhibited in Fig. 10 demonstrates that the phrase "carry more digits" does not always mean that greater overall accuracy will follow, and such cliches should not be used as a substitute for a true understanding of the formal structure of floating-point number systems and base conversion.

References

[1] D. W. Matula, "Base conversion mappings," *1967 Spring Joint Computer Conf., AFIPS Proc.*, vol. 30. Washington, D. C.: Thompson, 1967, pp. 311–318.
[2] ——, "The base conversion theorem," *Proc. Amer. Math. Soc.*, vol. 19, no. 3, pp. 716–723, June 1968.
[3] ——, "In-and-out conversions," *Commun. ACM*, vol. 11, no. 1, pp. 47–50, January 1968.
[4] ——, "Towards an abstract mathematical theory of floating-point arithmetic," *1969 Spring Joint Computer Conf., AFIPS Proc.*, vol. 34. Montvale, N. J.: AFIPS Press, 1969, pp. 765–772.
[5] N. Metropolis and R. L. Ashenhurst, "Radix conversion in an unnormalized arithmetic system," *Math Computation*, vol. 19, pp. 435–441, 1965.
[6] L. Collatz, *Functional Analysis and Numerical Mathematics* (translated by H. Oser). New York: Academic Press, 1966.
[7] M. Urabe, "Roundoff error distribution in fixed-point multiplication and a remark about the rounding rule," *SIAM J. Numer. Anal.*, vol. 5, no. 2, pp. 202–210, 1968.
[8] G. H. Hardy and E. M. Wright, *An Introduction to the Theory of Numbers*, 3rd ed. Oxford, England: Clarendon Press, 1954, p. 373.
[9] I. B. Goldberg, "27 bits is not enough for 8-digit accuracy," *Commun. ACM*, vol. 10, pp. 105–106, February 1967.

CADAC: A Controlled-Precision Decimal Arithmetic Unit

MARTY S. COHEN, T. E. HULL, AND V. CARL HAMACHER, SENIOR MEMBER, IEEE

Reprinted from *IEEE Transactions on Computers*, Volume C-32, Number 4, April 1983, pages 370-377.

***Abstract*—This paper describes the design of an arithmetic unit called CADAC (clean arithmetic with decimal base and controlled precision). Programming language specifications for carrying out "ideal" floating-point arithmetic are described first. These specifications include detailed requirements for dynamic precision control and exception handling, along with both complex and interval arithmetic at the level of a programming language such as Fortran or PL/I.**

CADAC is an arithmetic unit which performs the four floating-point operations add/subtract/multiply/divide on decimal numbers in such a way as to support all the language requirements efficiently. A three-level pipeline is used to overlap two-digit-at-a-time serial processing of the partial products/remainders. Although the logic design is relatively complex, the performance is efficient, and the advantages gained by implementing programmer-controlled precision directly in the hardware are significant.

***Index Terms*—Arithmetic unit, language requirements, numerical computation, pipelined multiplier, variable precision.**

I. INTRODUCTION

THERE has been a great deal of recent interest in the detailed specification of floating-point arithmetic, as evidenced, for example, by the work of the IEEE subcommittee on floating-point standards [1]. The goal of such work is to make scientific computing convenient from the point of view of the ultimate user. It should be powerful and flexible, and at the same time simple and easy to use. It is therefore appropriate to begin, in this section, by specifying a possible set of user needs. This is followed by a description of how these needs can be met, in an efficient way, through the development of suitable software and hardware. This paper is an updated version of recent work by the authors [2]–[5].

Existing programming languages tend to restrict our ability to describe the numerical processes we would like to implement. We believe that the situation would improve if a number of specific capabilities could be provided in the language in a straightforward and convenient way. These capabilities will be outlined in some detail in this section. An arithmetic unit, called CADAC, has been designed to support these capabilities in an efficient way. It has been implemented as an attached processor, and is currently undergoing extensive testing and evaluation. CADAC will be described in later sections.

We assume that all floating-point numbers have a normalized sign-and-magnitude decimal representation, and that all arithmetic operations on them are properly rounded (ties being broken by rounding to the nearest even). The main reason for this assumption is simply that it is easy and natural from the user's point of view. It also avoids I/O conversion errors.

Precision is to be under the control of the programmer and will be denoted by p, the number of decimal digits in the significand. The exponent range will be under the control of the programmer as well. Somewhat arbitrarily we assume that the exponent range is $[-10p, 10p]$ when the precision is p. Floating-point numbers can therefore be viewed by the user as being of the form

$$\underbrace{\pm 0 \cdot dd \cdots d}_{p \text{ decimal digits}} \times 10^{\overbrace{\pm ee \cdots e}^{\text{exponent range is } [-10p,\, 10p]}}.$$

We also assume that real variables can have two "values" in addition to the floating-point numbers, namely "null" and "indeterminate." Null is the uninitialized value, and wraparound results (see Sterbenz [6]) are left after overflow/underflow, except that indeterminate is the result of zerodivide.

Next, we assume that both complex and interval arithmetic are to be available. They can be provided directly as a part of the language, or, alternatively, through the specifications of user-supplied modules. In either event, their efficient implementation requires an arithmetic unit that can round up or round down (algebraically), as well as perform proper rounding and control of both precision and exponent range.

Precision control is now described in more detail. We identify two requirements which are not currently met by scientific programming languages.

1) The First Requirement: A means to enable the user to carry out intermediate stages of a calculation in precisions which differ from that currently being used in the program.

To consider one simple illustration, suppose that x and y are two arrays, and that we wish to calculate their dot product in higher precision in order to reduce the cumulative effect of roundoff error.

Manuscript received April 23, 1982; revised October 11, 1982. This work was supported in part by NSERC (Canada) Grants A-0055 and A-5192.

M. S. Cohen was with the Department of Electrical Engineering, University of Toronto, Toronto, Ont., Canada M5S 1A4. He is now with MSC Consulting, Toronto, Ont., Canada.

T. E. Hull is with the Departments of Computer Science and Mathematics, University of Toronto, Toronto, Ont., Canada M5S 1A4.

V. C. Hamacher is with the Departments of Electrical Engineering and Computer Science, University of Toronto, Toronto, Ont., Canada M5S 1A4.

Here is an example of what might appear in the program:

```
real(8) x(1:100), y(1:100), dotprod
  ⋮
begin precision(16)
    real(16) temp
    integer i
    temp = 0
    for i = 1, 2, ···, 100
        temp = temp + x(i)*y(i)
    end for
    dotprod = temp
end begin
  ⋮
```

The arrays x and y, and the result "dotprod," are in precision 8, but the intermediate calculations and the intermediate result "temp" are in precision 16. The value of "temp" is rounded to precision 8 on being assigned to "dotprod." The begin-end block is known as a *precision block*.

There are many other situations in which one might want to perform intermediate stages of a calculation in higher precision. These include calculating "residual" vectors in linear algebra, calculating Euclidean norms, and both determining and solving the normal equations associated with least-squares problems. See [3] and [4] for other examples. In some situations, the precision is increased in order to take advantage of the corresponding increase in exponent range.

2) *The Second Requirement:* A means to enable the user to carry out one part of a calculation two or more times in different precisions; for example, the calculation could be carried out in higher and higher precision until some error estimate becomes small enough.

To illustrate this second capability, we consider a program for solving a nonlinear equation within some prescribed accuracy, say "tol." We suppose that we have a procedure (such as Newton's method) for finding an approximation to the solution. And we also suppose that we have a way of determining a guaranteed bound on the error in an approximate solution, once that approximation has been found. The following informal program outline shows the essential features of what we have in mind:

```
initialize precision
flag = true
while (flag = true)
    begin precision p
        find approximate solution
        find error bound
        if bound ≤ tol
            root = approximation
            flag = false
        end if
    end begin
    increase precision
end while.
```

Here, the precision block is executed repeatedly, in higher and higher precision, until the error requirement is satisfied. Specific examples are given in [3] and [4].

The main implication of these two requirements, as far as the hardware is concerned, is that the precision of the data (which is concerned with storage) is chosen independently of the precision of the operations carried out on the data (which is concerned only with the arithmetic unit).

One of the principle requirements of any design is that the user know exactly what to expect under all circumstances. For example, the precision (of, say, eight decimal digits) should be exact, not "at least" (eight decimal digits).

The final topic to be discussed in connection with language facilities is that of exception handling.

3) *The Third Requirement:* A means to allow the user to have control over what is done in case of exceptions such as underflow and overflow.

The following example shows how a user might be allowed to specify what action is to be taken in case such exceptions arise:

```
begin
    on(underflow)
        result = 0
      (overflow)
        action to be taken
        ⋮
    end on
        ⋮} scope
end.
```

In this example, underflow anywhere in the scope of the block would lead to a value of 0 being assigned to "result" (a reserved word) in place of the wraparound value provided by the arithmetic unit before trapping, and then the calculation would resume from the point at which the trap took place.

Such a facility enables the user to choose alternative actions, such as terminating the calculation after output of an appropriate message, or perhaps repeating the calculation in a higher precision.

The difficulties with this third requirement are mostly in the software, and in connection with verification techniques. The essential hardware features are relatively easy to provide, since they only involve the provision of appropriate trapping on exceptions.

II. Overview of CADAC

The machine program view of CADAC is that it is a single-accumulator arithmetic unit with a specifiable precision. A decimal floating-point number with declared precision is stored in the main memory as a packed binary coded decimal (BCD) string with a short header that specifies sign, exponent, and length (precision of the significand).

Let us consider a typical sequence of operations involving CADAC. Suppose the Fortran expression

$$(B + C)/A$$

is to be evaluated and assigned to F. An assembly language code sequence corresponding to this task is

```
LOAD  B
ADD   C
DIV   A
STORE F.
```

Suppose the precisions of the significands of A, B, and C are declared to be eight decimal digits, and that the significand precision of F is six. If the current significand precision of CADAC is ten, then the above code executes as follows:

- the value in B is padded out with two zeros when it is loaded from main memory into the accumulator (ACC) in CADAC;
- the value in C is padded out with two zeros as it is moved from the main memory into CADAC and added to the contents of ACC, generating a properly rounded ten-digit sum in ACC;
- the value from A is padded out with two zeros when loaded into CADAC where it is divided into the contents of ACC, generating a rounded ten-digit quotient;
- the contents of ACC are rounded to six digits and stored at location F in the main memory.

A CADAC unit prototype has been built using high-speed SSI and MSI logic in the main data flow paths that process the significands. The unit is microprogram controlled by an AMD2910 microsequence generator, and exponent arithmetic is performed in three AMD2903 slices. The complete unit consists of two 7 × 16 in boards, each containing about 100 IC's, including all registers and buffers needed to interface the unit to a DMA controller. There are about 1K 88-bit microinstructions in the controller ROM. CADAC is interfaced to a PDP-11/34 minicomputer as an attached processor I/O device, and is currently undergoing extensive testing and evaluation. More details of this prototype will be provided in later sections. A number of data format details in the CADAC design, as reported below, are influenced by the 16-bit wordlength of the PDP-11 architecture. However, it should be clear that the design generalizes easily to other machine environments.

III. Number Representation

A main memory representation (suitable for a 16-bit word-length machine) for controlled-precision decimal floating-point numbers is shown in Fig. 1. The first word, termed the attribute, contains the sign, extend flag, exponent, and significand length descriptor. Succeeding words contain the fractional digit-normalized significand, low-order digits first. We have determined, from a consideration of typical user computations, that it is convenient to set the exponent range as $[-10p, 10p]$ when significand precision is p decimal digits.

The exponent E' is a binary integer ranging from 1 to 1023 and represents the true exponent E in excess-$(10p + 1)$ notation. Therefore, $E = E' - (10p + 1)$. A maximum symmetric range of $-511 \leq E \leq 511$ is available, reserving $E' = 0$ to indicate exact 0 and other special values, as described later.

The 4-bit length descriptor L is a binary integer that ranges from 0 to 15 and specifies the length p of the significand as $2L + 2$. Therefore, the significand length ranges from 2 to 32 decimal digits, in two-digit increments. For $2 \leq p \leq 32$, the exponent ranges are $[-20, 20]$-$[-320, 320]$, which can be accommodated in the 10-bit E'-field specified above.

A significand length p in the range [2, 32] along with an exponent in the range $[-10p, 10p]$ is sufficient for practically

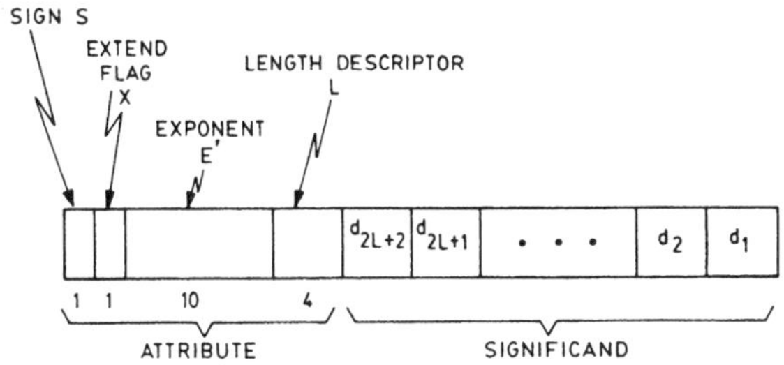

Fig. 1. Standard floating-point number representation.

all computational requirements. However, a longer significand and exponent range may be needed. This can be accommodated in an extended representation, indicated by setting the X bit to 1. (It is 0 in the above standard representation.) In extended representation, a second attribute word can be used to contain the length descriptor and additional exponent bits. The significand digits follow this second attribute word. We will not discuss the details of such a representation here because it will not be needed in this paper.

Let us now consider the representation of the special values: exact 0, null (or unassigned), "zero-divide," and indeterminate. As mentioned earlier, this class of values is indicated by an $E' = 0$ field. Exact 0 is represented by also setting $S = 0$. When $S = 1$ (with $E' = 0$) the operand is either null, zero-divide, or indeterminate. These three conditions are distinguished by the first digit of the significand following the attribute word, labeled d_{2L+2} in Fig. 1. If this digit is hexadecimal E, then the operand is null; if it is hexadecimal F, then the operand is indeterminate; and if it is hexadecimal D, then the operand is zero-divide. When space for a floating-point number is allocated to main memory, its attribute word is loaded with the appropriate L-field and X-field values. At this time, $S = 1$, $E' = 0$, and the first digit of the significand is set to hexadecimal E. A later assignment will change this special value. It should be noted that CADAC provides more than required by the language specifications of the preceding section in terms of special values. CADAC distinguishes between zero-divide ($A/0$, with $A \neq 0$) and indeterminate, while the language specification assigns indeterminate to either case.

IV. The Environment

CADAC is a single-accumulator arithmetic unit. It performs operations based on opcodes and operand addresses sent to its device input registers from the host. CADAC moves numbers between main memory and itself in DMA mode. A status register in CADAC indicates exceptions that occur as a result of instruction execution. This register can be read by the host. A control register is available in association with a maskable interrupt facility.

A. Precision

The floating-point number precision that is a part of the main memory number representation is independent of the current precision of the CADAC unit. Both the significand precision and the exponent range to be used in calculations can be independently set by appropriate commands to the CADAC unit. The particular case of $[-10p, 10p]$ for the exponent range, which should be sufficient for most applications, is thus easily supported by CADAC. Exponent range checking of main memory numbers can be done either by CADAC or by host software or firmware. For example, suppose CADAC precision p' is greater than the precision p of a result variable V. In association with rounding and storing from ACC to V, a check could be made in CADAC or the host for overflow/underflow with respect to the $[-10p, 10p]$ exponent range.

Suppose that the CADAC unit is set to operate with a given significand precision and exponent range. A number loaded into the unit is automatically rounded or padded to the current precision. If the exponent of the number exceeds the current CADAC range, an overflow/underflow will occur. When a result is returned to the main memory, the number is rounded or padded to the precision specified by the destination, and exponent range checking can also be performed. Since CADAC is a single accumulator machine, the high level language statement $X \leftarrow X$ is translated to a LOAD operation with appropriate rounding or padding, followed by a STORE operation. By setting the precision of the unit to some value less than that of a number, we have a convenient method for performing rounding. Thus we do not need a separate ROUND instruction. In normal operation we expect the precision of the unit to be greater than or equal to the precision of the operands.

B. Rounding

CADAC performs proper rounding by implementing "round to nearest, or, in case of a tie, to nearest even." The unit supports interval arithmetic with algebraic round up and round down. In addition to these rounding modes, CADAC also supports round up magnitude and round down magnitude (truncation). Rounding can occur:

- as part of an operation (rounding of results)
- when loading operands
- when storing results.

To properly support interval arithmetic, the rounding mode must be a property of the operation, not of the environment.

V. Instruction Set and Exception Handling

A. Instructions

This section introduces a basic instruction set consisting of instructions for hardware control and for basic arithmetic operations. Several additional instructions are included to support floating-point integer arithmetic. The instruction set consists of the following.

INIT—Initialize the unit and set the default precision and exponent range. Reset the exception flags. The default precision is 16 digits and the default exponent range is ± 160.

RESET—Reset the exception flags.

SET—Set the exponent range and significand precision.

LOADADDR—The unit uses temporary storage areas in the host main memory. These are required when extended precision numbers are processed. The starting addresses of these areas must be provided to the unit before extended precision processing can be requested.

LOAD—Load the accumulator from main memory.

STORE—Store the accumulator in main memory.

ADD—Add from main memory to the accumulator.

SUB—Subtract the main memory operand from the accumulator.

MUL—Multiply the accumulator by the main memory operand.

DIV—Divide the accumulator by the main memory operand.

NEG—Change the sign of the number in the accumulator.

FLOOR—Load the accumulator with the largest floating point integer less than or equal to the current contents of the accumulator.

CEIL—Load the accumulator with the smallest integer greater than or equal to the current contents of the accumulator.

COMPARE—Condition codes are set based on a comparison of the contents of the accumulator with the main memory operand. If the signs of these two values are the same, an actual subtraction ([ACC]-memory operand) is performed. Underflow/overflow exceptions are not set and the accumulator remains unaltered. If the signs differ, the condition codes can be set directly.

B. Floating-Point Integers

Our floating-point number representation can be used to represent large integers. These can be generated by the host or by the FLOOR or CEIL operators. Let NZ be the number of digit positions from the decimal point to the least-significant nonzero digit in a floating-point number. The condition for this number to be an integer is that $NZ \leq E$ where E is the exponent value. If it is known that a number is an integer, the programmer can test for a rounding "error." This can occur during any rounding operation.

C. Exception Handling

A number of exception conditions are detected. They are indicated by setting appropriate bits in the status register. An exception condition can generate an interrupt if its corresponding bit in the control register is set, and interrupts are enabled. The status register can always be read after an operation to determine if an exception occurred. The following exceptions are implemented.

1) OVERFLOW: This flag is set if exponent overflow or a zero-divide error occurs. On zero-divide (or indeterminate) the appropriate special value is loaded into the accumulator; otherwise, the wraparound exponent result [6] will be stored in the accumulator.

2) UNDERFLOW: Underflow is handled similar to overflow, with the wraparound result always being stored in the accumulator.

3) ZERO DIVIDE: A zero-divide error occurs when performing $A/0$ where A is a nonzero floating-point number. Both the overflow and zero-divide flags are set and the zero-divide code is stored in the accumulator.

4) INDETERMINATE: If an attempt is made to perform $0/0$, both zero-divide and indeterminate flags are set, and the indeterminate code is stored in the accumulator.

5) ILLEGAL OPERATION: This flag is set when the unit attempts to operate on a null, indeterminate, or zero-divide number. The indeterminate code is stored in the accumulator.

6) *INVALID FLOATING-POINT:* This flag is set if an invalid floating-point number is fetched. An invalid number may be an unnormalized number or a number containing a non-BCD digit. The indeterminate code is stored in the accumulator.

7) *NOT EXACT:* This flag is set whenever a rounding error occurs. If floating-point integer numbers are involved in an operation, any rounding operation produces an error. The rounded result is returned.

VI. The Pipeline

CADAC uses a hardware multiplier to achieve high performance. The multiplier includes a three-stage pipeline capable of running at 10 MHz. At this rate, the pipeline can multiply two 32-digit significands in approximately 30 μs. The design chosen for the multiplier unit uses BCD ALU's in a two-digit × two-digit structure. This structure multiplies two multiplicand digits by two multiplier digits and adds overlapped digits to produce two digits of partial product and two overlapped digits that are internally fed back. The first cycle produces two final product digits and two overlapped digits. All the multiplication is done in digit pairs and does not require shifting during product formation. The pipeline requires $17 \times 16 + 4$ iterations to multiply two 32-digit significands.

A. Operation of the pipeline

To understand the operation of the pipeline, consider the multiplication of two-digit numbers $a_1a_0 \times b_1b_0$ as illustrated in Fig. 2. Four two-digit product terms are required, $a_0 \times b_0$, $a_1 \times b_0$, $a_0 \times b_1$, and $a_1 \times b_1$. In the pipeline, these are obtained in parallel from look-up tables. Now note the significance of the two-digit product pairs. Pairs $a_1 \times b_0$ and $a_0 \times b_1$ have the same significance. The least significant (even) and most significant (odd) digits of these pairs are added together in parallel with two BCD ALU's. The even sum digit is added to the most significant digit of $a_0 \times b_0$. Similarly, the odd sum digit is added to the least significant digit of $a_1 \times b_1$. These two additions are done in parallel. A four-digit product is obtained.

Now consider how the pipeline would multiply $a_3a_2a_1a_0 \times b_1b_0$. In the first cycle the pipeline produces $a_1a_0 \times b_1b_0 = \alpha_{3_0}\alpha_{2_0}\alpha_{1_0}\alpha_{0_0}$. In the second cycle the pipeline produces $a_3a_2 \times b_1b_0 = \alpha_{3_1}\alpha_{2_1}\alpha_{1_1}\alpha_{0_1}$. Now the digit pairs $\alpha_{3_0}\alpha_{2_0}$ and $\alpha_{1_1}\alpha_{0_1}$ have the same significance and must be added together in the pipeline. In general, the digit pairs $\alpha_{1_i}\alpha_{0_i}$ emerge from the pipe. The pairs $\alpha_{3_i}\alpha_{2_i}$ are fed back to be added to $\alpha_{1_{i+1}}\alpha_{0_{i+1}}$.

The pipe must also add the previous partial product, generated with the previous multiplier digit pair, to the pairs $\alpha_{1_i}\alpha_{0_i}$. This partial product is fed into the pipe in pairs p_1p_0. The complete set of additions performed in the pipe is illustrated in Fig. 3. The eight required additions are indicated with brackets.

The three-stage pipeline is designed with as much parallelism as possible. Note that α_{3_i} is the odd digit of $a_1 \times b_1$. Call this digit a'_{1_i}. It must be added to the α_1 column during cycle

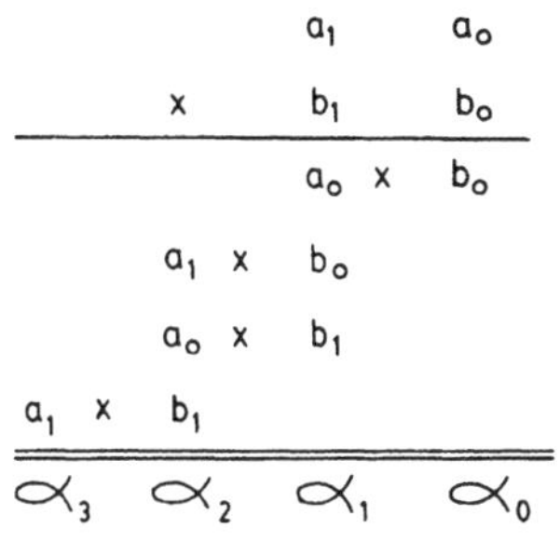

Fig. 2. A 2 × 2 multiply operation.

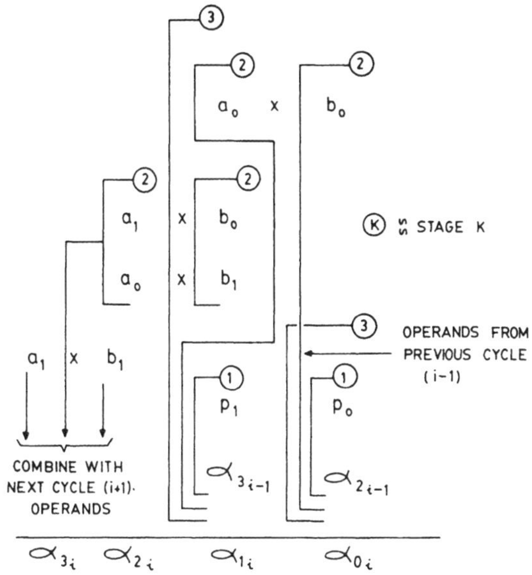

Fig. 3. Pipeline stage operations.

$i + 1$. It can be added to any of the digits in that column at any convenient time. It is most opportune to add it to p_1 in the first stage of the pipeline, as illustrated by the bracket labelled 1 in Fig. 3. Note that α_{2_i} consists of three terms, of which one is the even digit of $a_1 \times b_1$. Call this digit b'_{1_i}. It can be added to the α_0 column of cycle $i + 1$ in parallel with the addition of a'_{1_i}.

In the second stage of the pipe, the terms $a_1 \times b_0$ and $a_0 \times b_1$ can be added together. We can also add together the $a_0 \times b_0$ term and p_1p_0 (to which $a'_{1_{i-1}} \times b'_{1_{i-1}}$ has already been added).

We now consider the final stage of the pipe. The odd digit of $a_0 \times b_0$ (to which other terms have now been added) can be added to the even digit of $a_1 \times b_0 + a_0 \times b_1$. And lastly, the odd digit of $a_1 \times b_0 + a_0 \times b_1$ from the previous cycle, which is the remaining component of $\alpha_{2_{i-1}}$, can now be added to the least significant column.

The complete pipeline structure is illustrated in Fig. 4 and its interconnection with the other components of CADAC is shown in Fig. 5. The inputs to the pipeline are:

- the multiplier in R_{11}, R_{12}
- the multiplicand in R_{13}, R_{14}
- the previous partial product in R_{x1}, R_{x2}, denoted R_x.

The outputs from the pipeline are R_{41} and R_{42}.

Notice that all additions are arranged in pairs. Since there

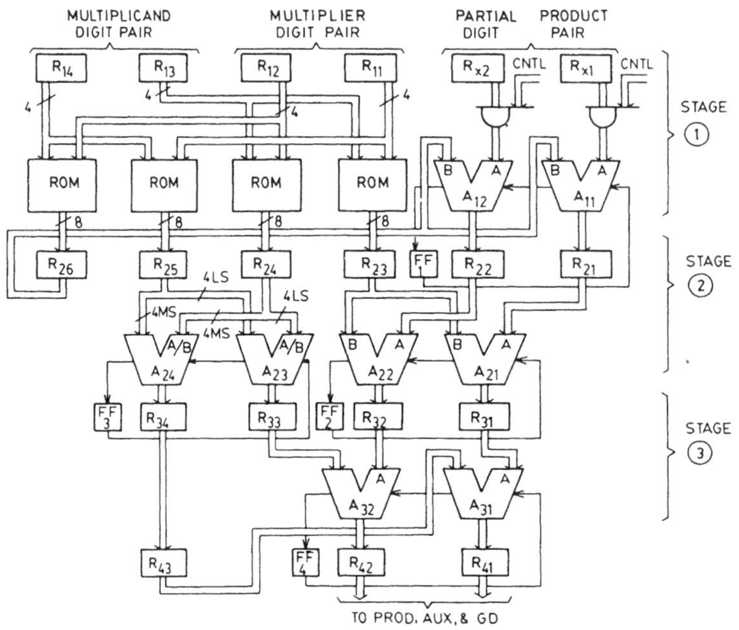

Fig. 4. Pipeline organization.

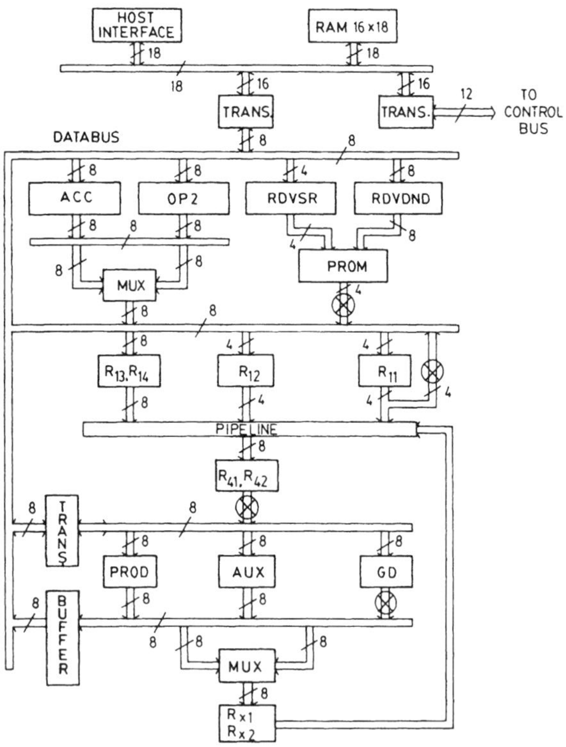

Fig. 5. CADAC dataflow paths.

is only one stage of carry propagation in each pair, there is no need for carry-lookahead. The carry output of the most significant adder is latched and fed back to the carry input of the least significant adder in the next iteration.

This is a convenient place to note that, with the exception of division, all arithmetic operations are performed with the least significant digit pair first. This is the reason that significands are stored least significant digit first. (Since we have decided to represent true zero by a special code, and not by a zero-most-significant digit, this storage arrangement poses no difficulty in zero detection.) This storage representation presents only a minor inconvenience in division, where slower execution times can be tolerated.

VII. Memory Requirements

A. Partial Product Storage (PROD)

The partial product digits emerging from the pipe must be buffered and fed back into the pipe at the appropriate time. A dual port memory, PROD, with 32-digit capacity, is used for this purpose. With this device a new partial product can be written into one port while the old partial product is read from the other. Separate counters are used to supply addresses to the two ports. See Fig. 5.

B. Operand Storage

To complement the 32-digit limit, we choose an accumulator (ACC) size of 32 digits. The second operand of an operation (*OP*2) is also stored. If a significand extends beyond 32 digits, it is not stored in CADAC but is buffered in main memory, in areas allocated to the unit. Regions are required for storing 1) the accumulator, 2) the second operand, and 3) the partial product. Data are loaded into the machine and written back in blocks of 32 digits as required by a calculation.

VIII. Standard (Nonextended) Arithmetic

A. Multiplication

We will consider the multiplication process in detail since the pipe has been optimized for this operation. See Fig. 5 while following this discussion. Consider the multiplication of two 32-digit numbers. The first multiplier pair is loaded into registers R_{11} and R_{12}. Now 16 pairs of multiplicand digits are loaded into R_{13} and R_{14} while partial product pairs appear at registers R_{41} and R_{42}. A zero is now loaded into R_{13}, R_{14} to obtain the 17th product pair. Of these 17 pairs, the first is a final product pair and the others are partial product pairs (stored in the dual-port memory PROD).

If we required a 64-digit final product, the final product pairs would be written into an additional 32-digit memory. However, all we require is a properly rounded 32-digit result. This requires the storage of a guard digit and the provision of a sticky bit. We provide a register (GD) to store a pair of potential guard digits. During the multiplication process, the first product pair produced is written to GD and the others are written to PROD.

While the zero is loaded into R_{13}, R_{14}, the next multiplier pair is loaded into R_{11}, R_{12}. This process is repeated until all digit pairs have been processed. Zeros are now fed into R_{13}, R_{14} until the pipe is flushed. There will now be a 32-digit product and a pair of digits in GD. If the most significant product digit is nonzero, the odd digit in GD is the guard digit and is used for rounding. If the most significant product digit is zero, the product is normalized by shifting left, with the odd digit in GD becoming the least significant product digit. The even digit in GD becomes the guard digit and is used for rounding.

Achievable execution times for processing significands for all four arithmetic operations are shown in Table I. These figures assume a 10 MHz clocking of the pipeline, which can be achieved. However, the AMD 2910 microsequence generator and the ROM used in the current prototype do not support this rate.

B. Division

Division is more complex than multiplication. We provide only an overview here. A straightforward long-division implementation is used. A PROM look-up table is used to estimate the successive quotient digits, by considering the first divisor digit and the first two partial remainder digits. The division process can be described as follows. The first quotient estimate is loaded into R_{11}, with R_{12} set to zero. The dividend is written into PROD. Now divisor pairs are loaded into R_{13}, R_{14} where they are multiplied by the quotient, and subtracted from the dividend pairs (fed into R_x). The partial remainder (PR) is stored in PROD. If the PR is negative, the quotient estimate is decremented and the divisor is added to the PR. This is repeated until the PR is positive. The quotient digit is written to ACC and the entire process repeated. One extra quotient digit is generated for a guard digit and the final remainder determines whether or not the sticky bit is set.

C. Addition/Subtraction

We will not discuss these operations in any detail. They are sequenced through part of the hardware described above for multiplication in a reasonably straightforward fashion. For both add and subtract instructions on signed operands, a true subtract may be required. If the difference is negative (10's complement), another pass through the ALU's is required to complement the result.

IX. Extended Precision Arithmetic

Extended precision arithmetic presents some problems not previously encountered. Most important, we discover that the

TABLE I
ACHIEVABLE EXECUTION TIMES* (μs)

DIGITS:	8	14	24	32
ADD/SUB	7	8	11	13
MULT	4	8	18	30
DIV	20	40	85	131

*execution times are for significand processing and they do not include operand fetching.

overlapped reading and writing of PROD during multiplication is not possible since the pairs involved may be in different blocks of 32 digits. Two separate RAM's are required. As before, we use PROD to contain the new partial product pairs. However, we add another RAM, which we call AUX, to hold the previous partial product. Arithmetic in extended precision is no different than before except that blocks of 32 digits are fetched from the host and written back as required.

The actual hardware cost of implementing extended precision arithmetic is therefore confined to extra buffers and ROM for the required additional microcode. This does not constitute a significant increase in cost of the basic CADAC unit.

X. Rounding

CADAC supports the five rounding modes discussed previously. These are:

- round to nearest
- round up/down algebraic
- round up/down magnitude.

Proper rounding requires a guard digit and a sticky bit. As was discussed previously, the unit retains two more digits than the precision of the calculation. This pair is stored in GD. This guarantees that the guard digit is retained even if a left shift is required (after multiplication or subtraction). The sticky bit is set to indicate that there are nonzero digits to the right of the guard digit. This bit is set automatically when overwriting a nonzero digit pair in GD. Rounding up is accomplished by applying a nonzero carry input to an ALU in the pipe.

XI. Zero Counting

It is important to know the number of leading and trailing zeros in a significand during various stages of processing in CADAC. The leading zero count is used:

- to skip over leading zeros in a partial remainder
- to normalize a product or sum/difference
- to determine if a partial remainder is zero.

The trailing zero count is used:

- to choose the multiplier from either ACC or $OP2$, picking the value whose significand has fewer nonzero significant digits
- to determine whether the sticky bit should be set when rounding a number
- to determine if a number is a floating-point integer
- to determine if a result is zero.

Leading zeros which are counted as digits emerge from the pipe and are transferred from R_{42}, R_{41} to PROD or GD. Trailing zeros which are counted as digits are written into ACC, $OP2$, or temporary storage.

XII. Control

So far we have not discussed the control of CADAC. The system uses an AMD2910 microsequence generator, ROM, and three AMD2903 4-bit ALU slices. The 2903's are used for:

- exponent arithmetic
- length calculation and handling
- provision of 12-bit registers to hold:
  - exponent limit
  - precision
  - operand lengths
  - trailing zero counts
  - other
- control logic.

XIII. Concluding Remarks

We have presented a possible set of requirements for a decimal arithmetic processor that performs clean arithmetic with specifiable precision. Such a machine should have sufficient rounding modes to support interval arithmetic as well as standard arithmetic. Such a machine should have a very large precision size and exponent range.

We have described the logic design and implementation of such a machine, named CADAC. It handles significands up to 32 digits very efficiently. In extended precision mode, computation is much slower but significand length is greatly increased. The final stages of testing the CADAC hardware are currently in progress at the University of Toronto.

Our primary goal in designing and implementing CADAC is to provide a thoroughly tested prototype that establishes the feasibility of supporting variable-precision decimal floating-point computations at a high execution speed. We do not claim that our programming environment requirements are the only reasonable set. However, they do provide powerful and flexible programming features, not currently supported in hardware, that are very useful in many practical applications.

The actual cost in integrated circuits, including ROM for microinstructions, of the CADAC prototype is not as interesting as the comparative cost of alternative implementations that either execute slower/faster or provide less/more functionality. For example, if only single (~8-digit) precision and double (~16-digit) precision is implemented in hardware, can a higher speed and a lower cost be achieved? Because of the significand alignment and normalization operations, our rough judgment is that both speed and cost cannot be significantly improved together by providing only single and double precision. Another interesting set of questions of the cost and speed type can be posed with respect to binary versus decimal implementations. It is clear that there are many more MSI and LSI components available for binary arithmetic, e.g., array multipliers. Therefore, it is tempting to conjecture that, with present technology, a binary implementation meeting similar, variable-precision functionality would be better. We have not carried out these or other comparative studies in quantitative detail. That type of study, interesting as it would be, requires significant effort; and it is outside of the scope of our present CADAC project.

To recap our project, we have demonstrated that it is feasible to build a decimal floating-point unit with very respectable performance, at reasonable cost, to meet very demanding requirements that focus on dynamically variable precision using small increments.

References

[1] "The proposed IEEE floating-point standard," four articles in *Comput.*, vol. 14, no. 3, Mar. 1981.

[2] M. Cohen, V. C. Hamacher, and T. E. Hull, "CADAC: An arithmetic unit for clean decimal arithmetic and controlled precision," in *Proc. 5th Symp. Comput. Arithmetic*, IEEE Publ. 81CH1630-3, May 1981, pp. 106–112.

[3] T. E. Hull, "The uses of controlled precision," in *Proceedings IFIP Working Conference on the Relationship Between Numerical Computation and Programming Languages*, J. K. Reid, Ed. North-Holland. Mar. 1982.

[4] ——, "Precision control, exception handling and a choice of numerical algorithm," in *Proc. Dundee Conf. on Numerical Analysis*, A. Watson, Ed. Springer-Verlag, to appear.

[5] ——, "Desirable floating-point arithmetic and elementary functions for numerical computation," in *Proc. 4th Symp. Comput. Arithmetic*, IEEE Publ. No. 78CH1412-6C, Oct. 1978, pp. 63–69.

[6] P. H. Sterbenz, *Floating-Point Computation.* Englewood Cliffs, NJ: Prentice-Hall, 1974.

Marty S. Cohen was born in Toronto, Canada, in 1954. He received the B.A.Sc. degree in engineering science and the M.A.Sc. degree in electrical engineering, specializing in computer design, from the University of Toronto in 1977 and 1980, respectively.

From 1977 to 1979 he was an Engineer with Ontario Hydro where he was involved in several software projects. In 1980 he was invited to join the CADAC development project where he was responsible for all aspects of design and implementation. He is currently President of MSC Consulting, a Toronto-based firm of consulting engineers.

Mr. Cohen is a member of the Association of Professional Engineers of Ontario.

T. E. Hull was born in Winnipeg, Canada. He received the M.A. degree in physics and the Ph.D. in applied mathematics, both from the University of Toronto, Canada.

He continued to work at the university, first in the Department of Mathematics and then served as the Computing Centre Director at UBC. Currently he is a Professor of Computer Science and Mathematics at the University of Toronto. His research interests include numerical software and language facilities for scientific computation.

Mr. Hull is active in AMS, MAA, SIAM, the Canadian Math Society, the Canadian Applied Math Society, CIPS and IFIP.

V. Carl Hamacher (SM'79) was born in London, Canada, in 1939. He received the B.A.Sc. degree in engineering physics in 1963 from the University of Waterloo, Waterloo, Canada; the M.Sc. degree in electrical engineering in 1965 from Queen's University, Kingston, Canada; and the Ph.D. degree in electrical engineering in 1968 from Syracuse University, Syracuse, NY.

Since 1968 he has been at the University of Toronto, Toronto, Canada, where he is a Professor in the Departments of Electrical Engineering and Computer Science, and the Associate Chairman of the Division of Engineering Science. His current research interests include local area computer networks, arithmetic processors, and real-time computer systems. During 1978–1979, he was a Visiting Scientist at the IBM Research Laboratory in San Jose, CA. He is a coauthor of the textbook *Computer Organization* (McGraw-Hill, 1978).

Dr. Hamacher is a member of ACM, Sigma Xi, and the Association of Professional Engineers of Ontario.

Finite Precision Rational Arithmetic: Slash Number Systems

DAVID W. MATULA AND PETER KORNERUP

Reprinted from *IEEE Transactions on Computers*, Volume C-34, Number 1, January 1985, pages 3-18.

***Abstract* — Fraction number systems characterized by fixed-slash and floating-slash formats are specified. The structure of arithmetic over such systems is prescribed by the rounding obtained from "best rational approximation." Multitiered precision hierarchies of both the fixed-slash and floating-slash type are described and analyzed with regards to their support of both exact rational and approximate real computation.**

***Index Terms* — Exact bit, finite precision, fixed-slash format, floating-slash format, fractions, hierarchical precision, number systems, precision-of-approximation, range/precision tradeoff, rounding, underflow-to-overflow range.**

I. Introduction and Summary

THERE are compelling reasons for investigating arithmetic in number systems composed of limited precision fractions. The simplicity and naturalness of specification of fraction number systems and arithmetic lead to a substantive body of number theory available to analyze and document properties of computation in such systems. To implement a fraction number system, consider that:

- a set of fractions $\{p/q\}$ can be represented as a set of integer pairs where the finite precision limitation characterizing the set can be specified simply by bounds imposed on the size of the numerator and denominator in each allowed pair, and
- any real value not exactly representable in such a finite precision fraction system can be *canonically* rounded to a value of the system by choosing what in number theory literature is called a "best rational approximation."

Our companion paper [9] describes an arithmetic unit capable of performing addition, subtraction, multiplication, and division incorporating this canonical rounding by best rational approximation on such fractional operands. Computation time is shown to be essentially equivalent to a floating-point divide operation of comparable precision per arithmetic operation. This speed is sufficiently competitive to make fraction number systems viable candidates for special application supplements to, or possibly alternatives for, a floating-point arithmetic system.

Manuscript received September 21, 1983; revised March 12, 1984. This work was supported in part by the National Science Foundation under Grant MCS 8012704 and by the Halliburton Foundation.

D. W. Matula is with the Department of Computer Science and Engineering, Southern Methodist University, Dallas, TX 75275.

P. Kornerup is with the Department of Computer Science, Aarhus University, Aarhus, Denmark.

Concern for numeric software portability has motivated several efforts [2], [6], [10] for defining unique and efficient finite precision computation systems with user specifiable precision levels available at the programming language level. To prescribe a particular floating-point arithmetic system recall that parameters for base, precision, and exponent range must be set. Other choices including rounding rule and possible allowance of denormalized numbers must be resolved to fully characterize the resultant number system and its attendant approximate real arithmetic. In contrast, fraction number systems may be fully characterized by a single (radix independent) integer valued parameter denoting the bound on numerator and denominator size. This parameter then serves quite naturally as a hierarchical precision specification variable. Approximate real arithmetic in fraction number systems is then uniquely prescribed by the canonical rounding obtained from best rational approximation of the exactly computed results. Lest this rounding rule appear esoteric consider the following.

The familiar process of rounding in radix representation illustrated by rounding the decimal representation $0.431464\cdots (= 277/642)$ to four digits yielding 0.4315, is elegantly simple as it corresponds to a *truncation of the radix representation*. Although less familiar, the "best rational approximation rounding" process, illustrated by rounding 277/642 to the four-digit fraction 22/51, can be shown to be equivalently simple as it results from *truncation of the corresponding (but not visible) continued fraction representation*.

The correspondence between fractions and continued fractions is fundamental to the characterization and implementation of finite precision rational arithmetic. A brief review of terminology will be useful here. A *fraction*, denoted p/q or $\frac{p}{q}$, is an ordered pair composed of a nonnegative integer *numerator* p, and a nonnegative integer *denominator* q, which are not both zero. The *quotient* (value) of p/q is the rational number determined by the ratio of p to q for $q \neq 0$, and is taken to be positive infinity when $q = 0$. The numerator and denominator of an *irreducible* fraction must have a greatest common divisor (gcd) of unity, other fractions being termed *reducible*. Two fractions are equal, denoted $p/q = r/s$, if $qr = ps$ ($p/q = r/s$ does not necessarily imply identical numerators and denominators). We do not always distinguish between "p/q" denoting a fraction (the ordered pair) or the quotient (a rational number).

Every nonnegative rational number x has both an irreducible fraction representation p/q and a finite *continued fraction* expansion

$$x = \frac{p}{q} = a_0 + \cfrac{1}{a_1 + \cfrac{1}{a_2 + \ddots \cfrac{1}{a_m}}} \qquad a_i \geqq 0\,, \qquad (1)$$

also denoted $p/q = [a_0/a_1/\cdots/a_m]$, where the *partial quotients* a_i are integral and unique (canonical) with the added requirements $a_0 \geqq 0$; $a_i \geqq 1$ for $1 \leqq i \leqq m-1$; and $a_m \geqq 2$ when $m \geqq 1$. The truncated continued fractions

$$\frac{p_i}{q_i} = a_0 + \cfrac{1}{a_1 + \ddots \cfrac{1}{a_i}} = [a_0/a_1/\cdots/a_i], \quad i = 0, 1, \cdots, m \qquad (2)$$

termed the *convergents* (or *best rational approximations*) of p/q, constitute a series of successively more accurate approximations.

Note then that

$$\frac{277}{642} = 0 + \cfrac{1}{2 + \cfrac{1}{3 + \cfrac{1}{6 + \cfrac{1}{1 + \cfrac{1}{3 + \cfrac{1}{3}}}}}} = [0/2/3/6/1/3/3].$$

The sequence of convergents to 277/642 are illustrated in continued fraction, fraction, and decimal radix representation along with the relative error of the approximations in Table I, from which the rounding of 277/642 to the four-digit fraction 22/51 is readily made visible.

The rounding rule prescribed by best rational approximation is simply to truncate the continued fraction representation at the largest index such that the corresponding fraction fits the format limitation. Importantly:

i) this process is unequivocally unique, i.e., we do not have to speculate on alternative rules for rounding "in the last place"; and

ii) this process is in fact radix independent, i.e., the sequence of best rational approximations does not depend on the radix employed for representation of the fractions, the final rounded value being determined simply by the last (largest indexed) convergent not exceeding the bound on size of numerator and denominator for the target fraction number system.

The canonical nature of the rounding is important not just in the fact that there is a resultant substantive body of number theory to analyze the properties of computation in these systems. For the computer architect, the availability of a *natural standard* to guide the number system and arithmetic specification is the ultimate key to effective design in support of numeric software portability.

TABLE I
THE SEQUENCE OF BEST RATIONAL APPROXIMATIONS TO 277/642 AND THE RELATIVE ERRORS OF THESE APPROXIMATIONS

continued fraction	fraction	decimal representation	relative error
[0]	0/1	0.0...	1
[0/2]	1/2	0.50...	0.15
[0/2/3]	3/7	0.428...	0.0067
[0/2/3/6]	19/44	0.4318...	0.00082
[0/2/3/6/1]	22/51	0.43137...	0.00021
[0/2/3/6/1/3]	85/197	0.431472...	0.000018
[0/2/3/6/1/3/3]	277/642	0.4314641...	0

In [15] we presented the foundations of finite precision rational arithmetic drawing heavily on classical material [5] from the number theoretic topics of Farey fractions and continued fractions. Basic results from these sources will be utilized (without proofs) as needed in our development here. Our companion paper [9] should be consulted for details on the efficient implementation of an arithmetic unit operating on rational operands.

In this paper, number systems composed of finite precision fractions in convenient binary formats facilitating exception handling features for underflow/overflow, infinity, exact zero, and not-a-number are prescribed. Innovations include a status bit for the exact/approximate status of each value. This bit is included in the word format to facilitate run time monitoring of the compatible exact-rational/approximate-real arithmetic that can be hosted by finite precision rational arithmetic. Accepting as a design goal the support of user specifiable precision at the programming language level, we show that finite precision rational arithmetic provides a conveniently parameterized hierarchical precision environment. Hierarchical precision is investigated both for support of exact rational and approximate real arithmetic.

We prescribe formats characterizing both fixed-slash and floating-slash number systems [12].

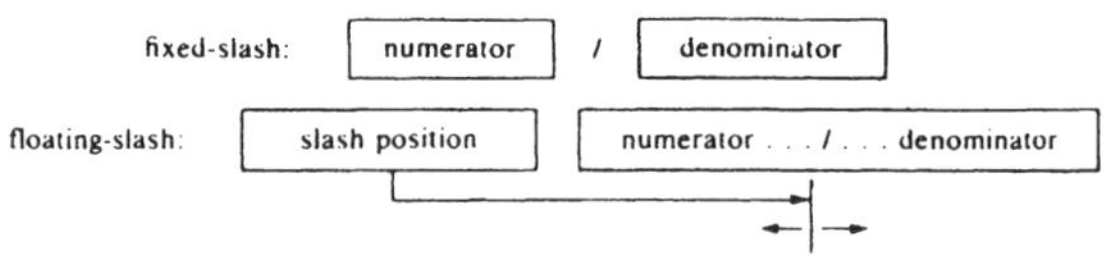

One format has an implicit slash "fixed" between equal sized numerator and denominator fields. The other has the "floating" slash position explicitly set by the value in an associated slash position field, with the sum of the number of bits in numerator and denominator fields prescribed.

In Section II we describe fixed-slash formats and analyze the resulting number systems and arithmetic. A major property of these systems is that the results of all arithmetic operations (+, −, ×, /) on single-word operands are exact in double-word representation, yielding a hierarchical exact rational subsystem within an approximate real computation environment. The feature of "graduated double-to-single precision rounding bias towards simple fractions" derived from best rational approximation in these systems is described. The benefits of this rounding bias for hosting "approximate rational" arithmetic (approximation of arithmetic over the rational field) are demonstrated. Exaggeration of precision degradation for "approximate real" arithmetic due to this rounding bias is examined in detail, with the result that a small loss of effective precision compared to format length should typically be realized.

Floating-slash formats and properties of the resulting number systems and arithmetic are described in Section III. Important features shared in common with fixed-slash systems are briefly summarized. Emphasis is placed on describing the two features distinguishing floating-slash from fixed-slash number systems:

i) the larger underflow-to-overflow range for comparable length formats, and

ii) the more uniform behavior of relative-error-of-approximation over the whole underflow-to-overflow range.

Of major importance is the characterization of extended range floating-slash systems and the identification of a precision fill feature through interpretation of "denormalized numbers." The extended range and precision fill features together allow specification of "extended floating-slash" systems having the traditional range and maximum relative gap size of comparable format length floating-point systems while containing an embedded "standard" floating-slash number system with "standard" floating-slash arithmetic as an accessible subsystem.

II. Fixed-Slash Number Systems

A. Format

The *(2k+2)-bit fixed-slash number system* is composed of 4-tuples (s, num, a, den) conveniently described by reference to the defining binary word format. The component fields illustrated in Fig. 1 are: the sign bit s, the k-bit integer field num, the exact bit a (for exact/approximate status), and the k-bit integer field den.

A fixed-slash number is termed *normalized* when $\gcd(num, den) = 1$, and so corresponds to an irreducible fraction. An unnormalized fixed-slash number corresponds to a reducible fraction.

The value v of the fixed-slash number is defined in the following.

a) If $num \neq 0$, $den \neq 0$, then $v = (-1)^s(num/den)$. [signed rational number]

b) If $num = 0$ and $den \equiv 1 \bmod 2$, then $v = (-1)^s 0$. [signed zero]

c) If $den = 0$ and $num \equiv 1 \bmod 2$, then $v = (-1)^s \infty$. [signed infinity]

d) If $den = 0$ and $num \equiv 0 \bmod 2$, or if $num = 0$ and $den \equiv 0 \bmod 2$, then v denotes "not-a-number." [not-a-number]

Standard *fixed-slash arithmetic* shall denote rounding (by best rational approximation to the fixed-slash format limit) of the exactly computed operation $(+, -, \times, /)$ on finite valued fixed-slash operands. When v is a number, $a = 0$

s	num	a	den
0	1 … k	k+1	k+2 … 2k+1

Fig. 1. $(2k+2)$-bit format fixed-slash number representation.

denotes that the value is exact; $a = 1$ denotes that the value is an approximation. The state $a = 1$ should be set when the corresponding represented value results from any of the following conditions.

1) Rounding error: The represented value is the result of rounding an otherwise exactly computed value that cannot be represented in the format provided.

2) Inherited error: A computed number where one of the arguments had $a = 1$.

3) Initial error: An initial number which is explicitly acknowledged to be not necessarily exact.

The exact bit and sign bit are unspecified when v is not-a-number.

A characterization of the numbers represented in a fixed-slash number system and an assessment of the space efficiency of the representation both follow from established number theory [5], [15], as summarized in the following observations.

Observation 1—Independence of Base: The set of representable extended real values of the $(2k+2)$-bit fixed-slash number system are the extended rational values of the Farey fractions F_{2^k-1} where the *order-n Farey fractions* F_n are defined by

$$F_n = \left\{ \pm\frac{p}{q} \,\middle|\, 0 \leq p, q \leq n, \ \gcd(p,q) = 1 \right\}. \tag{3}$$

Note further that representation of the order-n Farey fractions for any particular $n \leq 2^k - 1$ can be achieved by restricting the $(2k+2)$-bit fixed-slash numbers to $num \leq n, den \leq n$. □

Observation 2—Redundancy and Representation Efficiency: Redundancy in fixed-slash number systems entails a loss of less than one bit in storage efficiency independent of k.

Basis: Redundancy occurs in fixed-slash number systems since reducible as well as irreducible fractions can be represented. Dirichlet has shown (see [8, p. 324] for a proof) the following.

Theorem 1:

$$\lim_{n\to\infty} \frac{\left|\left\{ \frac{p}{q} \,\middle|\, 1 \leq p, q \leq n, \ \gcd(p,q) = 1 \right\}\right|}{n^2} = \frac{6}{\pi^2} = 0.6079\cdots. \tag{4}$$

Thus, approximately 60 percent of the representable fractions are irreducible for sufficiently large n, and direct computation [13] of the percentage of irreducible fractions for small n is in close agreement with this limit. For a numeric word format to attain 60 percent of the possible bit patterns as distinct values is to lose less than one bit (in this case $\log_2(\pi^2/6) = 0.718$ bits) per word. □

Suppose convenience at the user language interface

dictates prescribing fixed-slash number systems by the number of decimal digits allowed. By Observation 1 there is no need to employ binary coded decimal (BCD) representation to realize a faithful binary word implementation of such systems; e.g., imposing *num*, *den* $\leq 10^9 - 1$ in the 64 bit fixed-slash number system yields exactly those fractions having at most 9 decimal digits each in the numerator and denominator. In contrast, a 64 bit BCD string would contain only 16 decimal digits (with no sign). Note that BCD strings use only 10 of 16 possible patterns (losing $\log_2(16/10) = 0.68 \cdots$ bits) for every 4 bits, hence losing about 1/6 of the storage capacity. From Observation 2 it follows that faithful binary fixed-slash representation of "k-digit decimal" fixed-slash number systems will entail only a bounded loss (less than three bits) of storage capacity independent of k.

B. Exception Handling

A variety of implementation dependent exception handling procedures for nontrapping modes can be supported by the fixed-slash format described in Section II-A. Proceeding with a computation after exception by message passing is readily facilitated. The primary features are summarized in the following observations.

Observation 3—Extended Arithmetic: Signed infinity and zero are provided in a natural and symmetric manner. An implementation of the standard rules of arithmetic with fractions implicity ignores the signs of zero and/or infinity and correctly realizes extended arithmetic with infinity in the projective mode. Signed infinity is available for implementation of extended arithmetic in the affine mode. Note that the presence of a zero field in either numerator or denominator and a unit in the low order position of the other field is sufficient to characterize a zero or infinity, respectively. Then the leading $k - 1$ bits of that field are available to hold a message related to the corresponding zero or infinity value. □

Observation 4—Underflow/Overflow: The underflow and overflow thresholds are the reciprocals of each other. Underflow to either positive or negative zero and overflow to either positive or negative infinity may, by the exact/approximate bit, be distinguished from the occurrence of an exactly computed zero or infinity. The format then allows an implementation dependent message of $k - 1$ bits to be stored with the zero or infinity to describe the underflow/overflow situation (as noted in Observation 3). The message may be passed on, or otherwise becomes transparent to standard arithmetic on the zero or infinity value. □

Observation 5—Not-a-Number: When either the numerator field or denominator field is zero, it is sufficient that the other field have a low-order bit set to zero to determine that the value is not-a-number. Then the leading $k - 1$ bits of that field are available to hold an implementation dependent message that may be passed on for exception handling or as debugging information. Note that the null word is in the not-a-number class and is conveniently available to designate an "unassigned value." □

Observation 6—Denormalized Numbers: Standard fixed-slash arithmetic shall be expected to return normalized fixed-slash numbers, a feature readily obtained with the arithmetic unit described in [9]. A denormalized fixed-slash number shall refer to an unnormalized fixed-slash number where a specific meaning is associated with the value of gcd(*num*, *den*). Such a message will be transparent to standard fixed-slash arithmetic on the denormalized number, but can be visible in an enhanced environment to exception handling or extended arithmetic procedures. □

C. Exact Unary Operations

Fixed-slash number systems allow exact computation for the primary unary rational operators and certain conversion operations as summarized in the following observations.

Observation 7—Exact Additive and Multiplicative Inverses: Every member of a fixed-slash number system has an exact (efficiently computed) additive inverse (by changing sign) and multiplicative inverse (by swapping the contents of the *num* and *den* fields). The exact bit is not changed for these operations. □

Observation 8—Absolute Value: The absolute value of a fixed-slash number is exactly and efficiently computed by setting the sign bit to the positive state, leaving the remaining portion of the word unchanged. □

Observation 9—Integer and Fraction Parts: For the fixed-slash number x, floor(x), ceiling(x), and x mod 1 are all exactly computable each yielding a numerator and denominator no larger than the respective components of x. Recalling from (1)

$$x = \frac{p}{q} = [a_0/a_1/a_2/\cdots/a_m]$$

we obtain floor(x) and ceiling(x) using the value of a_0 or $a_0 + 1$; and x mod 1 using the value of $[a_1/a_2/\cdots/a_m]$, with proper modifications to account for the sign. Repeated application provides access to all of the partial quotients. The exact bit is not changed by these operations. □

Observation 10—Numerator, Denominator, gcd, and Normalization: The numerator and denominator of any fixed-slash number can be directly extracted. If unnormalized (corresponding to a reducible fraction) the rounding provided by the arithmetic unit described in [9] efficiently normalizes the fixed-slash number providing at the same time the value of gcd(*num*, *den*). □

Observation 11—Radix Represented Input Conversion: Floating-point and fixed-point numbers for any radix can be represented as fractions with integral numerator, and denominator a power of the radix. Such input can always be exactly represented in a $(2k+2)$-bit fixed-slash number system having sufficiently large k, and otherwise properly rounded by best rational approximation. □

Note that decimal data specified by a fixed-point format $F(i.j)$ denoting an i decimal digit field with $j \leq i$ places assumed to the right-hand side of the decimal point can always be exactly input into a 64-bit fixed-slash number system for any $i \leq 9$. Furthermore, data expressed in nondecimal or mixed radix units, e.g., feet/inches, weeks/days/hours/minutes, and degrees/seconds, are exactly representable in *any* of the measurement units in fixed-slash number systems.

Exact output conversion to mixed radix systems employing floor($r_i x$) and ($r_i x$) mod 1 is also conveniently available. These features provide added capacity beneficial to faithful hosting of data processing applications.

D. Hierarchical Precision Exact Rational Arithmetic

One functional goal of providing an arithmetic precision hierarchy in a programming language is that the user be able to control that the results of certain arithmetic operations shall be exactly represented. Hardware supported precision hierarchies are generally of the multitiered single/double or single/double/quad form. These tiers are usually defined simply to provide comparable arithmetic on operands whose format widths are corresponding multiples of a certain base word format width, rather than by attempting to determine the minimal most efficient hardware needed to realize particular functional arithmetic goals on base word operands.

For fixed-slash arithmetic the functional goal of realizing exact arithmetic on base word operands can be achieved with essentially optimal hardware efficiency by a convenient single/double-tiered hierarchy as shown in the following theorem.

Theorem 2: Let $n = 2k + 2 \geqq 4$. For any two n-bit (format) fixed-slash numbers with finite values x, y, the values

$$(x + y), \quad (x - y), \quad (x \times y), \quad \text{and} \quad (x/y) \tag{5}$$

are exactly representable as $2n$-bit (format) fixed-slash numbers. This result is best possible for addition and subtraction in that some values $(x + y), (x - y)$ are not representable as $2(n-1)$-bit fixed-slash numbers, and nearly best possible for multiplication and division in that some values $(x \times y)$, (x/y) are not representable as $2(n-2)$-bit fixed-slash numbers.

Proof: It is sufficient to consider nonnegative $x = p/q$ and $y = r/s$. By ordinary algebra of fractions:

$$x \pm y = \frac{p}{q} \pm \frac{r}{s} = \frac{ps \pm qr}{qs},$$

$$x \times y = \frac{p}{q} \times \frac{r}{s} = \frac{pr}{qs},$$

$$x/y = \frac{p}{q} \Big/ \frac{r}{s} = \frac{ps}{qr}.$$

For the $n = (2k+2)$-bit (format) fixed-slash numbers x, y we obtain $p, q, r, s \leq 2^k - 1$. Hence, $ps, pr, qs, qr < 2^{2k} - 1$, and $|ps \pm qr| < 2^{2k+1} - 1$. Thus, $(x \pm y)$ is exactly representable in a $2n = 2(2k + 2) = 2(2k + 1) + 2$ bit format, and $(x \times y)$, (x/y) are exactly representable in a $2(n - 1) = 2(2k + 1) = 2(2k) + 2$ bit format.

Choosing $x = 1/y = (2^k - 1)/(2^k - 2)$, we obtain

$$ps + qr = (2^k - 1)^2 + (2^k - 2)^2 > 2^{2k} \quad \text{for } k \geqq 3$$

where $\gcd(ps + qr, qs) = \gcd((2^k - 1) + (2^k - 2)^2, (2^k - 1)(2^k - 2)) = 1$. This confirms the necessity of a fixed-slash format of at least $2n$ bits for addition (similarly for subtraction). The necessity of a fixed-slash format of at least $2(n - 1)$ bits for multiplication follows readily for $x = y = (2^k - 1)/(2^k - 2)$, and for division for $x = 1/y = (2^k - 1)/(2^k - 2)$. □

Letting SINGLE and DOUBLE denote n-bit (word) and $2n$-bit (double word) formats of the same generic arithmetic type, the merger of functional goal with architectural convenience for fixed-slash arithmetic is succinctly expressed in the following observation.

Observation 12—Single-to-Double Exact Arithmetic: All arithmetic operations $(+, -, \times, /)$ on any pair of SINGLE fixed-slash operands yield exact DOUBLE fixed-slash results. □

Certain useful functional computations on SINGLE fixed-slash operands can also be shown to have exact DOUBLE or QUAD fixed-slash results where QUAD denotes the corresponding generic $4n$-bit (quadruple word) format. We state the following (noting that the proof follows in the same elementary manner as that of Theorem 2) as indicative of many that can be derived.

Lemma 3: Let $a_i, b_i, i = 0, 1, 2, 3$, be integer valued SINGLE, and x be a finite rational valued SINGLE, fixed-slash numbers. Then

$$y = \frac{a_0 + a_1 x}{b_0 + b_1 x} \tag{6}$$

is exactly representable as a DOUBLE fixed-slash result, and

$$z = \frac{a_0 + a_1 x + a_2 x^2 + a_3 x^3}{b_0 + b_1 x + b_2 x^2 + b_3 x^3} \tag{7}$$

is exactly representable as a QUAD fixed-slash result. □

It should be noted that the rational arithmetic unit described in [9] allows the computation of the full expression (6) in the same time as any one of the individual operations $+, -, \times, /$ (a time roughly equivalent to one floating-point divide of comparable precision).

Fixed-slash arithmetic can host exact rational arithmetic in a manner that may be viewed as an extension of "type INTEGER" arithmetic. At the same time, convenient support of traditional type INTEGER arithmetic as a subset of fixed-slash arithmetic (inserting the appropriate truncated integer divide) is feasible as summarized in the following observation.

Observation 13—Integer Compatibility: The $(2k+2)$-bit fixed-slash number system supports exact integer arithmetic over the full finite range $[-(2^k - 1), 2^k - 1]$ of the system. The format provides convenient architectural compatibility with sign-and-k-bit-magnitude integer representation by restriction to the leading $k + 1$ bits where $a = 0$ and $den = 1$ are maintained constant for all exact finite integer computations. □

Example: A four-tiered hierarchical precision fixed-slash arithmetic system is described in Table II. Word formats of 32, 64, and 128 bits are consistent with sizes of commercially available hardware supported floating-point arithmetic formats, e.g., the IBM370 system (where, however, SINGLE denotes 32 bits).

Designation of the 64-bit format as SINGLE width corresponds in size to the 64-bit format SINGLE width employed in the CDC CYBER 180 architecture. A 64-bit SINGLE width

TABLE II
FORMAT SIZES AND NUMERIC RANGES FOR A FOUR-TIERED FIXED-SLASH ARITHMETIC SYSTEM

Fixed-Slash Format	Word Width (in bits)	Numerator and Denominator Widths (in bits)	Underflow-to-Overflow Range (Decimal Equivalent)
HALF	32	15	$10^{\pm 4.5}$
SINGLE	64	31	$10^{\pm 9.3}$
DOUBLE	128	63	$10^{\pm 18.9}$
QUAD	256	127	$10^{\pm 38.2}$

for fixed-slash arithmetic is practically necessary both for sufficient range and accuracy of approximate arithmetic (accuracy is considered in Sections II-E and -F).

Provision of a QUAD precision of 256 bits is suggested primarily to achieve a range comparable to the range of several commercially available floating-point systems, e.g., the DEC Vax and Honeywell 6000 series both have ranges of approximately $10^{\pm 38}$ for SINGLE and DOUBLE width floating-point formats. The IEEE proposed standard [6] also has range $\sim 10^{\pm 38}$ for SINGLE, however a broader range for DOUBLE. For the body of scientific problems where a tradeoff favoring greater range to format size than that implicit in fixed-slash systems is desired, the floating-slash and extended floating-slash systems described in Section III are recommended.

Provision for HALF width allows memory saving in the storage of small integers and the relatively simple fractions frequently encountered as exact input, e.g., in many linear programming applications. Furthermore, provision of the four tiers HALF, SINGLE, DOUBLE, and QUAD provides a broader spectrum of user specifiable exact rational arithmetic control that could be efficiently supported at the programming language interface.

The fact that all fixed-slash arithmetic operations, including division, are exact in no more than twice the word length provides great utility to the feature of having an exact/approximate bit in the format of each fixed-slash number. This hardware facility can efficiently support a more comprehensive computation environment at the programming language level as noted in the following observation.

Observation 14 — Synergistic Exact Rational and Approximate Real Computation: Fixed-slash numeric representation with the exact/approximate bit associated with each value provides support for compatible exact rational and approximate real arithmetic. Thus, a synergistic dynamic precision controlled exact-rational/approximate-real computation environment could be accessible to the user at the programming language interface. □

Areas where a compatible exact-rational/approximate-real arithmetic computation environment could be beneficial include:

- symbolic computation;
- arithmetic for knowledge based systems; and
- combinatorial optimization, e.g., linear programming with sparse 0-1 constraint matrices.

Efficient hardware realization of compatible exact rational and approximate real arithmetic at relatively low cost in both computation time and architectural logic design complexity is an appealing dividend of fixed-slash representation. The strength of this exact-rational/approximate-real synergism also critically depends on the adequacy of support of approximate real arithmetic provided by fixed-slash computation, which is the subject of Sections II-E and -F.

E. Precision of Approximation

The parameter k implicitly determines the precision of a $(2k+2)$-bit fixed-slash approximation as noted in the following observation.

Observation 15 — Single Parameter Specification of Precision and Range: For the family of $(2k+2)$-bit fixed-slash number systems, the parameter k hierarchically specifies both the precision of approximate representation and the underflow-to-overflow range. Arbitrarily high precision over any finite region and an arbitrarily large range are achieved for sufficiently large k. There is a natural tradeoff between growth of precision-of-approximation and growth of underflow-to-overflow range in the family of fixed-slash number systems. □

The ability of any specific finite precision number system to host approximate real arithmetic is determined at the microscopic level by the spacing, or size of gaps, between representable values of the system. For floating-point systems the "gap function" [11], and/or equivalent "reciprocal-relative-spacing function" [1], are reasonably well behaved functions yielding the spacing at x (for x over the whole underflow-to-overflow range) in terms of the precision level of the system. Provision of an analogous "precision-of-approximation" function over the range of a fixed-slash system is accessible, but more complex. The analysis is aided by some pertinent facts from number theory.

Recalling that the values of the $(2k+2)$-bit fixed-slash numbers are just the order (2^k-1)-Farey fractions (3), we obtain the following [5], [15].

Theorem 4: Let $\frac{p}{q}, \frac{r}{s}$ be finite representable $(2k+2)$-bit fixed-slash numbers bounding the open interval $(\frac{p}{q}, \frac{r}{s})$, termed a *gap*, containing no other representable $(2k+2)$-bit fixed-slash numbers. Further, assume $0 \le \frac{p}{q} < \frac{r}{s}$ are irreducible fractions. Then

i) *absolute gap size:*

$$\frac{r}{s} - \frac{p}{q} = \frac{1}{sq}, \tag{8}$$

ii) *relative gap size:*

$$\frac{\frac{r}{s} - \frac{p}{q}}{\frac{p}{q}} = \frac{1}{sp}, \qquad \text{for } \frac{p}{q} \neq 0, \tag{9}$$

iii) *gap fill:* Letting $\text{FILL}(\frac{p}{q}, \frac{r}{s})$ denote the set of all irreducible fractions in the gap $(\frac{p}{q}, \frac{r}{s})$,

$$\text{FILL}\left(\frac{p}{q}, \frac{r}{s}\right) = \left\{ \frac{ip + jr}{iq + js} \,\middle|\, \gcd(i,j) = 1 \right\}. \tag{10}$$

□

Corollary 4.1: The extremes of gap size over the $(2k+2)$-bit fixed-slash number system are:

i) absolute gap sizes over $[0, 2^k - 1]$:

$$\text{max gap} = 1, \tag{11}$$

$$\text{min gap} = \frac{1}{(2^k - 1)(2^k - 2)} \sim 2^{-2k}, \tag{12}$$

where also

$$\text{max gap over } [0, 1] = \frac{1}{2^k - 1} \sim 2^{-k}, \tag{13}$$

ii) relative gap sizes over $[\frac{1}{2^k-1}, 2^k - 1]$:

$$\text{max rel gap} = 2^{-(k-1)}, \tag{14}$$

$$\text{min rel gap} = \frac{1}{(2^k - 1)(2^k - 3)} \sim 2^{-2k}. \tag{15}$$

□

Corollary 4.2: For any x in any gap $(\frac{p}{q}, \frac{r}{s})$ of the $(2k+2)$-bit fixed-slash numbers, the *gap size at* x, $\gamma(x)$, is given by

$$\gamma(x) = \frac{1}{sq} \begin{cases} \sim \dfrac{1}{\min\{q, s\}2^k} & \text{for } 0 < x < 1 \\ \sim \dfrac{x}{\min\{q, s\}2^k} & \text{for } x > 1. \end{cases} \tag{16}$$

Proof: Let x be in the gap $(\frac{p}{q}, \frac{r}{s})$ and assume without loss of generality that $q < s$. The fraction $(p + r)/(q + s)$ is in the gap $(\frac{p}{q}, \frac{r}{s})$, hence not representable as a $(2k+2)$-bit fixed-slash number. It follows

i) for $0 < x < 1$, that $s < 2^k \leq q + s < 2s$, so $2^{k-1} < s < 2^k$ and $qs \sim q2^k$,

ii) for $x > 1$, that $r < 2^k \leq p + r < 2r$, so $qs \sim qr/x \sim q2^k/x$

where the approximations are tight when $q << s$. □

The variation in gap sizes from $O(2^{-2k})$ to $O(2^{-k})$ over $[0, 1]$ noted in Corollary 4.1 for a $(2k+2)$-bit fixed-slash number system is very broad, being of the order "double-to-single" precision. This variation is, however, not capricious. By appropriate interpretation, Corollary 4.2 reveals a graduated precision hierarchy within a given fixed-slash number system that will be shown later to have its own merit for certain types of approximate computation.

Consider, for example, the 8-bit fixed-slash numbers (Farey fractions F_7) over the unit interval as illustrated in Fig. 2. Visualization of gap sizes in left-to-right order [Fig. 2(a)] yields an erratic (apparently chaotic) pattern. Alternatively, an enumeration of the gaps by decreasing size tagged only by the simpler fraction bound [Fig. 2(b)] reveals the graduated precision hierarchy of gap sizes. The hierarchy is made more explicit in the following where we restrict attention to the unit interval.

Corollary 4.3: Let j be the smallest integer (level) such that the given irreducible fraction $0 \leq p/q \leq 1$ is representable as a $(2j+2)$-bit fixed-slash number. Then the gap size on either side of p/q in the $(2k+2)$-bit fixed-slash number system for $k > j$ is approximately $2^{-(j+k)}$. □

Recall for binary floating-point systems that the addition

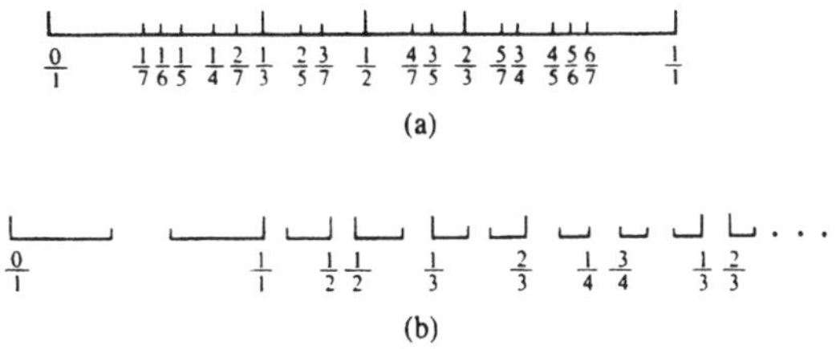

Fig. 2. Gaps between representable values of the 8-bit fixed-slash number system over [0, 1] arranged in (a) continuous order, and (b) in order by nonincreasing gap size.

of one more bit to the mantissa (significand) doubles the number of representable values by adding exactly one new representable value to each gap (bisecting each gap). There is an analogous but more complex situation for binary fixed-slash systems which we briefly summarize.

Corollary 4.4: Let F denote the $(2k+2)$-bit fixed-slash numbers, and F' the fixed-slash numbers obtained by allowing one more bit each in numerator and denominator, i.e., F' is the $(2(k+1)+2)$-bit fixed-slash numbers. Then we obtain [note: refer to Fig. 2 for illustration with F for $k = 2$ having $\{\frac{0}{1}, \frac{1}{3}, \frac{1}{2}, \frac{2}{3}, \frac{1}{1}\}$ representable over $[0, 1]$ (corresponding to raised tic marks), and F' for $k = 3$ having all values shown in the figure]:

i) density: F' has about four times the number of representable values as F;

ii) refinement: the *mediant* $\frac{p+r}{q+s}$, of each gap $(\frac{p}{q}, \frac{r}{s})$ of F is representable in F'.

Furthermore, for $\frac{p}{q} < \frac{u_1}{v_1} < \frac{u_2}{v_2} < \cdots < \frac{u_m}{v_m} < \frac{r}{s}$, with $\frac{u_i}{v_i}$, $1 \leqq i \leqq m$, giving all values representable in F' falling in the gap $(\frac{p}{q}, \frac{r}{s})$ of F,

iii) gap fill intensity: for $q < s$, $\lfloor \frac{s}{q} \rfloor \leqq m \leqq \lfloor \frac{s}{q} \rfloor + 3$,

iv) gap fill bias: for $q << s$, $\frac{u_1}{v_1}$ is approximately the midpoint of the gap $(\frac{p}{q}, \frac{r}{s})$, so the gap $(\frac{p}{q}, \frac{u_1}{v_1})$ of F' is still about one-half the size of the (previous) larger gap $(\frac{p}{q}, \frac{r}{s})$ of F where then the gaps $(\frac{u_1}{v_1}, \frac{u_2}{v_2}), (\frac{u_2}{v_2}, \frac{u_3}{v_3}), \cdots, (\frac{u_m}{v_m}, \frac{r}{s})$ of F' are each of the smaller (double-precision) size $\sim 1/s^2$, e.g., between $2^{-2(k-1)}$ and 2^{-2k} for p/q in the unit interval. □

The precision of approximation available for representation of a real number x is thus biased in the neighborhood of relatively simple fractions as summarized in the following observation.

Observation 16—Graduated Double-to-Single Precision Gaps: Gap sizes for the $(2k+2)$-bit fixed-slash numbers over the unit interval vary from a minimum of $\sim 2^{-2k}$ to a maximum of 2^{-k}. For fixed k, and any $j < k$, gaps of the $(2k+2)$-bit system, having as one bound a "simpler" fraction p/q with $2^j < q < 2^{j+1}$, have size about as large as $2^{-(j+k)}$. However, over $[0, 1]$ the "DOUBLE" $2n$-bit fixed-slash numbers have a maximum gap size still smaller than the minimum gap size of a "SINGLE" n-bit system. Relative gap sizes vary from a minimum of $\sim 2^{-2k}$ to a maximum of $2^{-(k-1)}$ over the whole underflow-to-overflow range, with bias towards larger relative gap sizes both from a "simpler" fraction gap bound and with distance of the gap from unity. □

F. Approximate Real Arithmetic

To examine the accuracy of approximate real arithmetic in any finite precision number system it is necessary to specify the rounding employed when the result of an arithmetic operation is not exactly representable in the system. For fixed-slash arithmetic the rounding is canonically specified utilizing the notion of "best rational approximation."

Recall that every nonnegative real number x has a *continued fraction* expansion

$$x = a_0 + \cfrac{1}{a_1 + \cfrac{1}{a_2 + \ddots}} = [a_0/a_1/a_2/\cdots] \qquad (17)$$

where the *partial quotients* a_i are integral and unique (canonical) with the added requirements $a_0 \geqq 0$; $a_i \geqq 1$ for $1 \leqq i$, and for terminating (rational) $x = [a_0/a_1/\cdots/a_m]$ also $a_m \geqq 2$ when $m \geqq 1$. The truncated continued fractions

$$\frac{p_i}{q_i} = [a_0/a_1/\cdots/a_i], \qquad i = 0, 1, \cdots,$$

termed the *convergents* (or *best rational approximations*) of x, constitute a series of successively more accurate approximations whose principal properties are (see [5], [7], or [15]):

i) recursive ancestry:

$$\text{with } p_{-2} = 0, p_{-1} = 1, q_{-2} = 1 \text{ and } q_{-1} = 0,$$
$$p_i = a_i p_{i-1} + p_{i-2},$$
$$q_i = a_i q_{i-1} + q_{i-2},$$

ii) irreducibility:

$$\gcd(p_i, q_i) = 1,$$

iii) adjacency:

$$q_i p_{i-1} - p_i q_{i-1} = (-1)^i,$$

iv) alternating convergence:

$$\frac{p_0}{q_0} < \frac{p_2}{q_2} < \cdots < \frac{p_{2j}}{q_{2j}} < \cdots \leqq x \leqq \cdots < \frac{p_{2j-1}}{q_{2j-1}} < \cdots < \frac{p_1}{q_1},$$

v) best rational approximation:

$$\frac{r}{s} \neq \frac{p_i}{q_i}, \qquad s \leqq q_i \Rightarrow \left|\frac{r}{s} - x\right| > \left|\frac{p_i}{q_i} - x\right|,$$

vi) quadratic convergence:

$$\frac{1}{q_i(q_{i+1} + q_i)} < \left|\frac{p_i}{q_i} - x\right| \leqq \frac{1}{q_i q_{i+1}}.$$

Let FXS_k denote the values of the $(2k+2)$-bit fixed-slash numbers (equivalently, FXS_k denotes the Farey fractions F_{2^k-1}). The rounding Φ_k: Reals $\rightarrow FXS_k$ is defined for every real number x where $p_0/q_0, p_1/q_1, p_2/q_2, \cdots$ are the convergents to $|x|$, by

$$\Phi_k(x) = \begin{cases} p_m/q_m & \text{if } x \geqq 0,\ x = p_m/q_m \\ & \quad \text{with } p_m, q_m \leqq 2^k - 1, \\ p_i/q_i & \text{if } x > 0,\ p_i, q_i \leqq 2^k - 1 \\ & \quad \text{and } \max\{p_{i+1}, q_{i+1}\} > 2^k - 1, \\ -\Phi_k(-x) & \text{if } x < 0. \end{cases} \qquad (18)$$

To illustrate the rounding suppose we are to round $x = 277/642$ into a 20 bit fixed-slash number, i.e., at most 9 bits in numerator and denominator. As noted in Table I,

$$\frac{277}{642} = [0/2/3/6/1/3/3],$$

$$\text{with convergents } \frac{0}{1}, \frac{1}{2}, \frac{3}{7}, \frac{19}{44}, \frac{22}{51}, \frac{85}{197}, \frac{277}{642}.$$

(These data may be used also to illustrate all the preceding properties i)–vi) of convergents.) We obtain as the rounded value the truncated continued fraction

$$[0/2/3/6/1/3] = \frac{85}{197} = \frac{001010101_2}{011000101_2}$$

corresponding to the last convergent representable as a 20-bit fixed-slash number.

Theorem 5: The rounding Φ_k: Reals $\rightarrow FXS_k$ satisfies the following four properties for all real x, y:

i) monotonic: $x < y \Rightarrow \Phi_k(x) \leqq \Phi_k(y)$,

ii) distributive sign: $\Phi_k(-x) = -\Phi_k(x)$,

iii) distributive inverse: $\Phi_k(1/x) = 1/\Phi_k(x)$,

iv) fixed points: $|x| = p/q \in FXS_k \Rightarrow \Phi_k(x) = x$. □

Theorem 6: Let $(\frac{t}{u}, \frac{p}{q})$ and $(\frac{p}{q}, \frac{r}{s})$ be the gaps on both sides of $\frac{p}{q}$ in the $(2k+2)$-bit fixed-slash number system. Then the interval of values rounding to $\frac{p}{q}$ includes all values between the mediants of the two gaps, i.e.,

$$\Phi_k(x) = \frac{p}{q} \qquad \text{for } \frac{t+p}{u+q} < x < \frac{p+r}{q+s} \qquad (19)$$

and also each mediant itself whenever $\frac{p}{q}$ is the "simpler" fraction bounding the gap, i.e.,

$$\Phi_k\left(\frac{p+r}{q+s}\right) = \frac{p}{q} \qquad \text{iff } q < s \text{ (hence also } p < r),$$

$$\Phi_k\left(\frac{t+p}{u+q}\right) = \frac{p}{q} \qquad \text{iff } q < u \text{ (hence also } p < t). \qquad (20)$$

□

Thus, the rounding Φ_k: Reals $\rightarrow FXS_k$ effectively satisfies the rule "round away from the mediant towards the boundary of each gap, rounding the mediant to the simpler fraction bounding the gap." Thus, we say Φ_k is *mediant rounding* into FXS_k.

Corollary 6.1: The mediant rounding error

$$|x - \Phi_k(x)| \qquad \text{for } \Phi_k(x) = p/q < \infty \text{ satisfies:}$$

absolute error bound:

$$|x - \Phi_k(x)| \leq \frac{1}{q 2^k}, \qquad (21)$$

relative error bound:

$$\left|\frac{x - \Phi_k(x)}{\Phi_k(x)}\right| \leq \frac{1}{p 2^k} \qquad \text{for } p \neq 0. \qquad (22)$$

□

It is important to note from Corollary 6.1 that most of the interval within a larger gap rounds to the simpler fraction bound. This provides then that the "simplicity" of the fractional result yields a measure of the graduated precision-of-approximation bias of the rounding. More specifically, we have the following.

Corollary 6.2: If $\Phi_k(x) = p/q \leq 1$, with $j \leqq k$ the smallest integer such that p/q is representable as a simpler $(2j+2)$-bit fixed-slash number, then

$$|x - \Phi_k(x)| < 2^{-(j+k)+1} \tag{23}$$

which is essentially a best possible bound in that

$$|x - \Phi_k(x)| \sim 2^{-(j+k)+1} \tag{24}$$

if x is the mediant of a gap in FXS_k. □

For our preceding example note that 277/642 is the mediant of the gap $(\frac{192}{445}, \frac{85}{197})$ in the 20 bit fixed-slash numbers. Hence, $\Phi_9(\frac{277}{642}) = \frac{85}{197}$ with a rounding error of $1/(642 \times 197)$ or $\sim 2^{-16.9}$, consistent with (24). The most important features of mediant rounding are summarized in the following observation.

Observation 17—Canonical Rounding: The rounding determined by truncating the continued fraction representation (best rational approximation) is unique and may be interpreted as the "round away from mediant" rule. This mediant rounding effects a graduated precision bias towards simpler fractions, i.e., for $j < k$, rounding to a $(2j+2)$-bit fixed-slash number within the set of $(2k+2)$-bit fixed-slash numbers is tolerated with error as large as $\sim 2^{-(j+k)+1}$ (over $[0, 1]$). □

A functional goal motivating the provision of a user controlled variable or multitiered precision hierarchy is to allow for appropriate portions of a computation sequence to be carried out in higher precision. Given that "graduated (downsizing) double-to-single precision rounding bias towards simpler fractions" is implicit without user request in a fixed-slash number system, we must ask if there is a significant class of problems for which this type of adaptive precision applies naturally to the appropriate portions of a computation sequence. We find an affirmative answer by distinguishing two types of approximate computation.

Type Q—Approximate Rational Computation: This corresponds to the class of finite sequences of rational arithmetic $(+, -, \times, /)$ operations on exact initial rational values (i.e., arithmetic over the rational number field) where approximation is employed whenever intermediate or final exact results would require too much storage. Examples of such computation occur in symbolic computation, combinatorial optimization, and operations on rational matrices.

Type R—Approximate Real Computation: This corresponds to the class of finite sequences of real arithmetic $(+, -, \times, /, \sqrt{\ }, \exp, \log, \sin, \cdots,)$ operations on real or approximate operands (exact reals, approximate values, intervals).

A benefit of the "graduated double-to-single precision rounding bias towards simple fractions" feature for hosting approximate rational arithmetic is summarized in the following observation and then illustrated by an example [14].

Observation 18—Recovery of Exactness: A moderate length approximate rational computation hosted by fixed-slash arithmetic will have an accumulated error governed with high probability by a near double precision error bound. If the exact result of the same rational computation is a rather simple fraction, the implicit single precision rounding interval associated with this simple fraction is very likely to contain the approximate near double precision computed result prior to the final rounding, so that the final rounding then recovers the exact simple fractional result. □

Example: Fig. 3 illustrates the computation of the determinant

$$D = \det\begin{vmatrix} \frac{10}{13} & \frac{20}{17} & \frac{1}{13} \\ \frac{11}{19} & \frac{7}{11} & \frac{77}{95} \\ \frac{69}{91} & \frac{4}{17} & \frac{56}{65} \end{vmatrix} = \frac{5}{13}$$

by evaluation of the rational expression

$$\left(\left(\left(\frac{10}{13} \times \frac{7}{11}\right)\frac{56}{65} + \left(\frac{11}{19} \times \frac{4}{17}\right)\frac{1}{13}\right) + \left(\frac{69}{91} \times \frac{20}{17}\right)\frac{77}{95}\right)$$
$$- \left(\left(\left(\frac{69}{91} \times \frac{7}{11}\right)\frac{1}{13} + \left(\frac{11}{19} \times \frac{20}{17}\right)\frac{56}{65}\right) + \left(\frac{10}{13} \times \frac{4}{17}\right)\frac{77}{95}\right)$$

in the order indicated by the parentheses. The fixed-slash approximate computation employed rounds all intermediate fractions by the mediant rounding Φ: Reals $\rightarrow F_{999}$, i.e., so as not to exceed 3 decimal digits in either numerator or denominator. The computed value along with the absolute accumulated error (compared to the corresponding exact value) is illustrated at each stage. Note that the final step of the computation involves the rounding of $\frac{320}{277} - \frac{84}{109} = \frac{11612}{30193}$, whose convergents are $0, \frac{1}{2}, \frac{1}{3}, \frac{2}{5}, \frac{3}{8}, \frac{5}{13}, \frac{1288}{3349}, \frac{2581}{6711}, \frac{11612}{30193}$. Thus, $\Phi(\frac{11612}{30193}) = \frac{5}{13}$, and the true result is recovered by the final rounding. □

For approximate real computation with no discernible preference for rational valued results, the graduated precision feature tends only to aggravate computational error. We show, however, that the degradation is not so severe as might be anticipated from the size of the maximum gap and maximum relative error. The analysis is aided by the following results from [15].

Theorem 7: For the random variable X chosen uniformly on $[0, 1]$ the expected value and the variance of the rounding error $|X - \Phi_k(X)|$, for rounding to $(2k+2)$-bit fixed-slash numbers, are given by

$$\mathrm{Exp}\left(\left|X - \Phi_k(X)\right|\right) = \frac{6 \ln 2}{\pi^2}\frac{k}{2^{2k}} + O\left(\frac{1}{2^{2k}}\right), \tag{25}$$

$$\mathrm{Var}\left(\left|X - \Phi_k(X)\right|\right) = \frac{c}{2^{3k}} + O\left(\frac{k^2}{2^{4k}}\right), \tag{26}$$

for c a constant. □

Theorem 8: If X is chosen uniformly on $[0, 1]$, then for any α, $1 \leqq \alpha \leqq 2$, with Φ_k denoting rounding to $(2k+2)$-bit fixed-slash numbers,

$$\mathrm{Prob}\left\{|X - \Phi_k(X)| > \frac{1}{(2^k - 1)^\alpha}\right\} \leqq \frac{2}{(2^k - 1)^{2-\alpha}}. \tag{27}$$

□

From Theorem 7 we note that the expected rounding error into $(2k+2)$-bit fixed-slash numbers is of an order much

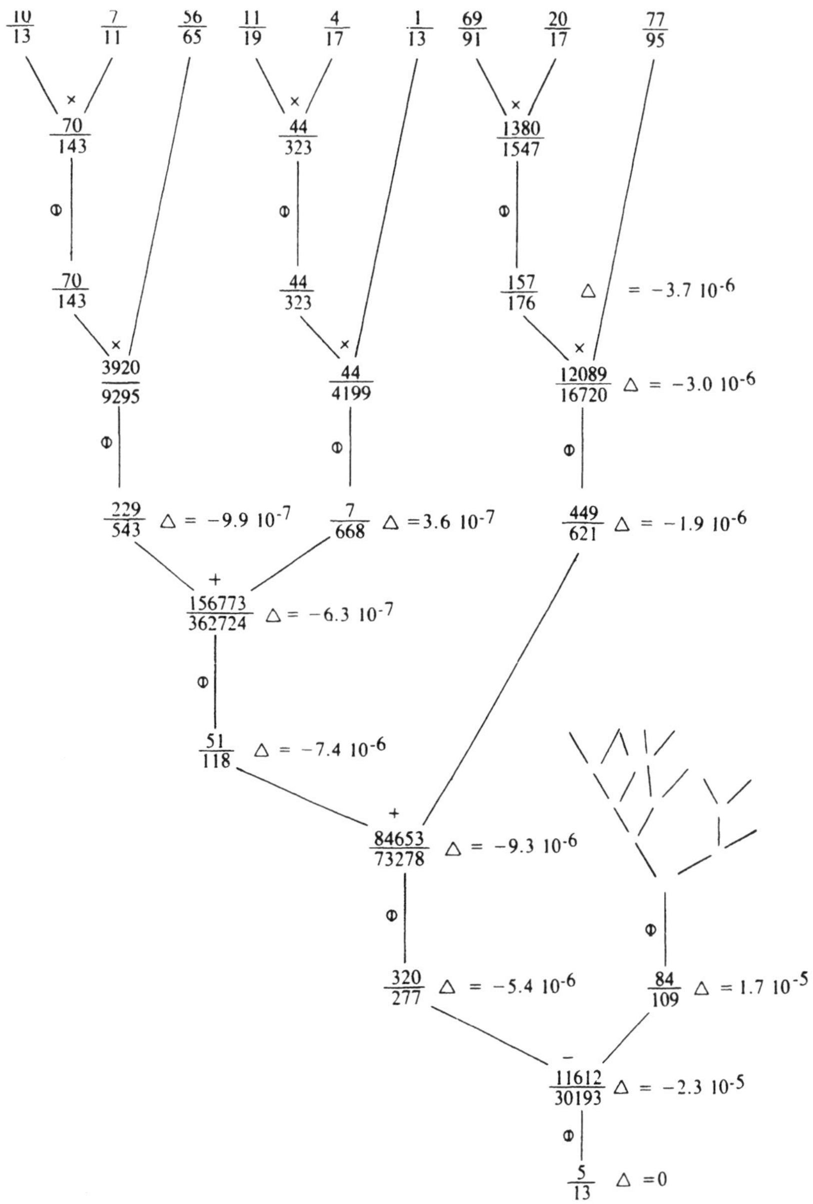

Fig. 3. A computation tree employing mediant rounding Φ: Reals → F_{999} showing the computed value and accumulated error Δ at each node.

closer to the double precision level $\sim 2^{-2k}$, than the single precision level $\sim 2^{-k}$, especially for large k. As the variance in rounding error is high, however, it is more important to ask for a bound on error size that will be violated with probability less than some very small tolerance, e.g., one-in-a-million (10^{-6}) or one-in-a-trillion (10^{-12}), which can be found from Theorem 8.

Table III shows a precision-of-approximation profile for each of the four tiers of the hierarchical precision fixed-slash arithmetic system of Table II. The profile gives, over the unit interval of each fixed-slash number system:

i) the minimum and maximum gap size,

ii) the average rounding error,

iii) a mediant rounding error bound (defined by one-half of all numbers chosen uniformly over [0, 1] should incur at most that error),

iv) a one-in-a-million rounding error bound (defined by at most one in a million numbers chosen uniformly over [0, 1] should incur and error larger than that value),

v) a one-in-a-trillion rounding error bound,

vi) the maximum possible error.

All entries are given in decimal exponent form so the values may be interpreted loosely to give equivalent numbers of "decimal digits of accuracy." Suppose, for example, we consider the rounding of a set of numbers assumed uniformly distributed on [0, 1] into DOUBLE (128-bit word) fixed-slash representation. Although we are only guaranteed the equivalent of at least 18 digit accuracy (max error bound) for each number, one-half of the values should have at least 37 digit accuracy, all but one-in-a-million should have at least 31 digit accuracy, and all but one-in-a-trillion should have at least 25 digit accuracy.

TABLE III
GAP SIZE AND ROUNDING ERROR BOUNDS OVER [0, 1] FOR CERTAIN FIXED-SLASH NUMBER SYSTEMS

Fixed-Slash Format	Word Width (in bits)	Gap Size over [0,1]		Precision-of-Approximation				
		min	max	Avg	Mediant	One-in-a-Million	One-in-a-Trillion	Max
HALF	32	$10^{-9.0}$	$10^{-4.5}$	$10^{-8.2}$	$10^{-9.0}$	$10^{-4.5}$	$10^{-4.5}$	$10^{-4.5}$
SINGLE	64	$10^{-18.6}$	$10^{-9.3}$	$10^{-17.5}$	$10^{-18.6}$	$10^{-12.3}$	$10^{-9.3}$	$10^{-9.3}$
DOUBLE	128	$10^{-37.9}$	$10^{-18.9}$	$10^{-36.5}$	$10^{-37.9}$	$10^{-31.6}$	$10^{-25.6}$	$10^{-18.9}$
QUAD	256	$10^{-76.4}$	$10^{-38.2}$	$10^{-74.7}$	$10^{-76.4}$	$10^{-70.1}$	$10^{-64.1}$	$10^{-38.2}$

For input of several thousand values the one-in-a-million bound would provide a conservative estimate of precision-of-approximation for all rounded inputs. For an extended computation on a machine performing 100 million (10^8) operations per second, the one-in-a-trillion error bound should be a conservative estimate of rounding precision for all results. Note that the compounded accumulated error of extended computation should probably dominate the infrequent larger errors introduced by the rounding precision bias of fixed-slash computation. Thus, the "average error" precision-of-approximation value could be used as a first-order estimate for comparison with accuracy of other finite precision number systems.

Primary effects of the graduated precision environment on approximate real computation are summarized in the following observation.

Observation 19—Precision of Approximate Real Computation: Rounding real values from [0, 1] to fixed-slash numbers with k-bit numerators and denominators yields an average error comparable to that of $(2k-\log_2 k)$-bit binary radix representation. Extended computation (assumed normalized to the unit interval) is thus hosted with (absolute) approximation errors much closer to the double precision level than the single precision (worst case guarantee) level of the single-to-double variable precision scale. A j-digit rounding error occurs with probability no larger than about one-in-10^j independent of word size. Thus, larger formats suffer proportionally less loss of precision on the single-to-double scale. The compounded accumulated error of extended computation should likely dominate subsequent precision loss due to the graduated precision environment. □

The net effect from Observations 18 and 19 is that the beneficial aspects of support of approximate rational arithmetic are achieved with no great degradation in support of approximate real computation for the larger format size (64, 128, 256)-bit fixed-slash number systems.

The graduated precision environment of fixed-slash number systems should not be considered an insurmountable obstacle for approximate computation with fractions if near uniform spacing of representable values is considered imperative for efficient use of resources. The use of denormalized numbers to yield precision fill in a manner analogous to that described for floating-slash representation in the next section is possible also for fixed-slash representation. We leave the outline of this feature to the floating-slash representation discussion where both precision fill *and* range extension become convenient extension options.

There are three implicit tradeoffs that are characteristic of fixed-slash number systems which should be recognized in comparison to other finite precision number systems.

Characteristic tradeoffs in fixed-slash number systems:

- greater precision-of-approximation is achieved at the expense of smaller numeric range for any format size;
- support for more uniform absolute error behavior over [0, 1] is achieved at the expense of relative error degrading with magnitudes away from unity;
- double-to-single downsizing precision rounding bias yields properties favorable to approximate rational computation at the expense of the attainable precision of approximate real computation.

In contrast to these tradeoffs, we show in the next section that floating-slash number systems provide a larger numeric range to format size tradeoff and more uniform control of relative-error-of-approximation over the whole underflow-to-overflow magnitude range.

III. FLOATING-SLASH NUMBER SYSTEMS

A. Format

The *$(k+\ell+1)$-bit floating-slash number system* is composed of 4-tuples (s, a, exs, f) conveniently described by reference to the fields of the defining binary word format. The component fields illustrated in Fig. 4 are: the sign bit s, the exact/approximate bit a, the ℓ bit integer field exs where $\ell = \lceil \log_2 k \rceil$, and the $(k-1)$-bit fraction field f containing the concatenated numerator and leading-bit-deleted denominator values. This representation uses an implicit leading denominator bit allowing the $(k-1)$-bit fraction field to represent a *k-bit fraction,* by which is meant any fraction with an i-bit integer numerator and j-bit integer denominator where $i + j \leqq k$, and $i, j \geqq 1$. The value $exs = 2^\ell - 1$ is reserved for encoding infinity and not-a-number. The value of exs, for $0 \leqq exs \leqq k - 2\ (\leqq 2^\ell - 2)$, is used to determine the slash position so that num is defined to be the integer in the leading $k - 1 - exs$ bits of the fraction field. den is composed by adjoining the leading bit (value 2^{exs}) to the remaining exs bits in the fraction field yielding an $(exs+1)$-bit integer. Thus,

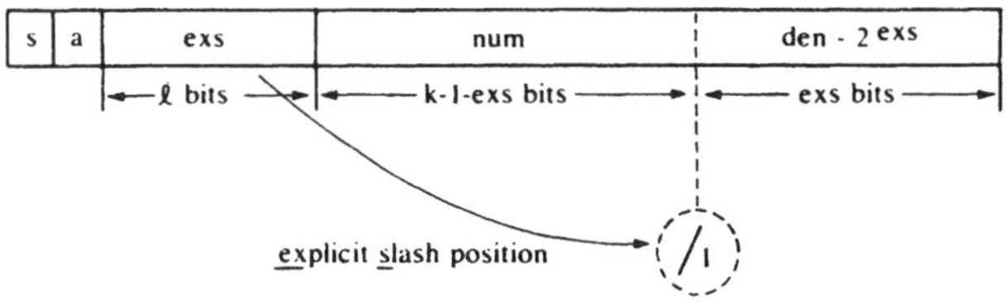

Fig. 4. Format for $(k+\ell+1)$-bit floating-slash number representation incorporating an implicit leading denominator bit.

$$\begin{aligned} num &= \lfloor f/2^{exs} \rfloor, \\ den &= (f \bmod 2^{exs}) + 2^{exs}. \end{aligned} \tag{28}$$

With these integer values for *exs*, *num*, and *den*, we then define the value v of a floating-slash number as follows.

a) If $exs = 0$, then $v = (-1)^s$ num. [integers including signed zero]

b) If $1 \leqq exs \leqq k - 2$, then $v = (-1)^s(num/den)$. [signed rational numbers]

c) If $exs = 2^\ell - 1$ and $f \equiv 0 \bmod 2$, then $v = (-1)^s\, \infty$. [signed infinity]

d) If $exs = 2^\ell - 1$ and $f \equiv 1 \bmod 2$, then v denotes "not-a-number." [not-a-number].

For a floating-slash number system with $k \leq 2^\ell - 1$ we shall have *exs* in the range $k - 1 \leqq exs \leqq 2^\ell - 2$ simply correspond to undefined values. For the *extended range* floating-slash number system we allow any *exs* field width $\ell \geqq \lceil \log_2 k \rceil$ and interpret *exs* in the range $k - 1 \leqq exs \leqq 2^\ell - 2$ as a biased exponent scaling either the numerator or denominator as shown in Fig. 5. In this interpretation we designate the $(k-1)$-bit fraction field to be a $(k-1)$-bit integer f which is augmented by a leading bit. The bit string then forms the leading k bits of either the numerator or denominator depending on whether the slash is interpreted to have "floated out" the right- or left-hand side of the word, respectively. The extended range floating-slash numbers corresponding to Fig. 5(a) are equivalent to k-bit floating-point numbers. The "small" floating-slash numbers of Fig. 5(b) are not equivalent to floating-point numbers, being rather the inverses of the "large" floating-slash (or floating-point) numbers.

The value v of an "extended floating-slash" number in the extended range is as follows.

e) If $\frac{k+2^\ell-3}{2} < exs \leqq 2^\ell - 2$, let $e = 2^\ell - 2 - exs$, and then $v = (-1)^s 2^e(2^{k-1} + f)$. [scaled numerators]

f) If $k - 1 \leqq exs < \frac{k+2^\ell-3}{2}$, let $e = exs - (k - 1)$, and then $v = (-1)^s/(2^e(2^{k-1} + f))$. [scaled denominators]

g) If $exs = \frac{k+2^\ell-3}{2}$, v is undefined.

A floating-slash number is termed *normalized* when $\gcd(num, den) = 1$ and so corresponds to an irreducible fraction. An unnormalized floating-slash number corresponds to a reducible fraction.

Standard *floating-slash arithmetic* shall denote rounding by best rational approximation to the standard (nonextended) floating-slash format limit of the exactly computed operation $(+, -, \times, /)$ on finite valued floating-slash operands. When v is a number, $a = 0$ denotes that the value is exact, and $a = 1$ denotes that the value is an approximation. The state $a = 1$ should be set corresponding to rounding error, inherited error, and/or initial error as noted for fixed-slash arithmetic.

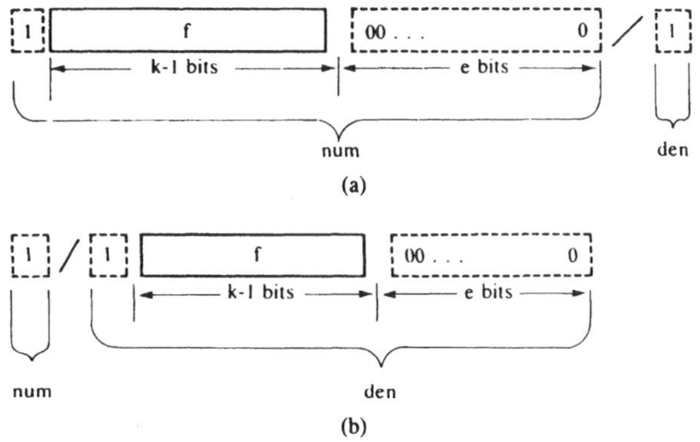

Fig. 5. Extended range interpretation of floating-slash number representation.

The values of the standard $(k+\ell+1)$-bit floating-slash numbers $(k \geqq 2)$ correspond to the values of all signed irreducible k-bit fractions, denoted FLS_k where

$$FLS_k = \left\{ \pm\frac{p}{q} \,\middle|\, p, q \geqq 1,\ \gcd(p,q) = 1, \right. \\ \left. \lfloor \log_2 p \rfloor + \lfloor \log_2 q \rfloor \leqq k - 2 \right\} \cup \left\{ \pm\frac{0}{1}, \pm\frac{1}{0} \right\}. \tag{29}$$

Binary floating-slash number systems thus have a dependence on the base implicit in their characterization. To determine if this is likely to cause any base dependent anomalies, consider the relation between FLS_k and the base independent set of *hyperbolic fractions* H_n defined by

$$H_n = \left\{ \pm\frac{p}{q} \,\middle|\, pq \leqq n,\ \gcd(p,q) = 1 \right\}. \tag{30}$$

It is readily determined that

$$FLS_{k-1} \subset H_{2^{k-1}-1} \subset FLS_k \tag{31}$$

and the implications for assessing the extent of base dependance and representation efficiency are summarized in the following observations.

Observation 20—Dependence on Base: The floating-slash numbers corresponding to FLS_k are characterized in a base dependent manner. The natural representation independent system $H_{2^{k-1}-1}$ is contained in FLS_k with less than one bit loss in representation capacity. Properties of $H_{2^{k-1}-1}$ can be analyzed to assure the avoidance of base dependent anomalies in FLS_k. Restriction to those fractions with $num \times den \leqq n$ in part (b) of the value specification given after Fig. 4 yields a realization of the values of the hyperbolic fractions H_n for any $n < 2^{k-1}$ if such a canonical "representation independent" system is desired. □

Dirichlet has shown the following (see [4, p. 283]).

Theorem 9:

$$\lim_{n\to\infty} \frac{|H_n|}{\left| \left\{ \pm\frac{p}{q} \,\middle|\, pq \leqq n, p \geqq 1, q \geqq 1 \right\} \right|} = \frac{6}{\pi^2} = 0.6079\cdots. \tag{32}$$

Thus, restriction to irreducible fractions in H_n loses only

$\log_2(\pi^2/6) = 0.718$ bits of representation capacity analogous to Observation 2 for fixed-slash representation. Further losses due to employing FLS_k rather than H_{2^k-1}, and due to possible unutilized values of the *exs* field are correspondingly small, so we have the following observation.

Observation 21 — Redundancy and Representation Efficiency: Floating-slash and extended range floating-slash number systems with the implicit leading denominator bit format of Figs. 4 and 5 incur a total loss in representation efficiency of only approximately one bit for any size format. □

B. Exception Handling

The important exception handling features of floating-slash systems applying to both standard and extended range formats are summarized here. Further details to explain and contrast these features with corresponding fixed-slash format features are available by reference to Observations 3–6.

- Signed infinities and signed zeroes are provided.
- Infinity may contain an implementation dependent message.
- Underflow and overflow thresholds are the reciprocals of each other.
- Not-a-number is provided.
- Denormalized numbers may be defined by giving meaning to the gcd of the reducible fraction corresponding to an unnormalized floating-slash number.

C. Unary Operators

The following unary rational operations on floating-slash and extended floating-slash numbers are exactly and efficiently computable and representable within the same $(k+\ell+1)$-bit format.

- Additive inverse.
- Multiplicative inverse.
- Absolute value.
- Integer part, fractional part (and successive partial quotient determination).
- Numerator, denominator.
- gcd, normalize.

Efficient numerator and denominator extraction is of great importance since to perform arithmetic it is necessary to move the numerator and denominator into separate registers. Since the unit position of the numerator can vary anywhere within the $(k-1)$-bit fraction field in the format of Fig. 4, shifting is required to extract the numerator. If the numerator bits were stored in reverse order left adjusted in the field, the numerator could possibly be extracted more efficiently by first reversing the full $(k-1)$-bit field (twisting the wires) and then masking. Alternatively, the numerator could be stored at the right-hand side of the field and the denominator at the left-hand side with the denominator bits reversed with unit bit left adjusted. Resolution of these architectural questions involving the most efficient procedure for ordering and moving numerator and denominator into registers would not alter the nature of the represented values.

Example: For floating-slash representation of the convergent approximation $\pi \sim 355/113$, which agrees with π to seven decimal places (rel. err. $\sim 10^{-7}$), note that $s = 0, a = 1$. Since $355 = 101100011_2$, $113 = 1110001_2 = 2^6 + 110001_2$, we must have $exs = 6 = 110_2$. For a 32-bit floating-slash format, $\ell = 5, k = 26$. Fig. 6 illustrates the representation of $\sim 355/113$ in the defining format (of Fig. 4) in (a), and in the twisted-denominator-left-adjusted format in (b).

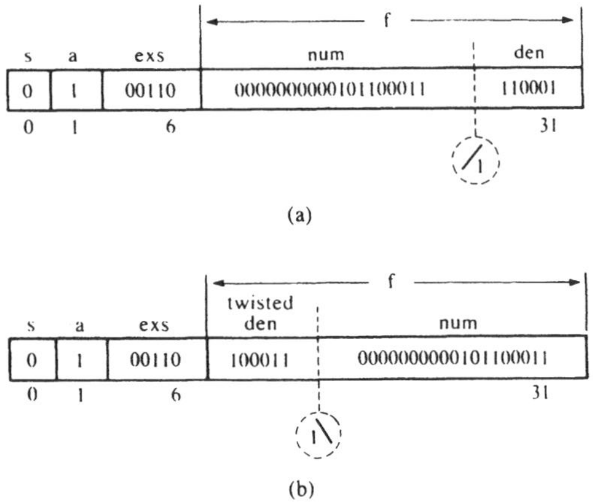

Fig. 6. The 32-bit floating-slash representation of ~355/113 in (a) standard format and (b) "twisted-denominator-low-order-bit-left-adjusted" format.

D. Hierarchical Precision Rational Arithmetic

Regarding the extent of exact simple rational computation in standard floating-slash number systems note [13]:

- Multiplication or division of an i-bit fraction by a j-bit fraction yields a value representable as an $(i+j)$-bit fraction.
- The sum or difference of an i-bit fraction and a j-bit $(j > i)$ fraction is representable as an $(i+2j-2)$-bit fraction; and furthermore, if both fractions have values greater than unity, then the result is representable as an $(i+j+1)$-bit fraction.
- Computing the difference $\frac{p}{q} - \frac{r}{s} = \frac{ps-qr}{qs}$ of two nearly equal fractions, a situation often termed "catastrophic cancellation" implies that $|ps - qr| << |ps + qr|$. Thus, the exact result will be representable without need for a great increase in precision, and possibly even with a smaller precision if $\frac{p}{q}, \frac{r}{s} >> 1$ (so that qs is relatively small). In particular, if the difference of two k-bit $(k = 3i)$ fractions of value greater than $2^{k/3}$ is less than $2^{-k/3}$, then the result is also a k-bit fraction.

Let SINGLE, DOUBLE, and TRIPLE denote n-bit (word), $2n$-bit (double word), and $3n$-bit (triple word) formats of the same generic type. We summarize primary aspects of the exact rational arithmetic hosted by such a multitiered standard floating-slash computation system in the following observation.

Observation 22 — Exact Rational Computation and Catastrophic Cancellation: Multiplication and division of standard SINGLE floating-slash operands yield exact DOUBLE floating-slash results. Addition and subtraction of standard SINGLE floating-slash operands yield exact TRIPLE floating-slash results. The occurrence of catastrophic cancellation, i.e., the difference of two nearly equal fractions, yields a great simplification in the resulting fraction, often allowing the difference of two SINGLE operands to be exactly represented in SINGLE format. □

Our utilization of an implicit leading denominator bit brings an important architectural property to the format beyond the savings of one bit in the representation.

Observation 23—Integer Compatibility: Floating-slash representation as provided by the format illustrated in Fig. 4 will have $a = 0, exs = 0$ for exact normalized representation of integers. The normalized representation of any integer in the range $[-2^{k-1} + 1, 2^{k-1} - 1]$ will then be identical to the sign-and-$(k+\ell)$-bit-magnitude representation of that integer. This same result also holds for the "twisted" format with the denominator at the left-hand side, illustrated in Fig. 6(b). □

A multitiered floating-slash system provides a synergistic integer/exact-rational/approximate-real computation environment with greater range than for corresponding fixed-slash format lengths.

Example: A four-tiered hierarchical floating-slash system using formats of size 32, 64, 128, and 256 bits is shown in Table IV, showing ranges of about twice the number of digits compared to the same size formats for fixed-slash in Table II.

Note that DOUBLE floating-slash representation with a 128-bit format achieves a range essentially equivalent to the DEC VAX and Honeywell 6000 series floating-point range (for their SINGLE or DOUBLE format). QUAD floating-slash (256 bits) gives essentially the range of the IBM 360 SINGLE, DOUBLE, or QUAD formats. Furthermore, QUAD contains as a subset all values representable in the proposed IEEE standard 32-bit single floating-point format. (In fact, all IEEE single x for $1 \leqq x \leqq 2^{56} - 1 = \sim 10^{16.8}$ are exactly representable in "SINGLE" 64-bit floating-slash representation, and all IEEE double x for $1 \leqq x \leqq 2^{119} - 1 = \sim 10^{35.8}$ are exactly representable in "DOUBLE" 128-bit floating-slash representation.)

Two approaches have been utilized for hierarchical multitiered floating-point range specification.

- Same underflow-to-overflow range for all floating-point precision tiers. This is the case for most commercially available large computer systems, including Cray-1, DEC VAX, Honeywell 6000, IBM 370, Interdata 8/32 (see [1]).
- Increased underflow-to-overflow range in higher floating-point precision tiers. This is the case for the proposed IEEE standard [6], that specified by the Ada language feature for parametrically declared FLOAT precision (see [3]), and that suggested for the CADAC arithmetic unit [2].

For one example of a multitiered hierarchical extended range floating-slash system we could have QUAD level remain a standard floating-slash system, with DOUBLE, SINGLE, and HALF being extended range floating-slash systems each having the same 8-bit slash position field. This provides downsizing of precision-of-approximation from QUAD to DOUBLE, SINGLE, or HALF, with a more moderate downsizing of range (e.g., for the 64-bit SINGLE format with $\ell = 8, k = 55$, the range is $\sim 10^{\pm 46.3}$, and for the 128 bit DOUBLE format $\ell = 8, k = 119$, the range is $\sim 10^{\pm 55.9}$, compared to QUAD range of $10^{\pm 74.0}$). Alternatively, following the more recent design trend, we may choose graduated broader ranges for extended range floating-slash in a manner comparable to those in existing and proposed floating-point systems, as suggested in Table V.

TABLE IV
FORMAT SIZES AND NUMERIC RANGES FOR A FOUR-TIERED FLOATING-SLASH ARITHMETIC SYSTEM

Floating-Slash Format	Word Width (in bits)	Slash Position Field Width	Fraction Field Width (in bits) Including Implicit Bit	Underflow-to-Overflow Range (Decimal) Equivalent)
HALF	32	5	26	$10^{\pm 7.5}$
SINGLE	64	6	57	$10^{\pm 16.8}$
DOUBLE	128	7	120	$10^{\pm 35.8}$
QUAD	256	8	247	$10^{\pm 74.0}$

TABLE V
FORMAT SIZES AND NUMERIC RANGES FOR A THREE-TIERED EXTENDED RANGE FLOATING-SLASH ARITHMETIC SYSTEM

Extended Floating-Slash Format	Word Width (in bits)	Slash Position Field Width	Fraction Field Width (in bits) Including Implicit Bit	Underflow-to-Overflow Range (Decimal Equivalent)
HALF	32	8	23	$10^{\pm 41.5}$
SINGLE	64	11	52	$10^{\pm 315}$
DOUBLE	128	14	113	$10^{\pm 2482}$

E. Precision of Approximation

The parameter k implicitly determines the precision-of-approximation as well as the underflow-to-overflow range for standard floating-slash systems. Essential features are summarized in the following observation.

Observation 24—Floating-Slash Range/Precision Specification: Floating-slash representation provides about twice the exponent range of fixed-slash representation for the same word size. The nested sequence $FLS_k \subset FLS_{k+1} \subset FLS_{k+2} \subset \cdots$ yields floating-slash systems having both greater range and precision governed naturally in their growth by a single parameter. Extended range floating-slash representation allows a separately parameterized range/precision specification if desired. The additional representable values of the extended range system all have absolute values outside the underflow-to-overflow range $[1/(2^{k-1} - 1), 2^{k-1} - 1]$ of the included subsystem FLS_k, preserving the integrity of computations and roundings within the underflow-to-overflow range of FLS_k. □

As it can be shown that the hyperbolic fractions H_n become log uniformly dense on the positive real line as $n \to \infty$, it follows that floating-slash representation yields "macroscopically" uniform relative error control for all parts of the range. This contrasts favorably with fixed-slash systems where relative errors consistently increase for approximations of quantities farther away from unity towards the overflow, or underflow, boundary.

At the "microscopic" level of gap size, we obtain a single-to-double precision variation in relative gap size graduated by the "simplicity" of the simpler fraction gap bound. Results for floating-slash relative gap size variation over the whole

TABLE VI
RELATIVE GAP SIZES AND BOUNDS ON RELATIVE ROUNDING ERROR FOR LOG UNIFORM DATA OVER THE WHOLE UNDERFLOW-TO-OVERFLOW RANGE FOR CERTAIN FLOATING-SLASH NUMBER SYSTEMS

Floating-Slash Format	Word Width (in bits)	Relative Gap Size		Precision-of-Approximation [Relative Error]			
		min	max	Avg	One-in-a-Million	One-in-a-Trillion	Max
HALF	32	$10^{-7.8}$	$10^{-3.9}$	$10^{-6.5}$	$10^{-3.9}$	$10^{-3.9}$	$10^{-3.9}$
SINGLE	64	$10^{-17.1}$	$10^{-8.5}$	$10^{-15.5}$	$10^{-10.4}$	$10^{-8.5}$	$10^{-8.5}$
DOUBLE	128	$10^{-36.1}$	$10^{-18.0}$	$10^{-34.2}$	$10^{-29.3}$	$10^{-23.4}$	$10^{-18.0}$
QUAD	256	$10^{-74.3}$	$10^{-37.1}$	$10^{-72.1}$	$10^{-67.4}$	$10^{-61.5}$	$10^{-37.1}$

underflow-to-overflow range are analogous to those of absolute gap size variation over $[0, 1]$ for fixed-slash representation. The following results adapted from [15] provide the relative gap size function and indicate the maximum and minimum bounds.

Theorem 10: For any x in any gap $(\frac{p}{q}, \frac{r}{s})$ of the $(k+\ell+1)$-bit floating-slash numbers the *relative gap size* at x, $\gamma^*(x) = (\frac{r}{s} - \frac{p}{q})/x$, is given by

$$\gamma^*(x) = \frac{1}{sqx} \sim \frac{1}{(\min\{sr, pq\})^{1/2} 2^{k/2}} \tag{33}$$

so that over $[\frac{1}{2^{k-1}-1}, 2^{k-1} - 1]$,

$$\text{max rel gap} \sim 2^{-k/2},$$
$$\text{min rel gap} \sim 2^{-k}. \tag{34}$$

□

F. Approximate Real Arithmetic

Rounding by best rational approximation into the $(k+\ell+1)$-bit floating-slash numbers is defined by Φ_k^*: Reals $\rightarrow FLS_k$ for every real number x where p_0/q_0, $p_1/q_1, p_2/q_2, \cdots$, are the convergents to $|x|$, by

$$\Phi_k^*(x) = \begin{cases} p_m/q_m & \text{if } x \geqq 0,\ x = \frac{p_m}{q_m} \in FLS_k \\ p_i/q_i & \text{if } x > 0,\ \frac{p_i}{q_i} \in FLS_k,\ \frac{p_{i+1}}{q_{i+1}} \notin FLS_k \\ -\Phi_k^*(-x) & \text{if } x < 0. \end{cases} \tag{35}$$

The rounding Φ_k^* satisfies the monotonic, distributive sign, distributive inverse, and fixed point properties just as noted for Φ_k in Theorem 5. Also Φ_k^*: Reals $\rightarrow FLS_k$ satisfies the rule "round away from the mediant to the boundary of each gap, rounding the mediant to the simpler fraction bounding the gap." Thus, we term Φ_k^*: Reals $\rightarrow FLS_k$ *mediant rounding to* FLS_k.

From [15] we obtain the following results on the average relative rounding error and on the distribution of relative error. For X chosen in a log uniform manner over the whole underflow-to-overflow range $[\frac{1}{2^{k-1}-1}, 2^{k-1} - 1]$

$$\operatorname{Exp}\left(\frac{|X - \Phi_k^*(X)|}{X}\right) < \sim \frac{k \ln 2}{2^k} \tag{36}$$

and for $\frac{1}{2} \leqq \alpha \leqq 1$,

$$\operatorname{Prob}\left\{\frac{|X - \Phi_k^*(X)|}{X} > \frac{1}{2^{(k-1)\alpha}}\right\} < \sim \frac{4\alpha}{2^{(k-1)(1-\alpha)}}. \tag{37}$$

These formulas are used to compute the precision-of-approximation profile for relative error of approximation in Table VI for the same four-tiered floating-slash hierarchy utilized in Table IV.

To avoid the graduated rounding precision bias of floating-slash representation note that for every "simple fraction" $\frac{p}{q}$ bounding a rather large gap in the $(k+\ell+1)$-bit floating-slash number system, the number of distinct unnormalized fractions equal to $\frac{p}{q}$ is proportional to the relative size of the gap. By interpreting the value of the gcd of such unnormalized fractions to imply a distinct real value of the gap (rather than the value $\frac{p}{q}$), we may obtain a "gap fill" through denormalized values in each gap so that the overall relative size of each gap throughout the extended system is $\sim 2^{-k}$.

Details of the encoding of such denormalized values and their decoding and use in an extension of the arithmetic unit of [9] are under investigation.

REFERENCES

[1] W. S. Brown and S. I. Feldman, "Environment parameters and basic functions for floating-point computation," *ACM Trans. Math. Software*, vol. 6, pp. 510–523, Dec. 1980.
[2] M. Cohen, T. E. Hull, and V. C. Hamacher, "CADAC: A controlled-precision decimal arithmetic unit," *IEEE Trans. Comput.*, vol. C-32, pp. 370–377, Apr. 1983.
[3] W. J. Cody, "Floating-point parameters, models and standards," in *The Relationship Between Numerical Computation and Programming Languages*, J. K. Reid, Ed. Amsterdam: North-Holland, 1982, pp. 51–65.
[4] L. E. Dixon, *History of the Theory of Numbers*, Vol. 1, (reprint). New York: Chelsea, 1971.
[5] G. H. Hardy and E. M. Wright, *An Introduction to the Theory of Numbers, 5th ed.* London: Oxford University Press, 1979.
[6] IEEE Task P754, "A proposed standard for binary floating-point arithmetic," *Computer*, vol. 14, pp. 51–62, Mar. 1981.
[7] A. Y. Khintchin, *Continued Fractions* P. Wynn, Trans. Grooningen: Noordhoff, 1963.
[8] D. E. Knuth, *The Art of Computer Programming, Vol. 2, Seminumerical Algorithms*, 2nd ed. Reading, MA: Addison-Wesley, 1981.

[9] P. Kornerup and D. W. Matula, "Finite precision rational arithmetic: An arithmetic unit." *IEEE Trans. Comput.*, vol. C-32, pp. 378–387, Apr. 1983.
[10] U. Kulisch and L. Miranker, *Computer Arithmetic in Theory and Practice*. New York: Academic, 1981.
[11] D. W. Matula, "A formalization of floating-point numeric base conversion," *IEEE Trans. Comput.*, vol. C-19, pp. 681–692, Aug. 1970.
[12] ——, "Fixed-slash and floating-slash arithmetic," in *Proc. 3rd Symp. Comput. Arith.*, 1975, pp. 90–91.
[13] D. W. Matula and P. Kornerup, "A feasibility analysis of binary fixed-slash and floating-slash number systems," in *Proc. 4th Symp. Comput. Arith.*, 1978, pp. 29–38.
[14] ——, "An approximate rational arithmetic system with intrinsic recovery of simple fractions during expression evaluation," in *Symbolic and Algebraic Computation*. E. W. Ng, Ed. New York: Springer-Verlag, 1979, pp. 383–397.
[15] ——, "Foundations of finite precision rational arithmetic," in *Computing, Suppl. 2*. New York: Springer-Verlag, pp. 85–111, 1980.

David W. Matula was born in St. Louis, MO, on November 6, 1937. He received the B.S. degree in engineering physics from Washington University, St. Louis, MO, in 1959 and the Ph.D. degree in engineering science from the University of California, Berkeley, in 1966.

He is now a Professor in the Department of Computer Science and Engineering, School of Engineering and Applied Science, Southern Methodist University, Dallas, TX, where he served as Department Head from 1974 to 1979. From 1966 to 1974 he was on the Faculty of the Applied Mathematics and Computer Science Department, Washington University, St. Louis, MO. He has been a Visiting Professor at Stanford University, Stanford, CA, the Naval Postgraduate School, Monterey, CA, the University of Texas at Austin, the University of Karlsruhe, Germany, and Aarhus University, Denmark. His research interests include computer arithmetic, analysis of algorithms, graph theory, and cluster analysis.

Prof. Matula is a member of the Association for Computing Machinery, the Society for Industrial and Applied Mathematics, the European Association for Theoretical Computer Science, the Mathematical Association of America, Sigma Xi, the Classification Society, and the Operations Research Society of America.

Peter Kornerup was born in Aarhus, Denmark, on April 29, 1939. He received the mag. scient. degree in mathematics (numerical analysis) from Aarhus University, Aarhus, in 1967.

After a period employed as a Consultant for the University Computing Center, from 1969 he was involved in establishing the computer science curriculum at Aarhus University, and founding the Computer Science Department in 1971. Throughout most of the 1970's he served as Chairman of the Department, and he has served as Head of the Institute of Mathematical Sciences at Aarhus University from 1981 to 1984. In 1975–1976 he spent a leave at the University of Southwestern Louisiana, Lafayette, LA, and in the Spring of 1979 another leave with Southern Methodist University, Dallas, TX. His research interests include microprogramming, computer architecture, and computer arithmetic.

Mr. Kornerup is a member of the Association for Computing Machinery.

FINITE PRECISION LEXICOGRAPHIC CONTINUED FRACTION NUMBER SYSTEMS

Peter Kornerup
Aarhus University
Aarhus, Denmark

David W. Matula
Southern Methodist University
Dallas, Texas 75275

Abstract

Lexicographic continued fraction binary (LCF) representation provides an order preserving bitstring representation of the non negative real numbers where every rational number has a finite length bitstring representation. We investigate the precision of k-bit LCF approximation. The maximum gap size over [0,1] for (k+1)-bit LCF representation is shown to be less than $2^{-.81k}$, comparable to binary coded decimal in worst case representation efficiency. The distribution of gap sizes for (k+1)-bit LCF representation over [0,1] is shown on a logarithmic scale to be bell shaped between $2^{-.81k}$ and $2^{-1.39k}$, becoming more peaked near the value corresponding to uniform spacing, 2^{-k}, with increasing k.

I. Introduction and Summary

For the purposes of computer arithmetic it is appropriate to view a number system as an abstract data type (ADT), where the specification of the set of arithmetic operations to be supported serve to characterize the ADT. The choice of a numeric representation and corresponding binary format then constitutes more than simply an encoding of numeric data. Numeric representation should be likened to the specification of a data structure to support the ADT. Questions of space efficiency of the representation and the time efficiency of procedures applied to the representation to realize the particular set of arithmetic operations characterizing the ADT are then criteria by which the suitability of the data representation must be judged.

The availability of VLSI and computer manufacturers' recent interest (e.g. [IEEE81], [KM84]) in providing firmware/hardware realizations of a variety of numeric representations (characterized by their ADT's) has brought renewed interest in the investigation of alternative numeric representations. In this paper we continue our investigation of a new binary numeric representation that ideally supports a number system with arithmetic operations synergistically allowing both exact rational arithmetic and approximate real arithmetic.

The lexicographic continued fraction binary (LCF) representation introduced in [MK83] provides an order-preserving one-to-one correspondence between the reals and all infinite bitstrings (excluding as in binary radix representation those bitstrings with a terminating infinite sequence of units). It was shown in [MK83] that LCF representation provides a space efficient finite bitstring encoding (with implied terminating infinite sequence of zeroes) for every rational number, and supports time efficient procedures for exact and/or approximate arithmetic (+, −, ×, ÷, <, =) as implicit from [KM83] and [Go72]. Our concern in this paper is to investigate the suitability of k-bit LCF representation in support of approximate real arithmetic, in particular, to study the uniformity of approximation error obtained by rounding to k-bit LCF numbers. For this purpose we first develop appropriate tools for characterizing and analysing the nature and size of gaps between representable k-bit LCF numbers. Given the variability of gap sizes between neighboring k-bit LCF representable numbers evident from enumerating the representable values for small k, we are then especially interested in determining the maximum gap size and distribution of gap sizes as a function of k for large k.

In Section II we review the definition and primary properties of LCF representation from [MK83]. We introduce the LCF tree as a convenient reference for displaying fractions in correspondence with their LCF representation, and for assessing the precision obtainable by k-bit LCF approximation.

*This research was supported in part by the National Science Foundation under grant DCR-8315289.

Our main result described in Section III derives from the formulation of an operation termed the "binary mediant" and provides for convenient development of the LCF tree and concomitant analysis of the nature and size of the gaps between neighboring k-bit LCF numbers. Specifically, if the irreducible fractions p/q and r/s are neighboring k-bit LCF numbers where say $p \leq r$, then there must exist a unique largest integer j such that $2^j p \leq r$ and $2^j q \leq s$. The irreducible fraction u/v, termed the binary mediant of p/q and r/s, which is the (k+1)-bit LCF representable number falling between p/q and r/s is then given by

$$\frac{u}{v} = \begin{cases} \dfrac{2^j p + r}{2^j q + s} & \text{if } \quad |ps - qr| = 1, \\ \\ \dfrac{(p+r)/2}{(q+s)/2} & \text{if } \quad |ps - qr| \neq 1. \end{cases} \tag{1}$$

Throughout this paper we use equality of fractions, a/b = c/d, in the stronger sense of ordered pair equality, i.e. to denote identical numerator values (a = c) and denominator values (b = d). When the numerator and/or denominator is itself an expression, the horizontal line format is used to clearly separate the fraction's numerator expression from the denominator expression as in the preceding formula for u/v.

In Section IV we employ the binary mediant and simulations to show that the gap size variability over the (k+1)-bit LCF numbers is reasonably narrow. For (k+1)-bit representation where the 2^k-1 representable values within [0,1] would effect a uniform spacing of size 2^{-k} if evenly spaced, we show that the maximum gap size is bounded by $2^{-.81k}$. By exhaustive enumeration through k = 24 and by simulation to k = 128, we show the distribution of gap sizes on the logarithmic scale to be bell shaped very narrowly peaked for large k near 2^{-k}, with vanishingly small tails truncated at points corresponding to the maximum gaps of size $2^{-.81k}$ and minimum gaps of size $2^{-1.39k}$. E.g. for k = 64 fully 99.9% of all gaps have size between $2^{-.89k}$ and $2^{-1.13k}$. Consider that binary coded decimal representation requires k = 4p bits to provide uniform spacing of size $1/10^p$ over [0,1]. Since $10^{-p} = 2^{-(\log_2 10)k/4} = 2^{-.83k}$, we note that the very infrequent largest gap size in k-bit LCF representation yields essentially no worse representation efficiency than that achieved throughout the unit interval for binary coded fixed-point decimal representation.

II. Lexicographic Binary Representation of Integers and Rationals

Utilizing the notation $[a_0/a_1/a_2/ \ldots]$ for the continued fraction

$$a_0 + \cfrac{1}{a_1 + \cfrac{1}{a_2 + \cfrac{1}{\ddots}}} \qquad a_i \geq 0$$

where the partial quotients a_i are assumed to be integral, it is known from the theory of continued fractions [HW79] that any non-negative rational number, denoted by the irreducible fraction p/q, has a finite expansion which is unique in terminal index even form

$$\frac{p}{q} = [a_0/a_1 \ldots /a_{2m}] \tag{2}$$

with the added requirements $a_0 \geq 0$, $a_i \geq 1$ for $1 \leq i \leq 2m$.

For the purpose of defining a continued fraction based binary representation of rationals we first provide representations of the positive integers (dealing separately with the case $a_0 = 0$) which are "self-delimiting", i.e. which implicitly contain end-markers when read left-to-right. The self-delimiting binary integer representation employed utilizes the standard binary representation preceded by a unary encoding of its length. Concatenation of such integer representations of the partial quotients of a continued fraction provides a binary representation where every rational number will have finite length. The LCF representation introduced in [MK83] further provides for the lexicographic order of the representations to correspond to numeric order of the rational values.

Formally, if the integer $p \geq 1$ has the (n+1)-bit base 2 radix representation $b_n b_{n-1} \ldots b_1 b_0$ with $b_i \in \{0,1\}$, $b_n = 1$, then*

$$\ell(p) = 1^n \circ 0 \circ b_{n-1} b_{n-2} \ldots b_1 b_0 \tag{3}$$

will be termed the lexicographic integer representation of p. This representation is order preserving in that the lexicographic ordering (leftmost bit first) of the bit strings corresponds to numeric ordering of the values. Table 1 contains the $\ell(\cdot)$ representations of the first 20 positive integers. For a discussion of various lexicographic order preserving binary encodings of the integers, see [Kn80] where similar representations are analysed.

---

*Here and in the following o denotes string-concatenation.

Integer	Standard Binary	Lexicographic	Integer	Standard Binary	Lexicographic
1	1_2	0	11	1011_2	1110011
2	10_2	100	12	1100_2	1110100
3	11_2	101	13	1101_2	1110101
4	100_2	11000	14	1110_2	1110110
5	101_2	11001	15	1111_2	1110111
6	110_2	11010	16	10000_2	111100000
7	111_2	11011	17	10001_2	111100001
8	1000_2	1110000	18	10010_2	111100010
9	1001_2	1110001	19	10011_2	111100011
10	1010_2	1110010	20	10100_2	111100100

Table 1. Right-adjusted standard binary representation and left-adjusted lexicographic bitstring representation of the integers 1, 2, ..., 20.

Note from the definition that the continued fraction expansion of $p/q = [a_0/a_1/a_2/ \dots /a_{2m}]$ is an increasing function of the even indexed partial quotients, and a decreasing function of the odd indexed quotients. Thus to obtain an order preserving representation of the rationals we simply represent the odd indexed quotients in complement form before concatenation. To be able to compare bit strings lexicographically from left to right it is assumed that any (finite length) representation is extended to the right with an arbitrary number of extra zeroes.

Formally, implicitly handling the case $a_0 = 0$ $(0 \le p/q < 1)$ by a leading zero bit and $a_0 \ge 1$ $(1 \le p/q)$ by a leading unit bit,

$$\mathrm{LCF}\left(\frac{p}{q}\right) = \begin{cases} 1 \circ \ell(a_0) \circ \overline{\ell(a_1)} \circ \ell(a_2) \circ \dots \circ \overline{\ell(a_{2m-1})} \circ \ell(a_{2m}) \circ \overline{\ell(\infty)} & \text{for } 1 \le \frac{p}{q}, \\ 0 \circ \overline{\ell(a_1)} \circ \ell(a_2) \circ \dots \circ \overline{\ell(a_{2m-1})} \circ \ell(a_{2m}) \circ \overline{\ell(\infty)} & \text{for } 0 \le \frac{p}{q} < 1, \end{cases} \quad (4)$$

where the finite bitstring form of LCF(p/q) is truncated at the last unit bit, except for LCF(0) whose finite form will be composed of a single 0.

<u>Example</u>

$\frac{22}{7} = [3/6/1]$ {terminal index even continued fraction form)

$= [3/6/1/\infty]$ {infinite extension}

$$\begin{aligned} \mathrm{LCF}\left(\frac{22}{7}\right) &= 1 \circ \ell(3) \circ \overline{\ell(6)} \circ \ell(1) \circ \overline{\ell(\infty)} \\ &= 1 \circ 101 \circ \overline{11010} \circ 0 \circ \overline{11\dots} \\ &= 1 \circ 101 \circ 00101 \circ 0 \circ 00\dots \\ &= 110100101 \quad \text{\{finite form\}} \end{aligned}$$ □

Observations concerning the LCF-representation given in [MK83] are summarized here:

— There is a one-to-one correspondence between all finite bitstrings terminating in a unit and all positive irreducible fractions.

— The LCF-representation extends naturally to the positive reals.

— The LCF-representation is order preserving over the non-negative reals, i.e. lexicographical ordering of LCF-strings corresponds to numerical ordering of the values.

— LCF(q/p) is the 2's complement of LCF(p/q) whenever $p > 0$, so the inverse of a k-bit LCF number is a k-bit LCF number.

— The bit pattern in LCF(p/q) may be interpreted as an encoding of the steps (transitions) in a finite automaton performing the Euclidean GCD-Algorithm (in binary) on p and q.

— LCF-representations may be used bit-by-bit left-to-right as input to on-line algorithms, for the direct computation of arithmetic expressions upon LCF-represented operands yielding LCF-represented results [KM83], [Go72].

— Signed LCF representation is given by:

$$\mathrm{SLCF}(p/q) = \begin{cases} 1 \circ \mathrm{LCF}(p/q) & \text{for } 0 \le p/q \\ 0 \circ \overline{\mathrm{LCF}(-p/q)} & \text{for } p/q < 0 \end{cases}$$

where $\overline{\mathrm{LCF}}$ denotes the 2's complement of the LCF bit string. Note that SLCF representation is then order preserving over the reals.

LCF representation is conveniently illustrated for all fractions hierarchically in terms of finite bitstring length by associating the positive irreducible fractions with the nodes of an infinite binary tree, termed the <u>LCF tree</u>, where the LCF bitstring denotes the path to the node containing the associated irreducible fraction. For LCF(p/q) given by the k-bit string $b_1 b_2 \dots b_{k-1} 1$, the fraction p/q is assigned to the node at depth k-1 reached by proceeding to the left child when $b_i = 0$ and to the right child when $b_i = 1$ for $i = 1, 2, \dots, k-1$.

The LCF tree illustrated to depth 4 in Figure 1 provides a convenient reference for interpreting the accuracy of finite precision LCF representation both with regard to the increasing precision obtained by a k-bit LCF approximation of a specific number as k increases, and with regard to the ordered enumeration of all k-bit LCF representable values for any fixed k. Specifically,

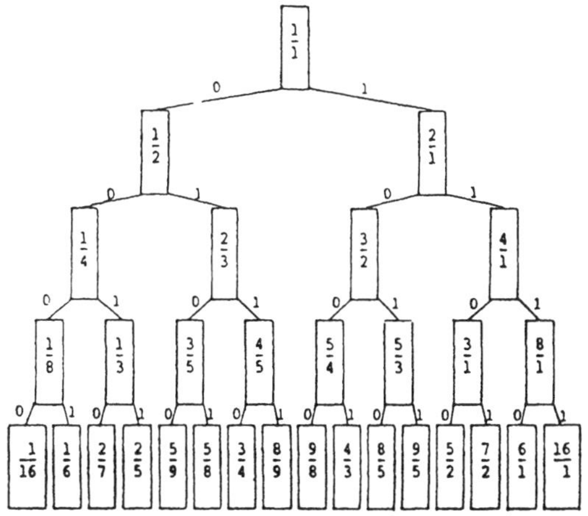

Figure 1. The LCF tree through depth four.

— Fractions in the nodes along a path determined by an LCF represented value provide successively tighter bounds on the value. E.g. the first five bits of the LCF number 01011... indicate a value in the interval [5/8, 2/3].

— Inorder enumeration of the fractions in the nodes of the LCF tree truncated at depth k-1 provides the ordered list of all positive rationals representable by k-bit LCF bitstrings. The seven positive 3-bit LCF numbers are then 1/4, 1/2, 2/3, 1/1, 3/2, 2/1, 4/1.

From the LCF tree it is further observed to the depth illustrated that:

— Each gap between representable positive k-bit LCF numbers is at most 2/3 the size of the gap between the (k-1)-bit LCF numbers in which it fell, at most 2/5 the size of the (k-2)-bit gap in which it fell, and at most 2/9 the size of the (k-3)-bit gap in which it fell.

— The irreducible fraction u/v associated with any node in the LCF tree can be conveniently determined as follows. If p/q, r/s are the irreducible fractions in the nearest left and right ancestor nodes where say $p \leq r$, and if j is the largest integer such that $2^j p \leq r$ and $2^j q \leq s$, then $\frac{u}{v} = \frac{2^j p+r}{2^j q+s}$ for $|ps - qr| = 1$, and $\frac{u}{v} = \frac{(p+r)/2}{(q+s)/2}$ for $|ps - qr| \neq 1$.

— If p/q is the fraction in the parent node of the node containing r/s in the LCF tree, then $.75\,p \leq r \leq 2p$ and $.75\,q \leq s \leq 2q$.

These latter observations hold for the whole LCF tree. Their formalization and application to the investigation of the accuracy of finite precision LCF representation are developed in the next sections.

III. The Binary Mediant and LCF Representation

The number theoretic concepts of the "mediant" of two fractions and the "adjacency relation" on fractions provided a foundation for our previous investigation and development of fixed-slash and floating-slash number systems [MK80, KM83, MK85]. The purpose of this section is to introduce and develop the analogous concept of "binary mediant" to provide a foundation for investigation of LCF representation.

Recall that fixed-slash number systems consist of Farey-sets

$$F_{2^n} = \left\{ \frac{p}{q} \;\middle|\; 0 \leq p,\, q < 2^n,\ \gcd(p,q) = 1 \right\}$$

whose members may be described by recursively introducing the mediant $(p+r)/(q+s)$ of two successive rationals $p/q < r/s$ already in the set, starting with 0/1 and 1/0. Restricting analysis to the interval $0 \leq p/q \leq 1$, it can be shown [MK80] that the size of gaps between successive members of F_{2^n} vary between 2^{-n} and 2^{-2n}, the larger gaps bounded on one side by rationals whose numerator and denominator are relatively small integers (intuitively termed simple fractions). It is easily seen that if say p/q is a very simple fraction, whereas r/s is not (e.g. $p \ll r$, $q \ll s$), then their mediant will be of numerical value very close to r/s but rather distant from p/q. If, however, p and q were both multiplied by some common factor c, chosen such that cp and cq were both of the same order of magnitude as r and s respectively, then the rational value $(cp+r)/(cq+s)$ would split the interval between p/q and r/s into two intervals of almost equal width.

This observation leads us to the introduction of an alternative form of mediant more nearly bisecting each gap which is conveniently related to binary representation. For any two non-negative fractions p/q and r/s we first define the partial ordering [MK80]:

$$\frac{p}{q} \lessdot \frac{r}{s} \quad \text{iff} \quad \frac{p}{q} \neq \frac{r}{s} \text{ and } p \leq r \text{ and } q \leq s, \qquad (5)$$

where the symbol $\lessdot$ denotes simpler than. For the irreducible fractions $p/q \lessdot r/s$, let j be the largest integer such that $2^j p/2^j q \lessdot r/s$. Then the binary mediant of p/q and r/s is the irreducible fraction u/v with value given by $(2^j p+r)/(2^j q+s)$. It is important to note that $\gcd(2^j p+r,\, 2^j q+s)$ is not necessarily unity, but the binary mediant is the reduced fraction u/v with the same value so then $\gcd(u,v) = 1$ by definition.

The relationship between the binary mediant and LCF representation is developed in the following lemmas which for brevity are given without proofs.

Lemma 1: If $p/q = [a_0/a_1/.../a_n]$ and $r/s = [a_0/a_1/.../a_n+2^i]$ where $i>0$ and $2^i \leq a_n$, then the binary mediant of p/q and r/s is

$$\frac{u}{v} = \frac{(p+r)/2}{(q+s)/2} = [a_0/a_1/.../a_n+2^{i-1}].$$

Lemma 2: If $p/q = [a_0/a_1/.../a_n]$, $r/s = [a_0/a_1/.../a_n/2^i]$, $i \geq 0$, then the binary mediant of p/q and r/s is

$$\frac{u}{v} = \frac{2^i p+r}{2^i q+s} = [a_0/a_1/.../a_n/2^{i+1}].$$

Lemma 3: If $p/q < r/s$ have LCF-representations

$$LCF(p/q) = \sigma\ o\ 0\ o\ 1^j \text{ and } LCF(r/s) = \sigma\ o\ 1\ o\ 0^j$$

for $j \geq 0$, and u/v is the binary mediant of p/q and r/s, then

$$LCF(\frac{u}{v}) = \sigma\ o\ 0\ o\ 1^{j+1}.$$

In Lemmas 1-3 we refer to a continued fraction expansion of a fraction without being specific about any canonical form. From continued fraction theory [HW79] any positive fraction will have exactly two continued fraction expansions under the assumption $a_i \geq 1$ for all $i \geq 1$, which are related as follows:

$$\frac{p}{q} = \begin{cases} [a_0/a_1/.../a_n] & (6) \\ & \text{for } a_n \geq 2. \\ [a_0/a_1/.../a_n-1/1] & (7) \end{cases}$$

Formula (6) is generally chosen as the canonical form, and we have earlier chosen between (6) and (7) to select that form whose terminal index is even simply to guarantee that the LCF representation has a terminal string of 0's. Note that utilizing the corresponding terminal index odd continued fraction and encoding into LCF form would yield a bit string differing only in that $LCF(p/q) = \sigma 100...$ would become $\sigma 011...$ for the appropriate substring σ. As in ordinary radix representation we take these bit strings as equivalent preferring to write $LCF(p/q) = \sigma 1$, assuming the finite sequence is extended to the right by an infinite string of zeros.

Lemma 4: The labeling of the nodes of the LCF-tree can be obtained by succesively generating the fraction occurring as label, as the binary mediant of its nearest left and right ancestors. The labeling of the edges (0 for left branch, and 1 for right branch) read as a path to a node provides the LCF representation of the fraction at that node, when completed by a unit-bit.

IV. The Distribution of Gap Sizes in the k-bit LCF System

To investigate the distribution of gap sizes between the k-bit LCF rationals, consider the following. The LCF rationals in the interval from one to infinity are the reciprocals of the LCF rationals in the interval zero to one. Hence we shall restrict our analysis to the unit interval. Also note that in the closed unit interval there are exactly 2^k+1 different rationals in the (k+1)-bit LCF system, the leading bit being used to distinguish that the number falls in the unit interval, rather than in the interval from one to infinity. Equivalent uniform spacing as in fixed-point binary would yield a uniform gap size of 2^{-k}, but would not provide for exact representation of all the relatively simple rationals.

For purposes of comparing the space efficiency of LCF representation we are concerned with the size k of the finite length k-bit LCF representation of p/q in comparison with the size of p and q. More critically for purposes of uniformity of approximation we are concerned with the extent of variability of the gap sizes in k-bit LCF representation about the ideal value 2^{-k} with increasing k.

Notice that the growth of the values of p and q, as a function of the position of the last non-zero bit in LCF(p/q), is dependent on the encoding of the partial quotients of the continued fraction expansion of p/q, i.e. on the encoding $\ell(\cdot)$ specified in Section II. It is known from the measure theory of continued fractions (e.g. [Kh63] or [Kn81]) that for p/q chosen uniformly on [0,1], the probability that any particular partial quotient a_i of p/q takes a specific value j depends only on j (for sufficiently large i) and is given by

$$P(a_i = j) = \log_2 \frac{(j+1)^2}{j(j+2)},$$

hence $P(a_i = 1) = 41.5\%$, $P(a_i = 2) = 17.0\%$,

$P(a_i = 3) = 9.3\%$, $P(a_i = 4) = 5.9\%$, etc.

From information theoretic considerations an object occurring with probability p is best encoded by using $-\log_2 p$ bits. Ideally then we should use 1.27 bits for encoding 1, 2.56 bits for encoding 2, 3.42 bits for 3, and 4.09 bits for 4. Since $\ell(1) = 0$, $\ell(2) = 100$, $\ell(3) = 101$ and $\ell(4) = 11000$, we observe that 1 has a rather short representation when compared to the frequency with which it occurs, whereas 2 and 4 have quite long representations. Hence we may expect that the numerator and denominator of $p/q = [0/1/1/.../1]$, where $LCF(p/q) = 0101...01$, will exhibit a fast growth rate as a function of k, whereas $r/s = [0/2/4/.../2/4]$, where $LCF(r/s) = 0011110000 1111...00001111$, will have a slow growth of r and s.

p and q will grow asymptotically as $2^{1.3885k}$, for LCF(p/q) = 0101...01, and similarly will be of the order $2^{0.8268k}$ for LCF(r/s) = 0011110000111l... 00001111. Based on these observations we may also expect asymptotically a minimal gap size between k-bit LCF-numbers to be at least as small as $2^{-1.3885k}$ and a maximal gap size to be at least as large as $2^{-0.8268k}$, where 2^{-k} would have been obtained if the gaps were all of the same width. Utilizing the notion of binary mediant from the preceding section we are able to derive an overall upper bound on the maximum gap size. For brevity we include here only the main idea of the proof.

Theorem 5: For any $k \geq 9$, the largest gap size over [0,1] between consecutive (k+1)-bit LCF numbers is less than $2^{-0.81k}$.

Sketch of Proof: Let the "worst case" effective precision α be defined by having $2^{-\alpha k}$ equal the largest gap size over [0,1] in (k+1)-bit LCF representation. By direct computation we obtain the following values of α for k = 1,2,...,20.

k	α	k	α
1	1.000	11	0.812
2	0.792	12	0.816
3	0.774	13	0.819
4	0.792	14	0.820
5	0.817	15	0.816
6	0.812	16	0.819
7	0.804	17	0.821
8	0.810	18	0.822
9	0.815	19	0.818
10	0.818	20	0.821

In general the LCF representation of any number $0 < x < 1$ may be given in the form

$$x = 0^{n_1}1^{n_2}0^{n_3}1^{n_4}\ldots$$

composed of blocks of $n_i \geq 1$ identical bits (0's or 1's) between reversals. Let r/s be the last node value on the path in the LCF tree corresponding to x before the ith block and let p/q be the preceding (parent) node value. Then x is known to be in a gap of size $|p/q - r/s|$ before interpreting the ith block of bits. For the case where $p/q \leqslant r/s$ it follows that $r/s \leqslant 2p/2q$. Then the value in the node corresponding to reading j of the next n_i bits is the irreducible fraction of value given by the not necessarily reduced fraction $((2^j-1)p+r)/((2^j-1)q+s)$ for $j = 1,2,\ldots,n_i$. After proceeding through the ith block x will then be known to fall in a gap of size

$$\left|\frac{(2^{n_i}-1)p+r}{(2^{n_i}-1)q+s} - \frac{(2^{n_i-1}-1)p+r}{(2^{n_i-1}-1)q+s}\right|$$

as compared to the previous gap size $|p/q-r/s|$. Taking the maximum ratio of new to old gap size for each value of $n=n_i$ over all $1 \leq s/q \leq 2$ yields 1/2 for n=1, $1/(2+\sqrt{3})$ for n=2, and $2^n/[(2^n+1)(2^{n-1}+1)]$ for $n \geq 3$. The effectiveness per bit of a string of n similar bits between reversals is then no worse than the negative base two logarithm of this ratio divided by n, giving an average effectiveness per bit of 0.8143 for n=4 as the worst case.

The details of handling the case where $r/s \leqslant p/q$ and of treating a terminal segment (no implied reversal) can be handled to verify the theorem. □

It should be noted that the argument of the preceding proof outline yields the LCF bit pattern 0000111100001111...00001111 as asymptotically achieving the minimum average reduction in gap size bounds per bit, consistent with the preceding heuristic argument that

LCF([0/2/4/.../2/4]) = 0011110000111l...00001111

would bound an interval of comparatively large gap size.

To further analyze the distribution of gap sizes a program was written to perform an exhaustive analysis of all gaps in the (k+1)-bit LCF systems up through k = 25. In each case it was found that the minimum size gap fell between two consecutive rationals of the form f(n-2)/f(n-1) and f(n-1)/f(n) where f(n) denotes the nth Fibonacci number, in correspondence with the preceding observation about LCF([0/1/1/.../1]). It was also found that the maximum size gap occurred next to a rational whose LCF representation contained replications of the bitpattern 00001111, consistent with the preceding discussions.

The main purpose of the program was however to obtain a bar-graph of the distribution of gap sizes, as shown in Figure 2. Here the frequencies (in percent) of the occurrence of a particular gap size is plotted, as a function of the gap size measured relative to the uniform gap size of 2^{-k} for k = 6, 12, 18 and 24, in a logarithmic scale and using a bin size of 0.01. The initially erratic distribution for k=6 settles to a smooth bell shape for k=24, and the variation from the minimum gap size to the maximum gap size indicated by the truncated tails of the bell shape is essentially the same for k = 12, 18 and 24.

To illustrate the narrowing of the "bell-shape," simulations were run for higher values of k. In these simulations a random bitpattern was generated, and the equivalent rational number as well as a neighbor was determined. For each k chosen, 2^{21} gap sizes were determined, and their distribution plotted in a bar-graph as before. Simulated bar-graphs for k = 32,64,128 are shown in Figure 3, overlayed over the family of exhaustively computed distributions for k = 15, 16,...,25, confirming that the distribution gets narrower with a higher peak for increasing values of k.

Table 2 gives "effective precision" bounds a,b such that, respectively, 95%, 99.9%, and 100% of the gap sizes fall in the interval $[2^{-bk}, 2^{-ak}]$ with symmetric cutoffs, summarizing the narrowing of the peak with increasing k from the data of the preceding figures.

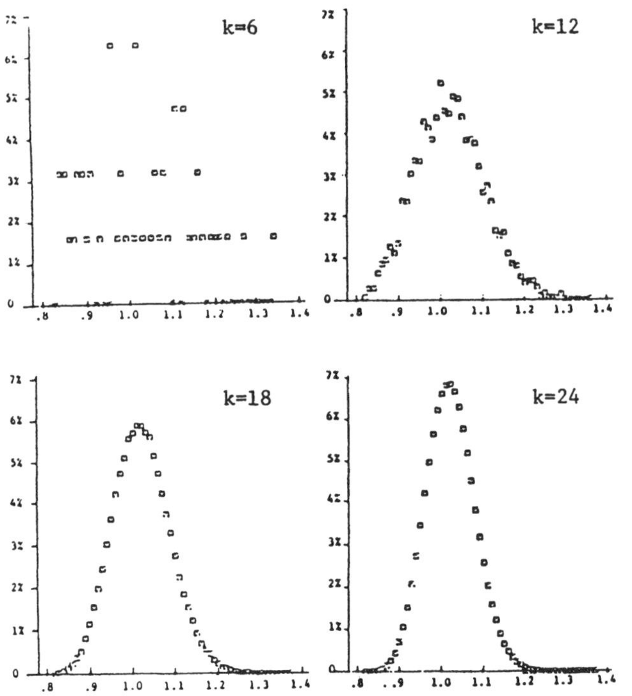

Figure 2: The percent of gaps in (k+1)-bit LCF representation over [0,1] that fall within the range $[2^{-ak}, 2^{-(a+.01)k}]$ for a = .80,.81,...,1.39 and k = 6,12,18,24, where values plotted for k = 12, 18 and 24 correspond exclusively to non zero percentages.

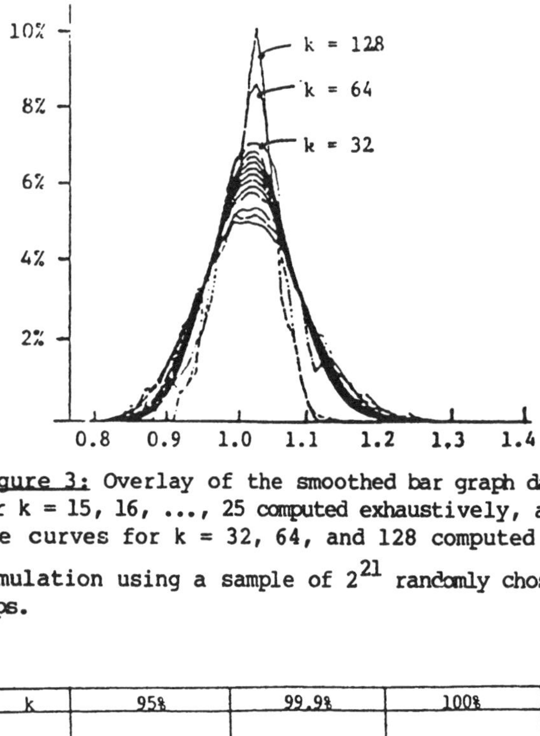

Figure 3: Overlay of the smoothed bar graph data for k = 15, 16, ..., 25 computed exhaustively, and the curves for k = 32, 64, and 128 computed by simulation using a sample of 2^{21} randomly chosen gaps.

k	95%	99.9%	100%
16	0.89 - 1.17	0.83 - 1.27	0.81 - 1.38
24	0.92 - 1.15	0.86 - 1.23	0.82 - 1.38
32	0.93 - 1.14	0.88 - 1.21	0.82 - 1.38
64	0.94 - 1.10	0.89 - 1.13	0.82 - 1.38
128	0.95 - 1.09	0.90 - 1.12	0.82 - 1.38

Table 2. Bounds a-b such that the indicated percentage of all gap sizes in (k+1)-bit LCF representation over [0,1] fall in the interval $[2^{-bk},2^{-ak}]$.

Overall our study shows that the gaps between LCF-representable values are subject to a variation in size corresponding to a worst case 18% precision loss (or equivalently an 18% storage capacity loss) and a best case 38% precision gain in comparison with an equivalent fixed-point binary system with uniform gap sizes over the unit interval. This appears to be a reasonably small price to pay in return for achieving exact representation of all simple rationals in a convenient format supported by the number theoretic foundations of continued fractions and "best rational approximation". It should be noted that the locations of the extreme larger gaps and smaller gaps within [0,1] correspond to continued fractions with particular sequences of partial quotients related to their chosen $\ell(\cdot)$ representation. Thus the gap size extremes may be regarded as base dependent anomalies of the representation and of no apparent intrinsic natural value other than to bound the precision variation of the representation.

Acknowledgments

The authors would like to express their gratitude to Tim Oglesby, who programmed the exhaustive and simulated studies of the gap distribution, and to the LTV Corporation, Dallas, who provided ample computing power for their execution during night-time. Also thanks are directed to Dorothy Volk for a patient job of typing the manuscript "on the fly," and the subsequent corrections and rearrangements.

References

[Go72] R. W. Gosper: Item 101 in HAKMEM, MIT-AIM 239, Feb. 1972, pp 37-44.

[HW79] G. H. Hardy and E. M. Wright: "An Introduction to the Theory of Numbers," 5th ed., Oxford University Press, London, 1979.

[IEEE81] IEEE Task P754: "A Proposed Standard for Binary Floating-Point Arithmetic," Computer, Vol. 14, March 1981, pp 51-62.

[Kh63] A. Y. Khintchin: "Continued Fractions," translated from Russian by P. Wynn. P. Noordhoff Ltd., Grooningen, 1963.

[Kn80] D. E. Knuth: "Supernatural Numbers", in Mathematical Gardner, D. A. Klarner, ed., Van Nostrand, New York, 1982, pp 310-325.

[Kn81] D. E. Knuth: "The Art of Computer Programming, Vol. 2, Seminumerical Algorithms," 2nd Ed., Addison-Wesley, Reading, 1981.

[KM83] P. Kornerup and D. W. Matula: "Finite Precision Rational Arithmetic: An Arithmetic Unit," IEEE TC, Vol. C-32, No. 4, April 1983, pp 378-87.

[KM84] U. W. Kulisch and W. L. Miranker (eds): A New Approach to Scientific Computation, Academic Press, New York, 1983. See also: High-Accuracy Arithmetic, Subroutine Library, General Information Manual (ACRITH), IBM Program Number 5664-185, 1984.

[MK80] D. W. Matula and P. Kornerup: "Foundations of Finite Precision Rational Arithmetic," in Computing, Suppl. 2, Springer-Verlag, pp 85-111, 1980.

[MK83] D. W. Matula and P. Kornerup: "An Order Preserving Finite Binary Encoding of the Rationals," Proc. 6th Sym. on Comp. Arith., IEEE Cat #83CH1892-9, 1983, pp 201-209.

[MK85] D. W. Matula and P. Kornerup: "Finite Precision Rational Arithmetic: Slash Number Systems," IEEETC, Vol. C-34, No. 1, January 1985, pp 3-18.

An Overflow/Underflow-Free Floating-Point Representation of Numbers*

SHOUICHI MATSUI** and MASAO IRI***

A new floating-point representation of numbers is proposed to cope systematically with the troubles in numerical computation due to exponent overflow/underflow in conventional floating-point representations.

The proposed representation resolves the phenomena of overflow and underflow and, at the same time, attains higher precision for numbers which are neither too large nor too small, by making mobile the boundary between the field for exponent and that for mantissa in a word.

A number system including "non-numbers" is also proposed which is closed with respect to the four arithmetic operations. The effectiveness of the proposed system is shown by numerical examples.

1. Introduction

For the numerical computation on a computer, a real number (or its approximation) is usually represented in a register of a fixed length in the form of a "floating-point" number. In the earlier days the fixed-point representation of numbers was commonly used, but, due to cumbersome problems related to "scaling", the floating-point representation, which can handle numbers over a wider range with the same relative precision at the sacrifice of a few significant digits to be allotted to the exponent, has superseded the fixed-point representation. Among a large variety of theoretically possible floating-point representation, the IBM-type (with the base 16, the mantissa represented by the sign and the magnitude, and 7-bit exponent biased by 64) is now prevalent. However, the more users experienced this type of representation of numbers for numerical computation, the more complaints arose. The exponent overflow/underflow is one of the most serious difficulties. S. Hitotumatu [7] discussed this issue. In general, as R. W. Hamming [5] points out, the range of the distribution of the exponents of numbers becomes broader and broader in the course of computation, so that the magnitudes of numbers are apt to overstride the pre-determined range, e.g. $(16^{-2^6}, 16^{2^6}) \fallingdotseq (10^{-77}, 10^{77})$. Careful scaling for avoiding this trouble would make the floating-point arithmetic less attractive, since it would invalidate the most important advantage of being scaling-free.

A floating-point number system was proposed by W. Kahan [10] to improve the undesirable properties of the IBM-type. His proposal consists basically in

(i) alloting a larger part of a word to the exponent, and

(ii) preparing "Not-A-Number (NAN)"s to guard against still possible overflow/underflow phenomena.

Other fundamental problems concerning the floating-point representation of numbers, such as the choice of the bases (=2, 4, 8, 16 etc.) [1, 2] (see also the references there) and representations different from the conventional ones [4, 6], have recently been given attention.

There is a claim [7] that the variable-length word would resolve all the difficulties, but, from the standpoints of the cost of memories and computation time, this kind of representation will not be practical except for extremely special purposes. Hence, we shall not discuss it here.

In this paper, we propose an ideal floating-point representation of real numbers under the condition that a number be represented with a fixed-length word. We discuss briefly also the characteristics of the proposed representation, its implementation, and examples of application to numerical computation.

The basic ideas of the proposed representation may be described informally as follows.

(i) When a number is represented in a fixed-length word, "bits" are allotted to the more important information with the higher priority. For example, if a number is represented by an exponent and a mantissa, a sufficient number of bits are assigned to the exponent first, and then the remaining bits are assigned to the mantissa. If the exponent is too large (or small) to be expressed in a word, then another form of representation is introduced, in which the exponent itself is represented by the "mantissa of the exponent" accompanied by the "exponent of the exponent", and a sufficient number of bits are assigned to the latter. This process will be repeated in a similar manner towards more complex representations.

(ii) A number system containing "non-number"s is established, which is closed under the four arithmetic operations, and which allows users to have complete

---

*This paper is a translation from Japanese of Transactions of IPSJ, Vol. 21, No. 4 (1980), which received an award as the 1980 winning paper of IPSJ.

**Information System Department, Economic Research Center, Central Research Institute of Electric Power Industry, Ohtemachi, Chiyoda-ku, Tokyo 100, Japan.

***Department of Mathematical Engineering and Instrumentation Physics, Faculty of Engineering, University of Tokyo, Hongo, Bunkyo-ku, Tokyo 113, Japan.

command of the numerical computation at a programming-language level (i.e. an overflow/underflow never interrupts the normal execution of computation), where the symmetry of the system with respect to the four arithmetic operations is taken account of.

The idea (i) is found, though incompletely, in R. Morris [13], and an exercise in D. E. Knuth [11] (Exercise 17 of Sec. 2.1 of Chap. 4) seems to suggest such a direction. In this paper, we shall extend the idea in full detail to propose an ideal floating-point number system. The idea (ii) which was ambiguously discussed in Kahan [10] is almost independent of (i). We shall show that, under some reasonable assumptions, the "non-number"s are determined uniquely.

2. Basic Ideas

2.1 Variable-length Exponent and Its Generalization

In a floating-point number system with the base β (=2, 4, 8, 16 etc.), a real number x is represented as

$$\text{(Level 0)} \qquad x=(-1)^{S_0}\times F\times \beta^{(-1)^{S_1}\times E_1}, \tag{2.1}$$

where s_0 and s_1 are either 0 or 1, E_1 is a nonnegative integer, and F is a real number which is normalized as

$$1>F\geqq 1/\beta \tag{2.2}$$

or

$$\beta>F\geqq 1. \tag{2.3}$$

For the normalization condition, the former (2.2) is widely used, but, for the reasons stated later, we adopt the latter (2.3) in the following. For any non-zero number x, the normalization condition determines its representation of the form (2.1) uniquely.

Since the number of bits to be allotted to F and E_1 is limited by the length of a word, it is obviously best to represent the exponent part E_1 exactly (if it is possible) and then to allot the remaining bits to the mantissa part F. In fact, the approximation error due to the rounding of the mantissa to a finite length is less significant than the error due to the variation of the exponent E_1 by ± 1.

By doing so, we can allocate a larger part of a word than in the conventional case to the mantissa if $|\log_\beta|x||$ is small, since the number of bits required for the exponent is small. As $|\log_\beta|x||$ becomes larger, the number of bits for E_1 increases and that for F decreases.

For those numbers, for which $|\log_\beta|x||$ is so large that E_1 may occupy most bits of a word, we are natually led to another kind of representation which we shall call the "level-1" representation:

$$\text{(Level 1)} \qquad x=(-1)^{S_0}\times F\times \beta^{(-1)^{S_1}\times E_1\times \beta^{E_2}}. \tag{2.4}$$

In this level, there is no room for the mantissa F so that we can say nothing more than that F lies between the upper and lower end of the normalization (2.3). E_1 and E_2 are determined so as to maximize the length of E_1 under the condition that E_1 and E_2 can be stored in a word and that they satisfy

$$E_1\times\beta^{E_2}\leqq E<(E_1+1)\times\beta^{E_2}, \tag{2.5}$$

where E is the exponent E_1 in the expression (2.1) when the number is expressed on level 0. We can introduce "higher levels" of representations in a similar manner. For example, for so large a number that the E_2 in (2.4) may occupy most bits of a word, the level 2 is introduced, where a number x is represented as follows:

$$\text{(Level 2)} \qquad x=(-1)^{S_0}\times F\times \beta^{(-1)^{S_1}\times E_1\times \beta^{E_2\times\beta^{E_3}}}. \tag{2.6}$$

In (2.6), F is a number which satisfies the condition (2.3), E_1 is a nonnegative integer which satisfies the condition

$$1\leqq E_1<\beta^{E_2\times\beta^{E_3}-1}, \tag{2.7}$$

and E_2 and E_3 are determined so as to maximize the length of E_2 under the condition that they can be stored in a word and they satisfy the condition

$$\beta^{E_2\times\beta^{E_3}}\leqq E<2\times\beta^{(E_2+1)\times\beta^{E_3}-1}, \tag{2.8}$$

where E is the E_1 in (2.1) when the number is expressed on level 0. Higher levels such as level 3, 4, 5, $\cdots$ are defined similarly.

For the actual implementation, additional informations, such as

(i) the level on which a number is represented,
(ii) a pointer to the boundary between the field for F and that for E_1 (in the case of the level-0 representation), or between E_i and E_{i+1} (in the case of the level-i representation with $i\geqq 1$),

are to be stored also in the same word.

2.2 Choice of the Base

The choice of the base of the number system is known to be important, and, judging from various theoretical arguments as well as experimental evidences, the best choice today should certainly be to take 2 for the base [1, 2] (see also the references there). Adopting the base 2 is advantageous also because we can suppress the first bit (i.e. the most significant bit) of the mantissa, which is always "1" in view of the normalization condition.

2.3 Sign-magnitude Notation

The representation of the mantissa as well as that of the exponent must be in the "sign-magnitude" notation, instead of "two's or one's complement" or "biased" notation, in order to realize the advantages (mentioned in §2.1 and §2.2) of the proposed representation. In passing, the sign-magnitude notation with the normalization condition (2.3) makes the number system "symmetric" with respect to the four arithmetic operations. This property of the symmetry is considered to be one of the desirable characteristics of the system [14]. Note that, for a number with the exponent of the largest absolute value, the mantissa is always equal to "1" in this system.

2.4 Rounding

The round-to-the-nearest is adopted.

2.5 "Non-numbers"

It is obvious that, within a single fixed-length word, only a finite number of real numbers can be distinguished from one another on the one hand, and on the other, that numbers larger (smaller) than the largest (smallest) representable number may possibly appear in the course of computation. Therefore, we must introduce into our number system "a number that is larger (smaller) than a certain number". It is compared to "$+\infty$" ("$-\infty$"). We shall take those "numbers which are not number in the usual sense" into our system, calling them "non-numbers". The number system containing non-numbers in addition to ordinary numbers should be closed with respect to the four arithmetic operations, so that the computation may not be interrupted by the operating system even when an overflow/underflow takes place. Then, we shall be able to describe every control of computational flow at a programming-language level, without using, e.g., the ON-statement in PL/I. Kahan also discusses such "non-number"s, "Not-A-Number (NAN)"s in his proposal [10], but they are not defined without ambiguity. We will determine such "non-number"s without ambiguity on the basis of the requirement of the symmetry of the number system. This is discussed in detail in Chap. 4.

3. Implementation of the Proposed Floating-point Number System

In practice, we can realize only a finite number of levels, so that we have to content ourselves up to a certain level. Here, we first consider the implementation of the "level 0". Higher levels may be implemented in a similar way.

3.1 Implementation of Level 0

For a real number x ($\neq 0$) expressed as in (2.1) under the normalization condition (2.3), we have the bit-strings of F and E looking like

$$F=(1.f_2f_3\cdots f_m), \tag{3.1}$$

$$E=(1e_{n-1}e_{n-2}\cdots e_1) \text{ (or } E=0). \tag{3.2}$$

Therefore, the $L=m+n$ bits

$$s_0f_2f_3\cdots f_me_{n-1}e_{n-2}\cdots e_1s_1$$

are necessary to represent x, where it is understood that $E=1$ if $n=1$ and $E=0$ (and no extra bit is needed for s_1) if $n=0$. L being fixed, n takes a value in the range from 0 to $L-1$. To hold the value of n which is the "pointer" to the boundary between the field for the mantissa and that for the exponent, $\lceil \log_2 L \rceil$ more bits are needed. Therefore, the L should be determined as the maximum integer which satisfies

$$L+\lceil \log_2 L \rceil \leqq W_L, \tag{3.3}$$

where W_L is the word length. Clearly, for any x, its representation in this form is unique, since we regard the ideal zero ("0") as a "non-number" in this system. (See Chap. 4.)

3.2 An Example of Implementation of Level 0

In this section we present an example of actual implementation of level 0. First, we assume that a word consists of 64 bits. Therefore, we have $L=58$ and $\lceil \log_2 58 \rceil=6$. The data structure for the level 0 is shown in Fig. 1. The length of the mantissa (without the sign bit) varies from 58 to 1 (including the suppressed first bit), and consequently, that of the exponent (with the sign bit) Varies from 0 to 57 (including the most significant bit, which is suppressed, when $B=n\neq 0$). Therefore, the numbers expressible on level 0 range widely, i.e. from about $10^{-10^{16}}$ to about $10^{10^{16}}$.

The values of B from 58 to 63 are meaningless, but they can be used to indicate (i) non-numbers, (ii) higher-level representations, etc.

The printing format of a level-0 number may be decided on the same principle as the partition of a word into the mantissa and the exponent part was based on. A specified field length of printing is divided into two parts, one for the exponent and the other for the mantissa. First, sufficient digits in the field are allocated to the exponent part, and then the remaining digits are allocated to the mantissa which is normalized in the form (X.XXX$\cdots$). (See Chap. 6 for an example.)

For the input formats, we allow all the formats which have been used in conventional computer systems.

4. A Floating-point Number System Including Non-numbers

A number system for numerical computation should desirably be closed at least with respect to the four arithmetic operations, so that, even if "an illegal pair of operands" appears, it should not disturb the normal progression of a programme. In general, a number represented within a fixed-length word is to be regarded as an interval rather than a point on the real line, regardless whether the width of the interval is explicitly expressed or not. Let us begin with the set of nonzero real numbers $R^+ \cup R^-$, where

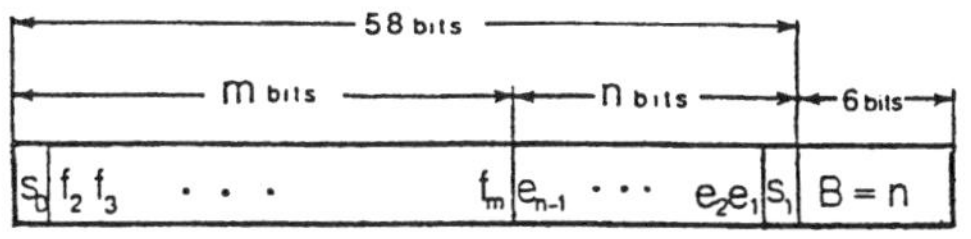

mantissa part $(-1)^{s_0}F=(-1)^{s_0}\times(1.f_2f_3\cdots f_m)_2$
exponent part $(-1)^{s_1}E=(-1)^{s_1}\times(1e_{n-1}\cdots e_2e_1)_2$
$B=n$ points to the boundary between the mantissa part and the exponent part.
($B=0, 1, \cdots, 57$; $E=0$ when $B=0$)

Fig. 1 Data structure for a floating-point number in the proposed representation (level 0)

$$R^+ = \{\text{positive real numbers}\}$$
and (4.1)
$$R^- = \{\text{negative real numbers}\}.$$

Only part of the elements of $R^+ \cup R^-$ can be represented in the form shown in Chap. 3. For the symmetry and closedness of the number system, the following 7 types of intervals (as well as those types of intervals which are defined with < in place of ≦) are necessary and sufficient:

(i) $\{x|a \leqq x \leqq b;\ a, b \in R^+\}$, $\{x|a \leqq x \leqq b;\ a, b \in R^-\}$,

(ii) $\{x|a \leqq x;\ a \in R^+\}$, $\{x|x \leqq a;\ a \in R^-\}$;

(iii) $\{x|0 < x \leqq a;\ a \in R^+\}$, $\{x|a \leqq x < 0;\ a \in R^-\}$;

(iv) $\{x|x \leqq a \text{ or } b \leqq x;\ a \in R^-, b \in R^+\}$;

(v) $\{x|a \leqq x \leqq b;\ a \in R^-, b \in R^+\}$;

(vi) R^+, R^-;

(vii) $R(= R^+ \cup R^- \cup \{0\})$.

In fact, starting from intervals of the basic type (i), all the other types of intervals can be produced by the four arithmetic operations. The following symbols are used to denote the above 7 types of intervals, respectively.

(i) +num, −num;

(ii) $+\infty$, $-\infty$;

(iii) +0, −0;

(iv) ∞;

(v) 0;

(vi) +?, −?;

(vii) ?.

The intervals of types (ii)~(vii) are called "non-numbers", which are schematically shown in Fig. 2.

We regard a number which is representable in the form shown in Chap. 3 as the representative point of an interval of type ±num, and assume the result of an arithmetic operation on two numbers of that kind to be one of the numbers and non-numbers of the above 7 kinds, as follows.

a. If the result is representable in the form shown in Chap. 3, then round it to an element of ±num, if necessary.
b. If the absolute value of the result is too large to be represented in the form of Chap. 3, then denote it by $\pm\infty$.
c. If the absolute value of the result is too small to be represented in the form of Chap. 3, then,
   c1. if the mantissa preserves information on the sign, then denote it by ±0;
   c2. if the mantissa carries no information on the sign, then denote it by 0.

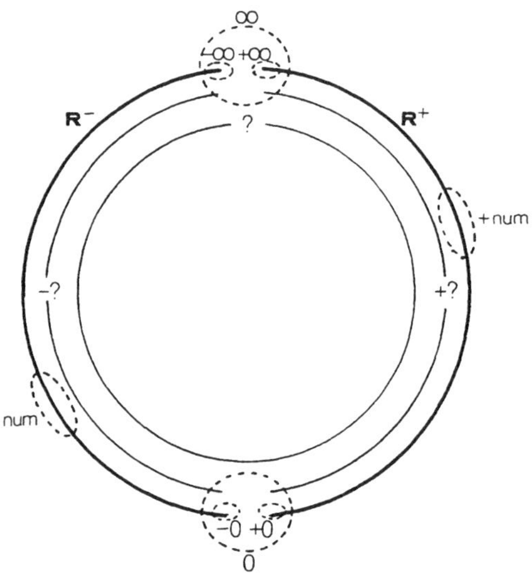

Fig. 2 Schematic diagram of the proposed number system including non-numbers.

The tables for the four arithmetic operations among numbers and non-numbers are given in Tables 1, 2, 3 and 4. These tables will explain how all the 7 types of intervals actually appear, and why those non-numbers are necessary and sufficient for the number system to be closed.

The order among numbers and non-numbers is defined in terms of the sign carried by the result of the subtraction operation as follows:

$X > Y$ iff the sign of $X - Y$ is "+";

$X < Y$ iff the sign of $X - Y$ is "−";

Table 1 Addition table for the number system including non-numbers.

op1 \ op2	+num	−num	+∞	−∞	+0	−0	+?	−?	∞	?	0
+num	+num +∞	±num ±0, 0	+∞	−∞	+num	+num	+?	?	∞	?	+num
−num	±num ±0, 0	−num −∞	+∞	−∞	−num	−num	?	−?	∞	?	−num
+∞	+∞	+∞	+∞	?	+∞	+∞	+∞	?	?	?	+∞
−∞	−∞	−∞	?	−∞	−∞	−∞	?	−∞	?	?	−∞
+0	+num	−num	+∞	−∞	+0	0	+?	?	∞	?	0
−0	+num	−num	+∞	−∞	0	−0	?	−?	∞	?	0
+?	+?	?	+∞	?	+?	?	+?	?	?	?	?
−?	?	−?	?	−∞	?	−?	?	−?	?	?	?
∞	∞	∞	?	?	∞	∞	?	?	?	?	∞
?	?	?	?	?	?	?	?	?	?	?	?
0	+num	−num	+∞	−∞	0	0	?	?	∞	?	0

Table 2 Subtraction table for the number system including non-numbers

op1 \ op2	+num	−num	+∞	−∞	+0	−0	+?	−?	∞	?	0
+num	±num ±0, 0	+num +∞	−∞	+∞	+num	+num	?	+?	∞	?	+num
−num	−num −∞	±num ±0, 0	−∞	+∞	−num	−num	−?	?	∞	?	−num
+∞	+∞	+∞	?	+∞	+∞	+∞	?	+∞	?	?	+∞
−∞	−∞	−∞	−∞	?	−∞	−∞	−∞	?	?	?	−∞
+0	−num	+num	−∞	+∞	0	+0	?	+?	∞	?	0
−0	−num	+num	−∞	+∞	−0	0	−?	?	∞	?	0
+?	?	+?	?	+∞	?	+?	?	+?	?	?	?
−?	−?	?	−∞	?	−?	?	−?	?	?	?	?
∞	∞	∞	?	?	∞	∞	?	?	?	?	∞
?	?	?	?	?	?	?	?	?	?	?	?
0	−num	+num	−∞	+∞	0	0	?	?	∞	?	0

Table 3 Multiplication table for the number system including non-numbers

op1 \ op2	+num	−num	+∞	−∞	+0	−0	+?	−?	∞	?	0
+num	+num +∞, +0	−num −∞, −0	+∞	−∞	+0	−0	+?	−?	∞	?	0
−num	−num −∞, −0	+num +∞, +0	−∞	+∞	−0	+0	−?	+?	∞	?	0
+∞	+∞	−∞	+∞	−∞	+?	−?	+?	−?	∞	?	?
−∞	−∞	+∞	−∞	+∞	−?	+?	−?	+?	∞	?	?
+0	+0	−0	+?	−?	+0	−0	+?	−?	?	?	0
−0	−0	+0	−?	+?	−0	+0	−?	+?	?	?	0
+?	+?	−?	+?	−?	+?	−?	+?	−?	?	?	?
−?	−?	+?	−?	+?	−?	+?	−?	+?	?	?	?
∞	∞	∞	∞	∞	?	?	?	?	∞	?	?
?	?	?	?	?	?	?	?	?	?	?	?
0	0	0	?	?	0	0	?	?	?	?	0

Table 4 Division table for the number system including non-numbers

op1 \ op2	+num	−num	+∞	−∞	+0	−0	+?	−?	∞	?	0
+num	+num +∞, +0	−num −∞, −0	+0	−0	+∞	−∞	+?	−?	0	?	∞
−num	−num −∞, −0	+num +∞, +0	−0	+0	−∞	+∞	−?	+?	0	?	∞
+∞	+∞	−∞	+?	−?	+∞	−∞	+?	−?	?	?	∞
−∞	−∞	+∞	−?	+?	−∞	+∞	−?	+?	?	?	∞
+0	+0	−0	+0	−0	+?	−?	+?	−?	0	?	?
−0	−0	+0	−0	+0	−?	+?	−?	+?	0	?	?
+?	+?	−?	+?	−?	+?	−?	+?	−?	?	?	?
−?	−?	+?	−?	+?	−?	+?	−?	+?	?	?	?
∞	∞	∞	?	?	∞	∞	?	?	?	?	∞
?	?	?	?	?	?	?	?	?	?	?	?
0	0	0	0	0	?	?	?	?	?	?	?

$X=Y$ iff X and Y have the same representation, i.e. they have the same bit-pattern.

With this definition, we have, for example,

$$-\infty < -\text{num} < -0 < +0 < +\text{num} < +\infty,$$

$$-\text{num} < 0 < +\text{num},$$

$$\text{if } x<y \text{ and } y<z, \text{ then } x<z.$$

However, it is not always true that

$$\text{if } x=y \text{ and } u<v, \text{ then } x+u<y+v.$$

5. Representation Error in the Proposed System

Usually, for a real number x, the representative point x^* of the interval containing x is obtained by rounding the next lower digit of the least significant representable digit of the mantissa of x. This convention is adopted for the level-0 numbers in our system, and the representative points of the numbers of higher levels are defined similarly.

The "representation error" of a number x is, then,

defined by

$$e_{\rm rep}(x) = \max_y \left\{ \frac{|x^* - y|}{|y|} \,\middle|\, y \text{ has the same representative point } x^* \text{ as } x \right\}. \qquad (5.1)$$

The "unit of rounding" of a number x is defined by

$$u(x) = \max_y \{ e_{\rm rep}(y) \mid y \text{ has the same exponent as } x \}. \qquad (5.2)$$

As is easily seen, for a floating-point number x with the base β and the L-digit mantissa, we have

$$u(x) = \frac{1}{2}\beta^{1-L} \qquad (5.3)$$

in the case where the round-to-the-nearest is adopted.

For the "non-number"s, it may be assumed that $e_{\rm rep}(x)$ is undefined, or defined to be ∞. In this respect, it is natural to take "0" and "± 0" for "non-number"s, rather than numbers.

The "representation error" as well as the "unit of rounding" is a good measure in terms of which to compare different manners of floating-point representation. We shall consider here the following four typical floating-point number systems for comparison. For convenience' sake, the comparison will be made using the 64-bit word.

(i) IBM-type:———The base is 16. The mantissa has 56 bits. The rounding is "chopping". The exponent has 7 bits ("biased" notation). The sign of the mantissa occupies 1 bit. (This is the so-called DOUBLE-PRECISION number).

(ii) Kahan-type:———The base is 2. The mantissa has 52 bits (substantially, 53 bits since the first bit is suppressed). The rounding is the round-to-the-nearest-to-even. The exponent has 11 bits ("biased" notation). The sign of the mantissa occupies 1 bit. ("DOUBLE" [8].) "Denormalized numbers" are also used to resist the underflow to some extent.

(iii) Morris-type:———The adaptation of his original proposal with the 36-bit word for the 64-bit word. The base is 2. The mantissa and the exponent use 61 bits. There is a pointer, G (3 bits), to the boundary between the field of the mantissa and that of the exponent. The rounding is the round-to-the-nearest. The length of the exponent field is specified by G as

(i) $G+1$ (Morris' type (i)),

or

(ii) $G+4$ (Morris' type (ii)).

(iv) Our system:———See Chap. 3 for the details. (We mainly consider the level 0. The rounding is the round-to-the-nearest as in Chap. 2.)

The normalization condition is (2.2) for (i) and (iii), and (2.3) for (ii) and (iv).

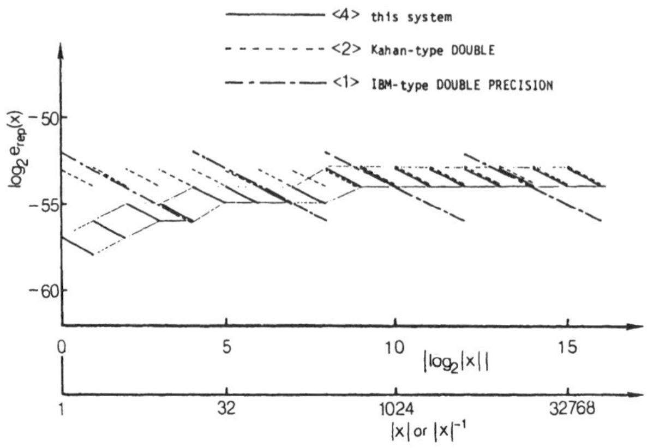

Fig. 3 Representation error $e_{\rm rep}(x)$ of various systems.

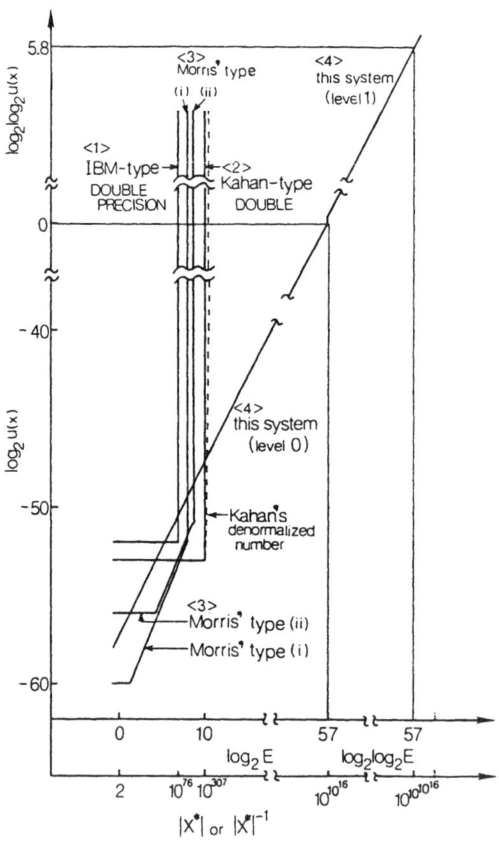

Fig. 4 Units of rounding $u(x)$ for various systems.

In Fig. 3 the representation error $e_{\rm rep}(x)$ for the numbers whose absolute value is neither too large nor too small is shown. In Fig. 4 the unit of rounding $u(x)$ is shown for a wider range of numbers.

As is seen in Fig. 3 and Fig. 4, the proposed representation of numbers has no "discontinuity" in precision, but the representation error "continuously" depends on the magnitude of a number. Furthermore, higher precision than the IBM-type or the Kahan-type is realized for numbers of moderate magnitude.

6. Numerical Examples Computed by Means of the Proposed System

6.1 Experimental Programmes

The proposed floating-point number system is implemented in the manner shown in Chap. 3 as a system of subroutines and functions written in FORTRAN on the HITAC 8800/8700 Computer Complex of the Computer Centre of the University of Tokyo. The experimental system of programmes consists of

(1) the subroutines to perform the four arithmetic operations (approximately 300 lines in total);
(2) the function subprogrammes to perform the input/output of numbers (approximately 200 lines in total);
(3) the function subprogrammes to perform the type conversion of the data in FORTRAN into those in our system and vice versa (approximately 200 lines in total);
(4) the function subprogrammes such as square root, exponential, and logarithm (approximately 200 lines in total);
(5) the auxiliary subroutines for comparison, rounding and pack/unpack operation of the mantissa and the exponent (approximately 300 lines in total).

All the non-numbers mentioned in Chap. 4 can be treated in the current version.

These subroutines and functions are merely for experimental use, so that they are not fast; e.g., for usual numerical computations, they run 600 through 1,000 times as slowly as the conventional FORTRAN statements on the HITAC system. A considerable speed-up, however, can be achieved by coding those subprogrammes in an assembly language. Moreover, it is theoretically possible that the proposed system runs as fast as the IBM-type or the Kahan-type number system if special-purpose arithmetic units are incorporated in the CPU.

6.2 Numerical Examples of Practical Computations

In this section, we show the effectiveness of the proposed number system by applying it to practical computational problems. There are so many difficult problems to solve by a simple program because both very large and small numbers appear in the course of computation. For example, the Graeffe method for algebraic equations is difficult to implement because coefficients may become very large or small quickly, even if the zeroes are neither very large nor very small. In order to solve these problems by a computer with a conventional floating-point number system such as that of the IBM-type, a highly sophisticated program equipped with a number of tricks to protect against overflow/underflow will be required.

The proposed system enables the users to solve those difficult problems by simple programs, and the solutions attained are generally no less accurate than those attained by complicated (or sophisticated) programs with conventional floating-point number systems.

6.2.1 Graeffe Method

Let us consider the solution of algebraic equations by the Graeffe method [18], where both very large and very small numbers often appear in the course of computation. For simplicity, we restrict our attention to the polynomials with all the zeroes positive real.

Let α_i $(i=1,\cdots,n)$ be the zeroes of a given polynomial

$$P(x)=x^n-a_{n-1}x^{n-1}+\cdots+(-1)^{n-1}a_1x+(-1)^na_0,$$

and let the polynomial with zeroes $\alpha_i^{2^\nu}$ $(i=1,\cdots,n)$ be

$$P^{(\nu)}(x)=x^n-a_{n-1}^{(\nu)}x^{n-1}+\cdots +(-1)^{n-1}a_1^{(\nu)}x+(-1)^na_0^{(\nu)}. \tag{6.1}$$

Then $a_i^{(\nu)}$ $(i=0,\cdots,n-1)$ satisfy the simple recurrence relation:

$$a_k^{(0)}=a_k, \tag{6.2}$$
$$a_k^{(\nu+1)}=\sum_{i+j=k}(-1)^{k+i}a_i^{(\nu)}a_j^{(\nu)}\quad(\nu=0,1,\cdots),$$

and the ν-th approximations to α_i are given by

$$\alpha_i\fallingdotseq(a_{i-1}^{(\nu)}/a_i^{(\nu)})^{1/2^\nu}. \tag{6.3}$$

Fig. 5 shows the iteration process using our system for the polynomial

$$\begin{aligned}P_1(x)&=x^8-11x^7+45.35x^6-88.55x^5+86.7542x^4\\&\quad-43.274x^3+10.984x^2-1.32x+0.0576\\&=(x-0.1)(x-0.2)(x-0.3)(x-0.4)(x-1)\\&\quad\times(x-2)(x-3)(x-4),\end{aligned} \tag{6.4}$$

and Fig. 6 shows that for the polynomial

$$\begin{aligned}&P_2(x)\\&=x^4-10.43857593020614x^3+40.58740567587410x^2\\&\quad-69.60408570545396x^2+44.36715614906059\\&=(x-2)(x-e)(x-\sqrt{7.4})(x-3)\\&\quad(e=2.718281828\cdots,\ \sqrt{7.4}=2.7202941017\cdots).\end{aligned} \tag{6.5}$$

In Fig. 5, all the zeroes are determined to a sufficient precision after 7 iterations ($\nu=7$). Note that the subsequent iterations do not deteriorate the accuracy. Iterations over the 6-th is impossible with the IBM-type representation of numbers.

In Fig. 6, it is seen that, due to the existence of nearly multiple zeroes, more iterations are required to get good approximations for zeroes, and that the final approximations are not very accurate. But, in our system, all the zeroes are determined after 16 iterations ($\nu=16$) with good accuracy, up to about 10 decimal digits. The IBM-type cannot proceed farther than the 6-th iteration, at which approximations are very poor—up to about 2 decimal places in the worst case. Neither can the Kahan-type proceed farther than the 7-th iteration at which it gives no better results than the IBM-type.

upper bound of the Kahan-type

```
v = 1
  3. 317760000000000 EE-3     =A( 1)
- 4. 770432000000000 EE-1     =A( 2)
  1. 639877247999999 EE 1     =A( 3)
- 1. 954100328000001 EE 2     =A( 4)
  8. 294775057599999 EE 2     =A( 5)
- 9. 027198200000001 EE 2     =A( 6)
  2. 820272999999999 EE 2     =A( 7)
- 3. 030000000000000 EE 1     =A( 8)
  8. 339569428721113 EE-2     =ROOT( 1)
  1. 705584280072372 EE-1     =ROOT( 2)
  2. 896891579541276 EE-1     =ROOT( 3)
  4. 853679799876546 EE-1     =ROOT( 4)
  9. 585743805733726 EE-1     =ROOT( 5)
  1. 7890847510520749 EE0     =ROOT( 6)
  3. 0508739212180358 EE0     =ROOT( 7)
  5. 5045435778091540 EE0     =ROOT( 8)
```

converged

```
v = 7
  2. 1581313729824 EE-159     =A( 1)
- 2. 15813137298240 EE-31     =A( 2)
  6. 34217809318084 EE 58     =A( 3)
- 5. 3792017006758 EE 125     =A( 4)
  4. 6455692578868 EE 176     =A( 5)
- 4. 6455692578870 EE 176     =A( 6)
  1. 3652101047498 EE 138     =A( 7)
- 1. 15792089237326 EE 77     =A( 8)
  9. 9999999999999997 EE-2    =ROOT( 1
  2. 0000000000000006 EE-1    =ROOT( 2
  2. 9999999999999984 EE-1    =ROOT( 3
  4. 0000000000000013 EE-1    =ROOT( 4
  9. 9999999999999996 EE-1    =ROOT( 5
  2. 0000000000000022 EE0     =ROOT( 6
  2. 9999999999999951 EE0     =ROOT( 7
  4. 0000000000000027 EE0     =ROOT( 8
```

```
v = 5
  2. 15535826671273 EE-40     =A( 1)
- 2. 15535826721458 9 EE-8    =A( 2)
  5. 01834615819841 EE 14     =A( 3)
- 2. 70846408510473 EE 31     =A( 4)
  1. 46811384664593 EE 44     =A( 5)
- 1. 46811384698750 EE 44     =A( 6)
  3. 41822664232871 EE 34     =A( 7)
- 1. 84485970981938 EE 19     =A( 8)
  9. 999999999927237 EE-2     =ROOT( 1)
  1. 999999855136427 EE-1     =ROOT( 2)
  2. 999990800367525 EE-1     =ROOT( 3)
  4. 000012555971386 EE-1     =ROOT( 4)
  9. 999999999927293 EE-1     =ROOT( 5)
  1. 9999998551364240 EE0     =ROOT( 6)
  2. 9999908003675364 EE0     =ROOT( 7)
  4. 0000125559713995 EE0     =ROOT( 8)
```

upper bound of the IBM-type

```
v = 8
  4. 6575310230509 EE-318     =A( 1)
- 4. 65753102305127 EE-62     =A( 2)
  4. 0223222965623 EE 117     =A( 3)
- 2. 8935810936553 EE 251     =A( 4)
  2. 1581313729823 EE 353     =A( 5)
- 2. 1581313729825 EE 353     =A( 6)
  1. 8637986301109 EE 276     =A( 7)
- 1. 3407807929945 EE 154     =A( 8)
  9. 9999999999999997 EE-2    =ROOT( 1
  2. 0000000000000006 EE-1    =ROOT( 2
  2. 9999999999999984 EE-1    =ROOT( 3
  4. 0000000000000013 EE-1    =ROOT( 4
  9. 9999999999999976 EE-1    =ROOT( 5
  2. 0000000000000022 EE0     =ROOT( 6
  2. 9999999999999951 EE0     =ROOT( 7
  4. 0000000000000027 EE0     =ROOT( 8
```

```
v = 6
  4. 64556925788690 EE-80     =A( 1)
- 4. 64556925788701 EE-16     =A( 2)
  2. 51836814092956 EE 29     =A( 3)
- 7. 33430419960579 EE 62     =A( 4)
  2. 15535826671270 EE 88     =A( 5)
- 2. 15535826671276 EE 88     =A( 6)
  1. 16842205763348 EE 69     =A( 7)
- 3. 40282370354638 EE 38     =A( 8)
  9. 9999999999999996 EE-2    =ROOT( 1)
  1. 999999999999838 EE-1     =ROOT( 2)
  2. 999999999527234 EE-1     =ROOT( 3)
  4. 000000000630681 EE-1     =ROOT( 4)
  9. 9999999999999996 EE-1    =ROOT( 5)
  1. 9999999999998341 EE0     =ROOT( 6)
  2. 9999999995272459 EE0     =ROOT( 7)
  4. 0000000006306702 EE0     =ROOT( 8)
```

```
v = 30
  2. 298526 EE-1330986224     =A( 1)
- 2. 2985271 EE-257244400     =A( 2)
  5. 4763732 EE 493268927     =A( 3)
- 2. 670771 EE 1054705705     =A( 4)
  1. 516089 EE 1481990536     =A( 5)
- 1. 516090 EE 1481990536     =A( 6)
  3. 612177 EE 1158762039     =A( 7)
- 1. 7616143 EE 646456993     =A( 8)
  9. 9999999999999999 EE-2    =ROOT( 1
  2. 0000000000000006 EE-1    =ROOT( 2
  2. 9999999999999984 EE-1    =ROOT( 3
  4. 0000000000000013 EE-1    =ROOT( 4
  9. 9999999999999994 EE-1    =ROOT( 5
  2. 0000000000000024 EE0     =ROOT( 6
  2. 9999999999999953 EE0     =ROOT( 7
  4. 0000000000000027 EE0     =ROOT( 8
```

Fig. 5 Iteration by the Graeffe method for $P_1(x)$.

upper bound of the Kahan-type

```
v = 1
  1. 968444544755123 EE 3     =A( 1)
- 1. 243233216278632 EE 3     =A( 2)
  2. 829367444181849 EE 2     =A( 3)
- 2. 778905609893071 EE 1     =A( 4)
  1. 5833268601423515 EE0     =ROOT( 1)
  2. 0961947379335459 EE0     =ROOT( 2)
  3. 1908602092232556 EE0     =ROOT( 3)
  5. 2715326138544105 EE0     =ROOT( 4)
```

```
v = 7
  6. 6662837510202 EE 210     =A( 1)
- 1. 9590447225817 EE 172     =A( 2)
  9. 6227697791204 EE 116     =A( 3)
- 1. 17902661948570 EE 61     =A( 4)
  1. 9999999999999518 EE0     =ROOT( 1)
  2. 7045783330773281 EE0     =ROOT( 2)
  2. 7340770775614733 EE0     =ROOT( 3)
  3. 0000001622446730 EE0     =ROOT( 4)
```

upper bound of the IBM-type

```
v = 5
  5. 08125451548211 EE 52     =A( 1)
- 1. 18320169534957 EE 43     =A( 2)
  3. 02539330106157 EE 29     =A( 3)
- 2. 01284256209763 EE 15     =A( 4)
  1. 9999931360575824 EE0     =ROOT( 1)
  2. 6592468559773728 EE0     =ROOT( 2)
  2. 7735140870227926 EE0     =ROOT( 3)
  3. 0077660744101009 EE0     =ROOT( 4)
```

converged

```
v = 16
  6. 7375701284 EE 107941     =A( 1)
- 3. 36284975669 EE 88213     =A( 2)
  4. 01365489311 EE 59751     =A( 3)
- 4. 15479228694 EE 31268     =A( 4)
  1. 9999999999999518 EE0     =ROOT( 1
  2. 7182818285951764 EE0     =ROOT( 2
  2. 7202941016100686 EE0     =ROOT( 3
  3. 0000000000009399 EE0     =ROOT( 4
```

```
v = 6
  2. 5819147451107 EE 105     =A( 1)
- 1. 39965879601065 EE 86     =A( 2)
  4. 38981722472593 EE 58     =A( 3)
- 3. 44645651957942 EE 30     =A( 4)
  1. 9999999998191536 EE0     =ROOT( 1)
  2. 6899448172248597 EE0     =ROOT( 2)
  2. 7487913512249693 EE0     =ROOT( 3)
  3. 0001740482911321 EE0     =ROOT( 4)
```

```
v = 30
  1. 577772 EE 1768518918     =A( 1)
- 3. 759240 EE 1445290421     =A( 2)
  2. 5108124 EE 978970272     =A( 3)
- 2. 0511859 EE 512305045     =A( 4)
  1. 9999999999999518 EE0     =ROOT( 1
  2. 7182818255951764 EE0     =ROOT( 2
  2. 7202941016100686 EE0     =ROOT( 3
  3. 0000000000009399 EE0     =ROOT( 4
```

Fig. 6 Iteration by the Graeffe method for $P_2(x)$.

In Fig. 7, the maximum relative errors of computed zeroes together with the maximum magnitudes of the coefficients are plotted against the iteration numbers.

6.2.2 Binomial Probability

Let us consider the problem of calculating the binomial probability

$$x_k=\binom{N}{k}p^kq^{N-k} \quad (p+q=1) \tag{6.6}$$

for a large N, say $N=2{,}000$ [16]. An N as large as 2,000 usually appears in practical problems of inventory control.

A simple program such as that shown in Fig. 8 will

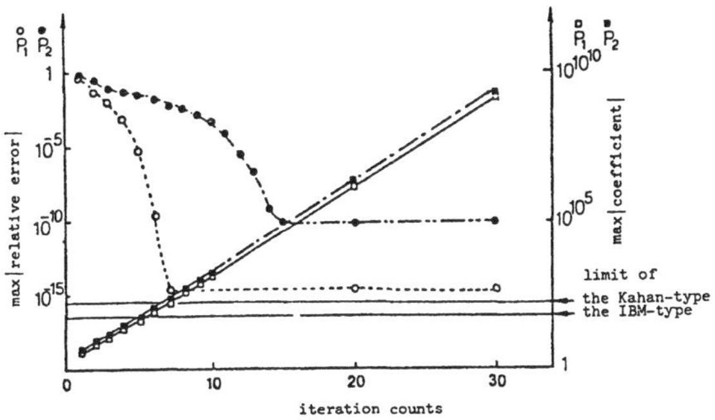

Fig. 7 Iteration by the Graeffe method for $P_1(x)$ and $P_2(x)$.

```
program 1
function Binomial(N,k:integer;p,q:real):real;
real: x; integer: i;
begin x:=1.0;
   for i:=1 to k do x:=x*(N-k+i)/i*p;
   for i:=1 to N-k do x:=x*q;
   return x
end
```

Fig. 8 A simple program to calculate binomial probability.

```
1.148130695274 EE-1398   = X(   0)
9.185045562194 EE-1395   = X(   1)
3.672181215765 EE-1391   = X(   2)
9.782690758799 EE-1388   = X(   3)

2.224107084496770 EE-2   = X(1599)
2.229667352208012 EE-2   = X(1600)
2.228274680532679 EE-2   = X(1601)
2.219929082478250 EE-2   = X(1602)
```

Fig. 9 Example of binomial probability computed by our system (N=2,000, p=0.8, q=0.2).

certainly fail because of overflow/underflow on a conventional system.

On a conventional system, a more sophisticated program which incorporates elaborate scaling techniques will be needed to get an exact answer avoiding overflow/underflow (see [16] for an example of complete program of that kind).

If we use the proposed floating-point number system, any simple program will work well, and even faster than the sophisticated one. Examples of binomial probabilities computed by the proposed system are shown in Fig. 9.

6.2.3 Divergence-convergence Boundary of a Quadratic Transformation

Let us consider the problem of calculating the divergence-convergence boundary of the quadratic transformation

$$x'^{\kappa}=\sum_{\lambda=1}^{n}\sum_{\mu=1}^{n}P_{\lambda\mu}^{\kappa}x^{\lambda}x^{\mu}\quad(\kappa=1,2,\cdots,n).\qquad(6.7)$$

This kind of transformation characterizes the asymptotic behaviour of the Newton-Raphson process near the solution (see [3] for details).

For $n=2$, the divergence-convergence boundary for the fixed direction θ is calculated by the algorithm shown in Fig. 10 [3]. The restart step, i.e., step 3.1, of the algorithm is used for avoiding overflow/underflow, and the presence of this step increases the total computing time.

```
1°    r0 ← 1
2°    x0 ← r0cos θ ;  y0 ← r0sin θ ;  m ← 1
3°    xm ← f^1(xm-1,ym-1)
      ym ← f^2(xm-1,ym-1)
      rm ← √(xm^2 + ym^2)
3.1°  if  rm > 10^30  or  rm < 10^-30
         then  r0 ← r0/rm^(2^-m) ;  goto 2°
3.2°  if  m=15  then  goto  4°
         else  m ← m+1;  goto  3°
4°    r*(divergence-convergence boundary) ← r0/r15^(2^-15)
```

Fig. 10 Procedure to calculate the divergence-convergence boundary of the quadratic transformation (n=2).

Table 5 Comparison of the performance between the IBM-type and our system for a quadratic transformation.

p_1	p_2	IBM-type		Our System	
		#restart	#total iterations	#restart	#total iterations
0.0	0.0	0	75,000	0	75,000
0.0	0.1	4,742	134,706	0	75,000
0.0	1.2	7,586	155,816	0	75,000
0.0	10.0	10,000	169,332	0	75,000
3.7	2.5	10,000	180,666	0	75,000

Since the proposed system can handle very wide ranges of numbers, no restart step is required, so that the total computing time is not increased.

In Table 5, comparison of the performances of the IBM-type number system and the proposed number system is shown for the quadratic transformation

$$x'=f^1(x,y)=-2xy+2/3x(p_1x+p_2y),$$
$$y'=f^2(x,y)=x^2+y^2+2/3y(p_1x+p_2y).\qquad(6.8)$$

The divergence-convergence boundary of the transformation (6.8) is calculated for $\theta=\theta_i=2\pi i/10{,}000$ ($i=1,\cdots,5{,}000$) to sufficient accuracy with the same stopping rule on the two systems. Table 5 indicates that the IBM-type system requires two restart steps for one θ_i on the average, while our system requires no restart step and the total iteration is nearly halved compared with the IBM-type.

6.2.4 6-j Symbols (Racah Symbols)

Many problems in quantum mechanics are formulated in terms of angular momenta and their combinations. 6-j symbols (Racah symbols), as defined below, are the coefficients used for combining angular momenta [9].

Definition. 6-j symbols $\begin{Bmatrix} j_1 j_2 j_3 \\ l_1 l_2 l_3 \end{Bmatrix}$:

$$\begin{Bmatrix} j_1 j_2 j_3 \\ l_1 l_2 l_3 \end{Bmatrix}\equiv\Delta(j_1j_2j_3)\Delta(j_1l_2l_3)\Delta(l_1j_2l_3)\Delta(l_1l_2j_3)w\begin{Bmatrix} j_1 j_2 j_3 \\ l_1 l_2 l_3 \end{Bmatrix},$$

$$\Delta(abc)\equiv\sqrt{\frac{(a+b-c)!(a-b+c)!(-a+b+c)!}{(a+b+c+1)!}},$$

$$w\begin{Bmatrix} j_1 j_2 j_3 \\ l_1 l_2 l_3 \end{Bmatrix} \equiv \sum_z \frac{(-1)^z (z+1)!}{F(z, j_1, j_2, j_3, l_1, l_2, l_3)},$$

$$\begin{aligned} F(z, j_1, j_2, j_3, l_1, l_2, l_3) & \\ \equiv (z-j_1-j_2-j_3)!(z-j_1-l_2-l_3)!(z-l_1-j_2-l_3)! & \\ \times (z-l_1-l_2-j_3)!(j_1+j_2+l_1+l_2-z)! & \\ \times (j_2+j_3+l_2+l_3-z)!(j_3+j_1+l_3+l_1-z)!, & \end{aligned}$$

where j'_*s and l'_*s are nonnegative half integers such that the three arguments appearing in one and the same $\Delta(a\,b\,c)$ may satisfy the triangular inequality, and the summation $\sum_z$ is taken for all integers such that the arguments of factorials may be nonnegative.

The calculation of the 6-j symbols has been tried by many researchers. There are also some explicit formulas for the 6-j symbols with relatively small momenta. Numerical tables are available for the angular momenta not exceeding eight [9, 15].

For nuclear physics, the 6-j symbols with small arguments are sufficient, and they are easy to compute directly from the definition. But, for some problems such as ion-beam scattering, we have to combine rather large angular momenta. In that case, a simple program does not work due to overflow and underflow if we use a conventional floating-point system. The proposed system enables us to run a simple program without any trick to get satisfactory results.

In Fig. 11, a few examples are shown of the computation of 6-j symbols with rather large angular momenta by means of the proposed system.

6.2.5 Newton-Raphson Iteration

The Newton-Raphson iteration for the solution of nonlinear equations is often embarrased by overflow/underflow. As is well known, this iteration process converges quadratically to a solution so long as the initial approximation is sufficiently close to the solution, and, in such a case, no overflow/underflow occurs. However, if the initial approximation is far from the solution, the convergence is slow, or, sometimes, a big jump takes place, which may cause overflow/underflow. However, the proposed system will work well even in those cases. Some discussions are made in [12].

$$\begin{Bmatrix} 10 & 10 & 10 \\ 10 & 10 & 10 \end{Bmatrix} = -2.919186780609217 \text{ EE-3}$$

$$\begin{Bmatrix} 20 & 20 & 20 \\ 20 & 20 & 20 \end{Bmatrix} = -5.029406456868522 \text{ EE-3}$$

$$\begin{Bmatrix} 30 & 30 & 30 \\ 30 & 30 & 30 \end{Bmatrix} = 4.102353215969380 \text{ EE-4}$$

$$\begin{Bmatrix} 40 & 40 & 40 \\ 40 & 40 & 40 \end{Bmatrix} = 1.828306972165276 \text{ EE-3}$$

$$\begin{Bmatrix} 50 & 50 & 50 \\ 50 & 50 & 50 \end{Bmatrix} = -1.121374316603310 \text{ EE-4}$$

$$\begin{Bmatrix} 60 & 60 & 60 \\ 60 & 60 & 60 \end{Bmatrix} = -1.006636256306322 \text{ EE-3}$$

Fig. 11 Examples of 6-j symbols for large angular momenta.

7. Concluding Remarks

In this paper, we proposed basic ideas of a new floating-point number system, compared it with the conventional ones, and gave examples of its implementation and numerical computations by means of it. The following problems are to be further investigated.

(i) Hardware implementation:———The advantages of our proposal will be more clearly proved by implementing it by hardware arithmetic circuits. The design and the actual implementation are being done. Theoretically, it is possible to realize as fast a speed as that realized by the conventional floating-point arithmetic circuits.

(ii) Rounding-error analysis:———We have discussed only the approximation error occurring when we represent a real number in a fixed-length word. The well-known analysis on the basis of the "fixed-length mantissa" [18] does not apply to our system. An analysis technique must be developed which can deal not only with numbers but also with non-numbers, and which is also capable of considering propagation, accumulation, and magnification of rounding errors in the process of numerical computation.

(iii) Practical implementation of higher levels and non-numbers:———Input/output format of higher-level numbers and non-numbers must be standardized. How many levels to consider will also be investigated from the theoretical as well as practical standpoint.

Acknowledgement

The authors are grateful to Professor Sigeiti Moriguti of the University of Electro-Telecommunication, Professor Emeritus of the University of Tokyo, and Professor Hideo Aiso of Keio University for their valuable suggestions and comments given to the authors at informal meetings. The authors are grateful also to the members of their research group for helpful discussions. Those suggestions, comments and discussions were very helpful to the authors in improving the present paper.

Reference

1. Brent, R. P. On the Precision Attainable with Various Floating-Point Number System. *IEEE Trans. on Computers*, **C-22**, 6 (1973), 601–607.
2. Cody, W. J., Jr. Static and Dynamic Numerical Characteristics of Floating-Point Arithmetic. *IEEE Trans. on Computers*, **C-22**, 6 (1973), 598–601.
3. Date, T. Properties of Divergence-Convergence Boundary of Quadratic Transformations (in Japanese). *Journal of Information Processing Society of Japan*, **19**, 6 (1978), 507–513.
4. Edger, A. D. and Lee, S. C. FOCUS: Microcomputer Number System. *Comm. ACM*, **22**, 3 (1979), 166–177.
5. Hamming, R. W. On the Distribution of Numbers. *The Bell System Technical Journal*, **40**, 8 (1970), 1609–1625.
6. Hehner, E. C. R. and Horspool, R. N. S. A New Representation of the Rational Numbers for Fast Easy Arithmetic. *SIAM Journal on Computing*, **8**, 2 (1979), 124–134.

7. HITOTUMATU, S. Proposal of a New Standard Floating Point Arithmetic System (in Japanese). *Journal of Information Processing Society of Japan*, **20**, 9 (1979), 793–797.
8. IDA, T. and GOTO, E. Overflow Free and Variable Precision Computing in FLATS. *Journal of Information Processing*, **1**, 3 (1978), 140–142.
9. JUDD, B. R. *Angular Momentum Theory for Diatomic Molecules*. Academic Press, New York, N.Y., 1975.
10. KAHAN, W. and PALMER, J. On a Proposed Floating-Point Standard. *ACM SIGNUM Newsletter*, Special Issue (Oct. 1979), 13–21.
11. KNUTH, D. E. *The Art of Computer Programming*, 2: *Seminumerical Algorithms*. Addison-Wesley, Reading, Mass., 1969.
12. MATSUI, S. *On Numerical Computation Oriented Floating-Point Number Systems* (in Japanese). Master Thesis, University of Tokyo, March 1981.
13. MORRIS, R. Tapered Floating Point: A New Floating-Point Representation. *IEEE Trans. on Computers*, **C-20**, 6 (1973), 1678–1679.
14. REINSH, C. H. Principle and Preference for Computer Arithmetic. *ACM SIGNUM Newsletter*, **14**, 1 (1979), 12–27.
15. ROTENBERG, et al. *The 3-j and 6-j Symbols*. Technology Press, MIT, Cambridge, Mass., 1959.
16. STERBENZ, P. H. *Floating-Point Computation*. Prentice-Hall, Englewood Cliffs, N.J., 1974.
17. SWARTZLANDER, E. E. and ALEXOPOULOS, A. G. The Sign/Logarithm Number System. *IEEE Trans. on Computers*, **C-24**, 12 (1975), 1238–1243.
18. WILKINSON, J. H. *Rounding Errors in Algebraic Processes*. Prentice-Hall, Englewood Cliffs, N.J., 1963.

Reprinted from *Proceedings of Eighth Symposium on Computer Arithmetic*, 1987, pages 139-143.

A CLOSED COMPUTER ARITHMETIC

F. W. J. OLVER

Institute for Physical Science and Technology, University of Maryland,
College Park, Maryland 20742, U.S.A.

ABSTRACT

Two closely related new systems of computer arithmetic are proposed. It is shown that both are closed under arithmetic operations in finite-precision arithmetic, thereby offering a permanent solution to the problems of overflow and underflow. Other advantages of the new systems pertaining to precision are described, and there is also a brief discussion of possible ways of hardware implementation.

1. Introduction and Summary

The floating-point (fℓp) system was adopted almost universally for computer arithmetic soon after the introduction of electronic computers in the early 1950's. Since then it has served the computing community remarkably well. Nevertheless there are persistent problems associated with the system. If this were not the case, then it would not have taken a large and distinguished committee many years to formulate the recent IEEE standard;[11] indeed, there may have been no need for such a standard.

Foremost among the problems associated with the fℓp system is failure caused by overflow or underflow. Anyone who has attempted to produce robust software is painfully aware of these problems; see, for example, references 1, 5. The reason overflow and underflow occur, of course, is that the fℓp system is not closed: starting from any two representable numbers it is always possible to generate numbers that lie outside the representable set by a finite number of additions, subtractions, multiplications or divisions, excluding division by zero. All modifications of the fℓp system, including for example those developed by Hull and his co-workers[9] and by Matsui and Iri,[14] also inevitably suffer from lack of closure.

Does a closed system of arithmetic exist? If it does, then we would expect it to be more complicated and costlier to execute than the fℓp system. As a result, the question has remained largely of academic interest. However, phenomenal advances in computer technology suggest that implementation of a closed system in an economic manner might now be feasible. Furthermore, although execution speed of arithmetic operations is always bound to be slower, some of the loss may be recoverable by the use of simpler algorithms. Human effort, too, would be saved by simplifications in the construction and debugging of programs.

In §2 we describe a system of computer arithmetic, called *level-index* (ℓi) arithmetic, together with a modification called *symmetric level-index* (sℓi) arithmetic. This section also outlines algorithms for implementing arithmetic operations in these systems.

In §3 we prove that the ℓi and sℓi systems are closed.

In §4 we discuss various aspects of the precision of the new systems, including some comparisons with the precision of the fℓp system. §5 supplies an illustrative example. In §6 we discuss briefly hardware implementation, and in the final section, §7, we present our main conclusions.

For fuller treatments of the topics in this paper, the reader may consult references 2, 3 in the case of §2, reference 13 in the case of §§3-5, and references 16, 19 in the case of §6.

2. The ℓi and sℓi Systems

The ℓi system is based on the mapping

$$X = \pm\phi(x), \tag{2.1}$$

and its inverse

$$x = \psi(|X|). \tag{2.2}$$

Here X is any real number, and its image (or "ℓi image") x is a nonnegative number that is stored internally in the computer (along with the sign in (2.1)) to represent X. The functions ϕ and ψ are "generalized" exponentials and logarithms, defined by

$$\phi(x) = x \quad (0 \le x < 1), \tag{2.3}$$

$$\phi(x) = e^{\phi(x-1)} \quad (x \ge 1), \tag{2.4}$$

$$\psi(X) = X \quad (0 \le X < 1), \tag{2.5}$$

$$\psi(X) = \psi(\ln X) + 1 \quad (X \ge 1). \tag{2.6}$$

Both $\phi(x)$ and $\psi(X)$ are increasing functions, with continuous first derivatives when $0 \le x < \infty$ and $0 \le X < \infty$.

The image x is stored in fixed absolute precision. Its integer part ℓ and fractional part f, say, are respectively the *level* and *index* of X. The reason for these names is that in consequence of (2.1), (2.3) and (2.4) X can be expressed explicitly in the form

$$X = \pm e^{e^{\cdot^{\cdot^{\cdot^{e^f}}}}},$$

where the exponentiations are carried out ℓ times. (When $\ell = 0$, we have $X = \pm f$.)

Suppose that we wish to subtract two numbers in the ℓi system. That is, we are given their images x and y, with $x \geq y \geq 0$, and we need to find z to satisfy the equation

$$\phi(z) = \phi(x) - \phi(y).$$

Let the result of decomposing x, y and z into their integer and fractional parts be denoted by

$$x = \ell + f, \quad y = m + g, \quad z = n + h.$$

If $\ell = m = 0$, then the calculation of n and h is trivial: $n = 0$ and $h = f - g$. If $\ell > 0$, then we compute three sequences $\{a_j\}$, $\{b_j\}$ and $\{c_j\}$, the members of which are defined by

$$a_j = 1/\phi(x-j), \quad b_j = a_j\phi(y-j), \quad c_j = a_j\phi(z-j).$$

The a_j are required for $j = \ell - 1, \ell - 2, \ldots, 0$. The b_j are required for $j = m - 1, m - 2, \ldots, 0$, except that we need b_0 also when $m = 0$. The sequence c_j, $j = 0, 1, \ldots$ is terminated when $c_j < a_j$ or $j = \ell - 1$, whichever comes first.

Recurrence relations and initial values are supplied by

$$a_{j-1} = e^{-1/a_j}, \quad a_{\ell-1} = e^{-f},$$

$$b_{j-1} = e^{-(1-b_j)/a_j},$$

$$b_{m-1} = a_{m-1}e^g \ (m \geq 1), \quad b_0 = a_0 g \ (m = 0),$$

$$c_j = 1 + a_j \ln c_{j-1}, \quad c_0 = 1 - b_0.$$

If $c_j (0 \leq j \leq \ell-1)$ is the first member of its sequence to satisfy $c_j < a_j$, then

$$n = j, \quad h = c_j/a_j,$$

However, if $c_j \geq a_j$ for $j = 0, 1, \ldots, \ell - 1$, then

$$n = \ell, \quad h = f + \ln c_{\ell-1}.$$

These relations are derivable from the definitions (2.1) to (2.6). It should be noted, incidentally, that the recurrence relations for the b_j and c_j are the same, but they are applied in opposite directions. Each of the quantities a_j, b_j and c_j lies in the interval $[0,1]$, and in reference 3 an error analysis is given to show that the algorithm can be executed in fixed-point arithmetic.

The algorithm for addition is similar to that for subtraction; the main change is to replace the equation $c_0 = 1 - b_0$ by $c_0 = 1 + b_0$. Multiplication and division are equivalent to addition and subtraction at a lower level. For example, if

$$\phi(z) = \phi(x)\phi(y), \tag{2.7}$$

with $x \geq 1$ and $y \geq 1$, then on taking logarithms we have

$$\phi(z-1) = \phi(x-1) + \phi(y-1). \tag{2.8}$$

Further details will be found in reference 3.

At level zero the ℓi system functions as a fixed-point system; compare (2.3). Often, however, we wish to preserve the relative precision of small numbers; in consequence an alternative way of representing numbers in the interval $[0,1]$ is needed. The $s\ell i$ system achieves this by using reciprocals of ℓi numbers, just as the $f\ell p$ system attains the same end by use of negative exponents. Thus in the $s\ell i$ system we have

$$X = \pm\{\phi(x)\}^{\pm 1},$$

the ambiguous sign in the exponent being positive or negative according as $|X| \gtrless 1$. Equivalently, the $s\ell i$ system can be represented by the mappings

$$X = \pm\Phi(x), \quad x = \Psi(|X|),$$

in which

$$\Phi(x) = 1/\phi(1-x) \quad (x < 0),$$

$$\Phi(x) = \phi(1+x) \quad (x \geq 0),$$

$$\Psi(X) = 1-\psi(1/X) \quad (0 < X < 1),$$

$$\Psi(X) = \psi(X) - 1 \quad (X \geq 1),$$

and $\phi(\cdot)$ and $\psi(\cdot)$ are defined as above. Again, both $\Phi(\cdot)$ and $\Psi(\cdot)$ are increasing functions, with continuous first derivatives, in their intervals of definition. Algorithms for arithmetic operations in the $s\ell i$ system are obtainable by modification of those for the ℓi system; see reference 4.

3. Proof of Closure

To fix ideas, let us suppose that the internal arithmetic base of the computer is r, and the fractional parts of the ℓi images are stored to d r-nary places. Then if we add any representable number $\phi(x)$ to itself the stored ℓi image of the sum will not exceed x, provided that

$$\phi(x+cr^{-d}) > 2\phi(x). \tag{3.1}$$

Here c is a positive constant: if the ℓi arithmetic processor computed to infinite precision, then we would have $c = 1$ for chopping, and $c = \frac{1}{2}$ for rounding. By using a sufficient number of guard digits in the processor we can approach these values arbitrarily closely, but the precise value of c is not really important. On taking logarithms the inequality (3.1) becomes

$$\phi(x-1+cr^{-d}) > \phi(x-1) + \ln 2.$$

If $x > 2$, then $\phi'(x-1)$ is increasing. Hence by application of the mean-value theorem we see that the last inequality is satisfied when

$$cr^{-d}\phi'(x-1) \geq \ln 2.$$

This is equivalent to

$$x \geq \theta(c^{-1}r^{d}\ln 2),$$

where $\theta(t)$ denotes the root of the equation

$$\phi'(x-1) = t.$$

Sample values of $\theta(t)$, computed with the aid of Newton's rule, are as follows:

$$\theta(2^{32}) = 5.06\ldots, \qquad \theta(2^{64}) = 5.26\ldots,$$

$$\theta(2^{1024}) = 5.63\ldots, \qquad \theta(2^{5,502,869}) = 6.00\ldots\ .$$

Consider now the set A of all numbers generated by the addition or subtraction of any two numbers, beginning with any pair of numbers whose ℓi images are less than $\theta(c^{-1}r^{d}\ln 2)$. Assume that the ℓi images of the members of A are generated by arithmetic operations in the ℓi system and stored to d r-nary places. In consequence of the result just proved, the absolute values of the members of A are bounded by $\phi\{\theta(c^{-1}r^{d}\ln 2)\}$. If we now extend the arithmetic processes to include multiplication and division, other than division by zero, then the set of numbers so obtained is bounded in absolute value by $\phi\{\theta(c^{-1}r^{d}\ln 2) + 1\}$. This is a consequence of the equivalence of multiplication and division to addition and subtraction at one level below; compare (2.7) and (2.8). See also reference 13.

We have therefore shown that when the ℓi system is used with any finite-precision arithmetic, there is a subset of its set of representable numbers that is closed under the operations of addition, subtraction, multiplication and division, excluding division by zero. Obviously, the same conclusion also applies to the $s\ell i$ system.

With the aid of the numerical values of $\theta(t)$ quoted above, we see that when $r = 2$ and $d = 32$, that is, with 32 bits assigned to the storage of the fractional part of the ℓi image, an upper bound for numbers generated by addition and subtraction is $\phi(5.07)$, whether the abbreviation mode be chopping or rounding. For multiplication and division the corresponding bound is $\phi(6.07)$. To raise the overall upper bound from $\phi(6.07)$ to $\phi(7)$, or more, we would need a wordlength in excess of 5,500,000 bits. Thus in practice levels beyond 6 will not be entered; in consequence *it will always suffice to allocate 3 bits to the storage of the integer part of the* ℓi *image.*

4. Precision

The main advantage of the ℓi and $s\ell i$ systems over other systems is the closure property established in the preceding section. However, there are also advantages pertaining to precision.

In the first place, because the mapping functions $\phi(x)$, $\psi(X)$, $\Phi(x)$ and $\Psi(X)$ have continuous derivatives, the systems are free from the annoying phenomenon of wobbling precision.[6]

Next, there is no need to introduce gradual underflow,[7,8] with its injurious effects on error analysis. In a sense, gradual underflow is already incorporated in a natural manner in the ℓi system at level zero, and in the $s\ell i$ system underflow is absent because of the closure property.

As for error analysis, the ℓi and $s\ell i$ systems have their own error measure, namely the absolute error in the ℓi and $s\ell i$ images. This measure is called *generalized precision*,[2] since it reduces to absolute error at level zero (in ℓi) and to relative error, or more precisely relative precision,[15] at level one (in ℓi). Moreover, in contrast to relative error, generalized precision has the elegant and useful property of being a metric.

Lastly, in order to compare directly local precisions* of the $f\ell p$ and ℓi (or $s\ell i$) systems for a given wordlength, we need to convert from generalized precision to relative precision. This comparison is carried out in some detail in reference 13, and we merely sketch the main results in the case of the ℓi system.

For a given mantissa length the relative precision of the $f\ell p$ system is constant, within a factor r (the internal base). For a given index length the relative precision of the ℓi system varies. Its measure is least for numbers $X = \phi(x)$ at level one, and increases steadily as X enters higher levels. At $X = 1$, the relative precision of the ℓi system is better than that of the IEEE $f\ell p$ system[11] by a factor of 32 in single precision and 256 in double precision. The relative precisions become approximately the same in the ranges $2^{11} < X < 2^{18}$ (single precision) and $2^{44} < X < 2^{70}$ (double precision). At higher values of X the $f\ell p$ system gains, up to a factor of about 37 just before overflow in single precision $(X = 2^{127})$, and of about 68 just before overflow in double precision $(X = 2^{1023})$. In consequence of these variations the comparative accuracy of results computed in the two systems will depend on the magnitude of these results as well as on the magnitudes of the numbers that appear during the intermediate steps. Unless most of these numbers happen to be in the upper part of the representable $f\ell p$ ranges (in which case the danger of overflow is increased),

*By "local precision" we mean the distance between consecutive representable numbers.

the ℓi system can be expected to yield the higher accuracy.

After X increases beyond the overflow limits, the fℓp system fails. The ℓi system still functions and the local relative precision continues to increase, becoming of order unity at $X = \phi\{\theta(c^{-1}r^{d}\ell n\ 2)\}$ that is, at "infinity" for the ℓi processes of addition and subtraction (§3). It might be argued that this gradual erosion of relative precision renders the ℓi system ineffective for very large numbers. However, relative precision is the appropriate error measure for the fℓp system, not the ℓi system. One might as well argue that the erosion of absolute precision that accompanied the change from the fixed-point system to the fℓp system would have rendered the fℓp system ineffective for large numbers.

5. Example

A good example of the power of the new arithmetics is afforded by the computation of the binomial probability distribution I(n,k,p), defined by

$$I(n,k,p) = \sum_{j=0}^{k} \binom{n}{j} p^{j}(1-p)^{n-j},$$

with $0 \le k \le n$ and $0 < p < 1$. Over six pages of Sterbenz's well-known book[18] are devoted to this problem. Sterbenz discusses failures caused by overflow and underflow in fℓp arithmetic, and provides an algorithm to overcome these difficulties, based on careful rescaling. The algorithm is for individual terms in the sum, and is quite complicated. In contrast, a simple recursive algorithm based on the relations $y_0 = (1-p)^n$,

$$y_j = \frac{n-j+1}{j}\,\frac{p}{1-p}\,y_{j-1} \qquad (j = 1,2,\dots k),$$

and

$$I(n,k,p) = \sum_{j=0}^{k} y_j,$$

can be executed straightforwardly in sℓi arithmetic and yields results of comparable accuracy to those generated by Sterbenz's algorithm.

The advantages of the sℓi algorithm are first that the construction and debugging of the program are less time-consuming. Secondly, the sℓi algorithm entails far fewer arithmetic operations, which offsets some of the increase in execution time needed for sℓi arithmetic.

Further details of this example, including computed results, will be found in reference 13.

6. Hardware Implementation

So far, implementations of the ℓi and sℓi systems have been made only in software. These implementations have been somewhat cumbersome because for convenience the steps of the arithmetic algorithms (§2) have been simulated in fℓp arithmetic, rather than the more natural fixed-point arithmetic.

The next stage will be the development of fast hardware processors for evaluating the requisite exponentials and logarithms. Given numbers a, b and c in fixed-point form we need to compute e^{-a}, $e^{-1/a}$, $e^{-(1-b)/a}$ and ℓn c, also in fixed-point form. On microcomputers algorithms of the CORDIC type (originated by J. E. Volder) might be useful; see, for example, references 17, 20. For mainframe computers, however, digital parallel algorithms of the kind used for fixed-point and fℓp arithmetic operations are likely to be more effective; see, for example, reference 10, §3.2, and reference 21, Chapter 3. Considerable progress on these lines has already been made by Johnsson and Krishnaswamy,[12] and Turner and Olver.[16,19]

The ultimate objective is to have the ℓi and sℓi systems encoded on a single silicon chip. Then at the outset of a program, programmers would be able to declare whether they wish the arithmetic to be executed in fℓp, ℓi, sℓi or some other arithmetic, e.g. that of reference 9, or even fixed-point arithmetic. This declaration would be analogous to those now made in FORTRAN pertaining to single or double precision, and real or complex variables.

Of course, it is unlikely that ℓi and sℓi arithmetic operations will ever be as fast as the corresponding operations in fℓp arithmetic. However, if the ratio of the speeds could be brought to within 10, or so, then the overall execution time of a program in ℓi or sℓi arithmetic might be only two or three times that of the same program in fℓp arithmetic. This is because only a part of program execution is spent on arithmetic operations, as a rule. Moreover, as indicated in §5, freedom from overflow and underflow means that it will be possible to implement simpler algorithms in the ℓi and sℓi systems than in the fℓp system. This advantage will help to offset further some of the speed loss of the arithmetic operations. It will also lighten human effort in constructing, testing and debugging programs and software.

7. Conclusions

We have described a new form of computer arithmetic, called the level-index system, together with a modification thereof, called the symmetric level-index system. In contrast to existing systems, the ℓi and sℓi systems are closed under arithmetic operations in finite-precision arithmetic. In consequence, these systems offer a permanent solution to the problems of overflow and underflow. There are also certain other advantages pertaining to precision. Execution speeds of arithmetic operations will be slower than for the floating-point system, but this loss may be offset to some extent by the ability to use simpler programs.

We have not entered into any discussion of error analysis in the new arithmetics. This subject is investigated in reference 13.

We can summarize briefly by stating that the advantages and challenges of the new systems are similar to those that were offered in the 1950's on the introduction of the floating-point system.

Acknowledgements

The work described in this paper is part of an ongoing project by several researchers, particularly C. W. Clenshaw and P. R. Turner at the University of Lancaster, D. W. Lozier at the National Bureau of Standards, and the author at the University of Maryland. Several helpful suggestions have been made by A. Feldstein of Arizona State University. Support has been provided, in part, by the U. S. Army Research Office, Durham under Contract DAAG-29-84-K-0022, and the National Science Foundation under Grant DMS-84-19820.

References

[1] W. S. Brown, "A simple but realistic model of floating-point computation," *ACM Trans. Math. Software,* v. 7, pp. 445-480, 1981.

[2] C. W. Clenshaw and F. W. J. Olver, "Beyond floating point," *J. Assoc. Comput. Mach.,* v. 31, pp. 319-328, 1984.

[3] C. W. Clenshaw and F. W. J. Olver, "Level-index arithmetic operations," *SIAM. J. Numer. Anal.,* 1987. [In press.]

[4] C. W. Clenshaw and P. R. Turner, "The symmetric level-index system." [Submitted for publication.]

[5] W. J. Cody, "An overview of software development for special functions." *In* Lecture Notes in Machematics, 506, *Numerical Analysis Dundee 1975,* G. A. Watson ed., Springer-Verlag, Berlin, pp. 38-48, 1976.

[6] W. J. Cody, "Basic concepts for computational software." *In* Lecture Notes in Computer Science, 142, *Problems and Methodologies in Mathematical Software Production,* P. C. Messina and A. Murli eds., Springer-Verlag, Berlin, pp. 1-23, 1982.

[7] J. T. Coonen, "Underflow and the denormalized numbers," *Computer,* v. 14, pp. 75-87, 1981.

[8] J. Demmel, "Underflow and the reliability of numerical software," *SIAM J. Sci. Statist. Comput.,* v. 5, pp. 887-919, 1984.

[9] T. E. Hull, "Precision control, exception handling and a choice of numerical algorithms." *In* Lecture Notes in Mathematics, 912, *Numerical Analysis,* G. A. Watson ed., Springer-Verlag, Berlin, pp. 169-178, 1982.

[10] K. Hwang and F. A. Briggs, *Computer Architecture and Parallel Processing,* McGraw-Hill, New York, 1984.

[11] IEEE Standard 754, *Binary Floating-Point Arithmetic,* The Institute of Electrical and Electronics Engineers, New York, 1985.

[12] S. L. Johnsson and V. Krishnaswamy "Floating-point Cordic." [Manuscript.]

[13] D. W. Lozier and F. W. J. Olver, "Closure and precision in computer arithmetic." [Manuscript.]

[14] S. Matsui and M. Iri, "An overflow/underflow-free floating-point representation of numbers," *J. Inform. Process.,* v. 4, pp. 123-133, 1981.

[15] F. W. J. Olver, "A new approach to error arithmetic," *SIAM J. Numer. Anal.,* v. 15, pp. 368-393, 1978.

[16] F. W. J. Olver and P. R. Turner, "Implementation of level-index arithmetic using partial table look-up," these proceedings.

[17] C. W. Schelin, "Calculator function approximation," *Amer. Math. Monthly,* v. 90, pp. 317-325, 1983.

[18] P. H. Sterbenz, *Floating-Point Computation,* Prentice-Hall, Englewood Cliffs, New Jersey, 1974.

[19] P. R. Turner, "Towards a fast implementation of level-index arithmetic," *Bull. Inst. Math. Appl.*, 1987. [In press.]

[20] J. S. Walther, "A unified algorithm for elementary functions," *AFIPS Conference Proc.*, v. 38, pp. 379-385, 1971.

[21] S. Waser and M. J. Flynn, *Introduction to Arithmetic for Digital Systems Designers,* Holt, Rinehart and Winston, New York, 1982.

Part VII:
IMPLEMENTATIONS

Editor's Comments on Papers 38 Through 40

38. White
Applications of Distributed Arithmetic to Digital Signal Processing: A Tutorial Review

39. Benini, De Micheli and Macii
Designing Low-Power Circuits: Practical Recipes

40. Coleman, Softley, Kadlec, Matuosek, Tichy, Pohl, Hermanek, and Benschop
The European Logarithmic Microprocessor

This part explores various aspects of implementations.

In Paper 38 White describes distributed arithmetic, which is attractive for its computational efficiency. Distributed arithmetic is basically a serial operation that computes sums of products. This is directly applicable to many operations such as digital filters that are widely used in digital signal processing applications.

In the next paper, Benini, De Micheli and Macii provide an overview of techniques for reducing the power dissipation of CMOS circuits. In CMOS there are three components of the power dissipation: switching power, short circuit power and leakage power. Switching power is due to the need to charge or discharge the capacitive load of the circuits, short circuit power is due to the brief period when the pairs of transistors from Vcc to ground are both on, and leakage power is due to sub-threshold effects. This paper shows methods for reducing all three components and provides over a dozen examples.

Finally, the last paper describes the European Logarithmic Processor, which represents data by their binary logarithms. While this obviously simplifies multiplication, division, square rooting, etc. it also can perform addition and subtraction as shown by Kingsbury and Rayner (1971). Basically addition is converted into multiplication of the largest operand by a non-linear factor that depends on the difference of the logarithms of the operands. The resulting logarithmic processor achieves attractive performance.

References

N. G. Kingsbury and P. J. W. Rayner, "Digital Filtering Using Logarithmic Arithmetic," *Electronics Letters*, vol. 7, pp. 56–58, 1971. (reprinted as Paper 32 in *Computer Arithmetic*, Volume I.)

Applications of Distributed Arithmetic to Digital Signal Processing: A Tutorial Review

Stanley A. White

DISTRIBUTED ARITHMETIC (DA) is so named because the arithmetic operations that appear in signal processing (e.g., addition, multiplication) are not "lumped" in a comfortably familiar fashion ("Aha, there's the multiplier over there," etc.), but are distributed in an often unrecognizable fashion. The most-often encountered form of computation in digital signal processing is a sum of products (or in vector analysis parlance, dot-product, or inner-product generation). This is also the computation that is executed most efficiently by DA.

Our motivation for using DA is its extreme computational efficiency. The advantages are best exploited in circuit design, but off-the-shelf hardware often can be configured effectively to perform DA. By careful design one may reduce the total gate count in a signal processing arithmetic unit by a number seldom smaller than 50 percent and often as large as 80 percent.

HISTORICAL PERSPECTIVE OF DISTRIBUTED ARITHMETIC

The first widely known description of DA was given at a presentation by Abraham Peled and Bede Liu on IIR digital filter mechanization in 1974 at the Arden House Workshop on Digital Signal Processing. Their work on both FIR and IIR digital filter mechanization was also published in the *IEEE ASSP Transactions* [1,2]. Earlier work (pre-1971) on DA had been performed in France by Croisier et al. [3]. The earliest documented work in the U.S. was done by Zohar [4,5,6], who had independently invented DA in 1968. Other early work in the United States was reported by Little [7]. From the Arden House description of DA, Bona and Terhan at Rockwell International designed an integrated-circuit DA compensator for control systems applications [8], and White generalized its control system application [9]. The April 1975 special digital-signal processing issue of *IEEE Proceedings* contained an article by Freeny on DA applications to the telephone system at Bell Laboratories [10] and an article by White on general vector dot-product formation using DA [11]. Later that year White et al. developed an AGM digital autopilot based on DA [12], Classen et al. at Philips in the Netherlands described communications systems applications [13], and Büttner and Schüessler at the University of Erlangen in Germany showed how to reduce the memory requirements [14]. In 1975, Kai-Ping Yiu at Hewlett-Packard showed a convenient rule for handling the sign bit [15]; H. Schröder of Siemens in Munich [16] and C. S. Burrus of Rice University [17] have given some suggestions and insight for speeding up the algorithms, and K. D. Kammeyer gave a survey/summary [18]. Mechanization studies on application of DA to digital filters were discussed by Burrus [19], Jenkins and Leon [20], Zeman and Nagel [21], Tam and Hawkins [22], Arjmand and Roberts [23] and White [24,25]. Kammeyer [26] and Taylor [27] have presented error analyses, Smith and White [28] have considered DA for coordinate transformations, and Burleson and Scharf have applied it to a rotator array [29]. Taylor applied DA to a Gray-Markel filter [30], and Zohar discussed a VLSI implementation of a correlator/filter [31]. DA is also discussed by Taylor in his text [32], in the text by Smith and Denyer [33], and by Mintzer in Elliott's handbook [34].

TECHNICAL OVERVIEW OF DISTRIBUTED ARITHMETIC

DA is basically (but not necessarily) a bit-serial computational operation that forms an inner (dot) product of a pair of vectors in a single direct step. The advantage of DA is its efficiency of mechanization. A frequently argued disadvantage has been its apparent slowness because of its inherent bit-serial nature. This disadvantage is not real if the number of elements in each vector is commensurate with the number of bits in each vector element, e.g., the time required to input eight 8-bit words one at a time in a parallel fashion is exactly the same as the time required to input (simultaneously on eight wires) all eight words serially. Other modifications to increase the speed may be made by employing techniques such as bit pairing or partitioning the input words into the most significant half and least significant half, the least significant half of the most significant half, etc., thereby introducing parallelism in the computation. This will be described in Section 4.

As an example of direct DA inner-product generation, consider the calculation of the following sum of products:

$$y = \sum_{k=1}^{K} A_k x_k . \tag{1}$$

The A_k are fixed coefficients, and the x_k are the input data words. If each x_k is a 2's-complement binary number scaled (for convenience, not as necessity) such that $|x_k| < 1$, then we may express each x_k as

$$x_k = -b_{k0} + \sum_{n=1}^{N-1} b_{kn} 2^{-n} \tag{2}$$

where the b_{kn} are the bits, 0 or 1, b_{k0} is the sign bit, and $b_{k,N-1}$ is the least significant bit (LSB).

Now let us combine Equations 1 and 2 in order to express y in terms of the bits of x_k:

$$y = \sum_{k=1}^{K} A_k \left[-b_{k0} + \sum_{n=1}^{N-1} b_{kn} 2^{-n} \right] . \tag{3a}$$

Equation 3a is the conventional form of expressing the inner product. Direct mechanization of this equation defines a "lumped" arithmetic computation. Let us instead interchange the order of the summations, which gives us:

$$y = \sum_{n=1}^{N-1} \left[\sum_{k=1}^{K} A_k b_{kn} \right] 2^{-n} + \sum_{k=1}^{K} A_k(-b_{k0}) . \tag{3b}$$

This is the crucial step: Equation 3b defines a distributed arithmetic computation. Consider the bracketed term in Equation 3b:

$$\sum_{k=1}^{K} A_k b_{kn} . \tag{3c}$$

Because each b_{kn} may take on values of 0 and 1 only, expression (3c) may have only 2^K possible values. Rather than compute these values on line, we may precompute the values and store them in a ROM. The input data can be used to directly address the memory and the result, i.e., the $\sum_{k=1}^{K} A_k b_{kn}$, can be dropped into an accumulator. After N such cycles, the memory contains the result, y.

As an example, let $K = 4$, $A_1 = 0.72$, $A_2 = -0.30$, $A_3 = 0.95$, and $A_4 = 0.11$. The memory must contain all possible combinations ($2^4 = 16$ values) *and* their negatives in order to accommodate the term

$$\sum_{k=1}^{K} A_k(-b_{k0}) \tag{3d}$$

which occurs at the sign-bit time. As a consequence, we need to use a $2 \cdot 2^K$ word ROM. Figure 1a shows the simple structure (with a $2 \times 2^4 = 32$-word ROM) that can be used to mechanize these equations; Table 1 shows the contents of the memory. The T_s signal is the sign-bit timing signal. We assume that the data on the x_1, x_2, x_3, and x_4 lines (which with T_s comprise the ROM address words) are serial, 2's-complement numbers. Each is delivered in a one-bit-at-a-time (1BAAT) fashion, with LSBs $\{b_{k,N-1}\}$ first. The sign bits $\{b_{k0}\}$ are the last bits to arrive. The clock period in which the sign bits all simultaneously arrive is called the "sign-bit time." During the sign-bit time the control signal $T_s = 1$, otherwise $T_s = 0$. For the moment we will assume essentially zero time delay between the time of arrival of the address pattern to the ROM and the availability of its output. The delay around the accumulator loop is assumed to be one clock cycle and is concentrated in the summer. Switch SWA remains in Position 1 except during the clock cycle that follows the sign-bit time, when it toggles for one clock cycle to Position 2, and the fully formed result is output.

We may reduce the memory size by half to a 2^K word ROM by modifying the adder to an adder/subtractor and using T_s as the add/subtract-control line as shown in Figure 1b. This configuration may now be mechanized with a 16-word ROM. The stored table is simply the upper half of Table 1.

The memory size may be halved again to $\frac{1}{2} 2^K$ words. In order to understand how this works, we shall interpret (*not* convert, but just interpret) the input data as being cast not in a (0, 1) straight binary code, but instead as being cast in a (−1, 1) offset binary code. Suppose that we think of x_k as

$$x_k = \frac{1}{2}[x_k - (-x_k)] , \tag{4}$$

and remember that in 2's-complement notation the negative of x_k is written as

$$-x_k = -\overline{b}_{k0} + \sum_{n=1}^{N-1} \overline{b}_{kn} 2^{-n} + 2^{-(N-1)} \tag{5}$$

where the overscore symbol indicates the complement of a bit. From Equations 2 and 5 we may rewrite Equation 4 as:

$$x_k = \frac{1}{2}\left[-(b_{k0} - \overline{b}_{k0}) + \sum_{n=1}^{N-1} (b_{kn} - \overline{b}_{kn}) 2^{-n} - 2^{-(N-1)} \right] . \tag{6}$$

In order to simplify our notation later, it is convenient to define the new variables

$$c_{kn} = b_{kn} - \overline{b}_{kn} \qquad n \neq 0 \tag{7}$$

and

$$c_{k0} = -(b_{k0} - \overline{b}_{k0}) \tag{8}$$

where the possible values of the c_{kn}, including $n = 0$, are ± 1. Now (6) may be rewritten as

$$x_k = \frac{1}{2}\left[\sum_{n=0}^{N-1} c_{kn} 2^{-n} - 2^{-(N-1)} \right] . \tag{9}$$

By substituting (9) into (1) we obtain

$$y = \frac{1}{2} \sum_{k=1}^{K} A_k \left[\sum_{n=0}^{N-1} c_{kn} 2^{-n} - 2^{-(N-1)} \right] \tag{10}$$

$$= \sum_{n=0}^{N-1} Q(b_n) 2^{-n} + 2^{-(N-1)} Q(0) \tag{11}$$

where

$$Q(b_n) = \sum_{k=1}^{K} \frac{A_k}{2} c_{kn} \quad \text{and} \quad Q(0) = \sum_{k=1}^{K} \frac{A_k}{2}. \tag{12}$$

Notice that $Q(b_n)$ has only $2^{(K-1)}$ possible amplitude values with a sign that is given by the instantaneous combination of bits. This is consistent with our earlier claim. The computation of y is mechanized using a $2^{(K-1)}$ word memory, a one-word initial condition register for $Q(0)$, and a single parallel adder/subtractor with the necessary control-logic gates. This is shown in Figure 1c, using the 8-word ROM, which contains the $Q(b_n)$.

Notice from the memory values of Table 2 that those values in the lower half under "Q" are the mirror image of the values in the upper half, but with the signs re-

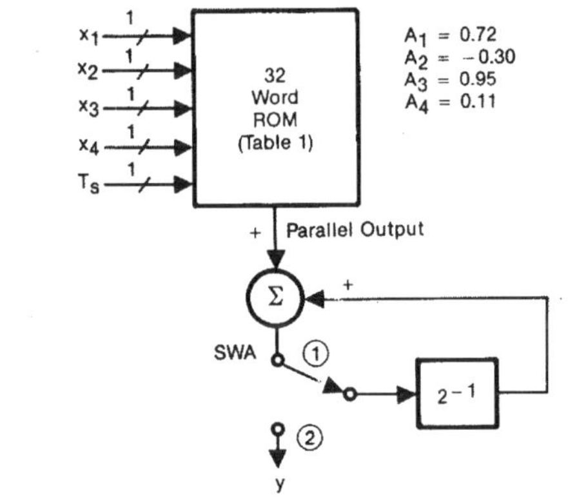

Figure 1a. Adder and Full Memory

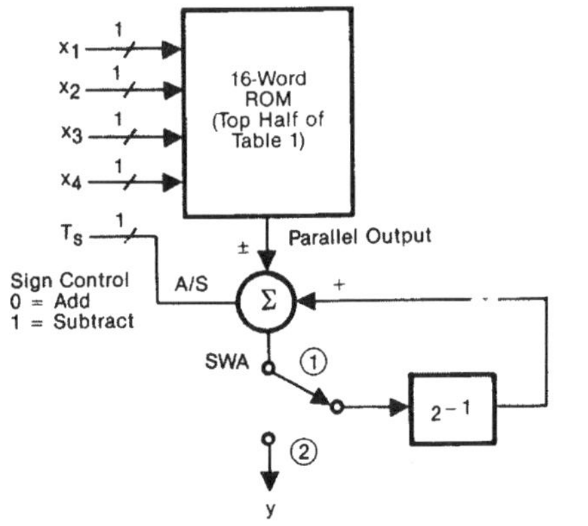

Figure 1b. Adder/Subtractor and Memory

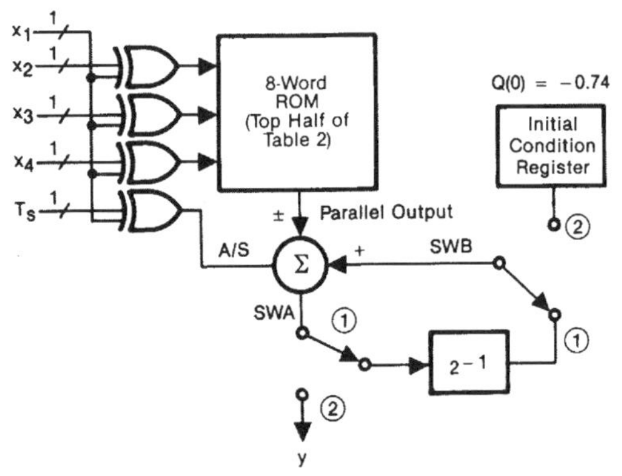

Figure 1c. Adder/Subtractor and Reduced Memory

Table 1

	Input Code					32-Word Memory Contents
	T_s	b_{1n}	b_{2n}	b_{3n}	b_{4n}	
$1 \le n \le N-1$	0	0	0	0	0	0
	0	0	0	0	1	A_4 = 0.11
	0	0	0	1	0	A_3 = 0.95
	0	0	0	1	1	A_3+A_4 = 1.06
	0	0	1	0	0	A_2 = −0.30
	0	0	1	0	1	A_2+A_4 = −0.19
	0	0	1	1	0	A_2+A_3 = 0.65
	0	0	1	1	1	$A_2+A_3+A_4$ = 0.75
	0	1	0	0	0	A_1 = 0.72
	0	1	0	0	1	A_1+A_4 = 0.83
	0	1	0	1	0	A_1+A_3 = 1.67
	0	1	0	1	1	$A_1+A_3+A_4$ = 1.78
	0	1	1	0	0	A_1+A_2 = 0.42
	0	1	1	0	1	$A_1+A_2+A_4$ = 0.53
	0	1	1	1	0	$A_1+A_2+A_3$ = 1.37
	0	1	1	1	1	$A_1+A_2+A_3+A_4$ = 1.48
$n = 0$	1	0	0	0	0	0
	1	0	0	0	1	$-A_4$ = −0.11
	1	0	0	1	0	$-A_3$ = −0.95
	1	0	0	1	1	$-(A_3+A_4)$ = −1.06
	1	0	1	0	0	$-A_2$ = +0.30
	1	0	1	0	1	$-(A_2+A_4)$ = +0.19
	1	0	1	1	0	$-(A_2+A_3)$ = −0.65
	1	0	1	1	1	$-(A_2+A_3+A_4)$ = −0.75
	1	1	0	0	0	$-A_1$ = −0.72
	1	1	0	0	1	$-(A_1+A_4)$ = −0.83
	1	1	0	1	0	$-(A_1+A_3)$ = −1.67
	1	1	0	1	1	$-(A_1+A_3+A_4)$ = −1.78
	1	1	1	0	0	$-(A_1+A_2)$ = −0.42
	1	1	1	0	1	$-(A_1+A_2+A_4)$ = −0.53
	1	1	1	1	0	$-(A_1+A_2+A_3)$ = −1.37
	1	1	1	1	1	$-(A_1+A_2+A_3+A_4)$ = −1.48

Table 2

Input Code				8-Word Memory Contents, Q
b_{1n}	b_{2n}	b_{3n}	b_{4n}	
0	0	0	0	$-1/2\,(A_1+A_2+A_3+A_4)$ = −0.74
0	0	0	1	$-1/2\,(A_1+A_2+A_3-A_4)$ = −0.63
0	0	1	0	$-1/2\,(A_1+A_2-A_3+A_4)$ = 0.21
0	0	1	1	$-1/2\,(A_1+A_2-A_3-A_4)$ = 0.32
0	1	0	0	$-1/2\,(A_1-A_2+A_3+A_4)$ = −1.04
0	1	0	1	$-1/2\,(A_1-A_2+A_3-A_4)$ = −0.93
0	1	1	0	$-1/2\,(A_1-A_2-A_3+A_4)$ = −0.09
0	1	1	1	$-1/2\,(A_1-A_2-A_3-A_4)$ = 0.02
1	0	0	0	$1/2\,(A_1-A_2-A_3-A_4)$ = −0.02
1	0	0	1	$1/2\,(A_1-A_2-A_3+A_4)$ = 0.09
1	0	1	0	$1/2\,(A_1-A_2+A_3-A_4)$ = 0.93
1	0	1	1	$1/2\,(A_1-A_2+A_3+A_4)$ = 1.04
1	1	0	0	$1/2\,(A_1+A_2-A_3-A_4)$ = −0.32
1	1	0	1	$1/2\,(A_1+A_2-A_3+A_4)$ = −0.21
1	1	1	0	$1/2\,(A_1+A_2+A_3-A_4)$ = 0.63
1	1	1	1	$1/2\,(A_1+A_2+A_3+A_4)$ = 0.74

Figure 1. DA mechanization of $y = A_1x_1 + A_2x_2 + A_3x_3 + A_4x_4$ for bit serial (1 BAAT) implementation.

versed. If we look at the bit patterns in the left-hand column, we discover that if we EXOR b_{in} with the remaining set of b_{2n}, b_{3n}, and b_{4n}, we properly address the 8-word memory to pull out the correct values of Q... except for the sign. By using the b_{1n} as the add/subtract control line for the accumulator input, we also now have the proper sign. During the sign-bit time the add/subtract command must be inverted. We therefore combine the b_{1n} and T_s signals through an EXOR gate in order to derive the proper add/subtract control signal.

The initial-condition memory that contains the value $Q(0)$ is shown on the extreme right side of Figure 1c. When the LSBs of the x_k are addressing the ROM, the value that is read out from the ROM must be corrected by the $Q(0)$ through switch SWB, which operates synchronously with switch SWA. This artifact of the binary-offset system can be seen in Equation 11. Subsequent values from the ROM are summed with the shifted previous sum. As before, we assume zero time delay between the application of the addressing bits and the availability of the contents of the ROM. There is a clock period of delay through the parallel adder, and the switches SWA and SWB are in Position 2 only for the clock cycle following the sign-bit time when $T_s = 1$. During the first clock cycle, the first output from the ROM, $Q(b_{N-1})$, is summed with $Q(0)$; during the second clock cycle it is right shifted and summed with $Q(b_{N-2})$ to produce $Q(b_{N-2}) + [Q(b_{N-1}) + Q(0)]2^{-1}$; during the third clock cycle it is again right shifted and summed with $Q(b_{N-3})$ to produce $Q(b_{N-3}) + Q(b_{N-2})2^{-1} + [Q(b_{N-1}) + Q(0)]2^{-2}$; up to the Nth clock cycle when we add $Q(b_0)$ to the right shifted previously computed quantity to produce $Q(b_0) + Q(b_1)2^{-1} + Q(b_2)2^{-2} + \cdots + Q(b_{N-2})2^{-(N-2)} + [Q(b_{N-1}) + Q(0)]^{2-(N-1)}$.

INCREASING THE SPEED OF DA MULTIPLICATION

One can see that ingesting the data serially, 1BAAT, results in a slow computation. If the input words are N bits in length, N clock cycles or periods are required in which to form the dot product. On the other hand, the equivalent of K separate products are being formed. If, therefore, $K > N$, the DA processor is faster than a single parallel multiplier/accumulator.

Additional speed may be bought in two ways; one at the expense of linearly increased memory plus more arithmetic operations, the other at the expense of exponentially increased memory. The speed may be increased by a factor of L by partitioning each input word into L subwords (L must be a factor of N). This effectively increases the dimension of the input vector by a factor of L. We can use L-times as many memories with an expanded-capacity accumulator for combining their results, or we can stay with a single memory, but its word capacity becomes $\frac{1}{2}2^{KL}$ and the lengths of the words grow by $\log_2 L$ bits. The first approach is obvious and is shown in Figure 2. The second is described below. Both are illustrated by example.

We seek a computationally simple means to introduce parallelism in the mechanization of Equation 11. The summation over n is next broken into two sums: the one over n sums from 0 to $(N/L) - 1$, and the second one sums l

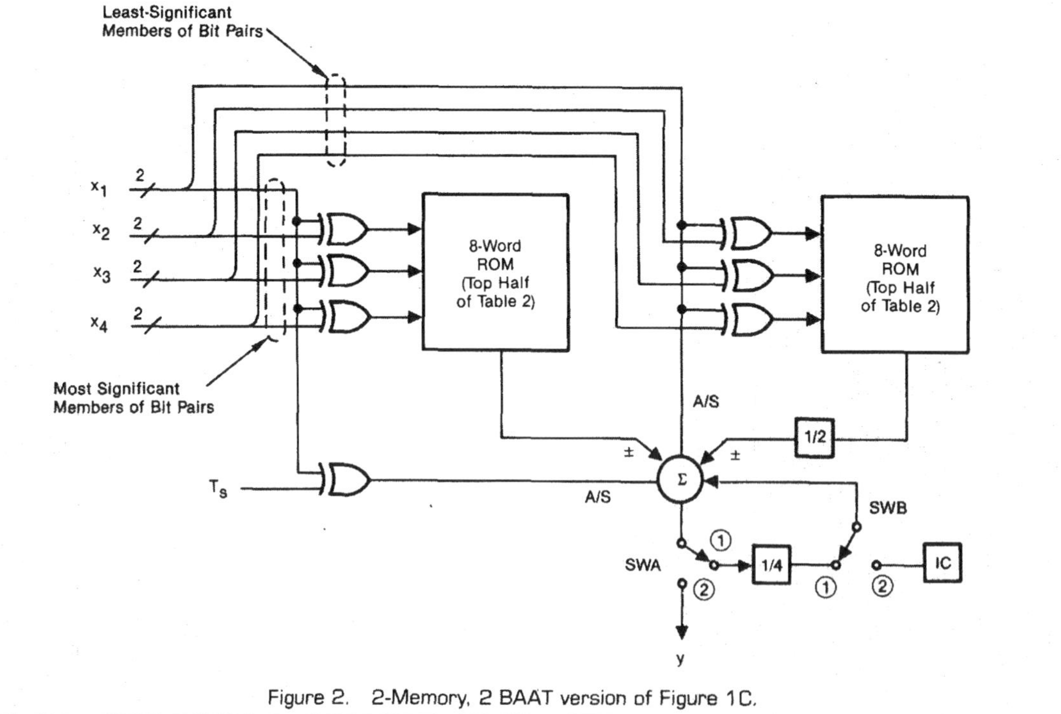

Figure 2. 2-Memory, 2 BAAT version of Figure 1C.

from 0 to $L - 1$. Now Equation 11 becomes:

$$y = \left[2^{-(N-L)}P_{IC} + \sum_{n=0}^{(N/L)-1} P(b_k)2^{-nL}\right] \tag{13}$$

where

$$P(b_k) = \sum_{k=1}^{K}\sum_{l=0}^{L-1} \frac{1}{2} A_k 2^{-l} c_{k,nL+l} \tag{14}$$

and

$$P_{IC} = -2^{-L}\sum_{k=1}^{K} A_k . \tag{15}$$

Only N/L rather than N clock times are required to form the inner product, therefore the reader can see that we have succeeded in increasing the speed of the computation by a factor of L.

We can parametrically explore the issue further. The total number of bits to be loaded is NK, and the number of clock periods required to read in these NK bits is N/L clocks. The number of input lines is

$$p = \frac{\text{total bits in}}{\text{number of clock periods}} = \frac{NK}{(N/L)} = KL . \tag{16}$$

There may be a relative cost

$$\begin{aligned} w &= \frac{\text{importance of minimizing pins}}{\text{importance of minimizing time}} \\ &= \frac{(N/L)}{KL} = \frac{N}{KL^2} \end{aligned} \tag{17}$$

We can solve this for L, the number of bits at a time (per input variable) that we are trying to load:

$$L = \left\lceil \sqrt{\frac{N}{wk}} \right\rceil . \tag{18}$$

The Gauss brackets tell us to round up to the next integer. DA is often most efficient when the number of input lines is commensurate with the number of clocks required to load the data, or equivalently, when $w = 1$. For our example

$$\sqrt{\frac{N}{wK}} = \sqrt{\frac{16}{1 \cdot 4}} = 2 ; \tag{19}$$

therefore, the data will be input 2 BAAT. Of each input bit pair, we identify the most significant bit (MSB) and the least significant bit (LSB). For all values of L, the gate at which the most significant bit appears receives special treatment. The T_s control signal is EXOR'd with the MSB for sign-bit time correction (because the sign bit is the MSB of the word). In Figure 3, we can see the configuration and also see that the LSB of the input pairs of x_1 controls the add/subtract line.

In order to demonstrate the validity of the approach, let us reconstruct by the same rules the structure for $L = 1$, as shown in Figure 4. Because for $L = 1$, a 1BAAT serial input line for each x_k uses the same line for the sign bit as for all other bits, all T_s correction must be EXOR'd with all x_k. The equivalence between Figures 1c and 4 should be easy for the reader to see because the ROM

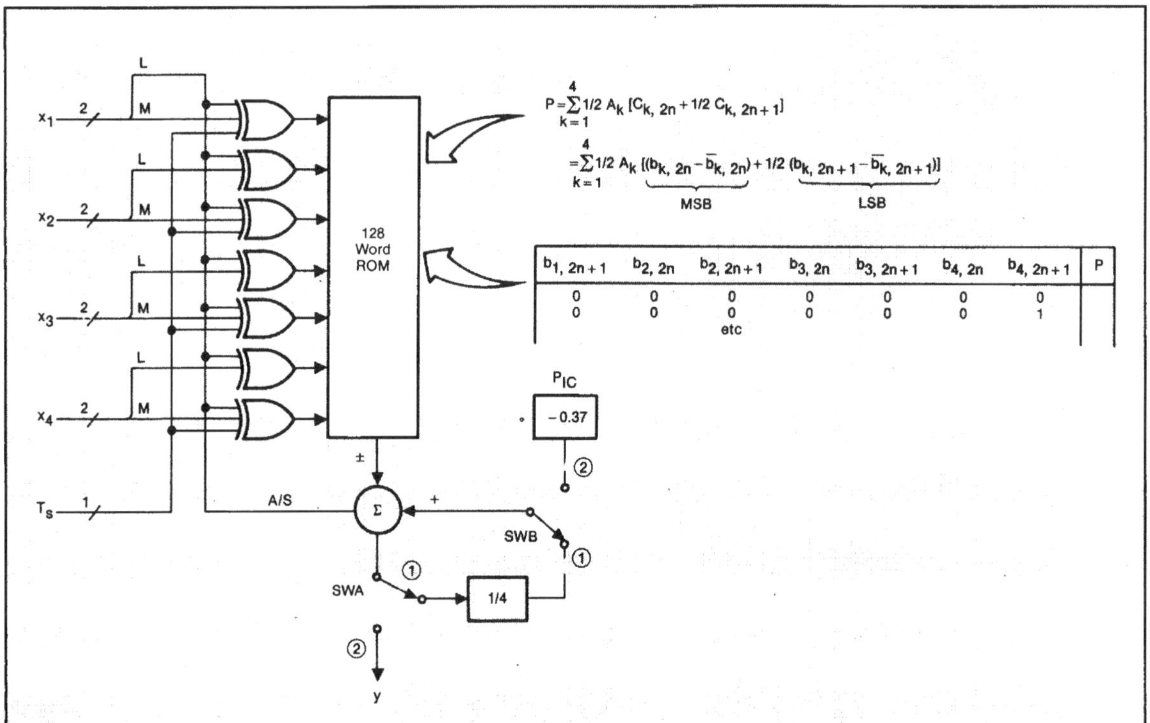

Figure 3. Single-memory, 2 BAAT version of Figure 1C.

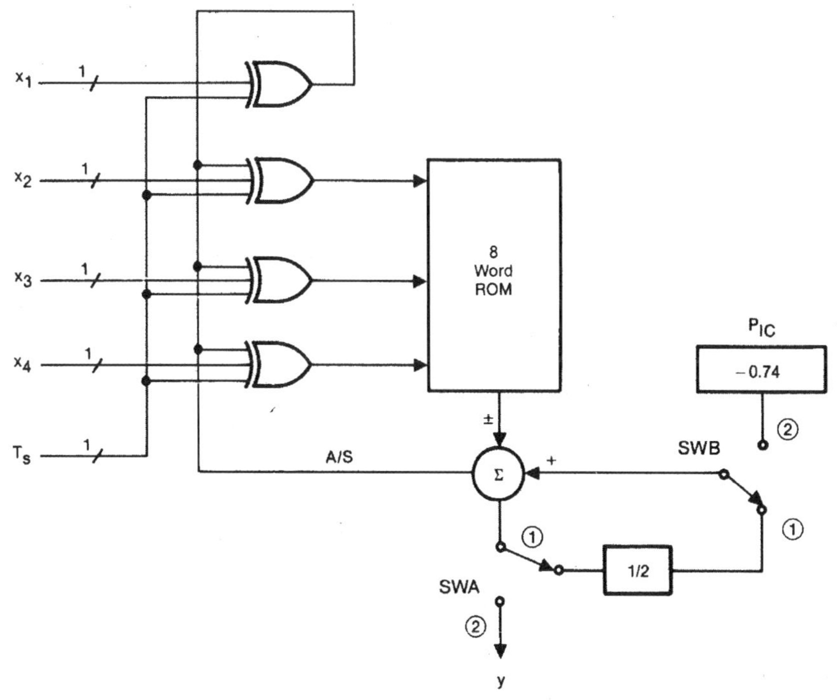

Figure 4. DA mechanization of $y = A_1x_1 + A_2x_2 + A_3x_3 + A_4x_4$ for 1 BAAT mechanization similar to Figure A3 showing alternate derivation of A/S (Add/Subtract) control.

address line in Figure 4 that is driven by x_k ($k = 2,3,4$) actually sees $x_k \oplus T_s \oplus (T_s \oplus x_1) = x_k \oplus x_1$. This is the same as in Figure 1c. The derivation of the A/S control lines is the same in both figures.

In Figure 5, we see a demonstration of an L = 4 design in order to show the structure. The memory cost is extremely great for such a small computation, but it is presented to illustrate the principle. Some more practical cases are shown below.

APPLICATION OF DA TO A BIQUADRATIC DIGITAL FILTER (AN EXAMPLE OF VECTOR DOT-PRODUCT AND VECTOR-MATRIX-PRODUCT MECHANIZATION)

A typical biquadratic digital filter has a transfer function of the form

$$\frac{Y(z)}{X(z)} = \frac{A_0 + A_1z^{-1} + A_2z^{-2}}{1 + B_1z^{-1} + B_2z^{-2}} \tag{20}$$

where the poles are determined by the B_1 and B_2, and the gain and zeros are determined by the A_0, A_1, and A_2. The time-domain description is

$$y_n = [A_0 A_1 A_2 B_1 B_2]^T [x_n x_{n-1} x_{n-2} y_{n-1} y_{n-2}] \tag{21}$$

where the coefficient vector is $[A_0A_1A_2B_1B_2]^T$ and the data vector is $[x_nx_{n-1}x_{n-2}y_{n-1}y_{n-2}]^T$. A direct DA mechanization of the filter is shown in Figure 6. Notice the extreme economy of this filter. This figure differs slightly from those that we have seen above. We have a 5-dimensional input vector to the addressing logic, but a scalar input, x_n, and a scalar output, y_n for the processor.

We have added a pair of delays (the z^{-1} blocks) in the input-signal path so that we could develop the delayed signals, x_{n-1} and x_{n-2} from the input, x_n; and we have added a third delay block so that we could develop the delayed output, y_{n-2}, from y_{n-1}. We obtain y_{n-1} from a parallel-to-serial register (P/S), since the output from the summer in the accumulator loop is a parallel word. The contents of the 16-word ROM are shown in Table 3. We may, of course, use parallelism to increase the speed, as was discussed in the previous section.

There is another important (because of low roundoff noise, low coefficient sensitivity and favorable limit-cycle behavior) form of digital filter (the normal form) that serves to illustrate the vector-matrix form of DA, and which we will now discuss. An excellent tradeoff study was recently presented by Barnes [35] to show how one could design a normal-form second order digital filter to meet prescribed performance criteria. In this section, an extremely efficient set of realizations is shown, one in which the speed and complexity can be effectively traded. (Much of the following is taken from References 36 and 37.)

A block diagram of the normal-form structure is shown in Figure 7. The multipliers b_1, b_2, c_1, and c_2 determine the pole locations of the filter. The input multipliers a_1 and a_2 are used to determine the input scaling and the multipli-

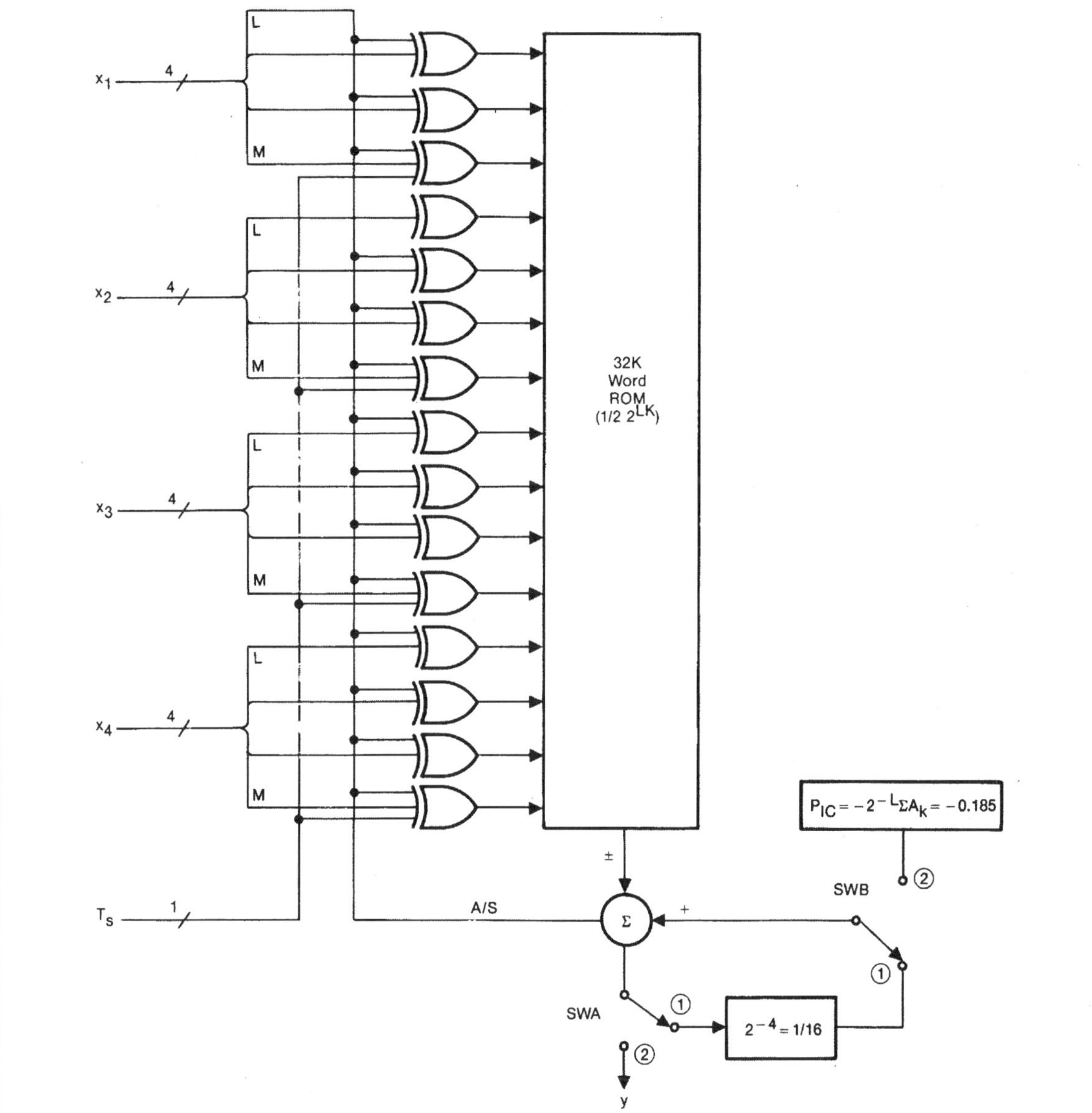

Figure 5. DA mechanization of $y = A_1x_1 + A_2x_2 + A_3x_3 + A_4x_4$ for 4 BAAT implementation using single Adder/Subtractor and minimum memory.

ers a_0, d_1, and d_2 to determine the zero locations of the filter. There are nine multipliers in total, as compared to the five that are required in the so-called direct mechanization. Barnes shows procedures whereby one may reduce this number of multipliers; however, we shall show how to eliminate them.

The vector matrix equation that describes the configuration of Figure 7 is given below:

$$\begin{bmatrix} u_n \\ v_n \\ y_n \end{bmatrix} = \begin{bmatrix} a_1 & b_1 & c_1 \\ a_2 & c_2 & b_2 \\ a_0 & d_1 & d_2 \end{bmatrix} \begin{bmatrix} x_{n-1} \\ u_{n-1} \\ v_{n-1} \end{bmatrix} \tag{22}$$

The relationships between Equations 20 and 22 are

$$\begin{aligned} A_0 &= a_0 \\ A_1 &= a_1d_1 + a_2d_2 - a_0(b_1 + b_2) \\ A_2 &= a_0(b_1b_2 - c_1c_2) + a_1(c_2d_2 - b_2d_1) \\ &\quad + a_2(c_1d_1 - b_1d_2) \\ B_1 &= b_1 + b_2 \\ B_2 &= -b_1b_2 + c_1c_2 . \end{aligned} \tag{23}$$

There are three common inputs to each of the "clumps" of Figure 7; so, for each output, we need to store $\frac{1}{2}2^3 = 4$

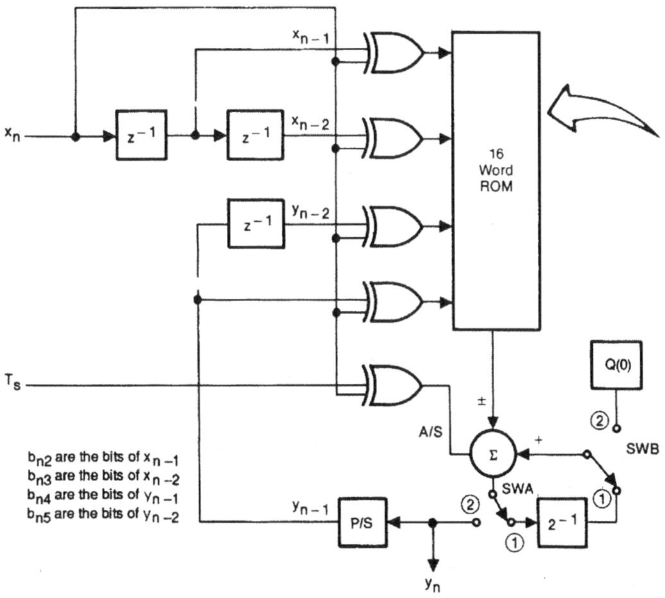

Table 3

Input Code b_{n2}	b_{n3}	b_{n4}	b_{n5}	Contents of 16-Word Memory, Q
0	0	0	0	$-1/2(A_0+A_1+A_2+B_1+B_2)$
0	0	0	1	$-1/2(A_0+A_1+A_2+B_1-B_2)$
0	0	1	0	$-1/2(A_0+A_1+A_2-B_1+B_2)$
0	0	1	1	$-1/2(A_0+A_1+A_2-B_1-B_2)$
0	1	0	0	$-1/2(A_0+A_1-A_2+B_1+B_2)$
0	1	0	1	$-1/2(A_0+A_1-A_2+B_1-B_2)$
0	1	1	0	$-1/2(A_0+A_1-A_2-B_1+B_2)$
0	1	1	1	$-1/2(A_0+A_1-A_2-B_1-B_2)$
1	0	0	0	$-1/2(A_0-A_1+A_2+B_1+B_2)$
1	0	0	1	$-1/2(A_0-A_1+A_2+B_1-B_2)$
1	0	1	0	$-1/2(A_0-A_1+A_2-B_1+B_2)$
1	0	1	1	$-1/2(A_0-A_1+A_2-B_1-B_2)$
1	1	0	0	$-1/2(A_0-A_1-A_2+B_1+B_2)$
1	1	0	1	$-1/2(A_0-A_1-A_2+B_1-B_2)$
1	1	1	0	$-1/2(A_0-A_1-A_2-B_1+B_2)$
1	1	1	1	$-1/2(A_0-A_1-A_2-B_1-B_2)$

Figure 6. DA implementation of direct form of biquadratic digital filter (1 BAAT).

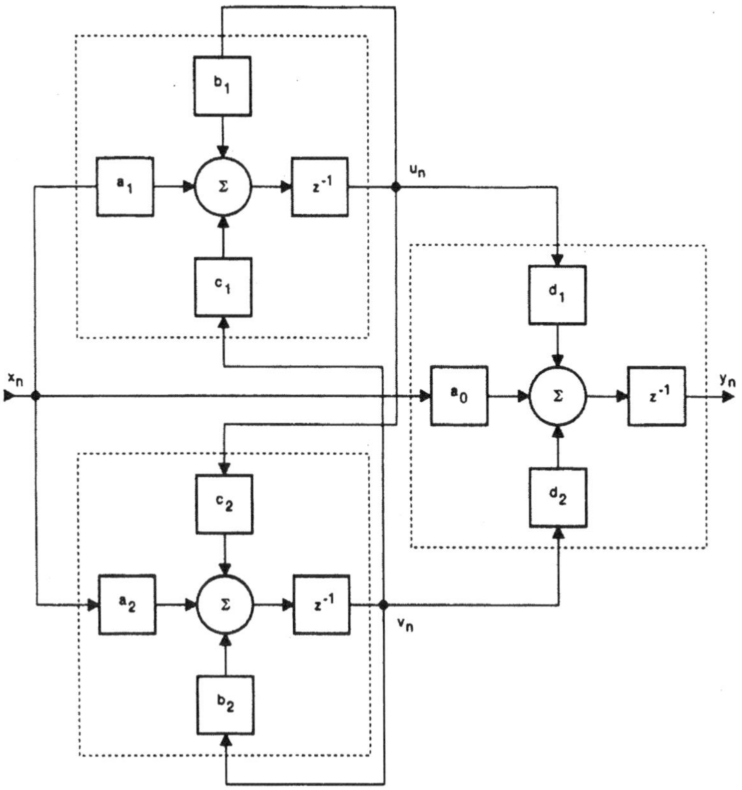

Figure 7. Normal form biquadratic filter organized into three clumps of identically driven inner products.

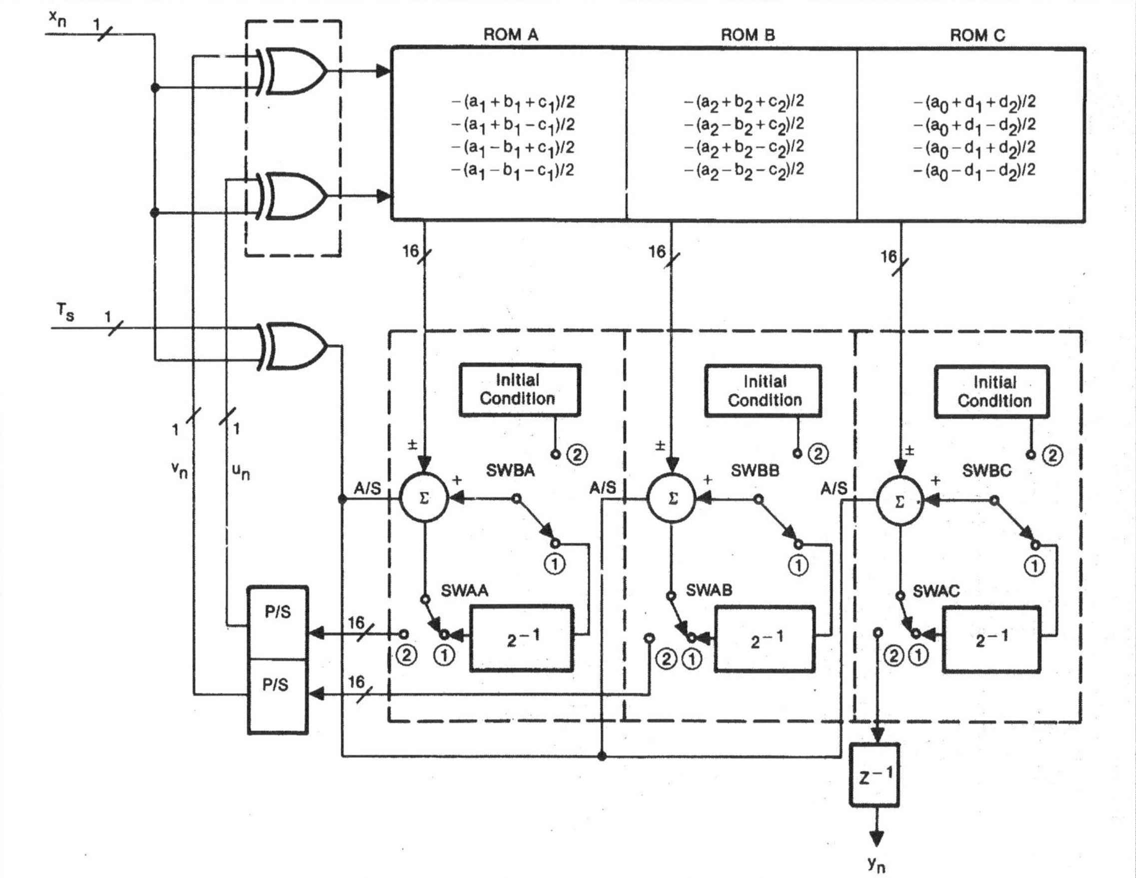

Figure 8. DA realization of state-space filter of Figure 7.

words. If the words that are stored are 16 bits long, then our three outputs together call for 3 × 16 = 48 bits per stored word. The total number of bits stored, however, is a modest 4 × 48 = 192. The somewhat detailed DA realization is illustrated in Figure 8. Notice how the addressing section has reduced to a pair of EXOR gates. The 4-word by 48-bit memory is shown for clarity as comprising three memories. In fact, the ROM may physically consist of three identically addressed 4-word by 16-bit memories or a single 4-word extended length (48-bit) memory. Each 16-bit output segment drives a separate accumulator loop, each with its own initial-condition register, just as we encountered earlier. The add/subtract control line is common, and the outputs of two of the accumulator loops are converted to serial form in their parallel-to-serial registers to be fed back to the memory addressing gates.

In order to simplify our subsequent development, it will be useful to redraw Figure 8 as shown in Figure 9. We can see the essence and utter simplicity of the structure, which is somewhat startling when one realizes that this is a realization of the 9-multiplier configuration of Figure 7.

Figure 10 shows a factor-of-four speedup over the circuit of Figure 9 by using the data 4 bits at a time (4BAAT). The parallel-to-serial registers shown do not output a serial data stream, but rather provide a sequence of four 4-bit wide segments. The time required to perform the filtering function has been reduced to 4 clock periods. This increase in speed demands that the ROM size be increased to $(\frac{1}{2})\ (2^3)^4 = 2048$ words. One would like to be able to make the throughput rate equal to the clock rate, rather than just a quarter of that rate. By quadrupling the memory of the configuration of Figure 10 and complicating the 3 adders, we can create a very fast but memory-hungry (8K word by 48 bit) filter structure that can perform the filtering function in a single clock period, as shown in Figure 11.

An alternate approach, which is shown in Figure 12, may be used in which eight ROMs are addressed by the three data streams, 2 bits at a time (2BAAT). Each memory is now a more modest $\frac{1}{2}\ (2^3)^2 = 32$ words × 48 bits for a total memory requirement of 8 × 32 × 48 = 12,288 bits. The three adders are each of a complexity less than that of a 16-bit modified-Booth multiplier. In this approach, we have reduced the memory size at the expense of complicating the adders.

The outputs from the memories are 16 (or whatever,

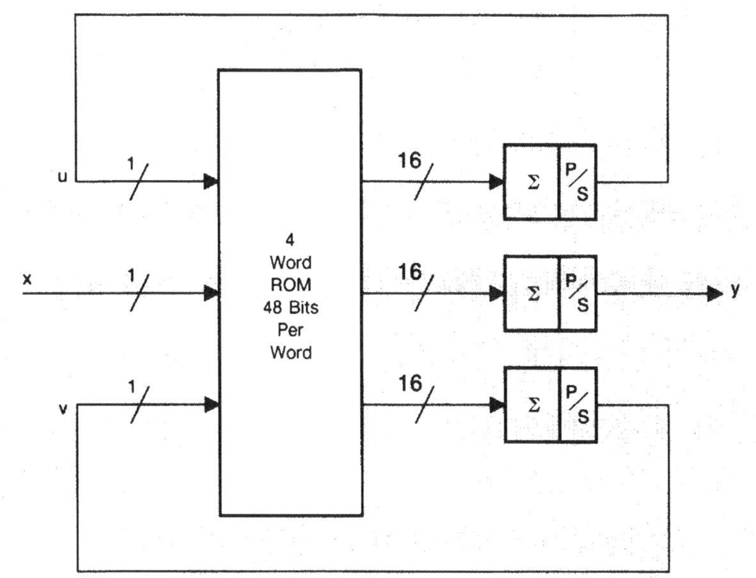

Figure 9. Alternate representation of Figure 8. Single memory 1 BAAT DA structure, 16 clock periods per filter function.

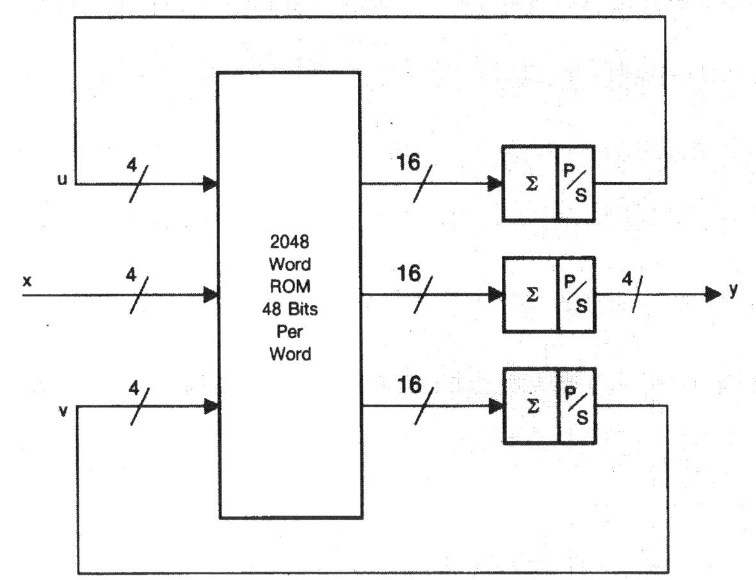

Figure 10. Single memory 4 BAAT DA structure, four clock periods per filter function.

say N, number of) bits. In the shift-and-add process that occurs in the accumulators, the accuracy of the results degrades because of the least-significant bits that are lost through the trauma of quantization. Figure 13 illustrates the problem; as the data circulates through the accumulator loop, LSB's are lost at the shift stage. (These lost bits are often modeled as an additive error.) From the point in the filter that this error is introduced (the accumulator outputs), we follow them around the recursive loop of the filter and see the reinforcement addition that is the cause of noise gain within recursive filters. We will create a parallel error path, now, with error subtractions that will nearly cancel the error additions.

Figure 14 shows the state-space filter of Figure 7 with

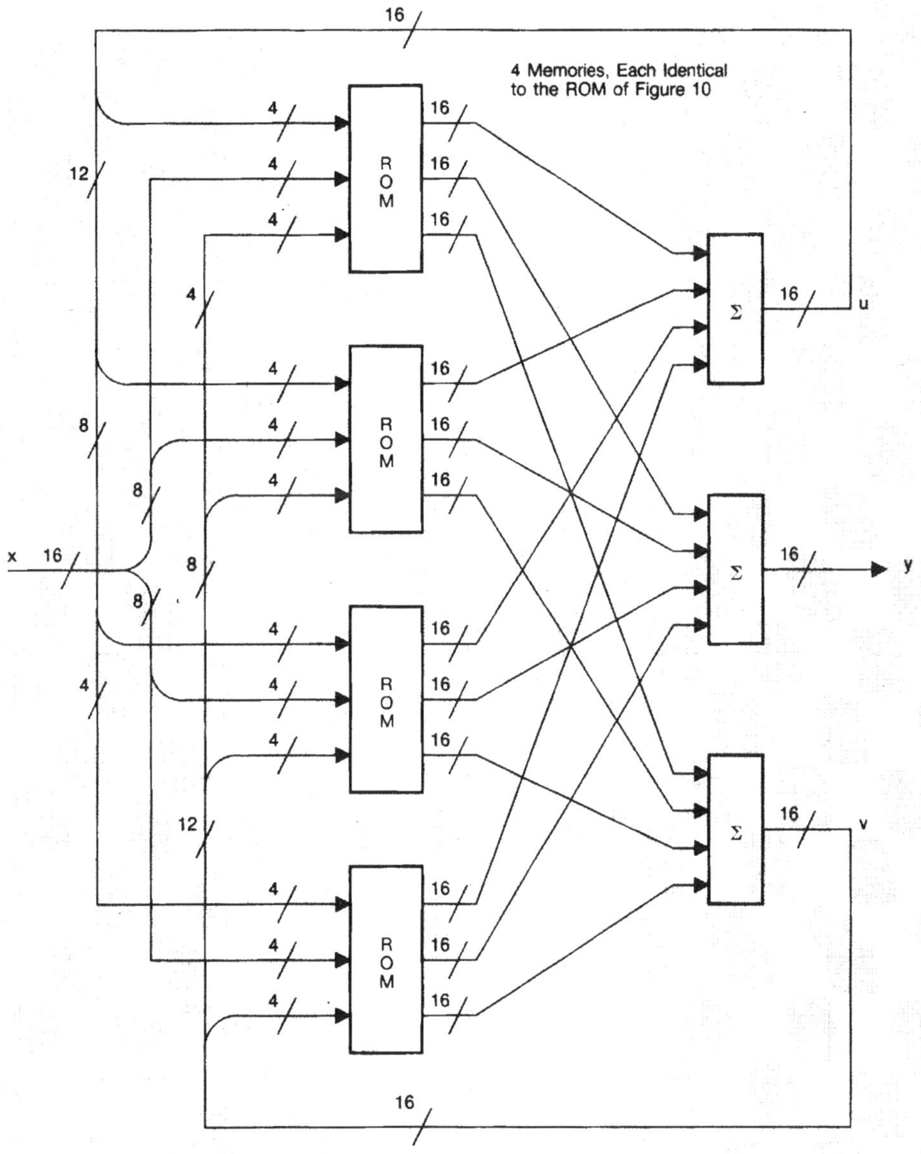

Figure 11. Four memory 4 BAAT DA structure, one clock period per filter function.

noise sources (quantization effects) ϵ_1, ϵ_2, and ϵ_3 added at the outputs of the accumulators. We have also added, within the dotted lines, noise-cancellation paths whose gains are shift approximations to the actual gains. This technique is known as error feedback and has long been used in DA filters [39]. We mechanize the noise cancellation paths by modifying Figure 8 as shown in Figure 15. Since at most the tilde-marked gains consist of a shift, those gains shown in Figure 15 are physically trivial. They feed a serial adder in which they are bit-by-bit combined with the initial conditions (ICs), then loaded into a serial-to-parallel register; that register now contains a chimerical initial condition, which includes the error feedback. If the entity within the dotted line of Figure 15 is redefined as the adder and parallel-to-serial register, then Figure 9 is valid for the error-feedback mechanization. If the adders that are fed by the ICs are L BAAT adders, then this structure can be generalized for use with structures that are shown in Figures 10–12.

APPLICATIONS IN TRANSFORMERS

DA has found important use in low-complexity, high-performance FFT structures [39,40,41]. The effective use

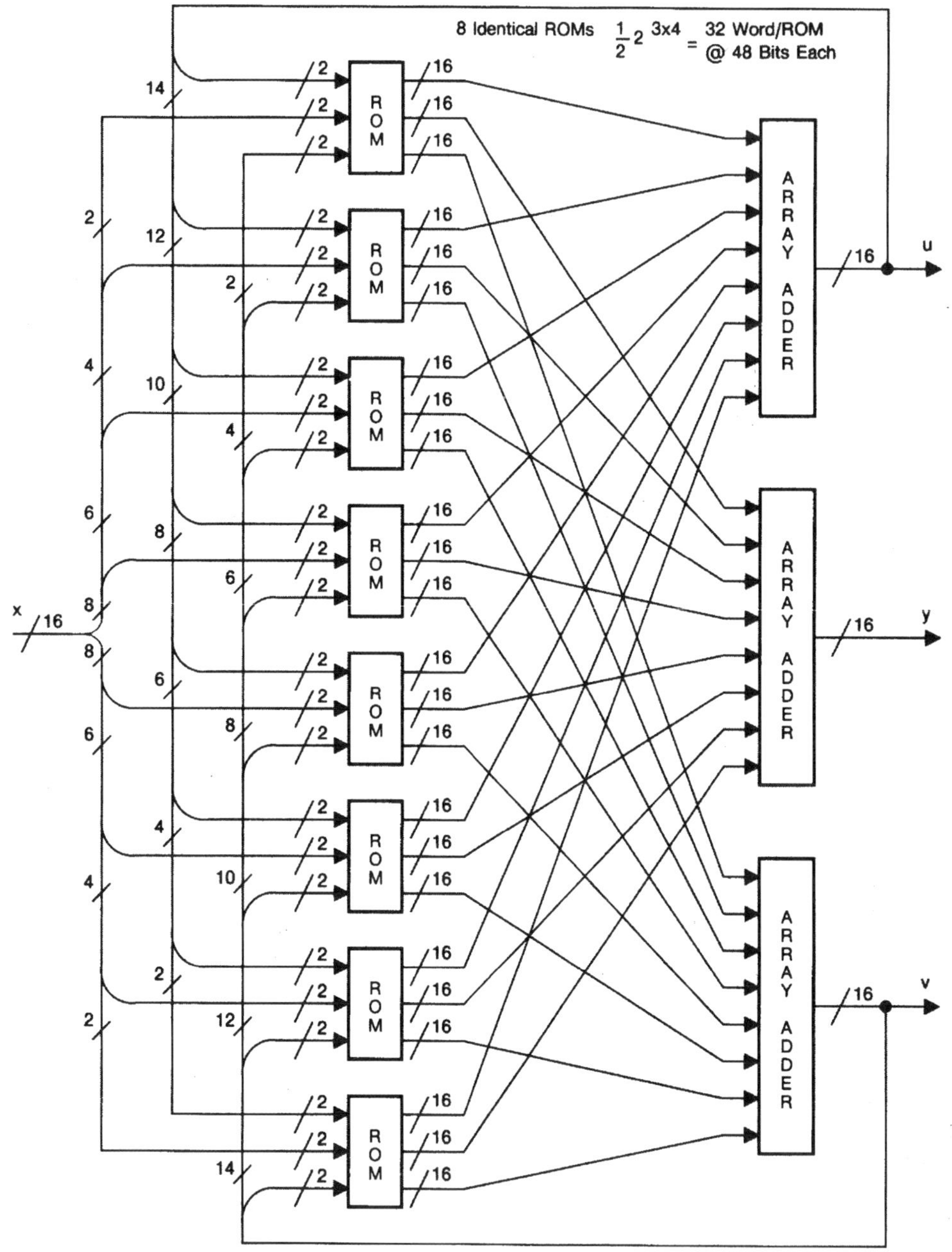

Figure 12. Eight memory 2 BAAT DA structure, one clock period per filter function.

of DA in the mechanization of a simple, direct, high-performance complex multiplier has been the reason for its success in FFT applications, and has spurred additional work in the development of efficient complex multipliers [42]. The development path for complex multipliers took an interesting twist with the advent of radix-3 and radix-6 FFT's [43, 44]. Their charm is that they give larger-radix multiplier-free building blocks that lead to increased efficiency in transform processing. By the simple expedient of turning to non-orthogonal coordinate systems for complex arithmetic, the resulting FFT structures [45] gave birth to a new complex multiplier [46] to perform the reduced number of "twiddles" on the interstage coupling.

Image processing has turned to the discrete-cosine

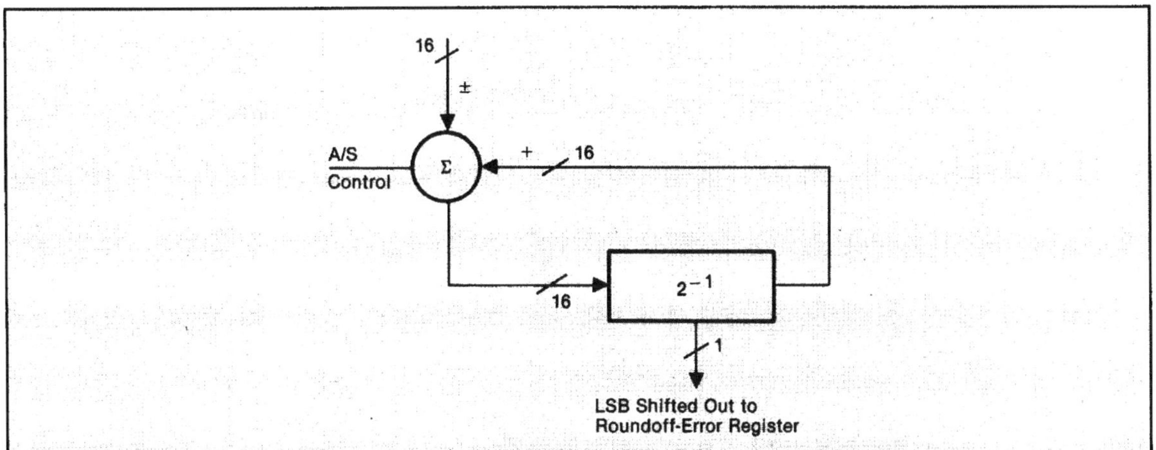

Figure 13. Accumulator loop showing LSB being dropped out on the right shift operation.

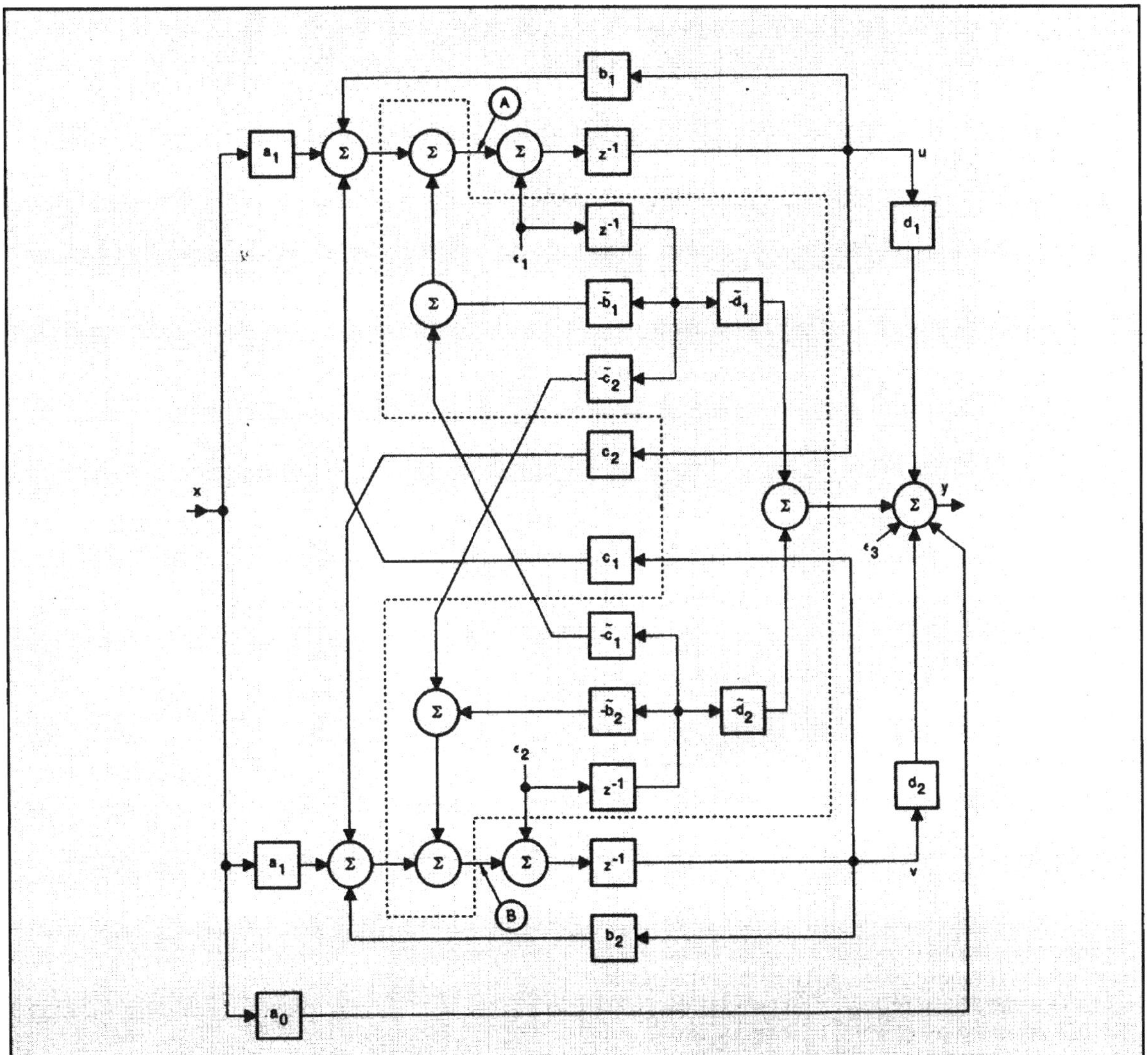

Figure 14. State space digital filter showing roundoff noise sources and cancellation loops.

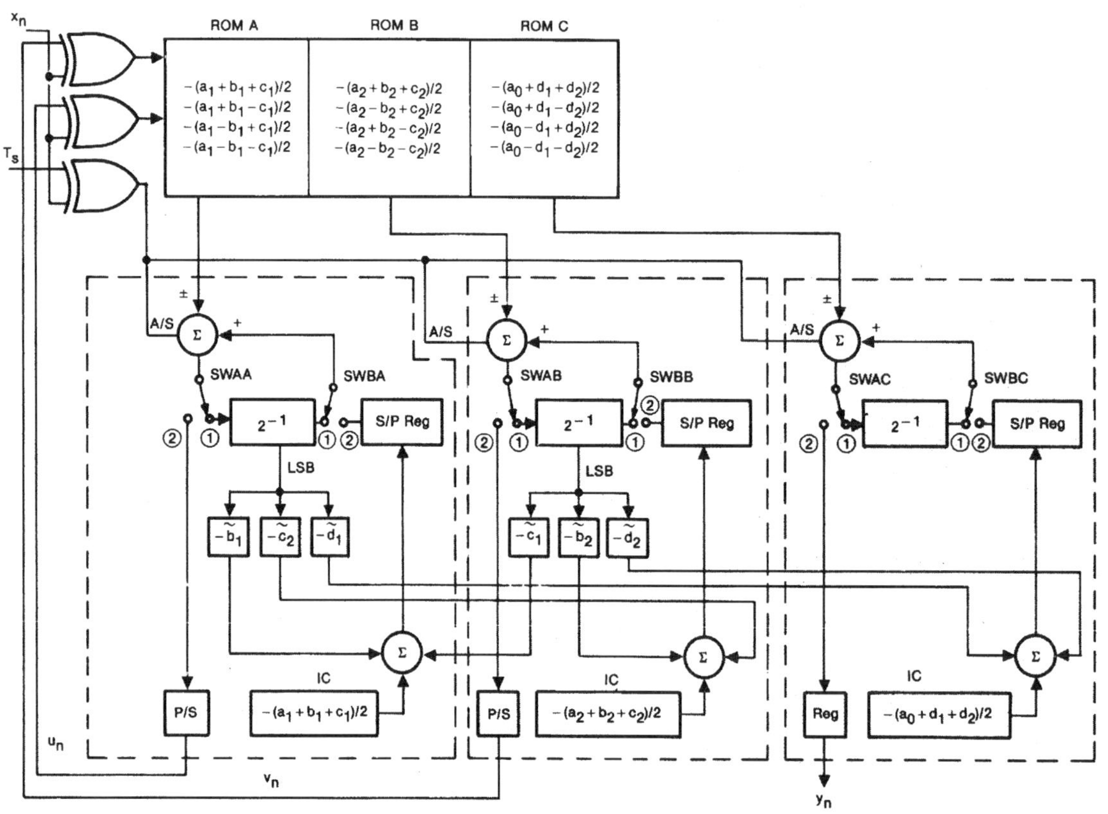

Figure 15. DA mechanization of Figure 14.

transform (DCT) for more efficient processing. There, again, DA has found a home [47,48].

NONLINEAR AND/OR NONSTATIONARY PROCESSING WITH DA

We have only considered the use of DA in linear, time-invariant systems. It is not so restricted. For variable coefficients, we may use RAM's rather than ROM's. In fact, one of the trailblazers in this approach was Schroder [16].

In 1981 Cowan and Mavor [49] described an 8-tap adaptive transversal filter that employed DA, and in 1983 an expanded version was published by Cowan, Smith, and Elliott [50]. Andrews [51] compared it favorably in a mechanization study involving traditional and nontraditional arithmetic mechanizations of adaptive filters. The capacity of the DA adaptive structure can be increased by using block-processing concepts [52,53].

The mechanization of nonlinear difference equations by DA was presented by Sicuranza in [54], who represented the filter by a truncated discrete Volterra series. Chiang et al. in [55] report the results of a mechanization trade-off study of various implementations of quadratic filters. They conclude that DA is as efficient as matrix decomposition implemented with systolic arrays, but that DA lacks the modularity.

Satisfactory DA adaptive nonlinear filters have been investigated and reported by Sicuranza and Ramponi [56], and by Smith et al. [57].

CONCLUSIONS

DA is a very efficient means to mechanize computations that are dominated by inner products. As we have seen, the coefficients of the equations can be time varying, and the equations themselves can be nonlinear. When a great many computing methods are compared, DA has always fared well, not always (but often) best, and never poorly. As a consequence, whenever the performance/cost ratio is critical (especially in custom designs), DA should be seriously considered as a contender.

ACKNOWLEDGMENTS

I want to express my thanks to the many workers in the field who have quickly and generously shared their results with me; to my coworkers who, over the years, have patiently simulated, analyzed, built, and tested endless DA concepts, and corrected my errors; and to the reviewers who offered very helpful suggestions and led me to references that were new and unfamiliar to me.

REFERENCES

[1] A. Peled and B. Liu, "A New Approach to the Realization of Nonrecursive Digital Filters," *IEEE Trans. Audio and Electroacoustics*, Vol. AU-21, No. 6, pp. 477–485, December 1973.

[2] A. Peled and B. Liu, "A New Hardware Realization of Digital Filters," *IEEE Trans. on A.S.S.P.*, Vol. ASSP-22, pp. 456–462, December 1974.

[3] A Croisier, D. J. Esteban, M. E. Levilion, and V. Rizo, "Digital Filter for PCM Encoded Signals," U.S. Patent 3,777,130, December 3, 1973.

[4] S. Zohar, "New Hardware Realization of Nonrecursive Digital Filters," *IEEE Trans. on Computers*, Vol. C-22, pp. 328–338, April 1973.

[5] S. Zohar, "The Counting Recursive Digital Filter," *IEEE Trans. on Computers*, Vol. C-22, pp. 338–347, April 1973.

[6] S. Zohar, "A Realization of the RAM Digital Filter," *IEEE Trans. on Computers*, Vol. C-25, pp. 1048–1052, October 1976.

[7] W. D. Little, "A Fast Algorithm for Digital Filters," *IEEE Trans. on Communications*, Vol. C-23, pp. 466–469, May 1974.

[8] B. E. Bona and F. C. Terhan, "A Special-Purpose Digital Control Compensator," *Proc. 8th Asilomar Conference on Circuits, Systems, and Computers*, Pacific Grove, California, pp. 454–457, December 1974.

[9] S. A. White, "Applications of Digital Signal Processing to Control Systems," *Proc. 8th Asilomar Conference on Circuits, Systems, and Computers*, Pacific Grove, California, pp. 278–284, December 1974.

[10] S. L. Freeny, "Special-Purpose Hardware for Digital Filtering," *Proc. IEEE*, Vol. 63, pp. 633–648, April 1975.

[11] S. A. White, "On Mechanization of Vector Multiplication," *Proc. IEEE*, Vol. 63, pp. 730–731, April 1975.

[12] S. A. White, W. P. Engler, J. P. Davis, S. L. Smith, J. V. Henning, "A Programmable Digital Servo Controller," Invention Disclosure PF75E145, October 16, 1975.

[13] T. R. C. M. Classen, W. F. G. Mecklenbrauker, and J. D. H. Peek, "Some Considerations on the Implementation of Digital Systems for Signal Processing," Philips Research Report 30, pp. 73–84, 1975.

[14] M. Buttner and H. W. Schuessler, "On Structures for the Implementation of the Distributed Arithmetic," *NTZ Communication Journal*, Vol. 6, June 1975.

[15] Kai-Ping Yiu, "On Sign-Bit Assignment for a Vector Multiplier," *Proc. IEEE*, Vol. 64, pp. 372–373, March 1976.

[16] H. Schroder, "High Word-Rate Digital Filters with Programmable Table Look-Up," *IEEE Trans. on Circuits and Systems*, Vol. CAS-24, No. 5, pp. 277–279, May 1977.

[17] C. S. Burrus, "Digital Filter Realization by Distributed Arithmetic," International Symposium on Circuits and Systems, Munich, April 1976.

[18] K. D. Kammeyer, "Digital Filter Realization in Distributed Arithmetic," *Proc. European Conf. on Circuit Theory and Design*, Genoa, Italy, September 1976.

[19] C. S. Burrus, "Digital Filter Structures Described by Distributed Arithmetic Filters," *IEEE Trans. on Circuits and Systems*, Vol. CAS-24, No. 12, pp. 674–680, December 1977.

[20] W. K. Jenkins, and B. J. Leon, "The Use of Residue Number System in the Design of Finite Impulse Response Filters," *IEEE Trans. on Circuits and Systems*, Vol. CAS-24, No. 4, April 1977.

[21] J. Zeman, and H. T. Nagle, Jr., "A High-Speed Microprogrammable Digital Signal Processor Employing Distributed Arithmetic," *IEEE Trans. on Computers*, Vol. C-29, No. 2, pp. 134–144, February 1980.

[22] B. S. Tam, and G. J. Hawkins, "Speed-Optimized Microprocessor Implementation of a Digital Filter," *IEEE Proc.*, Vol. 69, No. 3, pp. 85–93, May 1981.

[23] M. Arjmand and R. A. Roberts, "On Comparing Hardware Implementations of Fixed-Point Digital Filters," *IEEE Circuits and Systems Magazine*, Vol. 3, No. 2, 1981, pp. 2–8.

[24] S. A. White, "Architecture for a Digital Programmable Image Processing Element," *Proc. 1981 IEEE International Conference on Acoustics, Speech, and Signal Processing*, Atlanta, pp. 658–661, March 30–April 1, 1981.

[25] S. A. White, "An Architectural for a High-Speed Digital Signal Processing Device," *Proc. 1981 IEEE International Symposium on Circuits and Systems*, Chicago, pp. 893–896, April 28–29, 1981.

[26] K. D. Kammeyer, "Quantization Error of the Distributed Arithmetic," *IEEE Trans. on Circuits and Systems*, Vol. CAS-24, No. 12, pp. 681–689, December 1977.

[27] F. J. Taylor, "An Analysis of the Distributed-Arithmetic Digital Filter," *IEEE Trans. on A.S.S.P.*, Vol. ASSP-35, No. 5, pp. 1165–1170, Oct. 1986.

[28] S. G. Smith and S. A. White, "Hardware Approaches to Vector-Plane Rotation," *Proc. 1988 IEEE International Conference on Acoustics, Speech, and Signal Processing*, New York, pp. 212–213, April 11–14, 1988.

[29] W. P. Burleson and L. L. Scharf, "A VLSI Implementation of a Cellular Rotator Array," *Proc. 1988 IEEE Custom Integrated Circuits Conference*, pp. 8.1.2–8.1.4.

[30] F. J. Taylor, "A Distributed Gray-Markel Filter," *IEEE Trans. on A.S.S.P.*, Vol. ASSP-31, No. 3, pp. 761–763, June 1983.

[31] S. Zohar, "A VLSI Implementation of a Correlator/Digital Filter Based on Distributed Arithmetic," *IEEE Trans. on A.S.S.P.*, Vol. ASSP-37, No. 1, pp. 156–160, Jan. 1989.

[32] F. J. Taylor, *Digital Filter Design Handbook*, Marcel Dekker, Inc., pp. 678–697, June 1983.

[33] S. G. Smith and P. B. Deyer, *Serial-Data Computation*, Kluwer Academic Publishers, 1988.

[34] D. F. Elliott (ed.), *Handbook of Digital Signal Processing*, Academic Press, pp. 964–972, 1987.

[35] C. W. Barnes, "A Parametric Approach to the Realization of Second-Order Digital-Filter Sections," *IEEE*

Trans. on Circuits and Systems, Vol. CAS-32, No. 6, pp. 530–539, June 1985.

[36] S. A. White, "High-Speed Distributed-Arithmetic Realization of a Second-Order Normal-Form Digital Filter," *IEEE Trans. on Circuits and Systems,* Vol. CAS-33, No. 10, pp. 1036–1038, October 1986.

[37] S. A. White, "High-Speed Distributed-Arithmetic Realization of a Second-Order State-Space Digital Filter," *Proc. 20th Asilomar Conference on Signals, Systems, and Computers,* Pacific Grove, California, pp. 359–362, November 10–12, 1986.

[38] T. L. Chang and S. A. White, "An Error Cancellation Digital-Filter Structure and Its Distributed Arithmetic Implementation," *IEEE Trans. on Circuits and Systems,* Vol. CAS-28, No. 4, pp. 339–342, April 1981.

[39] S. A. White, "A Simple FFT Butterfly Arithmetic Unit," *IEEE Trans. on Circuits and Systems,* Vol. CAS-28, No. 4, pp. 352–366, April 1981.

[40] I. R. Mactaggart and M. A. Jack, "A Single Chip Radix-2 FFT Butterfly Architecture Using Parallel Data Distributed Arithmetic," *IEEE J. Solid State Circuits,* Vol. SC-19, pp. 368–373, June 1984.

[41] S. A. White, "An Architecture for a 16-Point FFT GaAs Building Block," *Proc. 21st Asilomar Conference on Signals, Systems, and Computers,* Pacific Grove, California, pp. 928–932, November 1987.

[42] S. G. Smith and P. B. Denyer, "Efficient Bit-Serial Complex Multiplication and Sum-of-Products Computation Using Distributed Arithmetic," *Proc. 1986 International Conference on Acoustics, Speech, and Signal Processing,* Tokyo, pp. 2203–2206, April 1986.

[43] E. Dubois and A. N. Venetsanopoulos, "A New Algorithm for the Radix-3 FFT," *IEEE Trans. on A.S.S.P.,* Vol. ASSP-26, No. 3, pp. 222–225, June 1978.

[44] S. Prakash and V. V. Rao, "A New Radix-6 FFT Algorithm," *IEEE Trans. on A.S.S.P.,* Vol. ASSP-29, No. 4, pp. 939–941, August 1981.

[45] S. A. White, "Results of a Preliminary Study of a Novel IC Arithmetic Unit for an FFT Processor," *Proc. 18th Asilomar Conference on Circuits, Systems, and Computers,* Pacific Grove, California, pp. 67–71, November 5–7, 1984.

[46] S. A. White, "A Complex Multiplier for Binary Two's-Complement Numbers," U.S. Patent No. 4,680,727, July 14, 1987.

[47] N. Demassieux, G. Concordel, J.-P. Durandeau, and F. Jutand, "An Optimized VLSI Architecture for a Multiformat Discrete Cosine Transform," *Proc. 1987 IEEE International Conference on Acoustics, Speech, Signal Processing,* pp. 547–550.

[48] A. M. Gottlieb, M. T. Sun, and T. C. Chen, "A Video Rate 16 × 16 Discrete Cosine Transform IC," *Proc. 1988 IEEE Custom Integrated Circuits Conference,* pp. 8.2.1–8.2.4.

[49] C. F. N. Cowan and J. Mavor, "New Digital-Adaptive Filter Implementation Using Distributed-Arithmetic Techniques," *IEE Proc.,* Vol. 128, Pt. F, No. 4, pp. 225–230, Aug. 1981.

[50] C. F. N. Cowan, S. G. Smith, and J. H. Elliott, "A Digital Adaptive Filter Using a Memory-Accumulator Architecture: Theory and Realization," *IEEE Trans. on A.S.S.P.,* Vol. ASSP-31, No. 3, pp. 541–549, June 1983.

[51] M. Andrews, "A Systolic SBNR Adaptive Signal Processor," *IEEE J. Solid State Circuits,* Vol. SC-21, No. 1, pp. 120–128, Feb. 1986.

[52] C.-H. Wei, and J.-J. Lou, "Multimemory Block Structure for Implementing a Digital Adaptive Filter Using Distributed Arithmetic," *IEE Proc.,* Vol. 133, Pt. G, No. 1, pp. 19–26, Feb. 1986.

[53] Y.-Y. Chiu and C.-H. Wei, "On the Realization of Multimemory Block Structure Digital Adaptive Filter Using Distributed Arithmetic," *J. Chinese Institute of Engineers,* Vol. 10, No. 1, pp. 115–122, 1987.

[54] G. L. Sicuranza, "Nonlinear Digital-Filter Realization by Distributed Arithmetic," *IEEE Trans. on A.S.S.P.,* Vol. ASSP-33, No. 4, pp. 939–1321, Aug. 1985.

[55] H.-H. Chiang, C. L. Nikias, and A. N. Venetsanopoulos, "Efficient Implementations of Quadratic Digital Filters," *IEEE Trans. on A.S.S.P.,* Vol. ASSP-34, No. 6, pp. 1511–1528, Dec. 1986.

[56] G. L. Sicuranza, "Adaptive Nonlinear Digital Filters Using Distributed Arithmetic," *IEEE Trans. on A.S.S.P.,* Vol. ASSP-34, No. 3, pp. 518–526, June 1986.

[57] M. J. Smith, C. F. N. Cowan, and P. F. Adams, "Nonlinear Echo Cancellers Based on Transpose Distributed Arithmetic," *IEEE Trans. on Circuits and Systems,* CAS-35, No. 1, pp. 6–18, Jan. 1988.

Stanley A. White (S'55, M'57, SM'69, F'82) was born in Providence, Rhode Island in 1931. He is Senior Scientist at Rockwell International Corporation's Autonetics Electronics Systems in Anaheim, California, and Adjunct Professor of Electrical Engineering at the University of California, Irvine. He received his B.S.E.E., M.S.E.E., and Ph.D. (with letter of commendation) degrees from Purdue University in 1957, 1959, and 1965, respectively. He is also a Fellow of the American Association for the Advancement of Science (AAAS), the New York Academy of Sciences (NYAS), and the Institute for the Advancement of Engineering (IAE). He is a member (and long-time Autonetics Chapter President) of Sigma Xi, Tau Beta Pi, the Audio Engineering Society, and has been International Director of Eta Kappa Nu. He received the Purdue University Distinguished Engineering Alumnus Award in 1988, the Leonardo Da Vinci Medallion in 1986, and was selected as Rockwell International Engineer of the Year in 1985. In 1984 he received both the Engineer of the Year Award from the Orange County (California) Engineering Council and the IEEE Centennial Medal. He has held several distinguished lecturerships and received the NEC Distinguished Lecturer Award.

Dr. White has over 100 publications in the open literature, holds 30 U.S. patents (with more pending issue), and is a registered Professional Engineer in both California and Indiana.

He is General Chairman of ISCAS '92; was General Chairman of ICASSP '84, Vice Chairman of ISCAS '83; was both General Chairman of the Asilomar Conference on Circuits, Systems, and Computers, and Technical Program Chairman of the IEEE Region 6 Conference in 1982; and was Technical Program Chairman of the Asilomar Conference in 1981. Dr. White is listed in *American Men and Women of Science, Who's Who in Biomedical Engineering, Who's Who in Engineering, Who's Who in America, Who's Who in The World,* and other standard biographical reference works.

Designing Low-Power Circuits: Practical Recipes

*by Luca Benini **
Giovanni De Micheli
Enrico Macii

1531-636X/03/$10.00©2001IEEE

Abstract—*The growing market of mobile, battery-powered electronic systems (e.g., cellular phones, personal digital assistants, etc.) demands the design of microelectronic circuits with low power dissipation. More generally, as density, size, and complexity of the chips continue to increase, the difficulty in providing adequate cooling might either add significant cost or limit the functionality of the computing systems which make use of those integrated circuits.*

In the past ten years, several techniques, methodologies and tools for designing low-power circuits have been presented in the scientific literature. However, only a few of them have found their way in current design flows.

The purpose of this paper is to summarize, mainly by way of examples,what in our experience are the most trustful approaches to low-power design. In other words, our contribution should not be intended as an exhaustive survey of the existing literature on low-power design; rather, we would like to provide insights a designer can rely upon when power consumption is a critical constraint.

We will focus solely on digital circuits, and we will restrict our attention to CMOS devices, this technology being the most widely adopted in current VLSI systems.

Introduction

Power dissipation has become a critical design metric for an increasingly large number of VLSI circuits. The exploding market of portable electronic appliances fuels the demand for complex integrated systems that can be powered by lightweight batteries with long times between re-charges (for instance, the plot in Fig. 1 shows the evolution of the world-wide market for mobile phones). Additionally, system cost must be extremely low to achieve high market penetration. Both battery lifetime and system cost are heavily impacted by power dissipation. For these reasons, the last ten years have witnessed a soaring interest in low-power design.

The main purpose of this paper is to provide a few insights into the world of low-power design. We do not intend to review the vast literature on the topic (the interested reader is referred to the many available surveys, e.g., [1–3]). Our objective is to give the readers a few basic concepts to help understanding the "nature of the beast", as well as to provide "silicon-proven recipes" for minimizing power consumption in large-scale digital integrated circuits.

Luca Benini

Giovanni De Micheli

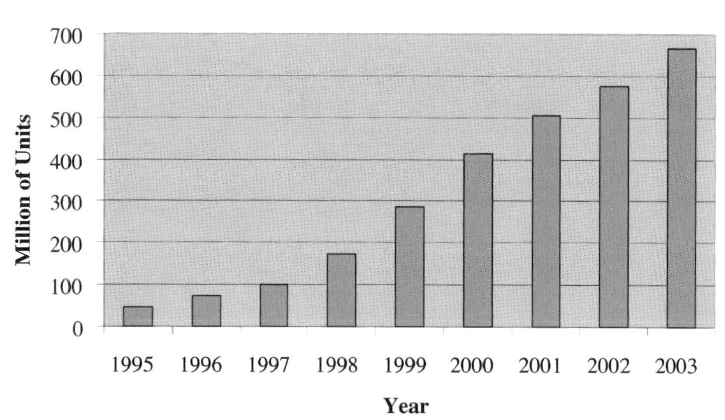

Figure 1. Global market for cellular phones.

The power consumed by a circuit is defined as $p(t) = i(t)v(t)$, where $i(t)$ is the instantaneous current provided by the power supply, and $v(t)$ is the instantaneous supply voltage. Power minimization targets maximum instantaneous power or average power. The latter impacts battery lifetime and heat

Enrico Macii

... continued on Page 8

* Contact Person: Luca Benini, Università di Bologna, DEIS, Viale Risorgimento 2, Bologna, ITALY 40136; Phone: +39-051-209.3782; Fax: +39-051-209.3073; E-mail: lbenini@deis.unibo.it

Low Power Circuits ... continued from Page 7

dissipation system cost, the former constrains power grid and power supply circuits design. We will focus on average power in the remainder of the paper, even if maximum power is also a serious concern. A more detailed analysis of the various contributions to overall power consumption in CMOS circuits (the dominant VLSI technology) is provided in the following section.

It is important to stress from the outset that power minimization is never the only objective in real-life designs. Performance is always a critical metric that cannot be neglected. Unfortunately, in most cases, power can be reduced at the price of some performance degradation. For this reason, several metrics for joint power-performance have been proposed in the past. In many designs, the *power-delay product* (i.e., energy) is an acceptable metric. Energy minimization rules out design choices that heavily compromise performance to reduce power consumption. When performance has priority over power consumption, the *energy-delay product* [4] can be adopted to tightly control performance degradation. Alternatively, we can take a *constrained optimization* approach. In this case, performance degradation is acceptable up to a given bound. Thus, power minimization requires optimal exploitation of the slack on performance constraints.

It is important to stress from the outset that power minimization is never the only objective in real-life designs. Performance is always a critical metric that cannot be neglected. Unfortunately, in most cases, power can be reduced at the price of some performance degradation. For this reason, several metrics for joint power-performance have been proposed in the past.

Besides power vs. performance, another key trade-off in VLSI design is *power vs. flexibility*. Several authors have observed that application-specific designs are orders of magnitude more power efficient than general-purpose systems programmed to perform the same computation [1]. On the other hand, flexibility (programmability) is often an indispensable requirement, and designers must strive to achieve maximum power efficiency without compromising flexibility.

The two fundamental trade-offs outlined above motivate our selection of effective low-power design techniques and illustrative cases. We selected successful low power design examples (real-life products) from four classes of circuits, spanning the flexibility vs. power-efficiency trade-off. To maintain uniformity, we chose designs targeting the same end-market, namely multimedia and Internet-enabled portable appliances, where power efficiency is not out-weighted by performance requirements. Our illustrative examples are (in order of decreasing flexibility):

1. Transmeta's Crusoe processor family [5]. This design is representative of the class of general-purpose *x86-compatible* processors for high-end portable appliances, and it features aggressive power optimizations as key market differentiator.

2. Intel's StrongARM family [6]. This design is representative of the class of low-power integrated core-and-peripherals for personal digital assistants and palmtop computers.

3. Texas Instrument TMS320C5x family [7]. This design is a typical very-low power digital signal processor for baseband processing in wireless communication devices (i.e., digital cellular phones).

4. Toshiba's MPEG4 Codec [8]. This is a typical application-specific system-on-chip for multimedia support, targeted for PDAs and digital cameras.

Needless to say, our selection neither implies any endorsement of the commercial products derived from the above mentioned designs, nor implies any form of comparison or benchmarking against competing products.

The remainder of this paper is organized as follows. In *Basic Principles*, we analyze in some detail the sources of power consumption in CMOS technology, and we introduce the basic principles of power optimization. In the section *Technology and Circuit Level Optimizations* we describe a few successful power optimization methods at the technology and circuit level. The section *Logic and Architecture Level Optimizations* covers logic and architecture-level optimizations, while *Software and System Level Optimizations* deals with software and system-level optimizations. Whenever possible, we will describe the optimization techniques with reference to the example designs.

Basic Principles

CMOS is, by far, the most common technology used for manufacturing digital ICs. There are 3 major sources of power dissipation in a CMOS circuit [9]:

$$P = P_{Switching} + P_{Short\text{-}Circuit} + P_{Leakage}$$

$P_{Switching}$, called *switching power*, is due to charging and discharging capacitors driven by the circuit. $P_{Short\text{-}Circuit}$, called *short-circuit power*, is caused by the short circuit currents that arise when pairs of PMOS/NMOS transistors are conducting simultaneously. Finally, $P_{Leakage}$, called *leakage power*, originates from substrate injection and subthreshold effects. For older technologies (0.8 μm and above), $P_{Switching}$ was predominant. For deep-submicron processes, $P_{Leakage}$ becomes more important.

Design for low-power implies the ability to reduce all three components of power consumption in CMOS circuits during the development of a low power electronic product. Optimizations can be achieved by facing the power problem from different perspectives: design and technology. En-

. . . continued on Page 10

Low Power Circuits ... continued from Page 9

hanced design capabilities mostly impact switching and short-circuit power; technology improvements, on the other hand, contribute to reductions of all three components.

Switching power for a CMOS gate working in a synchronous environment is modeled by the following equation:

$$P_{Switching} = \frac{1}{2} C_L V_{DD}^2 f_{Clock} E_{SW}$$

where C_L is the output load of the gate, V_{DD} is the supply voltage, f_{Clock} is the clock frequency and E_{SW} is the *switching activity* of the gate, defined as the probability of the gate's output to make a logic transition during one clock cycle. Reductions of $P_{Switching}$ are achievable by combining minimization of the parameters in the equation above.

Historically, supply voltage scaling has been the most adopted approach to power optimization, since it normally yields considerable savings thanks to the quadratic dependence of $P_{Switching}$ on V_{DD}. The major short-coming of this solution, however, is that lowering the supply voltage affects circuit speed. As a consequence, both design and technological solutions must be applied in order to compensate the decrease in circuit performance introduced by reduced voltage. In other words, speed optimization is applied first, followed by supply voltage scaling, which brings the design back to its original timing, but with a lower power requirement.

A similar problem, i.e., performance decrease, is encountered when power optimization is obtained through frequency scaling. Techniques that rely on reductions of the clock frequency to lower power consumption are thus usable under the constraint that some performance slack does exist. Although this may seldom occur for designs considered in their entirety, it happens quite often that some specific units in a larger architecture do not require peak performance for some clock/machine cycles. Selective frequency scaling (as well as voltage scaling) on such units may thus be applied, at no penalty in the overall system speed.

Optimization approaches that have a lower impact on performance, yet allowing significant power savings, are those targeting the minimization of the *switched capacitance* (i.e., the product of the capacitive load with the switching activity). Static solutions (i.e., applicable at design time) handle switched capacitance minimization through area optimization (that corresponds to a decrease in the capacitive load) and switching activity reduction via exploitation of different kinds of signal correlations (temporal, spatial, spatio-temporal). Dynamic techniques, on the other hand, aim at eliminating power wastes that may be originated by the the application of certain system workloads (i.e., the data being processed).

Static and dynamic optimizations can be achieved at different levels of design abstraction. Actually, addressing the power problem from the very early stages of design development offers enhanced opportunities to obtain significant reductions of the power budget and to avoid costly redesign steps. Power conscious design flows must then be adopted; these require, at each level of the design hierarchy, the exploration of different alternatives, as well as the availability of power estimation tools that could provide accurate feed-back on the quality of each design choice.

In the sequel, we will follow a bottom-up path to present power optimization solutions that have found their way in the development of realistic designs; on the other hand, issues related to power modeling and estimation will not be covered; the interested reader may refer to survey articles available in the literature (see, for example, [10–13]).

Technology and Circuit Level Optimizations

VLSI technology scaling has evolved at an amazingly fast pace for the last thirty years. Minimum device size has kept shrinking by a factor $k = 0.7$ per technology generation. The basic scaling theory, known as *constant field scaling* [14], mandates the synergistic scaling of geometric features and silicon doping levels to maintain constant field across the gate oxide of the MOS transistor. If devices are constant-field scaled, power dissipation scales as k^2 and power density (i.e., power dissipated per unit area) remains constant, while speed increases as k.

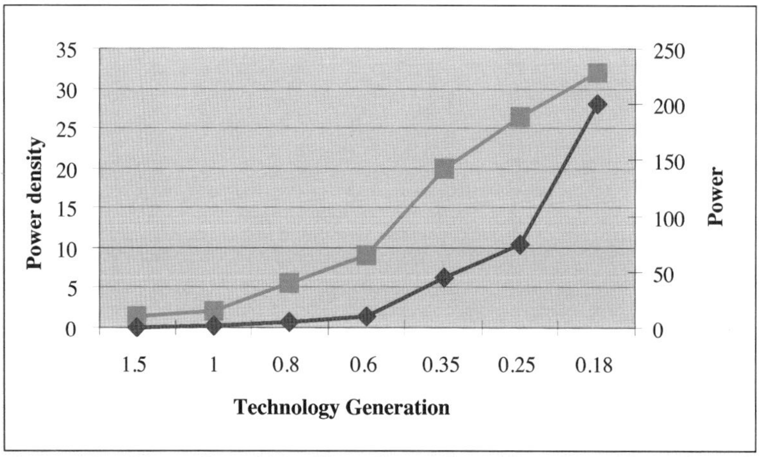

Figure 2. Power density (squares) and power (diamonds) vs. technology generation.

In a constant-field scaling regime, silicon technology evolution is probably the most effective way to address power issues in digital VLSI design. Unfortunately, as shown in Fig. 2, power consumption in real-life integrated circuits does not always follow this trend. Figure 2 shows average power consumption and power density trends for Intel Microprocessors, as a function of technology generations [14, 15]. Observe that both average power and power density increase as minimum feature size shrinks.

This phenomenon has two concomitant causes. First, die size has been steadily increasing with technology, causing an increase in total average power. Second, voltage supply has not been reduced according to the directives of constant-field scaling. Historically, supply voltage has scaled much more slowly than device size because: (i) supply voltage levels have been standardized (5 V, then 3.3 V), (ii) faster transistor operation can be obtained by allowing the electric field to raise in the device (i.e., "overdriving" the transistor). This approach is known as *constant voltage scaling*, and it has been adopted in older silicon technologies, up to the 0.8 μm generation.

... continued on Page 12

Designing Low-Power Circuits: Practical Recipes

Low Power Circuits ... continued from Page 11

Probably, the most widely known (and successful) power-reduction technique, known as *power-driven voltage scaling* [3], moves from the observation that constant-voltage scaling is highly inefficient from the power viewpoint, and that we can scale down the voltage supply to trade off performance for power. Depending on the relative weight of performance with respect to power consumption constraints, different voltage levels can be adopted. It is important to observe, however, that transistor speed does not depend on supply voltage V_{DD} alone, but on the *gate overdrive*, namely the difference between voltage supply and device threshold voltage ($V_{DD} - V_T$). Hence, several authors have studied the problem of jointly optimizing V_{DD} and V_T to obtain minimum energy, or minimum energy-delay product [1, 3].

Accurate modeling of MOS transistor currents is paramount for achieving acceptable scaled V_{DD} and V_T values. The simplistic quadratic model of CMOS ON-current $I_{DS} = S' \cdot (V_{GS} - V_T)^2$ leads to overly optimistic switching speed estimates for submicrometric transistors. In short-channel transistors, the *velocity saturation* of the electrons traveling between drain and source dictates a different current equation: $I_{DS} = S' \cdot (V_{GS} - V_T)^m$, with $1 \leq m < 2$ (e.g., $m = 1.3$). Furthermore, another important characteristic of CMOS transistors is sub-threshold conduction. When $V_{GS} < V_T$, the current is not zero, but it follows an exponential law: $I_{DS} = R' \cdot e^{V_T/V_o}$, V_o being the technology-dependent *sub-threshold slope*.

Velocity saturation favors supply down-scaling, because it implies diminishing returns in performance when supply voltage is too high. On the other hand, sub-threshold conduction limits down-scaling, because of increased static current leaking through nominally OFF transistors. Intuitively, optimization of threshold and supply voltage requires balancing ON-current and OFF-current, while at the same time maintaining acceptable performance. Optimizing V_{DD} and V_T for minimum energy delay product leads to surprisingly low values of both. V_{DD} should be only slightly larger than $2V_T$, with V_T a few hundred millivolts. This approach is known as ULP, or *ultra-low-power CMOS* [16].

ULP CMOS is not widely used in practice for two main reasons. First, threshold control is not perfect in real-life technology. Many transistors may have sub-threshold currents that are orders of magnitude larger than expected if their threshold is slightly smaller than nominal (remember that sub-threshold current is exponential in V_{GS}). Second, sub-threshold current is exponentially dependent on temperature, thereby imposing tight thermal control of ULP CMOS, which is not cost-effective. Nevertheless, aggressive voltage scaling is commonplace in low-power VLSI circuits: voltage supply and thresholds are not scaled to their ultimate limit, but they are significantly re-

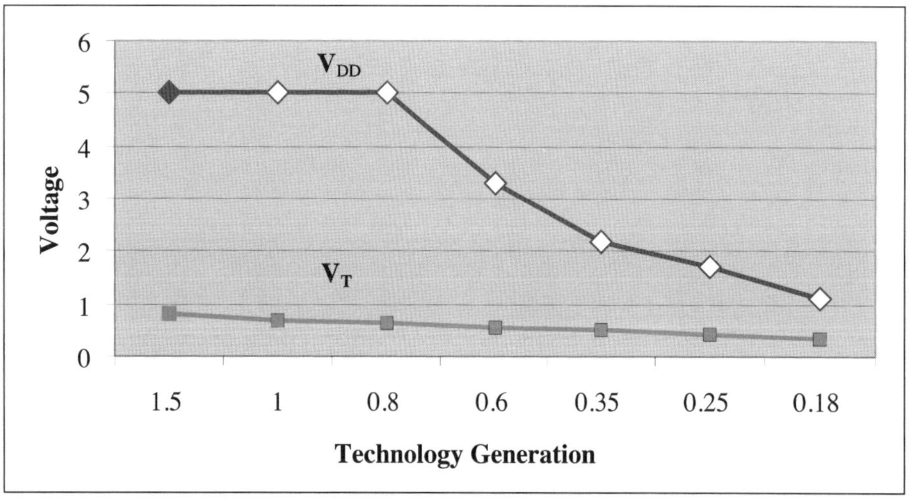

Figure 3. Supply voltage (diamonds) and threshold voltage (squares) vs. technology generation.

duced with respect to constant-voltage scaling. See Example 1.

Power-driven voltage scaling was a "low-hanging fruit" in older technologies. Unfortunately, starting from the 0.8 μm generation, the situation has changed. As shown in Fig. 3, supply voltage has started to decrease with shrinking device size even for high-performance transistors [15]. There are several technological reasons for this trend, which are outside the scope of this paper. From the point of view of power reduction, technology-driven voltage scaling has two important consequences. Aggressively scaled transistors with minimum channel length are becoming increasingly leaky in the OFF state (getting closer to ULP), and there is not much room left for further voltage scaling.

Leakage power is already a major concern in current technologies, because it impacts battery lifetime even if the circuit is completely idle. Quiescent power specifications tend to be very tight. In fact, CMOS technology has traditionally been extremely power-efficient when transistors are not switching, and system designers expect low leakage from CMOS chips. To meet leakage power constraints, *multiple-threshold* and *variable threshold* circuits have been proposed [3]. In multiple-threshold CMOS, the process provides two different thresholds. Low-threshold transistors are fast and leaky, and they are employed on speed-critical sub-circuits. High-threshold transistors are slower but exhibit low sub-threshold leakage, and they are employed in noncritical units/paths of the chip.

Unfortunately, multiple-threshold techniques tend to lose effectiveness as more transistors become timing-criti-

. . . continued on Page 14

Example 1. The StrongARM processor was first designed in a three-metal, 0.35 μm CMOS process, which had been originally developed for high-performance processors (the DEC Alpha family). The first design decision was to reduce supply voltage from 3.45 V to 1.5 V, with threshold voltage $VTN = |V_{TP}| = 0.35$ V, thus obtaining a 5.3x power reduction. The performance loss caused by voltage scaling was acceptable, because StrongARM has much more relaxed performance specifications than Alpha.

As a second example, the research prototype of the TMS320C5x DSP adopted a supply voltage $V_{DD} = 1$ V in a 0.35 μm technology. The aggressive V_{DD} value was chosen by optimizing the energy-delay product.

Example 2. The TMS320C5x DSP prototype adopted a dual-threshold CMOS process, in order to provide acceptable performance at the aggressively down-scaled supply voltage of $V_{DD} = 1$ V. The nominal threshold voltages were 0.4 V and 0.2 V for slow and fast transistors, respectively. Leakage current for the high-V_T transistors is below 1 nA/μm. For the low-V_T transistors, leakage current is below 1 μA/μm (a three-orders-of-magnitude difference!). The drive current of low-V_T transistors is typically twice that of the high-V_T devices.

The MPEG4 Video Codec prototype adopted the variable-threshold voltage scheme to reduce power dissipation. Substrate biasing is exploited to dynamically adjust the threshold: V_T is controlled to 0.2 V in active mode and to 0.55 V when the chip is in stand-by mode.

Low Power Circuits ... continued from Page 13

cal. Variable-threshold circuits overcome this shortcoming by dynamically controlling the threshold voltage of transistors through substrate biasing. When a variable-threshold circuit becomes quiescent, the substrate of NMOS transistors is negatively biased, and their threshold increases because of the well known *body-bias effect*. A similar approach can be taken for PMOS transistors (which require positive body bias). Variable-threshold circuits can, in principle, solve the quiescent leakage problem, but they require stand-by control circuits that modulate substrate voltage. Needless to say, accurate and fast body-bias control is quite challenging, and requires carefully designed closed-loop control [3]. See Example 2.

In analogy with threshold voltage, supply voltage can also be controlled to reduce power, albeit in the limited ranges allowed in highly scaled technologies. *Multiple-voltage* and *variable voltage* techniques have been developed to this purpose [3]. In multiple-voltage circuits two or more power supply voltages are distributed on chip. Similarly to the multiple-threshold scheme, timing-critical transistors can be powered at a high voltage, while most transistors are connected to the low voltage supply. Multiple voltages are also frequently used to provide standard voltage levels (e.g., 3.3 V) to input-output circuits, while powering on-chip internal logic at a much lower voltage to save power. The main challenges in the multiple-voltage approach are in the design of multiple power distribution grids and of power-efficient level-shifters to interface low-voltages with high-voltage sections.

Example 3. Supply voltage is closed-loop controlled in both the MPEG4 Codec core and the Crusoe Processor. Supply voltage control in the MPEG4 helps in compensating process fluctuations and operating environment changes (e.g., temperature changes). Variable-voltage operation in Crusoe is more emphasized: it automatically lowers the supply voltage when the processor is under-loaded. From the performance viewpoint, the processor gradually slows down when its full performance is not needed, but its supply voltage and clock speed are rapidly raised when the workload increases.

In both MPEG4-core and Crusoe, frequency and power supply are synergistically adjusted thanks to a clever feed-back control loop: a replica of the critical path is employed to estimate circuit slow-down in response to voltage down-scaling, and clock frequency is reduced using a PLL that locks on the new frequency.

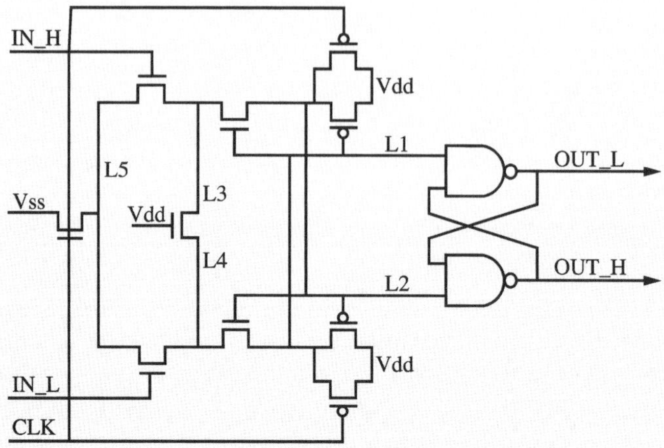

Figure 4. Low-power FF used in the StrongARM design.

Variable-voltage optimizations allow modulating the power supply dynamically during system operation. In principle, this is a very powerful technique, because it gives the possibility to trade off power for speed at run time, and to finely tune performance and power to non-stationary workloads. In practice, the implementation of this solution requires considerable design ingenuity. First, voltage changes require non-negligible time, because of the large time constants of power supply circuits. Second, the clock speed must be varied consistently with the varying speed of the core logic, when supply voltage is changed. Closed-loop feedback control schemes have been implemented to support variable supply voltage operation. See Example 3.

The techniques based on voltage scaling described above require significant process and system support, which imply additional costs that can be justified only for large-volume applications. Less aggressive circuit-level approaches, that do not require ad-hoc processes can also be successful. Among the proposals available in the literature, library cell design and sizing for low power have gained wide-spread acceptance. From the power viewpoint, probably the most critical cells in a digital library are the sequential primitives, namely, latches and flip-flops. First, they are extremely numerous in today's deeply pipelined circuits; and second, they are connected to the most active network in the chip, i.e., the clock. Clock drivers are almost invariantly the largest contributors to the power budget of a chip, primarily because of the huge capacitive load of the clock distribution network. Flip-flop (and latch) design for low power focuses on minimizing clock load and reducing internal power when the clock signal is toggled. Significant power reductions have been achieved by carefully designing and sizing the flip-flops [1].

Clock power optimization is effectively pursued at the circuit level also by optimizing the power distribution tree, the clock driver circuits and the clock generation circuitry (on-chip oscillators and phase-locked loops) [3]. Higher abstraction-level approaches to clock power reduction are surveyed in the next section. See Example 4.

Transistor sizing is also exploited to minimize power consumption in combinational logic cells. Rich libraries with many available transistor sizes are very useful in low-power design, because they help synthesis tools in achieving optimum sizing for a wide range of gate loads. Power savings can be obtained by adopting non-standard logic implementation styles such as pass-transistor logic, which can reduce the number of transistors (and, consequently the capacitive load), for implementing logic functions which are commonly found in arithmetic units (e.g., exclusive-or, multiplexers).

. . . continued on Page 16

Example 4. The StrongARM processor adopted the edge-triggered flip-flop of Figure 4 to reduce clock load with respect to the flow-through latches used in the Alpha processor. The flip-flop is based on a differential structure, similar to a sense amplifier. Notice that the internal structure of the FF is quite complex in comparison with the simple flow-through latches in Alpha. However,the FF has reduced clock load (only three transistors). This key advantage gave a 1.3 x overall power reduction over the latch-based design.

Example 5. Static power minimization was a serious concern for all our example designs. In the MPEG4 Codec, which adopted a dual supply voltage scheme, significant static power can be dissipated when high-supply gates are driven by a low-supply gate, because the driver cannot completely turn off the pull-up transistors in the fanout gates. For this reason, special level shifter cells were designed that exploited a regenerative cross-coupled structure similar to a sense amplifier to restore full swing in the high-supply section, and minimize static power.

Leakage power was the main concern in the StrongARM design, that could not benefit from multiple threshold or variable threshold control. To reduce leakage power in gates outside the critical path, non-minimum length transistors were employed in low-leakage, low-performance cells.

Example 6. The multiply-accumulate (MAC) unit of the StrongARM processor is based on a Wallace-tree multiplier coupled with a carry-lookahead adder. The Wallace-tree architecture was chosen because it is very fast, but also because it has low dynamic power consumption in the carry-save adder tree. The improvement comes from a sizable reduction in spurious switching, thanks to path delay balancing in the Wallace-tree. A 23% power reduction (as well as a 25% speed-up) is achieved by the Wallace-tree architecture with respect to the array multiplier.

Low Power Circuits ... continued from Page 15

While most traditional power optimization techniques for logic cells focus on minimizing switching power, circuit design for leakage power reduction is also gaining importance [17]. See Example 5.

Logic and Architecture Level Optimizations

Logic-level power optimization has been extensively researched in the last few years [11, 18]. Given the complexity of modern digital devices, hand-crafted logic-level optimization is extremely expensive in terms of design time and effort. Hence, it is cost-effective only for structured logic in large-volume components, like microprocessors (e.g., functional units in the data-path). Fortunately, several optimizations for low power have been automated and are now available in commercial logic synthesis tools [19], enabling logic-level power optimization even for unstructured logic and for low-volume VLSI circuits.

During logic optimization, technology parameters such as supply voltage are fixed, and the degrees of freedom are in selecting the functionality and sizing the gates implementing a given logic specification. As for technology and circuit-level techniques, power is never the only cost metric of interest. In most cases, performance is tightly constrained as well. A common setting is *constrained power optimization*, where a logic network can be transformed to minimize power only if critical path length is not increased. Under this hypothesis, an effective technique is based on *path equalization*.

Path equalization ensures that signal propagation from inputs to outputs of a logic network follows paths of similar length. When paths are equalized, most gates have aligned transitions at their inputs, thereby minimizing spurious switching activity (which is created by misaligned input transitions). This technique is very helpful in arithmetic circuits, such as adders of multipliers. See Example 6.

Glue logic and controllers have much more irregular structure than arithmetic units, and their gate-level implementations are characterized by a wide distribution of path delays. These circuits can be optimized for power by *resizing*. Resizing focuses on fast combinational paths. Gates on fast paths are down-sized, thereby decreasing their input capacitances, while at the same time slowing down signal propagation. By slowing down fast paths, propagation delays are equalized, and power is reduced by joint spurious switching and capacitance reduction. Resizing does not

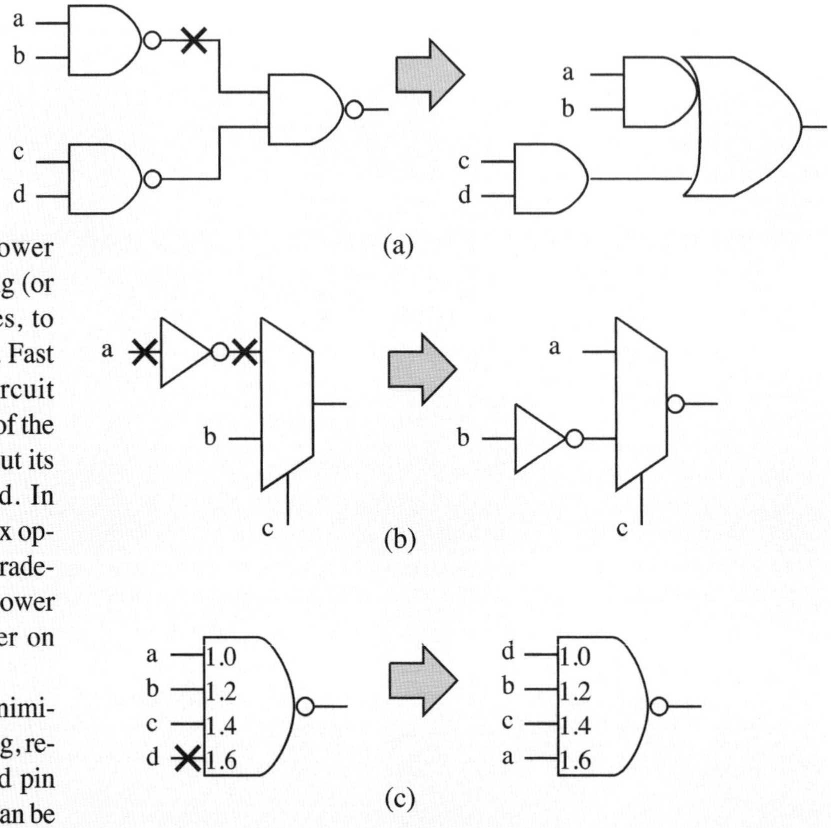

Figure 5. Local transformations: (a) re-mapping, (b) phase assignment, (c) pin swapping.

always imply down-sizing. Power can be reduced also by enlarging (or buffering) heavily loaded gates, to increase their output slew rates. Fast transitions minimize short-circuit power of the gates in the fanout of the gate which has been sized up, but its input capacitance is increased. In most cases, resizing is a complex optimization problem involving a trade-off between output switching power and internal short-circuit power on several gates at the same time.

Other logic-level power minimization techniques are re-factoring, re-mapping, phase assignment and pin swapping. All these techniques can be classified as *local transformations*. They are applied on gate netlists, and focus on nets with large switched capacitance. Most of these techniques replace a gate, or a small group of gates, around the target net, in an effort to reduce capacitance and switching activity. Similarly to resizing, local transformations must carefully balance short circuit and output power consumption. See Example 7.

Logic-level power minimization is relatively well studied and understood. Unfortunately, due to the local nature of most logic-level optimizations, a large number of transformations has to be applied to achieve sizable power savings. This is a time consuming and uncertain process, where uncertainty is caused by the limited accuracy of power estimation. In many cases, the savings produced by a local move are below the "noise floor" of the power estimation engine. As a consequence, logic-level optimization does

. . . continued on Page 18

Example 7. Figure 5 shows three examples of local transformations. In Fig. 5 (a) a re-mapping transformation is shown, where a high-activity node (marked with `x`) is removed thanks to a new mapping onto an `and-or` gate. In Fig. 5 (b), phase assignment is exploited to eliminate one of the two high-activity nets marked with `x`. Finally, pin swapping is applied in Fig. 5 (c) to connect a high-activity net with the input pin of the 4-input `nand` with minimum input capacitance.

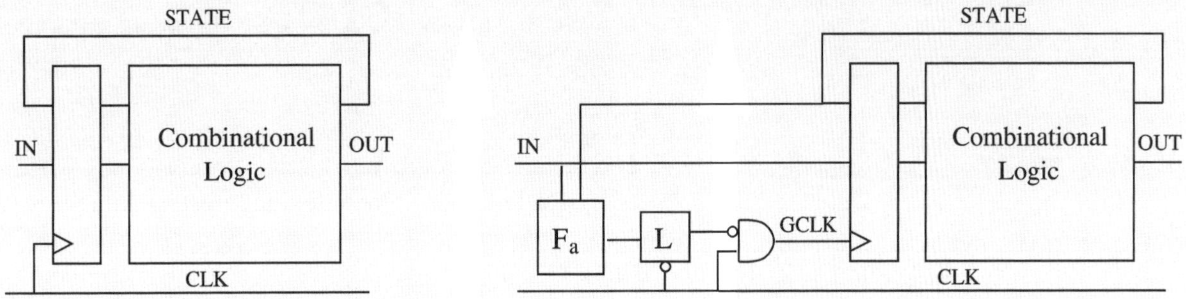

Figure 6. Example of gated clock architecture.

Low Power Circuits ... continued from Page 17

not result in massive power reductions. Savings are in the 10 to 20% range, on average. More sizable savings can be obtained by optimizing power at a higher abstraction level, as detailed next.

Complex digital circuits usually contain units (or parts thereof) that are not performing useful computations at every clock cycle. Think, for example, of arithmetic units or register files within a microprocessor or, more simply, to registers of an ordinary sequential circuit. The idea, known for a long time in the community of IC designers, is to disable the logic which is not in use during some particular clock cycles, with the objective of limiting power consumption. In fact, stopping certain units from making useless transitions causes a decrease in the overall switched capacitance of the system, thus reducing the switching component of the power dissipated. Optimization techniques based on the principle above belong to the broad class of *dynamic power management* (DPM) methods. As briefly explained in the section Basic Principles, they achieve power reductions by exploiting specific run-time behaviors of the design to which they are applied.

The natural domain of applicability of DPM is system-level design; therefore, it will be discussed in greater detail in the next section. Nevertheless, this paradigm has also been successfully adopted in the context of architectural optimization.

Clock gating [20] provides a way to selectively stop the clock, and thus force the original circuit to make no transition, whenever the computation to be carried out by a hardware unit at the next clock cycle is useless. In other words, the clock signal is disabled in accordance with the idle conditions of the unit.

As an example of use of the clock-gating strategy, consider the traditional block diagram of a sequential circuit, shown on the left of Fig. 6. It consists of a combinational logic block and an array of state registers which are fed by the next-state logic and which provide some feed-back information to the combinational block itself through the present-state input signals. The corresponding gated-clock architecture is shown on the right of the figure.

The circuit is assumed to have a single clock, and the registers are assumed to be edge-triggered flip-flops. The combinational block F_a is controlled by the primary inputs, the present-state inputs, and the primary outputs of the circuit, and it implements the activation function of the clock gating mechanism. Its purpose is to selectively stop the local clock of the circuit when no state or output transition takes place. The block named L is a latch, transparent when the global clock signal CLK is inactive. Its presence is essential for a correct operation of the system, since it takes care of filtering glitches that may occur at the output of block F_a. It should be noted that the logic for the activation function is on the critical path of the circuit; therefore, timing violations may occur if the synthesis of F_a is not carried out properly.

The clock management logic is synthesized from the Boolean function representing the idle conditions of the circuit. It may well be the case that considering all such conditions results in additional circuitry that is too large and power consuming. It may then be necessary to synthesize a simplified function, which dissipates the minimum possible power, and stops the clock with maximum efficiency. Because of its effectiveness, clock-gating has been applied extensively in real designs, as shown by the examples below, and it has lately found its way in industry-strength CAD tools (e.g., *Power Compiler* by Synopsys). See Examples 8 and 9.

Power savings obtained by gating the clock distribution network of some hardware resources come at the price of a global decrease in performance. In fact, resuming the operation of an inactive resource introduces a latency penalty that negatively impacts system speed. In other words, with clock gating (or with any similar DPM technique), performance and throughput of an architecture are traded for power. See Example 10.

Clock gating is not the only scheme for implementing logic shutdown; solutions ranging from register disabling and usage of dual-state flip-flops to insertion of guarding logic and gate freezing may be equally effective, although no industry-strength assessment of these design options has been done so far.

. . . continued on Page 20

Example 8. The TMS320C5x DSP processor exploits clock gating to save power during active operation. In particular, a two-fold power management scheme, local and global, is implemented. The clock signal feeding the latches placed on the inputs of functional units is enabled only when useful data are available at the units' inputs, and thus meaningful processing can take place. The gating signals are generated automatically by local control logic using information coming from the instruction decoder. Global clock gating is also available in the processor, and is controlled by the user through dedicated power-down instructions, `IDLE1`, `IDLE2`, and `IDLE3`, which target power management of increasing strength and effectiveness. Instruction `IDLE1` only stops the CPU clock, while it leaves peripherals and system clock active. Instruction `IDLE2` also deactivates all the on-chip peripherals. Finally, instruction `IDLE3` powers down the whole processor.

Example 9. The MPEG4 Codec exploits software-controlled clock gating in some of its internal units, including the DMA controller and the RISC core. Clock gating for a unit is enabled by setting the corresponding bit in a "sleep" register of RISC through a dedicated instruction. The clock signal for the chosen hardware resource is OR-ed with the corresponding output of the sleep register to stop clock activity.

Example 10. The latency for the CPU of the TMS320C5x DSP processor to return to active operation from the `IDLE3` mode takes around 50 μsec, due to the need of the on-chip PLL circuit to lock with the external clock generator.

Example 11. Several schemes [21–23] have been proposed to store compressed instructions in main memory and decompress them on-the-fly before execution (or when they are stored in the instruction cache). All these techniques trade off the efficiency of the compression algorithm with the speed and power consumption of the hardware de-compressor. Probably the best known instruction compression approach is the "Thumb" instruction set of the ARM microprocessor family [24]. ARM cores can be programmed using a reduced set of 16-bit instructions (in alternative to standard 32-bit RISC instructions) that reduce the required instruction memory occupation and required bandwidth by a factor of 2.

Low Power Circuits ... continued from Page 19

Software and System Level Optimizations

Electronic systems and subsystems consist of hardware platforms with several software layers. Many system features depend on the hardware/software interaction, e.g., programmability and flexibility, performance and energy consumption.

Software does not consume energy *per se*, but it is the execution and storage of software that requires energy consumption by the underlying hardware. Software execution corresponds to performing operations on hardware, as well as accessing and storing data. Thus, software execution involves power dissipation for computation, storage, and communication. Moreover, storage of computer programs in semiconductor memories requires energy (refresh of DRAMs, static power for SRAMs).

The energy budget for storing programs is typically small (with the choice of appropriate components) and predictable at design time. Thus, we will concentrate on energy consumption of software during its execution. Nevertheless, it is important to remember that reducing the size of program, which is a usual objective in compilation, correlates with reducing their energy storage costs. Additional reduction of code size can be achieved by means of compression techniques. See Example 11.

The energy cost of executing a program depends on its machine code and on the hardware architecture parameters. The machine code is derived from the source code from compilation. Typically, the energy cost of the machine code is affected by the back-end of software compilation, that controls the type, number and order of operations, and by the means of storing data, e.g., locality (registers vs. memory arrays), addressing, order. Nevertheless, some architecture-independent optimizations can be useful in general to reduce energy consumption, e.g., selective loop unrolling and software pipelining.

Software instructions can be characterized by the number of cycles needed to execute them and by the energy required per cycle. The energy consumed by an instruction depends weakly on the state of the processor (i.e., by the previously executed instruction). On the other hand, the energy varies significantly when the instruction requires storage in registers or in memory (caches).

The traditional goal of a compiler is to speed up the execution of the generated code, by reducing the code size (which correlates with the latency of execution time) and minimizing *spills* to memory. Interestingly enough, executing machine code of minimum size would consume the minimum energy, if we neglect the interaction with memory and we assume a uniform energy cost of each instruction. Energy-efficient compilation strives at achieving machine code that requires less energy as compared to a performance-driven traditional compiler, by leveraging the disuniformity in instruction energy cost, and the different energy costs for storage in registers and in main memory due to addressing and address decoding. Nevertheless, results are sometimes contradictory. Whereas for some architectures energy-efficient compilation gives a competitive advantage as compared to traditional compilation, for some others the most compact code is also the most economical in terms of energy, thus obviating the need of specific low-power compilers.

It is interesting to note that the style of the software source program (for any given function) affects the energy cost. Energy-efficient software can be achieved by enforcing specific writing styles, or by allowing source-code transformation to reduce the power consumption of a program. Such transformation can be automated, and can be seen as the front-end of compilation. See Example 12.

> Example 12. The IBM XL compilers transform source code (from different languages) into an internal form called W-code. The Toronto Portable Optimizer (TPO) [25] performs W-code to W-code transformations. Some transformations are directed to reducing power consumption. To this goal, switching activity is minimized by exploiting relations among variables that take similar values. Power-aware instruction scheduling can increase the opportunity of clock gating, by grouping instructions with similar resource usage.

> Example 13. Simultaneous multi-threading (SMT) [25] has been proposed as an approach for the exploitation of instruction-level parallelism. In SMT, the processor core shares its execution-time resources among several simultaneously executing threads (programs). Thus, at each cycle, instructions can be fetched from one or more independent threads, and passed to the issue and execution boxes. Since the resources can be filled by instructions from parallel threads, their usage increases and energy waste decreases. Compaq has announced that its future 21x64 designs will embrace the SMT paradigm.

Power-aware operating systems (OSs) trade generality for energy efficiency. In the case of embedded electronic systems, OSs are streamlined to support just the required applications. On the other hand, such an approach may not be applicable to OSs for personal computers, where the user wants to retain the ability of executing a wide variety of applications.

Energy efficiency in an operating system can be achieved by designing an energy aware *task scheduler*. Usually, a scheduler determines the set of start times for each task, with the goal of optimizing a cost function related to the completion time of all tasks, and to satisfy real time constraints, if applicable. Since tasks are associated with resources having specific energy models, the scheduler can exploit this information to reduce run-time power consumption. See Example 13.

Operating systems achieve major energy savings by implementing *dynamic power management* (DPM) of the system resources. DPM dynamically reconfigures an electronic system to provide the requested services and performance levels with a minimum number of active components or a

. . . continued on Page 22

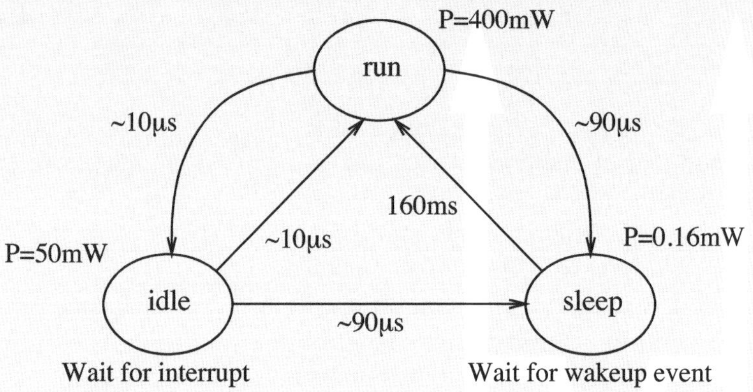

Figure 7. Power state machine for the StrongARM SA-1100 processor.

Low Power Circuits ... continued from Page 21

minimum load on such components [26]. Dynamic power management encompasses a set of techniques that achieve energy-efficient computation by selectively shutting down or slowing down system components when they are *idle* (or partially unexploited). DPM can be implemented in different forms including, but not limited to, clock gating, clock throttling, supply-voltage shut-down, and dynamically-varying power supplies, as described earlier in this paper.

We demonstrate here power management using two examples from existing successful processors. The former example describes the shut down modes of the StrongARM processor, while the latter describes the slow down operation within the Transmeta Crusoe chip. See Example 14.

An operating system that controls a power manageable component, such as the SA-1100, needs to issue commands to force the power state transitions, in response to the varying workload. The control algorithm that issues the commands is termed *policy*. An important problem is the computation of the policy for a given workload statistics and component parameter. Several avenues to policy computation are described in another tutorial paper [28]. We consider next another example, where power management is achieved by slowing down a component. See Example 15.

We mentioned before the important problem of computing policies, with guaranteed optimality properties, to control one or more components with possibly different management schemes. A related interesting problem is the design of components that can be effectively power managed. See Example 16.

Several system-level design trade-offs can be explored to reduce energy consumption. Some of these design choices belong to the domain of *hardware/software co-design*, and leverage the migration of hardware functions to software or vice versa. For example, the Advanced Configuration and Power Interface (ACPI) standard, initiated by Intel, Microsoft and Toshiba, provides a portable hw/sw interface that makes it easy to implement DPM policies for personal computers in software. As another example, the decision of implementing specific functions (like MPEG decoding) on specifically-dedicated hardware, rather than on a programmable processor, can significantly affect energy consumption. As a final interesting example, we would like to point out that *code morphing* can be a very powerful tool in reducing energy dissipation. See Example 17.

Conclusions

Electronic design aims at striking a balance between performance and power efficiency. Designing low-power applications is a multi-faceted problem, because of the plurality of embodiments that a system specification may have and the variety of

... continued on Page 24

Designing Low-Power Circuits: Practical Recipes

Example 14. The StrongARM SA-1100 processor [27] is an example of a power manageable component. It has three modes of operation: `run`, `idle`, and `sleep`. The `run` mode is the normal operating mode of the SA-1100: every on-chip resource is functional. The chip enters the `run` mode after successful power-up and reset. The `idle` mode allows a software application to stop the CPU when not in use, while continuing to monitor on-chip or off-chip interrupt requests. In the `idle` mode, the CPU can be brought back to the `run` mode quickly when an interrupt occurs. The `sleep` mode offers the greatest power savings and consequently the lowest level of available functionality. In the transition from `run` or `idle`, the SA-1100 performs an orderly shutdown of on-chip activity. In a transition from `sleep` to any other state, the chip steps through a rather complex wake-up sequence before it can resume normal activity.

The operation of the StrongARM SA-1100 can be captured by a power state machine model, as shown in Fig. 7. States are marked with power dissipation and performance values, edges are marked with transition times. The power consumed during transitions is approximately equal to that in the `run` mode. Notice that both `idle` and `sleep` have null performance, but the time for exiting `sleep` is much longer than that for exiting `idle` (10 μs versus 160 ms). On the other hand, the power consumed by the chip in `sleep` mode (0.16 mW) is much smaller than that in `idle` (50 mW).

Example 15. The Transmeta Crusoe TM5400 chip uses the Long Run^R power management scheme, that allows the OS to slow down the processor when the workload can be serviced at a slower pace. Since a slow down in clock frequency permits a reduction in supply voltage, then the energy to compute a given task decreases quadratically. From a practical standpoint, the OS observes the task queue and determines the appropriate frequency and voltage levels, which are transmitted to the phase-locked loop (PLL) and DC-DC converter. As a result, Transmeta claims that the Crusoe chip can play a soft DVD at the same power level used by a conventional x86 architecture in sleep state.

Example 16. The 160-Mhz implementation of the StrongARM described in [6] supports 16 possible operating frequencies generated by a PLL. Moreover, the PLL was designed within a power budget of 2 mW, so that it could be kept running in the `idle` state. Thus, the need for a low-power PLL is dictated by the power budget in the `idle` state. Note that the ability of keeping the PLL running in the `idle` state is key to achieving a fast transition to the `run` state.

Example 17. The Transmeta Crusoe chip executes x86-compatible binaries on a proprietary architecture, which is designed for low-power operation. The operating system performs run-time binary translation to this effect. The code morphing is key in reducing the energy cost associated with any given program.

Designing Low-Power Circuits: Practical Recipes

*by Luca Benini **
Giovanni De Micheli
Enrico Macii

Low Power Circuits … continued from Page 22

degrees of freedom that designers have to cope with power reduction.

In this brief tutorial, we showed different design options and the corresponding advantages and disadvantages. We tried to relate general-purpose low-power design solutions to a few successful chips that use them to various extents. Even though we described only a few samples of design techniques and implementations, we think that our samples are representative of the state of the art of current technologies and can suggest future developments and improvements.

References

[1] J. Rabaey and M. Pedram, *Low Power Design Methodologies*. Kluwer, 1996.

[2] J. Mermet and W. Nebel, *Low Power Design in Deep Submicron Electronics*. Kluwer, 1997.

[3] A. Chandrakasan and R. Brodersen, *Low-Power CMOS Design*. IEEE Press, 1998.

[4] T. Burd and R. Brodersen, "Processor Design for Portable Systems", *Journal of VLSI Signal Processing Systems*, vol. 13, no. 2–3, pp. 203–221, August 1996.

[5] D. Ditzel, "Transmeta's Crusoe: Cool Chips for Mobile Computing", *Hot Chips Symposium*, August 2000.

[6] J. Montanaro, *et al.*, "A 160-MHz, 32-b, 0.5-W CMOS RISC Microprocessor", *IEEE Journal of Solid-State Circuits*, vol. 31, no. 11, pp. 1703–1714, November 1996.

[7] V. Lee, *et al.*, "A 1-V Programmable DSP for Wireless Communications", *IEEE Journal of Solid-State Circuits*, vol. 32, no. 11, pp. 1766–1776, November 1997.

[8] M. Takahashi, *et al.*, "A 60-mW MPEG4 Video Coded Using Clustered Voltage Scaling with Variable Supply-Voltage Scheme", *IEEE Journal of Solid-State Circuits*, vol. 33, no. 11, pp. 1772–1780, November 1998.

[9] A. P. Chandrakasan, S. Sheng, and R. W. Brodersen, "Low-Power CMOS Digital Design", *IEEE Journal of Solid-State Circuits*, vol. 27, no. 4, pp. 473–484, April 1992.

[10] F. Najm, "A Survey of Power Estimation Techniques in VLSI Circuits", *IEEE Transactions on VLSI Systems*, vol. 2, no. 4, pp. 446–455, December 1994.

[11] M. Pedram, "Power Estimation and Optimization at the Logic Level", *International Journal of High-Speed Electronics and Systems*, vol. 5, no. 2, pp. 179–202, 1994.

[12] P. Landman, "High-Level Power Estimation", *ISLPED-96: ACM/IEEE International Symposium on Low Power Electronics and Design*, pp. 29–35, Monterey, California, August 1996.

[13] E. Macii, M. Pedram, F. Somenzi, "High-Level Power Modeling, Estimation, and Optimization", *IEEE Transactions on Computer-Aided Design of Integrated Circuits and Systems*, vol. 17, no. 11, pp. 1061–1079, November 1998.

[14] S. Borkar, "Design Challenges of Technology Scaling", *IEEE Micro*, vol. 19, no. 4, pp. 23–29, July-August 1999.

[15] S. Thompson, P. Packan, and M. Bohr, "MOS Scaling: Transistor Challenges for the 21st Century", *Intel Technology Journal*, Q3, 1998.

[16] Z. Chen, J. Shott, and J. Plummer, "CMOS Technology Scaling for Low Voltage Low Power Applications",

ISLPE-98: IEEE International Symposium on Low Power Electronics, pp. 56–57, San Diego, CA, October 1994.

[17] Y. Ye, S. Borkar, and V. De, "A New Technique for Standby Leakage Reduction in High-Performance Circuits", *1998 Symposium on VLSI Circuits*, pp. 40–41, Honolulu, Hawaii, June 1998.

[18] M. Pedram, "Power Minimization in IC Design: Principles and Applications", *ACM Transactions on Design Automation of Electronic Systems*, vol. 1, no. 1, pp. 3–56, January 1996.

[19] B. Chen and I. Nedelchev, "Power Compiler: A Gate Level Power Optimization and Synthesis System", *ICCD'97: IEEE International Conference on Computer Design*, pp. 74–79, Austin, Texas, October 1997.

[20] L. Benini, P. Siegel, and G. De Micheli, "Automatic Synthesis of Gated Clocks for Power Reduction in Sequential Circuits", *IEEE Design and Test of Computers*, vol. 11, no. 4, pp. 32–40, December 1994.

[21] Y. Yoshida, B.-Y. Song, H. Okuhata, T. Onoye, and I. Shirakawa, "An Object Code Compression Approach to Embedded Processors", *ISLPED-98: ACM/IEEE International Symposium on Low Power Electronics and Design*, pp. 265–268, Monterey, California, August 1997.

[22] L. Benini, A. Macii, E. Macii, and M. Poncino, "Selective Instruction Compression for Memory Energy Reduction in Embedded Systems", *ISLPED-99: ACM/IEEE 1999 International Symposium on Low Power Electronics and Design*, pp. 206–211, San Diego, California, August 1999.

[23] H. Lekatsas and W. Wolf, "Code Compression for Low Power Embedded Systems", *DAC-37: ACM/IEEE Design Automation Conference*, pp. 294–299, Los Angeles, California, June 2000.

[24] S. Segars, K. Clarke, and L. Goudge, "Embedded Control Problems, Thumb and the ARM7TDMI", *IEEE Micro*, vol. 15, no. 5, pp. 22–30, October 1995.

[25] D. Brooks, *et al.*, "Power-Aware Microarchitecture: Design and Modeling Challenges for Next-Generation Microprocessors", *IEEE Micro*, vol. 20, No. 6, pp. 26–44, November 2000.

[26] L. Benini and G. De Micheli, *Dynamic Power Management: Design Techniques and CAD Tools*. Kluwer, 1997.

[27] Intel, *SA-1100 Microprocessor Technical Reference Manual*. 1998.

[28] L. Benini, A. Bogliolo, and G. De Micheli, "A Survey of Design Techniques for System-Level Dynamic Power Management", *IEEE Transactions on VLSI Systems*, vol. 8, no. 3, pp. 299–316, June 2000.

Luca Benini received the B.S. degree *summa cum laude* in electrical engineering from the University of Bologna, Italy, in 1991, and the M.S. and Ph.D. degrees in electrical engineering from Stanford University in 1994 and 1997, respectively. Since 1998, he has been assistant professor in the Department of Electronics and Computer Science in the University of Bologna, Italy. He also holds visiting professor positions at Stanford University and the Hewlett-Packard Laboratories, Palo Alto, California.

Dr. Benini's research interests are in all aspects of computer-aided design of digital circuits, with special emphasis on low-power applications, and in the design of portable systems. On these topics, he has authored more than 100 papers in international journals and conferences, a book and several book chapters. He is a member of the technical program committee in several international conferences, including the Design and Test in Europe Conference and International Symposium on Low Power Design.

Giovanni De Micheli is professor of electrical engineering, and by courtesy, of computer science at Stanford University. His research interests include several aspects of design technologies for integrated circuits and systems, with particular emphasis on synthesis, system-level design, hardware/software co-design and low-power design. He is author of *Synthesis and Optimization of Digital Circuits*, McGraw-Hill, 1994, co-author and/or co-editor of four other books and of over 200 technical articles.

Dr. De Micheli is a fellow of ACM and IEEE. He received the Golden Jubilee Medal for outstanding contributions to the IEEE CAS Society in 2000. He received the 1987 *IEEE Transactions on CAD/ICAS* Best Paper Award and two best paper awards at the Design Automation Conference, in 1983 and in 1993.

He is editor-in-chief of the *IEEE Transactions on CAD/ICAS*. He was vice president (for publications) of the IEEE CAS Society in 1999–2000. Dr. De Micheli was program chair and general chair of the Design Automation Conference (DAC) in 1996–1997 and 2000 respectively. He was program and general chair of the International Conference on Computer Design (ICCD) in 1988 and 1989 respectively.

Enrico Macii holds the Dr. Eng. degree in electrical engineering from Politecnico di Torino, Italy, the Dr. Sc. degree in computer science from Università di Torino, and the Ph. D. degree in computer engineering from Politecnico di Torino. From 1991 to 1994 he was adjunct faculty at the University of Colorado at Boulder. Currently, he is associate professor of computer engineering at Politecnico di Torino.

His research interests include several aspects of the computer-aided design of integrated circuits and systems, with particular emphasis on methods and tools for low-power digital design. He has authored over 200 journal and conference articles in the areas above, including a paper that received the best paper award at the 1996 IEEE EuroDAC conference.

Enrico Macii is associate editor of the *IEEE Transactions on CAD/ICAS* (since 1997) and associate editor of the *ACM Transactions on Design Automation of Electronic Systems* (since 2000). He was technical program co-chair of the 1999 IEEE Alessandro Volta Memorial Workshop on Low Power Design, and technical program co-chair of the 2000 ACM/IEEE International Symposium on Low Power Electronics and Design (ISLPED). Currently, he is general chair of ISLPED'01.

The European Logarithmic Microprocessor

J. Nicholas Coleman, Chris I. Softley, *Member, IEEE*, Jiri Kadlec, Rudolf Matousek, Milan Tichy, *Member, IEEE*, Zdenek Pohl, Antonin Hermanek, *Member, IEEE*, and Nico F. Benschop

Abstract—In 2000 we described a proposal for a logarithmic arithmetic unit, which we suggested would offer a faster, more accurate alternative to floating-point procedures. Would it in fact do so, and could it feasibly be integrated into a microprocessor so that the intended benefits might be realized? Here, we describe the European Logarithmic Microprocessor, a device designed around that unit, and compare its performance with that of a commercial superscalar pipelined floating-point processor. We conclude that the experiment has been successful, and that for 32-bit work, logarithmic arithmetic may now be the technique of choice.

Index Terms—High-speed arithmetic, emerging technologies, instruction-set design, SIMD processors, design studies, logarithmic number system.

1 Introduction

A new microprocessor has been developed. Reliant for the execution of 32-bit real operations on a logarithmic arithmetic unit, such operations are performed substantially faster and also somewhat more accurately than with the floating-point (FLP) system. An operand is represented as its base-2 logarithm, itself a fixed-point value and so capable of multiplication, division, and square root in minimal time, and in the case of multiplication and division with no rounding error. Using an original approximation technique, addition and subtraction are carried out with speed and accuracy similar to that of FLP arithmetic. We described these techniques in [1], [2], to which reference should be made as the background to this work.

This device has been manufactured, integrated into a development system, and comprehensively evaluated. We envisage that applications for it will be in embedded systems of some numerical complexity, such as the more advanced digital filters and graphics systems. Inferences from [1] were that a significant reduction in cycle count may be expected in scalar codes, with about 6 dB better accuracy. However, a great many other factors will impinge in practice. Will the expected results be obtained in reality, and indeed will the new techniques work at all?

We begin, for completeness, with a short restatement of the techniques on which the device is based. It will be evident that a logarithmic ALU has little in common with an FLP one, and it is also true that conventional microprocessor architecture is adapted to deploy an FLP unit to best advantage. This raises a dilemma: does one simply take an FLP microprocessor and substitute a logarithmic ALU, forgoing some significant advantages that the logarithmic number system (LNS) may offer, or should one invest the substantial effort of reworking the microprocessor design accordingly? Among the few authors, including ourselves, who have considered this issue, the consensus is for the latter, and we next discuss the architectural issues relevant to the integration of the new ALU in the most favorable way.

With regard to accuracy, we had presented in [1] a series of simulations of the proposed ALU, in which we predicted its maximum and average error, and illustrated its typical performance on arithmetic kernels. We now revisit this experiment, this time running on the microprocessor itself, and show an almost identical set of results to that predicted.

Regarding speed, we had made crude predictions based on the cycle counts of the basic operations, but had not considered the many additional factors coming into play when the arithmetic unit is integrated into a processor. As we now have such a device, we are in a position to evaluate these practicalities, and for this purpose we chose an industry-standard superscalar pipelined FLP processor for comparison. Although the two devices had different arithmetic systems and supporting architectures, they were otherwise, so far as possible, similar. We present a simple analysis of the two devices to show that the predictions in [1] regarding scalar and peak vector performance are likely to hold in practice.

Theoretical processing rates, however, are an unreliable indicator of the behavior of a real device. We finish, therefore, with two case-study applications, both on a substantial scale and representative of real-world problems. Each application is programmed in optimized assembly language and run on the two processors. After measuring the speed and accuracy of each implementation, we analyze the dynamics of the architectural complex in order to isolate the contribution made by the different arithmetic systems to the results.

We show that the logarithmic arithmetic system used on the ELM has been instrumental in delivering an overall

- *J.N. Coleman is with the School of Electrical, Electronic and Computer Engineering, The University, Newcastle upon Tyne, NE1 7RU, UK. E-mail: j.n.coleman@ncl.ac.uk.*
- *C.I. Softley is with Photonfocus AG, Bahnhofplatz 10, CH-8853 Lachen, Switzerland. E-mail: softley@photonfocus.com.*
- *J. Kadlec, M. Tichy, Z. Pohl, and A. Hermanek are with the Institute of Information Theory and Automation, Academy of Sciences of the Czech Republic, Pod Vodárenskou věží 4, PO Box 18, 182 08 Praha 8, Czech Republic. E-mail: {kadlec, tichy, xpohl, hermanek}@utia.caz.cz.*
- *R. Matousek is with Acision, Villa J, Spielberg Office Centre, Holandská 5, 639 00 Brno, Czech Republic. E-mail: rudolf.matousek@acision.com.*
- *N.F. Benschop was with Philips Research, Eindhoven, The Netherlands. He is now retired and resides at Drossaardstraat 71, 5663 GJ Geldrop, The Netherlands. E-mail: n.benschop@chello.nl.*

Manuscript received 13 Feb. 2007; accepted 13 June 2007; published online 2 Aug. 2007.
Recommended for acceptance by B. Veeravalli.
For information on obtaining reprints of this article, please send e-mail to: tc@computer.org, and reference IEEECS Log Number TC-0058-0207.
Digital Object Identifier no. 10.1109/TC.2007.70791.

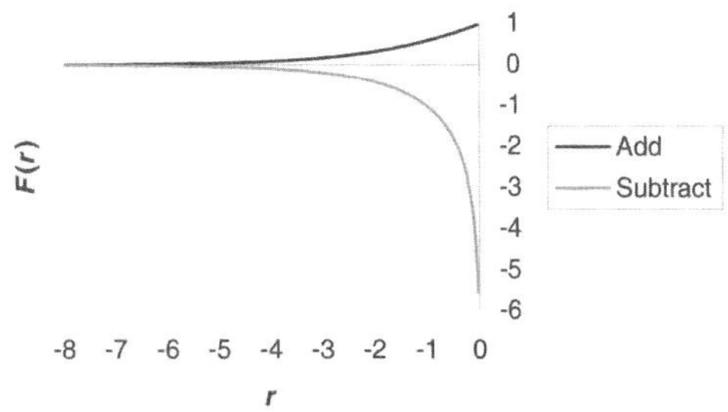

Fig. 1. LNS addition and subtraction functions.

increase in speed and improvement in accuracy. The higher speed has permitted a trade-off with reduced architectural complexity. The other relevant issues are silicon area and power dissipation. Although preliminary results suggest a broadly comparable area and a significant reduction in power, space does not permit an adequate treatment of these factors here, and in any case our results are not yet complete. We will therefore deal with these issues in a sequel to this paper. On the basis of the evidence currently presented, however, we conclude that, for 32-bit work, logarithmic arithmetic may now be the technique of choice.

2 LNS Arithmetic

In a logarithmic number system, a number x is represented as the fixed-point value $i = \log_2 x$, with a special arrangement to indicate zero x and an additional bit to show its sign. For $i = \log_2 x$ and $j = \log_2 y$, and assuming without loss of generality that, in dyadic operations, $j \le i$, LNS arithmetic involves the following computations:

$$\log_2(x+y) = i + \log_2(1 + 2^{j-i}),$$
$$\log_2(x-y) = i + \log_2(1 - 2^{j-i}),$$
$$\log_2(x \times y) = i + j,$$
$$\log_2(x \div y) = i - j,$$
$$\log_2(\sqrt{x}) = i \div 2.$$

Although multiplication, division, and square root are straightforward, and implemented with simple modifications to the fixed-point unit, addition and subtraction require the evaluation of a nonlinear function, $F = \log_2(1 \pm 2^r)$, as illustrated in Fig. 1 for $r = j - i$.

Early proposals for short-word-length systems, for example, [3] at 20 bits, suggested implementation with a lookup table containing all possible values of F. Beyond about 20 bits, the exponentially increasing storage requirements render this approach impractical and the function is better stored at intervals through r, with intervening values obtained by some interpolation procedure. This introduces inaccuracy and increases the delay time. Subsequent work therefore focused on finding an acceptable compromise between the two. The first interpolators had order-of-magnitude differences between multiplication and addition times. A first-order Taylor method [4] was fabricated in 3μ CMOS [5]. At 28-bits plus sign and accuracy within FLP error bounds, multiplications completed with 55 ns latency and additions with 1.4 μs. A higher order polynomial method at 31-bits plus sign [6] also offered better accuracy than FLP. A lower accuracy variant of it, which did not, was fabricated in 1.2μ CMOS [7]. Multiplication latency here was 13 ns and addition 158 ns. Both designs exposed a difficulty in subtractions when $r \to 0$, where the rapidly changing derivative demands successively narrower intervals and a vast increase in storage. To circumvent this problem, we described a cotransform in [2]. Applied to r and i when $-0.5 < r < 0$, it returns $r2$ and $i2$, where $r2 < -1$, the new values $r2$ and $i2$ being passed to the interpolator. A further development of the interpolator itself was then described in [2], [1]. Its errors were also within FLP limits, and with a significantly shorter critical speed path than hitherto, it raised the prospect of 32-bit LNS addition better than FLP in terms of both speed and accuracy. This design, together with the cotransform unit, forms the basis of the implementation described herein and is summarized in Section 2.1. Finally, ongoing work focuses not only on faster interpolation but also, for example, [13], on longer word lengths.

2.1 Addition and Subtraction Algorithms

For each function, the range of $r = j - i$ is partitioned into segments at increasing powers of 2. Each segment is divided into intervals, at each of which the function F and its derivative D are stored. By partitioning r, a linear Taylor interpolation produces an estimate of the function at intermediate points. The error in the estimation increases from zero when the required value lies on a stored point, to E when it falls immediately to the side of the next stored point. It was observed that the shape of this error curve is very similar in all intervals in both curves. It is therefore possible to calculate the error in any particular case by storing, alongside F and D for each interval, its value of E. A separate table stores the normalized shape of the common error curve, from 0 when the required value lies on an interpolation point, to 1. This table is known as the P (proportion) function. The error is calculated by multiplying E by P and is then added to the result of the interpolation. A full explanation is given in [1], the only modification to which was that the truncation of the interpolator multipliers was determined empirically, by successively trimming low-order bits from the multiplier in the hardware description model and simulating with random values while keeping within FLP error bounds.

The range shifter is inserted immediately after the selection of i and calculation of r. As described above, it is deployed selectively whenever r falls close to zero in the subtraction operation, transforming i and r into new values, where r is now in the linear region. Space does not permit an explanation of its operation here; it is fully explained in [2].

The entire unit is depicted in Fig. 2. Its accuracy was verified by exhaustive simulation of the hardware description model before fabrication, as described in Section 4.

2.2 Data Format, Range, and Precision

IEEE standard FLP representation uses a sign, an 8-bit biased exponent, and a 23-bit significand. The latter has an implied binary point immediately to its left and a hidden "1" to the left of the point. Extreme exponent values (0 and 255) are used for representing special cases; thus this format holds values in the range $\pm(1.0 \times 2^{-126})$ to $(1.111\ldots \times 2^{+127})$, $\approx \pm 1.2\mathrm{E}-38$ to $3.4\mathrm{E}+38$.

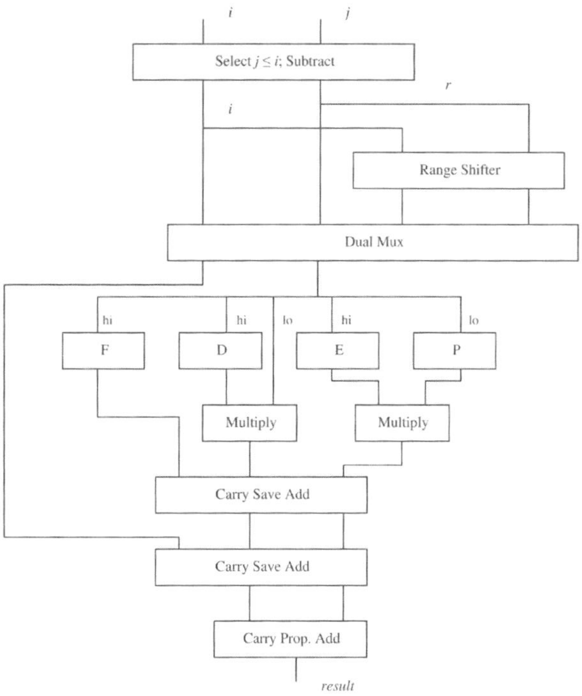

Fig. 2. Addition and subtraction hardware.

S	8-b Exponent	23-b Mantissa

In the equivalent LNS representation used throughout this work, the integer and fractional parts form a coherent two's complement fixed-point value in the range ≈ -128 to $+128$. The real numbers represented are in the range $\pm 2^{-128}$ to 2^{+128}, $\approx \pm 2.9\text{E}-39$ to $3.4\text{E}+38$. The smallest positive value, 40000000_{16}, is used for representing zero, while C0000000_{16} represents NaN.

S	8-b Integer	23-b Fraction

Underflows are taken to zero. Overflows, invalid operands, and any operation with NaN as an input return NaN.

3 Processor Philosophy and Architecture

As far as we were aware, there were no LNS-based microprocessor devices yet in existence. The motivation for this work was to develop one. However, a few authors had noted, as we did, that LNS ALUs have markedly different characteristics from their FLP equivalents and are therefore likely to require some reevaluation of the surrounding microprocessor design to deploy them to best advantage. Paliouras et al. [8] used the work in [6] as the basis for the addition unit in a proposed VLIW device, optimized for filtering and comprised of two independent ALUs, each with a 4-stage pipelined adder and a single-cycle multiplier. Simulations suggested an operating frequency of 12 MHz in 0.7μ CMOS. In an outline proposal for the ELM [9], we noted that the small size of the LNS multiplier-cum-integer unit encouraged its replication, leading to a short vector design with four single-cycle multipliers and two

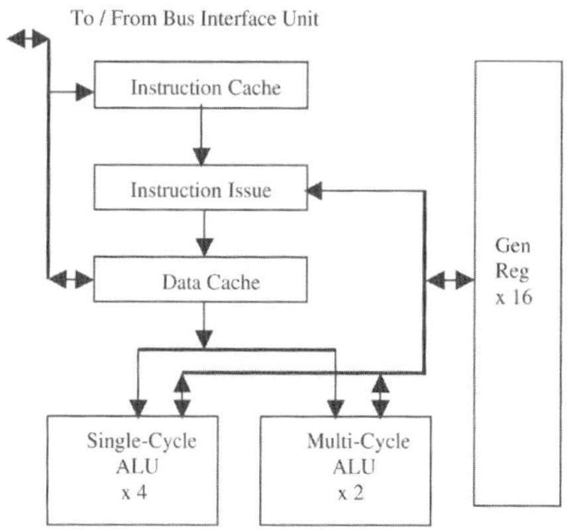

Fig. 3. ELM pipeline.

multicycle flowthrough adders. Arnold [10] instead proposed a VLIW device with only one single-cycle multiplier and one 4-stage pipelined adder, suggesting that large-scale replication of functional units could lead to difficulties with the complexity of the multiplexing paths leading to and from the registers. He then considered a number of issues related to suitable instruction set architectures (ISAs), but did not present a finished design or simulation results.

The following discussion is based on the characteristics of the LNS arithmetic unit at our disposal, and we show how these have influenced the design decisions relating to its integration into the ELM pipeline. Depicted in Fig. 3, this is based around 16 general-purpose registers.

Simulations of the hardware model had revealed that the basic addition/subtraction unit would incur about three times the delay of a fixed-point addition and that the range shifter, when invoked, would delay it by a further cycle. The first decision concerned the desirability of pipelining the arithmetic units. A number of factors seemed to weigh against:

- Whereas an FLP addition or multiplication circuit is complex enough that it is usually economic to emphasize its throughput by pipelining it, such an argument does not apply to fixed-point units and therefore not to the LNS multiplier. This leaves the LNS addition/subtraction unit as the only block of circuitry with a delay greater than a cycle —usually three and occasionally four—and it does not seem worthwhile to devise a pipelining scheme for this operation alone.
- One of the principal advantages of the LNS is the reduced latency of the multiplication operation. This implies that the device will be of particular utility in latency-sensitive applications, in which case it would seem consistent also to minimize the latency of the addition operation, but pipelining would inevitably increase it.
- The addition/subtraction unit has a variable delay time, which would complicate a pipeline control algorithm.

Accordingly, we have arranged a fully interlocked pipeline in which a single-cycle ALU executes all operations except logarithmic addition and subtraction. These are handled by the multicycle ALU, which executes with a three-cycle flowthrough time. The instruction issue unit continues to launch two more instructions while waiting for the addition to complete, provided that these are destined for the single-cycle ALU and do not otherwise conflict with the addition or subtraction in progress. This raises the possibility that both ALUs may complete at the same time, and the register file update paths are designed with sufficient bandwidth to accommodate this.

The second decision concerned the ISA. Mainstream computer architecture favors a register-register approach which decouples the data cache from the pipeline via a load/store unit in order to permit its independent concurrent operation. This is often combined with superscalar instruction processing in order that the loads can be issued in parallel with the arithmetic operations. Review of the likely application areas for this device, and the practicalities of designing it, suggested that this might not be the best approach and that a dual RR/RM architecture with a cache incorporated into the pipeline might be considered instead. Pertinent factors were:

- Time did not permit any consideration of superscalar design, so the device is limited to the issue of one instruction per cycle. Explicit load instructions would use up an issue slot.
- Complex numeric codes contain a high proportion of loads and stores as loaded data often have a short working life. Explicit loading would unavoidably increase the number of cycles per executed real instruction.
- A decoupled cache would inevitably have a longer access latency than a tightly integrated one, which would be inconsistent with the low latency emphasized elsewhere.

The drawback would be that an integrated cache could not operate independently of the pipeline. Potentially this could reduce performance in two ways:

- "Intelligent" features such as early data forwarding or predictive loading would be impossible. It would have to be accepted that, if required data were not in the cache, then the system would stall while the entire line was loaded. However, this could be minimized by making the cache as large as possible, with a relatively small line size.
- Disparities between the speed of the cache and that of the pipeline could not be accommodated so easily. However, since the LNS adder/subtractor uses several ROM devices itself, the pipeline speed will always be governed by memory speeds, so large disparities are unlikely to arise.

Consequently, the ISA allows for RR and RM addressing. Superficially similar to that of the IBM S/360, operands located in memory are addressed by a base register plus a 16-bit unsigned displacement. All instruction codes are 32 bits, and in total there are 35 distinct operations.

The final design question was more clear-cut. So far, we have emphasized low latency, sometimes at the expense of throughput. Functional unit replication is commonly employed to boost throughput on FLP devices. Unlike functional unit pipelining, this is a very appropriate strategy on the ELM, where the single-cycle unit is so small that it can easily be replicated four times. A vector capability is thereby provided, permitting the execution of four integer, logical, or LNS multiplication, division, or square root operations in one clock cycle. The vast majority of the silicon area used for the arithmetic circuitry can thus be reserved for the more substantial multicycle ALU. Two such units are provided, allowing two additions or subtractions to proceed in three cycles.

Accordingly, all instructions are available in either scalar or vector form, the former operating on only one memory location or register, and the latter on a set of four consecutive locations or registers (or two in the case of addition and subtraction). A hybrid mode is available to apply a scalar to all elements of a vector. The orthogonal ISA permits almost all meaningful combinations of scalarity and addressing mode.

Scalar integer and logical operations set a 2-bit condition code, according to the result being < 0, $= 0$, or > 0. Branch instructions may be conditional on any combination of the three. Because of the ambiguity, when a logarithmic and an integer instruction complete at the same time, logarithmic instructions do not set the condition code. Conditional tests on logarithmic values are instead made by applying an appropriate logical operation. For example, a register would be reloaded from itself to determine the state of the sign bit, which has the same significance in logarithmic values as in integers. A useful by-product of the data representation is that a one-place logical left shift brings the logarithmic part of a value into the same significance as that of an integer, transforming a sign comparison around zero into a magnitude comparison around unity. This might be used, for example, following a magnitude comparison between arbitrary values, which in the LNS would be best accomplished by division.

To maintain the required memory bandwidth, the instruction and data caches supply, respectively, one and four 32-bit words to the processor per cycle. Each cache is of 8-Kbyte capacity and is two-way set associative (until a late stage in the design they were to have been 16 Kbytes, which was within the timing tolerance but slightly over budget). A 64-bit asynchronous external bus interface is provided, with a minimum transaction time of three cycles per read and four per write, although these times can be increased with wait states if desired. Input and output devices may be mapped into the memory address space, and transfer 32 bits per operation. I/O devices may be accessed by polling, but also have recourse to a single-level nonmaskable interrupt.

Following verification on an FPGA, devices were fabricated in 0.18μ CMOS and packaged in 181-pin grid arrays for easy handling during development work. Sign-off simulations immediately before fabrication had indicated an operating frequency of 166 MHz. Fabricated devices ran on the tester at 150 MHz. The current devices have been installed on development boards and are functional at 125 MHz, although a batch of higher quality boards is ready for manufacture and it is expected that these will permit some increase in operating speed.

The existing development systems consist of a 7×8 in board containing a 125-MHz ELM1A, 4 Mbytes of 10-ns static RAM, basic peripherals, and serial interface to a host PC. The bus speed, both for read and write accesses, is set to four cycles. Software support is currently comprised of a relocating assembler, a linkage editor, and a mathematical

TABLE 1
Errors in ELM Addition and Subtraction Operations

	$\|e\|_{\text{max rel log}}$	$\|e\|_{\text{av rel log}}$	$e_{\text{max rel arith}}$	$e_{\text{min rel arith}}$	$e_{\text{av rel arith}}$	$\|e\|_{\text{av rel arith}}$
Add	0.61639037	0.24831750	0.37871610	−0.42724924	−0.010716149	0.17212057
Subtract	0.68221501	0.24811711	0.42695100	−0.47287539	−0.0051578344	0.17198167

link library. An implementation of one of the public versions of ANSI-C is in an advanced state of preparation.

4 Accuracy

In IEEE 754 arithmetic using the round-to-nearest option, a result is guaranteed to lie on the closest available quantization point. This is sometimes called "exact" rounding, and is difficult to achieve in the LNS. The logarithm itself might be rounded to the nearest point, but because of the nonlinear relationship between i and $x = 2^i$, this does not guarantee that x will also be exactly rounded when i lies near the midpoint of an interval. On the other hand, the LNS has the felicitous property of a significantly lower relative error than the maximum of that of FLP. Even with inexactly rounded i, this more than compensates for the inexact rounding of x, and consequently an LNS implementation will tend to have smaller worst-case relative errors than those of FLP. A detailed working out of this principle is presented by Arnold and Walter [11], who refer to this property of the LNS as "better than FLP" accuracy. The object of the experiments described herein is to determine whether this has happened in practice.

We evaluated the ELM addition and subtraction units in two ways. An exhaustive analysis determined the definitive error characteristics, and stochastic experiments illustrated their typical operational behavior. The units used for the measurement of accuracy are as defined in [1].

4.1 Definitive Analysis

ELM assembly language programs were developed to sweep through all representable values of the log. domain difference of the two operands to the logarithmic addition or subtraction, as far as a difference of 24×2^{23}. This corresponds to sweeping through the entire range of r up to the point at which the value of the relevant function is set to hardwired zero. In other words, the function approximation hardware in the logarithmic addition and subtraction unit has been tested for all representable values of its operand r.

These tests were carried out during the design verification stage prior to layout and fabrication. An FPGA prototype implementation of the ELM, logically identical to the ELM itself in all respects except cache size, was employed to speed up the test, and the test was split into 768 sections so that the data generated for each section would be small enough to fit into the memory available on the FPGA development board. A Perl script automated the generation and retrieval of a complete set of results for all 768 sections for both addition and subtraction.

Analysis of the results was automated by a Perl script, which read in the result saved from the FPGA run and evaluated the corresponding exact result using IEEE double-precision FLP. It then calculated the error in the ELM implementation and produced the maximum, minimum, mean, and mean-size figures for each section and for the entire data set, in terms of both log. domain LSBs and equivalent FLP LSBs. The script also binned the errors to form error histograms both for every section and for the entire data set.

The error figures for the entire range of r for LNS addition and subtraction on the ELM are given in Table 1.

The error distributions are shown in Fig. 4. Since the largest errors found for any value of r for both addition and subtraction were under 0.5 equivalent FLP bits, we conclude that the ELM implementation of LNS satisfies the required bounds for "better than FLP" arithmetic.

The final hardware implementation of the algorithms differs slightly from that already published [1, Table 1] and these results show its error performance to be marginally better. The only previous work to show the error distributions for LNS was Lewis' interleaved memory implementation of 32-bit LNS with a weak error model for subtraction of nearly equal quantities [6]. The distributions for normal operation of that unit agree in shape with the distributions presented here, except, of course, for the region in which the weak error model comes into play. No discussion of these distributions is given, however, and this is a possible area for further work.

4.2 Typical Behavior

In [1], we described an experiment in which we compared the accuracy of 32-bit FLP operations and kernels with that of their simulated LNS equivalents. Reference should be made to this work for a description of the method, and to [1, Fig. 8] for the results. The average processing error of the simulated LNS implementation had been compared with that of FLP, and the experiments were repeated several

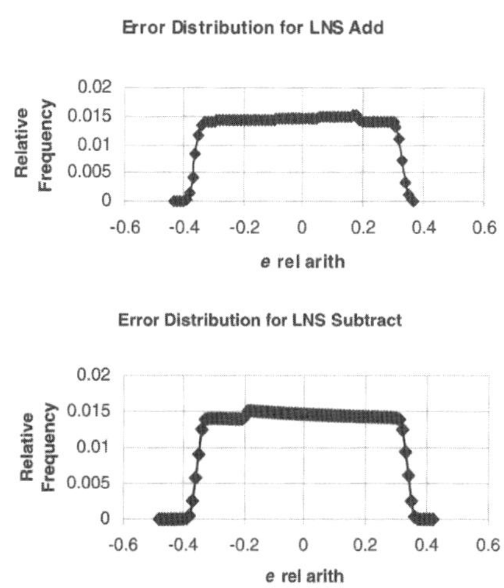

Fig. 4. Error distribution in ELM LNS addition and subtraction.

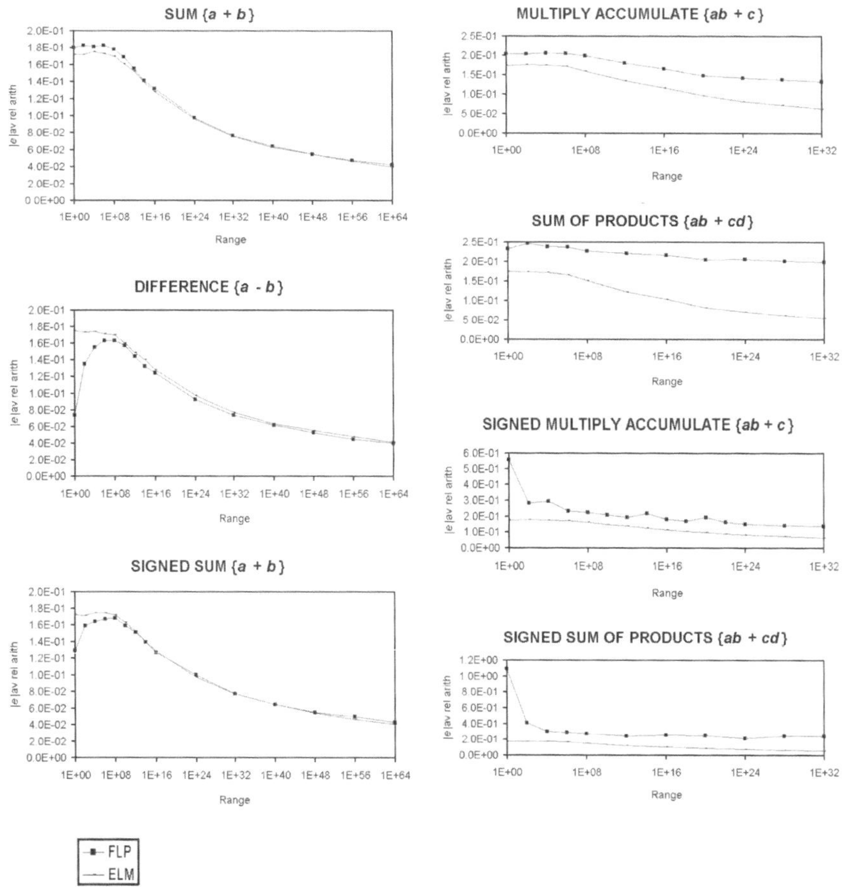

Fig. 5. Error in ELM execution of arithmetic kernels.

times for each kernel, with gradually increasing dynamic range.

We now repeated these experiments, this time running the LNS program on the ELM evaluation board rather than in simulation. The procedure was otherwise identical, although we had upgraded the Pascal compiler in the interim and were therefore reliant on a different randomization intrinsic and thus on different input data. We show these new results in Fig. 5. The results obtained from the ELM are virtually identical to the simulated LNS predictions. The FLP results are also similar, except that extremes, some of them anomalous, in the two signed multiplicative kernels seem to have varied. These presumably originate in some interference pattern between FLP methods and imperfections in the randomizers.

Overall, the LNS errors are generally less than those of FLP and are smoother throughout the dynamic range. They are less sensitive to the actual data values, although there might be some question as to whether this would be the case had the randomizer been implemented in LNS arithmetic.

5 Speed

The ELM is fabricated in 0.18μ technology, and all of the measurements reported in this work are taken from the existing 125-MHz systems. In order to discuss the speed, it is helpful to have an FLP standard for comparison. For this purpose, we looked for an industry-standard FLP DSP device, as similar as possible to the ELM in every respect apart from its arithmetic system. We chose the Texas Instruments TMS320C6711. Apart from DSP applications, this device is also used in graphics. It runs at 150 MHz, the fastest speed grade available in 0.18μ technology. Although a faster speed grade, the C6711D, has recently become available, it is fabricated at 0.13μ with copper interconnect and is therefore not directly comparable with the current implementation of the ELM. A variant, the C6701, is currently manufactured at 0.18μ and runs at 167 MHz, but is also not comparable because it has a simplified on-chip memory, having dispensed with the cache in favor of a RAM.

The TMS is a superscalar device with a VLIW arrangement and an RR instruction set. It has two independent arithmetic units, A and B, each with 16 registers $0..15$, parallel pipelined FLP add and multiply units, and integer and reciprocal-approximate units. Up to eight independent instructions may be issued per cycle (the symbol "||" in the assembly listings shown later designates that an instruction is issued in the same packet as the previous one). The add and multiply units have a latency of four cycles, the integer unit one. Divide and square root are implemented in software by using the Newton-Raphson method. Optimized assembly-language routines are provided by the manufacturer for this purpose and require about 30 and 40 cycles,

TABLE 2
Hardware Resources of TMS and ELM Devices

	TMS	ELM
Fabrication technology	0.18μ, 5-layer metal	0.18μ, 6-layer metal
Clock speed	150 MHz	125 MHz
Real add / subtract units	2	2
Real multiply units	2	
Real divide / sq-root units	2 (reciprocal / recip. sq-root approx.)	4
Integer / logical units	2	
Registers	32	16
Level-1 data cache size	4 Kbytes	8 Kbytes
Level-1 data cache mapping	2-way set associative	2-way set associative
Level-1 data cache policy	Read-allocate	Write-allocate, writeback
Level-1 data cache organisation	32 bytes x 2 ways x 64 lines, (dual ported)	32 bytes x 2 ways x 128 lines
Level-1 data cache read bandwidth	2.4 Gbytes / sec	2 Gbytes / sec
Level-1 data cache write bandwidth	1.2 Gbytes / sec	1 Gbyte / sec
Level-1 instruction cache size	4 Kbytes	8 Kbytes
Level-1 instr. cache mapping	Direct	2-way set associative
Level-1 instr. cache organisation	64 bytes x 64 lines	32 bytes x 2 ways x 128 lines
Level-2 cache size	64 Kbytes	-
Level-2 cache mapping	4-way set associative	-
Level-2 cache policy	Write-allocate, writeback	-
Level-2 cache organisation	128 bytes x 4 ways x 128 lines	-
External bus width	32 bits	64 bits
External bus protocol	Synchronous	Asynchronous
External bus bandwidth	340 Mbytes / sec	250 Mbytes / sec

respectively. Loads and stores enter the memory pipeline, which has a four-cycle latency. There is a 4-Kbyte level-1 (L1) data cache, 4-Kbyte L1 instruction cache, and a 64-Kbyte unified level-2 (L2) cache. The device has a 32-bit external bus with a configurable protocol. On the evaluation board used for these experiments, it used synchronous dynamic RAMs (SDRAMs), and was running at its maximum bandwidth of about 340 Mbytes/sec.

The ELM is a scalar device with an RM instruction set and 16 general registers. The device has a 64-bit asynchronous bus with a bandwidth on the current boards of 250 Mbytes/sec. The resources available to each device are detailed in Table 2.

Evidently the devices are comparable in terms of fabrication technology and clock speed. Each device has six principal ALUs. The TMS is comprised of two adders, two multipliers, and two integer units (the adders and multipliers can also be used for fixed-point work), while the ELM has two adders and four combined multiplier/integer units. The devices are also broadly comparable in terms of memory bandwidth, but do have different cache arrangements. The TMS has a unified L2 cache, which is not present on the ELM. There are also differences in the L1 caches, which to some extent are consequent on the different arithmetic techniques. The straightforward implementation of vector operations on the ELM, putting four identical functional units in parallel, requires a simple data cache organization in which four consecutive words are read out simultaneously. The TMS, with pipelined functional units launching operations in different cycles, is more likely to encounter bank conflicts and is therefore more likely to benefit from dual porting to the cache.

The TMS has an optimizing C compiler. It is not straightforward to program in assembly language because it does not have an interlocked pipeline, that is, there is no hardware mechanism for checking that one instruction has completed before a dependent one begins. The programmer must therefore construct a dependency chart from which the parallel operations are scheduled manually, calculating the state of the pipeline at each clock cycle and if necessary inserting no-ops to account for instruction latency. In vector operations, there might be several iterations of the loop in different stages of the pipeline simultaneously. The pipeline is brought into the correct state by a prologue before the start of the iterating kernel, and the terminating iterations are wound down by an epilogue at the end. To simplify programming, the manufacturer provides a "linear assembler," which determines the dependencies, schedule, and register allocation automatically. The preferred method of programming is then to use the compiler for the outer scalar sections of code, which can easily be scheduled efficiently by the compiler, and to write the inner loops in linear assembly language.

The ELM has an interlocked pipeline and also an RM ISA. The two features together make for a relatively straightforward assembly language interface, which is currently the only method of programming the device.

Latencies and vector throughput on the two devices are summarized in Table 3. Those for the discrete operations are shown first, followed by the values for multiply-accumulate (MAC) and sum-of-products (SOP) kernels. It is assumed in the two kernels that input data are taken from memory and results are returned there. The maximum vector execution rate on the TMS is two additions and two multiplications per

TABLE 3
Latency and Throughput of TMS and ELM Devices

Operation	TMS				ELM			
	Latency (cycles)	Latency (ns @ 150MHz)	Scalar Throughput (per cycle)	Vector Throughput (per cycle)	Latency (cycles)	Latency (ns @ 125MHz)	Scalar Throughput (per cycle)	Vector Throughput (per cycle)
Add	4	27	0.25	2	3	24	0.33	0.67
Subtract	4	27	0.25	2	3 (4)	24 (32)	0.33 (0.25)	0.67
Multiply	4	27	0.25	2	1	8	1	4
Divide	≈30	≈200	≈0.033		1	8	1	4
Square-root	≈40	≈267	≈0.025		1	8	1	4
Load	4	27	0.25	4	1	8	1	4
Store	4	27	0.25	2	2	16	0.5	2
MAC	16		0.0625	0.67	7		0.14	0.5
SOP	16		0.0625	0.67	7		0.14	0.5

A

```
LV      0,A0    * MAC, X = AB + C
MLV     0,B0
ALV     0,C0
ALV     2,C2
LV      4,A4
MLV     4,B4
STV     0,X0    * STORE 0, CYCLE 9
ALV     4,C4
ALV     6,C6
LV      0,A8
MLV     0,B8
STV     4,X4    * STORE 4, CYCLE 17
ALV     0,C8
ALV     2,C10
LV      4,A12
MLV     4,B12
STV     0,X8    * STORE 8, CYCLE 25
```

B

```
LV      0,A0    * SOP, X = AB + CD
MLV     0,B0
LV      4,C0
MLV     4,D0
ALRV    0,4
LV      8,A4
MLV     8,B4
ALRV    2,6
LV      4,C4
MLV     4,D4
STV     0,X0    * STORE 0, CYCLE 11
ALRV    8,4
LV      0,A8
MLV     0,B8
ALRV    10,6
LV      4,C8
MLV     4,D8
STV     8,X4    * STORE 4, CYCLE 19
ALRV    0,4
LV      8,A12
MLV     8,B12
ALRV    2,6
LV      4,C12
MLV     4,D12
STV     0,X8    * STORE 8, CYCLE 27
```

Fig. 6. ELM assembly language implementations of vector kernels. (A) MAC and (B) SOP.

cycle. The cache-processor bandwidth permits four loads or two stores per cycle. Its peak performance is therefore 0.67 MAC or SOP operations per cycle. These figures will only be achieved under best-case conditions with relatively long vectors, where the pipeline can be kept saturated with sequential elements. In purely scalar sequences, the loads, multiplications, addition, and store would execute in 16 cycles, a throughput of 0.0625 MAC or SOP operations per cycle.

On the ELM, it is possible to launch two logarithmic additions in one cycle and then to process four loads and four multiplications in the remaining two cycles before the addition completes. The four multiplications perform implicit loads, so all eight words are loaded in the two cycles. The maximum execution rate on this device is thus two additions, four multiplications, and eight loads in three cycles. The best schedule, again working from and to memory, is illustrated in Fig. 6, which also serves as an example of programming. The MAC calculates $x[n] = a[n]b[n] + c[n]$. It starts with the instruction "Load vector," which loads $a[0]..a[3]$ into R0..R3. "Multiply logarithm vector" multiplies each register by the corresponding element of b. Each instruction takes one clock cycle. "Add logarithm vector" then adds $c[0]..c[1]$ into R0..R1 in three cycles. Another ALV accumulates $c[2]..c[3]$ into R2..R3. This also requires three cycles, but during the last two, the products of the next four elements of a and b are prepared in R4..R7. With the addition now complete, the result is returned to $x[0]..x[3]$ via "Store vector," which takes two cycles. It is evident in Fig. 6A that the sequence can repeat every eight cycles, giving a peak vector rate of 0.5 MAC per cycle. Fig. 6B illustrates the SOP kernel. Here, the ab product is generated in R0..R3 and cd in R4..R7, the two products being added by "Add logarithm register vector." The second addition adds R6..R7 to R2..R3. Immediately after this starts, the former registers are finished with and R4..R7 are reused for the preparation of the next product. Again, the sequence maintains 0.5 SOP per cycle. Neither kernel makes any use of R12..R15, which are available for other purposes, for example, holding the base address of the vectors. The scalar MAC or SOP would execute in seven cycles, yielding a throughput of 0.14 per cycle.

Measuring by time, and recognizing the slightly slower clock speed of the current implementation of the ELM, additions and subtractions are still marginally faster than on the TMS, except when the range shifter is deployed. Allowing that this might occur in half of the subtractions, and that additions and subtractions might occur equally, the average speed of the ELM on these operations improves on that of the TMS by a few percent. Multiplications are

3.4 times the speed, and divisions and square roots complete in a small fraction of the time. It might be argued that additions and subtractions on the TMS are subject to the delay times of three more pipeline registers than on the ELM, so a strict calculation of flowthrough times would improve the TMS values. Against this, however, the entire design philosophy of the TMS device is based on pipelining. Its behavior would otherwise be radically different, and we must evaluate the device as it is.

Evidently the TMS is at an advantage in vector sequences. However, it will only attain this level of performance when it is possible to maintain a continuous pipeline flow. This will be difficult in code with data dependencies or with short vectors for which the start-up times dominate. In longer vector sequences, the TMS may exceed the performance of the ELM, particularly where fewer memory accesses are required, such as in running MAC sequences or weighted vector sums. However, in very long vector processing, other factors may come to dominate. In particular, it may become difficult to maintain the required memory bandwidth when the cache capacity is exceeded. At this point, the performance of the two devices is likely to converge as both become dominated by their external memory bandwidths, which are similar.

The ELM has a lower latency than the TMS in almost all cases, and hence a significantly higher scalar throughput. The figures suggest that, as predicted in [1], the ELM can be expected to offer around twice the performance of the TMS in scalar code and in less regular or short vector sequences. In code involving a significant proportion of division or square root operations, the ELM may exceed this.

6 Case Studies

Processor architecture, instruction scheduling, compiler efficiency, caching, memory access patterns, and other related factors all affect final performance. Two large-scale case studies were undertaken to determine whether the theoretical predictions hold in practice.

Material was drawn from two application areas: DSP and graphics. The first had a substantial proportion of vector code, the second far less, thus playing to the respective strengths of both processors. In both applications, all but a few percent of the processing workload consisted of relatively small kernels, so compiler effects were eliminated by writing these in assembly language. On the TMS we used the linear assembler, which was given every facility to optimize for speed, including speculative processing beyond array bounds and freedom from aliasing constraints. In most cases this produced an obviously near-optimal schedule, but in the few instances where this was not the case, the results were improved by assigning registers manually. On the ELM, we wrote standard assembly language and then optimized it manually by moving single-cycle instructions wherever possible to lie in the shadow of a preceding multicycle one. The outer mainlines were written in compiled C on the TMS, with options set to emphasize speed, and linked to the kernels. On the ELM, the mainlines were written in assembler, but were not subsequently optimized.

How efficient was the resulting TMS assembly programming? By considering the number of operations, the critical speed path between them, and the available functional units, we calculated the minimum possible latency for the real arithmetic processing in each of the scalar loop kernels. We show this as L_A. It is not realistically possible to avoid conditional branching latency in scalar codes, and taking this into account, a revised figure L_{AB} is also shown. This represents the minimum number of clock cycles in which the loop could possibly execute, assuming an unlimited supply of registers, perfect scheduling, loads, stores, and all housekeeping operations lying within the latency of arithmetic operations, and no overheads accruing from any other source such as from calling conventions. The latency achieved by the C compiler with the original high-level code is then given as L_C. Finally the latency actually achieved by the scheduled assembly program is shown as L. The ratios L_C/L_{AB} and L/L_{AB} therefore represent the quality of the result; the closer these values are to unity, the better. To avoid distortion, the latency of called arithmetic functions is shown separately, for example, as (+ div). For vector loops, the assembler will devise a schedule in which arithmetic operations and branches are both pipelined. The best possible schedule in this case, designated L_{ABV}, is determined simply by dividing whichever operation occurs most frequently by the number of functional units available to service it. For example, a loop containing six additions would require a minimum of three cycles per iteration with two pipelined addition units. As before, L/L_{ABV} grades the result.

For each application, we first describe the algorithm, mentioning any factors likely to influence execution speed. Excerpts of code are included where they serve to explain any significant differences in performance. Examples of input and output data are given. With regard to output, we have only shown that produced by the ELM, but where there are significant differences between it and that of the TMS, we have performed an analysis of accuracy against an 80-bit standard by using the same procedure as in Section 4. We measure the time taken by each processor, and by working through the assembly listings, counting the number of clock cycles for each kernel and multiplying by the number of passes through it, we analyze the number of clock cycles spent in each. The difference between the expected total clock cycles and the observed runtime is accounted for by stalling. This is predominantly because of cache activity, but also includes delays arising on the TMS from bank collisions and on the ELM from the subtractor range shifter (and to a negligible extent, it also includes the mainline processing). Finally, we discuss the results and show how the different arithmetic implementations have accounted for any differences in performance.

6.1 A Vector Application: Recursive Least Squares QR Filtering

This type of rapidly converging adaptive filter would be used at higher orders, such as N = 64, in precision control systems, such as that of a DVD player. At lower orders, it could be used in applications such as voice compression for Internet transmission. A preliminary version of this experiment and the algorithm used was described in [12].

The kernel is programmed as shown in Fig. 7A. The i loop processes the boundary cell to the left of each row and consists of a chain of serial dependencies, including a square root and a division. The inner j loop processes the remainder of the row and may be vectorized. Programming of the latter is illustrated in Figs. 7B and 7C. The calculated efficiency of the TMS version is shown in Table 4.

A

```
for (i = 0; i<= n; i++) {
  temp = sqrt (u [i] [i] * u [i] [i]
    + r [i][i] * r [i] [i]);
  den = 1 / temp;
  sfi = u [i] [i] * den;
  cfi = r [i] [i] * den;
  r [i] [i] = temp;
  z [i] = z [i] * cfi + d [i] * sfi;
  d [i + 1] = -z [i] * sfi + d [i] * cfi;
  p [i + 1] = p [i] * cfi;
  if (i != n) {
    for (j = i + 1; j <= n; j++) {
      r [i] [j] = r [i] [j] * cfi
        + u [i] [j] * sfi;
      u [i + 1] [j] = -r [i] [j] * sfi
        + u [i] [j] * cfi;
    }
  }
}
```

B

```
L2:      ; PIPED LOOP PROLOG
             LDW      .D2T2    *B5++,B7
             LDW      .D1T1    *A4++,A3
             MPYSP    .M2      B7,B4,B7
             MPYSP    .M1      A0,A6,A0
||           LDW      .D2T2    *B5++,B8
             LDW      .D1T1    *A4++,A0
||           MPYSP    .M2X     A0,B4,B7
             MPYSP    .M2      B7,B4,B8
             MPYSP    .M1      A3,A6,A3
||           LDW      .D2T2    *B5++,B7
             MPYSP    .M2X     A3,B4,B9
||           LDW      .D1T1    *A4++,A0
||           ADDSP    .L1X     B7,A0,A0
             MPYSP    .M2      B8,B4,B7
||           B        .S1      jloop
             SUB      .L2X     B12,A13,B0
||           SUB      .L1X     B12,A13,A1
||           MPYSP    .M1      A0,A6,A3
||           LDW      .D2T2    *B5++,B7
             MVK      .S1      0x3,A2
||           MV       .D2      B7,B8
||           MPYSP    .M2X     A0,B4,B
||           LDW      .D1T1    *A4++,A0
||           ADDSP    .L1X     B8,A3,A0
jloop:       ; PIPED LOOP KERNEL
   [ A2]     SUB      .L1      A2,1,A2
|| [ B0]     B        .S1      jloop
|| [ A1]     STW      .D1T1    A0,*A5++
||           MPYSP    .M1      A0,A6,A
||           MPYSP    .M2      B7,B4,B
             SUBSP    .L2X     B1,A3,B8
||           MV       .S2      B8,B1
||           MPYSP    .M1      A0,A6,A3
||           LDW      .D2T2    *B5++,B7
   [ A1]     SUB      .S1      A1,1,A1
|| [!A2]     STW      .D2T2    B8,*B6++
|| [ B0]     ADDK     .S2      0xffffffff,
||           MV       .L2      B9,B8
||           ADDSP    .L1X     B7,A3,A0
||           MPYSP    .M2X     A0,B4,B9
||           LDW      .D1T1    *A4++,A0
L4:      ; PIPED LOOP EPILOG
end:
             ADDK     .S1      0x1,A13
L6:
             CMPGT    .L1X     A13,B12,A1
   [!A1]     B        .S1      iloop
             MV       .D2      B12,B6
             ADDK     .S2      0x1,B6
             SHL      .S2      B6,0x2,B6
             ADDK     .S2      0x4,B6
||           ADDK     .S1      0x4,A11
             ADD      .L1X     B6,A10,A10
||           ADD      .L2      B6,B10,B10
||           ADDK     .S1      0x4,A12
||           ADDK     .S2      0x4,B11
|| [ A1]     LDW      .D2T2    *+SP(24),B3
             ; BRANCH OCCURS
```

C

```
         LV       0,R(14)
         MLSV     0,CFI
         LV       4,U(14)
         MLSV     4,SFI
JLOOP    ALRV     0,4
         LV       8,U(14)
         MLSV     8,CFI
         ALRV     2,6
         STV      0,R(14)
         MLSV     0,MSFI
         ALRV     8,0
         LV       4,U04(14)
         MLSV     4,SFI
         ALRV     10,2
         LV       0,R04(14)
         MLSV     0,CFI
         STV      8,U10(14)
         LA       14,16(14)
         LDA      12,1(12)
         BCL      NZ,JLOOP
```

Fig. 7. QR RLS algorithm. (A) C language version of i and j loops, (B) TMS assembly language, and (C) ELM assembly language implementation of the j loop.

TABLE 4
Efficiency of TMS Assembly Implementations of RLS Filter Loops

Loop	L_A	L_{AB}	L_{ABV}	L_C	L	$L_C / L_{AB(V)}$	$L / L_{AB(V)}$
i	29 (+ div, sqrt)	41 (+ div, sqrt)		74 (+ div, sqrt)	56 (+ div, sqrt)	1.80	1.37
j (not inc. prologue & epilogue)			2	23	3	11.5	1.5

The j loop, as implemented on the ELM, is shown in Fig. 7C. It may be largely understood by way of the explanation in Section 5, new points in this example being the addressing offset from R14 on line 1, "Multiply logarithm scalar vector" on line 2, which multiplies all four registers R0...R3 by the constant CFI from memory, and "Load address" and "Load decremented address" at the end, which respectively increment R14 by 16 and decrement R12 by 1. The latter is followed by a conditional branch back to the loop.

The experiment was repeated for filter order $\mathrm{N} = 8, 16, 32,$ 64, and 128. An input file spanned 2,000 time steps. Random noise, increasing in amplitude, was passed to the reference input, and $\frac{1}{2}\mathrm{N}$ time steps later was coupled into the signal. Inputs were quantized to 24-bit fixed-point values (23 bits plus sign) and thus took advantage of the entire 23-bit precision of the arithmetic. The same input data were presented to the TMS and the ELM, in each case being loaded into the memory of the respective development system before execution. In real life, the devices would be reading the data from an A-D converter, but this would take a similar length of time as that to access memory.

Both systems were timed (the TMS versions were run with and without the L2 cache), and the outputs were compared with that of an 80-bit Pentium to evaluate the SNR. An example of input and output is shown in Fig. 8. The SNR values for the results are presented in Fig. 9. The average TMS value is 108.4 dB and that of the ELM 117.5, so the ELM has delivered an average improvement of 9.1 dB or 1.5 bits. Since the 23-bit quantization has imposed a limit of

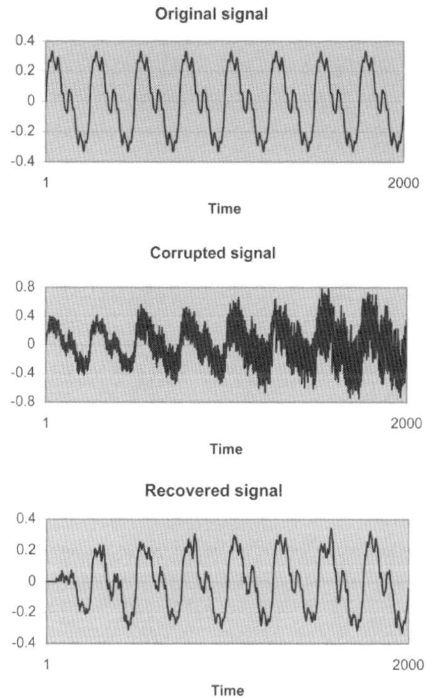

Fig. 8. Example of input and output data (output from ELM1A) for $N = 32$ filter.

138.5 dB, we can conclude that the TMS has added 30.1 dB of processing noise, whereas the ELM has added 21.0 dB. The generally smoother progression of the SNR values throughout the range of the ELM results is also noteworthy, and appears to be consistent both with the results in Fig. 5 and with the intuitive expectation that SNR will rise and fall around some optimum with N, although we make no attempt to explain it.

The runtimes are presented in Fig. 10, which shows an $O(N^2)$ increase consequent on the 2D nature of the arrays. A knee is present in each curve, where the total amount of data exceeds the capacity of the caches. For the TMS with only a 4-Kbyte L1 cache, this is at $N = 32$. For the ELM with its 8-Kbyte cache, it is at $N = 64$, and for the TMS with a 64-Kbyte L2 cache, at $N = 128$. Evidently there is little to choose between the ELM with one cache and the TMS with two, except for one specific instance at $N = 64$, where the

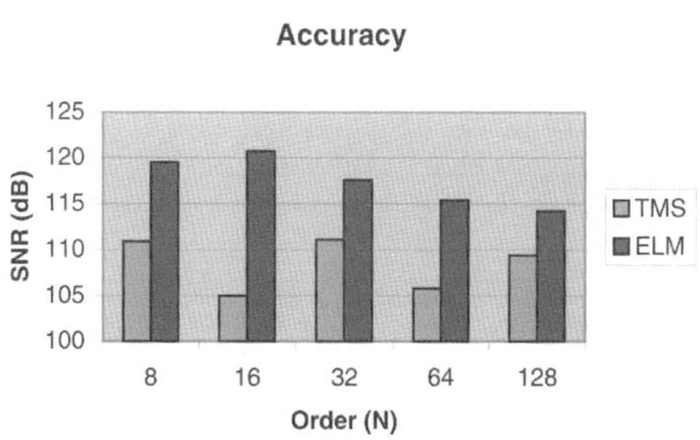

Fig. 9. Accuracy of TMS and ELM implementations of filter.

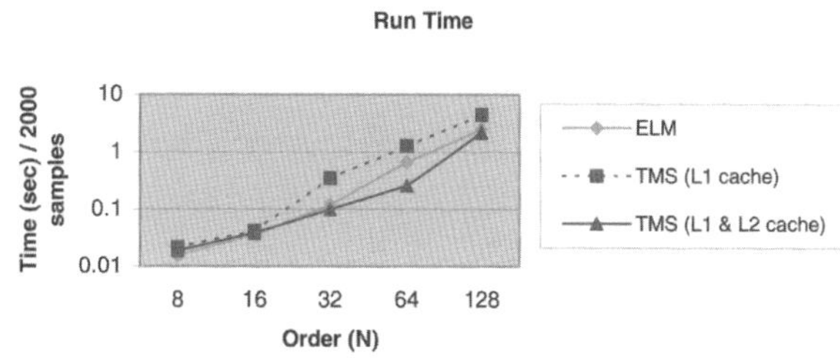

Fig. 10. Runtimes of TMS and ELM implementations of filter.

data are too large to fit into the cache on the ELM but just small enough to be cached at L2 on the TMS.

Figs. 7 and 11 together explain this result. With increasing N, the j loop increases in prominence as the i loop wanes. The j loop runs on the TMS in three cycles per iteration against an average of 5.25 on the ELM, the better performance on the TMS arising from the pipelining of the functional units. The i loop completes in 126 cycles on the TMS but 53 on the ELM, almost all of this difference being attributable to the division and square root operations. Furthermore, the ELM stalls for significantly longer than the TMS, obviously due to its lack of an L2 cache, except at $N = 128$ where this occurs on both processors. The improved performance of the i loop is, however, so significant that, except at $N = 64$, the ELM actually completes in slightly fewer cycles overall than the TMS does.

The longer cycle time but lower cycle count on the ELM result in the processors delivering a similar level of performance throughout. However, the ELM has done so despite what are evidently the major handicaps of having no L2 cache or vector pipeline. The dramatic simplifications inherent in the LNS division and square root operations have translated into an ability to dispense with a large measure of silicon complexity. The ELM has also achieved this result without running at its full design speed.

6.2 A Scalar Application: Recursive Ray Tracer

A demanding area of computational graphics is the ray tracer. An image is represented as a mesh of triangles into which a matrix of rays is projected from the viewer's eye. The first triangle on the path of each ray is visible to the viewer, and once identified its surface qualities are considered. If reflective, transparent, or both, then further rays are fired in the direction of reflection or refraction. A binary tree of recursive activations is built, and the color at each intersection point added to the total for that ray. To obviate testing each ray with each triangle, the latter are

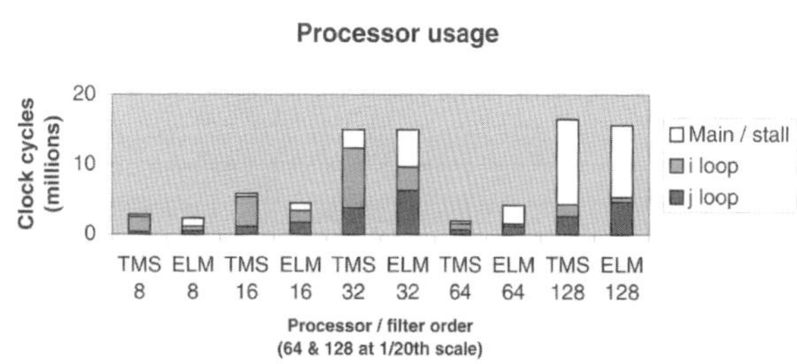

Fig. 11. Analysis of clock cycle usage in TMS and ELM implementations of filter.

TABLE 5
Efficiency of TMS Assembly Implementations of Ray Tracer Loops

Loop	L_A	L_{AB}	L_{ABV}	L_C	L	$L_C / L_{AB(V)}$	$L / L_{AB(V)}$
Ray-ellipsoid	32	44		82	51	1.86	1.16
Ray-plane	19 (+ div)	37 (+ div)		65 (+ div, fabsf)	45 (+ div)	1.76	1.22

grouped into a hierarchy of bounding axis-aligned ellipsoids. Only after a ray intersects an ellipsoid is it tested against the list of triangles within. There are two ray-ellipsoid intersection algorithms. The "algebraic" method treats the ellipsoid as a quadric, substitutes the ray parameters, and tests the discriminant. Programmatically, this involves a large number of multiterm SOPs. The "geometric" method rescales the ray into the ellipsoid's own space, effectively converting this into a sphere, and then makes a ray-sphere intersection test. This involves far fewer operations, but several are divisions or square roots. Intersection with the triangles is tested in two stages. The ray-plane algorithm rejects any rays that do not intersect the plane in which the triangle lies, and involves a division. Finally, Badouel's algorithm tests whether the intersection point lies within the triangle. This code is sparse and irregular but can be, and was, arranged so as to avoid significant use of division.

It would be very difficult to vectorize this application. The data consists of either scalars or three-element vectors. The control paths vary, even between successive iterations of the inner loops. Mapping a 3D model into a 1D memory results in fragmentation, so as the program will typically access randomly scattered blocks of between 8 and 32 words each, memory accesses are also difficult to vectorize.

A model was rendered at 256×256 pixels and comprised of 796 triangles within 86 ellipsoids in three levels, a total model size of 115 Kbytes. The TMS assembly language (Table 5) was highly optimal and a significant improvement on the compiled output.

There was no discernable difference between the outputs from the TMS and the ELM, which were similar to that shown in Fig. 13, except that this has random jitter on the refracted ray paths to simulate slightly opaque water. This increased the runtime unpredictably and was disabled during the timings to prevent discrepancies arising in the randomizers.

Runtimes were measured, excluding calls to initial runtime support on the TMS, for example, for storage allocation and cache initialization. The times were 14.9 s on the 150-MHz TMS using the algebraic algorithm and with both caches operational, and 10.9 s on the 125-MHz ELM with the geometric method. Restricted to the use of its L1 caches only, the TMS runtime increased to 241 s, probably because of thrashing in the program cache. Comparing the 125-MHz ELM with one cache to the 150-MHz TMS with two, the ratio of runtimes is thus 73 percent and that of clock cycles 61 percent. An analysis is presented in Fig. 14.

Unlike the digital filter, this application is intrinsically suited to the capabilities of the ELM, which runs with a lower cycle count than the TMS in every part of the algorithm. These gains have been achieved in different ways.

In the ray-triangle routine, the ratio of ELM/TMS clock cycles is 67 percent. The most heavily used part of this code is illustrated in Fig. 12, where the ratio for a pass through the complete sequence is 35/53. Broadly, the lower cycle count is attributable to the reduced latency of the multiplication and load operations, implicit loading, and the reduced branching latency.

In all its passes through the ray-plane loop, the TMS device consumed 999m clock cycles and the ELM 411m, a ratio of 41 percent. The processing included 11.9m divisions. Allowing 30 cycles for each on the TMS thus accounts for 357m cycles. Discounting the effect of these divisions, the ELM/TMS ratio would be 411/642 or 64 percent. The reduction in cycle count here stems from very similar factors to those just described for the ray-triangle routine.

In the ray-ellipsoid loop, the ratio of ELM/TMS cycles is 79 percent. As seen in Table 5, the TMS linear assembler has generated a particularly efficient schedule in this case, due to the fact that the code consists of a large number of independent SOPs which have allowed the processor to run at close to its peak vector rate. It is unlikely that the ELM would have improved on this directly. Instead, the algorithm was rewritten to use a smaller number of operations, despite several being divisions or square roots. Although this approach was counterproductive on the TMS, it exploited one of the main strengths of the LNS.

The data structure used in this algorithm was comprised of 115 Kbytes. This was far greater than the capacity of either L1 data cache, and since about 12 out of 64 Kbytes in the TMS L2 cache was occupied by frequently used program code, it was over twice the available size of the latter. Data accesses had little spatial locality, with random access across the entire structure. The small scattered accesses fit well into a fine-grained cache such as that on the ELM, which is eight words wide. In the TMS L2 cache with a 32-word line, more time is likely to be spent loading unnecessary data at the boundaries. Despite the availability of an L2 cache, the TMS device stalled for 21 percent of its overall runtime against 27 percent on the ELM.

In total, the ELM required only 61 percent of the clock cycles of the TMS, and did so while also dispensing with an L2 cache. It appears that, although some of this reduction is attributable to the low latency of division and square root, much also accrues from that of multiplication and from the streamlining of the load and control operations made in support of it.

7 Conclusion

As an alternative to FLP, we had proposed the use of the LNS for the representation and processing of real numbers. Although the advantages for multiplication, division, and square root were obvious, the question had been whether

A

```
if (td->u1 == 0) {
  /* uncommon case omitted */
}
else {
  beta = ((v0 * td->u1) - (u0 * td->v1))
    * td->recip1;
  if (beta < 0 || beta > 1) {
    goto quit;
  }
  alpha = (u0 - beta * td->u2)
    * td->recip2;
}
if (alpha < 0 || alpha + beta > 1) {
  goto quit;
}
```

B

```
                LDW     .D2T1   *+B13(8),A5
||              AND     .S1     A6,A5,A6
||              AND     .L1     A7,A5,A7
                LDW     .D1T2   *+A15[A6],B5
||              LDW     .D2T1   *+B13(4),A0
                LDW     .D2T1   *+B13(20),A3
                LDW     .D1T2   *+A15[A7],B4
                LDW     .D2T1   *B13,A4
||              ZERO    .S2     B0
                CMPEQSP .S2X    A5,B0,B0
        [!B0]   B       .S1     l1
||              SUBSP   .L1X    B5,A0,A0
                LDW     .D2T2   *+B13(12),B12
                STW     .D2T1   A3,*+SP(20)
                SUBSP   .L1X    B4,A4,A4
                STW     .D2T1   A0,*+SP(24)
                LDW     .D2T2   *+B13(16),B4
|| [ B0]        MVKL    .S2     _divf1,B5
                ; BRANCH OCCURS
; uncommon case omitted
l1:
                MPYSP   .M1     A5,A0,A12
                MPYSP   .M1X    B12,A4,A3
                LDW     .D2T1   *+B13(28),A0
                NOP             2
                SUBSP   .L1     A12,A3,A12
                MVKL    .S1     0x3f800000,A3
                MVKH    .S1     0x3f800000,A3
                NOP             1
                MPYSP   .M1     A12,A0,A0
                ZERO    .D1     A12
                NOP             2
                MV      .S2X    A0,B5
                CMPGTSP .S1     A0,A3,A1
||              CMPLTSP .S2X    B5,A12,B0
                OR      .S2X    B0,A1,B0
     [ B0]      B       .S1     L6
     [ B0]      ADDK    .S2     0xffffffff,DP
                STW     .D2T1   A0,*+SP(28)
     [ B0]      ADDK    .S1     0x10,A13
     [ B0]      MV      .D2     DP,B0
                NOP             1
                ; BRANCH OCCURS
                MPYSP   .M1X    B4,A0,A0
                LDW     .D2T2   *+B13(60),B5
                NOP             2
                SUBSP   .L1     A4,A0,A5
                NOP             3
                MPYSP   .M2X    A5,B5,B6
                NOP             3
                CMPLTSP .S2X    B6,A12,B0
L3:
     [ B0]      B       .S1     L6
     [!B0]      LDW     .D2T1   *+SP(28),A0
|| [ B0]        ADDK    .S2     0xffffffff,DP
     [ B0]      MV      .D2     DP,B0
|| [ B0]        ADDK    .S1     0x10,A13
                NOP             3
                ; BRANCH OCCURS
```

C

```
        L       0,U1(12)
        S       0,ZERO
        BC      NZ,TI1
* uncommon case omitted
TI1     LR      1,5
        ML      1,U1(12)
        LR      2,4
        ML      2,V1(12)
        SLR     1,2
        ML      1,RECIP1(12)
        LR      0,1
        BC      LT,TI9
        SLL     0,1
        BC      GT,TI9
        L       2,U2(12)
        MLR     2,1
        LR      0,4
        SLR     0,2
        DL      0,U1(12)
TI5     LR      3,0
        BC      LT,TI9
        ALR     3,1
        LR      2,3
        SLL     2,1
        BC      GT,TI9
```

Fig. 12. Excerpt from ray tracer: ray-triangle intersection. (A) Original HLL, (B) Equivalent in optimized TMS assembler, and (C) In optimized ELM assembler.

addition and subtraction could be carried out at least as fast and accurately as in FLP, not only in simulation but by working silicon. Anything less would be unlikely to gain acceptance. Once this were established, however, the LNS would become a viable alternative and the question would arise as to possible techniques for integrating it into a practical computer. How would the architecture of a new device need to be adapted to accommodate an LNS arithmetic unit? Would the theoretical gains actually translate into measurable improvements in performance?

Fig. 13. *Nymphaea alba*: 256×256 pixel ray traced image (with 512×512 adaptive supersampling) rendered in 12.0 s on 125-MHz ELM1A (available in color in the Computer Society Digital Library at http://

We developed a new microprocessor, the ELM, based on a 32-bit logarithmic arithmetic unit. We found that additive operators would in fact be marginally more accurate than

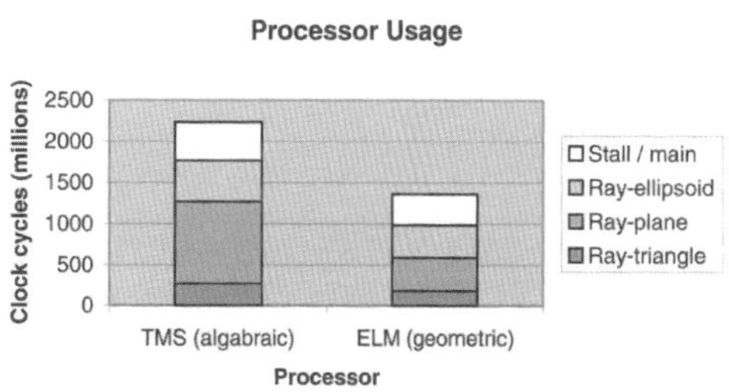

Fig. 14. Analysis of clock cycle usage in TMS and ELM implementations of ray tracer.

those of FLP, and as multiplicative operators return exact results, a typical kernel making equal use of each would incur roughly half the total error of FLP arithmetic. We had previously verified this expectation by using a simulation of the proposed unit. Running the same experiments again, this time on the fabricated microprocessor, we obtained an almost identical set of results. Clearly we may conclude that the LNS can indeed be implemented in silicon, with better accuracy than that of FLP arithmetic.

We evaluated its speed by comparison with a commercial FLP processor of similar fabrication technology. Each processor had six principal ALUs: two adders, two multipliers, and two integer units on the FLP processor, versus two adders and four combined multiplier/integer units on the LNS device. The FLP processor was the result of its designers' best efforts to produce a first-class arithmetic unit, in real silicon, integrated into a microprocessor in the most favorable way, and the LNS processor was the result of our own best efforts to do the same. We were, of course, comparing our first-off silicon and circuit board with the last of several revisions of the FLP device, but were nonetheless able to run the ELM at $\frac{5}{6}$ of the clock speed of the latter. Additive times were marginally better on the ELM, multiplications were 3.4 times the speed, and divisions and square roots were many times faster than the FLP software implementation. If, as we expect, a reworked PCB enables us to run the ELM a little closer to its design speed, which was similar to that of the FLP device, then we may simply compare clock cycles. In this case, the ELM would offer improvements to addition and multiplication of 1.3 and 4 times, respectively. In any event, LNS addition and subtraction can be implemented not only with "better than FLP" accuracy, but also with better speed. The criteria set out above have been satisfied, and we conclude that the LNS is a clear improvement on 32-bit FLP arithmetic.

In order to harness this new power, we had considered how the design of a microprocessor might be adapted to achieve the best synergy with desirable features of the arithmetic. The LNS is intrinsically of low latency, and this philosophy pervaded the design of the ELM. Functional unit pipelining was dispensed with, and an RM instruction set adopted to eliminate explicit loading. However, as the multiplier is effectively cost free, throughput was enhanced by functional unit replication, with a resulting short-vector capability.

Substantial case studies pinpointed how and where the use of LNS arithmetic had been of value. These studies also served to experiment with possible application areas for the ELM and to demonstrate its overall capability. Measuring by time, that is, despite the fact that the ELM was running at a slower clock speed, we found in one case that the inherent simplification of the division and square root operations almost entirely compensated for the lack of an L2 cache. In the other, we not only dispensed with the L2 cache but also obtained a reduction to 73 percent of the execution time. This arose through 1) the reduced latency of all real operations and of architectural decisions such as a tightly coupled L1 cache made in support of it, and 2) rewriting the algorithm to use fewer operations, despite several being divisions and square roots. In terms of clock cycles, this reduction was to 61 percent.

The ELM delivered an improvement in accuracy equivalent to 1.5 bits; in the application concerned, a reduction in processing noise to about $\frac{2}{3}$ of that of the FLP device.

In summary, the LNS delivered faster execution than FLP arithmetic, more accurate results, and did so with a significant reduction in architectural complexity. Power dissipation and silicon area are yet to be considered (although preliminary results are encouraging), but on the basis of the present results, the LNS may now be the preferred technique for 32-bit work. This is particularly so in emerging algorithms that are naturally cast in terms of divisions, but where a great deal of programmer time is currently spent in reworking to alternative formulations. A point that we made in [1] may be repeated here: the design of an LNS unit requires significant attention to only one block of circuitry. If the adder/subtractor could be reduced in latency to less than that of a 32×32 bit multiplier, then assuming an equal proportion of additions and multiplications, the LNS would be faster than *fixed-point* arithmetic. That is the next step.

Acknowledgments

The authors gratefully acknowledge the support of the European Strategic Programme for Research and development in Information Technology (ESPRIT) Long-Term Research Programme, which funded this work under Grants 23544 and 33544. Partners in the project were the University of Newcastle upon Tyne, University College Dublin, Academy of Sciences of the Czech Republic, Prague, Philips Research, Eindhoven, and Massana, Dublin. We also thank Philips Semiconductors, Nijmegen, where the ELM devices were fabricated. The initial design of the evaluation board was done by MEng student Steve Fishwick.

References

[1] J.N. Coleman, E.I. Chester, C. Softley, and J. Kadlec, "Arithmetic on the European Logarithmic Microprocessor," *IEEE Trans. Computers,* vol. 49, no. 7, pp. 702-715, July 2000, erratum, vol. 49, no. 10, p. 1152, Oct. 2000.

[2] J.N. Coleman and E.I. Chester, "A 32-Bit Logarithmic Arithmetic Unit and Its Performance Compared to Floating-Point," *Proc. 14th IEEE Symp. Computer Arithmetic,* 1999.

[3] F.J. Taylor, R. Gill, J. Joseph, and J. Radke, "A 20-Bit Logarithmic Number System Processor," *IEEE Trans. Computers,* vol. 37, pp. 190-200, 1988.

[4] D.M. Lewis, "An Architecture for Addition and Subtraction of Long Wordlength Numbers in the Logarithmic Number System," *IEEE Trans. Computers,* vol. 39, pp. 1325-1336, 1990.

[5] D. Yu and D.M. Lewis, "A 30-b Integrated Logarithmic Number System Processor," *IEEE J. Solid-State Circuits.,* vol. 26, pp. 1433-1440, 1991.

[6] D.M. Lewis, "Interleaved Memory Function Interpolators with Application to an Accurate LNS Arithmetic Unit," *IEEE Trans. Computers,* vol. 43, pp. 974-982, 1994.

[7] D.M. Lewis, "114 MFLOPS Logarithmic Number System Arithmetic Unit for DSP Applications," *IEEE J. Solid-State Circuits.,* vol. 30, pp. 1547-1553, 1995.

[8] V. Paliouras, J. Karagiannis, G. Aggouras, and T. Stouraitis, "A Very-Long Instruction Word Digital Signal Processor Based on the Logarithmic Number System," *Proc. Fifth IEEE Int'l Conf. Electronics, Circuits and Systems,* 1998.

[9] J.N. Coleman, C.I. Softley, J. Kadlec, R. Matousek, M. Licko, Z. Pohl, and A. Hermanek, "The European Logarithmic Microprocessor—A QR RLS Application," *Proc. 35th IEEE Asilomar Conf. Signals, Systems, and Computers,* 2001.

[10] M.G. Arnold, "A VLIW Architecture for Logarithmic Arithmetic," *Proc. Euromicro Symp. Digital System Design,* 2003.

[11] M.G. Arnold and C. Walter, "Unrestricted Faithful Rounding Is Good Enough for Some LNS Applications," *Proc. 15th IEEE Symp. Computer Arithmetic,* 2001.

[12] J.N. Coleman, C.I. Softley, J. Kadlec, R. Matousek, M. Licko, Z. Pohl, and A. Hermanek, "Performance of the European Logarithmic Microprocessor," *Proc. SPIE Ann. Meeting*, 2003.
[13] C.H. Chen, R.-L. Chen, and C.-H. Yang, "Pipelined Computation of Very Large Word-Length LNS Addition/Subtraction with Polynomial Hardware Cost," *IEEE Trans. Computers*, vol. 49, pp. 716-726, 2000.

J. Nicholas Coleman was initially educated in music, receiving the BA degree from York University. His first employment was in commercial applications and systems programming. Becoming increasingly interested in engineering, he joined Plessey Telecommunications Research, Poole, Dorset. Working initially as a software engineer and then as a hardware engineer, he was responsible for the design of a custom processor for real-time test of the System-X telephone exchange. At Brunel University, Uxbridge, Middlesex, he then began work on a PhD, for which he designed and built a special-purpose dataflow computer. After receiving the PhD degree, he joined Newcastle University, where he is a lecturer in computer engineering and the coordinator of the European Strategic Programme for Research and development in Information Technology (ESPRIT) Project on which this work is based. He has recently founded a spin-off company, Northern Digital, which, with research and development funding from the United Kingdom government, aims at commercializing the outcome of this project. His research interests include high-speed processor design and processor-intensive applications such as graphics.

Chris I. Softley received the BEng (hons) degree in microelectronics and software engineering from the University of Newcastle upon Tyne in 1999. His dissertation focused on arithmetic cells for a logarithmic number system (LNS) CPU. He was then employed by the university as a research assistant, seconded to the Philips Research Laboratories in Eindhoven, The Netherlands. Here, he worked primarily on the design of the European Logarithmic Microprocessor, under the European Strategic Programme for Research and development in Information Technology High-Speed Logarithmic Arithmetic Unit (ESPRIT HSLA) Project. Upon completing this work in 2002-2003, he moved to Switzerland and joined Photonfocus AG, becoming the head of sensor development in 2005. He currently develops high-performance CMOS image sensors for industrial applications which have a combined linear-logarithmic response for high dynamic range. He is also with the Hochschule für Technik, Rapperswil, where he supervises research on low-power low-noise readout circuits for fast CMOS imagers. His research interests include image sensors, energy-efficient analog circuits, computer arithmetic, number systems, and processor architecture, particularly as applied to vision and visualization systems. He is a member of the IEEE and the IEEE Computer Society.

Jiri Kadlec received the MSc degree from the Czech Technical University in Prague, in 1982 and the PhD degree from the Academy of Sciences of the Czech Republic in 1987. Since 1990, he has been with the Institute of Information Theory and Automation at the Academy of Sciences of the Czech Republic, where he is currently the head of the Department of Signal Processing. He was a researcher at Ruhr University, Bochum, Germany, in 1989, the University of Athens, Greece, in 1990, Queen's University of Belfast, United Kingdom, from 1992 to 1995, and the Katholieke Universiteit Leuven, Belgium, from 1995 to 1996. His research interests include recursive system identification algorithms suitable for FPGA, rapid prototyping of advanced signal processing algorithms, and scalable floating-point arithmetic for FPGA SoC designs.

Rudolf Matousek received the MSc degree in automation in transportation sciences from the Czech Technical University in Prague, in 2000. His research interests include computer arithmetic, FPGA implementation of DSP algorithms, and design methodology and tooling for the dynamic reconfiguration of FPGA devices.

Milan Tichy received the MSc and PhD degrees from the Czech Technical University in Prague, in 1999 and 2006, respectively. In 2003, he was awarded the two-year Marie Curie Fellowship, which he spent in the Department of Computer Science at Trinity College, Dublin, Ireland, from 2004 to 2006. He is currently a senior researcher in the Department of Signal Processing at the Institute of Information Theory and Automation of the Academy of Sciences of the Czech Republic. His research interests include reconfigurable systems, parallel algorithms and architectures, parallel adaptive algorithms, VLSI implementations, and embedded systems. He is a member of the IEEE and the IEEE Computer Society.

Zdenek Pohl received the bachelor's degree in electrical engineering from the Czech Technical University in Prague. He is currently working toward the PhD degree at the Czech Technical University. He is also a researcher in the Institute of Information Theory and Automation at the Academy of Sciences of the Czech Republic. His research interests include FPGA implementations of signal processing algorithms, speech coding, and rapid prototyping.

Antonin Hermanek received the MSEE degree from the Czech Technical University in Prague, in 1998 and the PhD degree from the Université Paris-Sud, Orsay, in 2005. He is a researcher in the Institute of Information Theory and Automation at the Academy of Sciences of the Czech Republic. His research interests include blind equalization, multiple-input, multiple-output (MIMO) and orthogonal frequency-division multiplexing (OFDM) communication systems, FPGA implementation of DSP algorithms, array processing, and rapid prototyping for signal processing. He has published more than 30 papers in these areas. He is a member of the IEEE.

Nico F. Benschop received the MSc degree in electronics from Delft University, The Netherlands, in 1966 and the PhD degree from Waterloo University, Ontario, Canada in 1971. From 1970 to 2002, he was with Philips Research Laboratories, Eindhoven, The Netherlands, researching design methods for digital VLSI in logic, arithmetic, and state machines. After retirement, he finished a book, *Associative Digital Network Theory*, combining these three main aspects under one heading of the finite associative algebra of function composition (semigroups) to improve structural insight into complex computer circuits and algorithms.

BIBLIOGRAPHY

Ulrich Ablassmeier, "Comparison of Various Binary Dividers in I^2L and CHIL Technology," *IEEE Journal of Solid-State Circuits*, Vol. SC-14, 1979, pp. 657-660

Dharma P. Agrawal, "Arithmetic Algorithms in a Negative Base," *IEEE Transactions on Computers*, Vol. C-24, 1975, pp. 998-1000

D. P. Agrawal, "Optimum Array-Like Structures for High-Speed Arithmetic," *Proceedings of the 3rd Symposium on Computer Arithmetic*, Dallas, TX, November 1975, pp. 208-219

Dharma P. Agrawal, "A Novel Technique for Computing Negabinary Squares," *IEEE Transactions on Computers*, Vol. C-27, 1978, pp. 266-270

D. P. Agrawal, "On Arithmetic Inter-Relationships and Hardware Interchangeability of Negabinary and Binary Systems," *Proceedings of the 4th Symposium on Computer Arithmetic*, Santa Monica, CA, October 1978, pp. 88-96

Dharma P. Agrawal, "On Negabinary-Binary Arithmetic Relationships and Their Hardware Reciprocity," *IEEE Transactions on Computers*, Vol. C-29, 1980, pp. 1032-1035

D. P. Agrawal and R. C. Joshi, "Negabinary Addition and Multiplication Using Binary Circuits," *Proceedings of the 5th Symposium on Computer Arithmetic*, Ann Arbor, MI, May 1981, pp. 270-273

D. P. Agrawal and T. R. N. Rao, "On Modular (2^n+1) Arithmetic Logic," *Proceedings of the 4th Symposium on Computer Arithmetic*, Santa Monica, CA, October 1978, pp. 104-108

Dharma P. Agrawal and T. R. N. Rao, "On Multiple Operand Addition of Signed Binary Numbers," *IEEE Transactions on Computers*, Vol. C-27, 1978, pp. 1068-1070

D. P. Agrawal, *et al.*, "On Design and Performance of VLSI Based Parallel Multiplier," *Proceedings of the 6th Symposium on Computer Arithmetic*, Aarhus, DENMARK, June 1983, pp. 17-21

Hassan M. Ahmed, "Efficient Elementary Function Generation with Multipliers," *Proceedings of the 9th Symposium on Computer Arithmetic*, Santa Monica, CA, September 1989, pp. 52-59

Akintunde Ibitayo Akinwande, *et al.*, "A 500-MHz 16 x 16 Complex Multiplier Using Self-Aligned Gate GaAs Heterostructure FET Technology," *IEEE Journal of Solid-State Circuits*, Vol. 24, 1989, pp. 1295-1300

I. Aleksander, "Array Networks for a Parallel Adder and Its Control," *IEEE Transactions on Electronic Computers*, Vol. EC-16, 1967, pp. 226-229

S. F. Anderson, *et al.*, "The IBM System/360 Model 91: Floating-Point Execution Unit," *IBM Journal of Research and Development*, Vol. 11, 1967, pp. 34-53

Michael Andrews, "A Systolic SBNR Adaptive Signal Processor," *IEEE Journal of Solid-State Circuits*, Vol. SC-21, 1986, pp. 120-128

M. Annaratone and R. Stefanelli, "A Multiplier with Multiple Error Correction Capability," *Proceedings of the 6th Symposium on Computer Arithmetic*, Aarhus, DENMARK, June 1983, pp. 44-51

Benjamin Arazi, "An Electrooptical Adder," *Proceedings of the IEEE*, Vol. 73, 1985, pp. 162-163

M. G. Arnold, T. A. Bailey, J. R. Cowles, and J. J. Cupal, "Redundant Logarithmic Number Systems," *Proceedings of the 9th Symposium on Computer Arithmetic*, Santa Monica, CA, September 1989, pp. 144-151

R. K. Arora and S. Kaushik, "Conversion Scheme in Residue Code," *Proceedings of the 5th Symposium on Computer Arithmetic*, Ann Arbor, MI, May 1981, pp. 152-157

Ehud Artzy, James A. Hinds, and Harry J. Saal, "A Fast Division Technique for Constant Divisors," *Communications of the ACM*, Vol. 19, 1976, pp. 98-101

Mohammad Javad Ashjaee and Sudhakar M. Reddy, "On Totally Self-Checking Checkers for Separable Codes," *IEEE Transactions on Computers*, Vol. C-26, 1977, pp. 737-744

David B. Aspinwall and Yale N. Patt, "Retrofitting the VAX-11/780 Microarchitecture for IEEE Floating Point Arithmetic—Implementation Issues, Measurements, and Analysis," *IEEE Transactions on Computers*, Vol. C-34, 1985, pp. 692-708

Daniel E. Atkins, "Higher-Radix Division Using Estimates of the Divisor and Partial Remainders," *IEEE Transactions on Computers*, Vol. C-17, 1968, pp. 925-932

Daniel E. Atkins, "Design of the Arithmetic Units of ILLIAC III: Use of Redundancy and Higher Radix Methods," *IEEE Transactions on Computers*, Vol. C-19, 1970, pp. 1085-1089

D. E. Atkins, "Higher Radix, Non-Restoring Division: History and Recent Developments," *Proceedings of the 3rd Symposium on Computer Arithmetic*, Dallas, TX, November 1975, pp. 158-160

D. E. Atkins and S. C. Ong, "A Comparison of Two Approaches to Multi-Operand Binary Addition," *Proceedings of the 4th Symposium on Computer Arithmetic*, Santa Monica, CA, October 1978, pp. 125-139

A. Avizienis, "Signed-Digit Representations for Fast Parallel Arithmetic," *IRE Transactions on Electronic Computers*, Vol. EC-10, 1961, pp. 389-400

Algirdas Avizienis, "Arithmetic Algorithms for Error-Coded Operands," *IEEE Transactions on Computers*, Vol. C-22, 1973, pp. 567-572

A. Avizienis, "Redundancy in Number Representations as an Aspect of Computational Complexity of Arithmetic Functions," *Proceedings of the 3rd Symposium on Computer Arithmetic*, Dallas, TX, November 1975, pp. 87-89

A. Avizienis, "Low-Cost Residue and Inverse Residue Error-Detecting Codes for Signed-Digit Arithmetic," *Proceedings of the 5th Symposium on Computer Arithmetic*, Ann Arbor, MI, May 1981, pp. 165-169

A. Avizienis, "Arithmetic Algorithms for Operands Encoded in Two-Dimensional Low-Cost Arithmetic Error Codes," *Proceedings of the 7th Symposium on Computer Arithmetic*, Urbana, IL, June 1985, pp. 285-292

A. Avizienis and C. S. Raghavendra, "Applications for Arithmetic Error Codes in Large, High-Performance Computers," *Proceedings of the 6th Symposium on Computer Arithmetic*, Aarhus, DENMARK, June 1983, pp. 169-173

Algirdas Avizienis and Chin Tung, "A Universal Arithmetic Building Element (ABE) and Design Methods for Arithmetic Processors," *IEEE Transactions on Computers*, Vol. C-19, 1970, pp. 733-745

Aqil M. Azmi and Fabrizio Lombardi, "On a Tapered Floating Point System," *Proceedings of the 9th Symposium on Computer Arithmetic*, Santa Monica, CA, September 1989, pp. 2-9

P. W. Baker, "Parallel Multiplicative Algorithms for Some Elementary Functions," *IEEE Transactions on Computers*, Vol. C-24, 1975, pp. 322-325

P. W. Baker, "More Efficient Radix-2 Algorithms for Some Elementary Functions," *IEEE Transactions on Computers*, Vol. C-24, 1975, pp. 1049-1054

P. W. Baker, "Suggestion for a Fast Binary Sine/Cosine Generator," *IEEE Transactions on Computers*, Vol. C-25, 1976, pp. 1134-1136

Gary L. Baldwin, *et al.*, "A Modular, High-Speed Serial Pipeline Multiplier for Digital Signal Processing," *IEEE Journal of Solid-State Circuits*, Vol. SC-13, 1978, pp. 400-408

P. T. Balsara and R. M. Owens, "Systolic and Semi-Systolic Digit Serial Multipliers," *Proceedings of the 8th Symposium on Computer Arithmetic*, Como, ITALY, May 1987, pp. 169-173

Nuno Bandeira, Ken Vaccaro, and James A. Howard, "A Two's Complement Array Multiplier Using True Values of the Operands," *IEEE Transactions on Computers*, Vol. C-32, 1983, pp. 745-747

S. Bandyopadhyay, S. Basu and A. K. Choudhury, "An Iterative Array for Multiplication of Signed Binary Numbers," *IEEE Transactions on Computers*, Vol. C-21, 1972, pp. 921-922

Dilip K. Banerji and Janusz A. Brzozowski, "On Translation Algorithms in Residue Number Systems," *IEEE Transactions on Computers*, Vol. C-21, 1972, pp. 1281-1285

D. K. Banerji, "A Novel Implementation Method for Addition and Subtraction in Residue Number Systems," *IEEE Transactions on Computers*, Vol. C-23, 1974, pp. 106-109

D. K. Banerji, "On the Use of Residue Arithmetic for Computation," *IEEE Transactions on Computers*, Vol. C-23, 1974, pp. 1315-1317

D. K. Banerji, "On Combinational Logic for Sign Detection in Residue Number Systems," *Proceedings of the 3rd Symposium on Computer Arithmetic*, Dallas, TX, November 1975, pp. 113-116

D. K. Banerji, T-Y. Cheung, and V. Ganesan, "A High-Speed Division Method in Residue Arithmetic," *Proceedings of the 5th Symposium on Computer Arithmetic*, Ann Arbor, MI, May 1981, pp. 158-164

D. K. Banerji and S. Kaushik, "Representation and Processing of Fractions in a Residue System," *Proceedings of the 6th Symposium on Computer Arithmetic*, Aarhus, DENMARK, June 1983, pp. 29-36

J. Bannur and A. Varma, "The VLSI Implementation of a Square Root Algorithm," *Proceedings of the 7th Symposium on Computer Arithmetic*, Urbana, IL, June 1985, pp. 159-165

Amnon B. Barak, "Multiplicative Algorithms for Ternary Arithmetic Using Binary Logic," *IEEE Transactions on Computers*, Vol. C-26, 1977, pp. 823-826

J. Barlow, "On the Distribution of Accumulated Roundoff Error in Floating Point Arithmetic," *Proceedings of the 5th Symposium on Computer Arithmetic*, Ann Arbor, MI, May 1981, pp. 100-105

G. Barrett, "A Formal Approach to Rounding," *Proceedings of the 8th Symposium on Computer Arithmetic*, Como, ITALY, May 1987, pp. 247-254

Ferruccio Barsi and Piero Maestrini, "Error Correcting Properties of Redundant Residue Number Systems," *IEEE Transactions on Computers*, Vol. C-22, 1973, pp. 307-315

Ferruccio Barsi and Piero Maestrini, "Error Detection and Correction by Product Codes in Residue Number Systems," *IEEE Transactions on Computers*, Vol. C-23, 1974, pp. 915-924

Ferruccio Barsi and Piero Maestrini, "Arithmetic Codes in Residue Number Systems with Magnitude Index," *IEEE Transactions on Computers*, Vol. C-27, 1978, pp. 1185-1188.

Dhruba Basu and Totadri Jayashree, "On a Single Postcorrection for Nonrestoring Division," *IEEE Transactions on Computers*, Vol. C-24, 1975, pp. 846-848

Gerard M. Baudet, Franco P. Preparata, and Jean E. Vuillemin, "Area-Time Optimal VLSI Circuits for Convolution," *IEEE Transactions on Computers*, Vol. C-32, 1983, pp. 684-688

C. R. Baugh and B. A. Wooley, "A Two's Complement Parallel Array Multiplication Algorithm," *IEEE Transactions on Computers*, Vol. C-22, 1973, pp. 1045-1047

M. A. Bayoumi, G. A. Jullien, and W. C. Miller, "Models for VLSI Implementation of Residue Number System Arithmetic Modules," *Proceedings of the 6th Symposium on Computer Arithmetic*, Aarhus, DENMARK, June 1983, pp. 174-183

Bernd Becker, "Efficient Testing of Optimal Time Adders," *IEEE Transactions on Computers*, Vol. C-37, 1978, pp. 1113-1121

O. J. Bedrij, "Carry Select Adders," *IRE Transactions on Electronic Computers*, Vol. EC-11, 1962, pp. 340-346

M. Benedek, "Developing Large Binary to BCD Conversion Structures," *Proceedings of the 3rd Symposium on Computer Arithmetic*, Dallas, TX, November 1975, pp. 188-196

W. Scott Bennett, "Quotient Generation with Conventional Binary Multipliers," *Proceedings of the IEEE*, Vol. 61, 1973, pp. 664-665

Bradley J. Benschneider, *et al.*, "A Pipelined 50-MHz CMOS 64-bit Floating-Point Arithmetic Processor," *IEEE Journal of Solid-State Circuits*, Vol. 24, 1989, pp. 1317-1323

R. O. Berg and Larry L. Kinney, "Serial Adders with Overflow Correction," *IEEE Transactions on Computers*, Vol. C-20, 1971, pp. 668-671

L. Bhuyan and D. P. Agrawal, "Multiple Addition and Parallel Counter in Generalized Binary and Negabinary Systems," *Proceedings of the 5th Symposium on Computer Arithmetic*, Ann Arbor, MI, May 1981, pp. 264-269

Laxmi N. Bhuyan and Dharma P. Agrawal, "On the Generalized Binary System," *IEEE Transactions on Computers*, Vol. C-31, 1982, pp. 335-338

R. L. Bivins and N. Metropolis, "Significance Arithmetic: Application to a Partial Differential Equation," *Proceedings of the 3rd Symposium on Computer Arithmetic*, Dallas, TX, November 1975, pp. 64-66

Robert L. Bivins and Nicholas C. Metropolis, "Significance Arithmetic: Application to a Partial Differential Equation," *IEEE Transactions on Computers*, Vol. C-26, 1977, pp. 639-642

P. E. Blankenship, "Comments on 'A Two's Complement Parallel Array Multiplication Algorithm,'" *IEEE Transactions on Computers*, Vol. C-23, 1974, pp. 1327

J. H. Bleher, A. E. Roeder, and S. M. Rump, "ACRITH: High-Accuracy Arithmetic--An Advanced Tool for Numerical Computation," *Proceedings of the 7th Symposium on Computer Arithmetic*, Urbana, IL, June 1985, pp. 318-321

G. Bohlender, "Floating-Point Computation of Functions with Maximum Accuracy," *Proceedings of the 3rd Symposium on Computer Arithmetic*, Dallas, TX, November 1975, pp. 14-23

Gerd Bohlender, "Floating-Point Computation of Functions with Maximum Accuracy," *IEEE Transactions on Computers*, Vol. C-26, 1977, pp. 621-632.

A. D. Booth, "A Signed Binary Multiplication Technique," *Quarterly Journal of Mechanics and Applied Mathematics*, Vol. 4, 1951, pp. 236-240

B. Bose, *et al.*, "Fast Multiply and Divide for a VLSI Floating-Point Unit," *Proceedings of the 8th Symposium on Computer Arithmetic*, Como, ITALY, May 1987, pp. 87-94

Ralph H. Brackert, Jr., M. D. Ercegovac, and Alan N. Willson, Jr., "Design of an On-Line Multiply-Add Module for Recursive Digital Filters," *Proceedings of the 9th Symposium on Computer Arithmetic*, Santa Monica, CA, September 1989, pp. 34-41

J. F. Brennan, "The Fastest Time of Addition and Multiplication," *IBM Research Reports*, Vol. 4, 1968.

R. Brent, "On the Addition of Binary Numbers," *IEEE Transactions on Computers*, Vol. C-19, 1970, pp. 758-759

Richard P. Brent, "On the Precision Attainable with Various Floating-Point Number Systems," *IEEE Transactions on Computers*, Vol. C-22, 1973, pp. 601-607

Richard P. Brent and H. T. Kung, "A Regular Layout for Parallel Adders," *IEEE Transactions on Computers*, Vol. C-31, 1982, pp. 260-264

R. P. Brent and H. T. Kung, "A Systolic Algorithm for Integer GCD Computation," *Proceedings of the 7th Symposium on Computer Arithmetic*, Urbana, IL, June 1985, pp. 118-125

Richard Brent, David Kuck, and Kiyoshi Maruyama, "The Parallel Evaluation of Arithmetic Expressions without Division," *IEEE Transactions on Computers*, Vol. C-22, 1973, pp. 532-534

C. L. Bridge, P. D. Fisher, and R. G. Reynolds, "Asynchronous Arithmetic Algorithms for Data-Driven Machines," *Proceedings of the 5th Symposium on Computer Arithmetic*, Ann Arbor, MI, May 1981, pp. 56-63

Bruce E. Briley, "Some New Results on Average Worst Case Carry," *IEEE Transactions on Computers*, Vol. C-22, 1973, pp. 459-463

Thomas J. Brosnan and Noel R. Strader, "Modular Error Detection for Bit-Serial Multiplication," *IEEE Transactions on Computers*, Vol. C-37, 1988, pp. 1043-1052

W. S. Brown, "A Simple but Realistic Model of Floating-Point Computation," *ACM Transactions on Mathematical Software*, Vol. 7, 1981, pp. 445-480

W. S. Brown and P. L. Richman, "The Choice of Base," *Communications of the ACM*, Vol. 12, 1969, pp. 560-561

W. S. Brown and C. S. Wetherell, "A Numeric Error Algebra," *Proceedings of the 6th Symposium on Computer Arithmetic*, Aarhus, DENMARK, June 1983, pp. 86-93

Thomas A. Brubaker and John C. Becker, "Multiplication Using Logarithms Implemented with Read-Only Memory," *IEEE Transactions on Computers*, Vol. C-24, 1975, pp. 681-690

Louis B. Bushard, "A Minimum Table Size Result for Higher Radix Nonrestoring Division," *IEEE Transactions on Computers*, Vol. C-32, 1983, pp. 521-526

Joaquín Bustoz, Alan Feldstein, and Richard Goodman, "Improving Trailing Digit Estimates Applied to Optimal Computer Arithmetic," *Journal of the ACM*, Vol. 26, 1979, pp. 716-730

V. Cantoni, *et al.*, "PAPIA: Pyramidal Architecture for Parallel Image Analysis," *Proceedings of the 7th Symposium on Computer Arithmetic*, Urbana, IL, June 1985, pp. 237-242

Peter R. Cappello and Willard L. Miranker, "Systolic Super Summation," *IEEE Transactions on Computers*, Vol. C-37, 1988, pp. 657-677

Maurus Cappa and V. Carl Hamacher, "An Augmented Iterative Array for High-Speed Binary Division," *IEEE Transactions on Computers*, Vol. C-22, 1973, pp. 172-175

P. Carmichael, *et al.*, "A New Method of Ultra-High-Speed Mod-2 Addition," *Proceedings of the IEEE*, Vol. 63, 1975, pp. 721-722

T. M. Carter, "Structured Arithmetic Tiling of Integrated Circuits," *Proceedings of the 8th Symposium on Computer Arithmetic*, Como, ITALY, May 1987, pp. 41-48

Tony M. Carter, "Cascade: Hardware for High/Variable Precision Arithmetic," *Proceedings of the 9th Symposium on Computer Arithmetic*, Santa Monica, CA, September 1989, pp. 184-191

J. R. Cavallaro and F. T. Luk, "CORDIC Arithmetic for an SVD Processor," *Proceedings of the 8th Symposium on Computer Arithmetic*, Como, ITALY, May 1987, pp. 113-120

J. R. Cavallaro and F. T. Luk, "CORDIC Arithmetic for an SVD Processor," *Journal of Parallel and Distributed Computing*, Vol. 5, 1988, pp. 271-290

Joseph J. F. Cavanaugh, *Digital Computer Arithmetic, Design and Implementation*, New York: McGraw-Hill Book Co., 1984.

S. Chakrabarti, "Divide-and-Correct Algorithm for Division in a Negative Base," *IEEE Transactions on Computers*, Vol. C-23, 1974, pp. 981-983

N. B. Chakraborti, John S. Soundararajan, and A. L. N. Reddy, "An Implementation of Mixed-Radix Conversion for Residue Number Applications," *IEEE Transactions on Computers*, Vol. C-35, 1986, pp. 762-764

Pak K. Chan and Martine D. F. Schlag, "Analysis and Design of CMOS Manchester Adder with Variable Carry-Skip," *Proceedings of the 9th Symposium on Computer Arithmetic*, Santa Monica, CA, September 1989, pp. 86-95

Chung Ho Chen, "An Algebraic Model of Arithmetic Codes," *IEEE Transactions on Computers*, Vol. C-31, 1982, pp. 318-321

Marina C. Chen, "The Generation of a Class of Multipliers: Synthesizing Highly Parallel Algorithms in VLSI," *IEEE Transactions on Computers*, Vol. C-37, 1988, pp. 329-338

Tien Chi Chen, "A Binary Multiplication Scheme Based on Squaring," *IEEE Transactions on Computers*, Vol. C-20, 1971, pp. 678-680

T. C. Chen, "Maximal Redundancy Signed-Digit Systems," *Proceedings of the 7th Symposium on Computer Arithmetic*, Urbana, IL, June 1985, pp. 296-300

P. W. Cheney, "A Digital Correlator Based on the Residue Number System," *IRE Transactions on Electronic Computers*, Vol. EC-10, 1961, pp. 67-70

Wu-Tung Cheng and Janak H. Patel, "A Minimum Test Set for Multiple Fault Detection in Ripple Carry Adders," *IEEE Transactions on Computers*, Vol. C-36, 1987, pp. 891-895

H. D. Cheng and K. S. Fu, "Algorithm Partition for a Fixed-Size VLSI Architecture Using Space-Time Domain Expansion," *Proceedings of the 7th Symposium on Computer Arithmetic*, Urbana, IL, June 1985, pp. 126-132

A. C. L. Chiang and Irving S. Reed, "Notes on the Arithmetic BN Modulo A Codes," *IEEE Transactions on Computers*, Vol. C-21, 1972, pp. 891-894

D. M. Chiarulli, W. G. Rudd, and D. A. Buell, "DRAFT: A Dynamically Reconfigurable Processor for Integer Arithmetic," *Proceedings of the 7th Symposium on Computer Arithmetic*, Urbana, IL, June 1985, pp. 309-317

Robert T. Chien and Se June Hong, "Error Correction in High-Speed Arithmetic," *IEEE Transactions on Computers*, Vol. C-21, 1972, pp. 433-438

Jean P. Chinal, "Some Comments on Postcorrections for Nonrestoring Division," *IEEE Transactions on Computers*, Vol. C-21, 1972, pp. 1385-1394

J. P. Chinal, "Minor Arithmetic," *Proceedings of the 3rd Symposium on Computer Arithmetic*, Dallas, TX, November 1975, pp. 98-107

J. P. Chinal, "The Logic of Modulo $2^k + 1$ Adders," *Proceedings of the 3rd Symposium on Computer Arithmetic*, Dallas, TX, November 1975, pp. 126-136

Jean P. Chinal, "Boolean Features of Sign-Flipping and Signed-Carry Logic in Base -2," *IEEE Transactions on Computers*, Vol. C-26, 1977, pp. 82-86

E. T. Chow and D. I. Moldovan, "Prime Factor DFT Parallel Processor Using Wafer Scale Integration," *Proceedings of the 7th Symposium on Computer Arithmetic*, Urbana, IL, June 1985, pp. 133-139

C. Y. Chow and J. E. Robertson, "Logical Design of a Redundant Binary Adder," *Proceedings of the 4th Symposium on Computer Arithmetic*, Santa Monica, CA, October 1978, pp. 109-115

P. Chow, Z. G. Vranesic, and J. L. Yen, "A Pipelined Distributed Arithmetic PFFT Processor," *Proceedings of the 5th Symposium on Computer Arithmetic*, Ann Arbor, MI, May 1981, pp. 198-206

L. Ciminiera, "Parallel Multipliers Based on Horizontal Compressors," *Proceedings of the 8th Symposium on Computer Arithmetic*, Como, ITALY, May 1987, pp. 63-69

L. Ciminiera and A. Serra, "Arithmetic Array for Fast Inner Product Evaluation," *Proceedings of the 5th Symposium on Computer Arithmetic*, Ann Arbor, MI, May 1981, pp. 207-215

L. Ciminiera and A. Serra, "Fast Iterative Multiplying Array," *Proceedings of the 6th Symposium on Computer Arithmetic*, Aarhus, DENMARK, June 1983, pp. 60-66

L. Ciminiera and A. Serra, "Efficient Serial-Parallel Arrays for Multiplication and Addition," *Proceedings of the 7th Symposium on Computer Arithmetic*, Urbana, IL, June 1985, pp. 28-35

L. Ciminiera, A. Serra, and A. Valenzano, "Fast and Accurate Matrix Triangularization Using an Iterative Structure," *Proceedings of the 5th Symposium on Computer Arithmetic*, Ann Arbor, MI, May 1981, pp. 215-221

G. W. Cobb, "The Impact of Parallelism on Software," *Proceedings of the 3rd Symposium on Computer Arithmetic*, Dallas, TX, November 1975, pp. 220-222

William J. Cody, Jr., "Static and Dynamic Numerical Characteristics of Floating-Point Arithmetic," *IEEE Transactions on Computers*, Vol. C-22, 1973, pp. 598-601

William J. Cody, Jr., "Analysis of Proposals for the Floating Point Standard," *Computer*, Vol. 14, March 1981, pp. 63-69

William J. Cody, Jr., "A Proposed Radix and Word Length Independent Standard for Floating-Point Arithmetic," *ACM SIGNUM Newsletter*, Vol. 20, January 1985, pp. 37-51

D. Cohen, "Mathematical Approach to Interative Computation Networks," *Proceedings of the 4th Symposium on Computer Arithmetic*, Santa Monica, CA, October 1978, pp. 226-238

M. Cohen, V. C. Hamacher, and T. E. Hull, "CADAC: An Arithmetic Unit for Clean Decimal Arithmetic and Controlled Precision," *Proceedings of the 5th Symposium on Computer Arithmetic*, Ann Arbor, MI, May 1981, pp. 106-112

A. Colagrossi and A. Miola, "A Normalization Algorithm for Truncated P-ADIC Arithmetic," *Proceedings of the 8th Symposium on Computer Arithmetic*, Como, ITALY, May 1987, pp. 212-216

M. Combet, H. Van Zonneveld, and L. Verbeek, "Computation of the Base Two Logarithm of Binary Numbers," *IEEE Transactions on Electronic Computers*, Vol. EC-14, 1965, pp. 863-867

J. T. Coonen, "An Implementation Guide to a Proposed Standard for Floating Point Arithmetic," *Computer*, Vol. 13, January 1980, pp. 68-79, Errata: *Computer*, Vol. 14, March 1981, p. 62

Jerome T. Coonen, "Underflow and the Denormalized Numbers," *Computer*, Vol. 14, March 1981, p. 75-87

J. Coonen, *et al.*, "A Proposed Standard for Binary Floating Point Arithmetic," *ACM SIGNUM Newsletter*, October 1979, pp. 4-12

Paolo Corsini and Graziano Frosini, "Uniform Shift Multiplication Algorithms without Overflow," *IEEE Transactions on Computers*, Vol. C-27, 1978, pp. 256-258

M. Cosnard, *et al.*, "The FELIN Arithmetic Coprocessor Chip," *Proceedings of the 8th Symposium on Computer Arithmetic*, Como, ITALY, May 1987, pp. 107-112

L. Dadda, "Some Schemes for Parallel Multipliers," *Alta Frequenza*, Vol. 34, 1965, pp. 349-356

L. Dadda, "On Parallel Digital Multipliers," *Alta Frequenza*, Vol. 45, 1976, pp. 574-580

L. Dadda, "Multiple Addition of Binary Serial Numbers," *Proceedings of the 4th Symposium on Computer Arithmetic, Santa Monica*, CA, October 1978, pp. 140-148

L. Dadda, "Some Schemes for Fast Serial Input Multipliers," *Proceedings of the 6th Symposium on Computer Arithmetic*, Aarhus, DENMARK, June 1983, pp. 52-59

L. Dadda, "Fast Multipliers for Two's-Complement Numbers in Serial Form," *Proceedings of the 7th Symposium on Computer Arithmetic*, Urbana, IL, June 1985, pp. 57-63

L. Dadda, "Squarers for Binary Numbers in Serial Form," *Proceedings of the 7th Symposium on Computer Arithmetic*, Urbana, IL, June 1985, pp. 173-180

Luigi Dadda, "Polyphase Convolvers," *Proceedings of the 9th Symposium on Computer Arithmetic*, Santa Monica, CA, September 1989, pp. 78-85

Luigi Dadda, "On Serial Input Multipliers for Two's Complement Numbers," *IEEE Transactions on Computers*, Vol. 38, 1989, pp. 1341-1345

Per E. Danielsson, "Serial/Parallel Convolvers," *IEEE Transactions on Computers*, Vol. C-33, 1984, pp. 652-667

Tich T. Dao, Marc Davio, and Colette Gossart, "Complex Number Arithmetic with Odd-Valued Logic," *IEEE Transactions on Computers*, Vol. C-29, 1980, pp. 604-611

Marc Davio, "Read-Only Memory Implementation of Discrete Functions," *IEEE Transactions on Computers*, Vol. C-29, 1980, pp. 931-934

D. F. Davis, "Elementary Functions on an Array Processor," *Proceedings of the 5th Symposium on Computer Arithmetic*, Ann Arbor, MI, May 1981, pp. 170-178

K. J. Dean, "Design for a Full Multiplier," *IEE Proceedings*, Vol. 115, 1969, pp. 1592-1594

Etienne Delhaye, *et al.*, "A 2.5-ns, 40-mW, 4 X 4 GaAs Multiplier in Two's Complement Mode," *IEEE Journal of Solid-State Circuits*, Vol. SC-22, 1987, pp. 409-414

J. Demmel, "Effects of Underflow on Solving Linear Systems," *Proceedings of the 5th Symposium on Computer Arithmetic*, Ann Arbor, MI, May 1981, pp. 113-120

J. W. Demmel, "On Error Analysis in Arithmetic with Varying Relative Precision," *Proceedings of the 8th Symposium on Computer Arithmetic*, Como, ITALY, May 1987, pp. 148-152

R. DeMori, "Suggestion for an IC Fast Parallel Multiplier," *Electronics Letters*, Vol. 5, 1969, pp. 50-51

R. DeMori and R. Cardin, "Design for a Recursive Parallel Multiplier," *Proceedings of the 7th Symposium on Computer Arithmetic*, Urbana, IL, June 1985, pp. 44-50

R. DeMori, M. Elia, and A. Serra, "Minimization Methods for Macrocellular Arithmetic Networks," *Proceedings of the 3rd Symposium on Computer Arithmetic*, Dallas, TX, November 1975, pp. 232-240

R. DeMori and A. Serra, "A Parallel Structure for Signed-Number Multiplication and Addition," *IEEE Transactions on Computers*, Vol. C-21, 1972, pp. 1453-1454

J. Deverrell, "Pipeline Iterative Arithmetic Arrays," *IEEE Transactions on Computers*, Vol. C-24, 1975, pp. 317-322

A. Dhurkadas, "Faster Parallel Multiplier," *Proceedings of the IEEE*, Vol. 72, 1984, pp. 134-136

Philip W. Diodato, *et al.*, "The Design of an IEEE Standard Math Accelerator Unit," *IEEE Journal of Solid-State Circuits*, Vol. SC-20, 1985, pp. 993-997

J. J. Dongarra and D. C. Sorensen, "A Fast Algorithm for the Symmetric Eigenvalue Problem," *Proceedings of the 7th Symposium on Computer Arithmetic*, Urbana, IL, June 1985, pp. 338-342

R. V. Donthi, M. Saleem, and H. Singh, "On Bit Sequential Multipliers," *Proceedings of the 6th Symposium on Computer Arithmetic*, Aarhus, DENMARK, June 1983, pp. 104-109

R. W. Doran, "Variants of an Improved Carry Look-Ahead Adder," *IEEE Transactions on Computers*, Vol. C-37, 1988, pp. 1110-1113

S. Dormido and M. A. Canto, "Synthesis of Generalized Parallel Counters," *IEEE Transactions on Computers*, Vol. C-30, 1981, pp. 699-703

S. Dormido and M. A. Canto, "An Upper Bound for the Synthesis of Generalized Parallel Counters," *IEEE Transactions on Computers*, Vol. C-31, 1982, pp. 802-805

Jean Duprat, Yvan Herreros, and Jean-Michel Muller, "Some Results about On-Line Computation of Functions," *Proceedings of the 9th Symposium on Computer Arithmetic*, Santa Monica, CA, September 1989, pp. 112-118

Jean Duprat and Jean-Michel Muller, "Hardwired Polynomial Evaluation," *Journal of Parallel and Distributed Computing*, Vol. 5, 1988, pp. 291-309

K. Efe, "Multi-Operand Addition with Conditional Sum Logic," *Proceedings of the 5th Symposium on Computer Arithmetic*, Ann Arbor, MI, May 1981, pp. 251-255

J. A. Eldon, "A Family of CMOS Floating Point Arithmetic Chips," *Proceedings of the 7th Symposium on Computer Arithmetic*, Urbana, IL, June 1985, pp. 101-109

K. M. Elleithy, M. A. Bayoumi, and K. P. Lee, "O(logN) Architectures for RNS Arithmetic Decoding," *Proceedings of the 9th Symposium on Computer Arithmetic*, Santa Monica, CA, September 1989, pp. 202-209

M. D. Ercegovac, "Radix-16 Evaluation of Certain Elementary Functions," *IEEE Transactions on Computers*, Vol. C-22, 1973, pp. 561-566

M. D. Ercegovac, "A General Method for Evaluation of Functions in a Digital Computer," *Proceedings of the 3rd Symposium on Computer Arithmetic*, Dallas, TX, November 1975, pp. 147-157

Milos D. Ercegovac, "A General Hardware-Oriented Method for Evaluation of Functions and Computations in a Digital Computer," *IEEE Transactions on Computers*, Vol. C-26, 1977, pp. 667-680

M. D. Ercegovac, "An On-Line Square Rooting Algorithm," *Proceedings of the 4th Symposium on Computer Arithmetic, Santa Monica*, CA, October 1978, pp. 183-189

M. D. Ercegovac, "A Higher-Radix Division with Simple Selection of Quotient Digits," *Proceedings of the 6th Symposium on Computer Arithmetic*, Aarhus, DENMARK, June 1983, pp. 94-98

Milos D. Ercegovac, "On-Line Arithmetic: An Overview," *SPIE Proceedings*, Vol. 495, 1984, pp. 86-93

M. D. Ercegovac and T. Lang, "A Division Algorithm with Prediction of Quotient Digits," *Proceedings of the 7th Symposium on Computer Arithmetic*, Urbana, IL, June 1985, pp. 51-56

M. D. Ercegovac and T. Lang, "On-Line Scheme for Computing Rotation Factors," *Proceedings of the 8th Symposium on Computer Arithmetic*, Como, ITALY, May 1987, pp. 196-203

Milos D. Ercegovac and Thomas Lang, "On-the-Fly Conversion of Redundant into Conventional Representations," *IEEE Transactions on Computers*, Vol. C-36, 1987, pp. 895-897

Milos D. Ercegovac and Tomas Lang, "Radix-4 Multiplication without Carry-Propagate Addition," *Proceedings of the IEEE International Conference on Computer Design*, Rye Brook, NY, October 1987, pp. 654-658

Milos D. Ercegovac and Tomas Lang, "On-Line Scheme for Computing Rotation Factors," *Journal of Parallel and Distributed Computing*, Vol. 5, 1988, pp. 209-227

Milos D. Ercegovac and Tomas Lang, "Radix-4 Square Root without Initial PLA," *Proceedings of the 9th Symposium on Computer Arithmetic*, Santa Monica, CA, September 1989, pp. 162-168

Milos D. Ercegovac and Tomas Lang, "Fast Radix-2 Division with Quotient-Digit Prediction," *Journal of VLSI Signal Processing*, Vol. 1, 1989, pp. 169-180

Milos D. Ercegovac and Tomas Lang, "On-the-Fly Rounding for Division and Square Root," *Proceedings of the 9th Symposium on Computer Arithmetic*, Santa Monica, CA, September 1989, pp. 169-173

M. D. Ercegovac and M. M. Takata, "An Arithmetic Module for Efficient Evaluation of Functions," *Proceedings of the 4th Symposium on Computer Arithmetic, Santa Monica*, CA, October 1978, pp. 190-199

M. D. Ercegovac *et al.*, "An Area-Time Efficient Binary Divider," *Proceedings of the IEEE International Conference on Computer Design*, Rye Brook, NY, October 1987, pp. 645-648

J. G. Earle, "Latched Carry-Save Adder," *IBM Technical Disclosure Bulletin*, Vol. 7, 1965, pp. 909-910.

Adly T. Fam, "Optimal Partitioning and Redundancy Removal in Computing Partial Sums," *IEEE Transactions on Computers*, Vol. C-36, 1987, pp. 1137-1143

J. Fandrianto, "Algorithm for High Speed Shared Radix 4 Division and Radix 4 Square-Root," *Proceedings of the 8th Symposium on Computer Arithmetic*, Como, ITALY, May 1987, pp. 73-79

Jan Fandrianto, "Algorithm for High Speed Shared Radix 8 Division and Radix 8 Square-Root," *Proceedings of the 9th Symposium on Computer Arithmetic*, Santa Monica, CA, September 1989, pp. 68-75

J. Fandrianto and B. Y. Woo, "VLSI Floating-Point Processors," *Proceedings of the 7th Symposium on Computer Arithmetic*, Urbana, IL, June 1985, pp. 93-100

Arnold S. Farber and Eugene S. Schlig, "Mathematical "Lower Bounds" and the Logic Circuit Designer," *IEEE Transactions on Computers*, Vol. C-19, 1970, pp. 80-81

P. M. Farmwald, "High Bandwidth Evaluation of Elementary Functions," *Proceedings of the 5th*

Symposium on Computer Arithmetic, Ann Arbor, MI, May 1981, pp. 139-143

Alan Feldstein and Richard Goodman, "Convergence Estimates for the Distribution of Trailing Digits," *Journal of the ACM,* Vol. 23, 1976, pp. 287-297

Alan Feldstein and Richard Goodman, "Loss of Significance in Floating Point Subtraction and Addition," *IEEE Transactions on Computers,* Vol. C-31, 1982, pp. 328-335

P. M. Fenwick, "Binary Multiplication with Overlapped Addition Cycles," *IEEE Transactions on Computers,* Vol. C-18, 1969, pp. 71-74

W. E. Ferguson, Jr. and D. W. Matula, "Rationally Biased Arithmetic," *Proceedings of the 7th Symposium on Computer Arithmetic,* Urbana, IL, June 1985, pp. 194-202

J. Ferguson and J. P. Shen, "The Design of Two Easily-Testable VLSI Array Multipliers," *Proceedings of the 6th Symposium on Computer Arithmetic,* Aarhus, DENMARK, June 1983, pp. 2-9

D. Ferrari, "A Division Method Using a Parallel Multiplier," *IEEE Transactions on Electronic Computers,* Vol. EC-16, 1967, pp. 224-226

Ivan Flores, *The Logic of Computer Arithmetic,* Englewood Cliffs, NJ: Prentice-Hall, 1963

Michael J. Flynn, "On Division by Functional Iteration," *IEEE Transactions on Computers,* Vol. C-19, 1970, pp. 702-706

C. C. Foster and F. D. Stockton, "Counting Responders in an Associative Memory," *IEEE Transactions on Computers,* Vol. C-20, 1971, pp. 1580-1583

C. Foster, *et al.*, "A Novel Multiply-by-Three Circuit," *Proceedings of the 3rd Symposium on Computer Arithmetic,* Dallas, TX, November 1975, pp. 185-187

D. L. Fowler and J. E. Smith, "An Accurate, High Speed Implementation of Division by Reciprocal Approximation," *Proceedings of the 9th Symposium on Computer Arithmetic,* Santa Monica, CA, September 1989, pp. 60-67

A. S. Fraenkel, "Systems of Numeration," *Proceedings of the 6th Symposium on Computer Arithmetic,* Aarhus, DENMARK, June 1983, pp. 37-43

K. S. Frankowski, "A Realistic Model for Error Estimates in the Evaluation of Elementary Functions," *Proceedings of the 4th Symposium on Computer Arithmetic,* Santa Monica, CA, October 1978, pp. 70-74

Gideon Frieder, "Algorithms for Binary Coded Balanced and Ordinary Ternary Operations," *IEEE Transactions on Computers,* Vol. C-24, 1975, pp. 212-215

N. Friedman, *et al.*, "Realization of a Multivalued Integrated Injection Logic (MI^2L) full adder," *IEEE Journal of Solid-State Circuits,* Vol. SC-12, 1977, pp. 532-534

A. Gabrielian, "Formal Systems of Numerals," *Proceedings of the 3rd Symposium on Computer Arithmetic,* Dallas, TX, November 1975, pp. 76-81

Terry G. Gaddess, "An Error-Detecting Binary Adder: A Hardware-Shared Implementation," *IEEE Transactions on Computers,* Vol. C-19, 1970, pp. 34-38

Nicolas Gaitanis, "Totally Self-Checking Checkers for Low-Cost Arithmetic Codes," *IEEE Transactions on Computers,* Vol. C-34, 1985, pp. 596-601

Daniel D. Gajski, "Parallel Compressors," *IEEE Transactions on Computers,* Vol. C-29, 1980, pp. 393-398

D. D. Gajski and L. P. Rubinfield, "Design of Arithmetic Elements for Burroughs Scientific Processor," *Proceedings of the 4th Symposium on Computer Arithmetic, Santa Monica,* CA, October 1978, pp. 245-256

Dragan Gamberger, "Incompletely Specified Numbers in the Residue Number System–Definition and Application," *Proceedings of the 9th Symposium on Computer Arithmetic,* Santa Monica, CA, September 1989, pp. 210-215

D. Gannon, "On the Structure of Parallelism in a Highly Concurrent PDE Solver," *Proceedings of the 7th Symposium on Computer Arithmetic,* Urbana, IL, June 1985, pp. 252-259

Gilles H. Garcia and William J. Kubitz, "Minimum Mean Running Time Function Generation Using Read-Only Memory," *IEEE Transactions on Computers,* Vol. C-32, 1983, pp. 147-156

O. N. Garcia, H. Glass, and S. C. Haines, "An Approximate and Empirical Study of the Distribution of Adder Inputs and Maximum Carry Length Propagation," *Proceedings of the 4th Symposium on Computer Arithmetic,* Santa Monica, CA, October 1978, pp. 97-103

Harvey L. Garner, "Generalized Parity Checking," *IRE Transactions on Electronic Computers*, Vol. EC-7, 1958, pp. 207-213

Harvey L. Garner, "The Residue Number System," *IRE Transactions on Electronic Computers*, Vol. EC-8, 1959, pp. 140-147

Harvey L. Garner, "Number Systems and Arithmetic," in *Advances in Computers*, Vol. 6., F. L. Alt and M. Rubinoff, eds., New York: Academic Press, 1965, pp. 131-194

Harvey L. Garner, "Error Codes for Arithmetic Operations," *IEEE Transactions on Computers*, Vol. C-15, 1966, pp. 763-770

Harvey L. Garner, "A Survey of Some Recent Contributions to Computer Arithmetic," *IEEE Transactions on Computers*, Vol. C-25, 1976, pp. 1277-1282

Harvey L. Garner, "Theory of Computer Addition and Overflows," *IEEE Transactions on Computers*, Vol. C-27, 1978, pp. 297-301.

George W. Gerrity, "Computer Representation of Real Numbers," *IEEE Transactions on Computers*, Vol. C-31, 1982, pp. 709-714

Clive Ghest, "Multiplying Made Easy for Digital Assemblies," *Electronics*, Vol. 44, November 22, 1971, 56-61

B. Gilchrist, J. H. Pomerene, and S. Y. Wong, "Fast Carry Logic for Digital Computers," *IRE Transactions on Electronic Computers*, Vol. EC-4, 1955, pp. 133-136

M. Ginsberg, "Numerical Influences on the Design of Floating-Point Arithmetic for Microcomputers," *Proceedings of the 1st Annual Rocky Mountain Symposium on Microcomputers*, August 1977, pp. 24-72

M. Ginsberg and D. J. Frailey, "The Design and Use of a Floating-Point (Software) Simulator for Testing the Arithmetic Behavior of Mathematical Software," *Proceedings of the 3rd Symposium on Computer Arithmetic*, Dallas, TX, November 1975, pp. 56-63

Patrick Glories, *et al.*, "A High-Performance N-MOS Adder Designed for Optimized Cryogenic Operation," *IEEE Journal of Solid-State Circuits*, Vol. SC-21, 1986, pp. 404-410

R. Gnanasekaran, "On a Bit-Serial Input and Bit Serial Output Multiplier," *IEEE Transactions on Computers*, Vol. C-32, 1983, pp. 878-880

R. Gnanasekaran, "A Fast Serial-Parallel Binary Multiplier," *IEEE Transactions on Computers*, Vol. C-34, 1985, pp. 741-744

Katsuaki Gonoi, *et al.*, "A GaAs 8 x 8 Bit Multiplier/Accumulator Using JFET DCFL," *IEEE Journal of Solid-State Circuits*, Vol. SC-21, 1986, pp. 523-529

A Gorji-Sinaki and M. D. Ercegovac, "Design of a Digit-Slice On-Line Arithmetic Unit," *Proceedings of the 5th Symposium on Computer Arithmetic*, Ann Arbor, MI, May 1981, pp. 72-80

J. B. Gosling, "A Review of High Speed Addition Techniques," *IEE Proceedings*, Vol. 118, 1971, pp. 29-35

J. B. Gosling, "Design of Large High-Speed Floating-Point-Arithmetic Units," *IEE Proceedings*, Vol. 118, 1971, pp. 493-498

J. B. Gosling, "Design of Large High-Speed Binary Multiplier Units," *IEE Proceedings*, Vol. 118, 1971, pp. 499-505

John B. Gosling, *Design of Arithmetic Units for Digital Computers*, New York: Macmillan Press Ltd., 1980

J. B. Gosling, "Some Tricks of the (Floating-Point) Trade," *Proceedings of the 6th Symposium on Computer Arithmetic*, Aarhus, DENMARK, June 1983, pp. 103-103

J. B. Gosling, J. H. P. Zurawski, and D. B. G. Edwards, "A Chip-Set for a High-Speed Low-Cost Floating-Point Unit," *Proceedings of the 5th Symposium on Computer Arithmetic*, Ann Arbor, MI, May 1981, pp. 50-55

L. N. Goyal, "Design of an Arithmetic Element for Serial Processing in an Iterative Structure," *Proceedings of the 3rd Symposium on Computer Arithmetic*, Dallas, TX, November 1975, pp. 223-231

Lakshmi N. Goyal, "A Note on Atrubin's Real-Time Iterative Multiplier," *IEEE Transactions on Computers*, Vol. C-25, 1976, pp. 546-548

R. T. Gregory, "Residue Arithmetic with Rational Operands," *Proceedings of the 5th Symposium on Computer Arithmetic*, Ann Arbor, MI, May 1981, pp. 144-145

R. T. Gregory and D. W. Matula, "Base Conversion in Residue Number Systems," *Proceedings of the 3rd Symposium on Computer Arithmetic*, Dallas, TX, November 1975, pp. 117-125

V. M. Gritsenko, "Comments on 'Error-Correcting Codes for Byte-Organized Arithmetic Processors,'" *IEEE Transactions on Computers*, Vol. C-27, 1978, pp. 189-190

A. L. Grnarov and M. D. Ercegovac, "On-Line Multiplicative Normalization," *Proceedings of the 6th Symposium on Computer Arithmetic*, Aarhus, DENMARK, June 1983, pp. 151-155

T. Gross, "Floating-Point Arithmetic on a Reduced-Instruction-Set Processor," *Proceedings of the 7th Symposium on Computer Arithmetic*, Urbana, IL, June 1985, pp. 86-92

C. C. Guest, M. M. Mirsalehi, and T. K. Gaylord, "Residue Number System Truth-Table Look-Up Processing—Moduli Selection and Logical Minimization," *IEEE Transactions on Computers*, Vol. C-33, 1984, pp. 927-931

L. Guibas and J. Vuillemin, "On Fast Binary Addition in MOS Technologies," *Proceedings of the 6th Symposium on Computer Arithmetic*, Aarhus, DENMARK, June 1983, pp. 22-23

H. H. Guild, "Fully Iterative Fast Array for Binary Multiplication and Addition," *Electronics Letters*, Vol. 5, 1969, p. 263

Alain Guyot, Bertrand Hochet, and Jean-Michel Muller, "A Way to Build Efficient Carry-Skip Adders," *IEEE Transactions on Computers*, Vol. C-36, 1987, pp. 1144-1152

Alain Guyot, Yvan Herreros, and Jean-Michel Muller, "JANUS, an On-Line Multiplier/Divider for Manipulating Large Numbers," *Proceedings of the 9th Symposium on Computer Arithmetic*, Santa Monica, CA, September 1989, pp. 106-111

A. Habibi and Paul A. Wintz, "Fast Multipliers," *IEEE Transactions on Computers*, Vol. C-19, 1970, pp. 153-157

Yoshimune Hagiwara, *et al.*, "A Single Chip Digital Signal Processor and its Application to Real-Time Speech Analysis," *IEEE Journal of Solid-State Circuits*, Vol. SC-18, 1983, pp. 91-99

C. Halatsis, N. Gaitanis, and M. Sigala, "Error-Correcting Codes in Binary-Coded Radix-r Arithmetic," *IEEE Transactions on Computers*, Vol. C-32, 1983, pp. 494-497

Ernest L. Hall, David D. Lynch, and Samuel J. Dwyer, III, "Generation of Products and Quotients Using Approximate Binary Logarithms for Digital Filtering Applications," *IEEE Transactions on Computers*, Vol. C-19, 1970, pp. 97-105

Thomas G. Hallin and Michael J. Flynn, "Pipelining of Arithmetic Function," *IEEE Transactions on Computers*, Vol. C-21, 1972, pp. 880-886

H. Hamada, "A New Real Number Representation and Its Operation," *Proceedings of the 8th Symposium on Computer Arithmetic*, Como, ITALY, May 1987, pp. 153-157

R. W. Hamming, "On the Distribution of Numbers," *Bell Systems Technical Journal*, Vol. 49, 1970, pp. 1609-1625

T. Han and D. Carlson, "Fast Area-Efficient VLSI Adders," *Proceedings of the 8th Symposium on Computer Arithmetic*, Como, ITALY, May 1987, pp. 49-56

Yoshihisa Harata, *et al.*, "A High-Speed Multiplier Using a Redundant Binary Adder Tree," *IEEE Journal of Solid-State Circuits*, Vol. SC-22, 1987, pp. 28-34

M. G. Harman, "An Attempt to Design an Improved Multiplication System," *IEEE Transactions on Computers*, Vol. C-17, 1968, pp. 1090-1092

Cornelis D. Hartgring, Binoy A. Rosario, and James M. Pickett, "High-Speed Low-Power Silicon MESFET Parallel Multipliers," *IEEE Journal of Solid-State Circuits*, Vol. SC-17, 1982, pp. 69-73

Mehdi Hatamian and Glenn L. Cash, "A 70-MHz 8-Bit x 8-Bit Parallel Pipelined Multiplier in 2.5-μm CMOS," *IEEE Journal of Solid-State Circuits*, Vol. SC-21, 1986, pp. 505-513

Yuji Hatano, *et al.*, "A 4-Bit X 4-Bit Multiplier and 3-Bit Counter in Josephson Threshold Logic," *IEEE Journal of Solid-State Circuits*, Vol. SC-22, 1987, pp. 606-612

Gene L. Haviland and Al A. Tuszynski, "A CORDIC Arithmetic Processor Chip," *IEEE Journal of Solid-State Circuits*, Vol. SC-15, 1980, pp. 4-15

W. Patrick Hays, *et al.*, "A 32-Bit VLSI Digital Signal Processor," *IEEE Journal of Solid-State Circuits*, Vol. SC-20, 1985, pp. 998-1004

Hartmut Henkel, "Improved Addition for the Logarithmic Number System," *IEEE Transactions on Acoustics, Speech, and Signal Processing*, Vol. 37, 1989, pp. 301-303

Dennis A. Henlin, *et al.*, "A 16 Bit X 16 Bit Pipelined Multiplier Macrocell," *IEEE Journal of Solid-State Circuits*, Vol. SC-20, 1985, pp. 542-547

L. C. Higbie, "Vector Floating-Point Data Format," *IEEE Transactions on Computers*, Vol. C-25, 1976, pp. 25-32

Irving T. Ho, "Multiple Addition by Residue Threshold Functions and Their Representation by Array Logic," *IEEE Transactions on Computers*, Vol. C-22, 1973, pp. 762-767

Irving T. Ho and Tien Chi Chen, "Multiple Addition by Residue Threshold Functions and Their Representation by Array Logic," *IEEE Transactions on Computers*, Vol. C-22, 1973, pp. 762-767

B. Hochet, P. Quinton, and Y. Robert, "Systolic Solution of Linear Systems over GF(p) with Partial Pivoting," *Proceedings of the 8th Symposium on Computer Arithmetic*, Como, ITALY, May 1987, pp. 161-168

R. N. Horspool and E. C. R. Hehner, "Exact Arithmetic Using a Variable-Length p-adic," *Proceedings of the 4th Symposium on Computer Arithmetic*, Santa Monica, CA, October 1978, pp. 10-14

D. Hough, "Applications of the Proposed IEEE-754 Standard for Floating Point Arithmetic," *Computer*, Vol. 14, March 1981, pp. 70-74

C. H. Huang, "A Fully Parallel Mixed-Radix Conversion Algorithm for Residue Number Applications," *IEEE Transactions on Computers*, Vol. C-32, 1983, pp. 398-402

T. E. Hull, "Desirable Floating-Point Arithmetic and Elementary Functions for Numerical Computation," *Proceedings of the 4th Symposium on Computer Arithmetic*, Santa Monica, CA, October 1978, pp. 63-69

T. E. Hull and M. S. Cohen, "Toward an Ideal Computer Arithmetic," *Proceedings of the 8th Symposium on Computer Arithmetic*, Como, ITALY, May 1987, pp. 131-138

G. Hunter, "A Quantitative Measure of Precision," *The Computer Journal*, Vol. 18, 1975, p. 231-233

John H. Huttenhoff and Richard R. Shively, "Arithmetic Unit of a Computing Element in a Global, Highly Parallel Computer," *IEEE Transactions on Computers*, Vol. C-18, 1969, pp. 695-698

Inseok S. Hwang and Aaron L. Fisher, "Ultrafast Compact 32-bit CMOS Adders in Multiple-Output Domino Logic," *IEEE Journal of Solid-State Circuits*, Vol. 24, 1989, pp. 358-369

K. Hwang, *Computer Arithmetic: Principles, Architecture, and Design*, New York: John Wiley & Sons, 1979

K. Hwang and T. P. Chang, "An International Rational/Radix Arithmetic System for High-Precision Computations," *Proceedings of the 4th Symposium on Computer Arithmetic*, Santa Monica, CA, October 1978, pp. 15-24

K. Hwang and Y-H. Cheng, "Partitioned Algorithms and VLSI Structures for Large-Scale Matrix Computations," *Proceedings of the 5th Symposium on Computer Arithmetic*, Ann Arbor, MI, May 1981, pp. 222-232

Kai Hwang and D. K. Panda, "Optical Arithmetic Using High-Radix Symbolic Substitution Rules," *Proceedings of the 9th Symposium on Computer Arithmetic*, Santa Monica, CA, September 1989, pp. 226-232

K. Hwang, H. C. Wang, and Z. Xu, "Evaluating Elementary Functions with Chebyshev Polynomials on Pipeline Nets," *Proceedings of the 8th Symposium on Computer Arithmetic*, Como, ITALY, May 1987, pp. 121-128

K. Hwang and Z. Xu, "Multiprocessors for Evaluating Compound Arithmetic Functions," *Proceedings of the 7th Symposium on Computer Arithmetic*, Urbana, IL, June 1985, pp. 266-275

Mary Jane Irwin and Robert Michael Owens, "Fully Digit On-Line Networks," *IEEE Transactions on Computers*, Vol. C-32, 1983, pp. 402-406

M. J. Irwin and D. R. Smith, "A Rational Arithmetic Processor," *Proceedings of the 5th Symposium on Computer Arithmetic*, Ann Arbor, MI, May 1981, pp. 241-245

T. Isobe, T. Nakamura, and G. Goto, "Application of Gunn Effect to Subnanosecond Full Adder and Full Subtracter," *Proceedings of the IEEE*, Vol. 61, 1973, pp. 792-793

D. Jacobsohn, "A Suggestion for a Fast Multiplier," *IEEE Transactions on Electronic Computers*, Vol. EC-13, 1964, p. 754

David H. Jacobsohn, "A Combinatoric Division Algorithm for Fixed-Integer Divisors," *IEEE Transactions on Computers*, Vol. C-22, 1973, pp. 608-610

Totadri Jayashree and Dhruba Basu, "On Binary Multiplication Using the Quarter Square Algorithm," *IEEE Transactions on Computers*, Vol. C-25, 1976, pp. 957-960

W. K. Jenkins, "A Highly Efficient Residue-Combinatorial Architecture for Digital Filters," *Proceedings of the IEEE*, Vol. 66, 1978, pp. 700-702

W. Kenneth Jenkins, "The Design of Error Checkers for Self-Checking Residue Number Arithmetic," *IEEE Transactions on Computers*, Vol. C-32, 1983, pp. 388-396

W. K. Jenkins and M. H. Etzel, "Special Properties of Complement Codes for Redundant Residue Number Systems," *Proceedings of the IEEE*, Vol. 69, 1981, pp. 132-133

Barry W. Johnson, James H. Aylor, and Haytham H. Hana, "Efficient Use of Time and Hardware Redundancy for Concurrent Error Detection in a 32-Bit VLSI Adder," *IEEE Journal of Solid-State Circuits*, Vol. SC-23, 1988, pp. 208-215

Everett L. Johnson, "A Digital Quarter Square Multiplier," *IEEE Transactions on Computers*, Vol. C-29, 1980, pp. 258-261

K. Johnsen, "An IEEE Floating Point Arithmetic Implementation," *Proceedings of the 6th Symposium on Computer Arithmetic*, Aarhus, DENMARK, June 1983, pp. 130-137

Paul Johnstone and Frederick E. Petry, "Higher Radix Floating Point Representations," *Proceedings of the 9th Symposium on Computer Arithmetic*, Santa Monica, CA, September 1989, pp. 226-232

G. A. Jullien, "Residue Number Scaling and Other Operations Using ROM Arrays," *IEEE Transactions on Computers*, Vol. C-27, 1978, pp. 325-336

G. A. Jullien, "Implementation of Multiplication, Modulo a Prime Number, with Applications to Number Theoretic Transforms," *IEEE Transactions on Computers*, Vol. C-29, 1980, pp. 899-905

G. A. Jullien and W. C. Miller, "Application of the Residue Number System to Computer Processing of Digital Signals," *Proceedings of the 4th Symposium on Computer Arithmetic*, Santa Monica, CA, October 1978, pp. 220-225

W. Kahan and E. LeBlanc, "Anomalies in the IBM ACRITH Package," *Proceedings of the 7th Symposium on Computer Arithmetic*, Urbana, IL, June 1985, pp. 322-331

Ralph Kallman, "A Faster 8-Bit Carry Circuit," *IEEE Transactions on Computers*, Vol. C-32, 1983, pp. 1209-1211

A. K. Kamal, Harpreet Singh, and D. P. Agrawal, "A Generalized Pipeline Array Procedure," *IEEE Transactions on Computers*, Vol. C-23, 1974, pp. 533-536

Ahmed A. Kamal and M. A. N. Ghannam, "High-Speed Multiplication Systems," *IEEE Transactions on Computers*, Vol. C-21, 1972, pp. 1017-1021

Dhirubhai V. Kanani and Kenneth H. O'Keefe, "A Note on Conditional-Sum Addition for Base-2 Systems," *IEEE Transactions on Computers*, Vol. C-22, 1973, pp. 626

Jack Kane, "A Low-Power Bipolar, Two's Complement Serial Pipeline Multiplier Chip," *IEEE Journal of Solid-State Circuits*, Vol. SC-11, 1976, pp. 669-678

Katsuyki Kaneko, *et al.*, "A VLSI RISC with 20-MFLOPS Peak, 64-bit Floating-Point Unit," *IEEE Journal of Solid-State Circuits*, Vol. 24, 1989, pp. 1331-1340

Toyohisa Kaneko and Bede Liu, "On Local Roundoff Errors in Floating Point Arithmetic," *Journal of the ACM*, Vol. 20, 1973, pp. 391-398

R.-S. Kao and F. J. Taylor, "Implementation of the Single Modulus Complex ALU," *Proceedings of the 8th Symposium on Computer Arithmetic*, Como, ITALY, May 1987, pp. 21-27

Alan H. Karp, "Exponential and Logarithm by Sequential Squaring," *IEEE Transactions on Computers*, Vol. C-23, 1984, pp. 462-464

S. Kaushik, "Sign Detection in Non-Redundant Residue Number Systems with Reduced Information," *Proceedings of the 6th Symposium on Computer Arithmetic*, Aarhus, DENMARK, June 1983, pp. 24-28

S. Kaushik, "Multiple Error Correction and Additive Overflow Detection," *Proceedings of the 7th Symposium on Computer Arithmetic*, Urbana, IL, June 1985, pp. 278-284

S. Kaushik and R. K. Arora, "Sign Detection in the Symmetric Residue Number System," *Proceedings of the 5th Symposium on Computer Arithmetic*, Ann Arbor, MI, May 1981, pp. 146-151

Shoji Kawahito, e*t al.*, "A 32 X 32-Bit Multiplier Using Multiple-Valued MOS Current-Mode Circuits," *IEEE Journal of Solid-State Circuits*, Vol. SC-23, 1988, pp. 124-132

T. H. Kehl and K. Burkhardt, "A Minicomputer Microprogrammable, Arithmetic Processor," *Proceedings of the 3rd Symposium on Computer Arithmetic*, Dallas, TX, November 1975, pp. 174-178

R. A. Keir, "Programmer-Controlled Roundoff and the Selection of a Stable Roundoff Rule," *Proceedings of the 3rd Symposium on Computer Arithmetic*, Dallas, TX, November 1975, pp. 73-75

R. A. Keir, "Compatible Number Representations," *Proceedings of the 3rd Symposium on Computer Arithmetic*, Dallas, TX, November 1975, pp. 82-86

J. G. Kent, "Comparison Sets: A Useful Partitioning of the Space of Floating Point Operand Pairs," *Proceedings of the 3rd Symposium on Computer Arithmetic*, Dallas, TX, November 1975, pp. 36-39

Jan G. Kent, "Highlights of a Study of Floating-Point Instructions," *IEEE Transactions on Computers*, Vol. C-26, 1977, pp. 660-666

S. Kent, "A High-Speed Threshold Gate Multiplier," *IEEE Transactions on Computers*, Vol. C-26, 1977, pp. 1279-1283

Juergen Kernhoff, *et al.*, "High-Speed CMOS Adder and Multiplier Modules for Digital Signal Processing in a Semicustom Environment," *IEEE Journal of Solid-State Circuits*, Vol. 24, 1989, pp. 570-575

T. Kilburn, D. B. G. Edwards, and D. Aspinall, "Parallel Arithmetic Unit Using a Saturated Transistor Fast Carry Circuit," *IEE Proceedings*, Vol. 107B, 1960, pp. 573-584

N. G. Kingsbury and P. J. W. Rayner, "Digital Filtering Using Logarithmic Arithmetic," *Electronics Letters*, Vol. 7, 1960, pp. 573-584

D. J. Kinniment and G. B. Steven, "Sequential State Binary Parallel Adder," *IEE Proceedings*, Vol. 117, 1970, pp. 1211-1218

Eisuke Kinoshita, Hideo Kosako, and Yoshiaki Kojima, "General Division in the Symmetric Residue Number System," *IEEE Transactions on Computers*, Vol. C-22, 1973, pp. 573-576

Eisuke Kinoshita, Hideo Kosako, and Yoshiaki Kojima, "Floating-Point Arithmetic Algorithms in the Symmetric Residue Number Systems," *IEEE Transactions on Computers*, Vol. C-23, 1974, pp. 9-20

John B. Kioustelidis and John K. Petrou, "A Piecewise Linear Approximation of $\text{Log}_2 x$ with Equal Maximum Errors in all Intervals," *IEEE Transactions on Computers*, Vol. C-24, 19735, pp. 858-861

R. Kirchner and U. Kulisch, "Arithmetic for Vector Processors," *Proceedings of the 8th Symposium on Computer Arithmetic*, Como, ITALY, May 1987, pp. 256-269

R. Kirchner and U. Kulisch, "Accurate Arithmetic for Vector Processors," *Journal of Parallel and Distributed Computing*, Vol. 5, 1988, pp. 250-270

R. Klatte and Ch. Ullrich, "Consequences of a Properly Implemented Computer Arithmetic for Periodicities of Iterative Methods," *Proceedings of the 3rd Symposium on Computer Arithmetic*, Dallas, TX, November 1975, pp. 33-35

Cornelis L. M. van der Klauw, "A Full Adder Using Junction Charge-Coupled Logic," *IEEE Journal of Solid-State Circuits*, Vol. SC-21, 1986, pp. 584-587

Donald E. Knuth, *The Art of Computer Programming, Vol. 2: Seminumerical Algorithms*, Reading, MA: Addison-Wesley, 1969

H. Kobayashi, "A Fast Multi-Operand Multiplication Scheme," *Proceedings of the 5th Symposium on Computer Arithmetic*, Ann Arbor, MI, May 1981, pp. 246-250

H. Kobayashi, "A Multioperand Two's Complement Addition Algorithm," *Proceedings of the 7th Symposium on Computer Arithmetic*, Urbana, IL, June 1985, pp. 16-19

H. Kobayashi and R. D. Bonnell, "Arithmetic for a High-Speed Adaptive Learning Network Element," *Proceedings of the 6th Symposium on Computer Arithmetic*, Aarhus, DENMARK, June 1983, pp. 164-168

Cetin Koc and Peter R. Cappello, "Systolic Arrays for Integer Chinede Remaindering," *Proceedings of the 9th Symposium on Computer Arithmetic*, Santa Monica, CA, September 1989, pp. 216-223

Shinji Komori, *et al.*, "A 40-MFLOPS 32-bit Floating-Point Processor with Elastic Pipeline Scheme," *IEEE Journal of Solid-State Circuits*, Vol. 24, 1989, pp. 1341-1347

I. Koren and Y. Maliniak, "A Unified Approach to a Class of Number Systems," *Proceedings of the 4th Symposium on Computer Arithmetic*, Santa Monica, CA, October 1978, pp. 25-28

Israel Koren and Yoram Maliniak, "On Classes of Positive, Negative, and Imaginary Radix Number Systems," *IEEE Transactions on Computers*, Vol. C-30, 1981, pp. 312-317

P. Kornerup, "A Unified Numeric Data Type in Pascal," *Proceedings of the 3rd Symposium on Computer Arithmetic*, Dallas, TX, November 1975, pp. 40-47

P. Kornerup and D. W. Matula, "A Feasibility Analysis of Fixed-Slash Rational Arithmetic," *Proceedings of the 4th Symposium on Computer Arithmetic*, Santa Monica, CA, October 1978, pp. 39-47

P. Kornerup and D. W. Matula, "An Integrated Rational Arithmetic Unit," *Proceedings of the 5th Symposium on Computer Arithmetic*, Ann Arbor, MI, May 1981, pp. 233-240

Peter Kornerup and David W. Matula, "Finite Precision Rational Arithmetic: An Arithmetic Unit," *IEEE Transactions on Computers*, Vol. C-32, 1983, pp. 378-388

P. Kornerup and D. W. Matula, "Finite-Precision Lexicographic Continued Fraction Number Systems," *Proceedings of the 7th Symposium on Computer Arithmetic*, Urbana, IL, June 1985, pp. 207-214

P. Kornerup and D. Matula, "A Bit-Serial Arithmetic Unit for Rational Arithmetic," *Proceedings of the 8th Symposium on Computer Arithmetic*, Como, ITALY, May 1987, pp. 204-211

Peter Kornerup and David W. Matula, "An On-Line Arithmetic Unit for Bit-Pipelined Rational Arithmetic," *Journal of Parallel and Distributed Computing*, Vol. 5, 1988, pp. 310-330

Peter Kornerup and David Matula, "Exploiting Redundancy in Bit-Pipelined Rational Arithmetic," *Proceedings of the 9th Symposium on Computer Arithmetic*, Santa Monica, CA, September 1989, pp. 119-126

Seigo Kotani, *et al.*, "Feasibility of an Ultra-High-Speed Josephson Multiplier," *IEEE Journal of Solid-State Circuits*, Vol. SC-22, 1987, pp. 98-103

N. Kouvaras, D. Lagoyannis, and L. Ponticopoulos, "A Digital System for Simultaneous Addition of Several Binary Numbers," *IEEE Transactions on Computers*, Vol. C-17, 1968, pp. 992-997

E. V. Krishnamurthy, "On Range-Transformation Techniques for Division," *IEEE Transactions on Computers*, Vol. C-19, 1970, pp. 157-160

E. V. Krishnamurthy, "On Optimal Iterative Schemes for High-Speed Division," *IEEE Transactions on Computers*, Vol. C-19, 1970, pp. 227-231

E. V. Krishnamurthy, "Complementary Two-Way Algorithms for Negative Radix Conversions," *IEEE Transactions on Computers*, Vol. C-20, 1971, pp. 543-550

E. V. Krishnamurthy, "Matrix Processors Using p-ADIC Arithmetic for Exact Linear Computations," *Proceedings of the 3rd Symposium on Computer Arithmetic*, Dallas, TX, November 1975, pp. 92-97

E. V. Krishnamurthy and H. Venkateswaran, "Multivariable Polynomial Processing–Applications to Interpolation," *Proceedings of the 4th Symposium on Computer Arithmetic*, Santa Monica, CA, October 1978, pp. 81-87

D. J. Kuck, D. S. Parker, and A. H. Sameh, "ROM Rounding: A New Rounding Scheme," *Proceedings of the 3rd Symposium on Computer Arithmetic*, Dallas, TX, November 1975, pp. 67-72

D. J. Kuck, D. S. Parker, Jr., and A. H. Sameh, "Analysis of Rounding Methods in Floating-Point Arithmetic," *IEEE Transactions on Computers*, Vol. C-26, 1977, pp. 643-650

Lars Kühnel and Hartmut Schmeck, "A Closer Look at VLSI Multiplication," *Integration*, Vol. 6, 1988, pp. 345-359

H. Kuki and W. J. Cody, "A Statistical Study of the Accuracy of Floating Point Number Systems," *Communications of the ACM*, Vol. 16, 1973, pp. 223-230

Ulrich Kulisch, "Mathematical Foundations of Computer Arithmetic," *Proceedings of the 3rd Symposium on Computer Arithmetic*, Dallas, TX, November 1975, pp. 1-13

Ulrich W. Kulisch and Willard L. Miranker, *Computer Arithmetic in Theory and Practice*, New York: Academic Press, 1981.

S. Kuninobu, *et al.*, "Design of High Speed MOS Multiplier and Divider Using Redundant Binary Representation," *Proceedings of the 8th Symposium on Computer Arithmetic*, Como, ITALY, May 1987, pp. 80-86

T. Kurokawa and H. Aiso, "Polynomial Transformer," *Proceedings of the 7th Symposium on Computer Arithmetic*, Urbana, IL, June 1985, pp. 153-158

Hung Chi Lai and Saburo Muroga, "Logic Networks of Carry-Save Adders," *IEEE Transactions on Computers*, Vol. C-31, 1982, pp. 870-882

A. L. Lang and B. D. Shriver, "The Design of a Polymorphic Arithmetic Unit," *Proceedings of the 3rd*

Symposium on Computer Arithmetic, Dallas, TX, November 1975, pp. 48-55

Eberhard Lange, "Implementation and Test of the ACRITH Facility in a System/370," *IEEE Transactions on Computers*, Vol. C-36, 1987, pp. 1088-1096

Corinna Lee, "Multistep Gradual Rounding," *IEEE Transactions on Computers*, Vol. 38, 1989, pp. 595-600

Frank S. Lee, *et al.*, "A High-Speed LSI GaAs 8 x 8-Bit Parallel Multiplier," *IEEE Journal of Solid-State Circuits*, Vol. SC-17, 1982, pp. 638-647

Hoon B. Lee and Robert O. Grondin, "A Comparison of Systolic Architectures for Matrix Multiplication," *IEEE Journal of Solid-State Circuits*, Vol. SC-23, 1988, pp. 285-289

Joseph Y. Lee, Hugh L. Garvin, and Charles W. Slayman, "A High-Speed High-Density Silicon 8 X 8-Bit Parallel Multiplier," *IEEE Journal of Solid-State Circuits*, Vol. SC-22, 1987, pp. 35-40

Samuel C. Lee and Albert D. Edgar, "The Focus Number System," *IEEE Transactions on Computers*, Vol. C-26, 1977, pp. 1167-1170

M. Lehman and N. Burla, "Skip Technique for High Speed Carry Propagation in Binary Arithmetic Units," *IRE Transactions on Electronic Computers*, Vol. EC-10, 1961, pp. 691-698

Claude P. Lerouge, Pierre Girard, and Joel S. Colardelle, "A Fast 16 Bit NMOS Parallel Multiplier," *IEEE Journal of Solid-State Circuits*, Vol. SC-19, 1984, pp. 338-342

D. Levan and J. C. Majithia, "Iterative Algorithms for Function Computations," *Proceedings of the IEEE*, Vol. 62, 1974, pp. 861-862

David M. Lewis and Lawrence K. Yu, "Algorithm Design for a 30 Bit Integrated Logarithmic Processor," *Proceedings of the 9th Symposium on Computer Arithmetic*, Santa Monica, CA, September 1989, pp. 192-199

Shuo-Yen Robert Li, "Fast Constant Division Routines," *IEEE Transactions on Computers*, Vol. C-34, 1985, pp. 866-869

X. Li and L. M. Ni, "A Pipeline Architecture for Computing Cumulative Hypergeometric Distributions," *Proceedings of the 7th Symposium on Computer Arithmetic*, Urbana, IL, June 1985, pp. 166-172

L. A. Liddiard, "Required Scientific Floating Point Arithmetic," *Proceedings of the 4th Symposium on Computer Arithmetic*, Santa Monica, CA, October 1978, pp. 56-62

Panos A. Ligomenides, "The Skip-and-Set Fast-Division Algorithm," *IEEE Transactions on Computers*, Vol. C-26, 1977, pp. 1030-1032

P. Ligomenides and R. Newcomb, "Complement Representations in the Fibonacci Computer," *Proceedings of the 5th Symposium on Computer Arithmetic*, Ann Arbor, MI, May 1981, pp. 6-9

S. L. Lillevik and P. D. Fisher, "Computational Design Alternatives with Microprocessor-Based Systems," *Proceedings of the 4th Symposium on Computer Arithmetic*, Santa Monica, CA, October 1978, pp. 267-273

H. Lin and H. J. Sips, "A Novel Floating-Point Online Division Algorithm," *Proceedings of the 8th Symposium on Computer Arithmetic*, Como, ITALY, May 1987, pp. 188-195

Haixiang Lin and Henk J. Sips, "On-Line CORDIC Algorithms," *Proceedings of the 9th Symposium on Computer Arithmetic*, Santa Monica, CA, September 1989, pp. 26-33

H. Ling, "High- Speed Binary Adder," *IBM Journal of Research and Development*, Vol. 25, 1981, pp. 156-166.

R. J. Linhardt and H. S. Miller, "Digit-by-Digit Transcendental-Function Computation," *RCA Review*, Vol. 30, 1969, pp. 209-247

G. J. Lipovski, "On Residue Number A/D and D/A Converters," *Proceedings of the 3rd Symposium on Computer Arithmetic*, Dallas, TX, November 1975, pp. 197-199

Tso-Kai Liu, *et al.*, "Optimal One-Bit Full Adders with Different Types of Gates," *IEEE Transactions on Computers*, Vol. C-23, 1974, pp. 63-70

W. Liu, J. C. Duh, and D. E. Atkins, "The Design of a Vector-Radix 2DFFT Chip," *Proceedings of the 7th Symposium on Computer Arithmetic*, Urbana, IL, June 1985, pp. 231-236

Chao-Kai Liuand and Tse Lin Wang, "Error-Correcting Codes in Binary-Coded-Decimal Arithmetic," *IEEE Transactions on Computers*, Vol. C-27, 1978, pp. 977-984

Jien-Chung Lo, Suchai Thanawastien, and T. R. N. Rao, "Concurrent Error Detection in Arithmetic and Logical Operations Using Berger Codes," *Proceedings of the 9th Symposium on Computer Arithmetic*, Santa Monica, CA, September 1989, pp. 233-240

R. Lohner and J. Wolff v. Gudenberg, "Complex Interval Division with Maximum Accuracy," *Proceedings of the 7th Symposium on Computer Arithmetic*, Urbana, IL, June 1985, pp. 332-336

Grazia Lotti and Francesco Romani, "Application of Approximating Algorithms to Boolean Matrix Multiplication," *IEEE Transactions on Computers*, Vol. C-29, 1980, pp. 927-928

Daniel W. Lozier, "The Use of Floating-Point and Interval Arithmetic in the Computation of Error Bounds," *IEEE Transactions on Computers*, Vol. C-32, 1983, pp. 411-417

R. I. MacDonald, "An Optoelectronic Adder," *Proceedings of the IEEE*, Vol. 74, 1986, pp. 1593-1595

O. L. MacSorley, "High-Speed Arithmetic in Binary Computers," *Proceedings of the IRE*, Vol. 49, 1961, pp. 67-91

I. Ross Mactaggart and Mervyn A. Jack, "A Single Chip Radix-2 FFT Butterfly Architecture Using Parallel Data Distributed Arithmetic," *IEEE Journal of Solid-State Circuits*, Vol. SC-19, 1984, pp. 368-373

Daniel J. Magenheimer, *et al.*, "Integer Multiplication and Division on the HP Precision Architecture," *IEEE Transactions on Computers*, Vol. C-37, 1988, pp. 980-990

Stanislaw Majerski, "On the Determination of Optimal Distribution of Carry Skips in Adders," *IEEE Transactions on Electronic Computers*, Vol. EC-16, 1967, pp. 45-58

S. Majerski, "Square-Root Algorithms for High-Speed Digital Circuits," *Proceedings of the 6th Symposium on Computer Arithmetic*, Aarhus, DENMARK, June 1983, pp. 99-102

Stanislaw Majerski, "Square-Rooting Algorithms for High-Speed Digital Circuits," *IEEE Transactions on Computers*, Vol. C-34, 1985, pp. 724-733

Stanislaw Majerski and Michael Wiweger, "NOR-Gate Binary Adder with Carry Completion Detection," *IEEE Transactions on Electronic Computers*, Vol. EC-16, 1967, pp. 90-92

J. C. Majithia, "Cellular Array for Extraction of Squares and Square Roots of Binary Numbers," *IEEE Transactions on Computers*, Vol. C-21, 1972, pp. 1023-1024

J. C. Majithia, "Some Comments Concerning Design of Pipeline Arithmetic Arrays," *IEEE Transactions on Computers*, Vol. C-25, 1976, pp. 1132-1134

J. C. Majithia and R. Kitai, "An Iterative Array for Multiplication of Signed Binary Numbers," *IEEE Transactions on Computers*, Vol. C-20, 1971, pp. 214-216

J. C. Majithia and D. Levan, "A Note on Base-2 Logarithm Computations," *Proceedings of the IEEE*, Vol. 61, 1973, pp. 1519-1520

David Mandelbaum, "Error Correction in Residue Arithmetic," *IEEE Transactions on Computers*, Vol. C-21, 1972, pp. 538-545

David M. Mandelbaum, "A Systematic Method for Division with High Average Bit Skipping," *IEEE Transactions on Computers*, Vol. 39, 1990, pp. 127-130

John D. Marasa and David W. Matula, "A Simulative Study of Correlated Error Propagation in Various Finite-Precision Arithmetics," *IEEE Transactions on Computers*, Vol. C-22, 1973, pp. 587-597

D. Marino, "New Algorithms for the Approximate Evaluation in Hardware of Binary Logarithms and Elementary Functions," *IEEE Transactions on Computers*, Vol. C-21, 1972, pp. 1416-1421

S. Markov, "On an Interval Arithmetic and Its Applications," *Proceedings of the 5th Symposium on Computer Arithmetic*, Ann Arbor, MI, May 1981, pp. 274-277

S. Markov, "On the Numerical Algorithms Formulated in Computer Arithmetic," *Proceedings of the 6th Symposium on Computer Arithmetic*, Aarhus, DENMARK, June 1983, pp. 82-85

Norman M. Martin and Stephen P. Hufnagel, "Conditional-Sum Early Completion Adder Logic," *IEEE Transactions on Computers*, Vol. C-29, 1980, pp. 753-756

David W. Matula, "In-and-Out Conversions," *Communications of the ACM*, Vol. 11, 1968, pp. 47-50

David W. Matula, "A Formalization of Floating-Point Numeric Base Conversion," *IEEE Transactions on Computers*, Vol. C-19, 1970, pp. 681-692

D. W. Matula, "Fixed-Slash and Floating-Slash Rational Arithmetic," *Proceedings of the 3rd Symposium on Computer Arithmetic*, Dallas, TX, November 1975, pp. 90-91

D. W. Matula, "Basic Digit Sets for Radix Representation of the Integers," *Proceedings of the 4th Symposium on Computer Arithmetic*, Santa Monica, CA, October 1978, pp. 1-9

D. W. Matula and P. Kornerup, "A Feasibility Analysis of Binary Fixed-Slash and Floating-Slash Number Systems," *Proceedings of the 4th Symposium on Computer Arithmetic*, Santa Monica, CA, October 1978, pp. 29-38

D. W. Matula and P. Kornerup, "An Order Preserving Finite Binary Encoding of the Rationals," *Proceedings of the 6th Symposium on Computer Arithmetic*, Aarhus, DENMARK, June 1983, pp. 201-209

T. G. McDaneld and R. K. Guha, "The Two's Complement Quasi-Serial Multiplier," *IEEE Transactions on Computers*, Vol. C-24, 1975, pp. 1233-1235

W. M. McKeeman, "Representation Error for Real Numbers in Binary Computer Arithmetic," *IEEE Transactions on Electronic Computers*, Vol. EC-16, 1967, pp. 682-683

Angelo R. Meo, "Arithmetic Networks and Their Minimization Using a New Line of Elementary Units," *IEEE Transactions on Computers*, Vol. C-24, 1975, pp. 258-280

R. D. Merrill, Jr., "Improving Digital Computer Performance Using Residue Number Theory," *IEEE Transactions on Electronic Computers*, Vol. EC-13, 1964, pp. 93-101

Nicholas C. Metropolis, "Analyzed Binary Computing," *IEEE Transactions on Computers*, Vol. C-22, 1973, pp. 573-576

Nicholas Metropolis and Robert L. Ashenhurst, "Radix Conversion in an Unnormalized Arithmetic System," *Mathematics of Computation*, Vol. 19, 1965, pp. 435-441

Veljko Milutinovic, Mark Bettinger, and Walter Helbig, "Multiplier/Shifter Design Tradeoffs in a 32-bit Microprocessor," *IEEE Transactions on Computers*, Vol. 38, 1989, pp. 874-880

A. Miola, "A Unified View of Approximate Rational Arithmetic and Rational Interpolation," *Proceedings of the 6th Symposium on Computer Arithmetic*, Aarhus, DENMARK, June 1983, pp. 210-220

P. S. Moharir, "Extending the Scope of Golub's Method beyond Complex Multiplication," *IEEE Transactions on Computers*, Vol. C-34, 1985, pp. 484-487

Paolo Montuschi and Luigi Ciminiera, "On the Efficient Implementation of Higher Radix Square Root Algorithms," *Proceedings of the 9th Symposium on Computer Arithmetic*, Santa Monica, CA, September 1989, pp. 154-161

Robert Morris, "Tapered Floating Point: A New Floating-Point Representation," *IEEE Transactions on Computers*, Vol. C-20, 1971, pp. 1578-1579

R. Morrison, *et al.*, "Experience with a High Level Language That Supports Interval Arithmetic," *Proceedings of the 6th Symposium on Computer Arithmetic*, Aarhus, DENMARK, June 1983, pp. 74-78

Jean-Michel Muller, "Discrete Basis and Computation of Elementary Functions," *IEEE Transactions on Computers*, Vol. C-34, 1985, pp. 857-862

Shinji Nakamura, "Algorithms for Iterative Array Multiplication," *IEEE Transactions on Computers*, Vol. C-35, 1986, pp. 713-719

Shinji Nakamura and Kai-Yu Chu, "A Single Chip Parallel Multiplier by MOS Technology," *IEEE Transactions on Computers*, Vol. C-37, 1988, pp. 274-282

Takashi Nakayama, *et al.*, "A 6.7-MFLOPS Floating-Point Coprocessor with Vector/Matrix Instructions," *IEEE Journal of Solid-State Circuits*, Vol. 24, 1989, pp. 1324-1330

Yoshiro Nakayama, *et al.*, "A GaAs 16 X 16 Bit Parallel Multiplier," *IEEE Journal of Solid-State Circuits*, Vol. SC-18, 1983, pp. 599-603

A. Naseem and P. D. Fisher, "The Modified Cordic Algorithm," *Proceedings of the 7th Symposium on Computer Arithmetic*, Urbana, IL, June 1985, pp. 144-152

Amos Nathan, "The Cascade Multiplier," *IEEE Transactions on Electronic Computers*, Vol. EC-14, 1965, pp. 243-247

T. -F. Ngai and M. J. Irwin, "Regular, Area-Time Efficient Carry-Lookahead Adders," *Proceedings of the 7th Symposium on Computer Arithmetic*, Urbana, IL, June 1985, pp. 9-15

D. D. Nguyen, "A Systematic Approach to the Design of Structures for Addition and Subtraction--Case of Radix r=mK," *Proceedings of the 5th Symposium on Computer Arithmetic*, Ann Arbor, MI, May 1981, pp. 50-55

L. M. Ni and K. Hwang, "Vector Reduction Methods for Arithmetic Pipelines," *Proceedings of the 6th Symposium on Computer Arithmetic*, Aarhus, DENMARK, June 1983, pp. 144-150

Lionel M. Ni and Kai Hwang, "Vector-Reduction Techniques for Arithmetic Pipelines," *IEEE Transactions on Computers*, Vol. C-34, 1985, pp. 404-411

Jean-Daniel Nicoud, "Iterative Arrays for Radix Conversion," *IEEE Transactions on Computers*, Vol. C-20, 1971, pp. 1479-1489

Dimitris Nikolos, Antonis M. Paschalis, and George Philokyprou, "Efficient Design of Totally Self-Checking Checkers for all Low-Cost Arithmetic Codes," *IEEE Transactions on Computers*, Vol. C-37, 1988, pp. 807-814

Takashi Nishimura, *et al.*, "A Bipolar 18K-Gate Variable Size Cell Masterslice," *IEEE Journal of Solid-State Circuits*, Vol. SC-21, 1986, pp. 727-732

Tobias Noll, *et al.*, "A Pipelined 330-MHz Multiplier," *IEEE Journal of Solid-State Circuits*, Vol. SC-21, 1986, pp. 411-416

Kenneth H. O'Keefe, "A Note on Fast Base Extension for Residue Number Systems with Three Moduli," *IEEE Transactions on Computers*, Vol. C-24, 1975, pp. 1132-1133

M. Ohhashi and R. E. Schneider, "High-Speed Computation of Unary Functions," *Proceedings of the 7th Symposium on Computer Arithmetic*, Urbana, IL, June 1985, pp. 82-85

L. Ohlsson and B. Svensson, "Matrix Multiplication on LUCAS," *Proceedings of the 6th Symposium on Computer Arithmetic*, Aarhus, DENMARK, June 1983, pp. 116-122

Norihiko Ohwada, Tadakatsu Kimura, and Masanobu Doken, "LSI's for Digital Signal Processing," *IEEE Journal of Solid-State Circuits*, Vol. SC-14, 1979, pp. 214-220

V. G. Oklobdzija and E. R. Barnes, "Some Optimal Schemes for ALU Implementation in VLSI Technology," *Proceedings of the 7th Symposium on Computer Arithmetic*, Urbana, IL, June 1985, pp. 2-8

Vojin G. Oklobdzija and Milos D. Ercegovac, "An On-Line Square Root Algorithm," *IEEE Transactions on Computers*, Vol. C-31, 1982, pp. 70-75

F. W. J. Olver, "A Closed Computer Arithmetic," *Proceedings of the 8th Symposium on Computer Arithmetic*, Como, ITALY, May 1987, pp. 139-143

F. W. J. Olver and P. R. Turner, "Implementation of Level-Index Arithmetic Using Partial Table Look-Up," *Proceedings of the 8th Symposium on Computer Arithmetic*, Como, ITALY, May 1987, pp. 144-147

S. Ong and D. E. Atkins, "Towards Quantitative Comparison of Computer Number Systems," *Proceedings of the 5th Symposium on Computer Arithmetic*, Ann Arbor, MI, May 1981, pp. 21-34

Shauchi Ong and Daniel E. Atkins, "A Basis for the Quantitative Comparison of Computer Number Systems," *IEEE Transactions on Computers*, Vol. C-32, 1983, pp. 359-369

S. Ong and D. E. Atkins, "A Comparison of ALU Structures for VLSI Technology," *Proceedings of the 6th Symposium on Computer Arithmetic*, Aarhus, DENMARK, June 1983, pp. 10-16

Yukihito Ooowaki, *et al.*, "A Sub-10ns 16 X 16 Multiplier Using 0.6μm CMOS Technology," *IEEE Journal of Solid-State Circuits*, Vol. SC-22, 1987, pp. 762-767

R. M. Owens, "Compound Algorithms for Digit On-Line Arithmetic," *Proceedings of the 5th Symposium on Computer Arithmetic*, Ann Arbor, MI, May 1981, pp. 64-71

Robert Michael Owens, "Techniques to Reduce the Inherent Limitations of Fully Digit On-Line Arithmetic," *IEEE Transactions on Computers*, Vol. C-32, 1983, pp. 406-411

R. M. Owens and M. J. Irwin, "Digit Online Networks," *Proceedings of the 6th Symposium on Computer Arithmetic*, Aarhus, DENMARK, June 1983, pp. 156-163

J. Palmer, "The Intel Standard for Floating Point Arithmetic," *Proceedings of COMPSAC*, Chicago, November 1977, pp. 1076-112

F. D. Parker, *The Structure of Number Systems*, Englewood Cliffs, NJ: Prentice-Hall, 1966

Christos A. Papachristou, "Direct Implementation of Discrete and Residue-Based Functions Via Optimal

Encoding: A programmable array logic approach," *IEEE Transactions on Computers*, Vol. C-32, 1973, pp. 961-968

C. A. Papachristou, "Algorithms for Parallel Addition and Parallel Polynomial Evaluation," *Proceedings of the 5th Symposium on Computer Arithmetic*, Ann Arbor, MI, May 1981, pp. 256-263

C. A. Papachristou, "Multi-Input Residue Arithmetic Utilizing Read-Only Associative Memory," *Proceedings of the 7th Symposium on Computer Arithmetic*, Urbana, IL, June 1985, pp. 182-188

B. Parhami, "Systolic Up/Down Counters with Zero and Sign Detection," *Proceedings of the 8th Symposium on Computer Arithmetic*, Como, ITALY, May 1987, pp. 174-178

Behrooz Parhami, "On the Complexity of Table Lookup for Iterative Division," *IEEE Transactions on Computers*, Vol. C-36, 1987, pp. 1233-1236

Behrooz Parhami, "Carry-Free Addition of Recorded Binary Signed-Digit Numbers," *IEEE Transactions on Computers*, Vol. C-37, 1988, pp. 1470-1476

Behrooz Parhami, "Generalized Signed-Digit Number Systems: A Unifying Framework for Redundant Number Representations," *IEEE Transactions on Computers*, Vol. 39, 1990, pp. 89-98

Janak H. Patel and Leona Y. Fung, "Concurrent Error Detection in Multiply and Divide Arrays," *IEEE Transactions on Computers*, Vol. C-32, 1983, pp. 417-422

Andrzej Peczalski, *et al.*, "A 6K GaAs Gate Array with Fully Functional LSI Personalization," *IEEE Journal of Solid-State Circuits*, Vol. SC-23, 1988, pp. 581-590

H. Peng, "Algorithms for Extracting Square Roots and Cube Roots," *Proceedings of the 5th Symposium on Computer Arithmetic*, Ann Arbor, MI, May 1981, pp. 121-126

V. Peng, S. Samudrala, and M. Gavrielov, "On the Implementation of Shifters, Multipliers, and Dividers in VLSI Floating Point Units," *Proceedings of the 8th Symposium on Computer Arithmetic*, Como, ITALY, May 1987, pp. 95-102

Ernesto H. Perea, *et al.*, "A GaAs Low-Power Normally-On 4 Bit Ripple Carry Adder," *IEEE Journal of Solid-State Circuits*, Vol. SC-18, 1983, pp. 365-369

Marko Petkovsek, "Continuous Digit Sets and Local Roundings," *Proceedings of the 9th Symposium on Computer Arithmetic*, Santa Monica, CA, September 1989, pp. 136-143

F. E. Petry, "Two's Complement Extension of a Parallel Binary Division by Ten," *Electronics Letters*, Vol. 19, 1983, pp. 718-720

Stylianos D. Pezaris, "A 40-ns 17-Bit by 17-Bit Array Multiplier," *IEEE Transactions on Computers*, Vol. C-20, 1971, pp. 442-447

V. Piuri, "Fault-Tolerant Systolic Arrays: An Approach Based upon Residue Arithmetic," *Proceedings of the 8th Symposium on Computer Arithmetic*, Como, ITALY, May 1987, pp. 230-238

Dhiraj K. Pradhan, "Fault-Tolerant Carry-Save Adders Design Reed-Muller Codes," *IEEE Transactions on Computers*, Vol. C-323, 1974, pp. 1320-1322

J. Prado and R. Alcantara, "A Fast Square-Rooting Algorithm Using a Digital Signal Processor," *Proceedings of the IEEE*, Vol. 75, 1987, pp. 262-264

Franco P. Preparata, "A Mesh-Connected Area-Time Optimal VLSI Multiplier of Large Integers," *IEEE Transactions on Computers*, Vol. C-32, 1983, pp. 194-198

Kamla K. Primlani and Jack L. Meador, "A Nonredundant-Radix-4 Serial Multiplier," *IEEE Journal of Solid-State Circuits*, Vol. 24, 1989, pp. 1729-17369

Carla Neaderhouser Purdy and George B. Purdy, "Integer Division in Linear Time with Bounded Fan-In," *IEEE Transactions on Computers*, Vol. C-36, 1987, pp. 640-644

Jesse T. Quatse and Roy A. Keir, "A Parallel Accumulator for General-Purpose Computer," *IEEE Transactions on Electronic Computers*, Vol. C-16, 1967, pp. 165-171

Jaan Raamot, "Microprocessor Applications of Integer Arithmetic," *Proceedings of the IEEE*, Vol. 66, 1978, pp. 238-244

C. S. Raghavendra and M. D. Ercegovac, "A Simulator for On-Line Arithmetic," *Proceedings of the 5th Symposium on Computer Arithmetic*, Ann Arbor, MI, May 1981, pp. 92-99

Jean-Luc Rainard and Yves-Jacques Vernay, "A 16-Bit Self-Testing Multiplier," *IEEE Journal of Solid-State Circuits*, Vol. SC-16, 1981, pp. 174-179

Vijaya Ramachandran, "Single Residue Error Correction in Residue Number Systems," *IEEE Transactions on Computers*, Vol. C-32, 1983, pp. 504-507

C. V. Ramamoorthy, J. R. Goodman, and K. H. Kim, "Some Properties of Iterative Square-Rooting Methods Using High-Speed Multiplication," *IEEE Transactions on Computers*, Vol. C-21, 1972, pp. 837-847

Thammavarapu R. N. Rao, "Error-Checking Logic for Arithmetic-Type Operations of a Processor," *IEEE Transactions on Computers*, Vol. C-17, 1968, pp. 845-849

Thammavarapu R. N. Rao, "Biresidue Error-Correcting Codes for Computer Arithmetic," *IEEE Transactions on Computers*, Vol. C-19, 1970, pp. 398-402

Thammavarapu R. N. Rao, "Error Correction in Adders Using Systematic Subcodes," *IEEE Transactions on Computers*, Vol. C-21, 1972, pp. 254-259

T. R. N. Rao, *Error Coding for Arithmetic Processors*, New York: Academic Press, 1974.

T. R. N. Rao, "Arithmetic of Finite Fields," *Proceedings of the 5th Symposium on Computer Arithmetic*, Ann Arbor, MI, May 1981, pp. 2-5

T. M. Rao and R. T. Gregory, "The Conversion of Hensel Codes to Rational Numbers," *Proceedings of the 5th Symposium on Computer Arithmetic*, Ann Arbor, MI, May 1981, pp. 10-20

T. R. N. Rao and K. Vathanvit, "A Class of A(N+C) Codes and Its Properties," *Proceedings of the 7th Symposium on Computer Arithmetic*, Urbana, IL, June 1985, pp. 293-295

G. R. Redinbo, "Protecting Convolution-Type Arithmetic Array Calculations with Generalized Cyclic Codes," *Proceedings of the 8th Symposium on Computer Arithmetic*, Como, ITALY, May 1987, pp. 219-225

I. S. Reed, *et al.*, "VLSI Residue Multiplier Modulo a Fermat Number," *Proceedings of the 7th Symposium on Computer Arithmetic*, Urbana, IL, June 1985, pp. 203-206

E. K. Reuter, *et al.*, "Some Experiments Using Interval Arithmetic," *Proceedings of the 4th Symposium on Computer Arithmetic*, Santa Monica, CA, October 1978, pp. 75-80

V. Thomas Rhyne, "A Simple Postcorrection for Nonrestoring Division," *IEEE Transactions on Computers*, Vol. C-20, 1971, pp. 213-214

Tom Rhyne, "Limitations on Carry Lookahead Networks," *IEEE Transactions on Computers*, Vol. C-33, 1984, pp. 373-374

Tom Rhyne and Noel R. Strader, II, "A Signed Bit-Sequential Multiplier," *IEEE Transactions on Computers*, Vol. C-35, 1986, pp. 896-901

R. K. Richards, *Arithmetic Operations in Digital Computers*, Princeton, NJ: D. Van Nostrand Co., 1955.

Naphtali Rishe, "Lexicographic Encoding of Numeric Data Fields," *Proceedings of the 9th Symposium on Computer Arithmetic*, Santa Monica, CA, September 1989, pp. 241-246

J. E. Robertson, "A New Class of Digital Division Methods," *IRE Transactions on Electronic Computers*, Vol. EC-7, 1958, pp. 218-222

James E. Robertson, "The Correspondence between Methods of Digital Division and Multiplier Recoding Procedures," *IEEE Transactions on Computers*, Vol. C-19, 1970, pp. 692-701

J. E. Robertson, "A Systematic Approach to the Design of Structures for Arithmetic," *Proceedings of the 5th Symposium on Computer Arithmetic*, Ann Arbor, MI, May 1981, pp. 35-42

J. E. Robertson, "Conditions for the Distributivity of Multiplication with Respect to Set Addition and Their Effect on the Design Array Multipliers," *Proceedings of the 6th Symposium on Computer Arithmetic*, Aarhus, DENMARK, June 1983, pp. 67-73

J. E. Robertson, "Error-Detection and Correction for Addition and Subtraction through Use of Higher Radix Extensions of Hamming Codes," *Proceedings of the 8th Symposium on Computer Arithmetic*, Como, ITALY, May 1987, pp. 226-229

James E. Robertson and Kishor S. Trivedi, "The Status of Investigations into Computer Hardware Design Based on the Use of Continued Fractions," *IEEE Transactions on Computers*, Vol. C-22, 1973, pp. 555-560

Albrecht Rothermel, *et al.*, "Realization of Transmission-Gate Conditional-Sum (TGCS) Adders with Low Latency Time," *IEEE Journal of Solid-State Circuits*, Vol. 24, 1989, pp. 558-561

S. M. Rump, "Higher Order Computer Arithmetic," *Proceedings of the 7th Symposium on Computer Arithmetic*, Urbana, IL, June 1985, pp. 302-308

R. A. Rutenbar and Y. E. Park, "Case Study of a VLSI Design Project: A Simple Inner Product Machine," *Proceedings of the 5th Symposium on Computer Arithmetic*, Ann Arbor, MI, May 1981, pp. 184-189

Takashi Saigo, *et al.*, "A 20K-Gate CMOS Gate Array," *IEEE Journal of Solid-State Circuits*, Vol. SC-18, 1983, pp. 578-584

Akito Sakurai and Saburo Muroga, "Parallel Binary Adders with a Minimum Number of Connections," *IEEE Transactions on Computers*, Vol. C-32, 1983, pp. 969-976

Mark R. Santoro, Gary Bewick, and Mark A. Horowitz, "Rounding Algorithms for IEEE Multipliers," *Proceedings of the 9th Symposium on Computer Arithmetic*, Santa Monica, CA, September 1989, pp. 176-183

Mark R. Santoro and Mark A. Horowitz, "SPIM: A Pipelined 64 x 64-bit Iterative Multiplier," *IEEE Journal of Solid-State Circuits*, Vol. 24, 1989, pp. 487-493

J. Santos and H. Arango, "Base 3 vs. Base 2 Synchronous Arithmetic Units," *IEEE Transactions on Electronic Computers*, Vol. EC-13, 1964, pp. 608-609

B. P. Sarkar and E. V. Krishnamurthy, "Economic Pseudodivision Processes for Obtaining Square Root, Logarithm, and Arctan," *IEEE Transactions on Computers*, Vol. C-20, 1971, pp. 1589-1593

Akio Sasaki, "Addition and Subtraction in the Residue Number System," *IEEE Transactions on Electronic Computers*, Vol. EC-16, 1967, pp. 157-164

Akio Sasaki, "The Basis for Implementation of Additive Operations in the Residue Number System," *IEEE Transactions on Computers*, Vol. C-17, 1968, pp. 1066-1073

Fumihiko Sato, *et al.*, "A 400 ps Bipolar 18 Bit RALU Using Advanced PSA," *IEEE Journal of Solid-State Circuits*, Vol. SC-15, 1980, pp. 802-808

J. Schaeffer and D. Makarenko, "Systolic Polynomial Evaluation and Matrix Multiplication with Multiple Precision," *Proceedings of the 7th Symposium on Computer Arithmetic*, Urbana, IL, June 1985, pp. 110-117

I. D. Scherson and Y. Ma, "Vector Computations on an Orthogonal Memory Access Multiprocessing Systems," *Proceedings of the 8th Symposium on Computer Arithmetic*, Como, ITALY, May 1987, pp. 28-37

I. Scherson and S. Ruhman, "Multi-Operated Associative Arithmetic," *Proceedings of the 6th Symposium on Computer Arithmetic*, Aarhus, DENMARK, June 1983, pp. 123-129

F. A. Schreiber and R. Stefanelli, "Two Methods for Fast Integer Binary-BCD Conversion," *Proceedings of the 4th Symposium on Computer Arithmetic*, Santa Monica, CA, October 1978, pp. 200-207

Jerry Schwarz, "Implementing Infinite Precision Arithmetic," *Proceedings of the 9th Symposium on Computer Arithmetic*, Santa Monica, CA, September 1989, pp. 10-17

M. Scott, "Fast Rounding in Multiprecision Floating-Slash Arithmetic," *IEEE Transactions on Computers*, Vol. 38, 1989, pp. 1049-1052

R. B. Seidensticker, "Continued Fractions for High-Precision and High-Accuracy Computer Arithmetic," *Proceedings of the 6th Symposium on Computer Arithmetic*, Aarhus, DENMARK, June 1983, pp. 184-193

O. A. Seriki and R. W. Newcomb, "Direct-Coupled MOS Squaring Circuit," *IEEE Journal of Solid-State Circuits*, Vol. SC-14, 1979, pp. 766-768

Z. Shaham and Z. Riesel, "A Note on Division Algorithms Based on Multiplication," *IEEE Transactions on Computers*, Vol. C-21, 1972, pp. 513-514

Naresh R. Shanbhag and Pushkal Juneja, "Parallel Implementation of a 4 X 4-Bit Multiplier Using Modified Booth's Algorithm," *IEEE Journal of Solid-State Circuits*, Vol. SC-23, 1988, pp. 1010-1013

R. Sharma, "Area-Time Efficient Arithmetic Elements for VLSI Systems," *Proceedings of the 8th Symposium on Computer Arithmetic*, Como, ITALY, May 1987, pp. 57-62

Ramautar Sharma, *et al.*, "A 6.75-ns 16 x 16-bit Multiplier in Single-Level-Metal CMOS Technology," *IEEE Journal of Solid-State Circuits*, Vol. 24, 1989, pp. 922-927

Robert F. Shaw, "Arithmetic Operations in a Binary Computer," *Review of Scientific Instruments*, Vol. 21, 1950, pp. 687-693

John J. Shedletsky, "Comment on the Sequential and Indeterminate Behavior of an End-Around-Carry Adder," *IEEE Transactions on Computers*, Vol. C-26, 1977, pp. 271-272

John Paul Shen and F. Joel Ferguson, "The Design of Easily Testable VLSI Array Multipliers," *IEEE Transactions on Computers*, Vol. C-33, 1984, pp. 554-560

Barry J. Shepherd, "Right Shift for Low-Cost Multiply and Divide," *IEEE Transactions on Computers*, Vol. C-20, 1971, pp. 1586-1589

Ryosaku Shimada, Yoshiteru Ohkura, and Junichi Aoe, "Nonbinary Arithmetic AN Codes Using Odd Radix Expression," *IEEE Transactions on Computers*, Vol. C-34, 1985, pp. 1050-1056

R. R. Shively, W. V. Robinson, Jr., and D. E. Orton, "Cascading Transmission Gates to Enhance Multiplier Performance," *IEEE Transactions on Computers*, Vol. C-33, 1984, pp. 677-679

B. Shriver and P. Kornerup, "The UNRAU–A Unified Numeric Representation Arithmetic Unit," *Proceedings of the 3rd Symposium on Computer Arithmetic*, Dallas, TX, November 1975, pp. 179-184

B. D. Shriver and E. K. Reuter, "A Bibliography on Computer Arithmetic," *Proceedings of the 3rd Symposium on Computer Arithmetic*, Dallas, TX, November 1975, pp. 241-249

H. C. Shyu, T. K. Truong, and I. S. Reed, "A Complex Integer Multiplier Using the Quadratic-Polynomial Residue Number System with Numbers of Form $2^{2n} + 1$," *IEEE Transactions on Computers*, Vol. C-36, 1987, pp. 1255-1258

Kripasindhu Sikdar, "Determination of Multipliers Mapping an Arbitrary Integer into a Range of Certain Type," *IEEE Transactions on Computers*, Vol. C-19, 1970, pp. 1221-1222

Shanker Singh and Ronald Waxman, "Multiple Operand Addition and Multiplication," *IEEE Transactions on Computers*, Vol. C-22, 1973, pp. 113-120

Bhabani P. Sinha and Pradip K. Srimani, "Fast Parallel Algorithms for Binary Multiplication and Their Implementations on Systolic Architectures," *IEEE Transactions on Computers*, Vol. 38, 1989, pp. 424-431

H. J. Sips, "Comment on 'An O(n) Parallel Multiplier with Bit-Sequential Input and Output,'" *IEEE Transactions on Computers*, Vol. C-31, 1982, pp. 325-327

Henk J. Sips, "Bit Sequential Arithmetic for Parallel Processors," *IEEE Transactions on Computers*, Vol. C-33, 1984, pp. 7-20

J. Sklansky, "An Evaluation of Several Two-Summand Binary Adders," *IRE Transactions on Electronic Computers*, Vol. EC-9, 1960, pp. 213-226

J. Sklansky, "Conditional-Sum Addition Logic," *IRE Transactions on Electronic Computers*, Vol. EC-9, 1960, pp. 226-231

A. G. Slekys and A. Avizienis, "A Modified Bi-Imaginary Number System," *Proceedings of the 4th Symposium on Computer Arithmetic*, Santa Monica, CA, October 1978, pp. 48-55

K. R. Sloan, Jr., "Comments on 'A Computer Algorithm for Calculating the Product AB Modulo M' by G. R. Blakley," *IEEE Transactions on Computers*, Vol. C-34, 1985, pp. 290-292

S. G. Smith, "Incremental Computation of Squares and Sums of Squares," *IEEE Transactions on Computers*, Vol. 38, 1989, pp. 1325-1328

S. G. Smith, "Comments on 'A Signed Bit-Sequential Multiplier'," *IEEE Transactions on Computers*, Vol. 38, 1989, pp. 1328-1330

S. G. Smith and P. B. Denyer, "Synthesis of Area-Efficient VLSI Architectures for Vector and Matrix Multiplication," *Proceedings of the 8th Symposium on Computer Arithmetic*, Como, ITALY, May 1987, pp. 13-20

S. G. Smith and P. B. Denyer, "Advanced Serial-Data Computation," *Journal of Parallel and Distributed Computing*, Vol. 5, 1988, pp. 228-249

S. P. Smith and H. C. Torng, "Design of a Fast Inner Product Processor," *Proceedings of the 7th Symposium on Computer Arithmetic*, Urbana, IL, June 1985, pp. 38-43

Michael A. Soderstrand, "A High-Speed Low-Cost Recursive Digital Filter Using Residue Number Arithmetic," *Proceedings of the IEEE*, Vol. 65, 1977, pp. 1065-1067

Michael A. Soderstrand, *et al.*, eds., *Residue Number System Arithmetic: Modern Applications in Digital Signal Processing*, New York: IEEE Press, 1986

Michael A. Soderstrand and Carmel Vernia, "A High-Speed Low-Cost Modulo Pi Multiplier with RNS Arithmetic Applications," *Proceedings of the IEEE*, Vol. 68, 1980, pp. 529-532

Mike Sousa and Fred Taylor, "Complex Integer to Complex Residue Encoding," *IEEE Transactions on Computers*, Vol. C-35, 1986, pp. 648-650

O. Spaniol, *Computer Arithmetic: Logic and Design*, New York: John Wiley & Sons, 1981

W. H. Specker, "A Class of Algorithms for Ln x, Sin x, Cos x, Tan^{-1} x, and Cot^{-1} x," *IEEE Transactions on Electronic Computers*, Vol. EC-14, 1965, pp. 85-86

Philip M. Spira, "Computation Times of Arithmetic and Boolean Functions in (d,r) Circuits," *IEEE Transactions on Computers*, Vol. C-22, 1973, pp. 552-555

D. G. Steer and S. R. Penstone, "Digital Hardware for Sine-Cosine Function," *IEEE Transactions on Computers*, Vol. C-26, 1977, pp. 1283-1286

Renato Stefanelli, "A Suggestion for a High-Speed Parallel Binary Divider," *IEEE Transactions on Computers*, Vol. C-21, 1972, pp. 42-55

Marvin L. Stein and William D. Munro, "Scaling Machine Arithmetic," *IEEE Transactions on Computers*, Vol. C-20, 1971, pp. 675-678

M. L. Steinard and W. D. Munro, *Introduction to Machine Arithmetic*, Reading, MA: Addison-Wesley, 1971

William J. Stenzel, William J. Kubitz, and Gilles H. Garcia, "A Compact High-Speed Parallel Multiplication Scheme," *IEEE Transactions on Computers*, Vol. C-26, 1977, pp. 948-957

Charles Stephenson, "Case Study of the Pipelined Arithmetic Unit for the TI Advanced Scientific Computer," *Proceedings of the 3rd Symposium on Computer Arithmetic*, Dallas, TX, November 1975, pp. 168-173

P. H. Sterbenz, *Floating Point Computation*, Englewood Cliffs, NJ: Prentice-Hall, 1974

D. Stevenson, "A Proposed Standard for Binary Floating-Point Arithmetic," *Computer*, Vol. 14, March 1981, pp. 51-62

Thanos Stouraitis and Fred J. Taylor, "Floating-Point to Logarithmic Encoder Error Analysis," *IEEE Transactions on Computers*, Vol. C-37, 1988, pp. 858-863

Noel R. Strader, II and V. Thomas Rhyne, "A Canonical Bit-Sequential Multiplier," *IEEE Transactions on Computers*, Vol. C-31, 1982, pp. 791-795

Masao Suzuki, *et al.*, "Design and Application of a 2500-Gate Bipolar Macrocell Array," *IEEE Journal of Solid-State Circuits*, Vol. SC-20, 1985, pp. 1025-1031

A. Svoboda, "An Algorithm for Division," *Information Processing Machines*, Vol. 9, 1963, pp. 25-32

A. Svoboda, "Adder with Distributed Control," *IEEE Transactions on Computers*, Vol. C-319, 1970, pp. 749-751

A. Svoboda, "Self Checking Adder for Large Scale Integration," *Proceedings of the 3rd Symposium on Computer Arithmetic*, Dallas, TX, November 1975, pp. 108-112

A. Svoboda, "Arithmetic Circuit Fault Detection by Modular Encoding," *Proceedings of the 4th Symposium on Computer Arithmetic*, Santa Monica, CA, October 1978, pp. 208-219

E. E. Swartzlander, Jr., "The Quasi-Serial Multiplier," *IEEE Transactions on Computers*, Vol. C-22, 1973, pp. 317-321

E. E. Swartzlander, Jr., "Parallel Counters," *IEEE Transactions on Computers*, Vol. C-22, 1973, pp. 1021-1024

E. E. Swartzlander, Jr., "Merged Arithmetic for Signal Processing," *Proceedings of the 4th Symposium on Computer Arithmetic*, Santa Monica, CA, October 1978, pp. 239-244

E. E. Swartzlander, Jr., ed., *Computer Arithmetic*, Stroudsburg, PA: Dowden, Hutchinson & Ross, 1980

E. E. Swartzlander, Jr., and A. G. Alexopoulos, "The Sign/Logarithm Number System," *IEEE Transactions on Computers*, Vol. C-24, 1975, pp. 1238-1242

E. E. Swartzlander, Jr. and J. Eldon, "Arithmetic for High Speed FFT Implementation," *Proceedings of the 7th Symposium on Computer Arithmetic*, Urbana, IL, June 1985, pp. 223-230

Earl E. Swartzlander, Jr. and Barry K. Gilbert, "Arithmetic for Ultra-High-Speed Tomography," *IEEE Transactions on Computers*, Vol. C-29, 1980, pp. 341-353

Earl E. Swartzlander, Jr., Barry K. Gilbert, and Irving S. Reed, "Inner Product Computers," *IEEE Transactions on Computers*, Vol. C-27, 1978, pp. 21-31

D. W. Sweeney, "An Analysis of Floating-Point Addition," *IBM Systems Journal*, Vol. 4, 1965, pp. 31-42

Nicholas S. Szabó and Richard I. Tanaka, *Residue Arithmetic and Its Applications to Computer Technology*, New York: McGraw-Hill Book Co.,1967

Haruyuki Tago, *et al.*, "A 6K-Gate CMOS Gate Array," *IEEE Journal of Solid-State Circuits*, Vol. SC-17, 1982, pp. 907-912

Naofumi Takagi, Hiroto Yasuura, and Shuzo Yajima, "High-Speed VLSI Multiplication Algorithm with a Redundant Binary Addition Tree," *IEEE Transactions on Computers*, Vol. C-34, 1985, pp. 789-796

Kazumitsu Takeda, *et al.*, "A Single-Chip 80-Bit Floating Point Processor," *IEEE Journal of Solid-State Circuits*, Vol. SC-20, 1985, pp. 986-992

Y. Takefuji, T. Kurokawa, and H. Aiso, "Fast Matrix Solver in GF(2)," *Proceedings of the 6th Symposium on Computer Arithmetic*, Aarhus, DENMARK, June 1983, pp. 138-143

Nabil Takla and Mark Hecker, "A Monolithic 64 Bit Floating-Point Coprocessor," *IEEE Journal of Solid-State Circuits*, Vol. SC-19, 1984, pp. 538-539

D. G. Tan, "The Theory and Implementation of High-Radix Division," *Proceedings of the 4th Symposium on Computer Arithmetic*, Santa Monica, CA, October 1978, pp. 154-163

K. Taniguchi, "Three-Dimensional IC's and an Application to High Speed Image Processor," *Proceedings of the 7th Symposium on Computer Arithmetic*, Urbana, IL, June 1985, pp. 216-222

Fred J. Taylor, "A VLSI Residue Arithmetic Multiplier," *IEEE Transactions on Computers*, Vol. C-31, 1982, pp. 540-546

Fred J. Taylor, "An Overflow-Free Residue Multiplier," *IEEE Transactions on Computers*, Vol. C-32, 1983, pp. 501-504

F. J. Taylor, "A More Efficient Residue Arithmetic Implementation of the FFT," *Proceedings of the 7th Symposium on Computer Arithmetic*, Urbana, IL, June 1985, pp. 243-250

Fred J. Taylor and William Dirr, Jr., "A New Residue to Decimal Converter," *Proceedings of the IEEE*, Vol. 73, 1985, pp. 378-380

Fred J. Taylor and Chao H. Huang, "An Autoscale Residue Multiplier," *IEEE Transactions on Computers*, Vol. C-31, 1982, pp. 321-325

F. J. Taylor, G. Papadourakis, A. Skavantzos, and A. Stouraitis, "A Radix-4 FFT Using Complex RNS Arithmetic," *IEEE Transactions on Computers*, Vol. C-34, 1985, pp. 573-576

Fred J. Taylor, *et al.*,"A 20 Bit Logarithmic Number System Processor," *IEEE Transactions on Computers*, Vol. C-37, 1988, pp. 190-200

G. S. Taylor, "Compatible Hardware for Division and Square Root," *Proceedings of the 5th Symposium on Computer Arithmetic*, Ann Arbor, MI, May 1981, pp. 127-134

G. S. Taylor, "Arithmetic on the ELXSI System 6400," *Proceedings of the 6th Symposium on Computer Arithmetic*, Aarhus, DENMARK, June 1983, pp. 110-115

G. S. Taylor, "Radix 16 SRT Dividers with Overlapped Quotient Selection Stages," *Proceedings of the 7th Symposium on Computer Arithmetic*, Urbana, IL, June 1985, pp. 64-71

G. S. Taylor and D. A. Patterson, "VAX Hardware for the Proposed IEEE Floating-Point Standard," *Proceedings of the 5th Symposium on Computer Arithmetic*, Ann Arbor, MI, May 1981, pp. 190-197

Jay J. Thomas and Sydney R. Parker, "Implementing Exact Calculations in Hardware," *IEEE Transactions on Computers*, Vol. C-36, 1987, pp. 764-768

J. E. Thornton, *Design of a Computer—The Control Data 6600*, Glenview, IL: Scott, Foresman and Co., 1970

K. D. Tocher, "Techniques of Multiplication and Division for Automatic Binary Computers," *Quarterly Journal of Mechanics and Applied Mathematics*, Vol. 11, 1958, pp. 364-384

K. S. Trivedi, "On the Use of Continued Fractions for Digital Computer Arithmetic," *Proceedings of the 3rd Symposium on Computer Arithmetic*, Dallas, TX, November 1975, pp. 137-146

K. S. Trivedi and M. D. Ercegovac, "On-Line Algorithms for Division and Multiplication," *Proceedings of the 3rd Symposium on Computer Arithmetic*, Dallas, TX, November 1975, pp. 161-167

Kishor S. Trivedi and Milos D. Ercegovac, "On-Line Algorithms for Division and Multiplication," *IEEE Transactions on Computers*, Vol. C-26, 1977, pp. 681-687

William W. Troutman, *et al.*, "Design of a Standard Floating-Point Chip," *IEEE Journal of Solid-State Circuits*, Vol. SC-21, 1986, pp. 396-399

Nai-Kuan Tsao, "Error Complexity Analysis of Algorithms for Matrix Multiplication and Matrix Chain Product,"

IEEE Transactions on Computers, Vol. C-29, 1980, pp. 662-665

Nai-Kuan Tsao, "A Simple Approach to the Error Analysis of Division-Free Numerical Algorithms," *IEEE Transactions on Computers*, Vol. C-32, 1983, pp. 343-351

P. K.-G. Tu and M. D. Ercegovac, "A Radix-4 On-Line Division Algorithm," *Proceedings of the 8th Symposium on Computer Arithmetic*, Como, ITALY, May 1987, pp. 181-187

Paul K.-G. Tu and Milos D. Ercegovac, "Design of On-Line Division Unit," *Proceedings of the 9th Symposium on Computer Arithmetic*, Santa Monica, CA, September 1989, pp. 42-49

Chin Tung, "Signed-Digit Division Using Combinational Arithmetic Nets," *IEEE Transactions on Computers*, Vol. C-19, 1970, pp. 746-748

Chin Tung, "A Division Algorithm for Signed-Digit Arithmetic," *IEEE Transactions on Computers*, Vol. C-17, 1968, pp. 887-889

Peter R. Turner, "Software Implementation of SLI Arithmetic," *Proceedings of the 9th Symposium on Computer Arithmetic*, Santa Monica, CA, September 1989, pp. 18-24

Silvio Tirrini, "Optimal Group Distribution in Carry-Skip Adders," *Proceedings of the 9th Symposium on Computer Arithmetic*, Santa Monica, CA, September 1989, pp. 96-103

Zenon D. Ulman, "Sign Detection and Implicit-Explicit Conversion of Numbers in Residue Arithmetic," *IEEE Transactions on Computers*, Vol. C-32, 1983, pp. 590-594

H. Umeo, "A Design of Time-Optimum and Register-Number-Minimum Systolic Convolver," *Proceedings of the 8th Symposium on Computer Arithmetic*, Como, ITALY, May 1987, pp. 5-12

Masaru Uya, Katsuyuki Kaneko, and Juro Yasui, "A CMOS Floating Point Multiplier," *IEEE Journal of Solid-State Circuits*, Vol. SC-19, 1984, pp. 697-702

Peter J. Varman, I. V. Ramakrishnan, and D. S. Fussell, "A Robust Matrix-Multiplication Array," *IEEE Transactions on Computers*, Vol. C-33, 1984, pp. 919-922

Stamatis Vassiliadis, David S. Lemon, and Michael Putrino, "S/370 Sign-Magnitude Floating-Point Adder," *IEEE Journal of Solid-State Circuits*, Vol. 24, 1989, pp. 1062-1070

Stamatis Vassiliadis, Eric M. Schwarz, and Don J. Hanrahan, "A General Proof for Overlapped Multiple-Bit Scanning Multiplications," *IEEE Transactions on Computers*, Vol. 38, 1989, pp. 172-183

Harry J. M. Veendrick and Leo C. Pfennings, "A 40 MHz Multiapplicable Digital Signal Processing Chip," *IEEE Journal of Solid-State Circuits*, Vol. SC-17, 1982, pp. 40-43

Jos R. Verjans, "A Serial-Parallel Multiplier Using the NENDEP Technology," *IEEE Journal of Solid-State Circuits*, Vol. SC-12, 1977, pp. 323-325

J. E. Volder, "The CORDIC Trigonometric Computing Technique," *IRE Transactions on Electronic Computers*, Vol. EC-8, 1959, pp. 330-334

Jean Vuillemin, "A Very Fast Multiplication Algorithm Algorithm for VLSI Implementation," *Integration*, Vol. 1, 1983, pp. 39-52

G. Walker, "Extension of the MC68000 Architecture to Include Standard Floating-Point Arithmetic," *Proceedings of the 5th Symposium on Computer Arithmetic*, Ann Arbor, MI, May 1981, pp. 179-183

C. S. Wallace, "A Suggestion for a Fast Multiplier," *IEEE Transactions on Electronic Computers*, Vol. EC-13, 1964, pp. 14-17

P. J. L. Wallis, "ADA Floating-Point Arithmetic as a Basis for Portable Numerical Software," *Proceedings of the 6th Symposium on Computer Arithmetic*, Aarhus, DENMARK, June 1983, pp. 79-81

J. S. Walther, "A Unified Algorithm for Elementary Functions," *Proceedings of the Spring Joint Computer Conference*, 1971, pp. 379-385

Tse Lin Wang and Chao-Kai Liu, "Weight-Preserved Single-Error-Correcting Scheme for Binary Adders," *IEEE Transactions on Computers*, Vol. C-23, 1974, pp. 1002-1007

Frederick A. Ware, *et al.*, "64 Bit Monolithic Floating Point Processors," *IEEE Journal of Solid-State Circuits*, Vol. SC-17, 1982, pp. 898-907

Katsuyoshi Washio, Kazuo Nakazato, and Tohru Nakamura, "2.7 ns 8 x 8-Bit Parallel Array Multiplier Using Sidewall Base Contact Structure," *IEEE Journal of Solid-State Circuits*, Vol. SC-22, 1987, pp. 613-614

S. Waser, "Survey of Arithmetic Integrated Circuits," *Proceedings of the 4th Symposium on Computer Arithmetic*, Santa Monica, CA, October 1978, pp. 257-266

S. Waser, "High Speed Monolithic Multipliers for Real-Time Digital Signal Processing," *Computer*, Vol. 11, October 1978

Shlomo Waser and Michael J. Flynn, *Introduction to Arithmetic for Digital Systems Designers*, New York: Holt, Rinehart and Winston, 1982

O. Watanuki and M. D. Ercegovac, "Floating-Point On-Line Arithmetic: Algorithms," *Proceedings of the 5th Symposium on Computer Arithmetic*, Ann Arbor, MI, May 1981, pp. 81-86

O. Watanuki and M. D. Ercegovac, "Floating-Point On-Line Arithmetic: Error Analysis," *Proceedings of the 5th Symposium on Computer Arithmetic*, Ann Arbor, MI, May 1981, pp. 87-91

Osaaki Watanuki and Milos D. Ercegovac, "Error Analysis of Certain Floating-Point On-Line Algorithms," *IEEE Transactions on Computers*, Vol. C-32, 1983, pp. 352-358

R. W. Watson and C. W. Hastings, "Self-Checked Computation Using Residue Arithmetic," *Proceedings of the IEEE*, Vol. 54, 1966, pp. 1920-1931

James W. Watterson and Jill J. Hallenbeck, "Modulo 3 Residue Checker: New Results on Performance and Cost," *IEEE Transactions on Computers*, Vol. C-37, 1988, pp. 608-612

A. Weinberger, "High-Speed Zero-Sum Detection," *Proceedings of the 3rd Symposium on Computer Arithmetic*, Dallas, TX, November 1975, pp. 200-207

A. Weinberger, "Parallel Adders Using Standard PLAs," *Proceedings of the 4th Symposium on Computer Arithmetic*, Santa Monica, CA, October 1978, pp. 116-124

A. Weinberger and J. L. Smith, "A One-Microsecond Adder Using One-Megacycle Circuitry," *IRE Transactions on Electronic Computers*, Vol. EC-5, 1956, pp. 65-73

A. Weinberger and J. L. Smith, "A Logic for High-Speed Addition," *National Bureau of Standards Circular*, No. 591, 1958, pp. 3-12

Charles W. Weller, "A High-Speed Carry Circuit for Binary Adders," *IEEE Transactions on Computers*, Vol. C-18, 1969, pp. 728-732

Stanley A. White, "On Mechanization of Vector Multiplication," *Proceedings of the IEEE*, Vol. 63, 1975, pp. 730-731

J. H. Wilkinson, *Rounding Errors in Algebraic Processes*, Engelwood Cliffs, NJ: Prentice-Hall, 1963

J. Williams and V. C. Hamacher, "A Linear-Time Divider Array," *Canadian Electrical Engineering Journal*, Vol. 6, 1981, pp.14-20

R. Willoner and I-N. Chen, "An Algorithm for Modular Exponentiation," *Proceedings of the 5th Symposium on Computer Arithmetic*, Ann Arbor, MI, May 1981, pp. 135-138

J. B. Wilson and R. S. Ledley, "An Algorithm for Rapid Binary Division," *IRE Transactions on Electronic Computers*, Vol. EC-10, 1961, pp. 662-670

S. Winograd, "On the Time Required to Perform Addition," *Journal of the ACM*, Vol. 12, 1965, pp. 277-285

S. Winograd, "On the Time Required to Perform Multiplication," *Journal of the ACM*, Vol. 14, 1967, pp. 793-802

Anthony S. Wojcik and Gernot Metze, "On the Cost of Base n Adders," *IEEE Transactions on Computers*, Vol. C-20, 1971, pp. 1196-1203

Gil Wolrich, *et al.*, "A High Performance Floating Point Coprocessor," *IEEE Journal of Solid-State Circuits*, Vol. SC-19, 1984, pp. 690-696

C. Wrathall and T. C. Chen, "Convergence Guarantee and Improvements for a Hardware Exponential and Logarithm Evaluation Scheme," *Proceedings of the 4th Symposium on Computer Arithmetic*, Santa Monica, CA, October 1978, pp. 175-182

Chwan-Chia Wu, "Time-Redundant Fault-Location in Bit-Sliced ALU's," *IEEE Transactions on Computers*, Vol. C-36, 1987, pp. 1387-1389

C.-C. Wu, L.-R. Tzeng and T.-S. Wu, "Time-Redundant Fault-Masking in ALUs," *Proceedings of the 8th Symposium on Computer Arithmetic*, Como, ITALY, May 1987, pp. 239-243

Ryuichiro Yamamoto, *et al.*, "Design and Fabrication of Depletion GaAs LSI High-Speed 32-Bit Adder," *IEEE Journal of Solid-State Circuits*, Vol. SC-18, 1983, pp. 592-599

Hironori Yamauchi, *et al.*, "10 NS 8 X 8 Multiplier LSI Using Super Self-Aligned Process Technology," *IEEE Journal of Solid-State Circuits*, Vol. SC-18, 1983, pp. 204-210

Stephen Sik-Sang Yau and Yu-Cheng Liu, "Error Correction in Redundant Residue Number Systems," *IEEE Transactions on Computers*, Vol. C-22, 1973, pp. 5-11

Kai-Ping Yiu, "On Sign Bit Assignment for a Vector Multiplier," *Proceedings of the IEEE*, Vol. 64, 1976, pp. 372-373

J. Michael Yohe, "Roundings in Floating-Point Arithmetic," *IEEE Transactions on Computers*, Vol. C-22, 1973, pp. 577-586

K. Yoshida, "Floating-Point Recurring Rational Arithmetic System," *Proceedings of the 6th Symposium on Computer Arithmetic*, Aarhus, DENMARK, June 1983, pp. 194-200

C. K. Yuen, "A Note on Base -2 Arithmetic Logic," *IEEE Transactions on Computers*, Vol. C-24, 1975, pp. 325-329

C. K. Yuen, "On the Floating Point Representation of Complex Numbers," *IEEE Transactions on Computers*, Vol. C-24, 1975, pp. 846-848

C. K. Yuen, "A New Representation for Decimal Numbers," *IEEE Transactions on Computers*, Vol. C-26, 1977, pp. 1286-1288

C. K. Yuen, "A Device for Finding the Largest N-Bit Integer in an N-Bit String," *Proceedings of the IEEE*, Vol. 69, 1981, pp. 130-132

D. Y. Y. Yun and C. N. Zhang, "Binary Paradigm and Systolic Array Implementation for Residue Arithmetic," *Proceedings of the 7th Symposium on Computer Arithmetic*, Urbana, IL, June 1985, pp. 189-193

R. J. Zaccone and J. L. Barlow, "Improved Normalization Results for Digit On-Line Arithmetic," *Proceedings of the 7th Symposium on Computer Arithmetic*, Urbana, IL, June 1985, pp. 20-27

Richard J. Zaccone and Jesse L. Barlow, "Eliminating the Normalization Problem in Digit On-Line Arithmetic," *IEEE Transactions on Computers*, Vol. C-36, 1987, pp. 36-46

W. Zadrozny, "Axiomatizations of Floating Point Arithmetics," *Proceedings of the 7th Symposium on Computer Arithmetic*, Urbana, IL, June 1985, pp. 74-81

Stanislaw H. Zak and Kai Hwang, "Polynomial Division on Systolic Arrays," *IEEE Transactions on Computers*, Vol. C-34, 1985, pp. 577-578

Jan Zeman and H. Troy Nagle, Jr., "A High-Speed Microprogammable Digital Signal Processor Employing Distributed Arithmetic," *IEEE Transactions on Computers*, Vol. C-29, 1980, pp. 134-144

Shalhav Zohar, "Rounding and Truncation in Radix (—2) Systems," *IEEE Transactions on Computers*, Vol. C-25, 1976, pp. 464-469

Dan Zuras and William H. McAllister, "Balanced Delay Trees and Combinatorial Division in VLSI," *IEEE Journal of Solid-State Circuits*, Vol. SC-21, 1986, pp. 814-819

J. H. P. Zurawski and J. B. Gosling, "Design of High-Speed Digital Divider Units," *IEEE Transactions on Computers*, Vol. C-30, 1981, pp. 691-699

J. H. P. Zurawski and J. B. Gosling, "Design of a High-Speed Square Root Multiply and Divide Unit," *IEEE Transactions on Computers*, Vol. C-36, 1987, pp. 13-23